ELECTRICAL POWER TECHNOLOGY

THEODORE WILDI
Professor of Electrical Engineering
Laval University

with the collaboration of
PERRY R. McNEILL
Professor and Chairman
Electronics/Electrical Power Technology
Oklahoma State University

JOHN WILEY & SONS

New York • Chichester • Brisbane • Toronto • Singapore

Library of Congress Cataloging in Publication Data:

Wildi, Theodore.
  Electrical power technology.

  Includes bibliographical references and index.
  1. Electric engineering.   2.   Electric power.
I. Title.

TK145.W489   621.31   80-39799
ISBN 0-471-07764-X

Printed in the United States of America

20 19 18 17 16 15 14 13 12

PREFACE

Electrical power technology has made enormous progress in recent years. Improvements in materials, transmission methods, and rotating machinery have been so great that electricity continues to be an all-pervasive source of energy.

This book is intended to give a broad overview of modern electrical power technology. It covers the basic principles of transformers and rotating machines, transmission and distribution systems, and power electronics associated with these fields. Toward this end, the subject matter has been divided into seven distinct parts: (1) Fundamentals; (2) Electrical Materials; (3) Alternating Current Circuits; (4) Transformers; (5) Rotating Machinery; (6) Electrical and Electronic Controls and (7) Electrical Energy. These major divisions, comprising 31 chapters, can be covered either separately or in sequence, in whole or part, depending on the particular program of study. Cross references to preceding chapters and equations make it easy to return to the original source of information, when required.

In order to convey the real-world aspects of machinery and systems, particular attention has been paid to such important elements as the inertia of revolving masses, the limitations of available materials, and the problems created by heat. It is felt that this approach also falls in line with the multidisciplinary programs of many colleges and technical institutes.

The subject matter requires a background in basic circuit theory, algebra and some trigonometry. Owing to its multidisciplinary approach, and its straightforward treatment of even complex topics, we believe this book will meet the needs of a broad range of readers. First, it is appropriate for students following a two-year electrical program in community colleges and technical institutes. It can also be used in those courses in an electronics program that deal with transformers or machinery and controls. The text can also be incorporated in a 4-year technology program, owing to its very broad coverage. Many of these programs are supported by hands-on laboratory work.

Students following industrial training programs will also find the book contains a great deal of practical information that can be directly applied in that greatest laboratory of all—the electrical industry itself. Finally, at a time when much effort is being devoted to continuing education, this book, with its many worked-out problems, is particularly well adapted for self-study.

The exercises at the end of each chapter are organized so that readers may solve problems appropriate to their particular level of interest, background and ability. They are divided into three levels—practical, intermediate and advanced. Furthermore, to encourage the reader in solving the exercises, the answers are given at the end of the book.

A quick glance through the text shows the importance given to photographs. All equipment and systems are illustrated by diagrams and pictures, showing them in various stages of construction, or in actual operation. Many students may not have had the opportunity to visit a generating station or to see at close hand the equipment used in the transmission and distribution of electrical energy. The photographs help convey the magnificent size of these devices and machines. Furthermore, many problems are directly related to the photographs, making them all the more real.

The International System of Units (SI), now widely adopted throughout the world, is used in this text. This decision is based upon the fact that textbooks in every field are rapidly adopting this new system of units. In this regard, special care has been taken to use units and symbols according to accepted standards. The reader not yet familiar with SI will be surprised by its basic simplicity. It may take a little time to get used to such units as weber, newton, pascal and so forth, but the simple method explained in Chapter 1 facilitates the conversion from one system to any other. Also, some of the exercises at the ends of the chapters use English units, and some use SI. Thus, students can become proficient at solving problems with both systems.

Power electronics is assuming an ever-growing place in the electrical power field. In recognition of this fact, we have included four important chapters on this subject. After introducing the principles of power electronics, the text goes on to cover the major aspects of ac and dc drives. They comprise the modern electronic methods of controlling medium and large-power machines.

Finally, a chapter is devoted to the dc transmission of electric power. Major installations of this kind are presently in operation throughout the world. It is expected that dc transmission lines and dc links will become increasingly important as they complement existing ac networks.

In covering such a broad range of subjects, a conscious effort has been made to establish cohesion so that the reader will see how they all fit together. Thus, we find that the terminology and equations for synchronous machines are similar to those relating to transmission lines. Transmission lines, in turn, bring up the question of reactive power. And reactive power is important in electronic converters. Thus, knowledge gained in one area is strengthened and broadened by using it in another. As a result, the learning of electrical power technology becomes a challenging, thought-provoking experience.

The explanation of the subject matter is straightforward and based on technical fact. Consequently, the reader who wishes to make an in-depth study of a particular topic will find that information given here agrees with knowledge revealed by more detailed or more advanced texts.

In summary, the book employs a theoretical, practical, multidisciplinary approach to give a broad understanding of electrical power technology. We hope it will stimulate renewed interest in this dynamic, exciting field.

It has been my good fortune to benefit from the comments and suggestions made by Robert Abrams of Alabama Technical College, Robert Cox of Wentworth Institute of Technology, A. J. Davidson of New Hampshire Vocational-Technical College at Claremont, Joseph M. DeGuilmo of Stevens Institute of Technology, R. Craig Hubele of Purdue University, James F. More of Ohio College of Applied Science-University of Cincinnati, James K. Shelton of Oklahoma State University and, especially, by Perry R. McNeill of Oklahoma State University. Their recommendations regarding the organization of the subject matter and their chapter-by-chapter analyses have been of inestimable help in making this book what it is. I wish to express my thanks for these very important contributions.

Finally, I want to thank my son, Karl, who made all the line drawings, and my wife, Rachel, for her encouragement and support.

Theodore Wildi

CONTENTS

LIST OF TABLES

XVI

PART I

FUNDAMENTALS

1

UNITS

Everything we see and feel and everything we buy and sell is measured and compared by means of units. Units enable us to measure the distance we walk, the land we own, the time of day, and the brightness of stars — and, in fact, every feature of the environment in which we live.

Some of these units have become so familiar that we often take them for granted, seldom stopping to think how they started, or why they were given the sizes they have.

Centuries ago, the foot was defined as the length of 36 barleycorns strung end to end, and the yard was the distance from the tip of King Edgar's nose to the end of his outstretched hand.

Since then, we have come a long way in defining our units of measure more precisely. Most units are now based upon the physical laws of nature, which are both invariable and reproducible. Thus, the meter and yard are measured in terms of the wavelength of light, and time by the duration of atomic vibrations. This improvement in our standards of measure went hand in hand with the advances in technology, and the one could not have been achieved without the other.

But if the basic standards of reference are recognized by all countries of the world, the units of everyday measure are far from being universal. For example, in measuring length, some people use the inch and yard, while others use the millimeter and meter. Astronomers employ the parsec, physicists use the angstrom, and surveyors measure with the rod and chain. But these units of length can be compared with great accuracy because the standard of length is based upon the wavelength of light.

Such standards of reference make it possible to compare the units of measure in one country, or in one specialty, with the units of measure in any other. Standard units of length, mass and time are the anchors that tie together the different systems of units used in the world today.

1-1 Systems of units

Over the years, systems of units have been devised to meet the needs of commerce, industry, and science. A system of units may be described as one in which the units bear a direct numerical relation-

1

ship to each other, usually expressed as a whole number. Thus, in the English system of units the inch, foot, and yard are related to each other by the numbers 12, 3, and 36. And again, in the same system, the ounce, pound, and short ton are related by the numbers 16, 2000, and 32 000.

The same correlation exists in metric systems, except that the units are related to each other by multiples of ten. Thus, the centimeter, meter and kilometer are related by the numbers 100, 1000, and 100 000. It is therefore easier to convert meters into centimeters than to convert yards into feet, and this decimal approach is one of the advantages of the metric system of units.*

The English system of units is very old and its origins may be traced back to the time of the Egyptians. By comparison, the metric system is a relative newcomer, having been started only some 150 years ago. As with every new system, it had to undergo important changes, during the course of which some units were added and others removed to meet the needs of the rapidly evolving electrical and mechanical technologies.

This process of evolution has yielded five important systems of units, some of which are gradually declining in use, having served the purpose for which they were designed. All, except the English system, are metric. The five systems include:

1. the English system;
2. the CGS (esu) or centimeter-gram-second, electrostatic unit system;
3. the CGS (emu) or centimeter-gram-second, electromagnetic unit system;
4. the MKS or meter-kilogram-second system;
5. the SI or International System of Units.

1-2   Metric systems

The CGS (esu) system was devised by scientists of the 1800 s to measure the phenomena associated with capacitors and static electricity. The same scientists devised the CGS (emu) system to measure magnetism and electric currents. For many years the two systems of units were used quite indepen-

dently of each other before it was suspected and then definitely established that they were related by — of all things — the speed of light.

The MKS system was then created to unify the CGS (esu) and CGS (emu) systems while retaining the practical units of electricity (volt, ampere, and watt), which had gained favor in the electrical industry. Although this was an important step, it did not quite solve the problem of coherence. Briefly, a system is said to be coherent when calculations involving one unit of A, one unit of B, one unit of C, and so forth, yield one unit of X.

To achieve coherence and to further simplify the system, the ampere (A) was subsequently added as a fourth base unit. This decision, taken in 1954, at the Tenth General Conference of Weights and Measures held at Sèvres, France, produced what was referred to as the MKSA system of units.

In 1960, at the Eleventh General Conference of Weights and Measures, the system of units proposed in 1954 was officially entitled "Système International d'Unités" for which the universal abbreviation is SI. The SI, or International System of Units, is a coherent system.

1-3   Getting used to SI

The official introduction of the International System of Units, and its adoption by most countries of the world, did not, however, eliminate the systems that were previously employed. Just like well-established habits, units become a part of ourselves, and we cannot readily let them go. In effect, it is not easy to switch overnight from yards to meters and from ounces to grams. Even scientists occasionally perform their everyday research in the CGS system of units, and convert to SI only when they publish their work. And this is quite natural, because long familiarity with a unit gives us an idea of its relative size, and how it relates to the physical world.

Nevertheless, the growing importance of SI (particularly in the electrical and mechanical fields) makes it necessary to be familiar with the essentials of this measurement system. Consequently, one

---

* The metric unit of length is spelled either *meter* or *metre*. In Canada, the official spelling is *metre*.

must be able to convert from one system to another in a simple unambiguous way. In this regard, the reader will discover that the conversion charts listed in the Appendix are particularly helpful.

1-4   Features of SI

The SI possesses a number of remarkable features shared by no other system of units.
1. It is a decimal system.
2. It employs many units commonly used in industry and commerce; for example, volt, ampere, kilogram, watt, etc.
3. It is a coherent system that expresses with startling simplicity some of the most basic relationships which occur in electricity, mechanics and electromechanics.
4. It can be used by the research scientist, the technician, the practicing engineer, and by the layman, thereby blending the theoretical and the practical world.

TABLE 1A   BASE UNITS

Quantity	Unit	Symbol
Length	meter	m
Mass	kilogram	kg
Time	second	s
Electric current	ampere	A
Temperature	kelvin	K
Luminous Intensity	candela	cd
Amount of substance	mole	mol

SUPPLEMENTARY UNITS

Plane angle	radian	rad
Solid angle	steradian	sr

Despite these advantages, SI is not the answer to everything. In specialized areas of atomic physics and even in day-to-day work, other units may be more convenient. Thus, we will continue to measure plane angles in degrees, even though the SI unit is the radian. Furthermore, "day" and "hour" will still be used, despite the fact that the SI unit of time is the second.

1-5   Base and derived units of the SI

The foundation of the International System of Units rests upon seven base units and two supplementary units. They are listed in Table 1A.

From these base units, we can derive other units to express quantities such as area, power, force, magnetic flux, and so on. There is really no limit to the number of units we can derive, but some occur so frequently that they have been given special names. Thus, instead of saying that the unit of pressure is the newton per square meter, we use the less cumbersome name, the pascal. Some of the derived units that have special names are listed in Table 1B. Definitions of the base and of the derived units are given in Secs. 1-6 and 1-7.

1-6   Definitions of base units

The following are official definitions of the SI base units. Although not required for an understanding of SI, they illustrate the extraordinary precision associated with this modern system of units. The text in italics is explanatory, and does not form part of the definition.

The **meter** (m) is the length equal to 1 650 763.73 wavelengths in vacuum of the radiation corresponding to the transition between the levels $2 p_{10}$ and $5 d_5$ of the krypton-86 atom.

The **kilogram** (kg) is the unit of mass; it is equal to the mass of the international prototype of the kilogram.

The international prototype of the kilogram is a

particular cylinder of platinum-iridium alloy that is preserved in a vault at Sèvres, France, by the International Bureau of Weights and Measures. Duplicates of the prototype exist in all important standards laboratories in the world. The platinum-iridium cylinder (90 percent platinum, 10 percent iridium) is about 4 cm high and 4 cm in diameter.

The **second** (s) is the duration of 9 192 631 770 periods of the radiation corresponding to the transition between the two hyperfine levels of the ground state of the cesium-133 atom.

A quartz oscillator, tuned to the resonant frequency of cesium atoms, produces a highly accurate and stable frequency.

TABLE 1B DERIVED UNITS WITH SPECIAL NAMES

Quantity	Unit	Symbol
Electric capacitance	farad	F
Electric charge	coulomb	C
Electric conductance	siemens	S
Electric potential	volt	V
Electric resistance	ohm	Ω
Energy	joule	J
Force	newton	N
Frequency	hertz	Hz
Illumination	lux	lx
Inductance	henry	H
Luminous flux	lumen	lm
Magnetic flux	weber	Wb
Magnetic flux density	tesla	T
Power	watt	W
Pressure	pascal	Pa

The **ampere** (A) is that constant current which, if maintained in two straight parallel conductors of infinite length, of negligible circular cross section, and placed 1 meter apart in a vacuum, would produce between these conductors a force equal to 2×10^{-7} newton per meter of length.

The **kelvin** (K) unit of thermodynamic temperature, is the fraction 1/273.16 of the thermodynamic temperature of the triple point of water.

Pure water in an evacuated cell is cooled until ice begins to form. The resulting temperature where ice, water and water vapor co-exist is called the triple point of water and is equal to 273.16 kelvin, by definition. The triple point is equal to 0.01 degree Celsius (°C). A temperature of 0°C is therefore equal to 273.15 kelvins, exactly.

The **candela** (cd) is the luminous intensity, in a given direction, of a source which emits monochromatic radiation of frequency 540×10^{12} hertz and whose radiant intensity in that direction is 1/683 watt per steradian.

The **mole** (mol) is the amount of substance of a system that contains as many elementary entities as there are atoms in 0.012 kilogram of carbon 12.

Note: When the mole is used, the elementary entities must be specified and may be atoms, molecules, ions, electrons, other particles or specified groups of such particles.

1-7   Definitions of derived units

Some of the more important derived units are defined as follows:

The **newton** (N) is that force which gives to a mass of 1 kilogram an acceleration of 1 metre per second per second. *(Hence 1 newton = 1 kilogram meter per second squared.)*

Although the newton is defined in terms of a mass and an acceleration, it also applies to stationary objects and to every application where a force is involved.

The **coulomb** (C) is the quantity of electricity transported in 1 second by a current of 1 ampere. *(Hence 1 coulomb = 1 ampere second.)*

The **farad** (F) is the capacitance of a capacitor between the plates of which there appears a difference of potential of 1 volt when it is charged by a quantity of electricity equal to 1 coulomb. *(Hence 1 farad = 1 coulomb per volt.)*

The **henry** (H) is the inductance of a closed circuit in which an electromotive force of 1 volt is produced when the electric current in the circuit varies uniformly at a rate of 1 ampere per second. *(Hence 1 henry = 1 volt second per ampere.)*

The **hertz** (Hz) is the unit of frequency equal to one cycle per second.

The **joule** (J) is the work done when the point of application of 1 newton is displaced a distance of 1 meter in the direction of the force. *(Hence 1 joule = 1 newton meter.)*

The **ohm** (Ω) is the electric resistance between two points of a conductor when a constant difference of potential of 1 volt, applied between these two points, produces in this conductor a current of 1 ampere, this conductor not being the source of

TABLE 1C   PREFIXES TO CREATE MULTIPLES AND SUBMULTIPLES OF SI UNITS

Multiplier	Exponent Form	Prefix	SI Symbol
1 000 000 000 000 000 000	10^{18}	exa	E
1 000 000 000 000 000	10^{15}	peta	P
1 000 000 000 000	10^{12}	tera	T
1 000 000 000	10^{9}	giga	G
1 000 000	10^{6}	mega	M
1 000	10^{3}	kilo	k
100	10^{2}	hecto	h
10	10^{1}	deca	da
0.1	10^{-1}	deci	d
0.01	10^{-2}	centi	c
0.001	10^{-3}	milli	m
0.000 001	10^{-6}	micro	μ
0.000 000 001	10^{-9}	nano	n
0.000 000 000 001	10^{-12}	pico	p
0.000 000 000 000 001	10^{-15}	femto	f
0.000 000 000 000 000 001	10^{-18}	atto	a

any electromotive force. *(Hence 1 ohm = 1 volt per ampere.)*

The **pascal** (Pa) is the unit of pressure or stress equal to one newton per square meter.

The **siemens** (S) is the unit of electric conductance equal to one reciprocal ohm.

The **tesla** (T) is the unit of magnetic flux density equal to one weber per square meter.

The **volt** (V) is the difference of electric potential between two points of a conducting wire carrying a constant current of 1 ampere, when the power dissipated between these points is equal to 1 watt. *(Hence 1 volt = 1 watt per ampere.)*

TABLE 1D	COMMON UNITS IN MECHANICS		
Quantity	**SI Unit**	**Symbol**	**Note**
Angle	radian	rad	1
Area	square meter	m^2	2
Energy (or work)	joule	J	
Force	newton	N	3
Length	meter	m	
Mass	kilogram	kg	
Power	watt	W	
Pressure	pascal	Pa	4
Speed	meter per second	m/s	
Speed of rotation	radian per second	rad/s	5
Torque	newton-meter	N·m	
Volume	cubic meter	m^3	
Volume	liter	L	6

1. Although the radian is the SI unit of angular measure, we use the degree almost exclusively in this book (1 rad $\cong 57.3°$).
2. Most countries, including Canada (as well as many organizations in the United States), use the spelling "metre" instead of "meter."
3. The newton is a very small force, roughly equal to the force needed to press a door bell.
4. The pascal is a very small pressure equal to 1 N/m^2.
5. In this book, we use the revolution per minute (r/min) to designate rotational speed (1 rad/s = 9.55 r/min).
6. This unit of volume is mainly used for liquids and gases. It is spelled "liter" or "litre." The official spelling in Canada is "litre."

The **watt** (W) is the power that gives rise to the production of energy at the rate of 1 joule per second. *(Hence 1 watt = 1 joule per second.)*

The **weber** (Wb) is the magnetic flux that, linking a circuit of one turn, produces in it an electromotive force of 1 volt as it is reduced to zero at a uniform rate in 1 second. *(Hence 1 weber = 1 volt second.)*

1-8 Multiples and submultiples of SI units

Multiples and submultiples of SI units are generated by adding appropriate prefixes to the units. Thus, prefixes such as kilo, mega, nano, centi, and so on multiply the value of the unit by factors listed in Table 1C. For example: 1 *kilo*ampere = 1000 amperes, 1 *nano*second = 10^{-9} seconds, and 1 *mega*watt = 10^6 watts.

1-9 Commonly used units

Tables 1D, 1E and 1F list some of the more common units encountered in mechanics, thermodynamics, electricity, and magnetism. They contain comments particularly useful to the reader who is not yet familiar with the SI.

1-10 Conversion charts and their use

The relative size of a unit is the key factor that gives us a "feel" for a unit. Familiar units can be converted to units we know less well by using standard conversion tables. But this is strictly an arithmetical process that often leaves us wondering if our calculations are correct.

The conversion charts in the appendix eliminate this problem because they show the relative size of

TABLE 1E	COMMON UNITS IN THERMODYNAMICS		
Quantity	**SI Unit**	**Symbol**	**Note**
Heat	joule	J	
Thermal power	watt	W	
Specific heat	joule per kilogram-kelvin	J/kg·K or J/kg·°C	1
Temperature	kelvin	K	2
Temperature difference	kelvin or degree Celsius	K or °C	1
Thermal conductivity	watt per meter-kelvin	W/m·K or W/m·°C	1

1. A temperature difference of 1 K is exactly equal to a temperature difference of 1°C. The °C is now a recognized SI unit and, in practical calculations, it is often used instead of the kelvin.
2. Thermodynamic, or absolute, temperature is expressed in kelvins. On the other hand, the temperature of objects is usually expressed in °C. The absolute temperature T is related to the Celsius temperature t by the equation: $T = t + 273.15$.

a unit by the position it occupies on the page. The largest unit is at the top, the smallest at the bottom, and intermediate units are ranked in between.

The units are connected by straight lines, each of which bears an arrow and an associated conversion factor. The conversion factor is the ratio of the larger to the smaller of the units so connected, and hence its value is always greater than unity. The arrow always points toward the smaller of the two units.

In Fig. 1-1, for example, five units of length —

the mile, meter, yard, inch and millimeter — are listed in descending order of size and the lines joining them bear an arrow that always points toward the smaller unit. The numbers show the relative size of the connected units; the yard is 36 times larger than the inch, the inch is 25.4 times larger than the millimeter, and so on. With this arrangement, we can convert from one unit to any other by a method we now explain.

Suppose we wish to convert from yards to millimeters. Starting from *yard* in Fig. 1-1, we have to

TABLE 1F	COMMON UNITS IN ELECTRICITY AND MAGNETISM		
Quantity	**SI Unit**	**Symbol**	**Note**
Capacitance	farad	F	
Conductance	siemens	S	1
Electric charge	coulomb	C	
Electric current	ampere	A	
Energy	joule	J	
Frequency	hertz	Hz	2
Inductance	henry	H	
Potential difference	volt	V	
Power	watt	W	
Resistance	ohm	Ω	
Resistivity	ohm-meter	$\Omega \cdot m$	
Magnetic field strength	ampere per meter	A/m	3
Magnetic flux	weber	Wb	
Magnetic flux density	tesla	T	4
Magnetomotive force	ampere	A	5

1. Formerly called "mho."
2. 1 Hz = 1 cycle per second.
3. 1 A/m = 1 ampere turn per meter.
4. 1 T = 1 Wb/m^2.
5. What was formerly called an ampere turn is now simply called ampere: 1 A = 1 ampere turn.

move downward and in the direction of the arrows for both lines (36 and 25.4) until we reach *millimeter*.

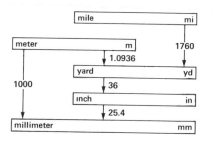

Figure 1-1   Units of length.

Conversely, if we wish to convert from millimeters to yards, we start at *millimeter* and move upward against the direction of the arrows until we reach *yard*. To convert from one unit to another, we apply the following rules:

1. *If, in going from one unit to another, we move in the direction of the arrow, we multiply by the associated conversion factor.*

2. *Conversely, if we move against the arrow, we divide.*

In moving from one unit to another, we can follow any path we please; the conversion result is always the same.

The rectangles bearing SI units extend slightly towards the left, to distinguish them from other units. Each rectangle bears the symbol for the unit, as well as the name of the unit written out in full.

Example 1-1:
Convert 2.5 yards to millimeters.

Solution:
Starting from *yard* and moving toward *millimeter*, we always move in the direction of the arrows. We must therefore *multiply* the numbers associated with each line:
2.5 yd = 2.5 x 36 x 25.4 millimeters
       = 2286 mm.

Example 1-2:
Convert 2000 meters into miles.

Solution:
Starting from *meter* and moving toward *mile*, we move once with, and once against the direction of the arrows. Consequently, we have:

$$2000 \text{ meters} = 2000 \times 1.0936 \div 1760 \text{ miles}$$
$$= \frac{200 \times 1.0936}{1760}$$
$$= 1.24 \text{ mi}$$

Example 1-3:
Convert 777 calories to kilowatthours.

Solution:
Referring to the ENERGY chart, we now move past four arrows, three of which point opposite to the path we are following. Consequently:
777 calories
$$= 777 \times 4.184 \div 1000 \div 1000 \div 3.6 \text{ kW·h}$$
$$= 9.03 \times 10^{-4} \text{ kW·h}.$$

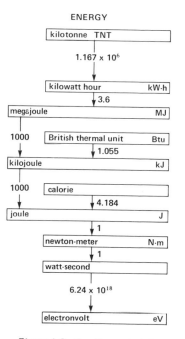

Figure 1-2   See Example 1-3.

QUESTIONS AND PROBLEMS

1-1   Name the seven base units of the International System of Units.

1-2   Name five derived units of the SI.

1-3   Give the symbols of the seven base units, with particular attention to capitalization.

1-4   Why are some derived units given special names?

1-5   What are the SI units of force, pressure, energy, power and frequency?

1-6   Give the appropriate prefix for the following multipliers: 100, 1000, 10^6, 1/10, 1/100, 1/1000, 10^{-6}, 10^{-9}, 10^{15}.

Express the following SI units in symbol form:

1-7	megawatt	1-21	millitesla
1-8	terajoule	1-22	millimeter
1-9	millipascal	1-23	revolution
1-10	kilohertz	1-24	megohm
1-11	gigajoule	1.25	megapascal
1-12	milliampere	1.26	millisecond
1-13	microweber	1.27	picofarad
1-14	centimeter	1.28	kilovolt
1-15	liter	1.29	megaampere
1-16	milligram	1.30	kiloampere
1-17	microsecond	1.31	kilometer
1-18	millikelvin	1-32	nanometer
1-19	milliradian	1-33	milliliter

State the SI unit for the following quantities, and write the symbol:

1-34	rate of flow	1-38	density
1-35	frequency	1-39	power
1-36	plane angle	1-40	temperature
1-37	magnetic flux	1-41	mass

Give the name of the SI units that correspond to the following units:

1-42	Btu	1-45	inch
1-43	horsepower	1-46	angstrom
1-44	line of flux	1-47	cycle per second

1-48	gauss	1-56	mho
1-49	line per square inch	1-57	pound force per square inch
1-50	°F		
1-51	bar	1-58	revolution
1-52	pound mass	1-59	degree
1-53	pound force	1-60	oersted
1-54	kilowatt-hour	1-61	ampere turn
1-55	gallon per minute		

The following problems can be **solved** by using the conversion charts:

1-62   10 square meters to square yards.

1-63   250 MCM to square millimeters.

1-64   1645 square millimeters to square inches.

1-65   13 000 circular mils to square millimeters.

1-66   640 acres to square kilometers.

1-67   81 000 watts to Btu per second.

1-68   33 000 foot pounds force per minute to kilowatts.

1-69   250 cubic feet to cubic meters.

1-70   10 foot pound force to microjoules.

1-71   10 pound force to kilogram force.

1-72   60 000 lines per square inch to teslas.

1-73   1.2 teslas to kilogauss.

1-74   50 ounces to kilograms.

1-75   76 oersteds to amperes per meter.

1-76   5 000 meters to miles.

1-77   80 ampere hours to coulombs.

1-78   25 pound force to newtons.

1-79   25 pound mass to kilogram.

1-80   3 tonnes to pounds mass.

1-81   100 000 lines of force to webers.

1-82   0.3 pounds per cubic inch to kilograms per cubic meter.

1-83   2 inches of mercury to millibars.

1-84   200 pounds per square inch to pascals.

1-85   70 pounds force per square inch to newtons per square meter.

1-86   15 revolutions per minute to radians per second.

1-87   A temperature of 120°C to kelvins.

1-88   A temperature of 200°F to kelvins.

1-89   A temperature difference of 120° Celsius to kelvins.

2

FUNDAMENTALS
OF ELECTRICITY
AND MAGNETISM

This chapter briefly reviews some of the fundamentals of electricity and magnetism. There is no particular order or sequence in the subject matter, and no attempt has been made to cover all fundamentals. Rather, we want to clarify the notation used throughout this book to designate voltages and currents. Some of the other topics treated here will also provide the reader with a sense of continuity for subjects covered in ensuing chapters.

2-1 Conventional and electron current flow

Consider the dry cell shown in Fig. 2-1, having one positive (+) and one negative (–) terminal. The difference of potential between them (measured in volts) is due to an excess of electrons at the negative terminal compared to the positive terminal.

If we connect a wire across the terminals, the potential difference causes an electric current to flow in the circuit. This current is composed of a steady stream of electrons that come out of the negative terminal, move along the wire, and reenter the cell by the positive terminal (Fig. 2-2).

Before the electron theory of current flow was fully understood, scientists of the 17th century arbitrarily decided that current in a conductor flows from the positive (higher) terminal to the negative (lower) terminal (Fig. 2-3). This so-called *conventional current flow* is still used today, and is the accepted direction of current flow in electrical power technology.

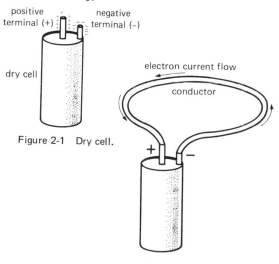

Figure 2-1   Dry cell.

Figure 2-2   Electron flow.

11

In this book, we use the conventional current flow, but it is worth recalling that the actual electron flow is opposite to the conventional current flow.

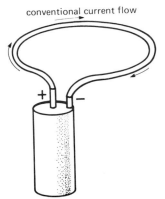

Figure 2-3   Conventional current flow.

2-2   Distinction between sources and loads

It is sometimes very important to identify the sources and loads in an electric circuit. By definition, a *source delivers* electrical power whereas a *load absorbs* it. Every circuit element that carries a current can be classified as either a source or a load. To distinguish the one from the other, consider two boxes A and B containing unknown devices and components. The boxes are connected by two conductors, as shown in Fig. 2-4. Using appropriate instruments, suppose we can find the *instantaneous* polarity of the voltage and the *instantaneous* direction of conventional current flow. How can we tell which box is the source and which box is the load?

We can identify them by means of the following rule:

- **Whenever current flows out of a positive terminal, the circuit element is a source;**
- **Whenever current flows into a positive terminal, the circuit element is a load.**

According to this rule, box A is a source and box B is a load.

2-3   Sign notation for voltages

In arithmetic, we use the symbols (+) and (−) to describe addition and substraction. In electricity and mechanics, we broaden the meaning to indicate the *direction* of an electric current, mechanical force, rotational speed, etc., compared to an arbitrary chosen direction. For example, if the speed of a motor changes from + 1000 r/min to − 400 r/min, it means that the direction of rotation has reversed. This interpretation of (+) and (−) is frequently met in the chapters that follow.

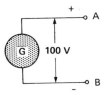

Figure 2-5   Double subscript notation.

We now describe a system of notation that enables us to indicate the polarity of voltages and the direction of currents. Figure 2-5 shows a source G having a positive terminal A and a negative terminal B. Terminal A is positive *with respect to* terminal B. Similarly, terminal B is negative *with respect to*

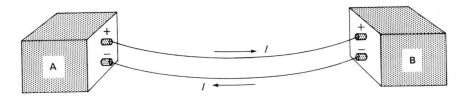

Figure 2-4   Distinction between a source and a load.

terminal A. Note that terminal A is not positive by itself: it is only positive with respect to B.

The potential difference and the relative polarities of terminals A and B may be designated by the *double-subscript* notation, as follows:

- E_{AB} =+ 100 V, which reads: **the voltage between A and B is 100 V and A is positive with respect to B**

- E_{BA} =– 100 V, which reads: **the voltage between A and B is 100 V and B is negative with respect to A**

For example, if we know that the generator in Fig. 2-6 has a voltage E_{21} = – 300 V, we know that the voltage between the terminals is 300 V and that terminal 2 is negative with respect to terminal 1.

During the interval from 1 to 2 seconds, E_{21} is negative, therefore, terminal 2 is *negative* with respect to terminal 1. The actual voltages and polarities at 0.5, 1.5, and 2.17 seconds are clearly shown by insets I, II, III of Fig. 2-7.

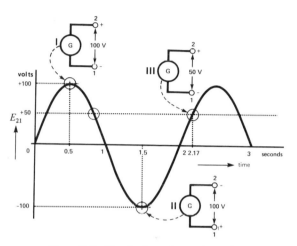

Figure 2-7   Graph of an alternating voltage having a peak of 100 V.

Figure 2-6   If E_{21} = – 100 V terminal 2 is negative with respect to terminal 1.

2-4   Graph of an alternating voltage

In the chapters that follow, we encounter sources whose voltages change polarity periodically. Such *alternating* voltages may be represented by means of a graph (Fig. 2-7). The vertical axis indicates the voltage at each instant, while the horizontal axis indicates the corresponding time. Voltages are positive when they are above the horizontal axis and negative when they are below. Figure 2-7 shows the voltage produced by the generator of Fig. 2-6.

Starting from zero, E_{21} gradually increases to + 100 volts and again falls to zero at the end of one second. During this one-second interval, terminal 2 is positive with respect to terminal 1, because E_{21} is positive (above the horizontal axis).

2-5   Positive and negative currents

We also make use of positive and negative signs to indicate the direction of current flow. The signs are allocated with respect to a reference direction given on the circuit diagram. For example, the cur-

Figure 2-8   Circuit element.

rent in a circuit element such as a resistor (Fig. 2-8) may flow from X to Y or from Y to X. If one of these two directions is considered to be positive (+), the other is negative (–).

The *positive* direction is shown *arbitrarily* by means of an arrow (Fig. 2-9). Thus, if a current of 2 A flows from X to Y, it flows in the positive direction and is designated by the symbol + 2 A. Conversely, if current flows from Y to X (direction opposite to that of the arrow), it is designated by the symbol − 2 A.

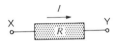

Figure 2-9   Circuit element showing positive direction of current flow.

Example 2-1:
The current in a resistor R varies according to the graph shown in Fig. 2-10. Interpret the meaning of this curve.

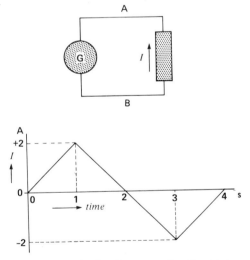

Figure 2-10   Graph representing the value and direction of current flow.

Solution:
According to the graph, the current increases from zero to + 2 A during the interval from 0 to 1 second. Because it is positive, the current flows from B to A in the resistor (direction of the arrow). During the interval from 1 to 2 seconds, the current decreases from + 2 A to zero, but it still circu-

lates from B to A in the resistor. Between 2 and 3 seconds, the current increases from zero to − 2 A and, because it is negative, it really flows in a direction opposite to that of the arrow, that is, from A to B in the resistor.

2-6   Another notation for voltages

Although we can represent the value and the polarity of voltages by the double-subscript notation (E_{12}, E_{ab}, etc.), we often prefer to use the *sign notation*. It consists of designating the voltage by a symbol (E_1, V, etc.) and identifying one of the terminals by a positive (+) sign. For example, Fig. 2-11 shows a source E_1 in which one of the terminals is *arbitrarily* marked with a positive (+) sign.

Figure 2-11   Sign notation.

With this notation, the following rules apply:
- **If we specify that E_1 = + 10 V, this signifies that the real polarity of the terminals corresponds to that indicated in the diagram. Furthermore, the voltage across the terminals is 10 V.**
- **Conversely, if E_1 = − 10 V, the real polarity of the terminals is the reverse of that shown on the diagram. The voltage across the terminals is 10 V.**

Example 2-2:
The circuit of Fig. 2-12 consists of three sources V_1, V_2 and V_3, each having a terminal marked with a positive (+) sign. Determine the actual value and polarity of the voltage across each source, knowing that V_1 = − 4 V, V_2 = + 10 V and V_3 = − 40 V.

Solution:
The true values and polarities are shown in Fig.

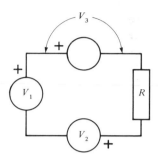

Figure 2-12   Circuit of Example 1-2.

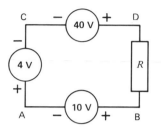

Figure 2-13   Solution of Example 2-2.

2-13. Directing our attention to terminal A, it seems impossible that it should be both positive (+) and negative (−). However, we must remember that point A is neither inherently positive or negative. It only has a polarity with respect to points B and C respectively. In effect, point **A** is *negative* with respect to point B and *positive* with respect to point C. This is why A carries both a positive and a negative sign.

2-7   Potential level

We have just described two ways of representing voltages in a circuit. We now introduce a third method that is particularly useful in circuits dealing with power electronics. The method is based upon the concept of potential levels.

To understand the operation of electronic cir-

cuits, it is useful to imagine that individual terminals have a potential level with respect to a reference terminal. The reference terminal is any convenient point chosen in a circuit; it is assumed to have zero electric potential. The potential level of all other points is then measured with respect to this zero reference terminal. In graphs, the reference level is shown as a horizontal line having a potential of 0 V.

Consider, for example, the circuit of Fig. 2-14, composed of an 80 V battery connected in series with an ac source having a crest voltage of 100 V. Of the three possible terminals, let us choose terminal 1 as the reference point. The potential level of this terminal is therefore shown by a horizontal line **1** (Fig. 2-15).

Consider now the potential level of terminal 2. The difference of potential between terminals 1

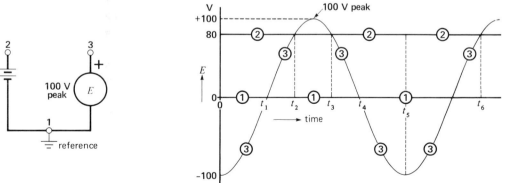

Figure 2-14   Potential level method of
representing voltages.

Figure 2-15   Potential levels of terminals 1, 2 and 3.

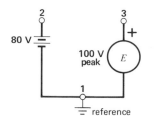

Figure 2-14   Potential level method of
representing voltages.

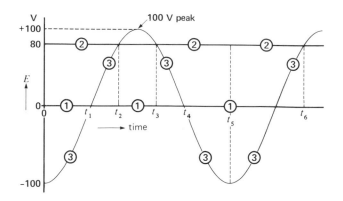

Figure 2-15   Potential levels of terminals 1, 2 and 3.

and 2 is always 80 V and terminal 2 is positive with respect to terminal 1. The level of this terminal is therefore indicated by a second horizontal line **2** placed 80 V above line **1**.

Now consider terminal 3. Voltage E between terminals 1 and 3 is alternating and we assume that its initial value is 100 V with terminal 3 negative with respect to terminal 1. Because E is alternating, the potential of terminal 3 is first negative, then positive, with respect to terminal 1. The changing level is shown by curve **3**. Thus, during the interval from 0 to t_1, the level of point **3** is below the level of point **1** which indicates that terminal 3 is negative with respect to terminal 1. During the interval t_1 to t_4, the polarity reverses so that the level of **3** is now above line **1**. Terminal 1 is now negative with respect to terminal 3, because line **1** is below curve **3**.

These simple explanations add nothing to what we already know. However, this potential-level method enables us to immediately determine the instantaneous voltages between any two terminals in a circuit, as well as their relative polarities. For example, during the interval from t_2 to t_3, terminal 3 is positive with respect to terminal 2, because curve **3** is above line **2**. The voltage between these terminals reaches a maximum of 20 V during this

interval. Then from t_3 to t_6, terminal 3 is negative with respect to terminal 2 and the voltage reaches a maximum value of 180 V at instant t_5.

We could have chosen another terminal as a reference terminal. Thus, in Fig. 2-16, we chose terminal 3 and, as before, we represent the zero potential of this reference terminal by a horizontal line (Fig. 2-17). Knowing that E is an alternating voltage and that terminal 1 is initially positive with respect to terminal 3 (Fig. 2-15), we can draw curve **1**.

To determine the level of terminal 2, we know that it is always 80 V positive with respect to terminal 1. Consequently, we draw curve **2**, so that it is always 80 V above curve **1**. By so doing, we automatically establish the level of terminal 2 with respect to terminal 3.

Figures 2-15 and 2-17 do not have the same appearance; however, at every instant, the relative polarities and potential differences between terminals are identical. From an electrical point of view, the two figures are identical and we invite the reader to check the truth of this statement.

In electronic circuits, we usually try to select a reference terminal so that it is easy to follow the voltages we are interested in.

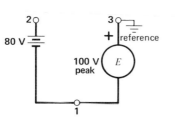

Figure 2-16   Changing the reference terminal.

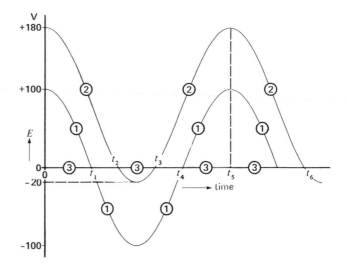

Figure 2-17   The relative potential levels are the same as in Fig. 1-15.

2-8   Faraday's law of electromagnetic induction

In 1831, while pursuing his experiments, Joseph Faraday made one of the most important discoveries in electromagnetism. Now known as Faraday's Law of electromagnetic induction, it revealed a fundamental relationship between the voltage and flux in a circuit. Faraday's law states that:

1. **If the flux linking a loop (or turn) varies as a function of time, a voltage is induced between its terminals.**
2. **The value of the induced voltage is proportional to the rate of change of flux.**

By definition, and according to the SI system of units, when the flux inside a loop varies at the rate of 1 weber per second, a voltage of 1 V is induced between its terminals. Consequently, if a flux varies inside a coil of N turns, the voltage induced is given by:

$$E = N \frac{\Delta \Phi}{\Delta t}$$ (2-1)

where

E = induced voltage [V]
N = number of turns in the coil

$\Delta \Phi$ = change of flux inside the coil [Wb]
Δt = time interval during which the flux changes [s]

Faraday's law of electromagnetic induction opens the door to a host of practical applications and establishes the basis of operation of transformers, generators and alternating current motors.

Example 2-3:
A coil of 2000 turns surrounds a flux of 5 mWb produced by a permanent magnet (Fig. 2-18). The magnet is suddenly drawn away causing the flux inside the coil to drop to 2 mWb in 1/10 of a second. What is the average voltage induced?

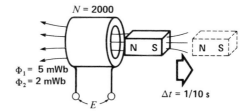

Figure 2-18   Voltage induced by a moving magnet.

Solution:

The flux change is:

$\Delta\Phi$ = (5 mWb – 2 mWb) = 3 mWb

Because this change takes place in 1/10 of a second (Δt), the average induced voltage is:

$$E = N\frac{\Delta\Phi}{\Delta t} = 2000 \times \frac{3}{1000 \times 1/10} = 60\ \text{V}$$

2-9 Voltage induced in a conductor

In many motors and generators, the coils move with respect to a constant flux. The relative motion produces a change in the flux linking the coils and, consequently, a voltage is induced, according to Faraday's law. However, in this special (although common) case, it is easier to calculate the induced voltage with reference to the *conductors*, rather than with reference to the coil itself. In effect, whenever a conductor "cuts" a magnetic field, voltage is induced across its terminals. The value of the induced voltage is given by:

$$\boxed{E = Blv} \qquad (2\text{-}2)$$

where

E = induced voltage [V]
B = flux density [T]
l = active length of the conductor in the magnetic field [m]
v = relative speed of the conductor [m/s]

Example 2-4:

The conductors of a large generator have a length of 2 m and are cut by a field of 0.6 teslas, moving at a speed of 100 m/s (Fig. 2-19). Calculate the voltage induced in each conductor.

Solution:

According to Eq. 2-2, we find:

$E = Blv$
  $= 0{,}6 \times 2 \times 100 = 120\ \text{V}$

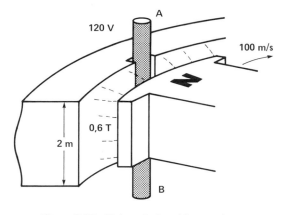

Figure 2-19   Voltage induced in a stationary conductor. Example 2-4.

2-10 Voltage induced in a rectangular conductor

Following the discovery of Faraday's law of electromagnetic induction, scientists and technicians of the 19th century did not wait long to invent and build all kinds of mechanical machines to generate electricity. The operating principle of these machines is always based upon the relative motion of a rectangular coil with respect to a magnetic field.

Consider a permanent magnet NS revolving at constant speed inside a stationary iron ring (Fig. 2-20). The ring (or stator) reduces the reluctance of the magnetic circuit; consequently, the flux density in the air gap is greater than if the ring were absent. A rectangular conductor ABCD is mounted inside the ring, but insulated from it (Fig. 2-21).

As the magnet turns, it sweeps across conductors AB and CD, inducing a voltage in them. The sum of the voltages appears across terminals AD. The terminal voltage E_{AD} is maximum when the poles are in the position of Fig. 2-22 because the flux cuts both conductors. On the other hand, the voltage is zero when the poles are in the position of Fig. 2-23 because flux does not cut the conductors at this moment.

If we plot E_{AD} as a function of the angle of ro-

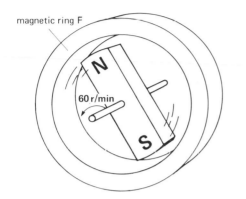

Figure 2-20 Elementary alternator with revolving field.

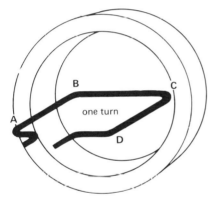

Figure 2-21 Rectangular conductor.

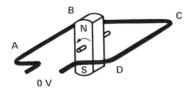

Figure 2-22 Induced voltage is maximum.

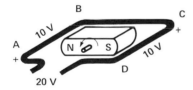

Figure 2-23 Induced voltage is zero.

tation, we obtain the curve shown in Fig. 2-24. This is an alternating voltage having a peak value of 20 V. Machines that produce such voltages are called alternating current generators or *alternators*.

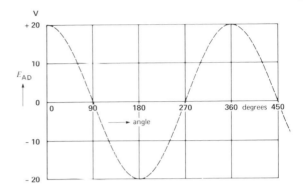

Figure 2-24 Voltage as a function of the angle of rotation.

2-11 Curve of induced voltage as a function of time

The magnet in our example revolves at uniform speed, therefore each angle of rotation corresponds to a specific interval of time. Because the magnet makes one turn per second, the angle of 360° in Fig. 2-24 corresponds to an interval of one second. Consequently, we can also represent the induced voltage as a function of time (Fig. 2-25).

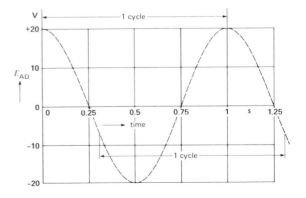

Figure 2-25 Voltage as a function of time.

2-12 Cycle and frequency

A *cycle* is defined as the interval of time between two consecutive voltages having the same polarity and changing at the same rate. Thus, referring to Fig. 2-25, the alternating voltage completes one cycle in 1 second. Note that a cycle may start anywhere on the volt-time curve.

The *frequency* of a periodic quantity designates the number of cycles it makes per second. Thus, if an ac voltage completes one cycle in 1/60th of a second, its frequency is 60 cycles per second.

The SI unit of frequency is the hertz (Hz). It is equal to one cycle per second. A frequency of 60 Hz corresponds therefore to a frequency of 60 cycles per second. The frequency of industrial voltages is fixed; it is 60 Hz in North America and 50 Hz in most other countries of the world.

2-13 Alternator having a revolving coil

Instead of using a revolving field, we can build an alternator in which the permanent magnet is fixed and the coil revolves (Fig. 2-26). As long as the relative motion of the coil with respect to the magnet

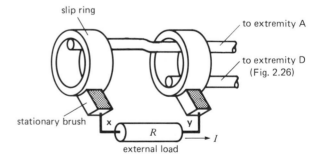

Figure 2-26   Elementary alternator with stationary field.

is the same as before, the value, polarity and waveshape of the induced voltage will be identical. However, we cannot make a direct connection between a stationary load and a revolving coil. Consequently, we have to use two stationary brushes

sliding on two revolving slip rings. The slip rings are connected to each end of the coil and are mounted on the shaft, but insulated from it (Fig. 2-27).

Although revolving-coil alternators are only built in small sizes (less than 10 kW), they help us understand the basic construction of direct current generators.

2-14 Direct current generator

If the brushes in Fig. 2-27 could be switched from one slip-ring to the other every time the polarity was about to change, we would obtain a voltage of constant polarity across the load. Brush X would always be positive and brush Y, negative. We can obtain this result by using a *commutator* (Fig. 2-28). A commutator in its most simple form is composed of a slip-ring that is cut in half, each half insulated from the other, as well as from the shaft. One segment is connected to coil-end A and the other to coil-end B. The commutator revolves with the coil and the induced voltage between the segments is picked up by two stationary brushes X and Y.

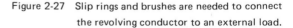

Figure 2-27   Slip rings and brushes are needed to connect the revolving conductor to an external load.

The voltage between brushes X and Y pulsates but never changes polarity (Fig. 2-29). In effect, the alternating voltage in the coil is *rectified* by the commutator, which acts as a mechanical reversing switch.

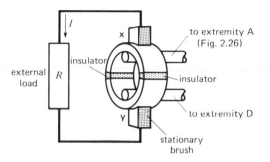

Figure 2-28   A commutator behaves like a mechanical reversing switch.

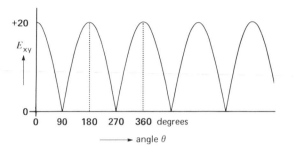

Figure 2-29   Pulsating dc voltage.

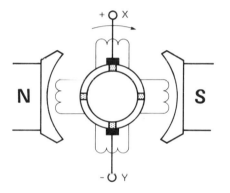

Figure 2-30   Dynamo having a four-segment commutator.

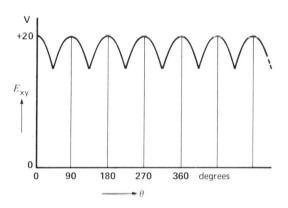

Figure 2-31   Output voltage of the dynamo.

Owing to the constant polarity between the brushes, the current in the external load always flows in the same direction. The machine represented in Fig. 2-28 is called a direct-current generator, or dynamo.

2-15   Improving the waveshape

We can improve the pulsating dc voltage by using four coils and four segments, as shown in Fig. 2-30. The resulting waveshape is given in Fig. 2-31. The voltage still pulsates, but it never falls to zero; it is much closer to a perfect dc voltage.

By increasing the number of coils and segments, we can obtain a dc voltage that is essentially constant. Modern dc generators produce voltages having a ripple of less than 5 percent.

2-16   Difference between an alternator and a dc generator

The elementary alternators and dc generators studied in the previous articles are essentially built the same way. In each case, a coil rotates between the poles of a magnet and an ac voltage is induced in the coil. The machines only differ in the way the coils are connected to the external circuit: alternators carry slip-rings while dc generators require a commutator (Fig. 2-32). We sometimes build small machines which carry both slip-rings and a commutator. Such machines can function simultaneously as alternators and as dc generators.

We study the construction and behavior of practical commercial alternators and dc generators in subsequent chapters.

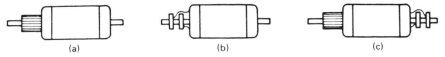

Figure 2-32   The same windings are used to make either an alternator or a dynamo.

2-17 Magnetic circuit

An *electric* circuit is composed of a group of interconnected elements that carry an electric current. By analogy, a *magnetic circuit* is composed of a group of elements which carry a magnetic flux. Figure 2-33 shows a magnetic circuit composed of iron, nickel and air carrying a flux Φ. The coil possesses N turns and carries a current I. It produces a magnetomotive force U given by:

$$U = NI \qquad (2\text{-}3)$$

where
   U = magnetomotive force of the coil [A]
   N = number of turns on the coil [1]
   I = current in the coil [A]

The SI unit of magnetomotive force is the ampere [A]; it is equal to the magnetomotive force (mmf) created when a current of 1 A flows in a coil of 1 turn.*

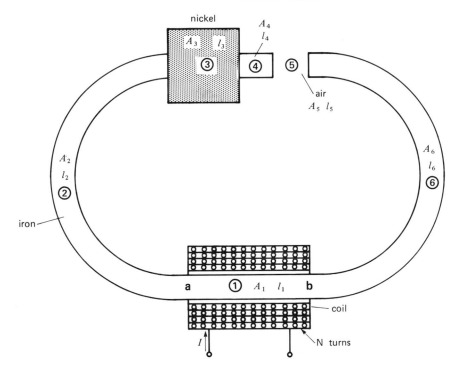

Figure 2-33   Magnetic circuit composed of iron, nickel, and air.

* Although the SI unit of mmf is the ampere, we shall continue to use the former designation "ampere turn" whenever it leads to better understanding.

The six components of the magnetic circuit have lengths $l_1, l_2, \ldots l_6$, and cross sections $A_1, A_2, \ldots A_6$, as shown in Figure 2-33. The mmf creates a magnetic flux (not shown) that passes in a closed loop through the iron, nickel and air. The flux density varies from point to point, depending on the cross-section of the flux path.

The SI unit of flux is the weber [Wb].

The SI unit of flux density is the tesla [T]. It is equal to 1 Wb/m². The flux density ranges from 1.6 T, in the core of large transformers, to 0.4 T, in the air gap of small motors.

2-18 Ampere's circuital law

The mmf developed by the coil produces a magnetic difference of potential, equal to NI ampere turns. This difference of potential acts upon the magnetic circuit, producing potential drops across each of the six component parts. Thus, there is a magnetic potential drop U_1 across component **1** another potential drop U_2 across component **2**, and so on until we come to a final potential drop U_6 across component **6**. According to *Ampere's circuital law*, the sum of these magnetic potential drops is equal to the mmf developed by the coil. Thus, we have:

$$U_1 + U_2 + U_3 + U_4 + U_5 + U_6 = U = NI \qquad (2\text{-}4)$$

We immediately recognize that this law is similar to Kirchhoff's law concerning voltage drops in a closed electric circuit. Indeed, we can draw the magnetic circuit as if it were an electrical circuit (Fig. 2-34). In this circuit, flux Φ is similar to a current, magnetomotive force NI is similar to the emf developed by a battery, and the resistances represent the so-called *reluctance* of each component.

We could, in fact, apply all the laws of electric circuits to solve magnetic circuits if the magnetic materials had a constant reluctance. Unfortunately, the reluctance varies with flux density. Because of this nonlinearity, we cannot apply standard circuit-solving techniques to magnetic circuits.

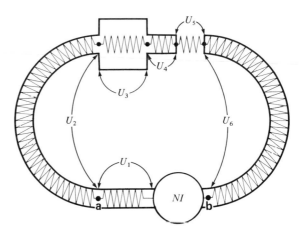

Figure 2-34   A magnetic circuit can be represented by an electric circuit.

2-19 Changing the reluctance of a magnetic circuit

The magnetic potential difference between the opposite faces **a**, **b** of the coil in Figure 2-33 is not NI, but rather $(NI - U_1)$, where U_1 is the potential drop in component **1** inside the coil. The reluctance of this component behaves like the internal resistance of a battery, causing its terminal voltage to be less than the generated emf.

We can reduce the potential drop in component **1** by using a material having a very high magnetic permeability. If this is done, the magnetic potential across the rest of the magnetic circuit will be almost exactly equal to NI. Finally, if we replace all the nickel and iron components by a material having infinite permeability, the entire mmf developed by the coil appears across the air gap. Consequently, by using materials of high permeability, we can concentrate the coil mmf so that it acts across a very short distance.

2-20 Magnetic field intensity H, flux density B

Each component of a magnetic circuit is subjected to a magnetic field intensity H, given by:

$$H = U/l \qquad (2\text{-}5)$$

where

H = magnetic field intensity [A/m]

U = difference of magnetic potential across the component [A] (or ampere turn)

l = length of the component [m]

For example, the magnetic field intensity acting across component **1** in Figure 2-33 is U_1/l_1 and that across component **5** is U_5/l_5.

As to the flux density, it is given by;

$$B = \Phi/A \qquad (2\text{-}6)$$

where

B = flux density [T]

Φ = flux in the component [Wb]

A = cross section of the component [m²]

Thus, the flux density in component **3** is $B = \Phi/A_3$.

A specific relationship always exists between the flux density (B) and the magnetic field intensity (H) of any magnetic material. This relationship is usually expressed graphically by the B-H curve of the material.

2-21 B-H curve of a vacuum

In a vacuum, the flux density B is proportional to the magnetic field H, and is expressed by the equation:

$$B = \mu_0 H \qquad (2\text{-}7)$$

where

B = flux density [T]

H = magnetic field [A/m]

μ_0 = magnetic constant [= $4\pi \times 10^{-7}$] *

---

* The complete expression for μ_0 is $4\pi \times 10^{-7}$ henry/meter.

In the SI, the magnetic constant is fixed, by definition. It has a numerical value of $4\pi \times 10^{-7}$ or 1/800 000 approximately. This enables us to write Eq. 2-7 in the approximate form:

$$H = 800\ 000\ B \qquad (2\text{-}8)$$

The B-H curve of a vacuum is a straight line which never saturates, no matter how great the flux density may be (Fig. 2-35). The curve shows that a magnetic field of 800 A/m produces a flux density of 1 millitesla.

Nonmagnetic materials such as copper, paper, rubber, air, and so forth, have B-H curves almost identical to that of a vacuum.

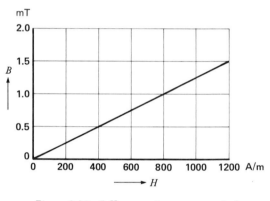

Figure 2-35 B-H curve of a vacuum and of nonmagnetic materials.

2-22 B-H curve of a magnetic material

The flux density in a magnetic material also depends upon the magnetic field intensity. Its value is given by:

$$B = \mu_0 \mu_r H \qquad (2\text{-}9)$$

where B, μ_0 and H have the same significance as before, and μ_r is the *relative permeability* of the material.

Unfortunately, μ_r is not constant, but varies with the flux density in the material. Consequently, the relationship between B and H is not linear, and this makes Eq. 2-9 rather impractical. We usually show the relationship by means of a B-H saturation curve. Thus, Fig. 2-36 shows typical saturation curves of three materials commonly used in electrical machines: silicon iron (1%), cast iron, and cast steel. The curves show that a magnetic field of 2000 A/m produces a flux density of 1.4 T and 0.5 T, in cast steel and cast iron, respectively.

2-23 Determining the relative permeability

The relative permeability μ_r of a material is the ratio of the flux density in the material to the flux density that would be produced in a vacuum, using the same magnetic field intensity H.

Given the saturation curve of a magnetic material, it is easy to calculate the relative permeability using the approximate equation:

$$\boxed{\mu_r \approx 800\,000\, B/H} \qquad (2\text{-}10)$$

where

B = flux density in the magnetic material [T]
H = corresponding magnetic field strength [A/m]

Example 2-5:
Determine the permeability of silicon iron (1%) at a flux density of 1.4 T.
Solution:
Referring to the saturation curve (Fig. 2-36), we see that a flux density of 1.4 T requires a magnetic field intensity of 1000 A/m. Consequently:

μ_r = 800 000 B/H
     = 800 000 x 1.4/1000 = 1120

At this flux density, the steel is 1120 times more permeable than a vacuum (or air).

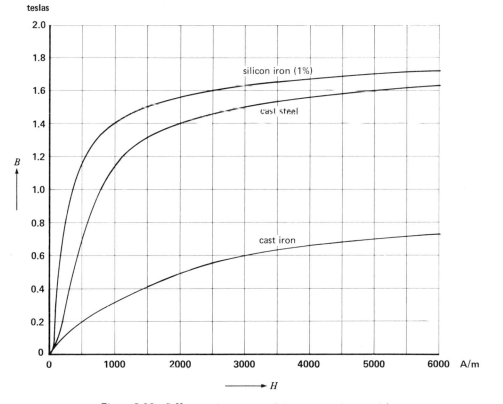

Figure 2-36   *B-H* saturation curves of three magnetic materials.

2-24 Solution of simple magnetic circuits

In practice, we usually have to solve two types of problems:

1. Calculate the mmf required to produce a given flux;
2. Given the mmf, calculate the flux produced in a given magnetic circuit.

Example 2-6:

Magnetomotive force U_g required for an air gap: We want to produce a flux density of 0.7 T in an air gap having a length of 2 mm (Fig. 2-37). Calculate the mmf required.

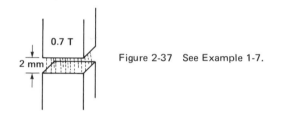

Figure 2-37   See Example 1-7.

Solution:

According to Eq. 2-8, we have:

$H = 800\,000\,B$

$= 800\,000 \times 0.7 = 560$ kA/m

Thus, to obtain a density of 0.7 T, we must produce a magnetic field of 560 kA/m. Because the air gap has a length of 2 mm, the mmf required is:

$U_g = Hl$

$= 560\,000 \times 2/1000 = 1120$ A

Example 2-7:

Magnetomotive force U_i required for an iron core: Consider a core made of silicon-iron (1%), whose dimensions are given in Figure 2-38. The average

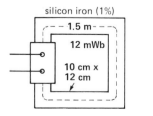

Figure 2-38
See Example 1-8.

length of the magnetic path is 1.5 m. Calculate the mmf of the coil to produce a flux of 12 mWb in the core.

Solution:

The cross section of the core is: $A = 10 \times 12 = 120$ cm^2 = 0.012 m^2.

$B = \Phi/A = 0.012/0.012 = 1$ T

Referring to the saturation curve (Fig. 2-36), we see that a magnetic field intensity of 300 A/m is required to produce this flux density. The coil must therefore produce a mmf of 300 A for every meter of length of the magnetic circuit. Because the average length of the circuit is 1.5 m, the mmf of the coil must be:

$U_i = 300 \times 1.5 = 450$ A or 450 ampere turns.

Example 2-8:

Magnetomotive force required for a series circuit: Calculate the mmf required to produce the same flux in the core of Example 2-7 if the circuit has an air gap of 1.5 mm (Fig. 2-39).

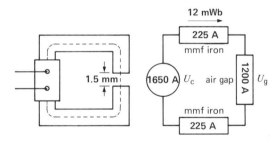

Figure 2-39   See Example 1-9.

Solution:

This is a series magnetic circuit because two paths of different reluctance (iron and air) carry the same flux (analogy with two resistors in series). The mmf of the coil is therefore equal to the sum of the mmf required (a) for the air gap, and (b) for the iron core.

a. U_g **for the air gap:** The flux density in the gap is the same as in the core: 1 T. Using Eq. 2-8,

we find that the mmf for the air gap is:
$U_g = Hl = 800\,000\,Bl$
$= 800\,000 \times 1\,T \times 0.0015\,m = 1200\,A$

b. U_i **for the iron core:** The length of the magnetic path in the iron is essentially the same as before; consequently, U_i is again 450 A. The coil must therefore develop a mmf:
$U_c = 1200 + 450 = 1650\,A$

Example 2-8 shows the great importance of air gaps in magnetic circuits. In effect, although the path is 1000 times longer through steel than through air, the air still requires more ampere turns. In rotating electrical machines, we try to keep the air gap between the rotor and stator as short as possible to keep the mmf of the poles to a minimum. By reducing the mmf, we can reduce the coil size, which enables us to reduce the dimensions and cost of the machine.

2-25 Lorentz force on a conductor

When a current-carrying conductor is correctly set in a magnetic field, it is subjected to a force which we call *electromagnetic force*, or Lorentz force. This force is of crucial importance because it constitutes the basis of operation of motors, of generator and of many electrical instruments. The magnitude of the force depends upon the orientation of the conductor with respect to the direction of the field. The force is greatest when the conductor is perpendicular to the field (Fig. 2-40) and zero

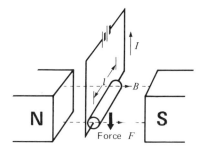
Figure 2-40   Force on a conductor.

when it is parallel to it (Fig. 2-41). Between these two extremes, the force has intermediate values.

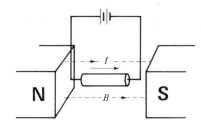
Figure 2-41   Force = 0.

The maximum force acting on a straight conductor is given by:

$$F = BlI$$ (2-11)

where
F = force acting on the conductor [N]
B = flux density of the field [T]
l = active length of the conductor [m]
I = current in the conductor [A]

Example 2-9:
A conductor 3 m long carrying a current of 200 A is placed in a magnetic field whose density is 0.5 T. Calculate the force on the conductor if it is perpendicular to the lines of force (Fig. 2-40).

Solution:
$F = BlI$
$= 0.5 \times 3 \times 200 = 300\,N$

2-26 Direction of the force acting on a straight conductor

We know that when a straight conductor carries a current, it is surrounded by a magnetic field. For a current flowing into the page of this book, the lines of force have the direction shown in Figure 2-42. On the other hand, Figure 2-43 shows the

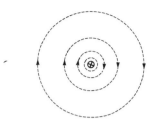

Figure 2-42   Magnetic field around a conductor.

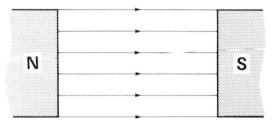

Figure 2-43   Magnetic field between two
permanent magnets.

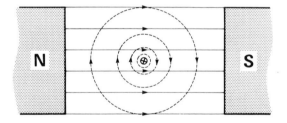

Figure 2-44   Magnetic field due to magnet and conductor.

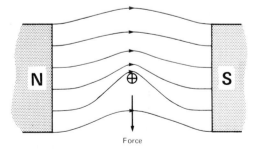

Force

Figure 2-45   Resulting magnetic field pushes
the conductor downwards.

uniform magnetic field created between the poles of a powerful permanent magnet. If we place the current-carrying conductor between the poles of the magnet, we obtain the superposed fields shown in Figure 2-44.

The resulting magnetic field does not, of course, have the shape shown in the figure because lines of force never cross each other. What, then, is the shape of the resulting field?

To answer the question, we observe that the lines of force created respectively by the conductor and by the permanent magnet act in the same direction above the conductor and in opposite directions below it. It follows that the number of lines above the conductor must be greater than the number below. Furthermore, the number of lines produced by the permanent magnet must remain the same because the strength of the magnet is unchanged. The resulting magnetic field therefore has the shape given in Figure 2-45.

Recalling that lines of flux act like stretched elastic bands, it is easy to visualize that a force acts upon the conductor, tending to push it downward.

QUESTIONS AND PROBLEMS

Practical level

2-1   Three dc sources G_1, G_2, G_3 (Fig. 2-46) generate voltages as follows:
$E_{12} = -100$ V; $E_{34} = -40$ V; $E_{56} = +60$ V.
Show the actual polarity (+) (−) of the terminals in each case.

2-2   In problem 2-1, if the three sources are connected in series, determine the voltage and polarity across the open terminals if the following terminals are connected together:
a. terminals 2-3 and 4-5;
b. terminals 1-4 and 3-6;
c. terminals 1-3 and 4-6.

2-3   Referring to Figure 2-47, show the voltage and the actual polarity of the generator terminals at instants 1, 2, 3, and 4.

Figure 2-46   See Problems 2-1 and 2-2.

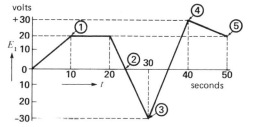

Figure 2-47   See Problem 2-3.

2-4   Explain the difference between a dynamo and an alternator.

2-5   What is the purpose of a commutator?

2-6   A conductor 2 m long moves at a speed of 60 km/h through a magnetic field having a flux density of 0.6 T. Calculate the induced voltage.

2-7   An alternator develops a peak voltage of 240 V at a frequency of 50 Hz. Calculate the new peak voltage and the frequency if:
   a. the turns per coil are doubled;
   b. the speed is reduced by half;
   c. the flux per pole is doubled.

2-8   A conductor cuts a flux of 3 Wb in 0,1 s. Calculate the induced voltage.

2-9   A coil having 200 turns links a flux of 3 mWb, produced by a permanent magnet. The magnet is moved, and the flux linking the coil falls to 1.2 mWb in 0.2 s. Calculate the average voltage induced.

2-10   What is the SI unit of:
   a. magnetic flux;
   b. magnetic flux density;
   c. magnetic field intensity;
   d. magnetomotive force.

2-11   Referring to Figure 2-36, calculate the relative permeability of cast iron at 0.2 T, 0.6 T and 0.7 T.

2-12   We want to produce a flux density of 0.6 T in an air gap having a length of 8 mm. Calculate the mmf required.

2-13   Conductor AB in Figure 2-19 carries a current of 800 A flowing from B to A.
   a. Calculate the force on the conductor.
   b. Calculate the force on the moving N pole.
   c. Does the force on the N pole act in the same direction as the direction of rotation?

3

FUNDAMENTALS
OF MECHANICS
AND HEAT

This introductory chapter gives certain fundamentals of mechanics and heat. These subjects are not immediately essential to an understanding of the chapters which follow, but they constitute a valuable reference source, which the reader may wish to consult from time to time. Consequently, we recommend a quick reading, followed by closer study of each section, as the various related subjects are met in the ensuing chapters.

3-1 Force

The most familiar force we know is the force of gravity. For example, whenever we lift a stone, we exert a muscular effort to overcome the gravitational force that continually pulls it downwards. There are other kinds of forces such as the force exerted by a stretched spring or the forces created by exploding dynamite. All these forces are expressed in terms of the newton (N), which is the SI unit of force.

The magnitude of the force of gravity depends upon the mass of a body, and is given by the approximate equation:

$$\boxed{F = 9.8\, m} \tag{3-1}$$

where

F = force of gravity acting on the body (also called its *weight*) [N]

m = mass of the body [kg]

9.8 = constant that applies when objects are relatively close to the surface of the earth (within 30 km)

3-2 Torque

Torque is produced when a force exerts a twisting action on a body, tending to make it rotate. Torque is equal to the product of the force times the perpendicular distance between the axis of rotation and the point of application of the force. For example, suppose a string is wrapped around a pulley having a radius r (Fig. 3-1). If we pull on the string with a force F, the pulley begins to rotate. The torque exerted on the pulley is given by:

$$T = Fr \qquad (3\text{-}2)$$

where

T = torque [N·m]
F = force [N]
r = radius [m]

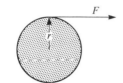

Figure 3-1   Torque $T = Fr$.

Example 3-1:
A motor develops a starting torque of 150 N·m. If the pulley has a diameter of 1 m, calculate the braking force needed to prevent the shaft from turning.

Solution:
The radius is 0.5 m; consequently, a braking force $F = T/r = 150/0.5 = 300$ N is required. If the radius were 2 m, a braking force of 75 N would be sufficient to prevent rotation.

3-3   Mechanical work

Mechanical work is done when a force F moves a distance d, in the direction of the force. The work is given by:

$$W = Fd \qquad (3\text{-}3)$$

where

W = work [J]
F = force [N]
d = distance the force moves [m]

Example 3-2:
A mass of 50 kg is lifted to a height of 10 m (Fig. 3-2). Calculate the work done.

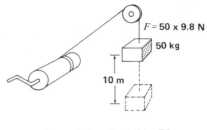

Figure 3-2   Work $W = Fd$.

Solution:
The force of gravity acting on the 50 kg mass is:
$F = 9.8 \times 50 = 490$ N   Eq. 3-1
The work done is:
$W = Fd = 490 \times 10 = 4900$ J

3-4   Power

Power is the rate of doing work and is given by:

$$P = W/t \qquad (3\text{-}4)$$

where

P = power [W]
W = work done [J]
t = time taken to do the work [s]

The SI unit of power is the watt (W). We often use the kilowatt (kW) which is equal to 1000 W. The power output of motors is sometimes expressed in horsepower (hp) which is an English unit equal to 746 W. It corresponds to the average power output of a dray horse.

Example 3-3:
An electric motor lifts a mass of 500 kg through a height of 30 m in 12 s (Fig. 3-3). Calculate the power developed by the motor, in kilowatts and in horsepower.

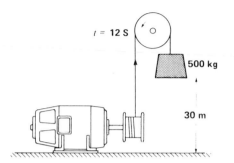

$t = 12$ S

500 kg

30 m

Figure 3-3   Power $P = W/t$.

We can measure the power output of a motor by means of a prony brake. It consists of a flat belt that rubs agains a pulley mounted on the motor shaft. The ends of the belt are connected to two spring scales, and the belt pressure may be adjusted by tightening screw V (Fig. 3-4). We can increase or decrease the power output by adjusting the tension of the belt. The mechanical power developed by the motor is entirely converted into heat by the belt rubbing against the pulley.

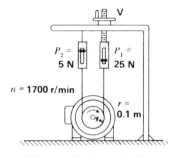

V

$P_2 = $ 5 N $P_1 = $ 25 N

$n = $ **1700 r/min**

$r = $ 0.1 m

Figure 3-4   Prony brake.

Solution:
The tension in the cable is:

$\quad F = 9.8\,m = 9.8 \times 500 = 4900$ N   Eq. 3-1

The work done is:

$\quad W = 4900 \times 30 = 147\,000$ J   Eq. 3-3

The power is:

$\quad P = W/t$

$\qquad = 147\,000/12 = 12\,250$ W $= 12.25$ kW

Expressed in horsepower:

$\quad P = 12\,250/746 = 16.4$ hp

When the motor is disconnected, the spring scales register the same pull and the resulting torque is zero. However, when the motor turns clockwise (as it does in Fig. 3-4), pull P_1 exceeds pull P_2. The resultant force acting on the circumference of the pulley is therefore $(P_1 - P_2)$ newtons. If the pulley has a radius r, the net torque $T = (P_1 - P_2)\,r$ newton-meters. Knowing the speed of rotation, we can calculate the power, using the above equation.

3-5   Power of a motor

The mechanical power output of a motor depends upon its rotational speed and the torque it develops. It is given by:

$$P = \frac{nT}{9.55} \qquad (3\text{-}5)$$

where

$\quad P = $ mechanical power [W]

$\quad T = $ torque [N·m]

$\quad n = $ speed of rotation [r/min]

$\quad 9.55 = $ a constant to take care of units (exact value $= 30/\pi$)

Example 3-4:
During a prony brake test on an electric motor, the spring scales indicate 25 N and 5 N, respectively (Fig 3-4). Calculate the power output if the motor turns at 1700 r/min and the radius of the pulley is 0.1 m.

Solution:

$\quad T = Fr$ Eq. 3-2

= (25 – 5) x 0.1 = 2 N·m

$P = nT/9.55$ Eq. 3-5

= 1700 x 2/9.55 = 356 W

The motor develops 356 W, or about 0.5 hp.

3-6   Transformation of energy

Energy may exist in one of the forms listed below:
1. Mechanical energy (energy in a waterfall, a coiled spring or a moving car);
2. Thermal energy (heat released by a stove, by friction, by the sun);
3. Chemical energy (energy contained in dynamite, coal, or an electric storage battery);
4. Electrical energy (energy produced by a generator, or manifested by lightning);
5. Atomic energy (energy released when the nucleus of an atom is split).

Although energy can be neither created nor destroyed, it can be converted from one form to another by means of appropriate devices or machines. For example, the chemical energy contained in coal may be transformed into thermal energy by burning the coal in a furnace. The thermal energy contained in steam can be transformed into mechanical energy by using a turbine. Finally, mechanical energy can be transformed into electrical energy by means of a generator.

In the above examples, the furnace, turbine and generator are the machines that do the energy transformation. Unfortunately, whenever energy is transformed, the ouput is always less than the input because all machines have losses. These losses appear in the form of heat, causing the temperature of the machine to rise. Thus, the electrical energy supplied to a motor is partly dissipated as heat in the windings. Furthermore, some of the mechanical energy is also lost, due to bearing friction and air turbulence created by the fan. The mechanical losses are also transformed into heat. Consequently, the useful mechanical power output is less than the electrical input.

3-7   Efficiency of a machine

The efficiency of a machine is given by:

$$\eta = \frac{P_o}{P_i} \times 100 \qquad (3\text{-}6)$$

where

η = efficiency [percent]
P_o = output power of the machine [W]
P_i = input power to the machine [W]

The efficiency is particularly low when thermal energy is converted into mechanical energy. Thus, the efficiency of steam turbines ranges from 25 to 40 percent, while that of internal combustion engines (automobile engines, diesel motors) lies between 15 and 30 percent. To realize how low these efficiencies are, we must remember that a machine having an efficiency of 20 percent loses 80 percent of the energy it receives.

Electric motors transform electrical energy into mechanical energy. Their efficiency ranges between 75 and 98 percent, depending on the size of the machine.

Example 3-5:
A 150 kW electric motor has an efficiency of 92 percent. Calculate the losses in the machine.

Solution:
The 150 kW rating always refers to the mechanical power *output* of the motor.
The input power is:
$P_i = P_o/\eta = 150/0.92 = 163$ kW;
The losses are:
$P_i - P_o = 163 - 150 = 13$ kW.

The losses are quite moderate, but they would still be enough to continuously heat a large home in the middle of winter.

3-8   Kinetic energy of linear motion

A falling stone or a swiftly moving automobile possess what is called kinetic energy, or energy due to motion. Kinetic energy is a form of mechanical energy given by:

$$E_k = \frac{1}{2} mv^2 \qquad (3\text{-}7)$$

where

E_k = kinetic energy [J]
m = mass of the body [kg]
v = speed of the body [m/s]

Example 3-6:

A bus having a mass of 6000 kg moves at a speed of 100 km/h. If it carries 40 passengers having a total mass of 2400 kg, calculate the total kinetic energy of the vehicle. What happens to this energy when the bus is braked to a stop?

Solution:

Total mass of the bus is:

m = 6000 + 2400 = 8400 kg.

v = 100 km/h = $\dfrac{100 \times 100 \text{ m}}{3600 \text{ s}}$ = 27.8 m/s.

E_k = ½ mv^2 = ½ x 8400 x 28.8²

   = 3 245 928 J = 3.25 MJ.

To stop the bus, the brakes are applied and the resulting frictional heat is entirely supplied by the decrease in kinetic energy. The bus will finally come to rest when all the kinetic energy (3.25 MJ) has been dissipated as heat.

3-9   Kinetic energy of rotation, moment of inertia

A revolving body also possesses kinetic energy. Its value depends upon the speed of rotation and upon the mass and shape of the body. The kinetic energy of of rotation is given by:

$$E_k = 5.48 \times 10^{-3} \, Jn^2 \qquad (3\text{-}8)$$

where

E_k = kinetic energy [J]
J = moment of inertia [kg·m²]
n = rotational speed [r/min]
5.48×10^{-3} = constant to take care of units [exact value = $\pi^2/1800$]

The moment of inertia J (sometimes simply called "inertia") depends upon the mass and shape of the body. Its value may be calculated for a number of simple shapes by using Eqs. 3-9 to 3-13 given in Table 3A. If the body has a more complex shape, it can always be broken up into two or more of the simpler shapes given in the table. The individual J's of these simple shapes are then added to give the total J of the body.

Inertia plays a very important part in rotating machines; consequently, it is worth our while to solve a few problems.

Example 3-7:

A solid 1400 kg steel flywheel has a diameter of 1 m and a thickness of 225 mm (Fig. 3-10). Calculate:

a. its moment of inertia.

b. the kinetic energy when the flywheel revolves at 1800 r/min.

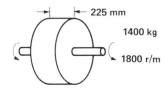

225 mm

1400 kg

1800 r/m

Figure 3-10   Flywheel in Example 3-7.

Solution:

a. referring to Table 3A, we find:

$$J = \frac{mr^2}{2} \qquad \text{Eq. 3-10}$$

TABLE 3A   MOMENT OF INERTIA J AROUND AN AXIS OF ROTATION 0

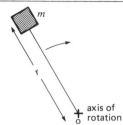

Figure 3-5   Mass m revolving at a distance r around axis 0.

$$J = mr^2 \qquad \text{(3-9)}$$

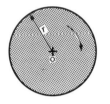

Figure 3-6   Solid disc of mass m and radius r.

$$J = \frac{mr^2}{2} \qquad \text{(3-10)}$$

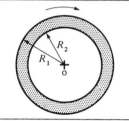

Figure 3-7   Annular ring of mass m having a rectangular cross section.

$$J = \frac{m}{2}(R_1{}^2 + R_2{}^2) \qquad \text{(3-11)}$$

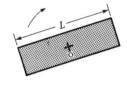

Figure 3-8   Straight bar of mass m pivoted on its center.

$$J = \frac{mL^2}{12} \qquad \text{(3-12)}$$

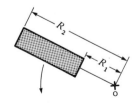

Figure 3-9   Rectangular bar of mass m revolving around axis 0.

$$J = \frac{m}{3}(R_1{}^2 + R_2{}^2 + R_1 R_2) \qquad \text{(3-13)}$$

$$= \frac{1400 \times 0.5^2}{2} = 175 \text{ kg·m}^2$$

b. The kinetic energy is:

$E_k = 5.48 \times 10^{-3} Jn^2$ Eq. 3-8

$= 5.48 \times 10^{-3} \times 175 \times 1800^2 = 3.1 \text{ MJ}$

Note that this relatively small flywheel possesses as much kinetic energy as the loaded bus mentioned in Example 3-6.

Example 3-8:

A flywheel having the shape given in Fig. 3-11 is composed of an annular ring supported by a rectangular hub. The ring and hub respectively have a mass of 80 kg and 20 kg. Calculate the moment of inertia of the flywheel.

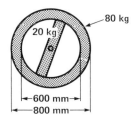

Figure 3-11   Flywheel in Example 3-8.

Solution:

For the annular ring:

$J_1 = m (R_1^2 + R_2^2)/2$ Eq. 3-11

$= 80 (0.4^2 + 0.3^2)/2 = 10 \text{ kg·m}^2$

For the hub:

$J_2 = mL^2/12$ Eq. 3-12

$= 20 \times (0.6)^2/12 = 0.6 \text{ kg·m}^2$

The total moment of inertia of the flywheel is:

$J = J_1 + J_2 = 10.6 \text{ kg·m}^2$

3-10 Torque, inertia and change in speed

There is only one way to change the speed of a revolving body and that is to subject it to a torque for a given period of time. The rate of change of speed depends upon the inertia, as well as on the torque. A simple equation relates these factors:

$$\boxed{\Delta n = 9.55\, T\Delta t/J} \qquad (3\text{-}14)$$

where

Δn = change in speed [r/min]

T = torque [N·m]

Δt = interval of time during which the torque is applied [s]

J = moment of inertia [kg·m^2]

9.55 = factor taking account of units [exact value = $30/\pi$]

If the torque acts in the direction of rotation, the speed rises. Conversely, if it acts against the direction of rotation, the speed falls. The term Δn may therefore represent either an increase or a decrease in speed.

Example 3-9:

The flywheel of Fig. 3-11 turns at 60 r/min. We wish to increase its speed to 600 r/min by applying a torque of 20 N·m. For how long must the torque be applied?

Solution:

Speed change, $\Delta n = (600 - 60) = 540$ r/min

Moment of inertia: $J = 10.6$ kg·m^2

$\Delta n = 9.55\, T\Delta t/J$ Eq. 3-14

$540 = 9.55 \times 20\Delta t/10.6$

therefore $\Delta t = 30$ seconds

3-11 Speed of a motor/load system

Consider a stationary load coupled to a motor by means of a shaft (Fig. 3-12). The load exerts a torque T_L and the motor develops an equal and opposite torque T_M. The system is at a standstill, and will remain so, as long as $T_M = T_L$. As a result of the opposing torques, the shaft twists and becomes slightly deformed, but otherwise nothing happens.

Suppose we want the load to turn clockwise at a speed n_1. To do this, we increase the motor current so that T_M exceeds T_L. The net torque on the system acts clockwise, and the speed increases progressively with time. However, as soon as the desired speed n_1 is reached, we reduce the motor current so that T_M is again exactly equal to T_L. The net torque acting on the system is now zero and the speed n_1 will neither increase or decrease any more (Fig. 3-13).

nitely at this new speed (Fig. 3-14).

In conclusion, torques T_M and T_L can be *identical* in Figs. 3-12, 3-13, and 3-14, and yet the speed may be zero, clockwise, or counterclockwise. The actual steady-state speed depends upon whether T_M was greater or less than T_L for a certain pe-

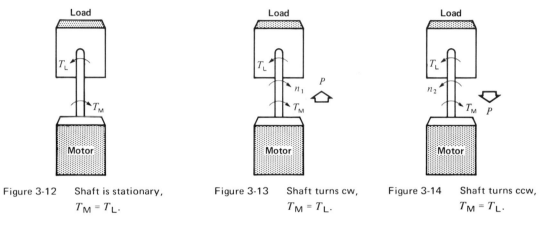

Figure 3-12   Shaft is stationary, $T_M = T_L$.

Figure 3-13   Shaft turns cw, $T_M = T_L$.

Figure 3-14   Shaft turns ccw, $T_M = T_L$.

This brings us to a very important conclusion. The speed of a mechanical load remains constant when the torque T_M developed by the motor is *equal* and *opposite* to the torque T_L exerted by the load. At first, this conclusion is rather difficult to accept, because we are inclined to believe that when $T_M = T_L$, the system should simply stop. But this is not so, as our reasoning (and reality) shows. We repeat: the speed of a motor remains constant when the motor torque is equal and opposite to the load torque. In effect, the motor/load system is then in a state of dynamic equilibrium.

With the load now running at speed n_1, suppose we reduce T_M so that it is less than T_L. The net torque on the system now acts counterclockwise. Consequently, the speed decreases and will continue to decrease as long as T_L exceeds T_M. If the imbalance between T_L and T_M lasts long enough, the speed will eventually become zero and then reverse. If we control the motor torque so that $T_M = T_L$ at the moment the reverse speed reaches a value n_2, the system will continue to run indefi-

riod of time *before* the actual steady-state condition was reached. The reader should ponder a few moments over this statement.

3-12 Power flow in a mechanically coupled system

Referring again to Fig. 3-13, we see that motor torque T_M acts in the same direction as speed n_1. This means that the motor *delivers* mechanical power to the shaft. Conversely, load torque T_L acts opposite to speed n_1. Consequently, the load *receives* mechanical power from the shaft. We can therefore state the following general rule:

When the torque developed by a motor acts in the same direction as the speed, the motor delivers power to the load. For all other conditions, the motor receives power from the load.

For example, referring to Fig. 3-14, the motor receives power from the load because T_M acts opposite to n_2. Although this is an unusual condition, it occurs for brief periods in electric trains and electric hoists. The behavior of the motor under these conditions will be examined in later chapters.

3-13 Motor driving a load possessing inertia

When a motor drives a mechanical load, the speed is usually constant. In this state of dynamic equilibrium, the torque T_M developed by the motor is exactly equal to the torque T_L imposed by the load. The inertia of the revolving parts does not come into play under these conditions. However, if the motor torque is raised so that it exceeds the load torque, the speed will increase, as we already know. Conversely, when the motor torque is less than that of the load, the speed drops. The increase or decrease in speed (Δn) is still given by the Eq. 3-14, except that torque T is now replaced by the *net* torque ($T_M - T_L$):

$$\Delta n = 9.55 \ (T_M - T_L) \ \Delta t / J \qquad (3\text{-}15)$$

where

Δn = change in speed [r/min]
T_M = motor torque [N·m]
T_L = load torque [N·m]
Δt = time interval during which T_M and T_L are unequal [s]
J = moment of inertia of all revolving parts [kg·m²]

Example 3-10:
A large reel of paper installed at the end of a paper machine has a diameter of 1.8 m, a length of 5.6 m and a moment of inertia of 4500 kg·m². It is driven by a directly coupled variable-speed dc motor turning at 120 r/min. The paper is kept at a constant tension of 6000 N.
a. Calculate the power of the motor when the reel turns at a constant speed of 120 r/min,
b. If the speed has to be raised from 120 r/min to 160 r/min in 5 seconds, calculate the torque that the motor must develop during this interval,
c. Calculate the power of the motor just before it attains 160 r/min,
d. Calculate the power of the motor after it has reached the desired speed of 160 r/min.

Solution:
a. 1 The torque exerted on the reel:

$$T = Fr = 6000 \times \frac{1.8}{2} = 5400 \text{ N·m}$$

a. 2 The power developed by the reel motor:

$$T = \frac{nT}{9.55} = \frac{120 \times 5400}{9.55} \qquad \text{Eq. 3-5}$$
$$= \quad 67.85 \text{ kW (equivalent to 91 hp)}$$

b. 1 The load torque (5400 N·m) stays constant because the tension in the paper remains unchanged as the speed increases from 120 r/min to 160 r/min. Let the required motor torque be T_M. It must be greater than the load torque in order that the speed may increase.

b. 2 We have:
Δn = 160 – 120 = 40 r/min
J = 4500 kg·m²
Δt = 5 s

b. 3
$$\Delta n = \frac{9.55 \ (T_M - T_L) \ \Delta t}{J}$$

$$40 = \frac{9.55 \ (T_M - 5400) \ 5}{4500}$$

$T_M - 5400 = 3770$
$T_M = 9170$

The motor must develop a torque of 9170 N·m during the acceleration period.

$$P = \frac{nT}{9.55} = \frac{160 \times 9170}{9.55}$$
$$= 153.6 \text{ kW (equivalent to 206 hp)}$$

c. As soon as the desired speed (160 r/min) is reached, the motor only has to develop a torque equal to the load torque (5400 N·m). The power of the motor is therefore reduced to:

$$P = \frac{nT}{9.55} = \frac{160 \times 5400}{9.55}$$
$$= 90.5 \text{ kW (equivalent to 121 hp)}$$

3-14 Electric motors driving linear motion loads

Rotating loads such as fans, pumps, machine tools, and so forth are well adapted for direct connection to electric motors because they also rotate. On the other hand, loads that move in a straight line, such as hoists, trains, wire-drawing machines, etc., must be equipped with a "motion converter" before they can be connected to a rotating machine. The motion converter may be a rope-pulley arrangement, a rack and pinion mechanism, or simply a wheel moving over a track. These converters are so utterly simple that we seldom think of the important part they play.

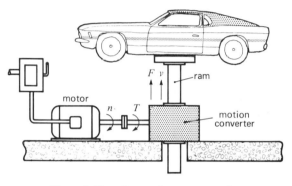

Figure 3-15   Converting rotary motion into linear motion.

Consider an electric jack driven by a motor that rotates at a speed n and exerts a torque T (Fig. 3-15). The jack is equipped with a motion converter that changes the rotary input into a linear output. The vertical ram exerts a powerful force F while moving at a linear speed v. The power supplied to the load is given by:

$$P_o = Fv$$

The power input to the converter is given by:

$$P_i = \frac{nT}{9.55} \qquad \text{Eq. 3-5}$$

Assuming no losses in the motion converter, we have:

$$P_i = P_o$$

Consequently:

$$\boxed{nT = 9.55\, Fv} \qquad (3\text{-}16)$$

where
n = rotational speed [r/min]
T = torque [N·m]
F = force [N]
v = linear speed [m/s]
9.55 = a constant [exact value = $30/\pi$]

Example 3-11:
A force of 25 kN is needed to pull a train at a speed of 90 km/h. The electric motor on board the locomotive turns at 1200 r/min. Calculate the load torque at the motor shaft.

Solution:
$$nT = 9.55\, Fv$$
$$1200\, T = 9.55 \times 25\,000 \times \frac{90\,000}{3600}$$
$$T = 4974 \text{ N·m} = 5 \text{ kN·m}$$

3-15 Heat and temperature

When we apply heat to a body, we supply it with thermal energy. Heat is therefore a form of energy, and the SI unit is the joule.

What happens when a body receives this type of energy? First, its temperature increases, a fact we can verify by touching it with the hand or by observing the reading of a thermometer. Second, the atoms of the body vibrate more intensely.

For a given amount of heat, the increase in temperature depends upon the mass of the body and the material of which it is made. For example, if we add 100 kJ of heat to 1 kg of water, the temperature rises by 24°C. The same amount of heat supplied to 1 kg of copper raises *its* temperature by 263°C. It is therefore obvious that heat and temperature are two quite different things.

If we remove heat from a body, its temperature drops. However, there is a lower limit to which the temperature can fall. This lower limit is called absolute zero. It corresponds to a temperature of 0 kelvin or – 273.15°C. At absolute zero, all atomic vibrations cease and the only motion which subsists is that of the moving electrons.

3-16 Temperature scales

The kelvin and the degree Celsius are the SI units of temperature. Figure 3-16 shows the relationship between the Kelvin, Celsius and Fahrenheit temperature scales.

m = mass of the body [kg]
c = specific heat capacity of the material making up the body [J/(kg·°C)]
Δt = change in temperature [°C]

The specific heat capacity of several materials is given in Table 5B, Sec. 5-8.

Example 3-12:

Calculate the heat required to raise the tempera-

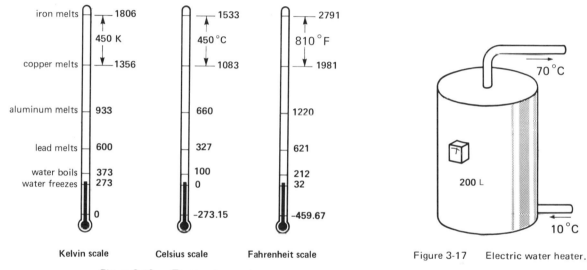

iron melts	1806	1533	2791
	450 K	450 °C	810 °F
copper melts	1356	1083	1981
aluminum melts	933	660	1220
lead melts	600	327	621
water boils	373	100	212
water freezes	273	0	32
	0	−273.15	−459.67
	Kelvin scale	Celsius scale	Fahrenheit scale

Figure 3-16   Temperature scales.

70 °C

200 L

10 °C

Figure 3-17   Electric water heater.

3-17 Heat required to raise the temperature of a body

We saw that the temperature rise of a body depends both upon the heat it receives and the nature of the material. The relationship between these quantities is given by the equation:

$$Q = mc\Delta t \qquad (3-17)$$

where
Q = quantity of heat added to (or removed from) a body [J]

ture of 200 L of water from 10°C to 70°C, assuming the tank is perfectly insulated (Fig. 3-17). The specific heat of water is 4180 J/kg·°C.

Solution:

Because 1 L of water has a mass of 1 kg, the heat required is:

$$Q = mc\Delta t$$
$$= 200 \times 4180 \times (70 - 10)$$
$$= 50.2 \text{ MJ}$$

Referring to the conversion table for Energy (see Appendix), we find that 50.2 MJ is equal to 13.9 kW·h.

3-18 Transmission of heat

Many problems in electric power technology are related to the adequate cooling of devices and machines. This, in turn, requires a knowledge of the mechanism by which heat is transferred from one body to another. In the sections that follow, we briefly review the elementary physics of heat transmission. We also include some simple but useful equations, enabling us to determine, with reasonable accuracy, the heat loss, temperature rise, etc., of electrical equipment.

Heat is transmitted from one body to another in three different ways: (1) conduction, (2) convection, and (3) radiation. We now examine these modes of heat transfer.

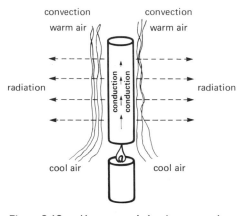

Figure 3-18   Heat transmission by convection, conduction and radiation.

3-19 Heat transfer by conduction

If we bring a hot flame near one end of an iron bar, its temperature rises owing to the increased vibration of its atoms (Fig. 3-18). This atomic vibration is transmitted from one atom to the next, to the other end of the bar. Consequently, the end opposite the flame also warms up, an observation we have all made at one time or another. In effect, heat is transferred along the bar by a process known as *conduction*.

The rate of heat transfer depends upon the *thermal conductivity* of the material. Thus, copper is a better thermal conductor than steel is, and insulators are especially poor conductors of heat.

The SI unit of thermal conductivity is the watt per meter degree Celsius [W/(m·°C)]. The thermal conductivity of several common materials is given in Table 4B, Sec. 4-11 and Table 5B, Sec. 5-8.

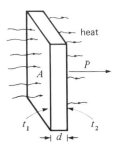

Figure 3-19   Heat transmission by conduction.

Referring to Fig. 3-19, we can calculate the rate of heat flow through a body by using the equation:

$$P = \frac{\lambda A (t_1 - t_2)}{d} \qquad (3\text{-}18)$$

where

P = power (heat) transmitted [W]

λ = thermal conductivity of the body [W/(m·°C)]

A = surface area of the body [m²]

$(t_1 - t_2)$ = difference of temperature between opposite faces [°C]

d = thickness of the body [m]

Example 3-13:
The temperature difference between two sides of a sheet of mica is 50°C (Fig. 3-20). Calculate the heat flowing through the sheet, in watts.

Solution:
According to Table 4B, the thermal conductivity of mica is 0.36 W/(m·°C). The heat transmitted is therefore:

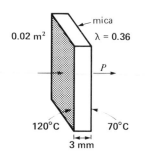

Figure 3-20   Mica sheet, Example 2-13.

$$P = \frac{\lambda A (t_1 - t_2)}{d}$$

$$= \frac{0.36 \times 0.02 \,(120 - 70)}{0.003} = 120 \text{ W}$$

3-20 Heat transfer by convection

In Fig. 3-18, the air in contact with the hot steel bar warms up and, becoming lighter, rises like smoke in a chimney. As the hot air moves upwards, it is replaced by cooler air which, in turn, also warms up. A continual current of air is therefore set up around the bar, removing its heat by a process called *natural convection*.

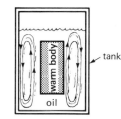

Figure 3-21   Convection currents in oil.

The convection process can be accelerated by employing a fan to create a rapid circulation of fresh air. Heat transfer by *forced convection* is used in most electric motors to obtain efficient cooling.

Natural convection also takes place when a hot body is immersed in a liquid, such as oil. The oil in

contact with the body heats up, creating convection currents which follow the path shown in Fig. 3-21. When the warm oil comes in contact with the cooler tank, it chills, becomes heavier, slides to the bottom, and moves upward again to replace the warmer oil now moving away. The heat dissipated by the body is therefore carried away by convection to the external tank. The tank, in turn, loses *its* heat by natural convection to the surrounding air.

3-21 Calculating losses by convection

The heat a body loses by natural convection in air is given by the following approximate equation:

$$P = 3A \,(t_1 - t_2)^{1.25} \qquad (3\text{-}19)$$

where
P = heat transferred by natural convection [W]
A = surface of the body [m]
t_1 = surface temperature of the body [°C]
t_2 = ambient temperature of the surrounding air [°C]

In the case of *forced* convection, the heat carried away is given by:

$$P = 1280 \, V_a \,(t_2 - t_1) \qquad (3\text{-}20)$$

where
P = heat transferred by forced convection [W]
V_a = volume of cooling air [m³/s]
t_1 = temperature of the incoming air [°C]
t_2 = temperature of the outgoing air [°C]

Equation 3-20 also applies if hydrogen cooling is used.

Example 3-14:
A totally enclosed motor has an external surface area of 1.2 m². When it operates at full load, the surface temperature rises to 60°C in an ambient of

20°C (Fig. 3-22). Calculate the heat loss by natural convection.

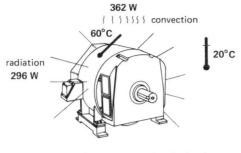

362 W
convection
60°C
20°C
radiation
296 W

Figure 3-22   Convection and radiation losses in a totally **enclosed motor**.

Solution:

$$P = 3A\ (t_1 - t_2)^{1.25}$$
$$= 3 \times 1.2\ (60 - 20)^{1.25} = 362\ W$$

Example 3-15:
A fan rated at 3.75 kW blows 240 m^3/min of air through a 750 kW motor to carry away the heat. If the inlet temperature is 22°C and the outlet temperature is 31°C, calculate the losses in the motor.

Solution:

$$P = 1280\ V_a\ (t_2 - t_1)$$
$$= 1280 \times \frac{240}{60}\ (31 - 22) = 46.1\ kW$$

3-22 Heat transfer by radiation

We have all basked in the heat produced by the sun's rays. This radiant heat energy possesses the same properties as light, and it readily passes through the empty space between the sun and the earth. Solar energy is only converted to heat when the sun's rays meet a solid body, such as the physical objects and living things on the surface of the earth. Scientists have discovered that all bodies radiate heat, even those that are very cold. The amount of energy given off depends upon the temperature of the body.

On the other hand, all bodies *absorb* radiant energy from the objects which surround them. The energy absorbed depends upon the temperature of the surrounding objects. There is, consequently, a continual exchange of radiant energy between material bodies, as if each were a miniature sun. Equilibrium sets in when the temperature of a body is the same as that of its surroundings. The body then radiates as much energy as it receives and the net radiation is zero. On the other hand, if the body is hotter than its environment, it will continually lose heat by radiation, even if it is located in a vacuum.

3-23 Calculating radiation losses

The heat a body loses by radiation is given by the equation:

$$\boxed{P = kA\ (T_1{}^4 - T_2{}^4)} \qquad (3\text{-}21)$$

where
P = heat radiated [W]
A = surface area of the body [m^2]
T_1 = absolute temperature of the body [K]
T_2 = absolute temperature of the surrounding objects [K]
k = a constant, which depends upon the nature of the body surface
Table 3B gives the values of k for surfaces commonly encountered in electrical equipment.

Example 3-16:
The motor in Example 3-14 is coated with a non metallic enamel. Calculate the heat lost by radiation, knowing that all surrounding objects are at ambient temperature (20°C).

Solution:
T_1 = surface temperature = 60°C
      or (273.15 + 60) = 333 K
T_2 = surrounding temperature = 20°C
      or (273.15 + 20) = 293 K

TABLE 3B   RADIATION CONSTANTS

Type of surface	Constant k $W/(m^2 \cdot K^4)$
polished silver	0.2×10^{-8}
bright copper	1×10^{-8}
oxidized copper	3×10^{-8}
aluminum point	3×10^{-8}
oxidized nichrome	2×10^{-8}
tungsten	2×10^{-8}
oxidized iron	4×10^{-8}
insulating materials	5×10^{-8}
paint or non metallic enamel	5×10^{-8}
perfect emitter (blackbody)	5.669×10^{-8}

From Table 3B, $k = 5 \times 10^{-8}$ W/(m$^2 \cdot$K^4)

$$P = kA\,(T_1^4 - T_2^4) \qquad \text{Eq. 3-21}$$
$$= 5 \times 10^{-8} \times 1.2\,(333^4 - 293^4)$$
$$= 296 \text{ W (see Fig. 3-22)}$$

QUESTIONS AND PROBLEMS

3-1 A cement block has a mass of 40 kg. What is the force of gravity acting on it? What force is needed to lift it?

3-2 How much energy is needed to lift a sack of flour weighing 75 kg to a height of 4 m?

3-3 Give the SI unit and the corresponding SI symbol for the following quantities:

force	work
pressure	area
mass	temperature
thermal energy	thermal power
mechanical energy	mechanical power
electrical energy	electrical power

3-4 In tightening a bolt, a mechanic exerts a force of 200 N at the end of a wrench having a length of 0.3 m. Calculate the torque he exerts.

3-5 An automobile engine develops a torque of 600 N·m at a speed of 4000 r/min. Calculate the power output in watts and in horsepower.

3-6 A crane lifts a mass of 600 lb to a height of 200 ft in 15 s. Calculate the power in watts and in horsepower.

3-7 An electric motor draws 120 kW from the line and has losses equal to 20 kW. Calculate:
a. the power output of the motor [kW] and [hp];
b. the efficiency of the motor;
c. the amount of heat released [Btu/h].

3-8 A large flywheel has a moment of inertia of 500 lb·ft^2. Calculate its kinetic energy when it rotates at 60 r/min.

3-9 The rotor of an induction motor has a moment of inertia of 5 kg·m^2. Calculate the energy needed to bring the speed:
a. from zero to 200 r/min;
b. from 200 r/min to 400 r/min;
c. from 3000 r/min to 400 r/min.

3-10 Name the three ways whereby heat is carried from one body to another.

3-11 A motor develops a cw torque of 60 N·m and the load develops a ccw torque of 50 N·m.
a. If this situation persists for some time, will the direction of rotation be cw or ccw?
b. Does the speed increase with time?
c. What value of motor torque is needed to keep the speed constant?

3-12 A motor drives a load at a cw speed of 1000 r/min. The motor develops a cw torque of 12 N·m, and the load exerts a ccw torque of 15 N·m.
a. Will the speed increase or decrease?
b. If this situation persists for some time, in what direction will the shaft eventually rotate?

3-13 Referring to Fig. 3-12, if T_M = 40 N·m, what is the power delivered by the motor?

3-14 Referring to Fig. 3-13, if T_M = 40 N·m and n_1 = 50 r/min, calculate the power delivered by the motor.

3-15 Referring to Fig. 3-14, if T_M = 40 N·m and n_2 = 50 r/min, calculate the power received by the motor.

Intermediate level

3-16 During a prony brake test on a motor (see Fig. 3-4), the following scale and speed readings were noted:
P_2 = 5 lbf, P_1 = 28 lbf, n = 1160 r/min.
If the diameter of the pulley is 12 inches, calculate the power output of the motor in kilowatts and horsepower.

3-17 A motor drives a flywheel having a moment of inertia of 5 kg·m². The speed increases from 1600 r/min to 1800 r/min in 8 s. Calculate:
a. the torque developed by the motor [N·m] ;
b. the energy in the flywheel at 1800 r/min [kJ] ;
c. the motor power at 1600 r/min;
d. the power input to the flywheel at 1750 r/min.

3-18 A dc motor drives a large grinder at a constant speed of 700 r/min and a power output of 120 hp. The moment of inertia of the revolving parts is equal to 2500 lb·ft².
a. Calculate the torque developed by the motor;
b. Calculate the motor torque needed so that the speed will increase to 750 r/min in 5 s.

3-19 The electric motor in a trolleybus develops a power output of 80 hp at 1200 r/min as the bus moves up an incline at a speed of 30 miles per hour. Assuming that the gear losses are negligible, calculate:
a. the torque developed by the motor [N·m] ;
b. the braking force acting on the bus [N] .

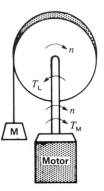

Figure 3-23   Electric hoist.

3-20 Calculate the amount of heat required to heat a 100 kg mass of copper from 20°C to 100°C.

3-21 Repeat Problem 3-20 for 100 kg of aluminum.

3-22 The motor in Fig. 3-23 drives a hoist, raising a mass m of 800 kg at a uniform rate of 5 m/s. The winch has a radius of 20 cm. Calculate the torque [N·m] and speed [r/min] of the motor.

3-23 If the hoisting rate is reduced to 1 m/s, calculate the new speed [r/min] and torque [ft·lbf] of the motor.

PART II

ELECTRICAL MATERIALS

4

INSULATING MATERIALS

The design and operation of electrical equipment to a large extent depends on the conducting and insulating materials that are available and used in construction. Both play a crucial, practical role. In this chapter, we only examine the properties of insulating materials, also known as *dielectrics*.

4-1 Types of insulators

Insulating materials may be classified into two main groups: *organic* insulators and *inorganic* insulators. In general, organic insulators such as rubber, paper, oil, cotton, thermoplastic materials, and so forth are composed of long molecular chains of carbon and hydrogen sometimes linked with other elements (oxygen, chlorine, etc.). Organic insulators deteriorate rapidly when the temperature exceeds about 150°C. On the other hand, inorganic insulators, such as mica, porcelain, air, etc., can function indefinitely in temperatures exceeding 1000°C.

Over the years, the number of available insulators has multiplied greatly, so that it is difficult to draw up a complete list. This diversity is due to the development of synthetic organic insulators (or plastics) invented and developed by chemists. Possessing thermal, electrical and mechanical properties far superior to natural insulators, these synthetic materials have greatly improved the design and manufacture of wires, cables, and electrical machines.

The variety of insulators sometimes appears even greater because competing manufacturers employ different trade names for essentially the same product. For example, polyurethane, a thin insulating film covering copper wires, is known in North America under seven different trade names, depending upon the manufacturer: Polysol (Canada Wire), Soldereze (Phelps Dodge), Isomelt (Pirelli Cables), Analac (Anaconda Copper), Solderex (Essex), Gendure (General Cable), and Beldure (Belden).

We often combine two or more insulators to create a new product háving the specific advantages of its several components. For example, we combine glass fibers with synthetic varnishes to produce an insulator that can resist both high temperatures and extreme vibration and shock.

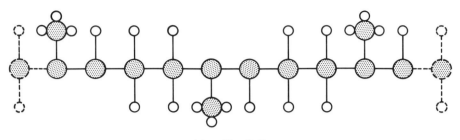

(a)   rubber $C_5 H_8$

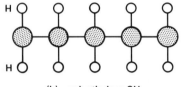

(b)   polyethylene CH_2

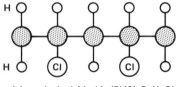

(c)   polyvinylchloride (PVC) $C_2 H_3 Cl$

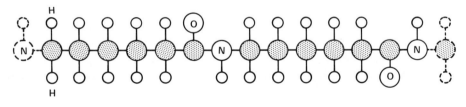

(d)   nylon $C_{12} H_2 O_2 H_{22}$

LEGEND

carbon C   oxygen O   hydrogen H   nitrogen N   chlorine Cl

Figure 4-1   Molecular structure of some natural and synthetic insulators.

4-2   Solid insulators

In one type of chemical reaction, called polymerization, some simple molecules are linked to form much larger molecules containing thousands of the original molecules (Fig. 4-1). A new substance formed in this way is said to be a *polymer* of the original molecule. All synthetic insulators are polymers. Natural rubber, resins, varnishes and bakelite are all polymers. Depending on their composition and molecular structure, polymers can be subdivided into many classes: polyvinyls, polyurethanes, polyesters, polyamides, polyimides, etc. For example, nylon is a polyamide, Dacron and Mylar® are polyesters and Kapton® is a polyimide.

Such synthetic materials are used to insulate the wires in the coils of motors, transformers, electromagnets, relays, etc., as well as the heavily-insulated wire and cable which distribute electric power in buildings. These polymers are sometimes combined with other insulating materials (cotton, glass, etc.) to produce sheets and plates possessing great mechanical strength, excellent resistance to high temperatures and superior electrical properties.

Although we tend to use more and more synthetic materials, natural insulators are still indispensable in many applications. Cotton is still widely used in the manufacture of insulating sheets, plates and cables. Paper is still one of the best materials to cover high voltage conductors. Asbestos, a natural inorganic material, is employed in the manufacture of high-temperature wire and fire-resistant control panels. Mica, a superb insulator, acts as support for the heating elements in toasters and as a high-temperature, uncrushable insulator in the commutators of direct current machines.

4-3   Liquid insulators

Mineral oil is used in big power transformers as both an insulator and heat-transferring agent. It also prevents oxidation of the insulating materials because it completely surrounds the windings and inhibits the access of air. Oxidation is a particularly important problem in high-voltage (HV) trans-

formers because they tend to produce corona discharges. In the presence of air, such HV discharges generate ozone, which is a very strong oxidizing agent. By immersing the windings in oil, we prevent the formation of ozone and, at the same time, heat dissipated by the windings is carried away to the surrounding tank. Because oil is a much better insulator than air, we can also reduce the size of the transformer.

Oil, however, has the disadvantage of being flammable and its ignition temperature is only about 150°C. Some higher-temperature synthetic liquid insulators get around this problem, but they are more expensive and often cannot be used with other insulators because they tend to attack them chemically.

4-4   Gaseous insulators

One of the best insulators known is the *air* which surrounds us. Its thermal properties are better than those of porcelain, it acts as a cooling agent, and costs absolutely nothing. However, at very high temperatures, air becomes a good conductor, owing to the phenomenon of ionization. For example, at 2000°C, the resistivity of air is still as high as that of porcelain, but when its temperature increases to between 5000°C and $50\,000^{\circ}$C, its conductivity approaches that of salt water.

Sulfur hexafluoride (SF_6) is another important insulating gas. Its molecules have the special ability to absorb free electrons, which accounts for its very high dielectric strength (10 times greater than air at a pressure of 400 kPa). Sulfur hexafluoride is used in high-voltage circuit breakers and enclosed transmission lines where space reduction is particularly important.

Hydrogen is another important insulating gas sometimes used to cool large rotating machines. Hydrogen has a much lower density and viscosity than air and, consequently, produces less friction at high rotational speeds. Furthermore, for a given temperature rise, it absorbs almost 14 times as much heat as air does. Finally, pure hydrogen prevents any oxidation of the insulating materials and thereby prolongs their life. Hydrogen cooling systems, however, are very complex and require constant maintenance; their use is only justified in very large machines.

From a safety point of view, hydrogen does not explode or burn, even in the presence of an electric arc, provided that the oxygen content is kept below 10 percent.

4-5   Deterioration of organic insulators

The factors that contribute most to the deterioration of insulators are: 1. heat; 2. humidity; 3. vibration; 4. acidity; 5. oxidation and 6. time (Fig. 4-2). Because of these various factors, the state of the insulators changes gradually; it slowly begins to crystallize and the transformation takes place more rapidly as the temperature rises.

In crystallizing, the organic insulator becomes hard and brittle. Eventually, the slightest shock or mechanical vibration will cause it to break. Under normal conditions, with respect to the above factors, most organic insulators have a life expectancy of eight to ten years provided that their temperature does not exceed 100°C. On the other hand

some synthetic polymers can withstand temperatures as high as 200°C for the same length of time.

Low temperatures are just as harmful as high temperatures are, because the insulation tends to freeze and crack. Special synthetic organic insulators have been developed, however, which retain their flexibility at temperatures even as low as –60°C.

4-6   Life expectancy of electric equipment

Apart from accidental electrical and mechanical failures, the life expectancy of electrical apparatus is limited by the temperature of its insulation: the higher the temperature, the shorter its life. Tests made on many insulating materials have shown that the useful life of electrical apparatus diminishes approximately by half every time the temperature increases by 10°C. This means that if a motor has a normal life expectancy of eight years at a temperature of 105°C, it will have a life expectancy of only four years at a temperature of 115°C, of two years at 125°C, and of only one year at 135°C!

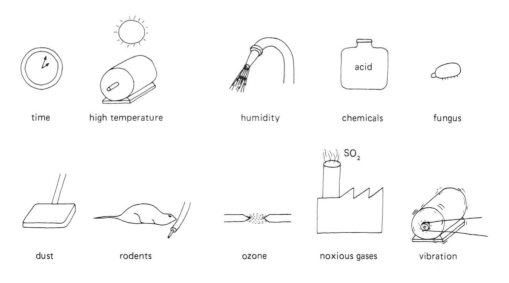

time   high temperature   humidity   chemicals   fungus

dust   rodents   ozone   noxious gases   vibration

Figure 4-2   Factors which may shorten the life expectancy of an insulator.

4-7   Thermal classification of insulators

Committees and organizations that set standards*
have grouped insulators into five classes, depending
upon their ability to withstand heat. These classes
correspond to the maximum temperature levels of:
105°C, 130°C, 155°C, 180°C and 220°C (formerly
represented by the letters A, B, F, H, and C). This
thermal classification (Table 4A)† is a cornerstone
in the design and manufacture of electrical apparatus. As we shall see in Chapter 5, a different
thermal classification is used for insulated wire and
cable employed in commercial and industrial
buildings and also homes.

4-8   Electrical resistivity of insulators

No insulator is perfect. When we apply a moderate
voltage to an insulator, a very small current flows,
in accordance with Ohm's law. Part of this current
filters through the volume of the insulator while
the remainder leaks over the surface. The surface
resistivity (and, consequently, the surface leakage
current) varies greatly with humidity and the
cleanliness of the surface. On the other hand, the
volume resistivity, usually expressed in teraohm-
meters (1 T$\Omega \cdot$m = 10^{12} $\Omega \cdot$m), is relatively constant.
Both resistivities become important when insula-
tors are subjected to very high voltages, such as in
lightning arrestors and transmission lines.

The resistance of an insulator is therefore composed of a "volume" resistance R in parallel with a
surface resistance R_s. These resistance are given by:

$$R = \rho \frac{l}{A} \qquad (4\text{-}1)$$

and

$$R_s = \rho_s \frac{l}{c} \qquad (4\text{-}2)$$

* Such as IEEE, Underwriters Laboratories, CSA.

† See Sec. 4-9.

where
R = volume resistance [Ω]
ρ = volume resistivity* [$\Omega \cdot$m]
l = length of insulator [m]
A = cross-section of insulator [m^2]
R_s = surface resistance [Ω]
ρ_s = surface resistivity [Ω]
c = perimeter of the cross-section [m]

Example 4-1:

An asbestos-base insulating spacer having a length
of 100 mm and a cross-section of 15 x 15 mm is
mounted between two flat bars operating at a potential difference of 120 kV (Fig. 4-3). The surface
and volume resistivities are respectively 0.5 TΩ
and 0.01 T$\Omega \cdot$m.

Calculate the value of:
a. the surface leakage current;
b. the current flowing through the spacer;
c. the power loss due to leakage current.

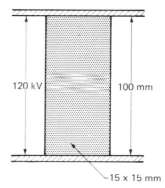

Figure 4-3   Voltage applied to a solid insulator.

Solution:
Volume resistance R is:

$$R = \rho \frac{l}{A} \qquad \text{Eq. 4-1}$$

$$= 0.01 \times 10^{12} \times \frac{0.100}{0.015 \times 0.015}$$

$$= 4.44 \times 10^{12}$$

$$= 4.44 \text{ T}\Omega$$

Surface resistance R_s is: Eq. 4-2

$$R_s = \rho_s \frac{l}{c}$$

$$= 0.5 \times 10^{12} \times \frac{0.10}{4 \times 0.015}$$

$$= 0.83 \text{ T}\Omega$$

a. Surface leakage current is:
$$I_s = E/R_s = \frac{120\,000}{0.83 \times 10^{12}}$$
$$= 0.14 \,\mu\text{A}$$

b. Volume leakage current is:
$$I_v = E/R = \frac{120\,000}{4.44 \times 10^{12}}$$
$$= 0.027 \,\mu\text{A}$$

c. Total leakage current is:
$$I = 0.14 + 0.027$$
$$= 0.167 \,\mu\text{A}$$

Power loss is:
$$P = EI = 0.167 \times 10^{-6} \times 120\,000$$
$$= 0.02 \text{ W} = 20 \text{ mW}$$

4-9   Dielectric strength - Insulation breakdown

The main purpose of a dielectric or insulator is to prevent significant current flow when subjected to a difference of potential. However, a dielectric cannot withstand higher and higher voltages. At a certain critical voltage, every dielectric material suddenly loses its insulating properties and breaks down. To explain such dielectric breakdown, consider a solid insulator placed between two metallic plates connected to a variable voltage source (Fig. 4-4).

When the applied voltage between the plates is zero, the electrons revolving around the nucleus of each atom of the insulator will follow a generally normal orbital path (Fig. 4-4a). As we increase the voltage, the electrons are drawn towards the positive plate and repelled by the negative plate so that the normal orbit becomes more flattened and elliptical (Fig. 4-4b). If we increase the voltage still more, the force of attraction becomes sufficiently great to tear the outermost valence electrons from their orbits (Fig. 4-4c). This separation occurs simultaneously in hundreds of millions of atoms, so that the insulator suddenly becomes filled with an avalanche of free electrons. The original feeble leakage current suddenly increases millions of times and the resulting intense heat (I^2R) decomposes the molecular structure, destroying the insulator.

The breakdown voltage required to produce this catastrophic failure depends upon the nature of the insulator and its thickness. The ratio of breakdown voltage to insulator thickness is called *dielectric strength*. It is generally expressed in kV/mm or in MV/m. Table 4B shows the dielectric strength of several insulators. Note that Mylar® possesses a dielectric strength 100 times greater than dry air. This synthetic material enables us to manufacture 400 V capacitors in which the thickness of the dielectric is only 0.006 mm, approximately one-tenth the thickness of a page of this book!

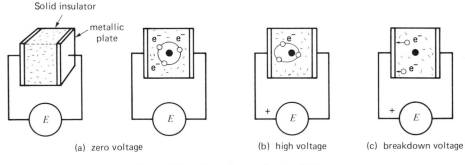

(a) zero voltage   (b) high voltage   (c) breakdown voltage

Figure 4-4   Breakdown of an insulator.

Example 4-2:
Ordinary window-pane glass has a dielectric strength of about 50 MV/m. Calculate the maximum voltage we could safely apply to a glass 3 mm thick.

Solution:
50 MV/m is equal to 50 kV/mm. The glass could therefore withstand a difference of potential of 3 x 50 = 150 kV before it punctures.

TABLE 4A   CLASSES OF INSULATION SYSTEMS

Class	Illustrative Examples
105°C A	Materials or combinations of materials such as cotton, silk, and paper when suitably impregnated or coated or when immersed in a dielectric liquid such as oil. Other materials or combinations of materials may be included in this class if by experience or accepted tests they can be shown to have comparable thermal life at 105°C.
130°C B	Materials or combinations of materials such as mica, glass fiber, asbestos, etc., with suitable bonding substances. Other materials or combinations of materials may be included in this class if by experience or accepted tests they can be shown to have comparable thermal life at 130°C.
155°C F	Materials or combinations of materials such as mica, glass fiber, asbestos, etc., with suitable bonding substances. Other materials or combinations of materials may be included in this class if by experience or accepted tests they can be shown to have comparable life at 155°C.
180°C H	Materials or combinations of materials such as silicone elastomer, mica, glass fiber, asbestos, etc., with suitable bonding substances such as appropriate silicone resins. Other materials or combinations of materials may be included in this class if by experience or accepted tests they can be shown to have comparable life at 180°C.
200°C N	Materials or combinations of materials which by experience or accepted tests can be shown to have the required thermal life at 200°C.
220°C R	Materials or combinations of materials which by experience or accepted tests can be shown to have the required thermal life at 220°C.
240°C S	Materials or combinations of materials which by experience or accepted tests can be shown to have the required thermal life at 240°C.
above 240°C C	Materials consisting entirely of mica, porcelain, glass, quartz, and similar inorganic materials. Other materials or combinations of materials may be included in this class if by experience or accepted tests they can be shown to have the required thermal life at temperatures above 240°C.

The above insulation classes indicate a normal life expectancy of 20 000 h to 40 000 h at the stated temperature. This implies that electrical equipment insulated with a class A insulation system would probably last for 2 to 5 years if operated continously at 105°C. Note that this classification assumes that the insulation system is not in contact with corrosive, humid, or dusty atmospheres.

For a complete explanation of insulation classes, insulation systems and temperature indices, see IEEE Std 1-1969, and the companion IEEE Standards Publications Nos 96, 97, 98, 99 and 101. See also IEEE Std 117-1974 and Underwriters Laboratories publication on insulation systems UL 1446, 1978.

4-10 Ionization of a gas

Most gases are excellent insulators, but they, too, break down in much the same way as solids do when the voltage becomes too high.

Consider two conducting plates separated by an insulating gas, such as air (Fig. 4-5a). Let us apply a voltage to the terminals of the plates, as we did before. The occasional free electrons, always found in a gas, are accelerated towards the positive plate and, in their hurried course, will bump into neutral atoms. If we gradually raise the voltage, the speed eventually becomes so great that the free electrons collide violently with the neutral atoms, knocking out one or more of their electrons. The electrons knocked out of orbit are accelerated in turn and strike even other atoms, liberating still more electrons. The gas is suddenly filled with a multitude of free electrons and a substantial current begins to flow. In effect, when the electric field (expressed in MV/m) attains this critical level, the gas becomes a conductor.

The electrons dislodged from their orbits leave behind atoms which now have a net positive charge. Such positively-charged atoms are called *positive ions*. Because these gaseous positive ions are free to move about, they will slowly drift towards the negative plate, while the free electrons move more rapidly towards the positive plate (Fig. 4-5a).

As soon as a positive ion touches the negative metallic plate, it captures one of the millions of free electrons available at the plate. The ion immediately becomes a simple atom of ordinary gas (electrically neutral). The electrons captured by the ions are immediately replaced by other free electrons arriving at the positive plate.

The circuit formed by the source, the plates and the gas now carries a moderate current, called *ionization current*. If we increase the voltage still more, the ionization current increases until it produces an electric arc. The luminous electric arc is the result of photon emission from electrons within the atoms of the gas. When these electrons drop back from outer to inner orbits, the atomic energy is released in the form of photons (visible light).

Such, ionization of a gas is always accompanied by the emission of radiant energy (visible, ultraviolet, or infra-red) and a hissing sound (mechanical energy). The effect is put to practical use in fluorescent lamps, as well as in mercury and sodium vapor lamps used in street lighting.

4-11 Thermal conductivity

All insulators are poor conductors of heat; their thermal conductivity is about 1000 times less than that of copper, for example. Stagnant air is the worst conductor of all: its thermal conductivity is

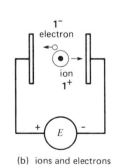
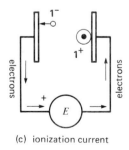

Figure 4-5   Ionization of a gas.
a. gas subjected to a potential difference;
b. collisions produce ions and electrons;
c. ionization current.

TABLE 4B	PROPERTIES OF INSULATING MATERIALS				
	Electrical Properties		**Thermal Properties**		**Mechanical Properties**	
Insulator	dielectric strength	dielectric constant	max operating temperature	thermal conductivity	density	notes
	MV/m or kV/mm	ϵ_r	°C	W/(m·°C)	kg/m³	
dry air	3	1	2000	0.024	1.29	gas density is at 0°C 101 kPa
hydrogen	2.7	1	-	0.17	0.09	
nitrogen	3.5	1	-	0.024	1.25	
oxygen	3	1	-	0.025	1.43	
sulfur hexafluoride (SF₆)	30 MV/m at 400 kPa	1	-	0.014	6.6	
solid asbestos	1	-	1600	0.4	2000	
asbestos wool	1	-	1600	0.1	400	
askarel	12	4.5	120	-	1560	synthetic liquid (restricted)
epoxy	20	3.3	130	0.3	1600 to 2000	
glass	100	5 to 7	600	1.0	2500	
magnesium oxide	3	4	1400	2.4	-	(powder)
mica	40 to 240	7	500 to 1000	0.36	2800	
mineral oil	10	2.2	110	0.16	860	
mylar®	400	3	150	-	1380	a polyester
nylon	16	4.1	150	0.3	1140	a polyamide
paper (treated)	14	4 to 7	120	0.17	1100	
polyamide	40	3.7	100 to 180	0.3	1100	
polycarbonate	25	3.0	130	0.2	1200	
polyethylene	40	2.3	90	0.4	930	
polyimide	200	3.8	180 to 400	0.3	1100	
polyurethane	35	3.6	90	0.35	1210	
polyvinylchloride (PVC)	50	3.7	70	0.18	1390	
porcelain	4	6	1300	1.0	2400	
rubber	12 to 20	4	65	0.14	950	
silicon	10	-	250	0.3	1800 to 2800	
teflon	20	2	260	0.24	2200	

16 000 times less than that of copper (Fig. 4-6). The poor conductivity of air is the reason we dip electrical windings in an insulating varnish. The liquid varnish fills and eliminates the air pockets. When it hardens, the windings become a solid mass of copper embedded in insulation. Heat generated within the impregnated windings moves out more freely to surrounding oil or air than when air-pockets are present.

Owing to the poor thermal conductivity of insulating materials, we tend to reduce their thickness to a strict minimum without, however, risking insulation breakdown or compromising mechanical strength. For example, a sheet of ordinary paper having a thickness of only 0.04 mm has sufficient dielectric strength to insulate the windings of a motor operating at 300 V. However, if used at this thickness, its weak mechanical strength renders it impractical. Consequently, we are obliged to use a sheet which is actually ten times thicker and produces a much higher dielectric strength.

5 m long is coated with a film of polyvinyl chloride (PVC) insulation 1 mm thick (Fig. 4-7). If the bus carries a current of 2000 A and dissipates I^2R losses of 450 W, calculate the difference of temperature between the copper and the external surface of the PVC.

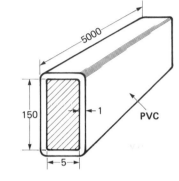

Figure 4-7
Insulation covering
a bus bar.

Solution:
Surface area A of the PVC next to the copper is:

$$A = \frac{(150 + 150 + 5 + 5)}{1000} \times 5 = 1.55 \text{ m}^2$$

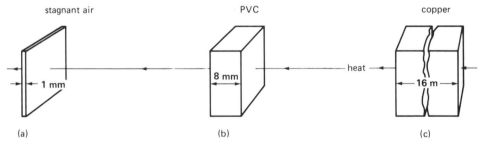

(a)   (b)   (c)

Figure 4-6   A film of stagnant air, 1 mm thick, offers as much resistance to the flow of heat, as does 8 mm of PVC, or 16 000 mm copper.

In high-voltage apparatus, it is the dielectric strength that mainly determines the thickness of the insulation used. The high dielectric strength of special polymers enables us to reduce the insulation thickness which, in turn, facilitates the cooling.

Example 4-3:
A copper bus bar 150 mm wide, 5 mm thick, and

Also from Table 4B, we have $\lambda = 0.18$

Using Eq. 3-5 we have:

$$P = \frac{\lambda A (t_1 - t_2)}{d} \qquad \text{Eq. 3-5}$$

$$450 = \frac{0.18 \times 1.55 (t_1 - t_2)}{0.001}$$

$$t_1 - t_2 = 1.6°C$$

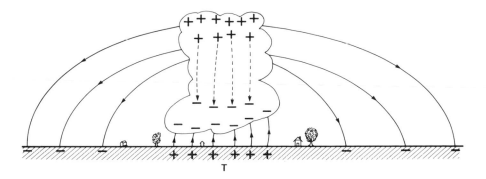

Figure 4-8   Electric fields created by a thundercloud.

4-12 Lightning strokes

During stormy weather, by a process not yet fully understood, a charge separation takes place inside certain clouds, so that positive charges move to the upper part of the cloud while negative charges stay below (Fig. 4-8). This transfer will obviously set up an electric field within the cloud. Furthermore, the negative charge at the base of the cloud repels the free electrons of the ground below. Consequently, region T becomes positively charged, by induction. It follows that an electric field and a difference of potential will also be established between the base of the cloud and the earth. Finally, yet another electric field exists between the electrons repelled from region T, and the positive charge at the top of the cloud.

As more and more positive charges move upward, the electric field below the cloud becomes more and more intense. Ultimately, it reaches the critical ionization level where air begins to break down. Ionization takes place first at the tips of church spires and the top of high trees, and may sometimes give rise to a bluish light. Mariners of old observed this light around the masts of their ships and called it St. Elmo's fire.

When the electric field becomes sufficiently intense, lightning will suddenly strike from cloud to earth. A single stroke may involve a charge transfer of from 0.2 to 20 coulombs, under a difference of potential of several hundred million volts. The current per stroke rises to peak in one or two microseconds and falls to half its peak value in about 40 μs. What is visually observed as a single stroke, is often composed of several strokes following each other in rapid succession. The total discharge time may last as long as 200 ms. Discharges also occur between positive and negative charges inside the cloud, rather than between the base of the cloud and ground. The thunder we hear is produced by a supersonic pressure wave. It is created by the sudden expansion of air around the intensely-hot lightning stroke.

4-13 Lightning arresters on buildings

The simplest lightning arresters are metallic rods which exceed the highest point of a building, channeling the lightning towards a ground electrode by means of a conducting wire. This prevents the high current from passing through the building itself, which might cause a fire, or endanger its occupants. A lightning arrester may be dangerous; during a discharge, it can momentarily create very high voltages between the conductor and ground.

Much more sophisticated lightning arresters are used on electrical utility systems. They divert lightning and high-voltage switching surges to ground before they damage costly and critical electrical equipment.

4-14 Lightning and transmission lines

When lightning makes a direct hit on a transmission line, it deposits a large electric charge, producing an enormous overvoltage between the line and ground. The dielectric strength of air is immediately exceeded and a flashover occurs. The line discharges itself and the overvoltage disappears in typically less than 5 μs.

point to point, between the line and ground (Fig. 4-9). The peak voltage (corresponding to the crest of the wave) may attain one or two million volts. Wave front **ab** is concentrated over a distance of about 300 m, while tail **bc** may stretch out over several kilometers.

The wave also represents the point-to-point value of the current flowing in the line. For most aerial lines, the ratio between surge voltage and

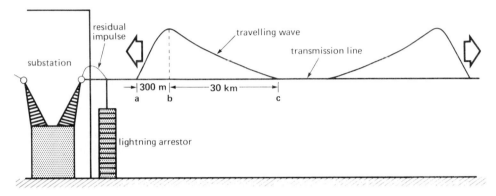

Figure 4-9   Flow of electric charge along a transmission line.

Unfortunately, the arc initiated by the lightning stroke produces a highly ionized path between the line and ground, which behaves like a conducting short-circuit. Consequently, the normal ac line voltage immediately delivers a large ac current which follows the ionized path. This follow-through current sustains the arc until the circuit-breakers open at the end of the line. The fastest circuit-breakers will trip in about 1/60th of a second, which is almost 16 000 μs after the lightning hit the line.

Direct hits on a transmission line are rare; more often, lightning will strike the overhead ground wire that shields the line. In the latter case, a local charge still accumulates on the line, producing a very high local overvoltage. This concentrated charge immediately divides into two waves that swiftly move in opposite directions at close to the speed of light (300 m/μs). The height of the impulse wave represents the voltage that exists from

surge current corresponds to a resistance of about 400 Ω. A surge voltage of 800 000 V at a given point is therefore accompanied by a surge current of 800 000/400 = 2000 A.

As the wave travels along the line, the I^2R and corona losses gradually cause it to flatten out, and the peak voltage decreases.

Should the wave encounter a line insulator, the latter will be briefly subjected to a violent overvoltage. The overvoltage period is equal to the time it takes for the wave to sweep past the insulator. The voltage rises from its nominal value to several hundred kilovolts in about 1 μs, corresponding to the length of wavefront **ab**. If the insulator cannot withstand this overvoltage, it will flash over, and the resulting follow-through current will cause the circuit-breakers to trip. On the other hand, if the insulator does not fail, the wave will continue to travel along the line until it eventually encounters a substation. It is here that the impulse wave can produce real havoc.

The windings of transformers, synchronous condensers, reactors, etc., are seriously damaged when they flash over to ground. Expensive repairs and even more costly shut-downs are incurred while the apparatus is out of service. The overvoltage may also damage circuit-breakers, switches, insulators, relays, etc., which make up a substation. To reduce the impulse voltage on station apparatus, lightning arresters are installed on all incoming lines.

Lightning arresters are designed to clip off the voltage peaks that exceed a certain level, say 400 kV. In turn, the apparatus within the substation is designed to withstand impulse voltages slightly higher than the arrester "clipping voltage," say 550 kV. Consequently, if a 1000 kV surge voltage enters a substation, the station arrester diverts a substantial part of the surge energy to ground. The residual impulse wave which travels beyond the arrestor then has a peak of only 400 kV. This impulse can easily be borne by station apparatus built to withstand an impulse of 550 kV.

4-15 Basic impulse insulation level (BIL)

How do insulating materials react to impulse voltages? Tests have shown that the withstand capability increases substantially when voltages are applied for very brief periods. To illustrate, suppose we wish to carry out an insulation test on a transformer, by applying a 60 Hz sinusoidal voltage between the windings and the tank. As we slowly raise the voltage, a point will be reached where breakdown occurs. Let us assume that the breakdown voltage is 46 kV (RMS) or 65 kV crest.

If we now apply a dc impulse between the windings and ground, we discover that it takes about twice the peak voltage (or 130 kV) before the insulation breaks down. The same is true of suspension insulators, bushings, spark gaps, etc., except that the ratio between impulse voltage and crest ac voltage is closer to 1.5.

In the interest of standardization, and to enable

a comparison between the impulse withstand capability of similar devices, standards organizations have precisely defined the shape and crest values of impulse waves. Figure 4-10 shows such a stan-

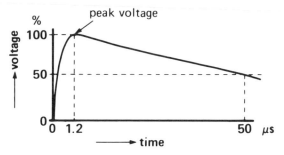

Figure 4-10   Standard shape of impulse voltage used to determine the BIL rating of electrical apparatus.

TABLE 4C   TYPICAL PEAK VOLTAGES FOR 1.2 x 50 μs BIL TESTS

Values are in kilovolts		
1550	825	250
1425	750	200
1300	650	150
1175	550	110
1050	450	90
900	350	30

dard impulse wave. It attains its peak after 1.2 μs and falls to one-half the peak in 50 μs. The peak voltage has a defined set of values that range from 30 kV to 2400 kV (see Table 4C).

The peak voltage is used to specify the basic impulse insulation levels (BIL) of equipment. Thus, a piece of equipment (transformer, insulator, capacitor, resistor, bushing, etc.) that can withstand a 1.2 x 50 microsecond wave of 900 kV, is said to possess a basic impulse insulation level (or BIL) of 900 kV.

The BIL of a device is always several times higher than its nominal ac operating voltage. For example, a 69 kV distribution transformer must have a BIL of 350 kV. However, there is no special relationship between BIL and nominal voltage. As the BIL rises, we must increase the amount of insulation which, in turn, increases the size and the cost of equipment.

In conclusion, the peak voltage across an arrester should never exceed the BIL of the apparatus it is intended to protect.

Figure 4-11   A 4 000 000 V impulse causes a flashover across an insulator string rated at 500 kV, 60 Hz. Such impulse tests increase the reliability of equipment in the field. The powerful impulse generator in the center of the photo is 24 m high and can deliver 400 kJ of energy at a potential of 6.5 MV (*IREQ*).

QUESTIONS AND PROBLEMS

Practical level

4-1 What determines whether a body is a conductor or an insulator?

4-2 Name two metals used for the transmission of electrical energy.

4-3 Do metals which are good conductors have a high or a low resistivity?

4-4 Name two metallic materials which are used as heating elements.

4-5 Why do we insulate certain conductors?

4-6 Name three insulators that can tolerate high temperatures without deteriorating appreciably.

4-7 Name some of the factors which contribute to the deterioration of organic insulators.

4-8 A transformer is built with class H insulation. What maximum temperature can it withstand?

4-9 Name three insulating gases and give the advantage of each.

4-10 Describe what happens when a voltage breakdown occurs:
   a. in a solid insulator;
   b. in a gaseous insulator.

4-11 What is meant by the abbreviation BIL?

Intermediate level

4-12 An electric motor has a normal life of eight years when the ambient temperature is 30°C. If brought to a location where the ambient temperature is 60°C, what is the new probable life of the motor?

4-13 An electromagnet (insulated class A) has a normal life of two years. What is its new life span if it is rewound using class F insulation? (Assume the temperature of the magnet remains the same).

4-14 A page of this book has a thickness of 80 μm. If its dielectric strength is 7 MV/m, what voltage can we safely apply across opposite faces before the page breaks down?

4-15 Estimate the breakdown voltage of the following:
   a. a window-pane having a thickness of 3 mm;
   b. two spheres in air separated by a distance of 25 mm;
   c. a polyimide film 0.025 mm thick which insulates a copper wire used in the coil of a relay.

4-16 We propose to insulate two adjacent electrical conductors with impregnated paper. What minimum thickness should be used if the voltage between the conductors is 200 kV?

4-17 Why do we impregnate windings with an insulating varnish? Give two reasons.

4-18 Why do we always try to minimize the quantity of insulation around electrical conductors?

4-19 Draw the graph of a 1.2 x 50 μs impulse having a peak of 900 kV.

4-20 A station lightning arrestor is designed to carry an impulse of 5000 A at a crest voltage of 600 kV. What BIL would you recommend for the station apparatus?

4-21 A lightning stroke delivers a charge of 6 coulombs at a potential difference of 200 MV.
   a. Calculate the energy, average current and average power of the stroke if it lasts for 200 μs;
   b. If it were possible to harness this energy to run a 7 kW electric stove, for how many hours would it heat?

4-22 In Example 4-3, calculate the difference of temperature across the insulator if the bus bar is insulated with 2 mm of epoxy.

4-23 A glass coffee pot is set upon a 1500 W stove element having a diameter of 150 mm. If the glass is 2 mm thick, calculate the temperature of the glass facing the element when the coffee is simmering at 80°C. Assume 80% of the heat given off by the element is available to warm the coffee.

5

CONDUCTING MATERIALS, WIRE AND CABLE

Most electrical conductors are either solid or liquid, but even gases become conductors when their temperature is high enough. In this chapter, we direct our attention mainly to the electrical, thermal, and mechanical properties of solid conductors.

5-1 Good conductors

Silver is an excellent conductor of electricity but, owing to its high cost, it is only used where its special properties are particularly important, such as in contacts which open and close electrical circuits. Copper is the most common material used because it conducts almost as well as silver, and costs much less. However, in the case of transmission lines, aluminum is often preferred because, for a given length and weight, its electrical resistance is about one-half that of copper. On the other hand, for a given *resistance*, the volume of aluminum is almost 1.7 times that of copper; consequently, it is far less attractive in the manufacture of electrical machines.

Phosphor bronze, an alloy of copper and tin, is used to make molded electrical parts, trolley wires, fuse clips, and flat springs.

5-2 Resistive conductors

The most important resistive materials are alloys of nickel and chromium, such as nichrome® and chromel® . They are used in rheostats, resistors and in heating elements of all kinds.

Tungsten is principally used to manufacture the filaments of incandescent lamps. Its high melting point permits a filament temperature sufficiently intense to emit a nearly white light.

There are at least thirty other resistive alloys of nickel, iron, cobalt and copper bearing names such as Nilvar®, Advance®, Karma®, Chromax®, Copel®, Ohmaloy®, etc. Their exceptional properties, such as high resistance to corrosion or great mechanical strength, render them useful for special applications.

5-3 Conductor shapes

Conductors are manufactured in a great variety of shapes and sizes to meet various applications. Wires are round, square, or flat; cables may be solid or stranded, and some heavy conductors are simply bare, rectangular bars (bus bars).

Most conductors are made of wire having a circular cross section, obtained by drawing the wire through successive dies (Fig. 5-1a).

Conductors having a square or rectangular cross section (Figs. 5-1b and 5-1c) are found in the windings of large transformers as well as in motors and generators. Rectangular conductors enable us to put almost 20 percent more copper in a given space, as compared to round conductors.

Stranded conductors are made up of several wires (called strands) that are twisted together to produce a more flexible wire or cable (Fig. 5-1d). The use of stranded conductors makes it easier to lay heavy cable, and to string transmission lines. Some stranded-aluminum conductors contain a central core composed of a stranded steel cable to increase the mechanical strength. Such cables are commonly designated by the term A.C.S.R. (aluminum cable steel reinforced).

called the "Brown & Sharp Gauge" (B & S).

According to this standard, each wire bears a gauge number, and the diameter of the wire diminishes as the gauge number increases: for instance, a #6 gauge wire is smaller than a #4 gauge wire. Table 5A gives the properties of round wires corresponding to the respective gauge numbers. The reader will find it useful to memorize the following rules which apply to the AWG system:*

1. **A conductor which has twice the cross section of another has a gauge number which is three numbers smaller.** Example: the cross section of a #15 wire is double that of a #18 wire.

2. **A conductor which has ten times the cross section of another has a gauge number which is ten numbers smaller.** Example: a #4 gauge wire has the same cross section as ten #14 gauge wires.

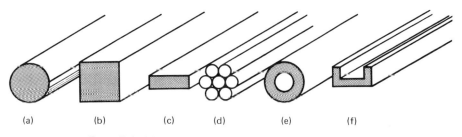

Figure 5-1   Various commercially available conductor shapes.

The high currents encountered in substations and in heavy industry are often carried by bare conductors having special shapes to facilitate the dissipation of heat and to increase mechanical strength (Figs. 5-1e and 5-1f).

Finally, we use resistive conductors having a circular or square cross section in the manufacture of electric heaters, motor starters, and rheostats.

5-4   Round conductors

In the United States and Canada, the diameters of round conductors are standardized according to what is known as the "Standard American Wire Gauge" (abbreviation AWG). This standard is also

Figure 5-2 shows how to measure the conductor size by means of a slotted wire gauge. After stripping all insulation (including the varnish covering), the bare wire is inserted into the slot that gives the best fit. The slot number corresponds to the wire size.

---

*   The AWG system follows a geometric progression in which the diameters for wire gauges #4/0 and #36 are fixed at 460 mils and 5 mils, respectively. Because there are 39 wire sizes between #4/0 and #36, the ratio between two successive diameters is
$$\sqrt[39]{460/5} = 1.1229 \ldots$$

TABLE 5A   PROPERTIES OF ROUND COPPER CONDUCTORS

Gauge number AWG/ B & S	Diameter of bare conductor		Cross section		Resistance mΩ/m or Ω/km		Weight g/m or kg/km	Typical diameter of insulated magnet
	mm	mils	mm^2	cmils	25°C	105°C		wire used in relays magnets motors trans- formers etc.
250MCM	12.7	500	126.6	250 000	0.138	0.181	1126	
4/0	11.7	460	107.4	212 000	0.164	0.214	953	
2/0	9.27	365	67.4	133 000	0.261	0.341	600	
1/0	8.26	325	53.5	105 600	0.328	0.429	475	
1	7.35	289	42.4	87 700	0.415	0.542	377	
2	6.54	258	33.6	66 400	0.522	0.683	300	
3	5.83	229	26.6	52 600	0.659	0.862	237	
4	5.18	204	21.1	41 600	0.833	1.09	187	
5	4.62	182	16.8	33 120	1.05	1.37	149	
6	4.11	162	13.30	26 240	1.32	1.73	118	
7	3.66	144	10.5	20 740	1.67	2.19	93.4	
8	3.25	128	8.30	16 380	2.12	2.90	73.8	mm
9	2.89	114	6.59	13 000	2.67	3.48	58.6	3.00
10	2.59	102	5.27	10 400	3.35	4.36	46.9	2.68
11	2.30	90.7	4.17	8 230	4.23	5.54	37.1	2.39
12	2.05	80.8	3.31	6 530	5.31	6.95	29.5	2.14
13	1.83	72.0	2.63	5 180	6.69	8.76	25.4	1.91
14	1.63	64.1	2.08	4 110	8.43	11.0	18.5	1.71
15	1.45	57.1	1.65	3 260	10.6	13.9	14.7	1.53
16	1.29	50.8	1.31	2 580	13.4	17.6	11.6	1.37
17	1.15	45.3	1.04	2 060	16.9	22.1	9.24	1.22
18	1.02	40.3	0.821	1 620	21.4	27.9	7.31	1.10
19	0.91	35.9	0.654	1 290	26.9	35.1	5.80	0.98
20	0.81	32.0	0.517	1 020	33.8	44.3	4.61	0.88
21	0.72	28.5	0.411	812	42.6	55.8	3.66	0.79
22	0.64	25.3	0.324	640	54.1	70.9	2.89	0.70
23	0.57	22.6	0.259	511	67.9	88.9	2.31	0.63
24	0.51	20.1	0.205	404	86.0	112	1.81	0.57
25	0.45	17.9	0.162	320	108	142	1.44	0.51
26	0.40	15.9	0.128	253	137	179	1.14	0.46
27	0.36	14.2	0.102	202	172	225	0.908	0.41
28	0.32	12.6	0.080	159	218	286	0.716	0.37
29	0.29	11.3	0.065	128	272	354	0.576	0.33
30	0.25	10.0	0.0507	100	348	456	0.451	0.29
31	0.23	8.9	0.0401	79.2	440	574	0.357	0.27
32	0.20	8.0	0.0324	64.0	541	709	0.289	0.24
33	0.18	7.1	0.0255	50.4	689	902	0.228	0.21
34	0.16	6.3	0.0201	39.7	873	1140	0.179	0.19
35	0.14	5.6	0.0159	31.4	1110	1450	0.141	0.17
36	0.13	5.0	0.0127	25.0	1390	1810	0.113	0.15
37	0.11	4.5	0.0103	20.3	1710	2230	0.091	0.14
38	0.10	4.0	0.0081	16.0	2170	2840	0.072	0.12
39	0.09	3.5	0.0062	12.3	2820	3690	0.055	0.11
40	0.08	3.1	0.0049	9.6	3610	4720	0.043	0.1

Figure 5-2   Measuring the size of a conductor with a wire gauge.

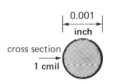

Figure 5-3   Circular mil.

5-5   Mils, circular mils

The diameter of a conductor is sometimes expressed in *mils* instead of in millimeters. The mil is a unit of length equal to one thousandth of an inch, or 0.0254 mm.

Similarly, the cross section of round wires is sometimes expressed in *circular mils* instead of in square millimeters. The circular mil (cmil or CM) is a unit of area equal to the area of a circle having a diameter of 1 mil (Fig. 5-3). A circular mil is therefore equal to 0.00 506 707 mm^2. The cross section of a round conductor expressed in circular mils is equal to the square of its diameter expressed in mils.

Example 5-1:
Calculate the cross section of a round wire having a diameter of 0.102 inches.

Solution:
The diameter is equal to 102 thousandths of an inch, or 102 mils. The cross section is therefore:

A = 102 mils x 102 mils = 10 404 circular mils

5-6   Stranded cable

The cross section of stranded cable is equal to the sum of the cross sections of all the strands, and does not include the area of the spaces between them. Thus, a #10 stranded wire possesses the same net cross section (and the same resistance) as a #10 solid wire.

Conductors bigger than #0000 (or 4/0) are generally identified by their cross section in thousands of circular mils (MCM).* Thus, a 250 MCM conductor has a cross sectiion of 250 000 cmil; it does not bear a gauge number.

5-7   Square wires

Square wires have gauge numbers that correspond to those of round wires. Here is the rule: If the side of a square wire has the same dimensions as the diameter of a round wire, the two bear the

---

* There is no standard symbol for this unit. Other common abbreviations are kCM and kcmil.

same gauge number (Fig. 5-4). It follows that the cross section of a square wire is about 25 percent greater than that of a round wire having the same gauge number.

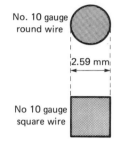

No. 10 gauge round wire

2.59 mm

No 10 gauge square wire

Figure 5-4   Square and round wires having the same gauge number.

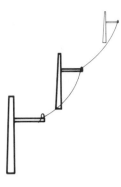

Figure 5-5   See Example 5-2.

5-8   Resistance of a conductor

At a given temperature, the resistance of a conductor depends upon its length, its cross section, and the nature of the material. The relationship is given by:

$$R = \rho \frac{l}{A} \qquad (5\text{-}1)$$

where

R = resistance of the conductor $[\Omega]$
ρ = resistivity of the material $[\Omega\cdot m]$
l = length of the conductor $[m]$
A = cross section of the conductor $[m^2]$

Resistivity is a basic property of a substance, representing its opposition to current flow. The SI unit of resistivity is the ohm meter $[\Omega\cdot m]$. However, owing to the low resistivity of metallic conductors, we prefer to use a submultiple, the nanohm-meter $[n\Omega\cdot m]$, equal to $10^{-9}\ \Omega\cdot m$. The resistivity of some common metals is given in Table 5B.

Example 5-2:
Calculate the resistance of copper conductor having a length of 2 km and a cross section of 22 mm^2 (Fig. 5-5). Assume the resistivity is 18 nΩ·m.

Solution:
Knowing that:
l = 2 km = 2000 m
A = 22 mm^2 = 22 x 10^{-6}m^2
ρ = 18 nΩ·m = 18 x 10^{-9} Ω
We find:

$$R = \rho \frac{l}{A}$$

$$= \frac{18}{10^9} \times \frac{2000}{22 \times 10^{-6}} = 1.64\ \Omega$$

Example 5-3:
Calculate the resistance of 25 m of #20 AWG nichrome wire at a temperature of 0°C.

Solution:
According to Table 5A, the cross section of #20 gauge round wire is 0.517 mm^2. Furthermore, the resistivity of nichrome is 1080 nΩ·m (Table 5B). The resistance of conductor is theretore:

$$R = \rho \frac{l}{A}$$

$$= \frac{1080}{10^9} \times \frac{25}{0.517 \times 10^{-6}} = 52.2\ \Omega$$

The resistance given by Equation 5-1 is the so-called the resistance of the conductor. The resistance may be considerably higher when the con-

TABLE 5B

ELECTRICAL, MECHANICAL AND THERMAL PROPERTIES OF SOME COMMON CONDUCTORS (AND INSULATORS)

Material	Chemical symbol or composition	Electrical properties			Mechanical properties			Thermal properties		
		resistivity ρ		temp coeff	density	yield strength	ultimate strength	specific heat	thermal conductivity	melting point
		0°C $n\Omega \cdot m$	20°C $n\Omega \cdot m$	at 0°C $(\times 10^{-3})$	kg/m³ or g/dm³	MPa	MPa	J/kg.°C	W/m.°C	°C
aluminum	Al	26.0	28.3	4.39	2703	21	62	960	218	660
brass	≈ 70% Cu, Zn	60.2	62.0	1.55	≈ 8300	124	370	370	143	960
carbon/ graphite	C	8000 to 30 000	—	≈ −0.3	≈ 2500	—	—	710	5.0	3600
constantan	54% Cu, 45% Ni, 1% Mn	500	500	−0.03	8900	—	—	410	22.6	1190
copper	Cu	15.88	17.24	4.27	8890	35	220	380	394	1083
gold	Au	22.7	24.4	3.65	19 300	—	69	130	296	1063
iron	Fe	88.1	101	7.34	7900	131	290	420	79.4	1535
lead	Pb	203	220	4.19	11 300	—	15	130	35	327
manganin	84% Cu, 4% Ni, 12% Mn	482	482	± 0.015	8410	—	—	—	20	1020
mercury	Hg	951	968	0.91	13 600	—	—	140	8.4	~ 39
molybdenum	Mo	49.6	52.9	3.3	10 200	—	690	246	138	2620
monel	30% Cu, 69% Ni, 1% Fe	418	434	1.97	8800	530	690	530	25	1360
nichrome	80% Ni, 20% Cr	1080	1082	0.11	8400	—	690	430	11.2	1400
nickel	Ni	78.4	85.4	4.47	8900	200	500	460	90	1455
platinum	Pt	9.7	10.4	3.4	21 400	—	—	131	71	1773
silver	Ag	15.0	16.2	4.11	10 500	—	—	230	408	960
tungsten	W	49.6	55.1	5.5	19 300	—	3376	140	20	3410
zinc	Zn	55.3	59.7	4.0	7100	—	70	380	110	420
air	78% N₂, 21% O₂	—	—	—	1.29	—	—	994	0.024	—
hydrogen	H₂	—	—	—	0.09	—	—	14 200	0.17	—
pure water	H₂O	—	2.5 × 10¹⁴	—	1000	—	—	4180	0.58	0.0

ductor carries an alternating current. The reason is that the ac flux created by the current produces an uneven current distribution in the conductor, thus raising its effective resistance. The ratio of ac to dc resistance depends upon the frequency, conductor size and permeability of the conductor and its surroundings. Thus, a large conductor embedded in iron may have an ac resistance that is 20% higher than the dc resistance. However, at 60 Hz, the ac resistance of most conductors is very close to the dc resistance.

5-9   Resistance as a function of temperature

The resistance of most metallic conductors increases with temperature. It may be calculated by using the equation:

$$\boxed{R_t = R_o (1 + \alpha t)}$$ (5-2)

where
R_t = resistance of the conductor at t °C [Ω]
R_o = resistance at 0°C [Ω]
α = temperature coefficient [1/°C] *
t = temperature [°C]
Table 5B gives the temperature coefficient of some common metals. Thus, for copper, α = 0.004 27.

Example 5-4:
A transmission line made of copper has a resistance of 100 Ω at 0°C. Calculate the change in resistance between summer and winter, knowing that the temperature varies from + 35°C to – 30°C.

Solution:
The resistance of the line at – 30°C is:
R = 100 {1 + (0.004 27 x – 30)}
= 100 (1 – 0.13) = 100 (0.87) = 87 Ω

At 35°C, the resistance is:
R = 100 {1 + (0.004 27 x 35)}
= 100 (1 + 0.15) = 115 Ω

* α is a greek letter pronounced "alpha".

The resistance of the transmission line varies from 87 Ω to 115 Ω, which represents a total change of 28 Ω (or 32 percent of 87 Ω). Consequently, for the same line current, the copper losses during the summer are 32 percent higher than during the winter.

Example 5-5:
The filament of a 60-watt incandescent lamp possesses a cold resistance of 17.6 Ω at 20°C (Fig. 5-6). The lamp draws a current of 0.5 A when connected to a 120 V source. Calculate the temperature of the hot filament. Take α = 0.0055/°C.

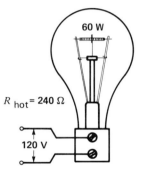

Figure 5-6 See Example 5-5.

Solution:
First, we calculate R_o, the resistance of the filament at 0°C:
R_t = R_o (1 + αt)
17.6 = R_o (1 + 0.0055 x 20)
17.6 = R_o (1.11)
from which
R_o = 15.85 Ω
Hot resistance of the filament is:
R_t = E/I = 120/0.5 = 240 Ω
Using Eq. 5-2, and letting t be the operating temperature, we obtain:
240 = 15.85 (1 + 0.0055 t)
from which
t = 2571°C

The resistance of some alloys, such as constantan and maganin, changes very little with temperature; consequently, we use these materials to make resistance standards and ammeter shunts.

Other alloys, such as nichrome and chromel, possess a high resistivity as well as a low temperature coefficient. They are used in heating elements and commercial resistors whose resistance should change only slightly with temperature. For example, the resistance of nichrome V increases by only 7 percent when its temperature increases from $20°C$ to $1000°C$.

5-10 Melting point

The melting point is one of the chief factors that determines the choice of metals used in furnace elements, fuses, and incandescent lamps.

Most metals soften and melt at around $1200°C$ and, consequently, they are not suitable as light-emitting materials. At this temperature, they emit a yellowish-orange color which is far from the white light emitted by the sun. Tungsten is a unique exception because it melts at $3410°C$ and has good mechanical strength up to $2800°C$; at this temperature the quality of light is good.

Zinc, which melts at $420°C$, is mainly used in fuses. Although lead melts at even lower temperatures, it is seldom used for this purpose.

Finally, molybdenum, which melts at $2610°C$, is employed as a resistive element in high-temperature electric furnaces. A hydrogen atmosphere is sometimes used to prevent oxidation of the heating elements.

5-11 Fusing current

If the current carried by a conductor is high enough, the conductor will melt. The fusing current depends upon the material and upon the diameter of the conductor. For bare, round conductors in free air, the current is given by the following approximate equation:

TABLE 5C	FUSING COEFFICIENT
Metal	k
copper	69
silver	62
aluminum	30
chromel	29
zinc	14
lead	6

$$I = kd^{3/2} \qquad (5\text{-}3)$$

where

I = fusing current [A]
d = diameter of the wire [mm]
k = coefficient depending upon the metal (see Table 5C)

Example 5-6:
Calculate the fusing current for #12 copper wire, in air.

Solution:
The diameter of #12 copper wire is 2.05 mm and $k = 69$.

$$I = kd^{3/2}$$
$$= 69 (2.05)^{3/2} = 203 \text{ A}$$

5-12 Fuses

The melting point of a conductor is put to practical use in the construction of fuses. These devices usually comprise a fuse link enclosed in a fiber tube (Fig. 5-7). Fiber is very popular because of its tough resistance to arcs, its mechanical strength, and its excellent insulating properties.

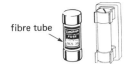

fibre tube

Figure 5-7 Cartridge fuse (non renewable) and fuse-holder, rated 30 A, 250 V.

(Bussmann)

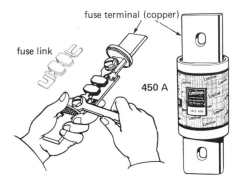

fuse terminal (copper)

fuse link

450 A

Figure 5-8 Renewable link fuse rated 450 A, 600 V.

15 A

Figure 5-9 Plug fuse rated 15 A, 240 V.

The fuse link is indented at one, two, or three places along its length so as to create short, narrow bridges of relatively high resistance (Fig. 5-8). When the current exceeds the nominal value, the bridges melt thereby interrupting the circuit. Plug fuses found in homes are designed along the same principles as industrial fuses (Fig. 5-9).

When a severe short-circuit occurs, the current becomes very high and the tremendous heat causes the fusible element to literally explode. The fiber tube must withstand the high internal pressure and special precautions must be taken to prevent the arc from being sustained by the vaporized metal.

To meet these requirements, we increase the length of the fuse as the operating voltage increases. Furthermore, the amount of fusible metal is kept to a strict minimum. Fuses are usually designed to carry the nominal current in an ambient temperature of 50°C.

Manufacturers of industrial fuses specify not only the nominal current, but also the maximum current the fuse may have to interrupt. For example, the 30 A, 250 V fuse shown in Figure 5-7 can interrupt a 250 V circuit even if the source has the ability to deliver a short-circuit current of 200 000 A.

High rupturing capacity (HRC) fuses often have a fusible element made of a thin copper or silver wire.

5-13 Contact resistance

When a fuse is pushed into the fuse holder, the contact with the stationary parts is never perfect. This may give rise to appreciable contact resistance and substantial I^2R losses. The heat may cause the fusible element to melt, even if the current is below its rated value. A poor contact sometimes becomes so hot that the insulating supports become carbonized and a massive, dangerous short-circuit can result. Poor contacts *anywhere* are always a potential hazard.

Example 5-7:
The terminals of a circuit-breaker are bolted to a bus-bar and the contact resistance is 0.0001 Ω. If the current is 6000 A, calculate the heat dissipated.

Solution:
The heat released is:
$$P = I^2R$$
$$= 0.0001 \times (6000)^2 = 3600\,W$$

This very high power loss will soon carbonize any surrounding insulation and will soften up and oxidize the metallic parts. This, in turn, increases the contact resistance still more until a catastrophic failure occurs. Note that the apparently "low resistance" is no guarantee that a joint will not overheat.

5-14 Nonlinear resistors

Most resistors are linear, which means that for a given temperature the current is proportional to the applied voltage. However, some materials are available in which the current does not increase in proportion to the voltage; they are called non-linear resistors. *Thermistors*, whose resistance is very sensitive to temperature changes, and *varistors* whose resistance drops with applied voltage are two examples of nonlinear resistors.

5-15 Thermistors

Figure 5-10 indicates that the resistance of a thermistor falls very rapidly as the temperature increases. The resistance decreases progressively from 4000 Ω to 3 Ω as its temperature varies from −50°C to +150°C. At a temperature of 25°C, the resistance falls at a rate of 4 percent per degree Celsius, which makes the thermistor useful as a temperature detector.

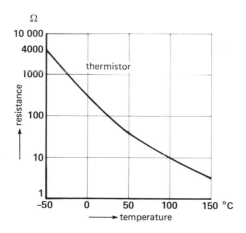

Figure 5-10   Typical thermistor characteristic.

5-16 The varistor (Thyrite)

The varistor is also a nonlinear resistor whose resistance decreases *instantaneously* as the voltage increases. It is made of a silicon-carbide powder and

built in the shape of a disc, sometimes as big as a hockey puck. The *E-I* characteristic of a varistor shows that the current increases dramatically with increasing voltage. Thus, when the voltage increases from 1.5 kV to 10 kV, the current rises from 1 mA to 100 A, an increase by a factor of 100 000 (Fig. 5-11).

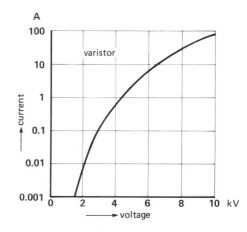

Figure 5-11   Typical varistor characteristic.

Varistors are placed in parallel with critical components which might be damaged by transient high voltages. They are also used in lightning arresters which protect very large electrical installations. Arresters are discussed in greater detail in Chapter 30.

5-17 Resistance of the ground

The resistance of the ground has a very important influence on electrical installations. First, for safety reasons, most electrical installations are connected to ground (earthed). Second, some high tension dc transmission lines use the ground as a return conductor to carry currents of several hundred amperes. Third, because the ground is a relatively good conductor, it offers a path for leakage currents of all kinds, which can corrode metallic pipes and structures buried in the ground.

The resistivity of the ground ranges between 5 Ω·m and 5000 Ω·m depending on its composition (clay, sand, granite, etc.) and the degree of moistness. For example, in the spring the resistivity of wet soil may be 50 $\Omega \cdot$m, and during the summer it may reach 300 $\Omega \cdot$m as the soil dries. Soil resistivity also varies with temperature. It increases slowly as temperatures approach the freezing point and then increases very rapidly as the temperature decreases further.

5-18 Resistance between two electrodes

In spite of its high resistivity, the ground is an excellent conductor. This is due to the enormous cross section it offers to current flow. For example, if we apply a voltage E between two electrodes driven into the ground, we note that the current flows throughout the entire volume of the earth, following a path similar to that shown in Fig. 5-12.

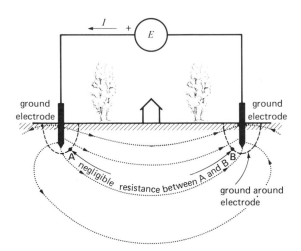

Figure 5-12   Ground current between two electrodes.

The resistance between the electrodes remains quite low, even though they may be many kilometers apart. We distinguish four resistances in the circuit of Fig. 5-12:

1. Resistance of the metal electrodes - negligible;
2. Contact resistance between the electrodes

and the soil - negligible;
3. Resistance of the soil around each electrode - important (this resistance depends mainly upon the nature of the soil and the depth of the electrodes);
4. Resistance of the ground between the electrodes - negligible.

Experience has shown that the resistance is mainly concentrated in a 10-meter radius around each electrode. Beyond this circle, the resistance is negligible. Consequently, the distance between electrodes does not change the resistance between them unless the electrodes are very close together.

We can reduce the resistance by driving the electrodes deeper into the ground or by impregnating the surrounding soil with chemicals, such as copper sulfate. In general, the electrode resistance diminishes by half whenever the depth is increased by a factor of 1.7. For example, if a depth of 1 m gives a resistance of 80 Ω, a depth of 1.7 m reduces the resistance to 40 Ω.

5-19 Measuring the resistance
of a ground electrode

We can easily measure the resistance of one electrode by using the set-up shown in Fig. 5-13. A voltage E_1 is applied between two electrodes A and B and we measure the resulting current I. We then measure the voltage E between electrode A and a small metallic probe placed at 60 percent of the distance d between the two electrodes. The ratio

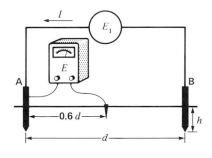

Figure 5-13   Measuring the resistance of a grounding electrode.

E/I gives the ground resistance of electrode A. It is simply called the resistance of electrode A. Note that in making the test the distance between the electrodes must be at least ten times the depth h.

Example 5-10:
Referring to Figure 5-13, two electrodes A and B, 12 m apart, are driven to a depth of 80 cm in a field. A source E_1 causes a current of 0.2 A to flow in the circuit. A voltage of 19 V is measured at a distance 60% x 12 = 7.2 m from one of the electrodes. The resistance of electrode A is, therefore:

$$R = E/I = 19/0.2 = 95\ \Omega^*$$

Special instruments are available to measure ground resistance (Fig. 5-14) but the simple method described above is sufficiently accurate in most cases.

Figure 5-14   Earth tester for measuring ground resistance; range: 0.01 to 9990 ohms. *(James G. Biddle Co.)*

5-20   Resistance of liquid conductors, electrolytes

Metals such as mercury, lead, iron, and so forth contain as many free electrons in their liquid state as they do when they are solid. It follows that such metallic liquids are excellent conductors of electricity.

* Both the National Electrical Code and the Canadian Electrical Code require a maximum ground resistance of 25 Ω per electrode.

Pure water, on the other hand, is an excellent insulator whose resistivity is about 10^{12} times greater than that of copper. Similarly, ordinary table salt is an excellent insulator. However, if we add a tablespoon of salt to a pail of water, the conductivity of the water increases dramatically. How can a liquid insulator and a solid insulator together produce a solution that conducts electricity? The following simplified explanation shows what happens when salt and water are mixed.

Each molecule of salt is composed of an atom of sodium (Na) and an atom of chlorine (Cl). In the presence of water, each molecule splits in two, to create a positive ion of sodium and a negative ion of chlorine. The sodium ion is simply an atom of sodium which has lost an electron, whereas the chlorine ion is an atom of chlorine which has one extra electron. These ions move about in random fashion within the water solution, very much like free electrons do inside a metal. Because the ions carry equal positive and negative charges, the solution (called electrolyte) is itself electrically neutral (Fig. 5-15).

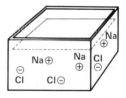

Figure 5-15   Positive and negative ions in an electrolyte.

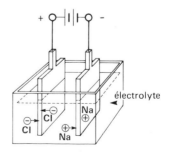

Figure 5-16   Migration of ions.

If we immerse two electrodes in the solution and connect them to a battery, the positive and negative ions migrate slowly towards the electrode of opposite polarity (Fig. 5-16). As soon as a positive sodium ion (Na$^+$) touches the negative plate, it captures a free electron from the plate and becomes a neutral sodium atom. In the same way, when a negative chlorine ion (Cl$^-$) touches the positive plate, it *loses* its extra electron to the plate and becomes a neutral atom of chlorine. Thus, the positive electrode continually receives electrons while the negative electrode continually gives them up. An electric current is therefore set up in the circuit formed by the electrolyte, electrodes and battery. However, the current in the electrolyte is not an electron flow but rather a migration of positive and negative ions which move respectively to the right and to the left. However, the current in the battery and the electrodes is a conventional electron current flow.

The same phenomenon takes place in a solution of water and any other salt. A solution of water and an acid also produces the same effect. In each case, the resulting electrolyte is composed of positive and negative ions, each representing an atom (or molecule) that has lost or gained one or more electrons. When current is made to pass through such a solution, a chemical reaction usually takes place at both electrodes. In effect, the atoms created when the ions are neutralized may react with the water or with the plate they touch. A gas is often released at one or both electrodes and sometimes the plates simply dissolve. In other cases, the atoms adhere to the plate to produce a smooth coating whose thickness depends upon the duration of current flow. This process is called electroplating.

Example 5-9:
Two electrodes having a surface of 1.2 m^2 and spaced 10 cm apart are plunged into sea water whose resistivity is 0.3 Ω·m. If we connect a 12 V battery to the electrodes, calculate the approximate value of the resulting current.

Solution:
The resistance of the electrolyte is:

$$R = \rho \frac{l}{A}$$
$$= 0.3 \times \frac{0.1}{1.2} = 0.025 \ \Omega$$

The current is $I = E/R = 12/0.025 = 480$ A

Note that even though the resistivity of salt water is twenty million times greater than that of copper, the current flow is still very large despite the relatively low voltage.

Electrolytes play a crucial role in electric batteries and fuel cells. Consequently, the interest in these liquid conductors goes far beyond the simple question of resistance.

MECHANICAL PROPERTIES OF CONDUCTORS

So far, we have directed our attention to the electrical and thermal properties of conductors, but in some cases we also have to consider their mechanical properties. Mechanical strength, for example, plays an important role in transmission lines and in coil-winding operations. First, the tractive force must not exceed the breaking strength of the conductor. Second, we must prevent any excessive stretching, otherwise the conductor may not return to its original length and shape when the tension is removed. These two considerations require a knowledge of the forces which change the length of a conductor permanently.

5-21 Tractive force and elongation

Let us gradually increase the pull (or tractive force) F on a metallic conductor, while observing the change in length d (Fig. 5-17). We obtain a series of readings which can be shown in the form of a graph (Fig. 5-18). On this graph, we indicate three reference points, **a**, **b**, and **c**, corresponding respectively to tractive forces F_1, F_2, and F_3.

As we increase the force from zero to F_1, the wire stretches uniformly from zero to d_1. During this straight portion **oa** of the curve, the increase

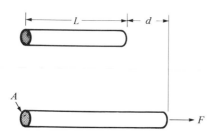

Figure 5-17   A conductor stretches when
subjected to a pull F.

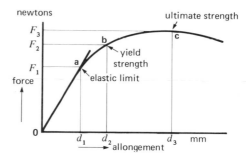

Figure 5-18   Relationship between pull
and increase in length.

in length d is proportional to the tractive force, and the wire acts like a spring. It returns to its original shape when the tractive force is removed.

For tractive forces greater than F_1 but less than F_2, the wire loses some of its elasticity and does not quite come back to its original length when the tension is removed. However, the permanent deformation is not serious enough to be important. The force F_2 at point **b** corresponds to the *yield strength* of the conductor. In practice, we must keep the tractive force below this limit.

If we raise the tractive force still more, the wire lengthens rapidly until we reach point **c** situated at the top of the curve. This corresponds to the *ultimate strength* of the conductor. Beyond point **c**, even a smaller force than F_3 may continue to stretch the conductor until it breaks.

We can draw a force-elongation curve for any conductor, but the actual forces F_2 and F_3 depend upon the type of metal and its cross section, and whether the conductor is hard-drawn or annealed. The shape of the curve also depends upon the operating temperature; the forces become smaller as the temperature increases.

5-22 Elastic limit, yield strength and ultimate strength

Referring to Figure 5-18, and assuming the conductor has a cross section A, we define three char-

acteristic "strengths"* of the conductor material:

$$\text{elastic limit} = F_1/A \qquad (5\text{-}4)$$

$$\text{yield strength} = F_2/A \qquad (5\text{-}5)$$

$$\text{ultimate strength†} = F_3/A \qquad (5\text{-}6)$$

$$\text{percent elongation} = d_3/l \qquad (5\text{-}7)$$

where

$F_1, F_2, F_3 =$ tractive forces as defined in Figure 5-18 [N]

$A =$ cross section of the material [m^2]

$d_3 =$ increase in length at fracture [m]

$l =$ original unstressed length [m]

Manufacturers can usually supply information regarding elastic limit, yield strength, and tensile strength of various materials (in megapascals, (MPa), as well as the percent elongation. Referring to Table 5D, we note that the yield strength of annealed copper is 35 MPa and that the ultimate strength is 220 MPa. Furthermore, the wire stretches by 60 percent before rupturing. Table 5D also reveals the great difference between annealed

---

* The technical term is "stress".

† Ultimate strength is often called "tensile strength".

TABLE 5D   MECHANICAL STRENGTH OF CONDUCTORS

Conductor material	Yield strength	Ultimate strength	Elongation
	MPa	MPa	%
pure aluminum	21	62	50
annealed copper	35	220	60
hard-drawn copper	410	470	14
steel	1170	1300	15

and hard-drawn copper. The first is used to wind coils while the second is usually found in aerial transmission lines. Finally, it is easy to understand why aluminum conductors used in transmission lines must be reinforced with a steel core: aluminum alone does not have the strength to support itself over long spans.

Example 5-8:
a. Calculate the maximum pull we can exert on a #12 annealed copper wire without over-stretching it.
b. What tractive force will cause the wire to break?
c. If the wire has an initial length of 5 m, calculate the length when it breaks. (The cross section of #12 AWG wire is 3.31 mm^2.)

Solution:
a. The yield strength is 35 MPa; consequently:
F_2 = (yield strength) x (cross section)
   = (35 x 10^6) x 3.31 x 10^{-6} = 116 N
b. The pull at the breaking point is:
F_3 = (ultimate strength) x (cross section)
   = (220 x 10^6) x 3.31 x 10^{-6} = 728 N
c. Because the elongation is 60 percent, the wire will lengthen by 5 m x 60% = 3 m before it snaps.

CONDUCTOR INSULATION

5-23 Types of insulation

Most conductors are insulated to prevent contact with other conductors, and to prevent short-circuits to ground.

Conductors used in electrical machines are often insulated with cotton, silk, or paper, and impregnated with special varnishes. These organic materials can tolerate moderate temperatures. Today, however, we tend to use synthetic insulating materials for machines of medium and large power.

Conductors that distribute electricity in houses and buildings are wrapped with flexible insulation, such as rubber, paper, cotton, cambric, and thermoplastic products. These conductors must operate at relatively low temperatures in order to ensure long life.

Conductors exposed to high temperatures, such as in electric stoves, must be covered with insulating materials having a mineral base. Glass, asbestos, porcelain, and mica can tolerate high temperatures without deteriorating appreciably.

5-24 Thermal capacity of conductors

Even the best conductors possess some resistance;

consequently, they all heat up when they carry an electric current. The heat produced in insulated conductors must be transmitted through the insulating layers and finally dissipated to the surrounding air. The greater the current in the wire, the higher the temperature of the insulation will be. On the other hand, to ensure a reasonable life, the temperature of the insulation must not be too high. We therefore arrive at the following important conclusion:

The maximum current an insulated wire can carry depends upon the maximum temperature its insulation can withstand, for an acceptable period of time.

Although bare conductors present no insulation problem, we must still keep the current and temperature within reasonable limits. A high temperature may produce excessive oxidation and flaking of the conductor and may represent a potential fire hazard for surrounding objects.

5-25 National Electrical Code and electrical installations*

The life expectancy of the electrical wiring in factories, buildings and homes must be particularly long because we cannot afford to replace the conductors every ten years. For this reason, the National Electrical Code specifies rather low maximum temperatures for wire and cable used in electrical installations. Depending on the type of insulation, the Code typically recognizes maximum temperatures of 60°C, 75°C, and 90°C. These are much lower than the maximum temperatures permitted in electrical apparatus using similar insulation.

A conductor must ultimately dissipate its heat to the surrounding air and, in order to establish its current-carrying capacity, the electrical code specifies that the standard ambient temperature shall be 30°C. Using this standard, and knowing the temperature rating of the particular insulation used,

---

* The general remarks made in this section also apply to the Canadian Electrical Code.

wire manufacturers can specify the current-carrying capacity (ampacity) of their products. For example, a #6 wire whose insulation is rated at 60°C can carry a current of 80 A in free air at an ambient temperature of 30°C (Fig. 5-19). On the other hand, a #6 wire whose insulation is rated at 90°C can carry a current of 100 A under the same ambient conditions (Fig. 5-20). Both conductors have the same life expectancy under the specified conditions.

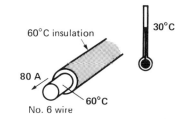

Figure 5-19 A No. 6 conductor having 60°C insulation can carry 80 A.

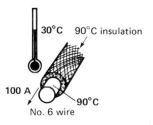

Figure 5-20 A No. 6 conductor having 90°C insulation can carry 100 A.

Table 5E, taken in part from the National Electrical Code, gives an idea of the ampacities of various conductors in free air at an ambient temperature of 30°C.

When several insulated conductors are placed in the same conduit, the heat dissipated by each raises the temperature of the others. The ampacity of each conductor must therefore be reduced as the number of conductors increases so as not to exceed the maximum permissible temperature. For example, the National Electrical Code specifies that when three #6 conductors rated at 60°C are

TABLE 5E

ALLOWABLE AMPACITIES FOR INSULATED CONDUCTORS, RATED 0-2000 V
SINGLE CONDUCTORS IN FREE AIR,
BASED ON AMBIENT TEMPERATURE OF 30°C

Gauge Number	Temperature rating of insulation		
	60°C	90°C	200°C
12	25	40 (25)	55
10	40	55 (40)	75
8	55	70	100
6	80	100	135
4	105	135	180
2	140	180	240
1/0	195	245	325
3/0	260	330	430
250 MCM	340	425	–

Note: The above values are derived from the National Electrical Code. The Canadian
Electrical Code gives identical values except for those values shown in paren-
thesis.

Figure 5-21

(a) Two-conductor rubber-covered
cable protected by a paper and
woven-fabric sheath.

(b) Asbestos-covered, two-conductor
cable.

(c) Two-conductor armored cable.

(d) Three-conductor lead-covered
armored cable.

(e) Three-conductor cable with a
lead sheath.

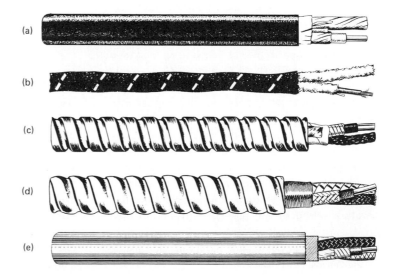

placed in a conduit, the ampacity per conductor is only 55 A compared to 80 A when they are suspended separately in free air.

5-26 Comparison of various conductors

The various cables shown in Figures 5-21 to 5-28 illustrate how their construction is affected by voltage rating, type of conductor and application. The cross sections are shown full-size so that the reader can visualize the actual physical dimensions of the cables. A detailed examination of these figures reveals the following information:

Figures 5-22a and 5-22b:
For the same size conductor, an increase in the voltage rating from 5 kV to 30 kV requires more insulation, therefore a larger cable. However, the ampacity is only slightly reduced.

Figures 5-22b and 5-22c:
A 30 kV cable with a 250 MCM aluminum conductor has an ampacity of 343 A; a similar cable made

of copper can carry 440 A.

Figures 5-22b and 5-23:
A 30 kV cable composed of three 250 MCM copper conductors has an ampacity of 359 A per conductor as compared with 440 A for a single conductor.

Figures 5-24a and 5-24b:
For the same voltage rating (600 V), a type R90 cross-linked polyethylene insulation is thinner than a standard-type RW60 insulation. Consequently, the ampacity of the conductor is also greater (425 A versus 340 A). However, the higher current rating is principally due to the 90°C versus the 60°C rating of the two insulating materials.

Figures 5-23 and 5-25:
The oil-cooled submarine cable rated at 138 kV, 630 A is no bigger than a standard 3-phase, 30 kV, 359 A cable, despite the fact that it is protected by a heavy steel wire sheath and carries almost five times as much power.

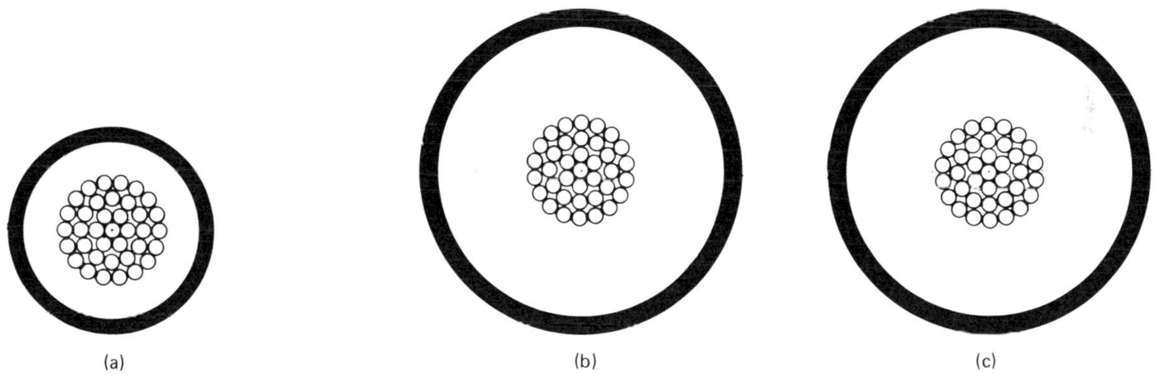

Figure 5-22   Ampacity in free air of three different conductors having a cross section of 250 MCM (full size).

    (a) Copper: 5 kV, 444 A.

    (b) Copper: 30 kV, 440 A.

    (c) Aluminum: 30 kV, 343 A.

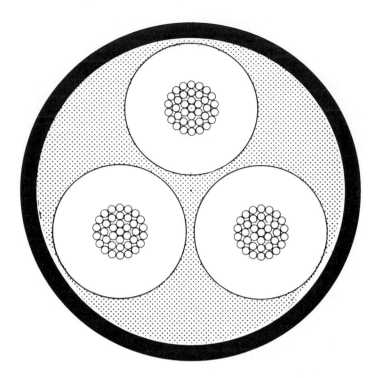

Figure 5-23 Three-phase cable composed of three 250 MCM conductors (full size) made of copper (30 kV, 359 A); ambient temperature, 40°C; maximum conductor temperature 90°C.

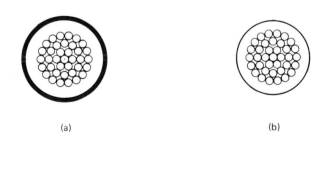

(a)   (b)

Figure 5-24 Conductors insulated for 600 V, using different insulation.

(a) Copper: RW 60; 600 V, 340 A.

(b) Copper: R 90; 600 V, 425 A.

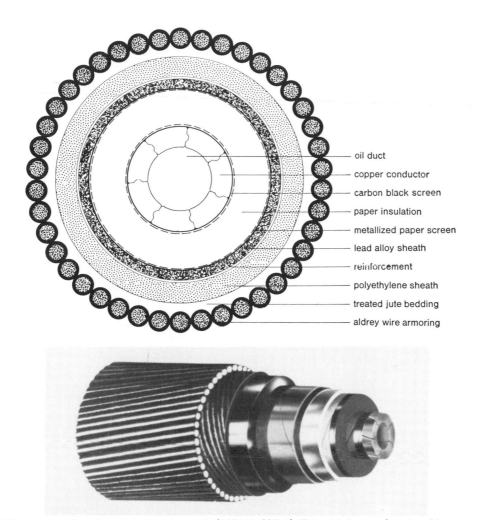

— oil duct

— copper conductor

— carbon black screen

— paper insulation

— metallized paper screen

— lead alloy sheath

— reinforcement

— polyethylene sheath

— treated jute bedding

— aldrey wire armoring

Figure 5-25   Cross section view of a submarine cable (138 kV, 630 A). This cable is one of seven cables submerged in Long Island Sound between Northport (Long Island) and Norwalk (Connecticut). *(Pirelli Cables Limited)*.

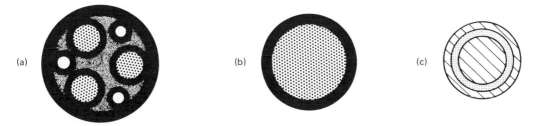

(a)   (b)   (c)

Figure 5-26   Cable for use in a mine, to feed a submersible pump. It is composed of 3 power conductors (No. 6, 133 strands), 2 control conductors (No. 10, 104 strands) and one grounding conductor (No. 8, 133 strands).

Figure 5-27   Ultra-flexible 600 A, 250 MCM conductor for arc welding (No. 34, 6834 strands-polyethylene sheath).

Figure 5-28   Single-conductor cable (250 MCM, 425 A) with copper sheath ground return, for use in hot locations (250°C).

QUESTION AND PROBLEMS

Practical level

5-1 Why can a conductor insulated with asbestos carry a larger current than one insulated with rubber?

5-2 What is a circular mil?

5-3 Upon what does the resistance of a round conductor depend?

5-4 Does the resistance of copper diminish with temperature?

5-5 What is meant by the term 500 MCM?

5-6 What determines the ampacity of a bare wire? of an insulated wire?

5-7 Why is aluminum seldom used in electrical machines?

5-8 When several insulated conductors are placed in the same conduit, the individual ampacity is less than that of a single conductor. Why?

5-9 What determines the insulation thickness of a conductor?

5-10 Why do we sometimes prefer to use a #10 stranded wire instead of a #10 solid wire?

5-11 If a #12 stranded wire is replaced by a #12 solid wire, does the resistance change?

5-12 Explain what is meant by elastic limit, yield strength and tensile strength.

5-13 Is a #10 AWG wire smaller than a #20 AWG? What is the cross section of these wires in cmils?

5-14 A motor is wound using two #12 wires in parallel. What is the size of the single wire that could replace them?

5-15 A cable is composed of four #16 AWG wires. What is its gauge size?

5-16 Calculate the resistance of 210 m of #14 wire at a temperature of 25°C. (Use the values given in Table 5A).

5-17 Express 500 MCM in mm².

5-18 Does a #4 square wire have a larger cross section than a #4 round wire? If so, by how much?

5-19 We have to select a cable capable of carrying a current of 95 A. What is the minimum size required, if the insulation is rated for 60°C? (See Table 5E).

5-20 Explain what is meant by an ion; by an electrolyte.

Intermediate level

5-21 A round conductor has a diameter of 0.0172 inches. Calculate its cross section in cmils.

5-22 A coil wound with #22 AWG wire has a resistance of 400 Ω at 25°C. Calculate the length of the wire and the weight of the coil. (Use the values given in Table 5A).

5-23 Using Table 5A and the rules which apply to the AWG system, determine the cross section in circular mils of #43 and #48 wires.

5-24 A coil made of copper wire has a resistance of 25 Ω when immersed in ice water (0°C). What would its resistance be in boiling water (100°C)? Assume the temperature coefficient is 0.004/°C.

5-25 A transmission line composed of two #4 copper wires has a length of 800 m. Calculate:
   a. the total resistance of the line at 25°C.
   b. the losses in the line if the current is 120 A.

5-26 Using Eq. 5-1, calculate the resistance of 1500 m of #6 aluminum wire at a temperature of 38°C. (See Tables 5A and 5B).

5-27 What is the maximum pull we can safely exert on a #40 copper wire (annealed) without overstretching it? What force would cause the wire to break?

5-28 Compare the electric power which each of the three cables of Figure 5-22 can carry.

Advanced level

5-29 A wire made of lead has a length of 2 km and a diameter of 2 mm. Calculate its resistance at a temperature of 130°C.

5-30 Referring to the 3-phase cable shown in Figure 5-23, calculate the maximum tractive force it can tolerate without stretching excessively when it is pulled through an underground duct. What tractive force might cause it to break?

5-31 The coil of an electromagnet, made of copper wire, has a resistance of 4 Ω at a temperature of 22°C. After operating for 2 days, the coil current is 42 A at a terminal voltage of 210 V. Calculate the average temperature of the coil at that time.

5-32 The resistivity of dry sand (and of granite) is about 1000 Ω·m. Calculate the resistance (between opposite faces) of a cube of this material having dimensions 10 m x 10 m x 10 m.

5-33 A 3-phase 15 kV, 750 MCM, 90°C cable, similar to the one shown in Figure 5-23 can carry a current of 545 A when laid in a conduit. Each conductor is composed of 61 strands of copper. Calculate:

a. the diameter of each strand;

b. the heat dissipated (in W/km) at 90°C.

5-34 In Problem 33, if the conductors are made of aluminum, what maximum current can each conductor carry without exceeding the maximum permissible temperature limits?

6
MAGNETIC MATERIALS AND PERMANENT MAGNETS

Many magnetic materials are available to the electrical equipment designer. They are the outcome of years of research in metallurgy in which certain magnetic properties were enhanced and others suppressed, depending upon the particular application.

In this chapter, we cover the basic properties of magnetic materials when they are subjected to dc and ac fields. We also include permanent magnets because of their commercial importance, and because their behavior is a logical extension of the properties of all magnetic materials.

6-1 B-H saturation curves

We discussed the B-H curves of several materials in Sections 2-20 to 2-23. We now add Figure 6-1 to show even a broader range. Owing to its logarithmic scale, the figure includes the B-H curves of the most permeable materials known to the least permeable of all, a vacuum.

As the magnetic field intensity increases, the relative permeability of all magnetic materials falls,

in accordance with Eq. 2-10:

$$\mu_r = 800\,000\,B/H \qquad \text{Eq. 2-10}$$

However, the permeability can never be less than 1. This is why all the B-H curves in Figure 6-1 ultimately follow the B-H curve of a vacuum when the magnetic field intensity H is high enough.

6-2 Saturation density

As we increase H in a magnetic material, more and more domains line up. We ultimately reach a state where *all* the domains are oriented in the same direction. The flux density at which this occurs is called the *saturation flux density*. If we increase H beyond this point, the flux density continues to rise, but only at a rate corresponding to the B-H curve of a vacuum. The reason is that the domains cannot contribute any further increase in B once they are totally lined up.

The saturation flux density for most magnetic

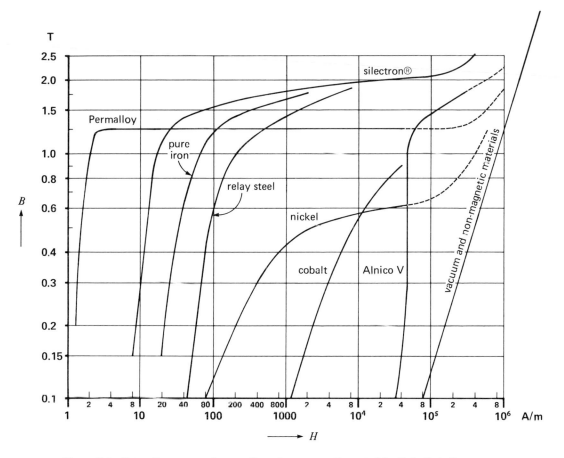

Figure 6-1 Saturation curves of magnetic and non magnetic materials. Note that all curves become asymptotic to the B-H curve of a vacuum when H is high.

materials is about 2 T.

6-3   Other magnetic properties

The B-H curve tells us how a magnetic material behaves as we increase H, beginning with the unmagnetized state. This information is useful when we deal with dc magnets and the dc fields of rotating machines. However, magnetic materials are often subjected to alternating fields. Under these conditions, the relationship between B and H is quite different from that given in Figure 6-1. Furthermore, the properties of permanent magnets cannot be explained with such a B-H curve. Consequently, we now direct our attention to the more general behavior of magnetic materials.

6-4   Residual flux density and coercive force

Consider the coil of Figure 6-2 which surrounds a magnetic material bent in the shape of a ring. The coil carries a current whose value and direction can be changed at will. Starting from zero, we gradually increase I, so that H and B increase. This increase follows curve **oa** of Figure 6-3. The flux density reaches a value B_m for a magnetic field H_m.

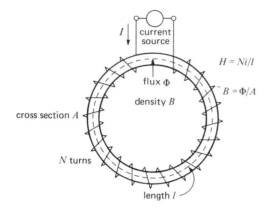

Figure 6-2   Method of determining the B-H properties of a magnetic material.

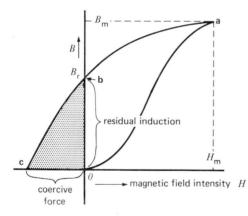

Figure 6-3   Residual induction and coercive force.

If the current is now gradually reduced to zero, the flux density B does not follow the original curve, but moves along a curve **ab** situated above **oa**. In effect, as we reduce the field intensity, the domains that were lined up under the influence of field H_m tend to retain their original orientation. This phenomenon is called *hysteresis*. Consequently, when H is zero, a substantial flux density

remains. It is called residual flux density or *residual induction* (B_r).

If we wish to eliminate this residual flux, we have to reverse the current in the coil and gradually increase H in the opposite direction. As we do so, we move along curve **bc**. The domains gradually change their previous orientation, until the flux density becomes zero at point **c**. The magnetic field required to reduce the flux to zero is called *coercive force* (H_c).

In reducing the flux density from B_r to zero, we also have to furnish energy. This energy is used to overcome the "frictional" resistance of the domains as they oppose the change in orientation. The energy supplied is dissipated as heat in the material. In effect, if we had a very sensitive thermometer, it would indicate a slight temperature rise as we demagnetize the ring.

We can prove that the energy required to demagnetize a cubic meter of material is exactly equal to the cross-hatched area under curve **bc**. For example, if area (Fig. 6-3) is equal to 72 tesla-amperes per meter, we have to furnish 72 J/m³ to force the flux to zero. For a "soft" material such as the steel used in transformer cores, the energy required is rather small: of the order of 10 J/m³. But for "hard" materials, such as Alnico, the energy is about 50 000 J/m³. It is precisely this high energy which distinguishes permanent magnet materials from other magnetic materials. Materials used in the manufacture of permanent magnets should have both a high B_r and a large H_c so that the required demagnetizing energy is as great as possible.

6-5   Hysteresis loop

Transformers and most electric motors operate on alternating current. In such devices, the flux in the iron changes continuously both in value and direction. The domains are therefore oriented first in one direction, then the other, at a rate that depends upon the frequency. Thus, if the flux has a frequency of 60 Hz, the domains describe a complete cycle every 1/60 of a second, passing successively through peak densities $+B_m$ and $-B_m$ as the peak magnetic field alternates between $+H_m$ and $-H_m$. If we draw the flux density B as a function of H, we obtain a closed curve called a *hysteresis loop* (Fig. 6-4). The residual induction B_r and coercive force H_c have the same significance as before.

Solution:
Volume of core is:
$$V = 20 \times 6 = 120 \text{ cm}^3 = 0.012 \text{ m}^3$$
Area of hysteresis loop is:
$$A = (1.2 \times 20) \times 4 = 96 \text{ T·A/m} = 96 \text{ J/m}^3$$
Energy loss per cycle is:
$$W = 96 \times 0.012 = 1.152 \text{ J}$$

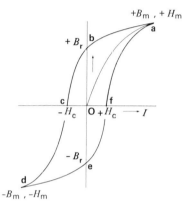

Figure 6-4   Hysteresis loop.

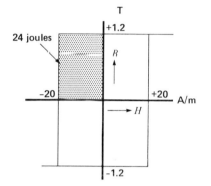

Figure 6-5   Hysteresis loop of Example 6-1.

6-6   Hysteresis loss

In describing a hysteresis loop, the flux moves successively from $+B_m$, $+B_r$, 0, $-B_m$, $-B_r$, 0 and $+B_m$, corresponding respectively to points **a**, **b**, **c**, **d**, **e**, **f**, and **a**, of Figure 6-4. The magnetic material absorbs energy during each cycle and this energy is dissipated as heat. We can prove that the amount of heat released (expressed in J/m^3) is equal to the area of the hysteresis loop.

Example 6-1:
A magnetic core made of Deltamax® has the rectangular hysteresis loop shown in Figure 6-5. If the core has a length of 20 cm and a cross section of 6 cm^2, calculate the hysteresis loss at a frequency of 400 Hz.

Energy loss per second is:
$$P = 1.152 \times 400 = 461 \text{ J/s}$$
Hysteresis loss is:
$$P = 461 \text{ W}$$

To reduce hysteresis losses, we select magnetic materials having a narrow hysteresis loop, such as the grain-oriented steel used in the cores of alternating-current transformers.

6-7   Hysteresis losses caused by rotation

Hysteresis losses are also produced when a piece of iron rotates in a constant magnetic field. Consider, for example, an iron armature AB that revolves in a field created by permanent magnets N, S (Fig. 6-6). The domains in the armature tend to line up with the magnetic field, irrespective of the position of the armature. Consequently, as the armature rotates, the N poles of the domains point first towards A and then towards B. A complete reversal occurs therefore every half revolution, as can be seen in Figures 6-6a and 6-6b. Consequently, the domains in the armature reverse periodically,

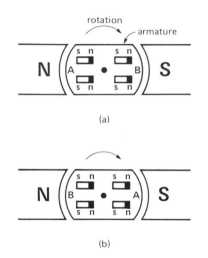

(a)

(b)

Figure 6-6   Hysteresis losses due to rotation.

even though the magnetic field is constant. Hysteresis losses are produced, just as they are in an ac magnetic field.

6-8   Eddy currents

Consider an ac flux Φ that links a rectangular-shaped conductor (Fig. 6-7). According to Faraday's law, an ac voltage E_1 is induced across its terminals.

If the conductor is short-circuited, a heavy ac current I_1 will flow, causing the conductor to heat up. If a second conductor is placed inside the first, a smaller voltage is induced because it links a smaller flux. Consequently, the short-circuit current I_2 is less than I_1 and so, too, is the dissipated power. Figure 6-8 shows four such concentric conductors carrying currents I_1, I_2, I_3, I_4. The currents are progressively smaller as the conductor area surrounding the flux decreases.

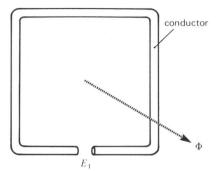

Figure 6-7   An ac flux Φ induces voltage E_1.

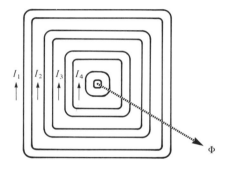

Figure 6-8   Concentric conductors carry ac currents due to ac flux Φ.

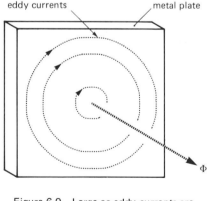

Figure 6-9   Large ac eddy currents are induced in a solid plate.

In Figure 6-9, the ac flux passes through a solid metal plate. It is basically equivalent to a whole series of rectangular conductors touching each other. Currents swirl back and forth inside the plate, following the paths shown in Figure 6-9. These so-called *eddy currents* (or Foucault currents) can be very large owing to the low resistance of the plate. Consequently, a metal plate that is pierced by an ac flux can become very hot. In this regard, special care has to be taken in transformers so that stray leakage fluxes do not cause sections of the enclosing tank to overheat.

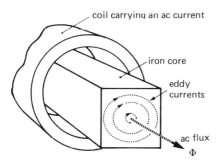

Figure 6-10   Solid iron core carrying an ac flux.

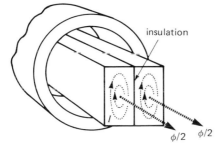

Figure 6-11   Eddy currents are reduced by splitting the core in half.

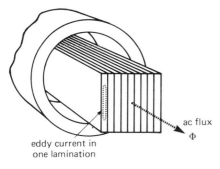

Figure 6-12   Core built up of thin, insulated laminations.

6-9   Eddy currents in a stationary iron core

The eddy current problem becomes crucially important when iron has to carry an ac flux. This is the case in all ac motors and transformers. Figure 6-10 shows a coil carrying an ac current which produces an ac flux in a solid iron core. Eddy currents are set up as shown, and they flow throughout the entire length of the core. A large core could eventually become red hot (even at a frequency of 60 Hz) owing to the tremendous I^2R losses.

We can reduce the losses by splitting the core in two along its length, taking care to insulate the two sections from each other (Fig. 6-11). The voltage induced in each section is one-half of what it was before, with the result that the eddy currents, and the corresponding losses, are considerably reduced.

If we continue to subdivide the core, we find that the losses decrease progressively. In practice, the core is composed of stacked laminations, usually a fraction of a millimeter thick. Furthermore, 3 to 4 percent silicon is added to increase the resistivity of the steel, thereby reducing the losses still more (Fig. 6-12).

The cores of ac motors and generators are therefore always laminated. A thin coating of insulation covers each lamination to prevent electrical contact between them. The laminations are stacked on top of each other and are tightly held in place by bolts and appropriate end-pieces.

The eddy current losses are given by:

$$P = kB^2f^2m \qquad (6\text{-}1)$$

where

P =   eddy current losses [W]

k =   constant depending upon the lamination thickness and the resistivity [typical values: 10^{-4} to 8×10^{-4}]

B =   peak flux density in the core [T]

f =   frequency of the ac flux [Hz]

m =   mass of iron which carries the flux [kg]

Example 6-2:

Calculate the eddy current loss in a transformer core weighing 150 kg if the peak flux density is 1.6 T at a frequency of 60 Hz. Take $k = 1.3 \times 10^{-4}$.

Solution:

$$P = kB^2f^2m$$
$$= 1.3 \times 10^{-4} \times 1.6^2 \times 60^2 \times 150 = 180 \text{ W}$$

6-10 Eddy current losses in a revolving core

The stationary field in direct current motors and generators produces a constant dc flux. This constant flux induces eddy currents in the revolving armature. To understand how they are induced, consider a solid cylindrical iron core that revolves between the poles of a magnet (Fig. 6-13). As it turns, the core cuts flux lines and, according to Faraday's law, a voltage is induced along its length having the polarities shown. Owing to this voltage,

large eddy currents flow in the core because its resistance is very low. These eddy currents produce large I^2R losses which are immediately converted into heat. The power loss is proportional to the square of the speed and the square of the flux density.

To reduce the eddy current losses, we laminate the armature as we do in the case of a stationary core. The laminations are tightly stacked with the flat side running parallel to the flux lines (Fig. 6-14).

6-11 Total iron losses

No losses are produced in a core that carries a constant flux. For example, if an electromagnet carries a dc current, copper losses are produced in the winding, but there are no iron losses at all.

On the other hand, hysteresis and eddy current losses are produced in iron whenever the flux

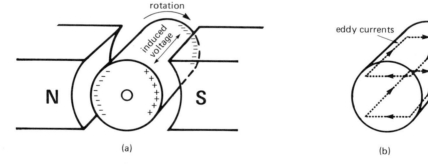

Figure 6-13 (a) Voltage induced in a revolving armature and (b) resulting eddy currents.

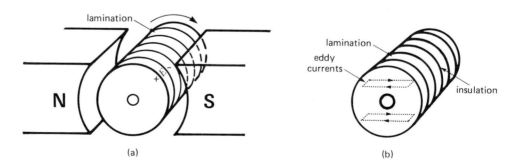

Figure 6-14 (a) Armature built up of thin laminations and (b) resulting eddy currents.

varies in magnitude or in direction.

Because the total losses depend upon several factors, such as frequency, flux density, thickness and quality, manufacturers of steel laminations usually show the losses by means of a set of curves. Thus, Figure 6-15 shows the losses as a function of flux density for various types of laminations at a frequency of 60 Hz. The losses include both hysteresis and eddy current losses.

affected by changes in lamination thickness.

6-12 Apparent power in iron

When an iron core carries an ac flux, it absorbs both active and reactive power. The active power supplies the hysteresis and eddy current losses, while the reactive power creates the ac flux. Consequently, the apparent power needed to excite an

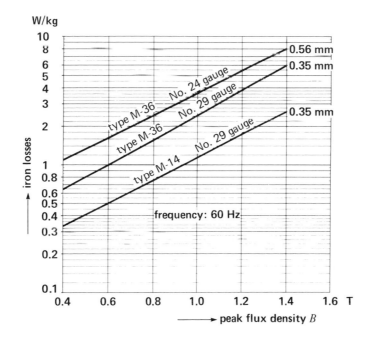

Figure 6-15 Typical iron losses in laminations operating at 60 Hz.

Low cost, type M-36 silicon-iron laminations are used in small armatures, whereas high-quality, type M-14 laminations are used in large rotating machines and transformers where high efficiency is essential. Note that for a given flux density, thin laminations (0.35 mm) produce substantially smaller losses than thicker laminations (0.56 mm) do. This decrease is entirely due to lower eddy current losses because hysteresis losses are not

iron core is given by:

$$S = \sqrt{P^2 + Q^2} \qquad (6\text{-}2)$$

where

S = apparent power to excite the core [VA/kg]

P = hysteresis and eddy current losses [W/kg]

Q = reactive power to create the flux [var/kg]

Figure 6-16 gives typical curves for a high-quality type M-6 grain-oriented steel used in transformers. Note that the apparent core loss S increases very rapidly when the flux density exceeds 1.6 T.

6-13 Hysteresis and permanent magnets

We have already seen that randomly oriented domains in a magnetic material tend to resist lining

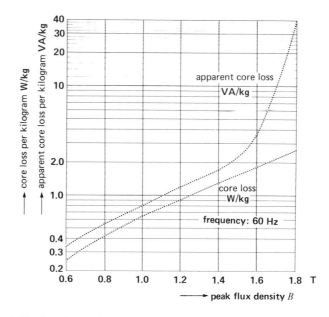

Figure 6-16   Core loss and apparent core loss in type M-7, 0.35 mm grain-oriented steel laminations containing 3 percent silicon.

Example 6-3:
A transformer core, composed of type M-6 laminations, has a mass of 150 kg. The flux alternates at a frequency of 60 Hz and the peak flux density is 1.6 T. Calculate the core loss and the apparent power needed to excite the core.

Solution:
Referring to Figure 6-16, the respective losses are:
   a. core loss = 1.8 W/kg
   b. apparent core loss = 3.7 VA/kg
The total core loss is:
   P = 1.8 x 150 = 270 W
The apparent power to excite the core is:
   S = 3.7 x 150 = 555 VA

up under the influence of an external magnetic field. However, once the domains are oriented, they try to stay in line, and again oppose further change to their orderly arrangement. This inherent opposition to change in domain orientation is called hysteresis. It is a property all magnetic materials possess to a greater or lesser degree. Although hysteresis produces losses when magnetic materials carry an ac flux, it also enables us to create permanent magnets.

Modern permanent magnets produce very strong magnetomotive forces so that they are often smaller than electromagnets of equal strength. Because no energy is required to sustain the magnetic field, permanent magnets enable us to manufacture devices having high efficiency and relatively small dimensions.

6-14 Magnetic energy stored in air

The "strength" of a permanent magnet depends exclusively upon the amount of energy it can store in its external magnetic field. It is therefore important to know how to calculate the energy.

6-15 Magnetic energy in a magnetic material

Surprising as it may seem, the energy stored in a magnetic material is usually much smaller than that in an air gap, even if the flux is the same. The reason is that the mmf needed to produce a given flux in a magnetic material is always very small owing to its excellent permeability. Consequently, in a magnetic circuit, we can usually neglect the energy in the iron compared to that in air.

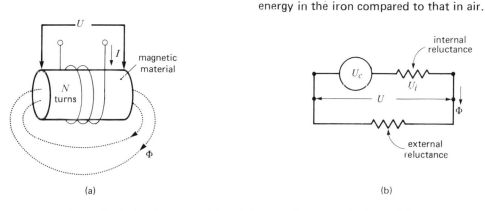

(a)   (b)

Figure 6-17   Electromagnet (a) and the equivalent magnetic circuit (b).

Consider a standard electromagnet that produces a constant flux Φ in air (Fig. 6-17). The dc coil develops a mmf given by $U_c = NI$.* However, the effective mmf U acting across the air gap is less than U_c owing to the magnetic potential drop U_i in the iron core. Thus, according to Sec. 2-19

$$U = U_c - U_i \qquad (6\text{-}3)$$

It can be proven that the magnetic energy stored in the air is given by:

$$\boxed{W = \frac{1}{2} U\Phi} \qquad (6\text{-}4)$$

where

W = magnetic energy in the air gap [J]
U = difference of magnetic potential across the air gap [A]
Φ = flux in the air gap [Wb]

6-16 Basic principle of a permanent magnet

Let us magnetize an iron rod by placing it inside a coil that develops a powerful mmf. Upon removing the rod, we discover that the rod itself produces a weak magnetic field, enabling it to pick up small tacks and iron filings. However, to produce such a field, the rod must develop a mmf of some kind. Because there is no coil, this mmf can only come from the domain structure within the iron rod itself. In turn, the magnetomotive forces produced by the individual domains are due to electron spin at the atomic level. In effect, the sum total of all the electron spins produces an internal mmf U_e. This mmf produces a flux Φ, whose value depends upon the internal reluctance of the rod and the

---

* U is the recognized ISO symbol for magnetic potential difference.

external reluctance of the magnetic circuit. If we assume an internal magnetic potential drop U_i inside the iron rod, we obtain an external mmf U given by:

$$U = U_e - U_i$$

The equivalent circuit of this so-called *permanent magnet* is shown in Figure 6-18b. Comparing it with a standard electromagnet, we find that the two have the same equivalent circuit (see Figs. 6-17b and 6-18b).

6-17 Strength of a permanent magnet

Consider a permanent magnet having a length l and a cross section A. It produces a flux Φ and an external mmf U. Because the external magnetic circuit is air, the energy stored in the magnetic field is given by Eq. (6-4):

$$W_a = \frac{1}{2} U\Phi \qquad \text{Eq. 6-4}$$

Flux Φ is equal to the product of the flux density

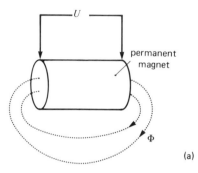

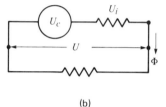

Figure 6-18   Permanent magnet (a) and the equivalent magnetic circuit (b).

This brings us to an important conclusion: permanent magnets are similar to ordinary electromagnets. There is one important difference, however. Whereas with a coil we know exactly what the driving mmf U_c is, we cannot measure the driving mmf U_e of a permanent magnet, nor can we isolate its internal potential drop U_i. However, this does not matter as long as we know the value of the *external* mmf U developed by the magnet. Fortunately, this is a quantity we can easily measure. From a practical standpoint, if we know the relationship between U and Φ, we can solve any circuit involving permanent magnets.

This is not to say that what goes on inside a magnet is unimportant. Such a conclusion would be false because most improvements in permanent magnet have been made precisely by looking "inside." However, from an application standpoint, it is the external features U and Φ that are important.

B in the magnet and its cross section A:

$$\Phi = BA \qquad (6\text{-}5)$$

Furthermore, the mmf U developed by the magnet is equal to the product of the magnetic field intensity it produces and the length l of the magnet:

$$U = Hl \qquad (6\text{-}6)$$

If we substitute Eqs. 6-5 and 6-6 in Eq. 6-4, we obtain the important relationship:

$$W = BHAl/2$$

$$\boxed{W = \frac{1}{2} (BH) V} \qquad (6\text{-}7)$$

In the above equations:

Φ = flux in the air [Wb]

B = flux density in the magnet [T]

H = magnetic field intensity produced by the magnet [A/m]

A = cross section of the magnet [m²]

l = length of the magnet [m]

V = volume of the magnet [m³]

Equation 6-7 leads us to the important conclusion that the energy stored in an external field is proportional (a) to the volume of the magnet and (b) to the product BH of the material of which it is made.

6-18 *B-H* relationship

Figure 6-19 shows the typical $B\text{-}H$ curves of three permanent magnet materials. Carbon steel has an acceptable residual induction, but its coercive force is very small. On the other hand, Indox® produces a low B_r, but a very high H_c. Alnico V produces both a high B_r and a strong H_c.

The residual flux density B_r is that obtained when the material forms a closed magnetic circuit. However, a flux imprisoned inside a solid is useless: to make it accessible, we have to open up an air gap. In order to see what happens under these con-

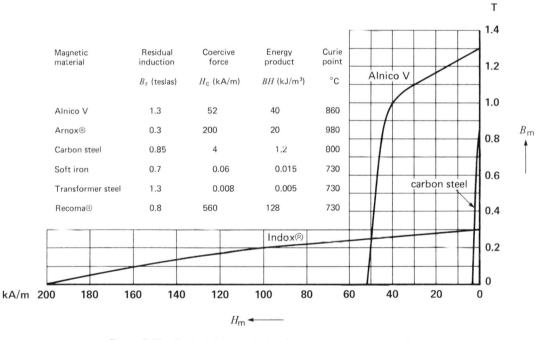

Magnetic material	Residual induction	Coercive force	Energy product	Curie point
	B_r (teslas)	H_c (kA/m)	BH (kJ/m³)	°C
Alnico V	1.3	52	40	860
Arnox®	0.3	200	20	980
Carbon steel	0.85	4	1.2	800
Soft iron	0.7	0.06	0.015	730
Transformer steel	1.3	0.008	0.005	730
Recoma®	0.8	560	128	730

Figure 6-19   Typical characteristics of permanent magnet materials.

The values of B and H are a property of the magnetic material itself. They can be read off the **bc** portion of the hysteresis loop shown in Figure 6-3.

ditions, let us start with a closed magnetized ring of Alnico V and gradually pry it open until it becomes a straight bar. The air gap starts therefore from zero and increases progressively until the magnet is straight. As the length of the gap increases, the flux density in the magnet decreases gradually, following the *B-H* curve of Figure 6-20. This curve is identical to the original Alnico V curve given in Figure 6-19. The maximum gap is

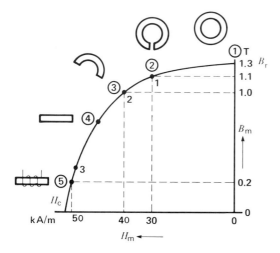

Figure 6-20   Effect of an increasing air gap on a permanent magnet.

obtained when the magnet is completely straightened out (point 4). If we want to reduce *B* still more, we have to add a counter mmf using an external coil (point 5).

   For every value of *B*, the magnet produces a corresponding magnetic field *H*. For example, referring to Figure 6-20, we obtain the following typical values:

B	1.3	1.0	0.2	[T]
H	0	40	50	[kA/m]

Knowing the dimensions of the magnet and the values of *B* and *H*, we can readily calculate the flux and the mmf the magnet produces. In effect, flux Φ = *B* x (cross section of the magnet) and *U* = *H* x (length of the magnet).

Example 6-4:
A permanent magnet made of Alnico V has the dimensions given in Figure 6-21. The pole pieces are made of soft iron and the flux in the air gap is

16 mWb. Calculate:
   a. the flux density in the magnet
   b. the flux density in the air gap
   c. the mmf developed by the magnet
   d. the energy in the air gap

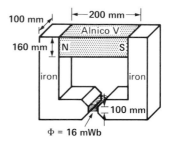

Figure 6-21   Permanent magnet of Example 6-4.

Solution:
a. The flux density in the magnet is:

$$B = \frac{16 \text{ mWb}}{100 \text{ mm} \times 160 \text{ mm}} = \frac{16 \times 10^{-3}}{0.1 \times 0.16} = 1 \text{ T}$$

b. The flux density in the air gap is:

$$B = \frac{16 \text{ mWb}}{100 \text{ mm} \times 100 \text{ mm}} = \frac{16 \times 10^{-3}}{0.1 \times 0.1} = 1.6 \text{ T}$$

Note that the density in the air gap is even greater than the residual flux density (1.3 T). We obtain this result by simply reducing the cross section of the poles facing the air gap.

c. Referring to the *B-H* curve of Alnico V, we note that *H* = 40 kA/m when *B* = 1 T. Because the magnet has a length of 200 mm, it develops a mmf:
U = 40 kA/m x 0.2 m
   = 8000 amperes, or 8000 ampere turns
This mmf forces the flux through the iron pole pieces and across the air gap. Because iron has a high permeability, the magnetic potential drop is negligible, and most of the mmf is available to produce the flux in the air gap.

d. The energy in the air gap is given by Eq. 6-4:

$$W = \frac{1}{2} U\Phi$$

$$= \frac{1}{2} \times 8000 \times \frac{16}{1000} = 64 \text{ J}$$

We could manufacture an electromagnet of equal strength by replacing the permanent magnet by an iron slug and winding a 1000-turn coil on top of it, carrying a current of 8 A (Fig. 6-22). Needless to say, such an electromagnet is both larger and more costly than the permanent magnet arrangement.

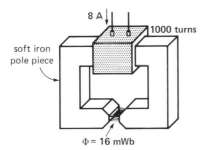

Figure 6-22   Electromagnet having the same strength.

6-19 Energy product

In most practical applications of permanent magnets, we have to produce a given flux Φ in an air gap of given length. These two requirements automatically establish the energy W of the magnetic field. The question then arises, how big a magnet do we need to create the energy required? Equation (6-7) provides the clue. If we rewrite it, we obtain:

$$\boxed{V = \frac{2W}{BH}} \qquad (6\text{-}8)$$

For a given energy W, we are naturally interested in using the least amount of magnetic material V. Consequently, the product BH should be as large as possible. This product is equal to the area of the rectangle shown in Figure 6-23. As we move along the curve, there is obviously one point $(B_{\rm d}, H_{\rm d})$

where the rectangle has the largest area. The corresponding $B_{\rm d}H_{\rm d}$ product is called *energy product* of the magnetic material.

Once we have selected a given magnetic material, we always design the magnetic circuit so that the flux density in the material is $B_{\rm d}$. This also fixes the value of $H_{\rm d}$, with the result that the amount of material is automatically minimized.

The energy product is an important way of comparing the relative merits of one magnetic material over another. In effect, the higher the energy product, the smaller the magnet will be. For example, the energy product of Alnico V is 40 kJ/m^3 whereas that of carbon steel (used in permanent magnets before 1935) is 1.2 kJ/m^3. Finally, it is easy to see why silicon steel is never used for permanent magnets; its energy product is only 0.005 kJ/m^3.

6-20 Types of permanent magnets

To manufacture powerful permanent magnets, we use various alloys made of iron, aluminum, cobalt, copper, platinum yttrium, oxygen, carbon, etc. Magnets having a metallic base have been used for many years. Carbon steel (1% carbon, 0.5% manganese, 98.5% iron) has been replaced by alloys such as Alnico V (8% aluminum, 14% nickel, 24% cobalt, 3% copper, 51% iron). A permanent magnet made of Alnico V is almost 30 times smaller than one made of carbon-steel.

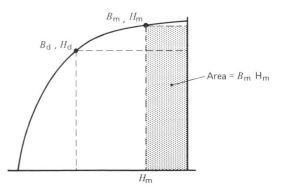

Figure 6-23   Energy product = $B_{\rm d} \times H_{\rm d}$.

Alnico was discovered in 1932 by the Japanese physicist I. Mishima. This invention revolutionized the industry of magnetic devices because, for the first time, it enabled the manufacture of permanent magnets that were smaller than electromagnets of equal strength. Alnico is extremely hard and brittle; it can only be shaped by grinding the cast parts.

Ceramic magnets constitute another class of magnets that are much lighter and have resistivities equivalent to that of good insulators. Known by trade names such as Indox®, Arnox®, Vectolite®, and Ferroxdure®, these magnets are essentially ferrites composed of an alloy of ferric oxide (Fe_2O_3), of barium oxide ($BaO)6(Fe_2O_3$), or oxides of zinc, cobalt, etc. Ceramic magnets have a lower B_r than Alnico has, but they develop much higher coercive forces. Consequently, the energy product of ceramic magnets is of the same order magnitude (Fig. 6-19). Ceramic magnets are used in the pole pieces of dc motors, in door-sealing magnetic rubber strips, and in the manufacture of magnetic tapes.

Rare-earth magnetic materials such as Recoma® have also been developed, that yield energy products almost three times that of Alnico.

6-21 Magnetizing and demagnetizing a magnet

To magnetize a permanent magnet, we place it in a very strong magnetic field. The domains line up almost at once so that the field has to be applied for only a few milliseconds. This short magnetizing period enables us to use a relatively small conductor to carry enormous currents (up to 200 000 A). The magnet is excited by a coil consisting of a single turn of wire (Fig. 6-24). A soft iron keeper ensures that the entire mmf developed by the coil is exerted on the magnet.

To demagnetize a permanent magnet, we place it in a strong ac field and gradually reduce the field to zero. In a manner similar to that of Fig. 6-24, we can use a single turn of wire carrying a high ac current, which is gradually reduced to zero.

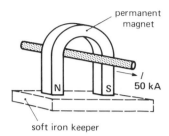

Figure 6-24   Magnetizing a permanent magnet.

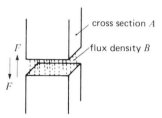

Figure 6-25 $F = 400\ 000\ B^2A$.

6-22 Change of magnetic field with time, Curie point

The magnetic fields produced by modern permanent magnets are very stable. Although any permanent magnet will tend to lose its strength with time, the fall-off is very slow and practically imperceptible. For example, a magnet made of Alnico V will retain 99 percent of its original strength after 100 years of use. By using special methods, we can stabilize the magnet even more. The now-obsolete carbon-steel magnets lost their magnetism much more quickly and so they had to be remagnetized periodically.

Temperature also plays an important part. As it increases, the flux density gradually decreases and, at a critical temperature called the Curie point, all magnetic materials lose their permeability as well as their permanent magnet properties. In effect, high temperatures produce intense atomic vibrations which destroy the orientation of the magnetic domains. The Curie point temperature depends upon the material, but usually lies between 700°C and 900°C.

6-23 Magnetic force of attraction

Consider two magnetic poles having a cross section A and separated by a relatively short air gap (Fig. 6-25). The force of attraction between them is given by:

$$F = 400\,000\,B^2A \qquad (6\text{-}9)$$

where

F = force of attraction [N]
B = flux density in the air gap [T]
A = cross section of the poles [m²]
400 000 = constant, taking account of units [exact value = $10^7/8\pi$]

Example 6-5:
Calculate the force of attraction between the electromagnet and the armature of Figure 6-26, knowing that the flux density in the air gap is 1.2 T.

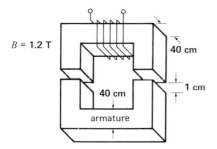

Figure 6-26   Magnet of Example 6-5.

Solution
Cross section per pole is:

A = 40 cm x 40 cm = 1600 cm² = 0.16 m²

The force is:

F = 400 000 B^2A

= 400 000 x (1.2)² x 0.16 = 92 160 N

Because there are two poles, the total force of attraction is 184 kN. The magnet can therefore lift a mass of 18.8 metric tons, a result that is found by applying Eq. 3-1.

Note that the force of attraction depends only upon the flux density and surface area of the piece being attracted. It is unaffected by the degree of saturation of the magnetic circuit. Furthermore, the source of flux is unimportant; it may be produced by the coil of an electromagnet or by a permanent magnet.

6-24 Converting mechanical energy into magnetic energy

When the north and south poles of a permanent magnet are bridged by a soft iron armature, the magnetic energy stored in the armature is negligible (Fig. 6-27). On the other hand, when we remove the armature, the energy stored in the resulting air gap may be very high (Fig. 6-28). The

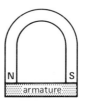

Figure 6-27   Negligible energy in the armature.

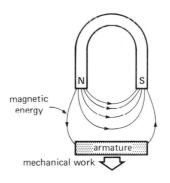

Figure 6-28   Mechanical work is converted into magnetic energy.

question now arises: Where did the energy come from? Did the magnet create the energy in the gap? If not, what did? To answer these questions, we recall that to remove the armature from the magnet, we did mechanical work. In effect, we had to exert a force F over a distance d. It is a remarkable fact that the mechanical work so expended is stored in the magnetic field of the air gap*

There is, consequently, a direct conversion of mechanical energy into magnetic energy as we pull the armature away from the poles. For example, if 6 J of mechanical work are needed to pull the armature away, then 6 J of magnetic energy are stored in the external field.

Conversely, as we slowly bring the armature back towards the magnet, we have to fight against the force of attraction. The magnetic energy in the air gap gradually decreases, as we approach the poles. The magnetic energy so released is directly converted into mechanical energy during the braking process†. When the armature is again stuck to the poles, the magnetic energy is zero.

Because it is mechanical energy that produces magnetic energy, we can understand why a permanent magnet does not lose its strength even if its armature is removed and replaced hundreds of times. It also explains why a magnetic nail collector does not lose its strength despite the fact that every nail attracted and removed becomes slightly magnetized. The remanent magnetism associated with every newly magnetized nail is entirely created by mechanical work. None of this remanent magnetism is "taken away" from the nail collector itself.

QUESTIONS AND PROBLEMS

Practical level

6-1 Referring to Figure 6-1, calculate the relative permeability of the following:
  a. relay steel at a flux density of 0.6 T;
  b. nickel at a density of 0.42 T.

---

* In practice, some of the mechanical energy is given up to the magnetic domains, producing a slight heating.
† The mechanical energy is ultimately converted to heat in the joints and muscles that are involved in the braking process.

6-2 Draw a complete hysteresis curve and show the coercive force and the residual flux density. Indicate the portion of the curve that is used for permanent magnets.

6-3 What means are used to reduce eddy current losses?

6-4 Referring to Figure 6-1, calculate the relative permeability of Permalloy for H values of 3 A/m, 100 A/m and 100 000 A/m.

6-5 As we pull an armature away from the poles of a magnet, we exert mechanical energy. What happens to this energy?

Intermediate level

6-6 Calculate the energy required to completely demagnetize 10 cm^3 of Alnico from an initial residual flux density of 1.3 T (use the curve of Fig. 6-19).

6-7 A laminated iron core made up of type M-36, 29 gauge laminations operates at a peak flux density of 10 kilogauss. If the core weighs 600 lb, calculate a) the resulting core loss [W], and b) the heat given off [Btu/h].

6-8 If the core in Problem 6-7 were made ot type M-36, 24 gauge laminations, calculate the new core loss [W]. Is this increased loss due to eddy currents or to hysteresis?

6-9 A sample of iron operating at a peak flux density of 1.2 T and a frequency of 60 Hz, has the following losses: hysteresis loss - 240 W; eddy current loss - 360 W.
  If the frequency is reduced to 50 Hz while maintaining the same flux density, calculate the new hysteresis and eddy current losses.

6-10 At what flux density do nickel and cobalt possess the same relative permeability and what is its value (see Fig. 6-1)?

6-11 The armature in Figure 6-6 rotates at a speed of 1200 r/min. Calculate the number of hysteresis loops that are completed in one second.

6-12 The electromagnet in Figure 6-17 has 2400 turns and carries a dc current of 2 A. It produces a total flux of 150 mWb, but the magnetic potential drop in the iron core is equal to 600 A. Calculate the magnetic energy stored in the air.

6-13 The permanent magnet pole of a dc motor is made of Indox® and operates at a flux density of 0.2 T. The magnet has a length of 2 cm and a cross section of 100 cm². Calculate:

    a. the MMF developed by the magnet;

    b. the flux produced by the magnet;

    c. the magnetic energy stored in the air gap.

6-14 Calculate the energy product for Alnico V at the following densities (see Fig. 6-19):

    a. 1.2 T;   b. 0.8 T;

    c. 64 500 lines per square inch.

Advanced level

6-15 A closed ring of Alnico V has a cross section of 4 in² and an average length of 20 in. A coil of 100 turns is wound around the ring in the manner shown in Figure 6-2. The remanent induction is 1.3 T, and the *B-H* curve is shown in Figure 6-19. Calculate the reverse current required to reduce the flux density to: a. 1 T,   b. 0 T.

6-16 Indox® has a maximum remanent induction of 0.3 T. How can we produce a flux density of 1.2 T in an air gap using this material?

6-17 We wish to produce a flux density of 0.5 T in an air gap whose length and cross section are 0.25 in and 9 in², respectively. If we decide to use Alnico V operating at a flux density of 1 T, calculate:

    a. the length of the magnet [in] ;

    b. the cross section of the magnet [in²] ;

    c. the volume of the magnet [in³] .

6-18 In Problem 6-17 we finally decide to use Indox® instead of Alnico V. Referring to Fig. 6-19, calculate:

    a. the optimum operating flux density in the magnet [T] ;

    b. the volume of the magnet [in³] .

PART III

ALTERNATING CURRENT CIRCUITS

7

ALTERNATING VOLTAGES AND CURRENTS

This chapter is essentially a review of ac voltages and currents, with particular emphasis on applications to power technology. Certain portions, dealing with harmonics and current flow in an inductor, also have a direct application in power electronics.

7-1 Sinusoidal voltage and current

The ac voltage generated by commercial alternators is very nearly a perfect sine wave. It may therefore be expressed by the equation:

$$e = E_m \sin \phi \qquad (7-1)$$

where

e = instantaneous voltage [V]
E_m = peak value of the sinusoidal voltage [V]
ϕ = angle, expressed in electrical degrees [°]

The electrical angle is related to time according to the equation:

$$\phi = 360\, ft \qquad (7-2)$$

where

ϕ = electrical angle [°]
f = frequency [Hz]
t = time [s]
360 = a constant, taking units into account

Sinusoidal voltages always produce sinusoidal currents, unless the circuit is nonlinear. Such currents are expressed in the same way as voltages are, yielding equations similar to Eq. 7-1. However, the current may lag or lead the voltage depending on the type of circuit used. For example, if a current lags behind the voltage by an angle θ, it is expressed by the equation:

$$i = I_m \sin (\phi - \theta) \qquad (7-3)$$

On the other hand, if the current *leads* the voltage, it is given by the equation:

$$i = I_m \sin (\phi + \theta) \qquad (7\text{-}4)$$

where

i = instantaneous current [A]
I_m = peak value of the sinusoidal current [A]
θ = phase angle between the voltage and current [°]

Figure 7-1 shows an ac source E_{ab} which generates a sinusoidal voltage having a peak value of 100 V. The instantaneous voltages and polarities are shown for electrical angles of 90° and 330°.

Why do we choose a sinusoidal wave rather than a "simple" curve such as a square or triangular wave? There are several good reasons for doing so:
a. in ac machines and transformers, sinusoidal voltages and currents respectively produce the least iron and copper losses for a given power

output. The efficiency is therefore better;
b. Sinusoidal voltages and currents produce less interference (noise) on telephone lines.
c. In electric motors, a sinusoidal voltage produces less noise, and the accelerating torque is smoother.

7-2   Instantaneous power

The instantaneous power supplied to a circuit is simply the product of the instantaneous voltage times the instantaneous current.

Instantaneous power is always expressed in watts, irrespective of the type of circuit used. The instantaneous power may be positive or negative. A positive value means that power flows from the source to the load. Consequently, a negative value indicates that power is flowing in reverse, from the load to the source.

Example 7-1:
A sinusoidal voltage having a peak value of 162 V and a frequency of 60 Hz is applied to the terminals of an ac motor. The resulting current has a

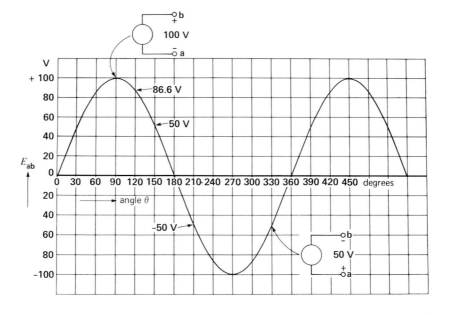

Figure 7-1   Sinusoidal voltage having a peak value of 100 V.

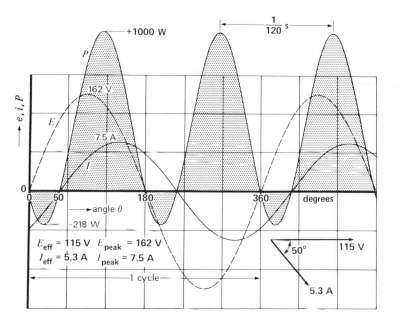

Figure 7-2   Instantaneous voltage, current and power in an ac circuit. (See Example 7-1).

peak value of 7.5 A and lags 50° behind the voltage.

a. Express the voltage and current in terms of the electrical angle ϕ;

b. Calculate the value of the instantaneous voltage and current at an angle of 120°;

c. Calculate the value of the instantaneous power at 120°;

d. Calculate the voltage at 5.55 ms.

Solution:

a. 1   From Eq. 7-1, we have:

$e = E_m \sin \phi = 162 \sin \phi$

a. 2   Furthermore, from Eq. 7-3, we have:

$i = I_m \sin (\phi - \theta) = 7.5 \sin (\phi - 50)$

b.   At $\phi = 120°$, we have:

$e = 162 \sin 120 = 162 \times 0.866$
$= 140.3 \text{ V}$

$i = 7.5 \sin (120 - 50) = 7.5 \sin 70$
$= 7.5 \times 0.94$
$= 7.05 \text{ A}$

c.   The instantaneous power at 120° is:

$p = ei = 140.3 \times 7.05 = +989 \text{ W}$

Because the power is positive, it flows at this instant from the source to the motor.

d. $e = E_m \sin \phi$ Eq. 7-1
$= E_m \sin 360 \, ft$ Eq. 7-2
$= 162 \sin (360) (60) (5.55 \times 10^{-3})$
$= 140.3 \text{ V}$

Figure 7-2 shows the waveshapes for e, i and p. Note that the instantaneous power is negative for brief intervals. During these moments, the motor returns power to the line.

7-3   Effective value of voltage and current

In power circuits, we sometimes encounter voltages that are periodic but not sinusoidal. For example, the waveshape may be rectangular, trapezoidal, or spiked. What voltage values should we assign to such peculiar waveshapes and how can the voltages be measured? In every case, we operate on a basis of comparison. We compare the peculiar voltage with a dc voltage. But how can a highly variable ac voltage be compared to a steady dc voltage? There is really only one meaningful way, and that is to compare their relative heating effects. We assume the two voltages are applied to two identical resistors. If the heat produced by the strange waveshape is equal to the heat produced by a dc voltage of, say 10 V, the strange waveshape is said to have an *effective value* of 10 V.

We can determine the effective value of a peculiar voltage by using two identical lamps and a variable dc source. The peculiar voltage is applied to one lamp, causing it to glow. The variable dc voltage is then connected to the second lamp and the voltage is adjusted so that both lamps become equally bright. The dc voltage is then measured; its value establishes the effective value of the peculiar voltage.

In practice, we can use a thermocouple voltmeter to measure the effective value of odd waveshapes. Such voltmeters contain a small resistor which heats up when a voltage is applied to the terminals. The heat, in turn, causes a tiny thermocouple to produce a small dc voltage, which is registered by a sensitive meter. The voltmeter is first calibrated by applying a dc voltage to its terminals. Moving-vane voltmeters and special electronic voltmeters can achieve the same result without using a resistor.

7-4 Determining the effective value of a voltage

The effective value of a periodic voltage can be found by calculating the thermal energy Q_e that the voltage would produce in a resistor of 1 Ω during one complete cycle. Let the cycle have a period T. This energy must be equal to the thermal energy which a dc voltage E_{eff} would produce in the same resistor, during the same interval of time. Consequently, we can write:

dc energy in time T = energy of periodic voltage in time T

dc power x T = Q_e

$$\frac{E^2_{eff}}{R} \times T = Q_e$$

$$\frac{E^2_{eff} \times T}{1\,\Omega} = Q_e$$

Consequently,

$$\boxed{E_{eff} = \sqrt{Q_e/T}} \qquad (7\text{-}5)$$

where

E_{eff} = effective value of the ac voltage [V]

Q_e = energy which the ac voltage dissipates in a hypothetical resistor of 1 Ω, during one complete cycle [J]

T = duration of one cycle [s]

If we know the waveshape of the voltage, we can calculate the value of Q_e, using mathematics. However, its value can also be found graphically. The graphical method is easier to understand and yields sufficiently accurate results. The following examples further clarify the meaning of effective voltage, and how its value can be found.

Example 7-2:
Calculate the effective value of the pulsating voltage shown in Fig. 7-3a.

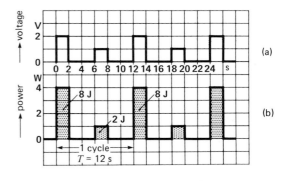

Figure 7-3   a. Pulsating voltage in Example 7-2.

b. Power and energy in Example 7-2.

Solution:
We assume the voltage is applied across the terminals of a 1 Ω resistor. The power dissipated is found from $p = e^2/R$. Because $R = 1\,\Omega$, the instantaneous power is simply equal to e^2. The graph of instantaneous power is shown in Fig. 7-3b. The cross-hatched area under the curve is equal to the energy dissipated in the resistor, expressed in joules. For the first 2 seconds the energy is 4 W

times 2 s (or 4 J/s times 2 s), yielding 8 J. The total energy Q_e dissipated during one cycle is obviously equal to 8 + 2 = 10 J. Furthermore, the duration T of one cycle is 12 s. The effective value of the pulsating voltage is:

$$E_{eff} = \sqrt{Q_e/T} = \sqrt{10/12} = \sqrt{0.83} = 0.91 \text{ V}$$

Thus a steady dc voltage of 0.91 V produces the same heating effect as the pulsating voltage of Fig. 7-3a.

Example 7-3:
Calculate the effective value of the ac voltage shown in Fig. 7-4a.

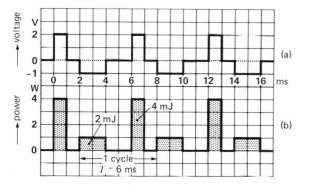

Figure 7-4   a. AC voltage in Example 7-3.

b. Power and energy in Example 7-3.

Solution:
Applying the same principles as in the previous example, we square the voltage to determine the instantaneous power in a hypothetical 1 Ω resistor. Note that the instantaneous power is always positive, even when the voltage is negative. The reason is that a negative voltage produces the same heating effect as a positive voltage does. The resulting energy is equal to the sum of the cross-hatched areas for one cycle (Fig. 7-4b). We have:

Q_e = 1 W x 2 ms + 4 W x 1 ms

= 2 mJ + 4 mJ = 6 mJ

also, T = 6 ms

Consequently, $E_{eff} = \sqrt{Q_e/T}$ Eq. 7-5

$$= \sqrt{6 \text{ mJ}/6 \text{ ms}}$$

$$= 1 \text{ V}$$

Consequently, a steady dc voltage of 1 V produces the same heating effect as the ac voltage of Fig. 7-4a.

Example 7-4:
Calculate the effective value of the sine wave shown in Fig. 7-5a, knowing that the peak voltage is 6 V.

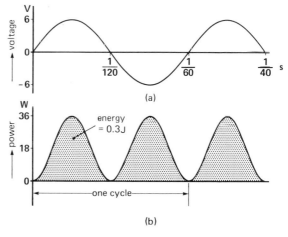

Figure 7-5   a. Voltage in Example 7-4.

b. Power and energy in Example 7-4.

Solution:
Squaring the voltage as before, we obtain the curve shown in Fig. 7-5b. The cross-hatched area for one complete cycle yields an average power of 18 W. Consequently,

Q_e = 18 W x 1/60 s = 0.3 J

since T = 1/60 s = 0.0167 s, we have

E_{eff} = $\sqrt{Q_e/T} = \sqrt{0.3/0.0167} = \sqrt{18}$

= 4.24 V

The previous examples show that the effective voltage is essentially equal to the square root of the average value of the square of the instantaneous voltages. For this reason, we sometimes use the term "root-mean-square" or "RMS" value instead of "effective" value.

7-5   Effective value of an ac current

The effective value of an ac current can be found in the same way as the effective value of a voltage. We again assume that the current flows in a 1 Ω resistor for one complete cycle. A quantity of heat Q_i is released during this interval T.

By definition, the effective value I_{eff} of a current is equal to that dc current which produces the same amount of heat in the same interval of time. Consequently,

$$\text{dc power} \times T = Q_i$$
$$I^2_{eff} R \times T = Q_i$$
$$I^2_{eff} \times 1 \, \Omega \times T = Q_i$$

whence $\boxed{I_{eff} = \sqrt{Q_i/T}}$ (7-6)

where

I_{eff} = effective value of the ac current [A]
Q_i = energy which the ac current dissipates in a hypothetical 1 Ω resistor during 1 complete cycle [J]
T = duration of one cycle [s]

7-6   Effective value of a sinusoidal current and voltage

Consider a sinusoidal current having a peak value I_m (Fig. 7-6a). If it flows in a hypothetical 1 Ω resistor, the instantaneous power p is given by the curve of Fig. 7-6b. The peak value of p is equal to I_m^2. It is obvious that the average power is equal to one half the peak power, or $I_m^2/2$. Consequently, the cross-hatched area under the power curve during one cycle is equal to $I_m^2/2 \times T$. But this area is equal to energy Q_i dissipated in the 1 Ω resistor. Thus, according to Eq. 7-6, we have:

$$I_{eff} = \sqrt{Q_i/T} = \sqrt{I_m^2 \times T/2\,T} = \sqrt{I_m^2/2} = I_m/\sqrt{2}$$

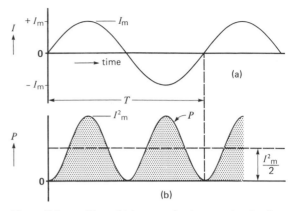

Figure 7 6   a. Sinusoidal current having a peak value I_m.

b. Power and energy dissipated in a 1 Ω resistor.

Consequently, $\boxed{I = \dfrac{I_m}{\sqrt{2}} = 0.707\,I_m}$ (7-7)

where

I = effective current [A]
I_m = peak value of sinusoidal current [A]
$\sqrt{2}$ = a constant, which applies only to sinusoidal waveshapes

In the same way, the effective value E of a sinusoidal voltage is given by:

$$\boxed{E = E_m/\sqrt{2} = 0.707\,E_m}$$ (7-8)

Most alternating current instruments are calibrated to show the *effective value* of voltage or current and *not the peak value*. When the value of an alternating voltage or current is given it is understood that it is the effective value. Furthermore, the subscript in E_{eff} and I_{eff} is usually dropped and the effective values of voltage and current are simply represented by the symbols E and I.

7-7   Phasors

We can represent the voltages and currents in an ac

circuit by drawing sine waves as shown in Fig. 7-2. This is a tedious process, and, consequently, a much simpler method was invented to yield the same information. It consists of replacing sine waves by *phasors*. A phasor is a straight line whose length is proportional to the effective voltage or effective current it represents. Thus an ac current of 10 A may be replaced by a straight line having a length of 10 cm.

Furthermore, in order to show the phase angle between voltages and currents, the phasors bear an arrow. The following rules apply to phasors:

1. Two phasors are said to be in phase when they point in the same direction (Fig. 7-7) The phase angle between them is then zero.

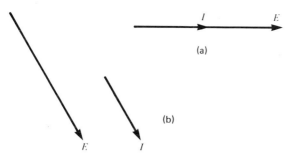

(a)

(b)

Figure 7-7

a. Phasors E and I are in phase, with common origin.

b. Phasors E and I are in phase, with no common origin.

2. Two phasors are said to be out of phase when they point in different directions. The phase angle between them is the angle through which one of the phasors has to be rotated to make it point in the same direction as the other. Thus, referring to Fig. 7-8, phasor I has to be rotated counterclockwise by an angle θ to make it point in the same direction as phasor E. Con-

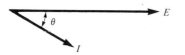

Figure 7-8   Phasor I lags behind phasor E.

versely, phasor E has to be rotated clockwise by an angle θ to make it point in the same direction as phasor I. Consequently, whether we rotate one phasor or the other, we have to sweep through the same angle to make them line up.

3. If a phasor E has to be rotated *clockwise* to make it point in the same direction as phasor I, then phasor E is said to *lead* phasor I. Conversely, a phasor I is said to lag behind phasor E if phasor I has to be rotated counterclockwise to make it point in the same direction. Thus, referring to Fig. 7-8, it is obvious that phasor E leads phasor I by θ degrees. It is also obvious that I lags behind E by θ degrees.

4. Referring now to Fig. 7-9, we *could* rotate

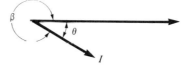

Figure 7-9   Phasor I leads E by β degrees.

phasor I clockwise by an angle β to make it point in the same direction as phasor E. We could then say that phasor I leads phasor E by β degrees. This is the same as saying that phasor I lags phasor E by θ degrees. In practice, we always select the smaller phase angle between two phasors to designate the lag or lead situation.

5. Phasors do not have to have a common origin, but may be entirely separate from each other, as shown in Fig. 7-10. By applying rule 3 above, we can easily show that phasor I leads phasor E_1 by $90°$ and that E_2 lags behind I by $135°$.

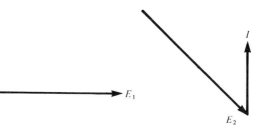

Figure 7-10   Phasor I leads E, by $90°$.

Example 7-5:

A 60 Hz source having an effective voltage of 240 V delivers an effective current of 10 A to a circuit. The current lags the voltage by 30°.

a. Draw the phasor diagram for the circuit;

b. Draw the corresponding waveshapes for E and I.

Solution:

a. We select any convenient direction for phasor E. Phasor I is drawn so that it lags 30° behind E (Fig. 7-11a).

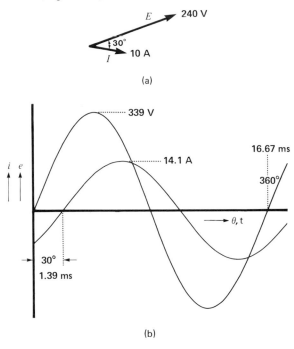

(a)

(b)

Figure 7-11

   a. Phasor I lags behind E by 30° (Example 7-5).

   b. Graph of E and I.

b. Peak voltage = $240\sqrt{2}$ = 339 V

   Peak current = $10\sqrt{2}$ = 14.1 V

   The resulting waveshapes are shown in Fig. 7-11b. The phase lag of 30° corresponds to a time interval t given by Eq. 7-2:

$$\theta = 360\,ft$$
$$30 = 360 \times 60\,t$$
$$t = 1.39 \text{ ms}$$

It is well known that the sum of two or more sinusoidal voltages (or currents) can be found by adding their respective phasors. The phasors are added by stringing them together in succession, joining head to tail. The resulting phasor extends from tail to head of the chain so formed.

Voltages and currents can also be subtracted. This is best done by first reversing the phasor to be subtracted and then adding it to the phasors that make up the string.

Example 7-6:

The three ac sources in Fig. 7-12a produce the following voltages.

$$E_{ab} = 100\ \underline{/0°}$$
$$E_{cd} = 60\ \underline{/+ 60°}$$
$$E_{12} = 40\ \underline{/- 150°}$$

If terminals a-c and d-2 are connected together, Fig. 7-12b), determine the magnitude of E_{1b} and its phase angle with respect to E_{ab}.

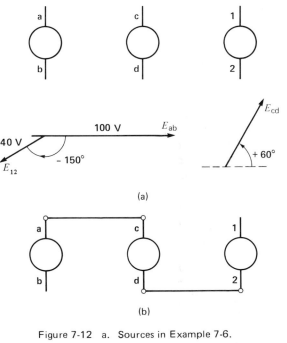

(a)

(b)

Figure 7-12 a. Sources in Example 7-6.

                  b. Sources connected in series.

Solution:

Referring to Fig. 7-12b, we can write:

$$E_{1b} = E_{12} + E_{dc} + E_{ab}$$

In this equation, phasor E_{dc} has the same magnitude as phasor E_{cd} but it points in the opposite direction. Phasors E_{12}, E_{dc} and E_{ab} are strung out head to tail as shown in Figure 7-12c. Phasor E_{1b} can be found either graphically or by trigonometry. Its value is

$$E_{1b} = 80.2 \; \underline{/ - 63.84°}$$

Consequently, E_{1b} lags 63.84° behind E_{ab} and its magnitude is 80.2 V.

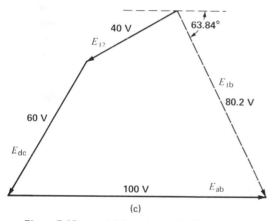

(c)

Figure 7-12   c. Adding phasors in Example 7-6.

7-8   Harmonics

Consider a set of sine waves in which the lowest frequency is f, and all other frequencies are integral multiples of f. By definition, the sine wave having the lowest frequency is called the *fundamental* and all the other waves are called *harmonics*. For example, a group of sine waves whose frequencies are 20, 40, 100, and 380 Hz is said to possess the following components:

fundamental frequency:	20 Hz
second harmonic:	40 Hz (2 x 20 Hz)
fifth harmonic:	100 Hz (5 x 20 Hz)
nineteenth harmonic:	380 Hz (19 x 20 Hz)

Let us now consider two sources e_1 and e_2 connected in series whose frequencies are respectively 60 Hz and 180 Hz (Fig. 7-13). The corresponding amplitudes are 100 V and 20 V. The fundamental and the third harmonic voltages pass through zero at the same time, and both are perfect sine waves.

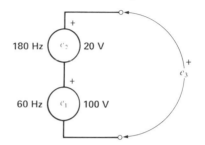

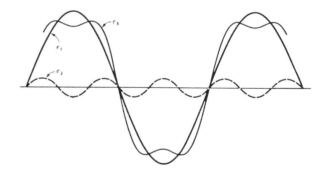

Figure 7-13   A fundamental and third harmonic voltage, together produce a flat-topped wave.

Because the sources are in series, the resulting terminal voltage is obviously equal to the sum of the instantaneous voltages produced by each source. The resulting terminal voltage e_3 is a distorted flat-topped wave whose frequency is equal to the fundamental frequency. We conclude that the sum of a fundamental voltage and a harmonic voltage yields a nonsinusoidal waveform whose frequency is equal to the fundamental frequency.

By using other frequencies and other amplitudes, we can produce a periodic voltage of any conceivable shape. All we have to do is to add together a fundamental voltage and an arbitrary set of harmonic voltages. For example, we can generate a square wave having an amplitude of 100 V and a frequency of 50 Hz by connecting the following sine wave sources in series:

TABLE 7A	100 V SQUARE WAVE		
Harmonic	**amplitude [V]**	**freq. [Hz]**	**relative amplitude**
fundamental	127.3	50	1
third	42.44	150	1/3
fifth	25.46	250	1/5
seventh	18.46	350	1/7
ninth	14.15	450	1/9
⋮	⋮	⋮	⋮
127th	1.00	6350	1/127
⋮	⋮	⋮	⋮
nth	127.3/n	50 n	1/n

A square wave is thus composed of a fundamental wave and an infinite number of harmonics. The higher harmonics have smaller and smaller amplitudes, and they are consequently less important. However, these high-frequency harmonics produce the steep sides and pointy corners of the square wave. In practice, square waves are not usually produced by adding sine waves, but the example does show that *any* wave can be built up from a fundamental wave and an appropriate number of harmonics.

Conversely, we can *decompose* a given periodic wave into its fundamental and harmonic components. However, the harmonic analysis of such waveforms is beyond the scope of this book.

Harmonic voltages and currents are usually undesireable, but in some ac circuits, they are also unavoidable. Harmonics are created by nonlinear loads, such as electric arcs and saturated magnetic circuits. They are also produced whenever voltages and currents are periodically switched, such as in power electronic circuits. All these circuits produce distorted waveshapes that are rich in harmonics.

7-9 Fundamental power and harmonic power

In ac circuits, the fundamental current and fundamental voltage together produce fundamental power. This fundamental power is the useful power that causes a motor to rotate, for example. On the other hand, the product of the harmonic voltage times the corresponding harmonic current produces a harmonic power. The latter is usually dissipated as heat in the ac circuit and consequently, does no useful work. Harmonic currents and voltages should therefore be kept as small as possible.

7-10 Effective value of a distorted periodic wave

The effective value of a distorted voltage is given by:

$$E = \sqrt{E_1{}^2 + E_2{}^2 + E_3{}^2 + \ldots + E_n{}^2} \qquad * \qquad (7\text{-}9)$$

where

E = effective value of the given waveform [V]

E_1 = effective value of the fundamental component [V]

$E_2, E_3, \ldots, E_n$ = effective value of the harmonics [V]

Example 7-7:
Calculate the effective value of the flat-topped ac voltage given in Fig. 7-13, knowing that the peak amplitudes e_1 and e_2 are respectively 100 V and 20 V.

---

* If the distorted voltage also contains a dc component E_0, the effective value is given by

$$E = \sqrt{E_0{}^2 + E_1{}^2 + E_2{}^2 + \ldots + E_n{}^2}$$

Solution:

a. 1 Effective value of the fundamental is:
$$E_1 = e_1/\sqrt{2} \qquad \text{Eq. 7-8}$$
$$= 100/\sqrt{2} = 70.7 \text{ V}$$

a. 2 Effective value of the third harmonic is:
$$E_2 - e_2/\sqrt{2}$$
$$= 20/\sqrt{2} = 14.14 \text{ V}$$

a. 3 Effective value of the distorted wave is:
$$e_3 = \sqrt{e_1{}^2 + e_2{}^2}$$
$$= \sqrt{70.7^2 + 14.14^2}$$
$$= \sqrt{5000 + 200}$$
$$= \sqrt{5200}$$
$$= 72.1 \text{ V}$$

Example 7-8:

A distorted wave has an effective value of 172 V. The fundamental has a frequency of 40 Hz and an effective value of 160 V. Calculate the effective value of all the harmonics combined.

Solution:
$$E = \sqrt{E_1{}^2 + E_2{}^2 + E_3{}^2 + \ldots + E_n{}^2} \qquad \text{Eq. 7-9}$$
$$= \sqrt{E_1{}^2 + E_h{}^2}$$

where

E_h = effective value of all the harmonics.
$$\therefore \quad 172 = \sqrt{160^2 + E_h{}^2}$$
$$172^2 = 160^2 + E_h{}^2$$
$$E_h{}^2 = 3984$$
$$E_h = 63.1 \text{ V}$$

7-11 Distorted waves and ac circuits

What kind of voltages and currents are produced when a distorted waveshape is applied to an ac circuit? The answer can be found quite easily by decomposing the distorted wave into its fundamental and harmonic components. Each component is then considered independently, and the currents and voltages *it* produces are determined by using standard circuit-solving techniques. The fundamental and harmonic voltages across each circuit element are then added together to determine the waveshape of the resulting voltage across

it. Similarly, the fundamental and harmonic currents flowing in a circuit element are added together to determine the waveshape of the current in that element. In effect, the resulting voltages and currents are found by applying the superposition theorem.

Finally, the total power associated with each circuit element is equal to the sum of the fundamental power and all the individual harmonic powers.

7-12 Current in an inductor

It is well know that in an inductive circuit the voltage and current are related by the equation:

$$e = L \frac{\Delta i}{\Delta t} \qquad (7\text{-}10)$$

where

e = instantaneous voltage induced in the circuit [V]

L = inductance of the circuit [H]

$\Delta i/\Delta t$ = rate of change of current [A/s]

This equation enables us to calculate the instantaneous voltage e, knowing the rate of change of current. However, it often happens that e is known and we want to calculate the resulting current I. We can use the same equation, but the solution requires a knowledge of advanced mathematics. To get around this problem, we can use a graphical solution, called the *volt-second method*. It yields the same results, and has the advantage of enabling us to visualize how the current increases and decreases with time, in response to a known applied voltage.

Consider, for example, Fig. 7-14, in which a voltage E is applied across an inductance L. We want to determine the resulting current I. According to the volt-second method, we must measure the area under the voltage curve E, during an interval of time T. For short intervals of time, the area is positive when the voltage is positive and negative when the voltage is negative. The *net* area

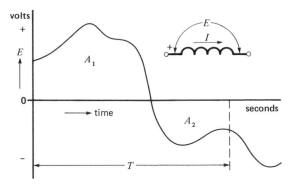

Figure 7-14 Voltage and current in an inductor.

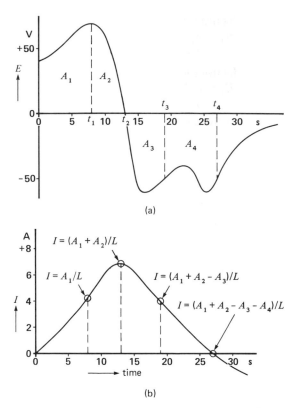

Figure 7-15 a. An inductor stores volt-seconds.

b. Current in the inductor.

after a "long" time interval T is found by subtracting the negative areas from the positive areas. Thus, in Fig. 7-14 the net area A after time interval T is equal to $(A_1 - A_2)$ volt-seconds. In general, the corresponding current is then given by:

$$\boxed{I = A/L} \qquad (7\text{-}11)$$

where

I = current after a time interval T [A]

A = net area under the volt-time curve during time T [V·s]

L = inductance [H]

Consider, for example, an inductor L having negligible resistance, connected to a source whose voltage varies according to the curve of Fig. 7-15a. If the initial current is zero, the value at instant t_1 is:

$$I = A_1/L$$

As time goes by, the area under the curve increases progressively, and so does the current. The current reaches its maximum value at time t_2 because at this moment the area under the voltage curve ceases to increase any more. Beyond t_2, the voltage becomes negative and, consequently, the net area begins to diminish. At instant t_3, for example, the net area is equal to $(A_1 + A_2 - A_3)$ and the corresponding current is:

$$I = (A_1 + A_2 - A_3)/L$$

At instant t_4, the negative area $(A_3 + A_4)$ is equal to the positive area $(A_1 + A_2)$. The net area is zero and the current is also zero. After instant t_4, the current becomes negative, in other words, it changes direction.

Another way of looking at the circuit is to consider that the inductor accumulates volt-seconds during the interval from 0 to t_2. As it becomes "charged up" with volt-seconds, the current increases in direct proportion to the volt-seconds received. Then during the "discharge" period from t_2 to t_4 the inductor loses volt-seconds and the current decreases accordingly. An inductor, therefore, behaves very much like a capacitor. However, instead of storing ampere-seconds or coulombs, an inductor stores volt-seconds.

Figure 7-15b shows the instantaneous current

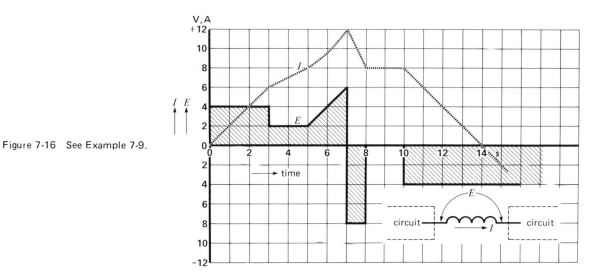

Figure 7-16   See Example 7-9.

obtained when the voltage of Fig. 7-15a is applied to an inductance of 100 H. The initial current is zero, and the current rises to a maximum of 6.9 A before again dropping to zero after a time interval of 27 s.

Important Note: If the current at the beginning of an interval T is not zero, we simply add the initial value to all the current values calculated by the volt-second method.

Example 7-9:

The voltage across the terminals of an inductor of 2 H varies according to the curve given in Fig. 7-16. a) Calculate the instantaneous current I in the circuit, knowing that the initial current is zero. b) Repeat the calculations for an initial current of 7 A.

Solution:

a. 1 **Interval from zero to 3 s:** During this interval, the area in volt-seconds increases uniformly and progressively. Thus, after one second, the area A is 4 V·s; after two seconds, it is 8 V·s; and so forth. Using the expression $I = A/L$, the current builds up to the following respective values: 2 A, 4 A, etc., attaining a final value of 6 A after three seconds.

a. 2 **Interval from 3 s to 5 s:** The area continues to increase, but at a slower rate, because voltage E is smaller than before. When $t = 5$ s, the total surface starting from the beginning is 16 V·s; therefore the current is 16 V·s/2 H = 8 A.

a. 3 **Interval from 5 s to 7 s:** The surface increases by 4 squares which is equivalent to 8 V·s. Consequently, the current increases by 4 A. Note that the current no longer follows a straight line because the voltage is not constant during this interval.

a. 4 **Interval from 7 s to 8 s:** The voltage suddenly changes polarity with the result that the 8 V·s during this interval subtract from the volt-seconds that were accumulated previously. The net area from the beginning is therefore 24 V·s – 8 V·s = 16 V·s. Consequently, the current at the end of this interval is $I = 16/2 = 8$ A.

a. 5 **Interval from 8 s to 10 s:** Because the terminal voltage is zero during this interval, the net volt-second area does not change and neither does the current (remember that we have assumed zero coil resistance).

a. 6 **Interval from 10 s to 14 s:** The negative volt-seconds continue to accumulate and at $t = 14$ s, the negative area is equal to the positive area, and so the net current is zero.

Beyond this point, the current changes direction.

b.   With an initial current of + 7 A, we would have to add 7 A to each of the currents calculated previously. The new current wave is simply 7 A above the curve shown in Fig. 7-16. Thus, the current at t = 11 s is 6 + 7 = 13 A.

7-13 Energy in an inductor

It is well known that a coil stores energy in its magnetic field when it carries a current I. The energy is given by:

$$W = \frac{1}{2} LI^2 \qquad (7\text{-}12)$$

where
W = energy stored in the coil [J]
L = inductance of the coil [H]
I = current [A]

If the current varies, the stored energy rises and falls in step with the current. Thus, whenever the current increases, the coil absorbs energy and whenever the current falls, energy is released. In a sense, the inductor does the job of a flywheel. Thus, a flywheel stores energy when its speed increases, whereas an inductor stores it when the current increases. Furthermore, in the same way that a flywheel tends to oppose a change in speed, an inductor tends to oppose a change in the current it carries.

Example 7-10:
a.   Referring to Fig. 7-16, calculate the energy stored in the inductor when t = 11 s.
b.   Is power flowing into or out of the coil at this instant, and what is its magnitude?

Solution:
a. 1 The inductance is 2 H, and the current at t = 11 s is 6 A. The energy stored is therefore:

$$W = \frac{1}{2} LI^2 \qquad \text{Eq. 7-12}$$
$$= \frac{1}{2} \times 2 \times 6^2$$
$$= 36 \text{ J}$$

b. 1 The current is decreasing; consequently, the stored energy must be falling. The coil is in the process of releasing its stored energy. The power flowing out of the coil is:
$$p = EI$$
$$= 4 \times 6$$
$$= 24 \text{ W}$$

7-14 Some useful equations

We end this chapter by listing (Table 7A) some useful equations that are frequently required when solving ac circuits. The equations are given without proof on the assumption that the student already possesses a knowledge of ac circuits in general.

QUESTIONS AND PROBLEMS

Practical level

7-1   a. Draw the waveshape of a sinusoidal voltage having a peak value of 200 V and a frequency of 5 Hz.
      b. If the voltage is zero at t = 0, what is the voltage at t = 5 ms? 75 ms? 150 ms?

7-2   A sinusoidal current has an effective value of 50 A. Calculate the peak value of current.

7-3   A sinusoidal voltage of 120 V is applied to a resistor of 10 Ω. Calculate:
      a. the effective current in the resistor;
      b. the peak voltage across the resistor;
      c. the power dissipated by the resistor;
      d. the peak power dissipated by the resistor.

7-4   A distorted voltage contains an 11th harmonic of 20 V, 253 Hz. Calculate the frequency of the fundamental.

7-5   The current in a 60 Hz single-phase motor lags 36 degrees behind the voltage. Calculate

TABLE 7A	IMPEDANCE OF SOME COMMON AC CIRCUITS

Circuit diagram	Impedance

$X_L = 2\pi f L$ (7-13)

$X_C = \dfrac{1}{2\pi f C}$ (7-14)

$Z = \sqrt{R^2 + X_L{}^2}$ (7-15)

$Z = \sqrt{R^2 + X_C{}^2}$ (7-16)

$Z = \sqrt{R^2 + (X_L - X_C)^2}$ (7-17)

$Z = \dfrac{R\,X_L}{\sqrt{R^2 + X_L{}^2}}$ (7-18)

$Z = \dfrac{R\,X_C}{\sqrt{R^2 + X_C{}^2}}$ (7-19)

$Z = \dfrac{X_C\,\sqrt{R^2 + X_L{}^2}}{\sqrt{R^2 + (X_L - X_C)^2}}$ (7-20)

the time interval between the positive peaks of voltage and current.

7-6 Referring to Fig. 7-17, determine the phase angle between the following phasors and, in each case, indicate which phasor is lagging.
   a. I_1 and I_3;
   b. I_2 and I_3;
   c. E and I_1.

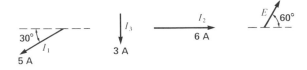

Figure 7-17   See Problem 7-6.

7-7 Referring to Fig. 7-17, determine the magnitude of the following phasors graphically by drawing the phasors to scale. Use a scale of 1 inch = 1 A.
   a. $I_a = I_1 + I_3$
   b. $I_b = I_2 + I_3$
   c. $I_c = I_1 - I_3$
   d. $I_d = I_3 - I_1$

7-8 In Problem 7-7, determine the position of phasors I_a, I_b, I_c and I_d with respect to phasor I_2. Use a protractor to measure the phase angles.

7-9 Referring to Fig. 7-18, determine the magnitude of the resulting phasor, and its lag or lead position with respect to phasor E_2, when the following additions are made. Use a graphic solution and determine the phase angles with a protractor.
   a. $E_1 + E_2 + E_3 + E_4$
   b. $E_1 + E_2 + E_3 - E_4$
   c. $E_1 - E_2 + E_3 - E_4$
   d. $-E_1 + E_4$

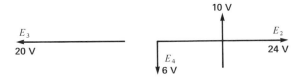

Figure 7-18   See Problem 7-9.

7-10 A 6 V battery is connected to an inductor of negligible resistance whose inductance is 2 H. If the initial current is zero,
   a. How many volt-seconds has the coil accumulated after 2 s?
   b. What is the current after 1 s? after 2 s?

Intermediate level

7-11 The voltage applied to an ac magnet is given by the expression $E = 160 \sin \phi$ and the current is $I = 20 \sin (\phi - 60)$, all angles being expressed in degrees.
   a. Draw the phasor diagram for E and I, using effective values.
   b. Draw the waveshape of E and I as a function of ϕ.
   c. Calculate the peak positive power and the peak negative power in the circuit.

7-12 A periodic current has the waveshape given in Fig. 7-19. If the peak current is 10 A, calculate the effective value.

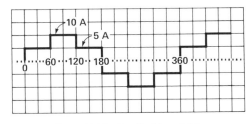

Figure 7-19   See Problem 7-12.

7-13 a. Referring to Fig. 7-13, draw the waveshape of the distorted sine wave, if the third harmonic voltage is reversed.
   b. Calculate the peak voltage of the resulting waveshape, as well as its effective value.

7-14 a. Referring to Fig. 7-16, calculate the power supplied to the coil at $t = 5$ s and the energy stored in the coil at that instant.
   b. Repeat the same calculations for $t = 9$ s.

8

ACTIVE,
REACTIVE AND
APPARENT POWER

The concept of active, reactive, and apparent power plays a major role in electric power technology. In effect, the transmission of electrical energy and the behavior of ac machines are often easier to understand by working with power, rather than by dealing with voltages and currents. The reader is therefore encouraged to pay particular attention to this chapter. He may also find it helpful to review the ac circuits discussed in Chapter 2, Sec. 2-2 to 2-6, and Chapter 7.

The terms active, reactive, and apparent power apply to steady-state alternating current circuits in which the voltages and currents are sinusoidal. We cannot use them to describe transient-state behavior, nor can we apply them to dc circuits.

8-1 Apparent power

Figure 8-1 shows a single-phase device operating at an effective voltage E and carrying an effective current I. The positive direction of current flow is designated by an arrow and, in a similar arbitrary way, the positive side of voltage is designated by a (+).

By definition, the *apparent power S* associated with the device is equal to the arithmetic product EI, expresssed in volt-amperes. The two conductors leading to the device carry the apparent power, but unlike current flow, power does not flow down one conductor and return by the other. Power flows over *both* conductors and, consequently, as far as power is concerned, we can replace the conductors by a single line, as shown in Fig. 8-2.

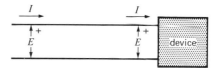

Figure 8-1 Arbitrary polarity and current flow in a device.

Figure 8-2 From a power standpoint a transmission line can be represented by a single line.

Clearly, power flows either towards the device or away from it. Unfortunately, if we only know the value of apparent power, we cannot state whether it flows in one direction or the other. The reason is that the power flow depends upon the phase angle between current phasor I and voltage phasor E. Thus, referring to Fig. 8-3, phasor I may occupy any position $i_1, i_2, .., i_8$ with respect to phasor E, but if the currents have the same magnitude, the apparent power EI is the same in every case. We shall shortly establish rules enabling us to

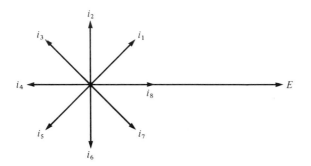

Figure 8-3   Apparent power is independent of the phase angle between voltage and current.

get a better understanding of power flow in ac devices and circuits. Specifically, they involve breaking up the apparent power into two components: an active power P and a reactive power Q.

As a final note, the maximum apparent power to be carried by a transmission line has a direct influence upon its physical size. Thus, as the line current I increases, we must employ larger conductors and, as the line voltage E becomes higher and higher, we have to increase the spacing between

them. Consequently, the size of a transmission line depends directly upon the product EI, which, of course, is equal to the apparent power S.

8-2   Active power*

The simple ac circuit of Fig. 8-4a consists of a resistor connected to an ac generator. The effective voltage and current are designated E and I respectively and, as we would expect in a resistive circuit, phasors E and I are in phase (Fig. 8-4b). If we connect a wattmeter into the line, it will give a reading $P = EI$ watts.

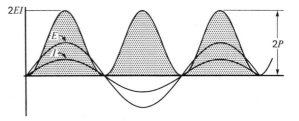

Figure 8-5   Active power is composed of a series of positive power pulses.

To get a better picture of what goes on in such a circuit, we have drawn the sinusoidal curves of E and I (Fig. 8-5). The peak values are respectively $\sqrt{2}E$ volts and $\sqrt{2}I$ amperes because, as we stated at the outset, E and I are *effective* values. By mul-

---

*   Many persons refer to active power as "real power" or "true power" considering it to be more descriptive. In this book, we use the term "active power", because it conforms to the IEEE designation.

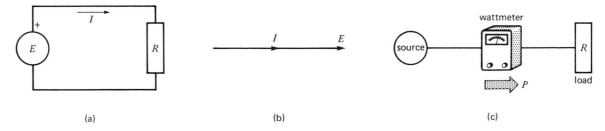

(a)   (b)   (c)

Figure 8-4   Circuit composed of a resistor connected to an ac generator.

tiplying the instantaneous values of voltage and current, we obtain the instantaneous power in watts.

The power wave consists of a series of positive loops that vary from zero to a maximum of $2EI$ watts. The fact that power is always positive indicates that it always flows from the generator to the resistor. This is one of the basic properties of what is called *active power*: although it pulsates between zero and maximum, it never changes direction. The direction of power flow is shown by an arrow P (Fig. 8-4c). As mentioned previously, power flows over *both* conductors and, consequently, we can replace them by a single line. In general, the line represents any transmission line connecting two devices, irrespective of the number of conductors it may have.

The generator is an *active source* and the resistor is an *active load*.

8-3 Definition of active sources and active loads

Many other devices, besides resistors, can function as active loads. To enable us to positively identify an active source or active load, consider Fig. 8-6 in which a device A is one of several devices making up an ac circuit. In solving such a circuit, we assign

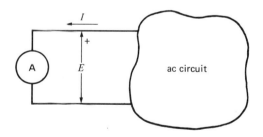

Figure 8-6 Device A may be an active source or active load depending upon the phasor relationship between E and I.

an arbitrary polarity to the voltage E across each device, and an arbitrary direction for the corresponding current flow I. This enables us to write the circuit equations based on Kirchhoff's voltage

and current laws. Naturally, to write these equations, we have to know the precise nature of all the devices making up the circuit. The equations give us a solution to E and I for each device and this solution can be represented by a phasor diagram. This diagram, together with the arbitrary polarity and current flow, permit us to state whether a device is an active load or active source. The following rule applies:

1. **A device is an** *active load* **when:**
   a. **voltage** E **and current** I **are in phase** *and*
   b. **current** I **is shown as entering the (+) terminal.**

The following additional rules follow naturally from the above basic rule:

2. A device is an *active source* when E and I are in phase and I is shown as leaving the (+) terminal.
3. A device is an active source when E and I are $180°$ out of phase and I is shown as entering the (+) terminal.
4. A device is an active load when E and I are $180°$ out of phase and I is shown as leaving the (+) terminal.

Example 8-1:
Two devices A and B are connected by a transmission line and an arbitrary polarity is assigned to the line voltage E and an arbitrary direction for line current I (Fig. 8-7). After solving the circuit, the equations show that phasors E and I are $180°$ out of phase. Which device is the active load?

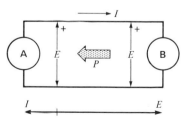

Figure 8-7 See Example 8-1.

Solution:

We could refer to the four rules listed above and select the one that applies to our problem. However, it is easier to remember only rule 1 and reason as follows: If E and I were in phase, B would obviously be the load. However, because they are 180° out of phase B must be the active source. It follows that device A is the active load, and the direction of active power flow is shown by arrow P.

Note that the phasor diagram applies simultaneously to device A, to device B and to the transmission line connecting them. In other words, a single phasor diagram suffices for all three elements of the circuit.

8-4   Reactive power

The circuit of Fig. 8-8a is identical to the original resistive circuit except that the resistor is now replaced by an inductor. After writing the equations, we discover that I lags 90° behind voltage E.

To see what really goes on in such a circuit, we have drawn the waveforms for E and I and, by again multiplying the instantaneous values, we obtain the curve of instantaneous power (Fig. 8-8b). The instantaneous power wave consists of a series of positive and negative loops: the positive loops correspond to instantaneous power delivered from the generator to the inductor and the negative loops represent instantaneous power delivered from the inductor to the generator. The dotted

area under each loop is the energy, in joules, transported in one direction or the other. Clearly, the energy is delivered in a continuous series of pulses of very short duration, every positive pulse being immediately followed by a negative pulse.

Energy moves back and forth between the generator and the inductor without ever being used up. Power which oscillates back and forth in this manner is called *reactive power*, to distinguish it from the unidirectional active power mentioned before. The reactive power in Fig. 8-8 is also given by the product EI. However, to distinguish this power from active power, another unit is used - the var. A larger unit, the kvar, is equal to 1000 var.

Special instruments, called varmeters, are available to measure the reactive power in a circuit (Fig. 8-9). A varmeter multiplies the effective line voltage E by the effective line current I times $\sin \theta$ (the angle between them) and gives a readout of the product. A reading is only obtained when E and I are out of phase; if they are in phase, the varmeter reads zero.

Example 8-2:

A voltmeter and ammeter connected into the circuit of Fig. 8-8a give readings of 140 V and 20 A, respectively. Calculate:

a. the apparent power of the circuit;

b. the reactive power of the circuit;

c. the active power of the circuit.

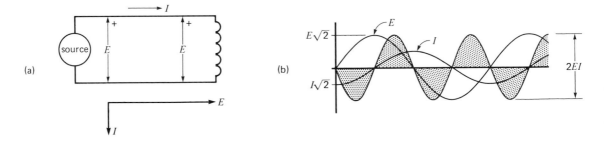

(a)

(b)

Figure 8-8   a.   Circuit composed of an inductor connected to an ac generator.

b.   Reactive power is composed of a sequence of identical positive and negative power pulses.

Solution:

a. The apparent power is:

$$S = EI = 140 \times 20$$
$$= 2800 \text{ VA} = 2.8 \text{ kVA}$$

b. The reactive power is:

$$Q = EI = 140 \times 20$$
$$= 2800 \text{ var} = 2.8 \text{ kvar.}$$

Note that if a varmeter were connected into the circuit, it would give a reading of 2800 var.

c. The active power is zero.

If a wattmeter were connected into the circuit it would read zero.

To recapitulate, the apparent power in Fig. 8-8a is EI voltamperes, but, because the current is 90° out of phase with the voltage, it is given the special name reactive power and the corresponding special unit, the var.

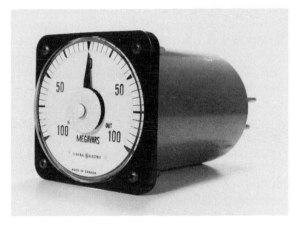

Figure 8-9   Varmeter. *(Canadian General Electric)*

8-5   Reactive sources and reactive loads

Reactive power involves energy that oscillates back and forth over a transmission line; consequently, it is really impossible to say whether it originates at one end of the line or the other. Nevertheless, it is very useful to *assume* that some devices generate reactive power while others absorb it. In other words, some devices behave like *reactive sources* and others like *reactive loads*. In order to distinguish between the two, we use the same line of

reasoning as in the case of active loads. Thus, suppose a device X is part of an ac circuit and that the voltage E and current I have respectively been assigned the arbitrary polarity and direction shown in Fig. 8-10. After solving the circuit equations

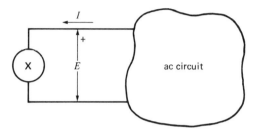

Figure 8-10   Device X may be a reactive source or reactive load depending on the phasor relationship between E and I.

suppose we find that the voltage and current are 90° out of phase. The following rule then applies*:

1. A device is a *reactive load* when;
   a. **current I lags 90° behind voltage E** and
   b. **current I is shown as entering the (+) terminal.**

The following additional rules follow naturally from the above basic rule:

2. A device is a reactive source when I lags behind E by 90° and I is shown as leaving the (+) terminal.
3. A device is a reactive source when I leads E by 90° and I is shown as entering the (+) terminal.
4. A device is a reactive load when I leads E by 90° and I is shown as leaving the (+) terminal.

Example 8-3:

Two devices X and Y are connected by a transmission line and an arbitrary polarity is assigned to the line voltage E and an arbitrary direction to

---

* These rules conform to definitions adopted by the IEEE.

line current I. (Fig. 8-11). After solving the circuit, the equations show that phasor E lags 90° behind phasor I. Which device is the reactive source?

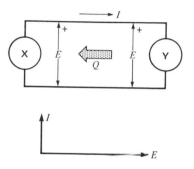

Figure 8-11   See Example 8-3.

Solution:

We could scan through the four rules given above and select the one that applies to our problem. However, it is easier to remember only rule 1 and reason as follows: If the current lagged behind the voltage in Fig. 8-11, then Y would be the reactive load. But since the current is actually leading the voltage, Y must be a reactive source. It follows that X is reactive load. The direction of reactive power flow is therefore from Y to X, as indicated by arrow Q.

The reader should note that the same phasor diagram applies to devices X and Y as well as to the transmission line connecting them.

8-6   The capacitor as a reactive source

Let us now study the behavior of a capacitor connected to an ac generator (Fig. 8-12a). Upon solving the circuit, we discover that the current leads the voltage by 90°. It follows from our definitions that the capacitor is a reactive source. Consequently, it delivers reactive power to the very generator to which it is connected! For most people, this takes a little time to swallow because how, we may ask, can a passive device like a capacitor possibly produce any power? The answer is that reactive power really represents energy which, like a pendulum, swings back and forth without ever doing any useful work. The capacitor acts as a temporary energy-storing device continually accepting energy for brief periods and releasing it again.

If we connect a varmeter into the circuit (Fig. 8-12b), it will give a negative reading of EI vars, showing that reactive power is indeed flowing from the capacitor to the generator. The generator is actually a reactive load, but we sometimes prefer to call it a *receiver* of reactive power, which, of course, amounts to the same thing.

8-7   Distinction between active and reactive power

There is a basic difference between active and reactive power and perhaps the most important thing to remember about them is that one cannot be converted into the other. Active and reactive power function independently of each other and, consequently, they can be treated as separate quantities in electric circuits.

Both place a burden on the transmission line that carries them, but, whereas active power eventually produces a tangible result (heat, mechanical power, light, etc.), reactive power only represents energy that oscillates back and forth.

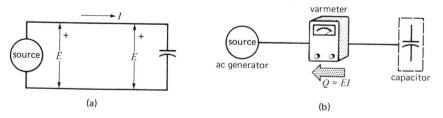

(a)   (b)

Figure 8-12   A capacitor is a source of reactive power.

8-8 Combined active and reactive loads

Most industrial loads absorb both active power P and reactive power Q. Consider, for example, the circuit of Fig. 8-13a in which a resistor and inductor are connected to a source E. The resistor draws

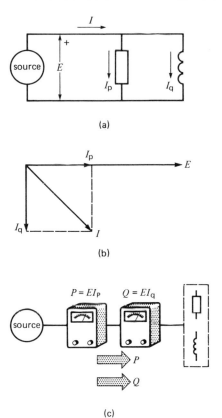

(a)

(b)

(c)

Figure 8-13 Active and reactive power flowing in the same direction.

a current I_p, while the inductor draws a current I_q. The phasor diagram is self-explanatory and, according to our definitions, the resistor is an active load while the inductor is a reactive load. The active and reactive power components P and Q both flow in the same direction as shown by the arrows in Fig. 8-13c. If we connect a wattmeter and a varmeter into the circuit, the readings will both be positive, indicating $P = EI_p$ watts and $Q = EI_q$ vars, respectively.

Let us now consider Fig. 8-14 in which a resistor and capacitor are connected to the source. The situation is similar to Fig. 8-13 except that the capacitor is a reactive source. The active and reactive power components now flow in opposite directions over the transmission line. A wattmeter con-

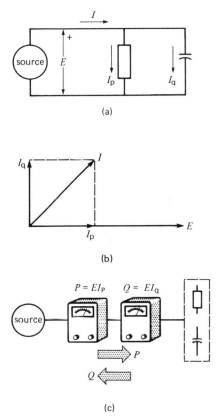

(a)

(b)

(c)

Figure 8-14 Active and reactive power flowing in opposite directions.

nected into the circuit will give a positive reading $P = EI_p$ watts but a varmeter will give a negative reading $Q = EI_q$. The so-called "source" G delivers active power P, but receives reactive power Q. Thus, it is both an active source and a reactive load. It may seem unusual to have two powers flowing in opposite directions over the same transmission line, but we must remember that active power P is not the same as a reactive power Q and that each flows independently of the other.

All ac inductive devices such as magnets, transformers, ballasts, induction motors, and so forth, absorb reactive power because one component of the current they draw lags 90° behind the voltage. The reactive power plays a very important role because it produces the ac magnetic field in these devices. The devices simultaneously behave as active loads because they must absorb active power to supply the losses and to do useful mechanical or electrical work.

A building, shopping center, or city may be considered to be an active and reactive load connected to an electric utility system. Such load centers contain thousands of induction motors and other electromagnetic devices which require reactive power (to sustain their magnetic fields) and active power (to do the useful work).

livers reactive power, the receptacle will receive it. In other words, a receptacle outlet is at all times ready to accept or deliver either active power P or reactive power Q to meet the requirements of the group of devices connected to it.

8-9   Relationship between P, Q, S

Consider the single-phase circuit of Fig. 8-15 composed of a source, a load and appropriate meters. Let us assume that:

- the voltmeter indicates E volts;
- the ammeter indicates I amperes;
- the wattmeter indicates $+P$ watts;
- the varmeter indicates $+Q$ vars;
- the current lags behind the voltage by θ degrees.

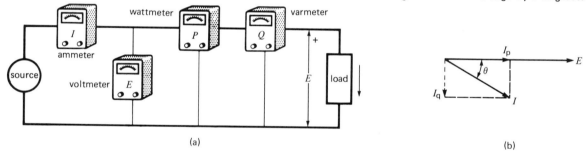

(a)

(b)

Figure 8-15   Relationship between P, Q, and S.

Speaking of sources and loads, the deceptively simple electrical outlet also deserves our attention. Such an outlet may be the 120 V receptacle in a home, a 3-phase 480 V service entrance to a factory, or the terminals of a high-power 345 kV transmission line. All such outlets are ultimately connected to the huge alternators that power the electrical transmission and distribution systems. Strange as it may seem, an outlet can act not only as an active or reactive source (as we would expect), but it may also behave as an active or reactive load. What factors determine whether it will behave in one way or the other? It depends entirely upon the type of device (or devices) connected to the receptacle. If the device requires active power, the receptacle will provide it; if the device de-

Current I can be decomposed into two components I_p and I_q, respectively in phase, and in quadrature, with phasor E. The numerical values of I_p and I_q can be found directly from the meter readings:

$$I_p = P/E \qquad (8\text{-}1)$$

$$I_q = Q/E \qquad (8\text{-}2)$$

Furthermore, the apparent power S transmitted over the line is given by $S = EI$, from which:

$$I = S/E \qquad (8\text{-}3)$$

Referring to the phasor diagram, it is obvious that:

$$I^2 = I_p^2 + I_q^2$$

so that:

$$\left[\frac{S}{E}\right]^2 = \left[\frac{P}{E}\right]^2 + \left[\frac{Q}{E}\right]^2$$

or

$$\boxed{S^2 = P^2 + Q^2}\ ^* \tag{8-4}$$

in which

S = apparent power [VA]
P = active power [W]
Q = reactive power [var]

Example 8-1:
An alternating current motor absorbs 40 kW of active power and 30 kvar of reactive power. Calculate the apparent power supplied to the motor.

Solution:

$$\begin{aligned}
S &= \sqrt{P^2 + Q^2} \qquad \text{Eq. 8-4}\\
&= \sqrt{40^2 + 30^2}\\
&= 50 \text{ kVA.}
\end{aligned}$$

Example 8-2:
A wattmeter and varmeter are connected into a 120 V single-phase line feeding an ac motor. They respectively read 1800 W and 960 var. Calculate:
a. the in-phase and quadrature components I_p and I_q;
b. the line current I;
c. the apparent power supplied by the source.

Solution:
Referring to Fig. 8-15, we have:

---

* This equation may be represented by a right-angle "power triangle" having sides P, Q, and hypotenuse S. The relationship between the three sides is the well-known pythagorean equation $S^2 = P^2 + Q^2$. The concept of the power triangle is particularly useful when solving ac circuits using complex notation.

a. 1 $I_p = P/E = 1800/120 = 15$ A   Eq. 8-1

a. 2 $I_q = Q/E = 960/120 = 8$ A   Eq. 8-2

b.   From the phasor diagram:

$$I = \sqrt{I_p^2 + I_q^2} = \sqrt{15^2 + 8^2} = 17 \text{ A}$$

c.   The apparent power is:
$S = EI = 120 \times 17 = 2040$ VA

8-10 Power factor

The power factor of an alternating current circuit is given by the equation:

$$\boxed{\cos \theta = P/S} \tag{8-5}$$

where
$\cos \theta$ = power factor, expressed as a simple number, or as a percentage
P = active power delivered or absorbed by the circuit [W]
S = apparent power of the circuit [VA]

Because the active power P can never exceed the apparent power S, it follows that the power factor can never be greater than unity (or 100 percent). The power factor of a resistor is 100 percent because the apparent power it draws is equal to the active power. On the other hand, the power factor of an ideal coil having no resistance is zero because it does not consume any active power.

To sum up, the power factor of a circuit or a device is simply a way of stating what fraction of its apparent power is real, or active, power.

In a single-phase circuit the power factor is also a measure of the phase angle θ between the voltage and current. Thus, referring to Fig. 8-15:

$$\begin{aligned}
\text{power factor} &= P/S\\
&= EI_p/EI\\
&= I_p/I\\
&= \cos \theta
\end{aligned}$$

Consequently,

$$\boxed{\text{power factor } = \cos \theta} \qquad (8\text{-}6)$$

where

power factor = power factor of a single-phase circuit

θ = phase angle between the voltage and current

If we know the power factor, we know the cosine of the phase angle between E and I and, hence, we can calculate the angle. A power factor is said to be *lagging* if the current lags behind the voltage. Conversely, the power factor is said to be *leading* if the current leads the voltage.

Example 8-3:

Calculate the power factor of the motor in Example 8-2 and the phase angle between the line voltage and line current.

Solution:

$$\begin{aligned}
\text{power factor} &= P/S \\
&= 1800/2040 \\
&= 0.882 \text{ or } 88.2 \text{ percent lagging} \\
\cos \theta &= 0.882 \\
\therefore \theta &= 28°
\end{aligned}$$

Example 8-4:

A single-phase motor draws a current of 5 A from a 120 V, 60 Hz line. The power factor of the motor is 65 percent. Calculate:

a. the active power absorbed by the motor;
b. the reactive power supplied by the line.

Solution:

a. 1 The apparent power drawn by the motor is:

$$\begin{aligned}
S &= EI = 120 \times 5 \\
&= 600 \text{ VA}
\end{aligned}$$

a. 2 The active power absorbed is:

$$P = S \cos \theta \qquad \text{Eq. 8-5}$$
$$= 600 \times 0.65 = 390 \text{ W}$$

b. The reactive power is:

$$Q = \sqrt{S^2 - P^2} \qquad \text{Eq. 8-4}$$
$$= \sqrt{600^2 - 390^2}$$

$$= 456 \text{ var}$$

Note that the motor draws more reactive power from the line than active power. This is unfortunate, because it burdens the line with non-productive power.

Example 8-5:

A 50 μF paper capacitor is placed across the motor terminals in Example 8-4. Calculate:

a. the reactive power generated by the capacitor;
b. the active power absorbed by the motor;
c. the reactive power absorbed from the line;
d. the new line current.

Solution:

a. 1 The impedance of the capacitor is:

$$X_C = 1/2\pi f C \qquad \text{Eq. 7-14}$$
$$= 1/2\pi \times 60 \times 50 \times 10^{-6}$$
$$= 53 \ \Omega$$

a. 2 The current in the capacitor is:

$$I = E/X_C = 120/53$$
$$= 2.26 \text{ A}$$

a. 3 The reactive power generated by the capacitor is:

$$Q_C = EI = 120 \times 2.26$$
$$= 271 \text{ var}$$

b. 1 The motor continues to draw the same active power because it is still fully loaded.

$$P_m = 390 \text{ W}$$

b. 2 Furthermore, the motor also draws the same reactive power as before because nothing has taken place to change its magnetic field.

Consequently:

$$Q_m = 456 \text{ var}$$

c. 1 The motor draws 456 var from the line, but the capacitor furnishes 271 var to the same line. The net reactive power drawn from the line is therefore:

$$\begin{aligned}
Q_L &= Q_m - Q_C \\
&= 456 - 271 \\
&= 185 \text{ var}
\end{aligned}$$

c. 2 The active power drawn from the line is obviously:

$$P_L = P_m = 390 \text{ W}$$

d. 1 The apparent power drawn from the line is:
$$S = \sqrt{P_L{}^2 + Q_L{}^2}$$
$$= \sqrt{390^2 + 185^2}$$
$$= 432 \text{ VA}$$

d. 2 The new line current is:
$$I_L = S_L/E = 432/120$$
$$= 3.6 \text{ A}$$

Thus, the line current drops from 5 A to 3.6 A by placing the capacitor in parallel with the motor.

8-11 Systems comprising several loads

The concept of active and reactive power enables us to simplify the solution of some rather complex circuits. Consider, for example, a group of loads connected to a 380 V receptacle (Fig. 8-16a). We wish to calculate the apparent power of the system as well as the current supplied by the receptacle.

is inductive, it absorbs reactive power; consequently, the 5 kvar arrow flows from the source to the load. On the other hand, because load C represents a capacitor, it delivers reactive power to the system. The 6 kvar arrow is directed accordingly.

The distinct nature of active and reactive power enables us to add all the active powers in a circuit to obtain the total active power P. In the same way, we can add all the reactive powers to obtain the total reactive power Q. The resulting total apparent power S is then found by Eq. (8-4):
$$S = \sqrt{P^2 + Q^2}.$$

When adding reactive powers, we assign a positive value to those that are absorbed and a negative value to those that are generated (such as by a capacitor). In the same way, we assign a positive value to active powers that are absorbed and a negative value to those that are generated (such as by an alternator).

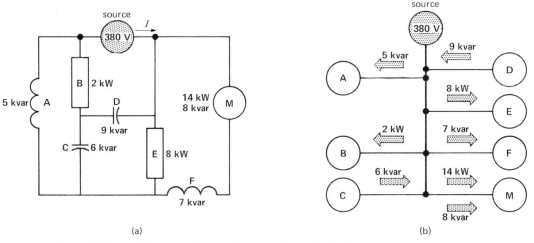

Figure 8-16   a. Example of grouped active and reactive loads.

b. All loads are assumed to be directly connected to the 380 V receptacle.

Using the power approach, we do not have to worry about the way the loads are interconnected. We simply draw a block diagram of the individual loads, indicating the direction of active and reactive power flow (Fig. 8-16b). Thus, because load A

Note that usually we *cannot* add the apparent powers in various parts of a circuit to obtain the total apparent power S. We can only add them if their power factors are identical.

Let us now analyze the circuit of Fig. 8-16:

1. active power absorbed by the system:
$$P = (2 + 8 + 14) = + 24 \text{ kW}$$
2. reactive power absorbed by the system:
$$Q_1 = (5 + 7 + 8) = + 20 \text{ kvar}$$
3. reactive power supplied by the capacitors:
$$Q_2 = (- 9 - 16) = - 25 \text{ kvar}$$
4. net reactive power Q absorbed by the system:
$$Q = (+ 20 - 25) = - 5 \text{ kvar}$$
5. apparent power of the system:
$$S = \sqrt{P^2 + Q^2} = \sqrt{24^2 + (- 5)^2} = 24.5 \text{ kVA}$$
6. because the receptacle furnishes the apparent power, the line current is:
$$I = S/E = 24\,500/380 = 64.5 \text{ A}$$

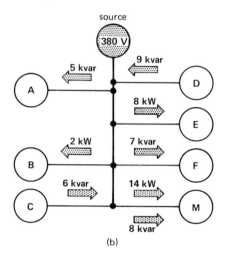

(b)

Figure 8-16b.

The receptacle delivers 24 kW of active power, but it *receives* 5 kvar of reactive power. This reactive power flows into the receptacle and out over the local distribution system of the electrical utility company where it becomes available to create magnetic fields. The magnetic fields may be associated with distribution transformers, transmission lines or even relays and electromagnets of other customers connected to the same distribution system.

8-12 Reactive power without magnetic fields

We sometimes encounter situations where loads absorb reactive power without creating any magnetic field at all. This can happen in electronic power circuits when we delay the current flow by means of a rapid switching device, such as a thyristor.

Consider, for example, the circuit of Fig. 8-17 in which a 100 V, 60 Hz source is momentarily connected to a resistive load of 10 Ω by means of a synchronous switch. The switch opens and closes its contacts so that current only flows during the latter part of each half cycle. We can see, almost by intuition, that this forced delay causes the current to lag behind the voltage. Indeed, if we con-

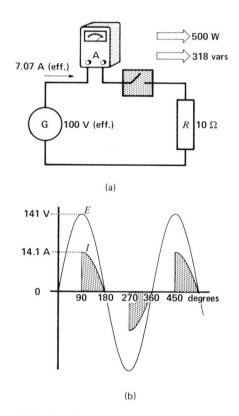

Figure 8-17   Reactive power in switched loads having no magnetic field.

nected a wattmeter and varmeter between the source and the switch, they would respectively read + 500 W and + 318 var, which corresponds to a lagging power factor of 84 percent. The reactive power is associated with the rapidly operating switch rather than with the resistor itself. Nevertheless, reactive power is consumed just as surely as if an inductor were present in the circuit.

8-13 Solving AC circuits

We have seen that active and reactive powers can be added algebraically. This enables us to solve some rather complex ac circuits without ever having to draw a phasor diagram. We calculate the active and reactive powers associated with each circuit element and deduce the corresponding voltages and currents.

2. $I_R = 60/12 = 5$ A, from which
$$P = 5 \times 60 = 300 \text{ W}.$$

3. apparent power supplied to terminals 1 - 3:
$$S = \sqrt{P^2 + Q^2} = \sqrt{300^2 + (-720)^2}$$
$$= 780 \text{ VA}.$$

4. current $I_L = S/E_{31} = 780/60 = 13$ A.

5. voltage across the inductance:
$$E_{23} = 13 \times 8 = 104 \text{ V}.$$

6. reactive power absorbed by the inductance:
$$Q_L = E_{23} \times I_L = 104 \times 13$$
$$= +1352 \text{ var}.$$

7. total reactive power absorbed the circuit:
$$Q = +1352 - 720 = +632 \text{ var}.$$

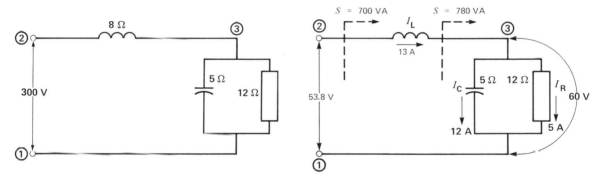

Figure 8-19   Solving ac circuits by the active/reactive power method.

Example 8-6:
Calculate the impedance of the circuit shown in Fig. 8-19a and determine the current flowing in the resistor if $E_{21} = 300$ V.

Solution:
The impedance of a circuit is independent of the voltage applied to its terminals. We can therefore assume a voltage across the terminals of an appropriate circuit element and, by successive deduction, calculate the voltages and the currents in all the other elements. Thus, let us suppose a voltage of 60 V between terminals 3 and 1 (Fig. 8-19b). We then proceed in logical steps, as follows:

1. $I_C = 60/5 = 12$ A, from which
$$Q_C = 12 \times 60 = -720 \text{ var}.$$

8. total active power absorbed by the circuit:
$$P = 300 \text{ W}.$$

9. apparent power absorbed by the circuit:
$$S = \sqrt{300^2 + 632^2} = 700 \text{ VA}.$$

10. voltage $E_{21} = S/I_L = 700/13 = 53.8$ V.

11. impedance between terminals 2-1:
$$Z = E_{21}/I_L = 53.8/13 = 4.14 \text{ } \Omega.$$

The actual current in the resistor can be found by simple proportion. Because 53.8 V between terminals 2-1 produces a current of 5 A in the resistance, it follows that a voltage of 300 V will yield:
$$I_R = 5 \text{ A} (300/53.8) = 27.9 \text{ A}$$

QUESTIONS AND PROBLEMS

Practical level

8-1   What is the unit of active power? reactive power? apparent power?

8-2   A capacitor of 500 kvar is placed in parallel with an inductor of 400 kvar. Calculate the apparent power of the group.

8-3   Name a static device which can generate reactive power.

8-4   Name a static device which absorbs reactive power.

8-5   What is the approximate power factor, in percent, of a capacitor? of a coil? of an incandescent lamp?

Intermediate level

8-6   The current in a single-phase motor lags 50° behind the voltage. What is the power factor of the motor?

8-7   A large motor absorbs 600 kW at a power factor of 90 percent. Calculate the apparent power and reactive power absorbed by the machine.

8-8   A 200 μF capacitor is connected to a 240 V, 60 Hz source. Calculate the reactive power it generates.

8-9   A 10 Ω resistor is connected across a 120 V, 60 Hz source. Calculate:
   a. the active power absorbed by the resistor;
   b. the apparent power absorbed by the resistor;
   c. the peak power input to the resistor;
   d. the duration of each positive power pulse.

8-10   A 10 Ω reactance is connected to a 120 V, 60 Hz line. Calculate:
   a. the reactive power absorbed by the reactor;
   b. the apparent power absorbed by the reactor;
   c. the peak power input to the reactor;
   d. the peak power output of the reactor;
   e. the duration of each positive power pulse.

8-11   Determine which of the devices in Figs. 8-20a through 8-20f acts as an active (or reactive) power source.

8-12   A single-phase motor draws a current of 12 A at a power factor of 60 percent. Calculate the in-phase and quadrature components of current I_p and I_q with respect to the line voltage.

8-13   A single-phase motor draws a current of 16 A from a 240 V, 60 Hz line. A wattmeter connected into the line gives a reading of 2765 W. Calculate the power factor and the reactive power absorbed by the machine.

8-14   If a capacitor having a reactance of 30 Ω is connected in parallel with the motor of problem 8-13, calculate:
   a. the active power reading of the wattmeter;
   b. the total reactive power absorbed by the capacitor and motor;
   c. the apparent power of the ac line;
   d. the line current;
   e. the power factor of the motor/capacitor combination.

8-15   Using only power concepts and without drawing any phasor diagrams, find the impedance of the circuits in Fig. 8-21.

8-16   An induction motor absorbs an apparent power of 400 kVA at a power factor of 80 percent. Calculate:
   a. the active power absorbed by the motor;
   b. the reactive power absorbed by the motor;
   c. what purpose does the reactive power serve?

8-17   A circuit composed of a 12 Ω resistor in series with an inductive reactance of 5 Ω carries an ac current of 10 A. Calculate:
   a. the active power absorbed by the resistor;
   b. the reactive power absorbed by the inductor;
   c. the apparent power of the circuit;
   d. the power factor of the circuit.

8-18   A coil having a resistance of 5 Ω and an inductance of 2 H carries a direct current of 20 A. Calculate:
   a. the active power absorbed;
   b. the reactive power absorbed.

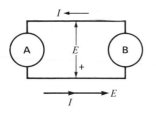

(a)

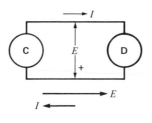

(b)

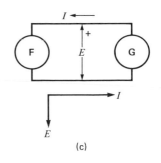

(c)

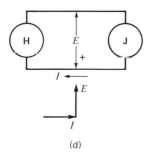

(d)

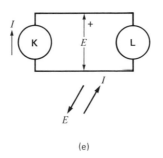

(e)

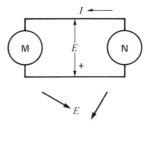

(f)

Figure 8-20   See Problem 8-11.

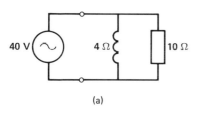

(a)

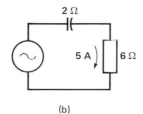

(b)

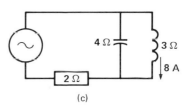

(c)

Figure 8-21   See Problem 8-15.

Advance level

8-19   A motor having a power factor of 0.8 absorbs an active power of 1200 W. Calculate the reactive power drawn from the line.

8-20   In Problem 8-13, if we place a capacitor of 500 var in parallel with the motor, calculate:
   a. the total active power absorbed by the system;
   b. the apparent power of the system;
   c. the power factor of the system.

8-21   A coil having a reactance of 10 Ω and a resistance of 2 Ω is connected in parallel with a capacitive reactance of 10 Ω. If the supply voltage is 200 V, calculate:
   a. the reactive power absorbed by the coil;
   b. the reactive power generated by the capacitor;
   c. the active power dissipated by the coil;
   d. the apparent power of the circuit;

8-22 The power factor at the terminals of a 120 V
      source is 0.6 lagging (Fig. 8-22). Without
      using phasor diagrams, calculate:
      a. the value of E;
      b. the impedance of the load Z.

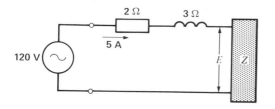

Figure 8-22   See Problem 8-22.

8-23 In Figs. 8-23a and 8-23b, indicate the magni-
      tude and direction of the active and reactive
      power flow. Hint: decompose I into I_p and
      I_q and treat them independently.

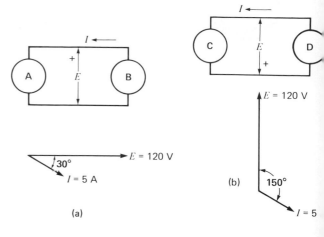

Figure 8-23   See Problem 8-23.

9

THREE-PHASE CIRCUITS

Electric power is generated, transmitted and distributed in the form of three-phase power. Homes and small establishements are wired for single-phase power, but this merely represents a tap-off from the basic three-phase system. Three-phase power is preferred over single-phase power for several important reasons:

a. 3-phase motors, generators and transformers are simpler, cheaper and more efficient;

b. 3-phase transmission lines can deliver more power for a given weight and cost;

c. the voltage regulation of 3-phase transmission lines is inherently better.

A knowledge of 3-phase power and 3-phase circuits is therefore essential to an understanding of power technology. Fortunately, the basic circuit techniques used to solve single-phase circuits can be directly applied to 3-phase circuits. Furthermore, we shall see that the vast majority of 3-phase circuits can be reduced to elementary single-phase diagrams. In this regard, we assume the reader is familiar with the previous chapters dealing with ac circuits and power.

9-1 Polyphase systems

We can gain an immediate preliminary understanding of polyphase systems by referring to the common gasoline engine. A single-cylinder engine having one piston is comparable to a single-phase machine. On the other hand, a two-cylinder engine is comparable to a 2-phase machine. The more common six-cylinder engine could be called a six-phase machine. In a six-cylinder engine, identical pistons move up and down inside identical cylinders, but they do not move in unison. They are staggered in such a way so as to deliver power to the shaft in successive pulses rather than at the same time. As the reader may know from personal experience, this produces a smoother running engine and a much smoother output torque.

Similarly, in a 3-phase electrical system, the three phases are identical, but they deliver power at different times. As a result, the power flow is very smooth. Furthermore, because the phases are identical, one phase may be used to represent the behavior of all three.

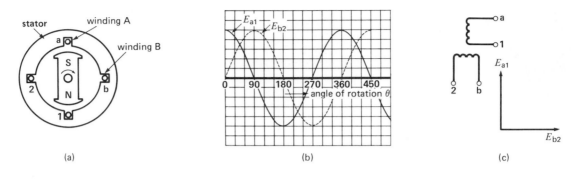

Figure 9-1   a. Two-phase alternator.   b. Voltages induced in a 2-phase alternator.   c. Phasor diagram

Although we must beware of carrying analogies too far, the above description reveals that a three-phase system is basically composed of three single-phase systems which operate in sequence. Once this basic fact is realized, much of the mystery surrounding 3-phase systems disappears.

9-2   Two-phase alternator

In reviewing the single-phase alternator in Section 2-10, we saw that an ac voltage is induced in the stator winding when it is cut by the flux of a revolving magnet. The magnet is driven by an external mechanical source, such as a turbine. Let us mount a second winding on the stator, identical to the first, but displaced from it by a mechanical angle of 90° (Fig. 9-1a).

As the magnet rotates, sinusoidal voltages are induced in each winding. They obviously have the same magnitude and frequency, but they do not reach their maximum value at the same time. In effect, at the moment when the magnet occupies the position shown in Fig. 9-1a, voltage E_{a1} passes through its maximum positive value, whereas voltage E_{b2} is zero. This is because the flux only cuts across the conductors in slots 1 and a at this instant. However, after the rotor has made one quarter turn (or 90°), voltage E_{a1} becomes zero and voltage E_{b2} attains its maximum positive value. The two voltages are therefore out of phase by 90°. They are represented as curves in Fig. 9-1b

and as phasors in Fig. 9-1c. Note that E_{a1} leads E_{b2} because it reaches its peak positive value before E_{b2} does.

This machine is called a *two-phase alternator*, and the stator windings are respectively called phase A and phase B.

Example 9-1:
The alternator shown in Fig. 9-1a rotates at 6000 r/min and generates an effective sinusoidal voltage of 170 V per winding. Calculate:

a. the peak voltage across each phase;

b. the output frequency;

c. the time interval corresponding to a phase angle of 90°.

Solution:

a. The peak voltage per phase is:
$$E_m = \sqrt{2}\,E = 1.414 \times 170 \qquad \text{Eq. 7-8}$$
$$= 240 \text{ V}$$

b. One cycle is completed every time the magnet makes one turn. The period of one cycle is:
$$T = 1/6000 \text{ min} = 60/6000 \text{ s} = 0.01 \text{ s}$$
$$= 10 \text{ ms}$$
The frequency is:
$$f = 1/T = 1/0.01 = 100 \text{ Hz}$$

c. A phase angle of 90° corresponds to a time interval of one quarter revolution or 10 ms/4 = 2.5 ms. Consequently, phasor E_{b2} lags 2.5 ms behind phasor E_{a1}.

Let us now connect two identical resistive loads across phases A and B (Fig. 9-2a). Currents I_a and I_b will flow in each resistor, respectively in phase with E_{a1} and E_{b2}. The currents are therefore 90° out of phase with each other (Fig. 9-2b). This means that I_a reaches its maximum value one quarter period before I_b does. Furthermore, the alternator now produces a power output.

one phase. In other words, the total power output of the 2-phase alternator is the same at every instant. As a result, the mechanical power needed to drive the alternator is also constant.

9-4   Three-phase alternator

A three-phase alternator is similar to a two-phase

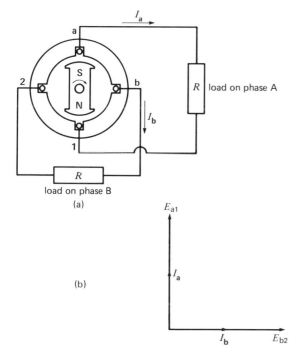

Figure 9-2   a.   Two-phase alternator under load.
b.   Phasor diagram of 2-phase alternator.

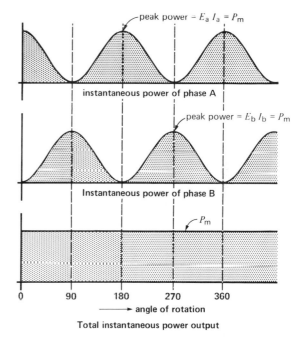

Figure 9-3   Power produced by a 2-phase alternator.

9-3   Power output of a 2-phase alternator

The instantaneous power supplied to each resistor is equal to the instantaneous voltage times the instantaneous current. This yields the power waves shown in Fig. 9-3. Note that when the power of phase A is maximum, that of phase B is zero, and vice-versa. If we add the instantaneous powers of both phases, we discover that the resultant power is constant, and equal to the peak power P_m of

alternator, except that the stator has three identical windings instead of two. The three windings a-1, b-2 and c-3 are placed at 120° to each other, as shown in Fig. 9-4a.

When the magnet is rotated at constant speed, the voltages induced in the three windings have the same effective values, but the peaks occur at different times. In effect, at the moment when the magnet occupies the position shown in Fig. 9-4a, only voltage E_{a1} is at its maximum position value.

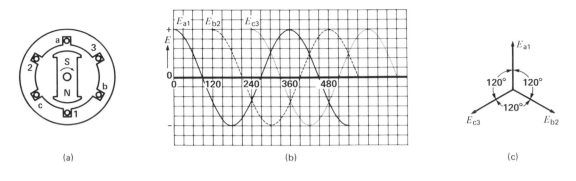

Figure 9-4 a. Three-phase alternator.   b. Voltages induced in a 3-phase alternator.   c. Phasor diagram

Voltage E_{b2} will reach its positive peak after the rotor has turned through an angle of 120° (or one third of a turn). Similarly, voltage E_{c3} will attain *its* positive peak after the rotor has revolved through 240° (or two-thirds of a turn) from its initial position.

Consequently, the three stator voltages, E_{a1}, E_{b2} and E_{c3} are respectively out of phase by 120°. They are shown as sine waves in Fig. 9-4b, and as phasors in Fig. 9-4c.

Let us connect the three windings of the alternator to three identical resistors. This arrangement requires six wires to deliver power to the individual single-phase loads (Fig. 9-5a). The resulting currents I_a, I_b and I_c are respectively in phase with voltages E_{a1}, E_{b2} and E_{c3}. Because the resistors are identical, the currents have the same effective values, but they are mutually out of phase by 120° (Fig. 9-5b). The fact that they are out of phase simply means that they reach their positive peaks at different times.

9-5   Power output of a 3-phase alternator

The instantaneous power supplied to each resistor is again composed of a wave which surges between zero and a maximum value P_m. However, the power peaks in the three resistors do not occur at the same time, owing to the phase angle between the voltages. If we add the instantaneous powers of all three resistors, we discover that the resulting power is constant, and has a magnitude of $1.5\,P_m$. As is the case of a 2-phase alternator, the instantaneous output of a 3-phase alternator is constant. This

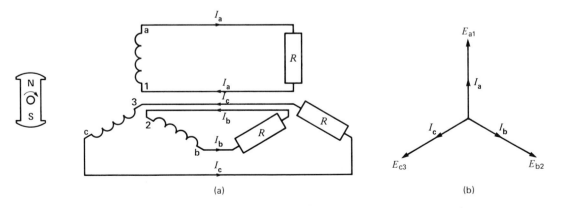

Figure 9-5   a. Three-phase, 6-wire system.   b. Corresponding phasor diagram.

also means that the power flow over the transmission line, connecting the alternator to the load, is constant. Finally, the mechanical power required to drive the rotor is also constant.

Example 9-2:
The 3-phase alternator shown in Fig. 9-5a is connected to three 20 Ω load resistors. If the voltage induced in each phase is 120 V, calculate:
a. the power dissipated in each resistor;
b. the power dissipated in the 3-phase load;
c. the peak power P_m dissipated in each resistor;
d. the total 3-phase power compared to P_m.

Solution:
a. Each resistor behaves as a single-phase load subjected to an effective voltage of 120 V. The power dissipated in each resistor is therefore:
$$P = E^2/R = 120^2/20$$
$$= = 720 \text{ W}$$
b. The power dissipated in the 3-phase load (all three resistors) is obviously:
$$P_3 = 3P = 3 \times 720$$
$$= 2160 \text{ W}$$
This power is constant and does not vary from instant to instant.
c. 1 The peak voltage across one resistor is
$$E_m = \sqrt{2}\,E = \sqrt{2} \times 120$$
$$= 169.7 \text{ V}$$
c. 2 The peak current in each resistor is:
$$I_m = E_m/R = 169.7/20$$
$$= 8.485 \text{ A}$$

c. 3 The peak power in each resistor is:
$$P_m = E_m I_m = 169.7 \times 8.485$$
$$= 1440 \text{ W}$$
d. 3 The ratio of P_3 to P_m is:
$$P_3/P_m = 2160/1440$$
$$= 1.5$$

Thus, whereas the power in each resistor pulsates between zero and a maximum of 1440 W, the total power for all three resistors is unvarying and equal to 2160 W.

9-6 Wye connection

The three single-phase circuits of Fig. 9-5 are electrically independent. Consequently, we can connect the three return conductors together to form a single return conductor (Fig. 9-6a). This reduces the number of transmission line conductors from 6 to 4. The return conductor, called *neutral* conductor (or simply, neutral), carries the sum of the three currents ($I_a + I_b + I_c$). At first, it seems that the cross section of the neutral conductor should be three times that of lines a, b, and c. However, the diagram of Fig. 9-6b. clearly shows that the *sum of the three return currents is zero at every instant.* For example, at the instant corresponding to 240°, $I_c = +I_{max}$ and $I_b = I_a - 0.5 I_{max}$, making $I_a + I_b + I_c = 0$.

We can therefore remove the neutral wire altogether without in any way affecting the voltages or currents in the circuit (Fig. 9-7). In one stroke, we accomplish a great saving because the number of

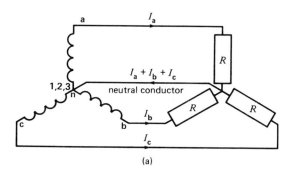

(a)

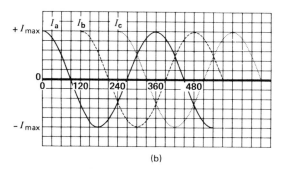

(b)

Figure 9-6   a. Three-phase, 4-wire system.   b. Line currents in 3-phase, 4-wire system.

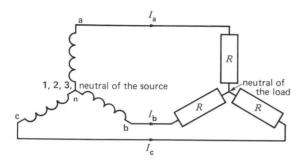

Figure 9-7   Three-phase, 3-wire system.

line conductors drops from six to three! However, the loads in Fig. 9-6a must be identical in order to remove the neutral wire. If the loads are not identical, the absence of the neutral conductor produces unequal voltages across the three loads.

The circuit of Fig. 9-7, composed of the alternator, transmission line and load, is called a *3-phase, 3-wire system*. The alternator, as well as the load, are said to be connected in *wye*, because the three branches ressemble the letter Y. For equally obvious reasons, some people prefer to use the term "connected in star".

The circuit of Fig. 9-6a is called a *3-phase, 4-wire system*. The neutral conductor in such a system is usually about the same size as the line conductors. Three-phase, 4-wire systems are widely used to distribute electric power in commercial and industrial buildings. The line conductors are often called *phases*, which is the same term applied to the alternator windings.*

---

* The term "phase" is used to designate different things. Consequently, it has to be read in context to be understood. The following examples show some of the ways in which the word "phase" is used.
1. the current is out of phase with the voltage (refers to phasor diagram);
2. the three phases of a transmission line (meaning the three conductors of the line);
3. the phase-to-phase voltage (meaning line voltage);
4. the phase sequence (the order in which the phasors follow each other);
5. the burned-out phase (the burned-out winding of a 3-phase machine);
6. the three-phase voltage (the line voltage of a 3-phase

9-7   Voltage relationships

Consider the wye-connected armature windings of a 3-phase alternator (Fig. 9-8a). The induced voltage in each winding has an effective value E_{LN} and the corresponding phasor diagram is shown in Fig. 9-8b. Knowing that the line-to-neutral voltages are E_{an}, E_{bn} and E_{cn} the question is, what are the line-to-line voltages E_{ab}, E_{bc} and E_{ca}? We can write the following equations, based on Kirchhoff's voltage law:

$$E_{ab} = E_{an} + E_{nb} \tag{9-1}$$
$$= E_{an} - E_{bn} \tag{9-1}$$

$$E_{bc} = E_{bn} + E_{nc} \tag{9-2}$$
$$= E_{bn} - E_{cn} \tag{9-2}$$

$$E_{ca} = E_{cn} + E_{na} \tag{9-3}$$
$$= E_{cn} - E_{an} \tag{9-3}$$

Referring first to Eq. 9-1, we draw phasor E_{ab} exactly as the equation indicates:

$$E_{ab} = E_{an} - E_{bn} = E_{an} + (-E_{bn})$$

The resulting phasor diagram shows that line voltage E_{ab} leads E_{an} by 30° (Fig. 9-8c). Using simple trigonometry, and based upon the fact that the length of the line-to-neutral phasors is E_{LN}, we have:

$$\text{length } E_L \text{ of phasor } E_{ab} = 2 \times E_{LN} \cos 30°$$
$$E_L = 2 \times E_{LN}/\sqrt{3}/2$$
$$= \sqrt{3}\, E_{LN}$$

system);
7. the 3-phase currents are unbalanced (the currents in a 3-phase line or machine are unequal and not displaced at 120°);
8. phase-shift transformer (a device which can change the phase angle of the output voltage with respect to the input voltage);
9. phase-to-phase fault (a short circuit between two line conductors);
10. phase-to-ground short (a short circuit between a line, or winding, and ground);
11. the phases are unbalanced (the line voltages, or the line currents, are unequal or not displaced at 120° to each other).

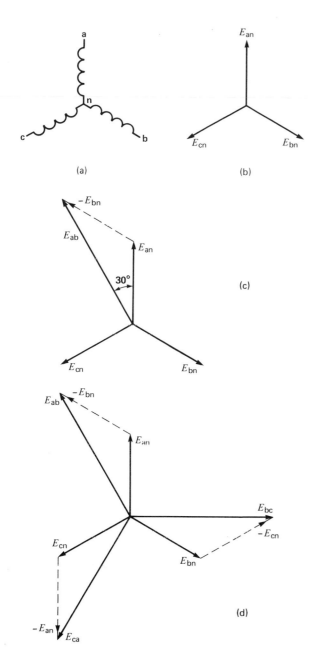

(a)

(b)

(c)

(d)

Figure 9-8 a. Wye-connected stator windings of a 3-phase alternator.
b. Line-to-neutral voltages.
c. Method to determine line voltage E_{ab}.
d. Line voltages E_{ab}, E_{bc}, E_{ca} are equal and displaced at 120°.

The line-to-line voltage (or line voltage) is therefore $\sqrt{3}$ times greater than the line-to-neutral voltage:

$$E_L = \sqrt{3}\, E_{LN} \qquad (9\text{-}4)$$

where

E_L = effective value of the line voltage [V]

E_{LN} = effective value of the line-to-neutral voltage [V]

$\sqrt{3}$ = a constant [approximate value = 1.73]

Owing to the symmetry of a 3-phase system, we conclude that the line voltage across *any* two alternator terminals is equal to $\sqrt{3}\,E_{LN}$. The truth of this can be seen by referring to Fig. 9-8d, which shows phasors E_{bc} and E_{ca}. The phasors are drawn according to Eqs. 9-2 and 9-3, respectively. The line voltages are equal in magnitude and mutually displaced by 120°.

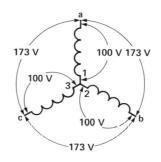

Figure 9-9   Voltages induced in a wye-connected alternator.

To further clarify these results, Fig. 9-9 shows the voltages between the terminals of a 3-phase alternator whose line-to-neutral voltage is 100 V. The line voltages are all equal to $100\sqrt{3}$ or 173 V. The voltages between lines a, b, c constitute a 3-phase system, but the voltage between any two lines (a and b, b and c, b and n, etc.) is nevertheless an ordinary single-phase voltage.

Example 9-3:
A 3-phase 60 Hz alternator, connected in wye, generates a line voltage of 23 900 V. Calculate:
a. the line-to-neutral voltage;
b. the voltage induced in the individual windings;

c. the time interval between the positive peak voltage of phase A and the positive peak of phase B;

d. the peak value of the line voltage.

Solution:

a. $E_{LN} = E_L/\sqrt{3} = 23\,900/1.73$ Eq. 9-4
    $= 13\,800$ V

b. The windings are connected in wye; consequently, the voltage induced in each winding is 13 800 V.

c. One complete cycle corresponds to 1/60 s, or 360°. Consequently, a phase angle of 120° corresponds to an interval of

$$T = \frac{120}{360} \times \frac{1}{60} = 1/180 \text{ s, or } 5.55 \text{ ms.}$$

The positive peaks are therefore separated by intervals of 5.55 ms.

d. The peak line voltage is:

$E_m = \sqrt{2}\, E_L = 1.414 \times 23\,900$ Eq. 7-8
    $= 33\,800$ V

The same voltage relationships exist in a wye-connected load, such as that shown in Figs. 9-6 and 9-7. In other words, the line voltage is $\sqrt{3}$ times the line-to-neutral voltage.

Example 9-4:
The alternator in Fig. 9-7 generates a line voltage of 865 V, and each load resistor has an impedance of 50 Ω. Calculate:

a. the voltage across each resistor;
b. the current in each resistor;
c. the total power output of the alternator.

Solution:

a. $E_{LN} = E_L/\sqrt{3} = 865/1.73$ Eq. 9-4
    $= 500$ V

b. Using Ohm's law:
    $I = E_{LN}/R = 500/50$
    $= 10$ A

All the line currents are therefore equal to 10 A.

c. Power absorbed by each resistor is:
    $P = E_{LN}\, I = 500 \times 10$
    $= 5000$ W

The power delivered by the alternator to all three resistors is:
    $P = 3 \times 5000 = 15$ kW

9-8   Delta connection

A 3-phase load is said to be *balanced* when the line voltages are equal and the line currents are equal. This corresponds to three identical impedances connected across the 3-phase line. This is the usual condition encountered in 3-phase circuits.

The three impedances may be connected in wye (as we already have seen) or in *delta* (Fig. 9-10a).

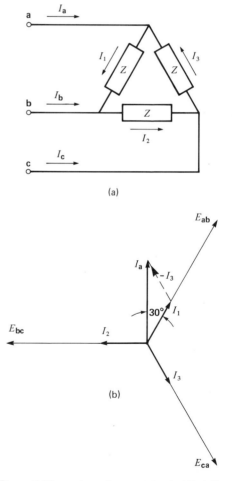

Figure 9-10   a. Impedances connected in delta.
               b. Current phasor relationships.

Let us now determine the voltage and current relationships in such a delta connection,* assuming a resistive load. The resistors are connected across the line; consequently, resistor currents I_1, I_2 and I_3 are in phase with the respective line voltages E_{ab}, E_{bc} and E_{ca}. The line voltages are produced by an external alternator (not shown). Furthermore, according to Kirchhoff's law, the line currents are given by:

$$I_a = I_1 - I_3 \qquad (9\text{-}5)$$

$$I_b = I_2 - I_1 \qquad (9\text{-}6)$$

$$I_c = I_3 - I_2 \qquad (9\text{-}7)$$

Referring first to Eq. 9-5, we draw phasor I_a exactly as the equation indicates. The resulting phasor diagram shows that I_a leads I_1 by 30° (Fig. 9-10b). Using simple trigonometry and letting I_z be the effective resistor current, and I_L the line current, we have:

$$\begin{aligned} I_L &= 2 \times I_z \cos 30° \\ &= 2 \times I_z \sqrt{3}/2; \\ &= \sqrt{3}\, I_z \end{aligned}$$

The line current is therefore $\sqrt{3}$ times greater than the current in each delta-connected load:

$$\boxed{I_L = \sqrt{3}\, I_z} \qquad (9\text{-}8)$$

where

I_L = effective value of the line current [A]
I_z = effective value of the current in a delta-connected load [A]
$\sqrt{3}$ = a constant [approximate value = 1.73]

The reader can easily determine the magnitude and position of phasors I_b and I_c, and observe that the three line currents are equal and displaced by 120°.

* The connection is so named because it resembles the Greek letter Δ.

Table 9A summarizes the basic relationships between the voltages and currents in wye-connected and delta-connected loads. The relationships are valid for any type of circuit element (resistor, capacitor, inductor, motor winding, alternator winding, etc.) as long as the elements in the three phases are identical. In other words, the relationships apply to any balanced 3-phase system.

Example 9-5:
Three identical impedances are connected in delta across a 3-phase 550 V line. If the line current is 10 A, calculate:
a. the current in each impedance;
b. the value of each impedance [Ω] .

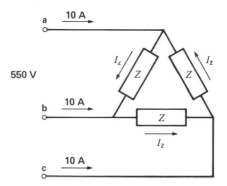

Figure 9-10 c. See Example 9-5.

Solution:
a. The current in each impedance is:
$$I_z = 10/\sqrt{3} - 5.78 \text{ A}$$

b. The voltage across each impedance is 550 V. Consequently,
$$\begin{aligned} Z &= E/I_z = 550/5.78 \\ &= 95 \ \Omega. \end{aligned}$$

9-9   Power transmitted by a 3-phase line

The apparent power supplied by a single-phase line is equal to the product of the line voltage E times the line current I. The question now arises: What is the apparent power supplied by a 3-phase line having a line voltage E and a line current I?

TABLE 9A   VOLTAGE AND CURRENT RELATIONSHIPS IN 3-PHASE CIRCUITS

Wye connection	**Delta connection**
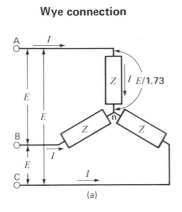	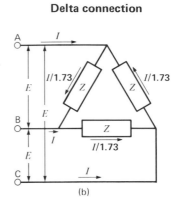
(a)	(b)

Figure 9-11   a.   Impedances connected in wye.   b.   Impedances connected in delta.

- The current in each element is equal to the line current I,

- The voltage across each element is equal to the line voltage E divided by $\sqrt{3}$

- The voltage across the elements are 120° out of phase,

- The currents in the elements are 120° out of phase.

- The current in each element is equal to the line current I divided by $\sqrt{3}$

- The voltage across each element is equal to the line voltage E,

- The voltages across the elements are 120° out of phase,

- The currents in the elements are 120° out of phase.

If we refer to the wye-connected load of Fig. 9-11a, the apparent power supplied to each impedance is:

$$S_z = \frac{E}{\sqrt{3}} \times I$$

The apparent power supplied to all three impedances is obviously three times as great.[*] Consequently,

$$S = \frac{E}{\sqrt{3}} \times I \times 3 = \sqrt{3}\, EI$$

The same result is obtained in the case of a delta-connected load (Fig. 9-11b). We therefore have:

$$\boxed{S = \sqrt{3}\, EI} \qquad (9\text{-}9)$$

where

S = total apparent power delivered by a 3-phase line [VA]

E = effective line voltage [V]

I = effective line current [A]

$\sqrt{3}$ = a constant [approximate value = 1.73]

The relationship between active power P, reactive power Q, and apparent power S is the same for balanced 3-phase circuits as for single-phase circuits. We therefore have:

$$S = \sqrt{P^2 + Q^2} \qquad (9\text{-}10)$$

and

$$\cos \theta = P/S \qquad (9\text{-}11)$$

where
S = total 3-phase apparent power [VA]
P = total 3-phase active power [W]
Q = total 3-phase reactive power [var]
$\cos \theta$ = power factor of the 3-phase load
θ = phase angle between the line current and the line-to-neutral voltage [°]

Example 9-6:
A 3-phase motor, connected to a 440 V line, draws a line current of 5 A. If the power factor of the motor is 80 percent, calculate:
a. the apparent power;
b. the active power, and
c. the reactive power absorbed by the machine.

Solution:
a. The apparent power is:
   $$S = \sqrt{3}\,EI = 1.73 \times 440 \times 5 = 3810 \text{ VA}$$
   $$= 3.81 \text{ kVA}$$
b. The active power is:
   $$P = S \cos \theta = 3.81 \times 0.80$$
   $$= 3.05 \text{ kW}$$
c. The reactive power is:
   $$Q = \sqrt{S^2 - P^2} = \sqrt{3.81^2 - 3.05^2}$$
   $$= 2.28 \text{ kvar}.$$

* In 3-phase balanced circuits, we can add the apparent powers of the three phases because they have identical power factors. If the power factors are not identical, the apparent powers cannot be added.

9-10 Solving 3-phase circuits

A *balanced* 3-phase load may be considered to be composed of three identical single-phase loads. Consequently, the easiest way to solve such a circuit is to consider only one phase. The following examples illustrate the method to be employed.

Example 9-7:
Three identical resistors are connected in wye across a 3-phase 550 V line, dissipating a total power of 3000 W (Fig. 9-12). Calculate:
a. the current in each line;
b. the value of each resistance.

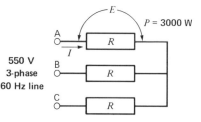

Figure 9-12 See Example 9-7.

Solution:
a. Power dissipated by each resistor:
   $$P = 3000 \text{ W}/3 = 1000 \text{ W}$$
   Voltage across the terminals of each resistor:
   $$E = 550 \text{ V}/1.73 = 318 \text{ V}$$
   Current in each resistor:
   $$I = P/E = 1000 \text{ W}/318 \text{ V} = 3.15 \text{ A}$$
   The current in each line is also 3.15 A.
b. Resistance of each element:
   $$R = E/I = 318/3.15 = 101 \ \Omega$$

Example 9-8:
In the circuit of Fig. 9-13, calculate:
a. the current in each line;
b. the voltage across the inductor terminals.

Solution:
a. Each phase is composed of an inductive react-

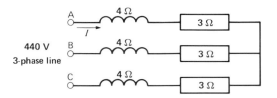

Figure 9-13 See Example 9-8.

ance X_L = 4 Ω in series with a resistance R = 3 Ω. Consequently, the impedance of each phase is:

$$Z = \sqrt{4^2 + 3^2} = 5 \ \Omega \qquad \text{Eq. 7-15}$$

The voltage across each phase is:

$$E_{LN} = 440 \ V/1.73 = 254 \ V$$

The current in each circuit element is therefore:

$$I = 254/5 = 50.8 \ A$$

This is also the line current.

b. The voltage across each inductor is:

$$E = I X_L = 50.8 \times 4$$
$$= 203.2 \ V$$

Example 9-9:

A 3-phase 550 V, 60 Hz line is connected to three identical capacitors connected in delta (Fig. 9-14). If the line current is 22 A, calculate the capacitance of each capacitor.

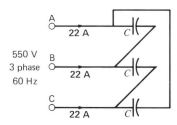

Figure 9-14 See Example 9-9.

Solution:

The current in each capacitor is:

$$I = 22 \ A/1.73 = 12.7 \ A$$

Voltage across each capacitor = 550 V

Capacitive reactance X_C of each capacitor:

$$X_C = 550/12.7 = 43.3 \ \Omega$$

The capacitance is:

$$C = 1/2\pi f X_C = 1/(2\pi \times 60 \times 43.3)$$
$$= 61.3 \ \mu F$$

9-11 Indeterminate loads

In many cases, we do not know whether a 3-phase load is connected in delta or in wye. For example, 3-phase motors, alternators, transformers, capacitors, etc., often have only three external terminals, and there is no way to tell how the internal connections are made. Under these circumstances, we simply *assume* that the connection is in wye. (A wye connection is slightly easier to handle than a delta connection.)

In a wye connection, the impedance per phase is understood to be the line-to-neutral impedance. The voltage per phase is simply the line voltage divided by $\sqrt{3}$. Finally, the current per phase is equal to the line current.

The assumption of a wye connection can be made not only for individual loads, but for entire load centers comprising motors, lamps, heaters, furnaces, and so forth. We simply assume that the load center is connected in wye, and proceed with the usual calculations.

Example 9-10:

A manufacturing plant draws 415 kVA from a 2400 V 3-phase line (Fig. 9-15a). If the plant power factor is 87.5 percent lagging, calculate:

a. the impedance of the plant, per phase;
b. the phase angle between the phase voltage and phase current;
c. the complete phasor diagram for the plant.

Solution:

a. 1 We assume a wye connection composed of three identical impedances Z (Fig. 9-15b).

a. 2 The voltage per phase is:

$$E = 2400/1.73$$
$$= 1390 \ V$$

a. 3 The current per phase is:

$$I = S/1.73 \ E = 415 \ 000/1.73 \times 2400$$
$$= 100 \ A \qquad \text{Eq. 9-9}$$

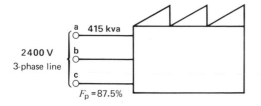

a 415 kva

2400 V
3-phase line

b

c

$F_p = 87.5\%$

(a)

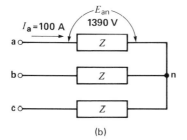

E_{an}
$I_a = 100$ A 1390 V

a o— Z

b o— Z —• n

c o— Z

(b)

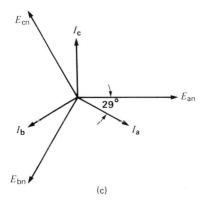

E_{cn} I_c

29° E_{an}

I_b I_a

E_{bn}

(c)

Figure 9-15 a. See Example 9-10.
b. Equivalent wye connection.
c. Phasor diagram.

a. 4 The impedance per phase is:

$$Z = E/I = 1390/100 = 13.9\ \Omega$$

b. The phase angle θ between the line-to-neutral voltage (1390 V) and the corresponding line current (100 A) is given by:

$$\cos\theta = 0.875 \qquad \text{Eq. 9-11}$$
$$\therefore\ \theta = 29°$$

The current in each phase lags 29° behind the respective phase voltage.

c. The complete phasor diagram is shown in Fig. 9-15c. In practice, we would show only one phase.

Example 9-11:

A 5000 hp wye-connected motor is connected to a 4000 V, 3-phase, 60 Hz line (Fig. 9-16). A delta-connected capacitor bank rated at 1800 kvar is also connected to the line. If the motor produces an output of 3594 hp at an efficiency of 93 percent and a power factor of 90 percent (lagging), calculate:

a. the active power absorbed by the motor;
b. the reactive power absorbed by the motor;
c. the reactive power supplied by the transmission line;
d. the apparent power supplied by the transmission line;
e. the transmission line current;
f. the motor line current.
g. Draw the complete phasor diagram for one phase.

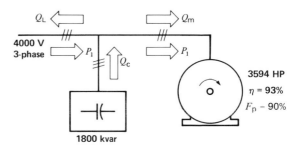

Figure 9-16 See Example 9-11.

Solution:

a. 1 Power output of 3594 hp is equivalent to:
$$P_2 = 3594 \times 0.746 = 2681\ \text{kW}$$

a. 2 Active power input to motor:
$$P_1 = P_2/\eta = 2681/0.93 \qquad \text{Eq. 3-6}$$
$$= 2883\ \text{kW}$$

b. 1 Apparent power absorbed by the motor:
$$S_m = P/\cos\theta = 2883/0.90$$
$$= 3203\ \text{kVA}$$

b. 2 Reactive power absorbed by the motor:
$$Q_m = \sqrt{S_m^2 - P_1^2} = \sqrt{3203^2 - 2883^2}$$
$$= 1395\ \text{kvar}$$

c. 1 Reactive power supplied by the capacitor bank:

Q_C = 1800 kvar

c. 2 Reactive power returned to the transmission line:

Q_L = $Q_C - Q_m$ = 1800 – 1395

= 405 kvar

d. Apparent power supplied by the line:

S_L = $\sqrt{Q_L{}^2 + P_1{}^2}$ = $\sqrt{405^2 + 2883^2}$

= 2911 kVA

e. Transmission line current:

I_L = $S_L/1.73\,E_L$ Eq. 9-9

= 2 911 000/(1.73 × 4000)

= 420 A

f. Motor line current:

I_m = $S_m/1.73\,E_L$

= 3 203 000/(1.73 × 4000)

= 462 A

g. 1 The line-to-neutral voltage for one phase:

E_{LN} = 4000/1.73 = 2312 V

g. 2 Phase angle θ between the motor current and the line-to-neutral voltage:

cos θ = 0.9

∴ θ = 25.8°

(The motor current lags behind the voltage as shown in Fig. 9-17a.)

g. 3 Line current drawn by the capacitor bank:

I_c = $Q_C/1.73\,E_L$

= 1 800 000/(1.73 × 4000)

= 260 A

Where should phasor I_c be located on the phasor diagram? The question is important because the capacitors are connected in delta, and we assumed a wye connection for the motor. This situation can create problems if we try to follow the actual currents in the capacitor bank. The solution is to recognize that if the capacitors *were* connected in wye (while generating the same reactive power) the line current of 260 A would lead E_{LN} by 90°. Consequently, we draw I_c 90° ahead of E_{LN}.

g. 4 Phase angle θ_L between the transmission line

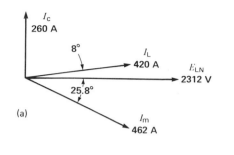

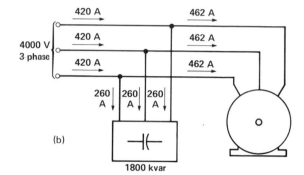

Figure 9-17 a. Phasor relationships for one phase.
b. Line currents, Example 9-11.

current and E_{LN} is:

cos θ = P_1/S_L = 2883/2911

= 0.99

θ_L = 8°

The line current (420 A) leads E_{LN} by 8° because the kvars supplied by the capacitor bank exceed the kvars absorbed by the motor.

g. 5 The phasor diagram for one phase is shown in Fig. 9-17a.

g. 6 The circuit diagram is shown in Fig. 9-17b.

We want to bring to the reader's attention the importance of assuming a wye connection, irrespective of the actual connection used. By assuming a wye connection for all circuit elements, we simplify the calculations and eliminate confusion.

As a final remark, the reader has no doubt noticed that the solution of a 3-phase problem involves active, reactive, and apparent power. Impedances such as resistors, inductors and capacitors do not often appear, and indeed, are not even needed to arrive at a solution. This is to be expected because most industrial loads involve electric motors, furnaces, lights, etc. which are seldom described in terms of resistance and reactances. They are usually represented as devices that draw a given amount of power at a given power factor.

The situation is somewhat different in the case of 3-phase transmission lines. Here, we can define resistances and reactances because the parameters are fixed. The same remarks apply to equivalent circuits describing the behavior of individual machines such as induction motors, synchronous machines, and so forth. In conclusion, the solution of 3-phase circuits may involve either active and reactive power or R, L, C elements - and sometimes both.

9-12 Phase sequence

Consider a 3-phase source having terminals a, b, c. Let us assume that three resistors are connected across the terminals as shown in Fig. 9-18. Three-phase currents I_a, I_b, I_c will flow in the wye-connected load, and each will attain a positive peak at successive intervals of time. If we use the positive peak of I_a as a starting point, it is clear that the positive peaks of the currents will follow each other in either the sequence

$$I_a - I_b - I_c - I_a - I_b - I_c \ldots \quad \text{(positive sequence)}$$

or the sequence

$$I_a - I_c - I_b - I_a - I_c - I_b \ldots \quad \text{(negative sequence)}$$

A 3-phase source can therefore have two sequences: one arbitrarily called positive, the other negative. The so-called *phase sequence* can be related to the terminals of the source rather than to the current flow in an imaginary 3-phase load. Thus, the positive sequence above may be described by the sequence a - b - c - a - b - c . . . while the negative sequence may be designated by a - c - b - a - c - b . . .

The sequence a - b - c . . . also means that the positive peaks of the line-to-neutral voltages follow each other in the sequence

$$E_{an} - E_{bn} - E_{cn} \ldots \quad \text{(positive sequence)}$$

Similarly, the sequence a - c - b . . . means that the positive peaks of the line-to-neutral voltages follow each other in the sequence

$$E_{an} - E_{cn} - E_{bn} \ldots \quad \text{(negative sequence)}$$

We can represent these two sequences by the phasor diagrams of Figs. 9-19a and 9-19b. The reader will recall that phasors are assumed to rotate counterclockwise; consequently, in Fig. 9-19a, the phasors do indeed cross the horizontal axis in the sequence $E_{an} - E_{bn} - E_{cn} \ldots$

The neutral n refers to the neutral of the imagi-

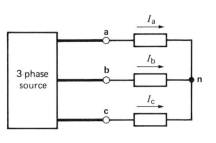

Figure 9-18   Phase sequence of a 3-phase source.

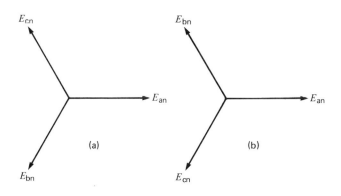

Figure 9-19   a. "Positive" phase sequence.
b. "Negative" phase sequence.

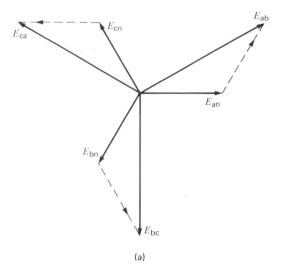

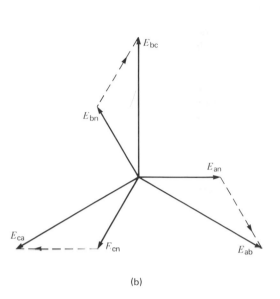

(a)   (b)

Figure 9-20   a. "Positive" phase sequence for line voltages.   b. "Negative" phase sequence for line voltages.

nary load in Fig. 9-18. If the source happens to have a neutral, then Fig. 9-19 applies equally well to the source. But how can we designate the phase sequence if the source has no neutral? To answer the question, let us consider Fig. 9-19a, in which the line-to-neutral phasors have a "positive" phase sequence. We can derive the corresponding phasors for the line voltages by using Equations 9-1 to 9-3. These line phasors (shown in Fig. 9-20a) will also have a positive phase sequence. As they sweep past the horizontal axis in the conventional counterclockwise direction, they follow the sequence

E_{ab} - E_{bc} - E_{ca} - E_{ab} - E_{bc} . . .

If we compare this with the positive sequence a - b - c - a . . ., we see that the first subscript follows the same sequence. As to the second subscript, it follows the sequence b - c - a - b . . ., but this is obviously the same as the sequence a - b - c - a . . . We can therefore state the following rule:

When using the double subscript notation, the sequence of the first subscript and of the second subscript corresponds to the phase sequence of the source.

Example 9-12:
In Fig. 9-12, the phase sequence of the source is known to be A - C - B. Draw the phasor diagram of the line voltages.

Solution:

a. The voltages follow the sequence A - C - B, which is the same as the sequence AC - CB - BA - AC . . . Consequently, the line voltage sequence is E_{AC} - E_{CB} - E_{BA} and the corresponding phasor diagram is shown in Fig. 9-21.

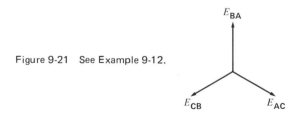

Figure 9-21   See Example 9-12.

Phase sequence is important because it determines a) the direction of rotation of 3-phase motors and b) whether one 3-phase system can be connected in parallel with another. Consequently, in 3-phase systems, phase sequence is as important as the frequency and voltage are.

Fortunately, we can reverse the phase sequence by simply interchanging any two conductors of a 3-phase line. Although this may appear to be a trivial change, it becomes a major problem when very large bus bars or high-voltage transmission lines have to be reversed. In practice, measures are taken so that such drastic changes do not have to be made at the last minute. The phase sequence of all major distribution systems is known in advance, and any future connections are planned accordingly.

9-13 Determining the phase sequence

Special instruments are available to determine phase sequence, but we can also determine it by using two incandescent lamps and a capacitor. The

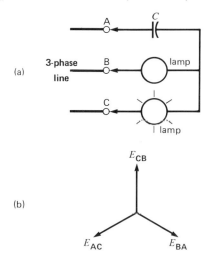

Figure 9-22 a. Determining phase sequence using two lamps and a capacitor.
b. Resulting phasor diagram.

devices are connected in wye. If we connect the circuit to a 3-phase line (without connecting the neutral), one lamp will always burn brighter than the other. The phase sequence is in the order: *bright lamp - dim lamp - capacitor*.

Suppose, for example, that the capacitor - lamp circuit is connected to a 3-phase line as shown in Fig. 9-22a. Because the lamp connected to phase C burns more brightly, the phase sequence is C-B-A. The line voltages follow each other in the sequence CB-BA-AC, which is to say in the sequence E_{CB}, E_{BA}, E_{AC}. The corresponding phasor diagram is given in Fig. 9-22b.

9-14 Unbalanced 3-phase loads

Most 3-phase loads are reasonably well balanced, and so we can usually employ the circuit solutions covered in the previous sections. However, to solve unbalanced circuits, it is common practice to write the circuit equations using complex numbers (a + jb). We can also solve the circuits by using simple trigonometry. In applying this method, we have to follow some simple, definite rules:

1. The phase sequence must be known. A change in sequence can produce entirely different line currents, even though the line voltages and loads are identical.
2. The current in each single-phase load is assumed to flow in the same sense as the phase sequence.
3. Three-phase loads that are balanced are broken up into three single-phase loads, having one-third the power rating, but possessing the same power factor. The single-phase loads are assumed to be connected in wye.
4. The currents flowing in the individual single-phase loads are calculated, and the lag or lead situation is determined from the power factor. The current phasors are then drawn on the phasor diagram, in proper relationship (lag/lead) to their respective voltages.
5. The line current in each phase is then found by applying Kirchhoff's current law.

Example 9-13:
Three different loads are connected across a 480 V, 3-phase line (Fig. 9-23). If the phase sequence is a - b - c, determine the value of the line currents and the phase angle between them.

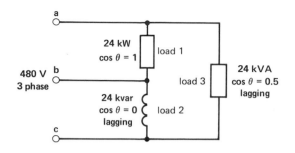

Figure 9-23 See Example 9-13.

Solution:

a. 1 The current flow in each load is drawn in the same sense as the phase sequence. Consequently, i_1 flows in the direction a - b, i_2 in the direction b - c, and i_3 in the direction c - a (Fig. 9-24). We *cannot* assign an arbitrary direction to the current flow.

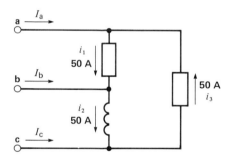

Figure 9-24 See Example 9-13.

a. 2 Next, we determine the phase angle of the current from the power factor of each load. Thus, referring to Fig. 9-24,

i_1 is in phase with E_{ab}
i_2 lags 90° behind E_{bc}
i_3 lags 60° behind E_{ca} (cos 60° = 0.5)

a. 3 The currents have the same magnitudes in all three loads. The value is obviously given by:

$$I = S/E = 24\ 000/480 = 50\ A$$

a. 4 We now draw the phasor diagram for the voltages, using the given sequence a - b - c (Fig. 9-25).

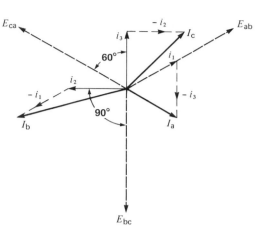

Figure 9-25 See Example 9-13.

a. 5 The respective load current phasors are then added to the phasor diagram. Note that:

i_1 is in phase with E_{ab}
i_2 lags 90° behind E_{bc}
i_3 lags 60° behind E_{ca}

The length of each phasor corresponds to a current of 50 A.

a. 6 All the line currents are shown as flowing towards the load (Fig. 9-24). Although we could assign an arbitrary direction to each, it is common practice to adopt the current flow as shown. Applying Kirchhoff's law, we find:

$$I_a = i_1 - i_3$$
$$I_b = i_2 - i_1$$
$$I_c = i_3 - i_2$$

By constructing the phasors graphically, according to these equations, we immediately find the magnitude and position of the line currents (Fig. 9-25). Then, using simple trigonometry, we find:

I_a = 50 A, lagging 60° behind E_{ab}
I_b = 86.6 A, lagging 60° behind E_{bc}
I_c = 70.7 A, lagging 105° behind E_{ca}

a. 7 From this information, we find that:

I_b lags 120° behind I_a
I_c leads I_a by 75°

The results are summarized in Fig. 9-26.

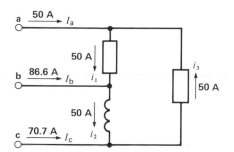

Figure 9-26 See Example 9-13.

Example 9-14:

A 3-phase delta-connected motor and a single capacitor are connected to a 480 V, 3-phase source (Fig. 9-27). The motor draws 200 kVA at a power factor of 86.6 percent lagging and the capacitor supplies 48 kvar of reactive power. Furthermore, a 23 kW lighting load is connected between phase a and neutral. If the phase sequence is a - b - c, determine the magnitude and phase angle of the line currents.

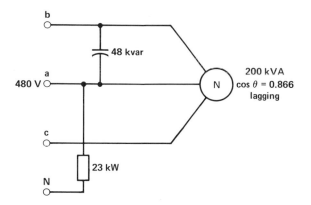

Figure 9-27 See Example 9-14.

Solution:

a. 1   To solve this problem, we essentially follow the same procedure as before. Current flow is drawn in the same sense as the phase sequence; consequently, i_4 flows in the direction a - b and i_5 in the direction a - n (Fig. 9-28).

However, the motor creates a problem because it is connected (internally) in delta. We simply ignore this information and assume the motor is connected in wye. The apparent power and power factor remain unchanged. This has the advantage of creating an artificial neutral N inside the motor. The line-to-neutral voltage in the motor is obviously the same as the line-to-neutral voltage of the source. As a result, currents i_1, i_2, i_3 must flow in the direction shown in Fig. 9-28.

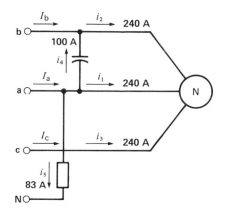

Figure 9-28 See Example 9-14.

a. 2   The magnitudes of the various currents is easily found:

$$I_{motor} = S/\sqrt{3}\,E$$
$$= 200\,000/(1.73 \times 480)$$
$$= 240\;A$$

Because cos θ = 0.866, the motor currents lag 30° behind the respective line-to-neutral voltages.

$$I_4 = Q/E = 48\,000/480$$
$$= 100\;A$$

Current i_4 leads E_{ab} by 90°.

$$I_5 = P/E = 23\,000/277$$
$$= 83\;A$$

Current i_5 is in phase with E_{an}.

a. 3   We now draw the phasor diagram for the voltages, using the given sequence a - b - c. In

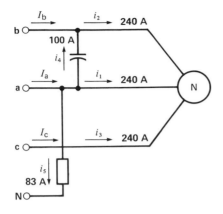

Figure 9-28 See Example 9-14.

this case, it is preferable to draw the line-to-neutral phasors first (Fig. 9-29). The line voltage phasors are then constructed according to Equations 9-1 to 9-3.

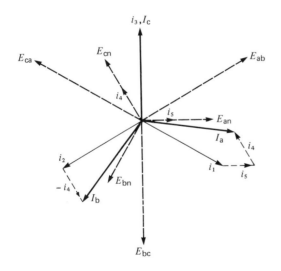

Figure 9-29 See Example 9-14.

a. 4 The current phasors are now added to the phasor diagram using the results given in a. 2. Thus,

i_1 lags E_{an} by $30°$

i_2 lags E_{bn} by $30°$

i_3 lags E_{cn} by $30°$

i_4 leads E_{ab} by $90°$

i_5 is in phase with E_{an}

a. 5 The line currents are found by applying Kirchhoff's law. Thus, referring to Fig. 9-28, we find:

$$I_b = i_2 - i_4$$
$$I_a = i_1 + i_4 + i_5$$
$$I_c = i_3$$

We can construct these phasors graphically, by drawing them to scale, or we can solve the problem by simple trigonometry. For example, to find I_a, the phasors are constructed as shown in Fig. 9-30. The horizontal component of I_a is:

$$|I_a|\,\underline{/0} = 240\cos 30 + 83 + 100\cos 120$$
$$= 240.8\text{ A}$$

The vertical component of I_a is:

$$|I_a|\,\underline{/90°} = 240\sin(-30) + 100\sin 120$$
$$= 33.4\text{ A}$$

Consequently, $|I_a| = \sqrt{249.8^2 + 33.4^2}$
$$= 243\text{ A}$$

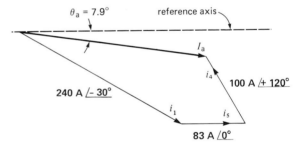

Figure 9-30 See Example 9-14.

The phase angle of I_a with respect to the reference (horizontal) axis is:

$$\theta_a = \arctan(-33.4/240.8)$$
$$= 7.9°$$

Using the same method to calculate I_b and I_c, we find:

$I_b = 260$ A, lagging $7.4°$ behind E_{bn}

$I_c = 240$ A, lagging $30°$ behind E_{cn}

a. 6 I_b lags almost exactly $120°$ behind I_a.

I_c lags $142.6°$ behind I_b.

QUESTIONS AND PROBLEMS

Practical level

9-1 A 3-phase wye-connected alternator generates 2400 V in each of its windings. Calculate the line voltage.

9-2 The alternator in Fig. 9-4 generates a peak voltage of 100 V per phase.
   a. Calculate the instantaneous voltage between terminals 1a at $0°$, $90°$, $120°$, $240°$ and $330°$;
   b. What is the polarity of terminal a with respect to terminal 1 at each of these instants?
   c. What is the instantaneous value of the voltage across terminals 2b at each of these same instants?

9-3 Referring to Fig. 9-4c, voltage E_{b2} is $120°$ behind voltage E_{a1}. Could we also say that E_{b2} is $240°$ ahead of E_{a1}?

9-4 The voltage between lines a - b - c of Fig. 9-7 is 620 V.
   a. What is the voltage across each resistor?
   b. If $R = 15 \Omega$, what is the current in each line?
   c. Calculate the power supplied to the 3-phase load.

9-5 Three resistors are connected in delta. If the line voltage is 13.2 kV and the line current is 1202 A, calculate:
   a. the current in each resistor;
   b. the voltage across each resistor;
   c. the power supplied to each resistor;
   d. the power supplied to the 3-phase load;
   e. the ohmic value of each resistor.

9-6 a. What is the phase sequence in Fig. 9-5?
   b. How can we reverse it?

9-7 A 3-phase motor connected to a 600 V line draws a line current of 25 A. Calculate the apparent power supplied to the motor.

9-8 Three incandescent lamps rated 60 W, 120 V are connected in delta. What line voltage is needed so that the lamps burn normally?

Intermediate level

9-9 Three 10 Ω resistors are connected in delta on a 208 V 3-phase line.
   a. What is the power supplied to the 3-phase load?
   b. If the fuse in one line burns out, calculate the new power supplied to the load.

9-10 If one line conductor of a 3-phase line is cut, is the load then supplied by a single-phase or a 2-phase voltage?

9-11 A 3-phase heater dissipates 15 kW when connected to a 208 V, 3-phase line.
   a. What is the line current if the resistors are connected in wye?
   b. What is the line current if the resistors are connected in delta?
   c. If the resistors are known to be connected in wye, calculate the resistance of each.

9-12 We wish to apply full load to a 100 kVA, 4 kV 3-phase alternator using a resistive load. Calculate the value of each resistance if the elements are connected;
   a. in wye,   b. in delta.

9-13 The windings of a 3-phase motor are connected in delta. If the resistance between any two terminals is 0.6 Ω, what is the resistance of each winding?

9-14 Three 24 Ω resistors are connected in delta across a 600 V, 3-phase line. Calculate the resistance of three elements connected in wye which would dissipate the same amount of power.

9-15 A 60 hp 3-phase motor absorbs 50 kW from a 600 V, 3-phase line. If the line current is 60 A, calculate:
   a. the efficiency of the motor;
   b. the apparent power absorbed by the motor;
   c. the reactive power absorbed by the motor;
   d. the power factor of the motor.

9-16 Three 15 Ω resistors and three 8 Ω reactors are connected as shown in Fig. 9-13. If the line voltage is 530 V, calculate:

a. the active, reactive and apparent power supplied to the 3-phase load;

b. the voltage across each resistor.

9-17 Two 60 W lamps and a 10 μF capacitor are connected in wye. The circuit is connected to the terminals X-Y-Z of a 3-phase 120 V outlet. The capacitor is connected to terminal Y, and the lamp which burns brighter is connected to terminal X.

a. What is the phase sequence?

b. Draw the phasor diagram for the line voltages.

Advanced level

9-18 Three 10 μF capacitors are connected in wye across a 2300 V, 60 Hz line. Calculate:

a. the line current;

b. the reactive power generated.

9-19 In Problem 9-17, if the capacitor is connected to terminal X, which lamp will be brighter?

9-20 Three delta-connected resistors absorb 60 kW when connected to a 3-phase line. If they are reconnected in wye, calculate the new power absorbed.

9-21 Three 15 Ω resistors (R) and three 8 Ω reactors (X) are connected in different ways across a 530 V, 3-phase line. Without drawing a phasor diagram, calculate the line current for each of the following connections:

a. R and X in series, connected in wye;

b. R and X in parallel, connected in delta;

c. R connected in delta, X connected in wye.

9-22 In Fig. 9-14, calculate the line current if the frequency is 50 Hz.

9-23 In Problem 9-15, assume that the motor is connected in wye, and that each branch can be represented by a resistance R in series with an inductive reactance X.

a. Calculate the value of R and X.

b. What is the phase angle between the line current and the corresponding line-to-neutral voltage?

9-24 An industrial plant draws 600 kVA from a 2.4 kV line at a power factor of 80 percent lagging.

a. What is the equivalent line-to-neutral impedance of the plant?

b. Assuming that the plant can be represented by an equivalent circuit similar to Fig. 9-13, determine the values of the resistance and reactance.

10

ALTERNATING CURRENT INSTRUMENTS

The most common alternating current meters and instruments are ammeters, voltmeters, wattmeters, and watthourmeters; they enable us to measure the current, voltage, power, and energy in a circuit. In this chapter, we study the construction and operating principles of these instruments, as well as the methods employed to measure power and energy in single-phase and three-phase circuits.

10-1 Alternating current ammeter

The pointer of a dc ammeter swings right or left depending upon the direction of current flow. It would be impossible to measure an alternating current with such an instrument because the d'Arsonval movement would be subjected to a rapid succession of opposing impulses. The pointer would simply vibrate, while remaining in the zero position. In ac circuits, we must therefore use ammeters whose construction differs from that of a dc ammeter.

One common type of ac ammeter is the *moving-vane ammeter*. Its principle of operation may be understood by referring to Fig. 10-1a. The current

I to be measured flows in a stationary coil, producing a pair of N, S poles. These poles induce a pair of weaker n, s poles in a soft iron lamination (the vane). The vane is inclined to the N-S axis by a spiral spring (not shown), but is otherwise free to pivot about its center. The magnetic attraction be-

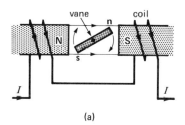

(a)

Figure 10-1 a. Elementary moving-vane ammeter.

tween the poles of opposite polarity produces a torque which tends to line up the vane in the direction of the lines of force created by the coils.

When the current reverses, all magnetic polarities reverse, with the result that the torque continues to act in the same direction and with the same intensity as before. The electromagnetic torque causes the vane to rotate against the restraining torque of the spring.

The scale of these instrument is not linear like that of a d'Arsonval movement, and it is difficult to take accurate measurements when the currents are less than 10 percent of full scale. However, these instruments are rugged, simple and precise (Fig. 10-1b).

read low, owing to the increase in the inductive reactance of the coil.

10-3 ac voltmeter with rectifier

We can build a linear scale ac voltmeter having good sensitivity by using a d'Arsonval movement and a rectifier. The rectifier is a device which only

Figure 10-1 b. Typical moving-iron (or moving vane) instruments.

10-2 Alternating current voltmeter

We can construct a moving-vane *voltmeter* by winding the stationary coil with many turns of fine wire and connecting an appropriate resistor in series with it. The value of the resistance depends upon the full-scale voltage of the instrument.

Because the current required to produce a full-scale deflection is of the order of 20 to 50 mA, these voltmeters are not very sensitive. Nevertheless, they are very accurate. They are generally used to measure alternating voltages in power circuits, where the frequency is less than 150 Hz. At higher frequencies, moving-vane voltmeters tend to

permits current to flow in one direction. In its simplest form, the rectifier consists of a single diode. It is represented by the symbol ——▶|—— and the arrow designates the permitted direction of conventional current flow.

If we use a single diode in series with a d'Arsonval movement (Fig. 10-2), the movement receives one impulse per cycle, that is whenever terminal A is (+) with respect to terminal B. On the other hand, if we use four diodes, mounted in a so-called bridge circuit, the movement receives two impulses per cycle (Fig. 10-3).* This enables us to double

* See Section 23-6.

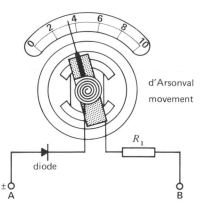

Figure 10-2 Rectifier-type ac voltmeter using one rectifier.

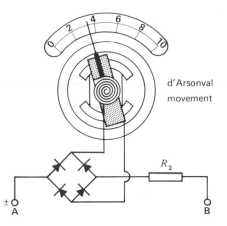

Figure 10-3 Rectifier-type ac voltmeter using a diode bridge circuit.

the value of the series resistor and consequently, to double the sensitivity (ohms per volt) of the voltmeter. These instruments are calibrated to indicate the effective ac voltage, based on a *sinusoidal* waveshape. We cannot use them to measure the non-sinusoidal voltages encountered in some electronic power circuits.

10-4 Wattmeter

To measure the active power P absorbed by a dc circuit, we simply multiply the voltage E by current I. In ac circuits, however, the EI product gives the *apparent power*, which is not necessarily equal to the *active power*. To measure active power

Figure 10-4

High-precision wattmeter rated 50 V 100 V 200 V; 1 A 5 A. The scale ranges from 0-50 W to 0-1000 W. *(Weston Instruments)*

(watts), we use an instrument called a *wattmeter*. (Fig. 10-4).

The construction of a wattmeter is similar to that of a d'Arsonval movement, except that the permanent magnet is replaced by an electromagnet (Fig. 10-5). The wattmeter is composed of the following parts:

crosses the potential coil B_p. The latter carries a small current I_p that is proportional to the line voltage E. The potential coil is therefore subjected to a torque whose instantaneous value is proportional to the flux Φ_c and current I_p, as in any d'Arsonval movement. Because the flux and current are alternating, the torque varies rapidly,

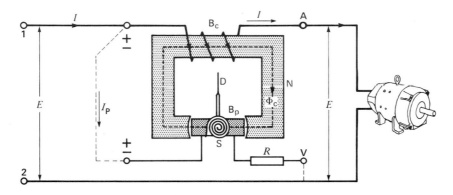

Figure 10-5   Circuit of a single-phase wattmeter.

B_c - a fixed coil having several turns of wire, large enough to carry the line current I. It is called *current coil*

N - a laminated iron core which offers an easy path for the alternating flux created by the current coil

B_p - a rectangular coil of fine wire mounted on a pivot. It is called *potential coil*

R - a high resistance in series with the potential coil

S - a spiral spring which produces a torque opposing the rotation of the coil

D - a pointer, mounted on the pivot, and which moves over a graduated scale

To measure active power, we connect the wattmeter into the circuit as shown in Fig. 10-5. Line current I flows through current coil B_c and line voltage E is applied to the potential coil circuit.

10-5 Operation of the wattmeter

Current I produces an alternating flux Φ_c which

producing a series of impulses at twice the line frequency. On a 60 Hz system, the vibrations created by these impulses are not visible because they are damped out by the inertia of the coil and the pointer.

When current I and voltage E are in phase, all the impulses act in the same direction, producing an average positive torque and a corresponding deflection. The deflection is proportional to the product EI, which is the active power supplied to the load.

When the load is inductive or capacitive, E and I are 90° out of phase. The potential coil receives the same impulses as before, but they act successively in opposite directions. The resulting torque is zero and the coil simply vibrates while the pointer stays at zero. The wattmeter registers only that component of current which is *in phase* with the voltage E and, consequently, it only reads the active power in a circuit. Figure 10-4 shows a high-precision wattmeter.

10-6 Power measurement in single-phase circuits

Owing to its external connections and the way it is built, a wattmeter may be considered to be a voltmeter and ammeter combined in the same box. The maximum voltage and current the instrument can tolerate are usually shown on the nameplate. In single-phase circuits, the pointer moves upscale when the ± terminal of the current coil is connected to the ± terminal of the potential coil (Fig. 10-6).

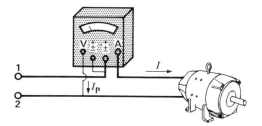

Figure 10-6 Method of connecting a single-phase wattmeter.

Example 10-1:

A wattmeter connected into the circuit of a single-phase motor gives a reading of 280 W. If the line current is 4.8 A and the line voltage 112 V, calculate the power factor of the motor.

Solution:

Apparent power to the motor:

S = 112 x 4.8 = 538 VA

Active power absorbed by the motor:

P = 280 W

Power factor = P/S = 280/538

= 0.52 (or 52 percent)

10-7 Power measurement in 3-phase, 3-wire circuits

The active power supplied to a 3-phase load may be measured by two wattmeters connected as shown in Fig. 10-7. The total power is equal to the sum of the two wattmeter readings. If the power factor of the load is less than 100 percent, the instruments will give different readings. Indeed, if the power factor is less than 50 percent, one of the wattmeters will actually give a negative reading. We must then reverse the connections of either the current coil or the potential coil, so as to obtain a numerical reading of this negative quantity. In this case, the power of the three-phase circuit is equal to the *difference* between the two wattmeter readings.

The two-wattmeter method gives the active power absorbed whether the load is balanced or unbalanced.[*]

[*] Proof of this statement may be found in basic texts on ac circuits.

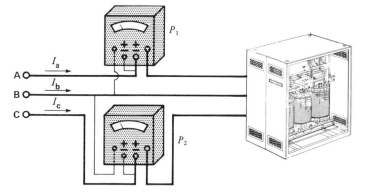

Figure 10-7 Measuring power in a 3-phase circuit using the two-wattmeter method.

Example 10-2:

A test on a three-phase motor yields the following results: P_1 = + 5950 W, P_2 = + 2380 W, the current in each of the three lines is 10 A and the line voltage is 600 V. Calculate the power factor of the motor.

Solution:

Apparent power supplied to the motor:

$$S = 1.73 \times E \times I = 1.73 \times 600 \times 10$$
$$= 10\ 380\ VA$$

Active power supplied to the motor:

$$P = 5950 + 2380$$
$$= 8330\ W$$
$$\cos \theta = P/S = 8330/10\ 380$$
$$= 0.80\ \text{or}\ 80\ \text{percent.}$$

10-8 Power measurement in 3-phase, 4-wire circuits

In 3-phase, 4-wire circuits, three wattmeters are needed to measure the total power. To make the connection, the current coil of one wattmeter is first connected in series with one of the live lines. The potential coil of the meter is then connected between the same live line and neutral. The other

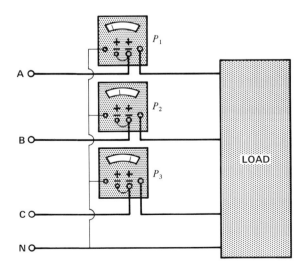

Figure 10-8 Measuring power in a 3-phase, 3-wire circuit.

instruments are connected in the same way to the remaining live lines (Fig. 10-8). The total power supplied to the load is equal to the sum of the three wattmeter readings. The three-wattmeter method gives the active power for both balanced and unbalanced loads.

10-9 Measuring instantaneous power

A wattmeter is an instrument which basically multiplies the instantaneous voltage by the instantaneous current and, by mechanical damping (inertia of the coil and pointer), gives the average value of

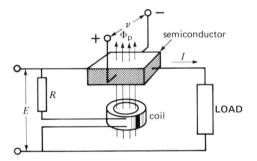

Figure 10-9 Output voltage v is proportional to the instantaneous power.

the product. However, in some circuits, we wish to obtain a non-mechanical readout of the active power; we then use solid-state multipliers, such as the Hall effect multiplier.

The Hall effect multiplier is composed of a special semiconductor material in the form of a six-sided wafer. Load current I flows through the semiconductor so as to lie in the path of flux Φ_p. This ac flux is proportional to ac line voltage E (Fig. 10-9). The remarkable feature of the Hall effect device is that a voltage v, proportional to the instantaneous product $I\Phi_p$, appears between the two remaining "unused" sides. This voltage is therefore proportional to the instantaneous EI product, or instantaneous power. If the voltage is applied to a sensitive dc voltmeter, the instrument indicates the active power consumed by the load. The voltage v may also be used as an electrical signal to monitor or control power in a circuit. Finally, by applying it to the terminals of an oscilloscope, we can actually observe instantaneous power curves such as the one shown in Fig. 7-2, chapter 7.

Figure 10-10 shows a megawatt-range wattmeter circuit used to measure power in a generating station. The current transformers (C T) and potential transformers (P T) step down the line currents and voltages to values compatible with the instrument rating. (see Secs. 12-4 and 12-5).

erating stations and the substations of electrical utilities and large industrial consumers.

In 3-phase, 3-wire circuits, we can deduce the reactive power from the two wattmeter readings (Fig. 10-7). We simply multiply the difference of the two readings by 1.73. For example, if two wattmeters indicate + 5950 W and + 2380 W respectively, the reactive power is (5950 – 2380) x 1.73 = 6176 vars. Note that this method of var measurement is only valid for *balanced* three-phase circuits.

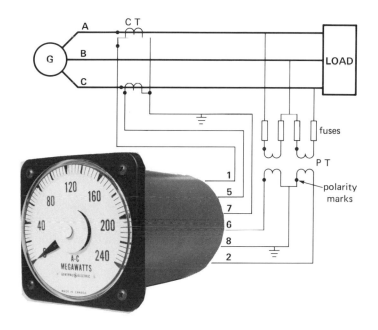

Figure 10-10 Power measurement in a high-power circuit.

10-10 Varmeter

A varmeter indicates the reactive power in a circuit. It is built the same way as a wattmeter is, but an internal circuit shifts the line voltage by 90° before it is applied to the potential coil. Varmeters are mainly employed in the control rooms of gen-

10-11 Watthourmeter

We have already seen that the SI unit of energy is the joule. However, for many years, power utilities have been using the kilowatthour to measure the energy supplied to industry and private homes. One kilowatthour (kW·h) is exactly equal to

3.6 MJ.

Meters which measure industrial and residential energy are called *watthourmeters*; they are designed to multiply power by time. The electricity bill is usually based upon the number of kilowatthours consumed during one month. Watthourmeters must therefore be very precise. *Induction* watthourmeters are practically the only types employed on ac circuits.

Figure 10-11 shows the principal parts of such a meter: a potential coil B_p wound with many turns of fine wire; a current coil B_c; an aluminum disc D supported on a vertical spindle; a permanent magnet A; and, finally, a gear mechanism that registers the number of turns made by the disc. When the meter is connected to a single-phase line, the disc is subjected to a torque which causes it to turn, like a high-precision motor.

(a)

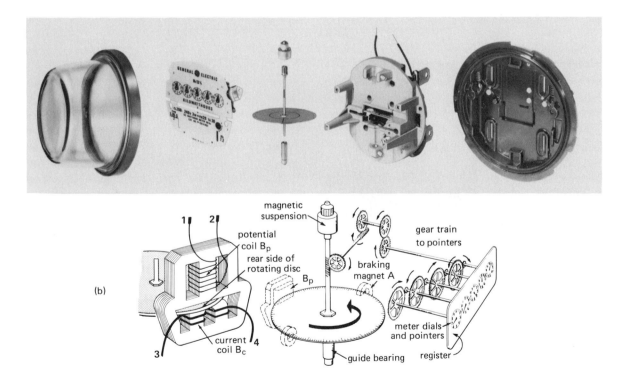

(b)

Figure 10-11 a. Complete watthourmeter. *(General Electric)*
b. Components making up the meter. *(General Electric)*

10-12 Operation of the watthourmeter

The operation of a watthourmeter can be understood by referring to Fig. 10-12. Load current I produces an alternating flux Φ_c which crosses the aluminum disc, inducing in it a voltage and, consequently, eddy currents I_f. On the other hand, potential coil B_p produces an alternating flux Φ_p which intercepts current I_f. The disc is therefore subjected to a torque which causes it to rotate.

latter, we have seen, is proportional to the active power supplied to the load. Consequently, the number of turns per second is proportional to the number of joules per second. It follows that the number of turns of the disc is proportional to the number of joules (energy) supplied to the load.

10-13 Meter readout

In addition to other details, the nameplate of a

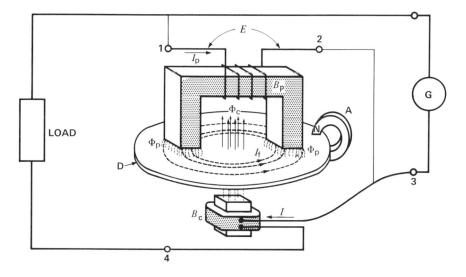

Figure 10-12 Principle of operation of a watthourmeter.

The torque is proportional to flux Φ_p and current I_f. Because these two quantities depend respectively. upon voltage E and load current I, the torque is proportional to the active power delivered to the load.

As in the case of a wattmeter, the average torque on the disc is zero when the E and I are $90°$ out of phase. This, however, is only part of the story.

As the disc moves between the poles of permanent magnet A, a braking torque is produced whose value is proportional to the *speed* of the disc. Because the motor torque is always equal to the braking torque (see Sec. 3-11), it follows that the speed is proportional to the motor torque. The

watthourmeter lists the rated voltage, current and frequency, and a metering constant K_h. Constant K_h is the amount of energy, in watthours, which flows through the meter for each turn of the disc. Consequently, we can calculate the amount of energy which flows through a meter in a given time by counting the number of turns. Then, dividing energy by time, we can calculate the active power supplied to the load.

Example 10-3:
The nameplate of a watthourmeter shows $K_h = 3.0$. If the disc makes 17 turns in two minutes, calculate the energy consumed by the load during this interval, and the average power of the load.

Solution:

Each turn corresponds to an energy consumption of 3.0 W·h. Energy consumed during the 2-minute interval is:

E_h = K_h x number of turns = 3.0 x 17
    = 51 W·h

Average power absorbed by the load during this interval is:

P = E_h/t = 51/2 min = 51/(1/30 h)
   = 51 x 30
   = 1530 W

Most watthourmeters have four dials to indicate the amount of energy consumed. The dials are read from left to right and the number so obtained is the number of kilowatthours consumed since the meter was first put in service. In reading the individual dials, we always take the number the pointer swept over last. For example, in Fig. 10-13,

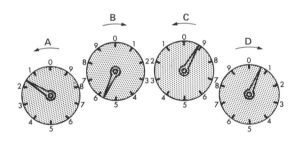

Figure 10-13 Reading the dials of a watthourmeter.

the reading is 1-5-9-0, or 1590 kW·h. Obviously, to measure the energy consumed during one month, we must subtract the readings at the beginning and end of the month. Some modern watthourmeters give a digital readout which, of course, is much easier to read.

10-14 Measuring three-phase energy and power

The energy consumed by a 3-phase load (three-wire system) can be measured with two single-phase watthourmeters. The two meters are often combined into one by mounting two discs on the same spindle and using a single register (Fig. 10-14). The current and potential coils are con-

Figure 10-14 Watthourmeter for a 3-phase, 3-wire circuit *(General Electric)*.

nected to the line in the same way as those of two wattmeters.

Figure 10-15 is a 3-phase solid-state watthourmeter having a precision exceeding that of induction-type watthourmeters.

Special ac meters such as demand meters are covered in Chapter 27. Many other instruments are available to measure power factor, phase angle, frequency and so forth. Details on their construction can be found in manufacturer's bulletins and specialized texts (Fig. 10-16).

QUESTIONS AND PROBLEMS

Practical level

10-1 A moving-vane ammeter can be used to measure a dc current. Explain why.

10-2 Referring to Fig. 10-13, if the pointer of dial B is between 1 and 2, what is the new meter reading?

10-3 Draw a simple diagram of a wattmeter, and explain how it works.

10-4 Describe the construction of a watthour-meter. Explain why the disc rotates.

10-5 Draw the circuit of a wattmeter connected into a single-phase circuit.

10-6 Show how to connect two wattmeters into a 3-phase, 3-wire line.

10-7 A voltmeter, ammeter and wattmeter are connected into the line leading to a single-phase motor. If the respective readings are 114 V, 5 A and 360 W, calculate the power factor of the motor.

10-8 In Problem 10-7, calculate the mechanical power developed if the motor has an efficiency of 60 percent.

Intermediate level

10-9 Two wattmeters connected into a 3-phase, 3-wire 220 V line indicate 3.5 kW and 1.5 kW, respectively. If the line current is 16 A, calculate:

a. the apparent power;

b. the power factor of the load.

10-10 An electric motor having a cos θ of 82 percent draws a current of 25 A from a 600 V 3-phase line.

a. Calculate the active power supplied to the motor;

b. If the motor has an efficiency of 85 percent, calculate the mechanical power output;

c. How much energy does the motor consume in 3 h?

10-11 We want to determine the power of an electric heater, installed in a home, by means of a watthourmeter. All other loads are shut off and it is found that the disc makes ten complete turns in one minute. If K_h = 3.0, calculate the power of the heater.

10-12 A wattmeter having a scale 0-3 kW is rated 300 V and 10 A. It is used to measure power in a 200 V single-phase circuit where the power factor is 10 percent. The pointer reads 1.7 kW, but after a few minutes,

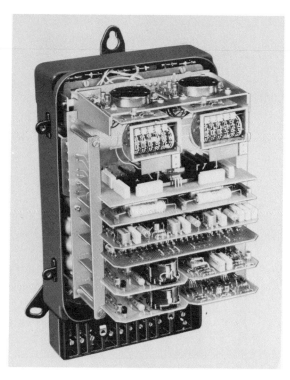

Figure 10-15 This high-precision electronic watthour-meter gives a numerical readout of the energy delivered by a 3-phase transmission line. It has an accuracy of 0.2 percent, which compares favorably with the 0.5 percent accuracy of some of the best induction-type watthourmeters. This meter is used on high-power lines where the monthly consumption exceeds 10 GW·h. *(Siemens)*

smoke is seen to come out of the instrument. Explain.

10-13 A domestic watthourmeter has a precision of 0.7 percent. Calculate the maximum possible error if the monthly consumption is 800 kW·h.

Advanced level

10-14 The wattmeters in Fig. 10-7 register + 35 kW

Figure 10-16   a. Power factor meter. *(General Electric)*
b. Frequency meter. *(Canadian General Electric)*

and – 20 kW, respectively. If the load is balanced, calculate:

a. the load power factor;

b. the line current if the line voltage is 630 V.

10-15   A sine wave has an effective value of 120 V. Determine the average voltage during a positive half-cycle and show that it is 90 percent of the effective value.

10-16   Using the results of Problem 10-15, calculate the average current flowing in the coil of Fig. 10-2, knowing that $R_1 = 60$ kΩ and sinusoidal voltage $E_{AB} = 120$ V (effective value).

10-17   A 50 Ω d'Arsonval movement gives a full-scale deflection at a current of 1 mA. We wish to convert it into an ac voltmeter, using the circuit of Fig. 10-2.

a. Calculate the value of R_1, so that the scale range is 0-150 Vac.

b. If we apply a sinusoidal voltage of 50 V between terminals AB, what will the instrument read? If we reverse the terminals, what is the new reading?

c. If we apply a dc voltage of 50 V across terminals AB, with A positive with respect to B, what will the meter read? If we reverse the terminals what is the new reading?

10-18   In Problem 10-17, calculate the voltmeter reading if a triangular ac voltage having a peak of 50 V is applied across the terminals.

10-19   The disc in Fig. 10-12 turns at 10 r/min for a load of 10 kW. If a 5 kvar capacitor is connected in parallel with the load, what is the new rate of rotation?

10-20   a. The flux created by the permanent magnet in Fig. 10-12 decreases by 0.5% in 10 years. What is the effect on the speed of rotation and the precision of the meter?

b. The resistance of coil B_c changes with temperature. Does this affect the speed of rotation if the active load remains fixed?

PART IV

IV

TRANSFORMERS

11
TRANSFORMERS

The transformer is probably one of the most useful electrical devices ever invented. It can raise or lower the voltage or current in an ac circuit, it can isolate circuits from each other, and it can increase or decrease the value of a capacitor, an inductor, or a resistor. Finally, the transformer enables us to transmit electrical energy over great distances and to distribute it safely in factories and homes.

We will study some of the basic properties of transformers in this chapter. It will help us understand not only the special transformers covered in later chapters but, also the basic operating principle of induction motors, alternators, and synchronous motors. All these devices are based upon the laws of electromagnetic induction. Consequently, we encourage the reader to pay particular attention to the subject matter covered here.

11-1 Voltage induced in a coil

Consider the coil of Fig. 11-1 which surrounds (or "links") a variable flux Φ. The flux alternates sinusoidally at a frequency f, periodically reaching positive and negative peaks Φ_{max}. The alternating

flux induces a sinusoidal ac voltage in the coil,

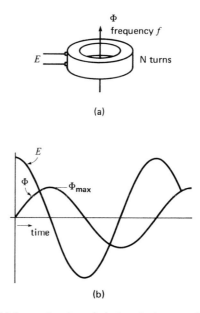

(a)

(b)

Figure 11-1 a. A voltage is induced when a coil links a variable flux.
b. A sinusoidal flux induces a sinusoidal voltage.

whose value is given by:

$$E = 4.44\, fN\Phi_{max} \quad *$$ (11-1)

where

E = effective voltage induced [V]

f = frequency of the flux [Hz]

N = number of turns on the coil

Φ_{max} = peak value of the flux [Wb]

4.44 = a constant [exact value = $2\pi/\sqrt{2}$]

It does not matter where the ac flux comes from: It may be created by a moving magnet, a nearby ac coil, or even by an ac current which flows in the coil itself.

Example 11-1:

The coil in Fig. 11-1 possesses 4000 turns and links an ac flux having a peak value of 2 milli-webers. If the frequency is 60 Hz, calculate the value and frequency of the induced voltage E.

Solution:

E = 4.44 $fN\Phi_{max}$ Eq. 11-1

= 4.44 × 60 × 4000 × 0.002

= 2131 V

The induced voltage has an effective value of 2131 V and a frequency of 60 Hz. The peak voltage is $2131\sqrt{2}$ = 3013 V.

11-2 Applied voltage and induced voltage

Figure 11-2a shows a coil of N turns connected to a sinusoidal ac source E_g. The coil has a reactance X_m and draws a current I_Φ. If the resistance of the coil is negligible, the effective current is given by:

$$I_\Phi = E_g/X_m$$

The detailed behavior of the circuit can be explained as follows:

The sinusoidal current I_Φ produces a sinusoidal

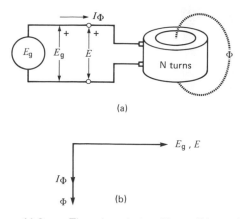

(a)

(b)

Figure 11-2 a. The voltage induced in a coil is equal to the applied voltage.

b. Phasor relationships.

mmf NI_Φ which, in turn, creates a sinusoidal flux Φ. The peak value of this ac flux is Φ_{max}. The flux induces a voltage E across the terminals of the coil, whose value is given by Eq. 11-1. On the other hand, the *applied* voltage E_g and the *induced* voltage E *must be identical* because they appear between the same pair of conductors. Because $E_g = E$, we may write:

$$E_g = 4.44\, fN\Phi_{max}$$

from which

$$\Phi_{max} = \frac{E_g}{4.44\, fN}$$ (11-2)

This equation indicates that for a given frequency and a given number of turns, Φ_{max} varies in proportion to the applied voltage E_g. This means that if E_g is kept constant, the peak flux *must remain* constant.

For example, suppose we gradually insert an iron core into the coil while keeping E_g fixed (Fig. 11-3). The peak value of the ac flux will remain absolutely constant during this operation, retaining its original value Φ_{max} even when the core is completely inside the coil. In effect, if the flux increased (as we would expect), the induced voltage E would also increase. But this is impossible because $E = E_g$ at every instant, and E_g is fixed.

---

* This equation is deduced from Faraday's law $E = N\Delta\Phi/\Delta t$ (Eq. 2-1).

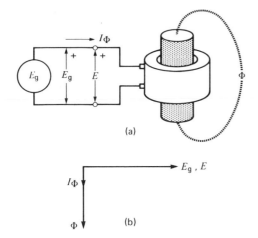

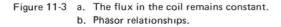

Figure 11-3   a. The flux in the coil remains constant.
b. Phasor relationships.

For a given supply voltage E_g, the ac flux in Figs. 11-2 and 11-3 is therefore the same. However, the so-called *magnetizing current* I_Φ is much smaller when the iron core is inside the coil. In effect, to produce the same flux, a smaller magnetomotive force is needed with an iron core than with an air core. Consequently, the magnetizing current in Fig. 11-3 is smaller than in Fig. 11-2. As in any inductive circuit, I_Φ lags 90° behind E_g, and Φ is in phase with the current (Figs. 11-2b and 11-3b).

Example 11-2:
A coil having 90 turns is connected to a 120 V, 60 Hz source. If the effective value of the magnetizing current is 4 A, calculate:
a. the peak value of flux;
b. the peak value of the mmf;
c. the inductive reactance of the coil;
d. the inductance of the coil.

Solution:
a. $\Phi_{max} = E_g/(4.44\,fN)$ Eq. 11-2
$= 120/(4.44 \times 60 \times 90)$
$= 0.005 = 5\text{ mWb}$
b. 1 The peak current is:
$I_m = 1.41\,I = 1.41 \times 4$
$= 5.64\text{ A}$

b. 2 The peak mmf is:
$U = NI_m = 90 \times 5.64$ Eq. 2-3
$= 507.6\text{ A}$
The flux is equal to 5 mWb when the coil mmf is 507.6 ampere-turns.
c.   The inductive reactance is:
$X_m = E_g/I_\Phi = 120/4$
$= 30\ \Omega$
d.   The inductance is:
$L = X_m/2\pi f$ Eq. 7-12
$= 30/(2\pi \times 60) = 0.0796$
$= 79.6\text{ mH}$

11-3 Elementary transformer

In Fig. 11-4, a coil having an air core is excited by an ac source E_g producing a total flux Φ. If we bring a second coil close to the first, it will surround a portion Φ_{m1} of the total flux. An ac voltage E_2 is therefore induced in the second coil and its value can actually be measured with a voltmeter. The combination of the two coils is called a *transformer*. The coil connected to the source is called the *primary winding* (or "primary") and the one connected to the load is called the *secondary winding* (or "secondary").

A voltage exists only between primary terminals 1-2 and secondary terminals 3-4. No voltage exists between the primary and secondary terminals. The secondary is therefore electrically isolated from the primary.

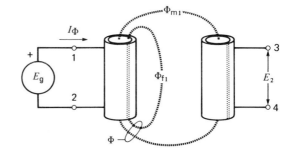

Figure 11-4 Voltage induced in a secondary winding.

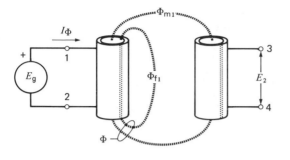

Figure 11-4

The flux Φ created by the primary is composed of two parts: 1) a *mutual flux* Φ_{m1}, which links the turns of both coils, and 2) a *leakage flux* Φ_{f1}, which links only the turns of the primary. If the coils are far apart, the mutual flux is very small compared to the total flux Φ; we then say that the coupling between the two coils is weak. We can obtain a better coupling (and a higher secondary voltage E_2) by bringing the two coils closer together. However, even if we bring the secondary right up to the primary so that the two coils touch, the mutual flux will still be small compared to the total flux Φ. When the coupling is weak, voltage E_2 is relatively small and, worse still, it collapses almost completely when a load is connected across the secondary terminals. In most industrial transformers, the primary and secondary windings are wound on top of each other to improve the coupling between them.

11-4 Polarity of a transformer

In Fig. 11-4, fluxes Φ_{f1} and Φ_{m1} are both produced by current I_Φ. Consequently, the fluxes are in phase, both reaching their maximum values at the same instant. It follows that voltage E_2 will reach *its* peak value at the same instant as E_g does. Suppose, during one of these peak moments, that primary terminal 1 is positive with respect to primary terminal 2, and that secondary terminal 3 is positive with respect to secondary terminal 4 (Fig. 11-5). Terminals 1 and 3 are then said to possess the same *polarity*. This "sameness" can be shown

by placing a large dot beside primary terminal 1 and secondary terminal 3. The dots are called *polarity marks*.

The polarity marks in Fig. 11-5 could equally well be placed beside terminals 2 and 4 because, as the voltage alternates, they, too, become simultaneously positive. Consequently, the polarity marks may be shown beside terminals 1 and 3 *or* beside terminals 2 and 4.

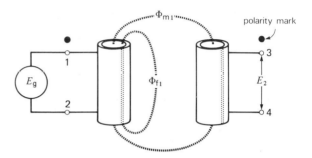

Figure 11-5   Terminals having the same instantaneous polarity are marked with a dot.

11-5 Properties of polarity marks

The following rules apply to polarity marks:
1. A current entering a polarity-marked terminal produces a flux in a "positive" direction* (Fig. 11-6). Conversely, a current flowing out of a polarity-marked terminal produces a flux in the "negative" direction. This rule applies to both primary and secondary windings.
2. If one polarity-marked terminal is momentarily positive with respect to its other terminal, then all polarity-marked terminals are positive with respect to their other terminals (Fig. 11-7).

11-6 Ideal transformer at no-load - voltage ratio

Before undertaking the study of a practical trans-

---

* "Positive" and "negative" are shown in quotation marks because we can rarely look inside a transformer to see in which direction the flux is actually circulating.

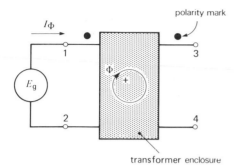

Figure 11-6   A current entering a polarity-marked terminal produces a flux in a "positive" direction.

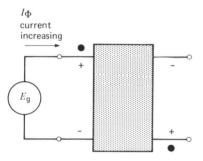

Figure 11-7   Instantaneous polarities when the magnetizing current is increasing.

former, we shall examine the properties of the so-called *ideal transformer*. An ideal transformer has no losses and its core is infinitely permeable. Furthermore, any flux produced by the primary is completely linked by the secondary, and vice versa. Consequently, an ideal transformer has no leakage flux of any kind.

Practical transformers have properties which approach those of an ideal transformer. Consequently, our study of the ideal transformer will help us understand the properties of transformers in general.

Figure 11-8a shows a transformer in which the primary and secondary respectively possess N_1 and N_2 turns. The primary is connected to a sinusoidal source E_g and the magnetizing current I_Φ creates a flux Φ_m. The flux is completely linked by the pri-

mary and secondary windings and, consequently, it is called a *mutual flux*. The flux varies sinusoidally, and reaches a peak value Φ_{max}. According to Eq. 11-1, we can therefore write:

$$E_1 = 4.44\, fN_1\Phi_{max} \qquad (11\text{-}3)$$

and

$$E_2 = 4.44\, fN_2\Phi_{max} \qquad (11\text{-}4)$$

From these equations, we deduce the expression for the *voltage ratio* of an ideal transformer:

$$\boxed{\dfrac{E_1}{E_2} = \dfrac{N_1}{N_2}} \qquad (11\text{-}5)$$

where
  E_1 = voltage induced in the primary [V]
  E_2 = voltage induced in the secondary [V]
  N_1 = number of turns on the primary
  N_2 = number of turns on the secondary

This equation shows that the ratio of the primary and secondary voltages is equal to the ratio of

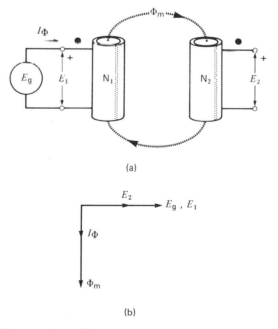

(a)

(b)

Figure 11-8   a. Primary and secondary linked by a mutual flux.
b. Phasor relationships.

the number of turns. Furthermore, because the primary and secondary voltages are induced by the same mutual Φ_m, they are necessarily in phase.

The phasor diagram at no load is given in Fig. 11-8b. If the transformer has fewer turns on the secondary than on the primary, phasor E_2 is shorter than phasor E_1. As in any inductor, current I_Φ lags 90° behind applied voltage E_g. The phasor representing flux Φ_m is obviously in phase with magnetizing current I_Φ which produces it.

In an *ideal* transformer, the magnetic circuit is infinitely permeable so that no magnetizing current is required to produce the flux Φ_m. Thus, under no-load conditions, the phasor diagram of such a transformer is identical to Fig. 11-8b except that phasor I_Φ does not appear. In effect, an ideal transformer requires no reactive power to set up its magnetic field (see Sec. 8-8).

Example 11-3:
A transformer having 90 turns on the primary and 2250 turns on the secondary is connected to a 120 V, 60 Hz source. The coupling between the primary and secondary is perfect, but the magnetizing current is 4 A. Calculate:
a. the effective voltage across the secondary terminals;
b. the peak voltage across the secondary terminals;
c. the instantaneous voltage across the secondary when the instantaneous voltage across the primary is 37 V.

Solution:

a. 1 The turns ratio is:
$$N_2/N_1 = 2250/90 \qquad \text{Eq. 11-5}$$
$$= 25$$
a. 2 The secondary voltage is therefore 25 times greater than the primary voltage because the secondary has 25 times more turns. Consequently:
$$E_2 = 25 \times E_1 = 25 \times 120$$
$$= 3000 \text{ V}$$
a. 3 Instead of reasoning as above, we can apply Eq. 11-5:

$$E_1/E_2 = N_1/N_2$$
$$120/E_2 = 90/2250$$
which yields $E_2 = 3000$ V
b. The voltage varies sinusoidally; consequently, the peak secondary voltage is:
$$E_m = \sqrt{2}\,E = 1.414 \times 3000 \qquad \text{Eq. 7-8}$$
$$= 4242 \text{ V}$$
c. The secondary voltage is *always* 25 times greater than E_1. Consequently,
$$E_2 = 25 \times 37 = 925 \text{ V}$$

11-7 Ideal transformer under load - current ratio

Pursuing our analysis, let us connect a load Z across the secondary of the ideal transformer (Fig. 11-9). A secondary current I_2 will immediately flow, given by:
$$I_2 = E_2/Z$$
Does E_2 change when we connect the load? To answer this question, we must remember that in an ideal transformer the primary and secondary windings are always linked by a mutual flux Φ_m, and by no other flux. In other words, an ideal transformer, by definition, has no leakage flux. Consequently, the voltage ratio under load is the same as at no-load, namely:

$$E_1/E_2 = N_1/N_2$$

If the supply voltage E_g is kept fixed, then the primary induced voltage E_1 remains fixed. Consequently, E_2 also remains fixed. We conclude that E_2 remains constant whether a load is connected or not. For the same reasons, mutual flux Φ_m also remains fixed.

Let us now examine the magnetomotive forces created by the primary and secondary windings. First, current I_2 produces a secondary mmf $N_2 I_2$. If it acted alone, this mmf would produce a profound change in the mutual flux Φ_m. But we just saw that Φ_m does not change under load. We conclude that flux Φ_m can only remain fixed if the primary develops a mmf which *exactly* counterbalances $N_2 I_2$. Thus, a primary current I_1 must

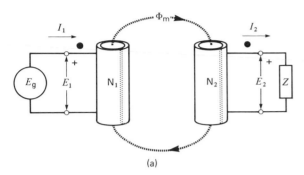

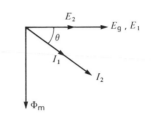

(a)

(b)

Figure 11-9   a.   Ideal transformer under load.

b.   Phasor relationships.

flow so that:

$$N_1 I_1 = N_2 N_2 \qquad (11\text{-}6)$$

To obtain the required instant-to-instant bucking effect, currents I_1 and I_2 must increase and decrease at the same time. In other words, they must be in phase. Furthermore, when I_1 flows *into* a polarity-mark on the primary side, I_2 must flow *out* of the polarity-mark on the secondary side (see Fig. 11-9).

Using these facts, we can now draw the phasor diagram of an ideal transformer under load (Fig. 11-9b). Assuming a resistive-inductive load, current I_2 lags behind E_2 by an angle θ. Flux Φ_m lags 90° behind E_g, but no magnetizing current I_Φ is needed because this is an ideal transformer. Finally, the primary and secondary currents are in phase, and according to Eq. 11-6, they are related by:

$$\boxed{I_1/I_2 = N_2/N_1} \qquad (11\text{-}7)$$

where

I_1 = primary current [A]
I_2 = secondary current [A]
N_1 = number of turns on the primary
N_2 = number of turns on the secondary

Comparing Eq. 11-5 and Eq. 11-7, we see that the transformer current ratio is the inverse of the voltage ratio. In effect, what we gain in voltage, we lose in current and vice versa. This is consistent with the requirement that the instantaneous power input $E_1 I_1$ to the primary must equal the instanta-

neous power output $E_2 I_2$ of the secondary. If the two were not identical, it would mean that the transformer itself absorbs power. By definition, this is impossible in an ideal transformer.

Example 11-4:
An ideal transformer having 90 turns on the primary and 2250 turns on the secondary is connected to a 200 V, 50 Hz source. The load across the secondary draws a current of 2 A at a power factor of 80 percent lagging (Fig. 11-10a). Calculate:
a. the effective value of the primary current;
b. the peak flux linked by the secondary winding;
c. the instantaneous current in the primary when the instantaneous current in the secondary is 100 mA;
d. draw the phasor diagram, based on Fig. 11-9.

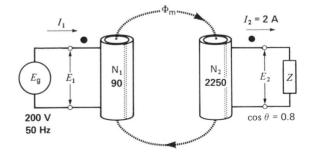

Figure 11-10   a.   See Example 11-4.

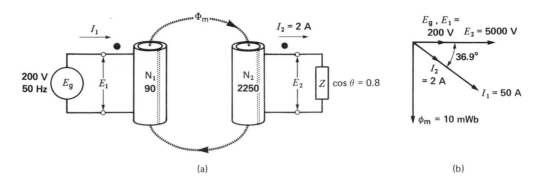

(a) (b)

Figure 11-10   a. See Example 11-4.   b. Phasor relationships.

Solution:

a. 1 The turns ratio is:
$$a = N_1/N_2 = 90/2250$$
$$= 1/25$$

a. 2 The current ratio is therefore 25 and because the primary has fewer turns, the primary current is 25 times greater than the secondary current. Consequently:
$$I_1 = 25 \times 2 = 50 \text{ A}$$

a. 3 Instead of reasoning as above, we can calculate the current directly by means of Eq. 11-6.
$$N_1I_1 = N_2I_2$$
$$90\,I_1 = 2250 \times 2$$
$$\therefore I_1 = 50 \text{ A}$$

b. In an ideal transformer, the flux linking the secondary is the same as that linking the primary.
$$\Phi_{max} = E_g/(4.44\,fN_1)$$
$$= 200/(4.44 \times 50 \times 90) = 0.01$$
$$= 10 \text{ mWb}$$

c. The instantaneous current in the primary is *always* 25 times greater than the instantaneous current in the secondary.
$$I_{1\,instantaneous} = 25\,I_{2\,instantaneous}$$
$$= 25 \times 0.1$$
$$= 2.5 \text{ A}$$

d. 1 Secondary voltage is:
$$E_2 = 25 \times E_1 = 25 \times 200$$
$$= 5000 \text{ V}$$

d. 2 Phase angle between E_2 and I_2 is:
$$\text{power factor} = \cos \theta$$
$$0.8 = \cos \theta$$
$$\therefore \theta = 36.9°$$

d. 3 The phase angle between E_1 and I_1 is also 36.9°.

d. 4 The mutual flux lags 90° behind E_g (Fig. 11-10b).

11-8 Circuit symbol for an ideal transformer

To highlight the bare essentials of an ideal transformer, it is best to show it in symbolic form. Thus, instead of drawing the primary and secondary windings and the mutual flux Φ_m, we simply show a box having primary and secondary terminals (Fig. 11-11). Polarity marks are added, enabling us to indicate the direction of current flow and the polarities of voltages E_1 and E_2. For example, a current I_1 flowing into one polarity-mark terminal is *always* accompanied by a current I_2

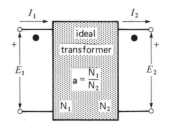

Figure 11-11 Symbol for an ideal transformer.

flowing out of the other polarity-mark terminal.

Furthermore, if we let the ratio of transformation N_1/N_2 = a, we obtain:

$$E_1 = aE_2$$

and $$I_1 = I_2/a$$

In an ideal transformer, and specifically referring to Fig. 11-11, E_1 and E_2 are *always* in phase, and so are I_1 and I_2.[*]

It is sometimes useful to make analogies between electrical and mechanical devices. For example, it is well known that a resistor is comparable to a mechanical brake, that a capacitor is similar to a mechanical spring and that an inductor behaves like a flywheel. Similarly, an ideal transformer may be compared to a gearbox. Thus, in the same way that a gearbox can increase or decrease the speed of rotation, a transformer can increase or decrease the current. The analogy between the two devices is best shown in tabular form:

Ideal transformer	Ideal gearbox
Primary turns	Teeth on the driving gear-wheel
Secondary turns	Teeth on the driven gear-wheel
Turn ratio	Gear ratio
Primary voltage	Input torque
Secondary voltage	Output torque
Primary current	Input rotational speed
Secondary current	Output rotational speed
Primary and secondary polarity marks	Arrows showing input and output direction of rotation

11-9 Impedance ratio

Although a transformer is generally used to change

[*] Some texts show the respective voltages and currents as being 180° out of phase. This situation can arise depending upon how the behavior of the transformer is described, or how the voltage polarities and current directions are assigned.

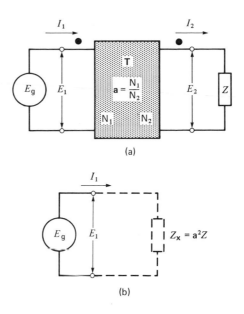

(a)

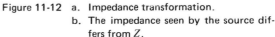

(b)

Figure 11-12   a. Impedance transformation.
b. The impedance seen by the source differs from Z.

a voltage or current, it also has the important ability to transform an impedance. Consider, for example, Fig. 11-12a in which an ideal transformer T is connected between a source E_g and a load Z. As far as the source Is concerned, it "sees" an impedance Z_x between the primary terminals given by:

$$Z_x = E_1/I_1$$

However,

$$\frac{E_1}{I_1} = \frac{aE_2}{I_2/a} = \frac{a^2E_2}{I_2} = a^2Z$$

Consequently, $$Z_x = a^2Z \qquad (11\text{-}8)$$

This means that the impedance "seen" by the source is a^2 times the actual impedance (Fig. 11-12b). Thus, an ideal transformer has the amazing ability to increase or decrease the value of an impedance "seen" in the primary, by a factor equal to the square of the turns ratio.

The impedance change is real, and not illusory like the image produced by a magnifying glass. An ideal transformer can modify the value of any component, be it a resistor, capacitor, or inductor. For example, if a 1000 Ω resistor is placed across the secondary of a transformer having a primary to secondary turn ratio of 1:5, it will appear across the primary as if it had a resistance of 1000 × $(1/5)^2$ = 40 Ω. Similarly, a capacitor having a reactance of 1000 Ω connected to the secondary, appears as a 40 Ω capacitor across the primary. However, because the reactance of a capacitor is inversely proportional to its capacitance (X_c = $1/2\pi fC$), the apparent capacitance between the primary terminals is 25 times greater than its actual value. We can therefore artificially increase (or decrease) the microfarad value by means of a transformer having an appropriate turns ratio.

11-10 Shifting impedances from primary to secondary and vice-versa

As a further illustration of the impedance-changing properties of an ideal transformer, consider the circuit of Fig. 11-13a. It is composed of a source E_g, a transformer T and of four impedances Z_1 to Z_4. The transformer has a turn ratio a. We can progressively shift one or more of the secondary impedances to the primary side, as shown in Figs. 11-13b to 11-13e. The circuit arrangement of the impedances shifted in this way remains the same, but all impedance values are multiplied by a^2.

If all the impedances are transferred to the primary side, the ideal transformer ends up at the extreme right-hand side of the circuit (Fig. 11-13d). In this position, the secondary of the transformer is on open-circuit. Consequently, both the primary and secondary currents are zero. We can therefore remove the ideal transformer altogether, yielding the equivalent circuit shown in Fig. 11-13e.

In some cases, it is useful to shift impedances from the primary side to the secondary side. The procedure is the same, but all impedances so transferred are now *divided* by a^2 (Fig. 11-14b). We can even shift the source E_g to the secondary side, where it becomes a source having a voltage E_g/a. The ideal transformer is now located at the ex-

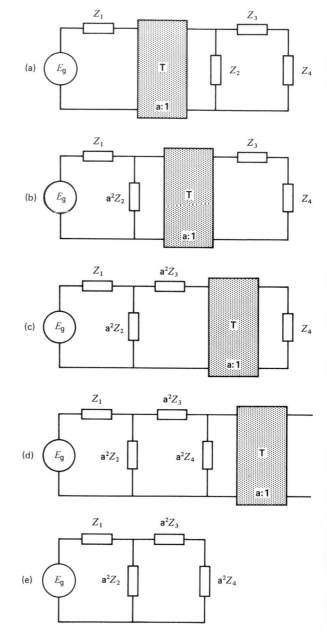

Figure 11-13   Effect upon the impedances as they are progressively transferred to the primary side.

treme left-hand side of the circuit (Fig. 11-14c). In this position, the primary of the transformer is on open-circuit. Consequently, both the primary and secondary currents are zero. As before, we can remove the transformer completely, leaving the equivalent circuit of Fig. 11-14d.

In comparing Figs. 11-13a and 11-13e, we may wonder how a circuit which contains a real transformer T can be reduced to a circuit which has no transformer at all. In effect, is there any useful relationship between the two circuits? And the same question applies to Fig. 11-14. The answer is yes - there is a useful relationship between the real circuit of Fig. 11-13a and the equivalent circuit of Fig. 11-13e.

Suppose that the real voltage across Z_4 is E_4 volts, and that the real current through it is I_4 amperes (Fig. 11-15). Then, in the equivalent circuit, the voltage across the equivalent a^2Z_4 impedance is equal to E_4 x a volts. On the other hand, the current through the impedance is equal to $I_4 \div a$ amperes (Fig. 11-16). In other words, whenever an impedance is transferred to the primary side, the real voltage across the impedance increases by a factor a. Simultaneously, the real current through it decreases by a factor a.

In general, whenever an impedance is transferred from one side of a transformer to the other, the real voltage across it changes in proportion to the turns ratio. If the impedance is transferred to the side where the transformer voltage is higher, the voltage across the transferred impedance will also be higher. Conversely, if the impedance is

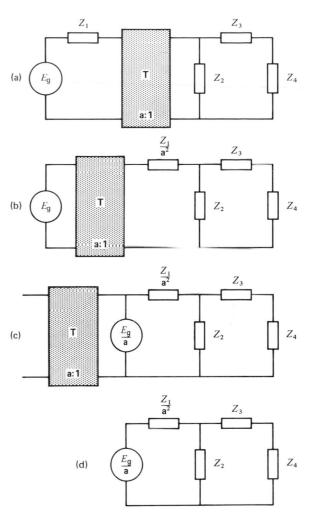

Figure 11-14   Shifting impedances.

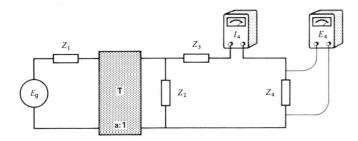

Figure 11-15   Actual voltage and current in an impedance.

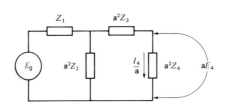

Figure 11-16   Equivalent voltage and current.

transferred to the side where the transformer voltage is lower, the voltage across the transferred impedance is lower than the real voltage - again, of course, in the ratio of the number of turns.

Example 11-5:

Calculate voltage E and current I in the circuit of Fig. 11-17, knowing that ideal transformer T has a primary to secondary turns ratio of 1:100.

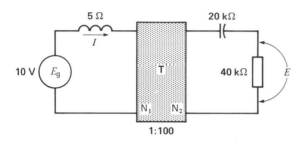

Figure 11-17   See Example 11-5.

Solution:

The easiest way to solve this problem is to shift all the impedances to either the primary or secondary side of the ideal transformer. In Fig. 11-18, we have shifted them to the primary side. Because the primary has 100 times fewer turns than the secondary, the impedance values are divided by 100^2, or 10 000. Voltage E become $E/100$, but current I remains unchanged because it is already on the primary side.

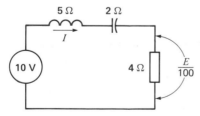

Figure 11-18   Equivalent circuit of Fig. 11-17.

a. 1   The impedance of the circuit in Fig. 11-18 is:

$$Z = \sqrt{R^2 + (X_L - X_c)^2} \qquad \text{Eq. 7-16}$$

$$= \sqrt{4^2 + (5 - 2)^2}$$

$$= \sqrt{16 + 9}$$

$$= 5 \ \Omega$$

a. 2 $I = E/Z = 10/5 = 2$ A

a. 3 $E/100 = IR = 2 \times 4 = 8$

$$\therefore E = 800 \text{ V}$$

11-11   Ideal transformer with an imperfect core

The ideal transformer of Fig. 11-11 has an infinitely permeable core. What happens if such a perfect core is replaced by an iron core having hysteresis and eddy-current losses, and whose permeability is rather low? We can represent these imperfections by two circuit elements R_m and X_m in parallel with the primary terminals of the ideal transformer (Fig. 11-19). The resistance R_m represents the iron losses and the heat they produce. To furnish these losses, a small current I_f is drawn from the line. The reactance X_m represents the permeability of the transformer core. Thus, if the permeability is low, X_m is low. The current I_Φ flowing through X_m represents the magnetizing current needed to create the flux Φ_m in the core.

The respective impedances R_m and X_m may be deduced from the following equations:

$$R_m = E_1^2/P_m \qquad (11\text{-}9)$$

$$X_m = E_1^2/Q_m \qquad (11\text{-}10)$$

where

R_m = resistance representing the iron losses $[\Omega]$

X_m = reactance of the primary winding $[\Omega]$

E_1 = Primary induced voltage [V]

P_m = iron losses [W]

Q_m = reactive power needed to set up the mutual flux Φ_m [vars]

The total current needed to produce the flux Φ_m in an imperfect core is equal to the sum of I_f and I_Φ. It is called the *exciting current* I_0. The phasor diagram at no-load for a less-than-ideal transformer is given in Fig. 11-20.

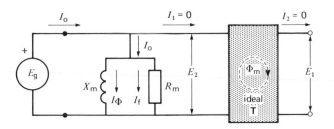

Figure 11-19   An imperfect core represented by a reactance and a resistance.

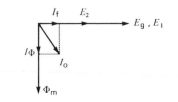

Figure 11-20   Phasor diagram.

Example 11-6:

A large transformer operating at no-load draws an exciting current of 5 A when the primary is connected to a 120 V, 60 Hz source (Fig. 11-21a). From a wattmeter test, we know that the iron losses are equal to 180 W.* Calculate:

a. the reactive power absorbed by the core;
b. the value of R_m and X_m;
c. the value of I_f and I_Φ.

b. 1 $R_m = E_1^2/P_m = 120^2/180$
            $= 80 \; \Omega$
b. 2 $X_m = E_1^2/Q_m = 120^2/572$
            $= 25.2 \; \Omega$
c. 1 $I_f = E_1/R_m = 120/80$
            $= 1.5 \; A$
c. 2 $I_\Phi = E_1/X_m = 120/25$
            $= 4.8 \; A$
c. 3   The phasor diagram is given in Fig. 11-21b.

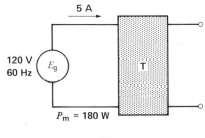

(a)

Figure 11-21   a. See Example 11-6.

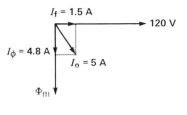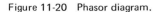

(b)

Figure 11-21   b. Phasor diagram, Example 11-6.

Solution:

a. 1   The apparent power is:
            $S_m = EI = 120 \times 5$
                 $= 600 \; VA$
a. 2 $P_m = 180 \; W$
a. 3   The reactive power is:

$$Q_m = \sqrt{S_m^2 - P_m^2} = \sqrt{600^2 - 180^2}$$

$$= 572 \; vars$$

*   Iron losses are discussed in Sections 6-6 to 6-12.

11-12 Elementary transformer under load

Now that we know the basic properties of an ideal transformer, we can begin a detailed analysis of a practical transformer. This analysis will give us a better understanding of transformers in general, and automatically leads us to the basic equivalent circuit of any power transformer.

Consider the transformer in Fig. 11-22 connected to a source E_g and operating at no-load. The transformer is ideal in every respect, except that the flux produced by the primary is not com-

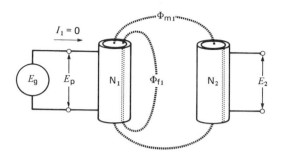

Figure 11-22   Transformer having leakage flux.

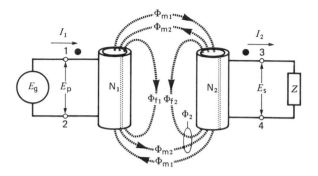

Figure 11-23   Mutual fluxes and leakage fluxes produced by a transformer under load.

pletely linked by the secondary. However, because the core is infinitely permeable, and because it has no losses, the no-load current $I_1 = 0$.

Let us now connect a load Z across the secondary, keeping the source voltage E_g fixed (Fig. 11-23). This simple operation sets off a train of events which we list as follows:

1. The original secondary voltage E_2 (Fig. 11-22) causes an ac current I_2 to flow in the load;
2. I_2 also flows in the secondary winding, producing a mmf N_2I_2;
3. N_2I_2 produces an ac flux Φ_2;
4. A portion of Φ_2 (Φ_{m2}) links with the primary winding;
5. Another portion of Φ_2 (Φ_{f2}) does not link with the primary.

The mmf produced by I_2 thus upsets the magnetic field which existed before I_2 began to flow. This modifies the previous situation so that the currents and voltages in Fig. 11-23 are quite different from those in Fig. 11-22.

The question is, how can we bring order to this rather confusing situation, where the secondary current and voltage affect the primary voltage and current, and vice-versa?

Referring to Fig. 11-23, we reason as follows:

First, the total flux produced by I_1 is still composed of two parts: a new mutual flux Φ_{m1} and a new leakage flux Φ_{f1}.

Second, the total flux produced by I_2 is composed of a mutual flux Φ_{m2} and a leakage flux Φ_{f2}.

Third, we combine Φ_{m1} and Φ_{m2} to show a single mutual flux Φ_m (Fig. 11-24). This mutual flux is created by the joint action of the primary and secondary mmfs. Fluxes Φ_{m1} and Φ_{m2} act in opposite directions because I_1 flows into a polarity mark (terminal 1) while I_2 flows out of a polarity mark (terminal 3).

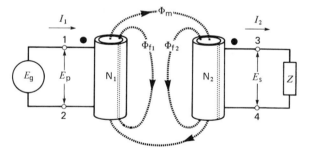

Figure 11-24   A transformer possesses two leakage fluxes and a mutual flux.

Fourth, we note that the primary leakage flux Φ_{f1} is created by N_1I_1, while the secondary leakage flux is created by N_2I_2. Consequently, leakage flux Φ_{f1} is in phase with I_1 and leakage flux Φ_{f2} is in phase with I_2.

Fifth, the secondary induced voltage E_s is now composed of two parts:

1. a voltage E_{f2} induced by leakage flux Φ_{f2} and given by:

$$E_{f2} = 4.44 \, fN_2\Phi_{f2} \qquad (11\text{-}11)$$

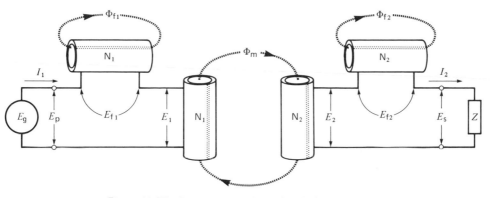

Figure 11-25   Separating out the various induced voltages.

2. a voltage E_2 induced by mutual flux Φ_m and given by:

$$E_2 = 4.44\,fN_2\Phi_m \qquad (11\text{-}12)$$

In general, E_{f2} and E_2 are not in phase.

Similarly, primary induced voltage E_p is composed of:

3. a voltage E_{f1} induced by leakage flux Φ_{f1} and given by:

$$E_{f1} = 4.44\,fN_1\Phi_{f1} \qquad (11\text{-}13)$$

4. a voltage E_1 induced by mutual flux Φ_m and given by:

$$E_1 = 4.44\,fN_1\Phi_m \qquad (11\text{-}14)$$

Sixth, induced voltage E_p = applied voltage E_g.

Using these six basic facts, we now proceed to develop the equivalent circuit of the transformer.

11-13 Primary and secondary leakage reactance

We can segregate the four voltages E_1, E_2, E_{f1} and E_{f2} by rearranging the transformer circuit as shown in Fig. 11-25. The secondary winding is drawn twice to show even more clearly that the N_2 turns are linked by both Φ_{f2} and Φ_m. This rearrangement obviously does not change the induced voltages, but it does make each voltage stand out by itself. Thus, it becomes clear that E_{f2} is really a voltage drop across a reactance. This so-called *secondary leakage reactance* X_{f2} is given by:

$$X_{f2} = E_{f2}/I_2 \qquad (11\text{-}15)$$

The primary winding is also shown twice, to separate E_1 from E_{f1}. Again, it is clear that E_{f1} is simply a voltage drop across a reactance. This so-called *primary leakage reactance* X_{f1} is given by:

$$X_{f1} = E_{f1}/I_1 \qquad (11\text{-}16)$$

The primary and secondary leakage reactances are shown in Fig. 11-26. We have also added the primary and secondary winding resistances R_1 and R_2 which, of course, act in series with the respective windings.

11-14 Equivalent circuit of a practical transformer

The circuit of Fig. 11-26 is composed of resistive and inductive elements (R_1, R_2, X_{f1}, X_{f2}, Z) coupled together by a mutual flux Φ_m which links the primary and secondary windings. The leakage-free magnetic coupling enclosed in the dotted square is actually an ideal transformer. It possesses the same properties and obeys the same rules as the ideal transformer we discussed in Secs. 11-6 to 11-11. For example, we can shift impedances to the primary side or the secondary side by multiplying (or dividing) their values by a^2, as we did before.

If we add circuit elements X_m and R_m to represent a practical core, we obtain the complete equivalent circuit of a practical transformer (Fig. 11-27). In this circuit, only the primary and sec-

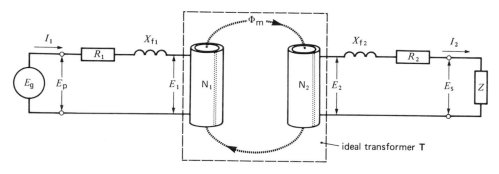

Figure 11-26   Resistance and leakage reactance of primary and secondary.

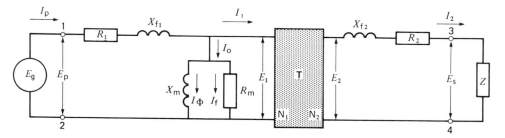

Figure 11-27   Complete equivalent circuit of a practical transformer.

ondary terminals 1-2 and 3-4 are accessible; all other components are "buried" inside the transformer itself. However, by appropriate measurements and tests, we can find the values of all circuit elements.

Example 11-7:

The secondary winding of a transformer possesses 180 turns. When the transformer is under load, the secondary current has an effective value of 18 A, 60 Hz. Furthermore, the mutual flux Φ_m has a peak value of 20 mWb. The secondary leakage flux Φ_{f2} has a peak value of 3 mWb. Calculate:

a. the voltage induced in the secondary winding by the leakage flux;
b. the value of the secondary leakage reactance;
c. the value of E_2.

Solution:

a. E_{f2} = 4.44 $fN_2\Phi_{f2}$ Eq. 11-11
   = 4.44 x 60 x 180 x 0.003
   = 143.9 V

b. The leakage reactance is:
   X_{f2} = E_{f2}/I_2 = 143.9/18   Eq. 11-15
   = 8 Ω
c. E_2 = 4.44 $fN_2\Phi_m$ Eq. 11-12
   = 4.44 x 60 x 180 x 0.02
   = 959 V

11-15 Construction of a power transformer

We usually design a power transformer so that it approaches the characteristics of an ideal transformer. Thus, to attain high permeability, the core is made of iron (Fig. 11-28a). The resulting magnetizing current I_m is at least 5000 times smaller than it would be if an air core were used. Furthermore, to keep the iron losses down, the core is laminated, and high resistivity, high-grade silicon steel is used. Consequently, iron loss current I_f is usually 2 to 10 times smaller than I_m, and the resulting exciting current I_0 is small.

Leakage reactances X_{f1} and X_{f2} are made as small as possible by winding the primary and sec-

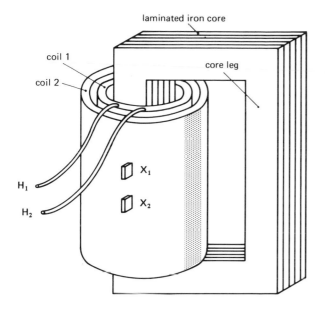

Figure 11-28   a. Construction of a transformer.

Figure 11-28   b. Inserting laminations.

ondary coils on top of each other, and as closely together as insulation considerations will permit. The coils are carefully insulated from each other and from the core. Such tight coupling between the coils yields the highest possible secondary voltage at no-load. It also guarantees good voltage regulation when a load is connected to the secondary terminals.

Winding resistances R_1 and R_2 are also minimized both to reduce the I^2R loss and resulting heat and to ensure high efficiency.

Figure 11-28a is a simplified version of a power transformer. In practice, the primary and secondary coils are distributed over both core legs in order to reduce the amount of copper. For the same reason, in larger transformers the iron core cross section is not square (as shown) but is built up to be nearly circular.

Figure 11-28b shows how the laminations of a small transformer are stacked to build up the core. Figure 11-28c shows the primary windings of a very big transformer.

A transformer is reversible in the sense that either coil can be used as the primary winding.

Figure 11-28   c. Primary winding of a large transformer; rating 128 kV, 290 A. *(ASEA)*

The number of turns on the primary and secondary windings depends upon their respective voltages. A high-voltage winding has far more turns than a low-voltage winding does. On the other hand, the current in a HV winding is much smaller, enabling us to use a smaller size conductor. The result is that the amount of copper in the primary and secondary windings is about the same. In practice, the outer coil (coil 2, in Fig. 11-28a) weighs more because the length per turn is greater. Aluminum or copper conductors are used.

11-16 Losses, and nominal voltage, current and power

Like any electrical machine, a transformer has losses. They are composed of:
1. I^2R losses in the windings;
2. hysteresis and eddy-current losses in the core;
3. stray losses due to currents induced in the tank and metallic supports, by the primary and secondary leakage fluxes.

The losses appear in the form of heat and produce 1) an increase in temperature and 2) a drop in efficiency. Under normal operating conditions, the efficiency of transformers is very high; it may reach 99.5 percent for very big power transformers. The heat produced by the iron losses depends upon the peak value of the mutual flux Φ_m, which, in turn, depends upon the applied voltage. On the other hand, the heat dissipated in the windings depends upon the current they carry. Consequently, to keep the transformer temperature at an acceptable level, we must set limits to both the applied voltage and the current drawn by the load. These two limits determine the *nominal voltage* and *nominal current* of the transformer.

The *power rating* of a transformer is equal to the product of the nominal voltage times nominal current. However, the result is not expressed in watts, because the phase angle between the voltage and current may have any value at all, depending on the nature of the load. Consequently, the power-handling capacity of a transformer is expressed in volt-amperes (VA), in kilovolt-amperes

(kVA) or in megavolt-amperes (MVA), depending on the size of the transformer. In effect, the temperature rise of the transformer is directly related to the *apparent* power which flows through it. This means that a 500 kVA transformer will get just as hot feeding a 500 kvar inductive load as a 500 kW resistive load.

The rated kVA, frequency and voltage are always shown on the nameplate. In large transformers, the corresponding rated currents are also shown.

Example 11-8:
The nameplate of a distribution transformer indicates 250 kVA, 60 Hz, primary 4160 V, secondary 480 V.
a. Calculate the nominal primary and secondary currents.
b. If we apply 2000 V to the 4160 V primary, can we still draw 250 kVA from the transformer?

Solution:
a. Nominal current of the 4160 V winding is:

$$I_p = \frac{\text{nominal } S}{\text{nominal } E_p} = \frac{250 \times 1000}{4160} = 60 \text{ A}$$

Nominal current of the 480 V winding is:

$$I_s = \frac{\text{nominal } S}{\text{nominal } E_s} = \frac{250 \times 1000}{480} = 521 \text{ A}$$

b. By applying 2000 V to the primary, the flux and the iron losses will be lower than normal and the core will be cooler. However, the load current should not exceed its nominal value, otherwise the windings will overheat. Consequently, the maximum power output using this lower voltage is:

$$S = 2000 \text{ V} \times 60 \text{ A} = 120 \text{ kVA}$$

11-17 No-load saturation curve

Let us apply a gradually increasing ac voltage E_p to the primary of a transformer, with the secondary open-circuited. As the voltage rises, the mutual flux Φ_m increases in accordance with Eq. 11-2. Exciting current I_0 will also increase but, when the iron begins to saturate, the current has to increase much more to produce the required flux. If we draw a graph of E_p versus I_0, we see the dramatic increase in current as we pass the saturation point (Fig. 11-29). Transformers are usually designed to operate at a peak flux density of about 1.5 T, which corresponds roughly to the knee of the saturation curve. Thus, when nominal voltage is

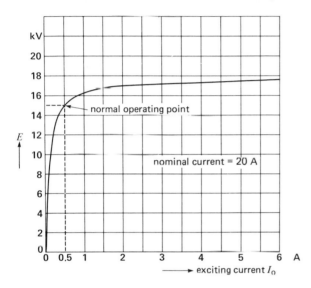

Figure 11-29   No-load saturation curve of a 167 kVA, 14.4 kV/480 V, 60 Hz transformer.

applied to a transformer, the corresponding flux density is about 1.5 T. We can exceed the nominal voltage by perhaps 20 percent, but if we apply twice the nominal voltage, the exciting current may exceed the nominal full-load current, even when the transformer is not loaded.

The non-linear relationship between E_p and I_0 shows that the exciting branch (composed of R_m and X_m) is not as constant as it appears. In effect, although R_m is reasonably constant, X_m decreases rapidly with increasing saturation. However, because transformers usually operate at close to rated voltage, R_m and X_m remain essentially constant. In any event, at full load, a slight change in voltage makes no difference because current I_0 is

swamped by the much larger load current I_1 (Fig. 11-27).

11-18   Nominal impedance of a transformer

The nominal impedance of a transformer is the value of the load impedance when the transformer delivers full power. It is equal to E_n/I_n, where E_n and I_n are the nominal voltage and nominal current on either the primary or secondary side. The nominal impedance may also be calculated by the equation:

$$\boxed{Z_n = E_n^2/S_n} \qquad (11\text{-}17)$$

where

Z_n = nominal impedance of the transformer [Ω]

S_n = nominal power rating of the transformer [VA]

E_n = nominal voltage of the transformer on either the primary or secondary side [V]

Because the primary and secondary voltages are usually different, it follows that a transformer has two nominal impedances, one for the primary and one for the secondary. Depending upon the respective voltages, the two impedances may be vastly different.

Example 11-9:
Calculate the nominal impedance of a 250 kVA, 4160 V/480 V, 60 Hz transformer, on the primary and secondary side.

Solution:
a. 1   Nominal primary impedance is:
$$Z_{np} = E_p^2/S_n = 4160^2/250\ 000$$
$$= 69\ \Omega$$
a. 2   Nominal secondary impedance is:
$$Z_{ns} = E_s^2/S_n = 480^2/250\ 000$$
$$= 0.92\ \Omega$$

11-19   Per-unit impedance

We can get a better idea of the relative magnitude

TABLE 11A   TYPICAL PER-UNIT VALUES OF TRANSFORMERS

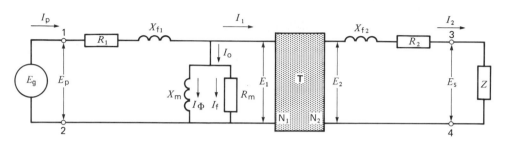

Figure 11-30   Equivalent circuit of a transformer.

Circuit Element (see Fig. 11-30)	Typical per-unit values	
	3 kVA to 250 kVA	**1 MVA to 100 MVA**
R_1 or R_2	0.009 - 0.005	0.005 - 0.002
X_{f1} or X_{f2}	0.008 - 0.025	0.03 - 0.06
X_m	20 - 30	30 - 50
R_m	20 - 50	100 - 500
I_0	0.05 - 0.03	0.03 - 0.02

---

of the various circuit elements of a transformer by comparing their impedances with the nominal impedance of the transformer. In making the comparison, elements located on the primary side are compared with the primary nominal impedance. Similarly, elements on the secondary side are compared with the secondary nominal impedance. The comparison can be made either on a percent or on a per-unit basis; we shall use the latter. Typical per-unit values are listed in Table 11A.

Example 11-10:
Using the information given in Table 11A, calculate the approximate values of the equivalent-circuit impedances of a 250 kVA, 4160 V/480 V, 60 Hz distribution transformer.

Solution:
We first determine the nominal impedances on the primary and secondary side. From the results of Example 11-9, we have: Z_{np} = 69 Ω and Z_{ns} = 0.92 Ω. We now calculate the various impedances by multiplying Z_{np} and Z_{ns} by the per-unit values given in Table 11A. Thus, referring to Fig. 11-30, we find:

R_1 = 0.005 x 69   = 0.35 Ω
R_2 = 0.005 x 0.92 = 4.6 mΩ
X_{f1} = 0.025 x 60   = 1.5 Ω
X_{f2} = 0.025 x 0.92 = 23 mΩ
X_m = 30 x 69   = 2070 Ω = 2 kΩ
R_m = 50 x 69   = 3450 Ω = 3.5 kΩ

This example shows the power of the per-unit method of estimating impedances. The equivalent

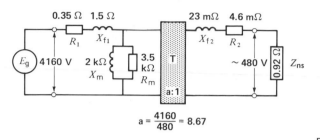

$$a = \frac{4160}{480} = 8.67$$

Figure 11-31   See Example 11-10.

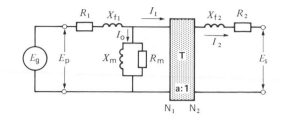

Figure 11-32   Complete equivalent circuit of a transformer at no-load.

circuit is shown in Fig. 11-31. The actual values may be 20 percent higher or lower than indicated.

11-20 Simplifying the equivalent circuit

The equivalent circuit of a transformer (Fig. 11-30) is relatively simple, but it is far more exact than needed in most practical applications. Consequently, we can often simplify it to make the calculations easier. To illustrate, let us try to simplify the circuit when the transformer operates a) at no-load and b) at full-load.

(a)   At no-load, I_2 is zero and so is I_1 because T is an ideal transformer (Fig. 11-32). Consequently, only the exciting current I_0 flows in R_1 and X_{f1}. These impedances are so small that the voltage drop across them is negligible. Furthermore, the voltage drop across R_2 and X_{f2} is zero. We can therefore neglect these impedances entirely, giving us the much simpler circuit of Fig. 11-33. The turn ratio, a = E_1/E_2, is obviously equal to the ratio of the primary to secondary voltages E_p/E_s measured across the terminals.

(b)   At full-load, I_p is at least 20 times larger than I_0. Consequently, we can neglect I_0 and the corresponding magnetizing branch. The resulting circuit is shown in Fig. 11-34. This simplified circuit may be used whenever the load exceeds 15 percent of the rated capacity of the transformer.

We can simplify the circuit still more by shifting everything to the primary side, thus eliminating transformer T (Fig. 11-35). By adding the respective resistances and reactances, we obtain

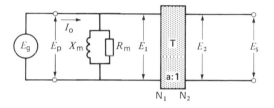

Figure 11-33   Simplified circuit at no-load.

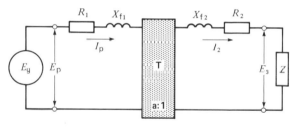

Figure 11-34   Simplified equivalent circuit of a transformer at full load.

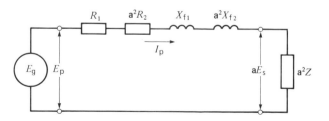

Figure 11-35   Equivalent circuit with impedances shifted to the primary side.

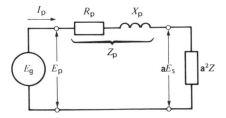

Figure 11-36   Total transformer resistance and reactance referred to the primary side.

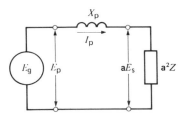

Figure 11-37   The internal impedance of large transformers is mainly reactive.

the circuit of Fig. 11-36. In this circuit,

$$R_p = R_1 + a^2 R_2 \qquad (11\text{-}18)$$
$$X_p = X_{f1} + a^2 X_{f2} \qquad (11\text{-}19)$$

where

R_p = equivalent transformer resistance referred to the primary side

X_p = equivalent transformer leakage reactance referred to the primary side

The combination of R_p and X_p constitutes the total transformer impedance Z_p referred to the primary side. From Eq. 7-15, we have:

$$Z_p = \sqrt{R_p^2 + X_p^2} \qquad (11\text{-}20)$$

Impedance Z_p is usually expressed as either a percent or a per-unit of the nominal primary impedance Z_{np}, mentioned before. The per-unit impedance is always given on the nameplate of the transformer.

Transformers above 500 kVA possess a leakage reactance X_p at least five times greater than R_p. In such transformers, we can neglect R_p, as far as voltage regulation is concerned.* The equivalent circuit is thus reduced to a simple reactance X_p between the source and the load (Fig. 11-37). In conclusion, it is really quite remarkable that the relatively complex circuit of Fig. 11-30 can be re-

---

* The full load regulation of a power transformer is the change in secondary voltage, expressed in percent of rated secondary voltage, which occurs when the rated kVA output at a specified power factor is reduced to zero, with the primary impressed terminal voltage maintained constant.

duced to a simple reactance in series with the load.

Example 11-11:
A single-phase transformer rated 3000 kVA, 69 kV/4.16 kV, 60 Hz has an impedance of 8 percent. Calculate:

a. the internal impedance of the transformer referred to the primary side;

b. the voltage regulation from no-load to full-load for a 2000 kW resistive load;

c. the primary and secondary currents if the secondary is accidentally short-circuited.

Solution:

a. 1 Nominal transformer impedance on primary side:
$$Z_{np} = E_n^2 / S_n = 69\,000^2 / 3\,000\,000$$
$$= 1587 \ \Omega$$

a. 2 $Z_p = 1587 \times 0.08 = 127 \ \Omega$

a. 3 Because the windings have negligible resistance compared to their leakage reactance, we can write:
$$X_p = 127 \ \Omega \ (\text{Fig. 11-38a})$$

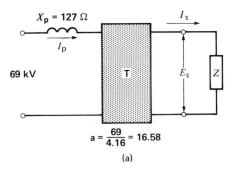

Figure 11-38   a. See Example 11-11.

b. 1 Approximate impedance Z of a 2000 kW load:

$$Z = E_s^2/P = 4160^2/2\,000\,000 = 8.65\ \Omega$$

b. 2 Impedance referred to primary side:

$$a^2Z = \left(\frac{69}{4.16}\right)^2 \times 8.65 = 2380\ \Omega$$

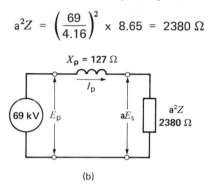

(b)

Figure 11-38 b. See Example 11-11.

b. 3 Referring to Fig. 11-38b, we have:

$$I_p = 69\,000/\sqrt{127^2 + 2380^2} = 28.95\ A$$

b. 4 $aE_s = (a^2Z)\,I_p = 2380 \times 28.95$
$= 68\,902\ V$
$E_s = 68\,902 \times (4.16/69) = 4154\ V$

b. 5 Secondary no-load voltage = 4160 V

b. 6 Voltage regulation in percent is:

$$\frac{\text{no-load voltage} - \text{full-load voltage}}{\text{no-load voltage}} \times 100$$

$$\frac{4160 - 4154}{4160} \times 100 = 0.14\ \text{percent}$$

The voltage regulation is excellent.

c. 1 Referring again to Fig. 11-38b, if the secondary is accidentally short-circuited, we find:

$$I_p = E_p/X_p = 69\,000/127$$
$$= 543\ A$$

c. 2 The corresponding current I_s on the secondary side:

$$I_s = aI_p = (69/4.16) \times 543$$
$$= 9006\ A$$

The short-circuit currents are 12.5 times greater than the rated values. The I^2R losses are therefore 156 times greater than normal and the circuit-breaker or fuse protecting the transformer must open immediately to prevent overheating. Very powerful electromagnetic forces

are also set up. They, too, are 156 times stronger than normal and, unless the windings are firmly braced and supported, they may be damaged or torn apart.

11-21 Measuring transformer impedances

We can determine all the internal impedances shown in Fig. 11-30 by means of an open-circuit test and a short-circuit test on the transformer.

During the *open-circuit test*, rated voltage is applied to the primary winding and current I_0, voltage E_p, and active power P_m are measured (Fig. 11-39). The secondary open-circuit voltage

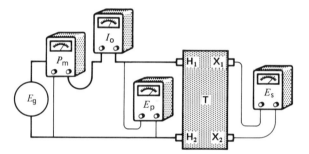

Figure 11-39 Open-circuit test and determination of R_m, X_m, and turns ratio.

E_s is also measured. We can make the following calculations:

active power absorbed by core $= P_m$
apparent power absorbed by core $= S_m = E_pI_0$
reactive power absorbed by core $= Q_m$
                                  $= \sqrt{S_m^2 - P_m^2}$

$$R_m = E_p^2/P_m \qquad \text{Eq. 11-9}$$

$$X_m = E_p^2/Q_m \qquad \text{Eq. 11-10}$$

$$a = N_1/N_2 = E_p/E_s \qquad \text{Eq. 11-5}$$

During the *short-circuit test*, the secondary winding is shorted and a much lower-than-normal voltage (ordinarily less than 5 percent of rated voltage) is applied to the primary. The primary current should not exceed its nominal value to prevent overheating and a consequent rapid change in winding resistance.

Voltage E_{sc}, current I_{sc} and power P_{sc} are measured as before (Fig. 11-40) and the following calculations made:

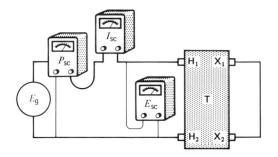

Figure 11-40 Short-circuit test to determine leakage reactance and winding resistance.

$$Z_p = E_{sc}/I_{sc} \qquad (11\text{-}21)$$

$$R_p = P_{sc}/I_{sc}^2 \qquad (11\text{-}22)$$

$$X_p = \sqrt{Z_p^2 - R_p^2} \qquad \text{Eq. 11-20}$$

Example 11-12:
During a short-circuit test on a 500 kVA, 69 kV/4.16 kV, 60 Hz transformer, voltage, current and power were measured as follows: E_{sc} = 2600 V; I_{sc} = 4 A; P_{sc} = 2400 W (see Fig. 11-40). Calculate the value of the reactance and resistance of the transformer, referred to the primary, as well as the percent impedance.

Solution:
Referring to the equivalent circuit of the transformer under short-circuit conditions (Fig. 11-41a), we find the following values:

$Z_p = E_{sc}/I_{sc}$ = 2600/4 = 650 Ω
$R_p = P_{sc}/I_{sc}^2$ = 2400/16 = 150 Ω
$X_p = \sqrt{650^2 - 150^2}$ = 632 Ω
Nominal primary voltage:
  E_p = 69 kV

Nominal primary current:
  I_p = 500 kVA/69 kV = 7.25 A
Nominal impedance referred to the primary:
  $Z_{np} = E_p/I_p$ = 69 000/7.25 = 9517 Ω

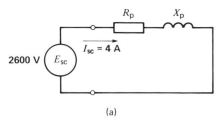

(a)

Figure 11-41 a. See Example 11-12.

Per-unit impedance = Z_p/Z_{np} = 650/9517
              = 0.0682
              (Percent impedance = 6.82)
Per-unit reactance = X_p/Z_{np} = 632/9517
              = 0.0664
              (Percent reactance = 6.64)
Per-unit resistance = R_p/Z_{np} = 150/9517
              = 0.0158
              (Percent resistance = 1.58)

Example 11-13:
An open-circuit test on the transformer given in Example 11-12 yields the following results when the *low voltage* winding is excited. (In some cases, such as in a repair shop, a 69 kV voltage may not be available and the open-circuit test has to be done by exciting the LV winding.)

E_s = 4160 V I_0 = 2 A P_m = 5000 W

Calculate the voltage regulation and efficiency of the transformer when it supplies a 250 kVA, 80 percent power factor (lagging) load.

Solution:
Industrial loads and voltages fluctuate all the time. Thus, when we state that a load is 250 kVA, with cos θ = 0.8, it is understood that the load is *about* 250 kVA and that the power factor is *about* 0.8. Furthermore, the primary voltage is *about* 69 kV.

Consequently, in calculating voltage regulation and efficiency, there is no point in arriving at a precise mathematical answer, even if we were able to give it. Knowing this, we can make certain assumptions

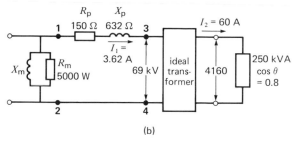

(b)

Figure 11-41 b. See Example 11-13.

that make it much easier to arrive at a solution.

The equivalent circuit may be represented by Fig. 11-41b. The values of R_p and X_p are already known, and so we only have to add the magnetizing branch. Let us assume that the voltage across the load is 4160 V. We now calculate the efficiency of the transformer.

a. 1 The load current is:

$$I_2 = S/E_s = 250\ 000/4160$$
$$= 60\ A$$

a. 2 $I_1 = I_2/a = 60 \times 4.16/69$
$$= 3.62\ A$$

a. 3 The copper loss is:

$$P_{copper} = I_1^2 R_p = 3.62^2 \times 150$$
$$= 1966\ W$$

a. 4 The iron loss is equal to that measured on the LV side of the transformer:

$$P_{iron} = 5000\ W$$

a. 5 Total losses are:

$$P_{losses} = 5000 + 1966$$
$$= 6966\ W = 7\ kW$$

a. 6 The active power to the load is:

$$P_0 = S \cos \theta = 250 \times 0.8$$
$$= 200\ kW$$

a. 7 The efficiency is:

$$\eta = P_0/P_i = 200/(200 + 7)$$
$$= 0.966\ or\ 96.6\ percent$$

Note that in making the above calculations, we only consider the active power. The reac-

tive power does not enter into efficiency calculations. Let us now determine the voltage regulation.

b. 1 In examining Fig. 11-41b, it is obvious that the presence of the magnetizing branch does not affect the voltage drop across R_p and X_p. Consequently, the magnetizing branch does not affect the voltage regulation.

b. 2 We could determine the regulation by drawing a phasor diagram, but we shall use the method explained in Sec. 8-13, based on active and reactive powers.

b. 3 The active power absorbed by the load is:

$$P = S \cos \theta = 250 \times 0.8$$
$$= 200\ kW$$

b. 4 The reactive power abosrbed by the load is:

$$Q = \sqrt{S^2 - P^2} = \sqrt{250^2 - 200^2}$$
$$= 150\ kvar$$

b. 5 The active and reactive power absorbed by "terminals" 3, 4 are:

$$P = 200\ kW, \quad Q = 150\ kvar$$

because the ideal transformer consumes no active or reactive power.

b. 6 Reactive power associated with X_p is:

$$Q_x = I_1^2 X_p = 3.62^2 \times 632$$
$$= 8.28\ kvar$$

b. 7 Active power associated with R_p is:

$$P_r = I_1^2 R_p = 3.62^2 \times 150$$
$$= 1.97\ kW$$

b. 8 Active power input to terminals 1-2 is:

$$P_1 = P + P_r = 200 + 1.97$$
$$= 202\ kW$$

b. 9 Reactive power input to terminals 1-2 is:

$$Q_1 = Q + Q_x = 150 + 8.28$$
$$= 158\ kvar$$

b. 10 Apparent power input to terminals 1-2 is:

$$S_1 = \sqrt{P_1^2 + Q_1^2} = \sqrt{202^2 + 158^2}$$
$$= 256.4\ kVA$$

b. 11 Voltage across terminals 1-2 is:

$$E_{12} = S_1/I_1 = 256\ 400/3.62$$
$$= 70\ 829\ V$$

b. 12 If the load were suddenly removed, the voltage drop across R_p and X_p would disappear, and the voltage across the open-circuit secondary terminals would rise from 4160 V to:

$E_2 = 4160 \times (70\ 829/69\ 000)$
$= 4270 \text{ V}$

b. 13 The voltage regulation is therefore:

Regulation = $(4270 - 4160)/4270$
$= 0.0264$
$= 2.64$ percent

11-22 Standard terminal markings

We saw in Sec. 11-4 that the polarity of a transformer can be shown by means of dots on the primary and secondary terminals. This type of marking is used on instrument transformers. On power transformers, however, the terminals are designated by the symbols H_1 and H_2 for the high-voltage (HV) winding and by X_1 and X_2 for the low-voltage (LV) winding. By convention, H_1 and X_1 have the same polarity.

Although the polarity is known when the symbols H_1, H_2, X_1 and X_2 are given, it is common practice to mount the four terminals in a standard way so that the transformer has either *additive or subtractive* polarity. A transformer is said to have *additive* polarity when terminal H_1 is diagonally

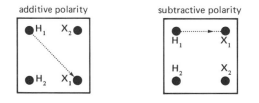

Figure 11-42   Additive and subtractive polarity depends on the location of the H_1 - X_1 terminals.

opposite terminal X_1. Similarly, a transformer has *subtractive* polarity when terminal H_1 is adjacent to terminal X_1 (Fig. 11-42). If we know that a transformer has additive (or subtractive) polarity, we do not have to identify the terminals by symbols.

Subtractive polarity is standard for all single-phase transformers above 200 kVA provided the high-voltage winding is rated above 8660 V. All other transformers have additive polarity.

11-23 Polarity tests

To determine whether a transformer possesses additive or subtractive polarity, we connect the high-voltage winding to an ac source E_g. A jumper J is connected between any two adjacent HV and LV terminals, and a voltmeter E_x is connected between the other two adjacent HV and LV terminals (Fig. 11-43a). Another voltmeter E_p is con-

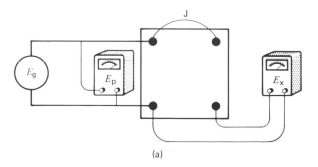

(a)

Figure 11-43   a. Determining the polarity of a transformer using an ac source.

nected across the HV winding. If E_x gives a higher reading than E_p, the polarity is additive. This tells us that H_1 and X_1 are diagonally opposite. On the other hand, if E_x gives a lower reading than E_p, the polarity is subtractive, and terminals H_1 and X_1 are adjacent.

In this polarity test, jumper J effectively connects the secondary voltage E_s in series with the primary voltage E_p. Consequently, E_s either adds to or subtracts from E_p. In other words, $E_x = E_p + E_s$ or $E_x = E_p - E_s$, depending on the polarity. We can now see how the terms "additive" and "subtractive" originated.

In making the polarity test, an ordinary 120 V, 60 Hz source can be connected to the HV winding, even though its nominal voltage may be several hundred kilovolts.

Example 11-14:
During a polarity test on a 500 kVA, 69 kV/600 V transformer (Fig. 11-43a), the following readings are obtained: $E_p = 118$ V, $E_x = 119$ V. Determine the polarity markings of the terminals.

Solution:

The polarity is additive because E_x is greater than E_p. Consequently, the HV and LV terminals connected by the jumper must respectively be labelled H_1 and X_2 (or H_2 and X_1).

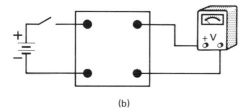

(b)

Figure 11-43 b. Determining the polarity of a transformer using a dc source.

Figure 11-43b shows another circuit which may be used to determine the polarity of a transformer. A dc source, in series with an open switch, is connected to the LV winding of the transformer. The terminal connected to the positive side of the source is marked X_1. A dc voltmeter is connected across the HV terminals. When the switch is closed, a voltage is momentarily induced in the HV winding. If, at this moment, the pointer of the voltmeter moves upscale, the transformer terminal connected to the (+) terminal of the voltmeter is marked H_1 and the other is marked H_2.

11-24 Transformer taps

Owing to the internal impedance Z_p of a transformer, the secondary voltage tends to fall with increasing load. The drop from no-load to full-load seldom exceeds 2 percent and may often be neglected. However, owing to additional voltage drops in transmission lines, the voltage on a system may consistently be lower than normal. Thus, a transformer having a ratio of 2400 V to 120 V may be connected to a transmission line where the voltage is never higher than 2000 V. Under these conditions, the voltage across the secondary is considerably less than 120 V. Incandescent lamps are dim, electric stoves take longer to cook food, and electric motors may stall under quite moderate overloads.

To correct this problem, *taps* are provided on the primary windings of distribution transformers (Fig. 11-44). Taps enable us to change the transformer ratio so as to raise the secondary voltage by 4½, 9 or 13½ percent. We can therefore maintain

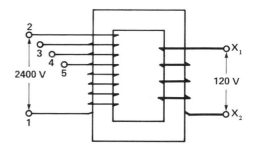

HV tap	percent	primary voltage	secondary voltage
1 – 2	0	2400 V	120 V
1 – 3	4½	2292 V	120 V
1 – 4	9	2184 V	120 V
1 – 5	13½	2076 V	120 V

Figure 11-44   Taps on a transformer.

a satisfactory secondary voltage, even though the primary voltage may be 4½, 9, or 13½ percent below normal. Thus, referring to the transformer of Fig. 11-44, we obtain 120 V across the secondary whether we apply 2400 V between taps 1 and 2, or 2076 V between taps 1 and 5.

Some transformers are designed to change the taps automatically, whenever the secondary voltage is above or below a preset level. Such tap-changing transformers help maintain the secondary voltage within ± 2 percent of its rated value throughout the day.

11-25 Transformers in parallel

When a growing load eventually exceeds the power rating of an installed transformer, we sometimes connect a second transformer in parallel with it. To ensure proper load-sharing between the two, they must possess:

a. the same primary and secondary voltages;

b. the same per-unit impedance.

Particular attention must be paid to the polarity of each transformer, so that only terminals having the same polarity are connected together (Fig. 11-45). An error in polarity produces a dead short-circuit as soon as the transformers are excited.

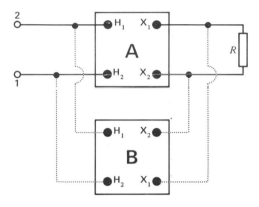

Figure 11-45   Connecting transformers in parallel to share a load.

Example 11-15:

A 100 kVA transformer is to be connected in parallel with an existing 250 kVA transformer to supply a load of 330 kVA. Both transformers are rated 7200 V/240 V, but the 100 kVA unit has an

impedance of 4 percent while the 250 kVA transformer has an impedance of 6 percent (Fig. 11-46a). Calculate:

a. the nominal primary current of each transformer;

b. the equivalent impedance of the load referred to the primary side;

c. the internal impedance of each transformer referred to the primary side;

d. the actual primary current in each transformer.

Solution:

a. 1 Nominal primary current of the 250 kVA unit:
$$I_{n1} = 250\ 000/7200 = 34.7\ \text{A}$$

a. 2 Nominal primary current of the 100 kVA unit:
$$I_{n2} = 100\ 000/7200 = 13.9\ \text{A}$$

b. 1 The equivalent circuit of the two transformers and the load, referred to the primary side, is given in Fig. 11-46b. Note that transformer impedances Z_{p1} and Z_{p2} are considered to be entirely reactive. This assumption is justified because the transformers are fairly big. The magnetizing branches (R_m and X_m) are neglected because they have a negligible effect on the load currents. The circuit elements can be rearranged to yield the circuit of Fig. 11-46c.

b. 2 Load impedance referred to the primary

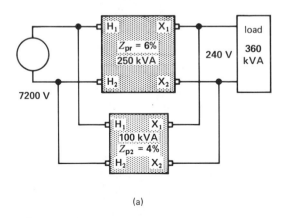

(a)

Figure 11-46   a. Actual transformer connections.

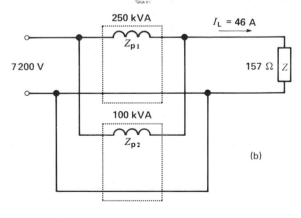

(b)

Figure 11-46   b. Circuit diagram.

side:
$$Z = E_p^2/S_{load} = 7200^2/330\,000$$
$$= 157\ \Omega$$

b. 3 The approximate load current:
$$I_L = S_{load}/E_p = 330\,000/7200 = 46\ A$$

c. 1 Nominal impedance of the 250 kVA unit:
$$Z_{np1} = 7200^2/250\,000 = 207\ \Omega$$
Internal impedance referred to the primary side:
$$Z_{p1} = 0.06 \times 207 = 12.4\ \Omega$$

c. 2 Nominal impedance of the 100 kVA unit:
$$Z_{np2} = 7200^2/100\,000 = 518\ \Omega$$
Internal impedance referred to the primary side:
$$Z_{p2} = 0.04 \times 518 = 20.7\ \Omega$$

d. 1 Referring to Fig. 11-46c, we easily find that the 46 A load current divides in the following way:
$$I_1 = 46 \times 20.7/(12.4 + 20.7) = 28.8\ A$$
$$I_2 = 46 - 28.8 = 17.2\ A$$

The 100 kVA transformer is seriously overloaded because it carries a primary current of 17.2 A, which is 25 percent above its nominal value of 13.9 A. The 250 kVA unit is not overloaded (compare 28.8 A vs 34.7 A).

The 100 kVA transformer is overloaded because of its low impedance (4 percent), compared to the impedance of the 250 kVA transformer (6 percent). A low-impedance transformer always tends to carry more than its proportionate share of the load. If the percent impedances were equal, the load would be shared between the transformers in proportion to their respective power ratings.

11-26 Cooling methods

To prevent rapid deterioration of the insulating materials inside a transformer, adequate cooling of the windings and core must be provided.

Low-power transformers below 50 kVA can be cooled by the natural flow of the surrounding air. The metallic housing is fitted with ventilating louvres so that convection currents may flow over the windings and around the core (Fig. 11-47). Larger

Figure 11-47 Single-phase transformer, type AA, rated 15 kVA, 600 V/240 V, 60 Hz, insulation class 150°C for indoor use. Height: 600 mm; width: 434 mm; depth: 230 mm; weight: 79.5 kg. *(Hammond)*

transformers can be built the same way, but forced circulation of clean air must be used. Such *dry-type* transformers are used inside buildings, away from hostile atmospheres.

Distribution transformers below 200 kVA are usually immersed in mineral oil and enclosed in a

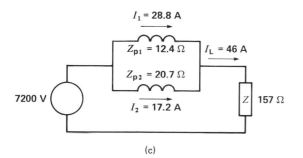

Figure 11-46 c. Equivalent circuit.

steel tank. Oil carries the heat away to the tank, where it is dissipated by radiation and convection to the outside air (Fig. 11-48). Oil is a much better insulator than air is; consequently, it is invariably used on high-voltage transformers.

Figure 11-48 Two single-phase transformers, type OA, rated 75 kVA, 14.4 kV/240 V, 60 Hz, 55°C temperature rise, impedance 4.2%. The small radiators at the side increase the effective cooling area.

Figure 11-49 Three phase, type OA grounding transformer, rated 1900 kVA, 26.4 kV, 60 Hz. The power of this transformer is 25 times greater than that of the transformers shown in Fig. 11-48, but it is still self-cooled. Note, however, that the radiators occupy as much room as the transformer itself.

As the power rating increases, we add external radiators to increase the cooling surface of the oil-filled tank (Fig. 11-49). Oil circulates around the transformer and moves through the radiators where the heat is again released to surrounding air. For still higher ratings, cooling fans blow air over the radiators (Fig. 11-50).

For transformers in the megawatt range, cooling may be effected by an oil-water heat exchanger. Hot oil from the transformer is pumped through a serpentine tube which is in contact with cool water. Such a heat exchanger is very effective, but also very costly, because the water itself has to be continuously cooled and recirculated.

Some big transformers are designed to have a variable rating, depending on the method of cooling used. Thus, a transformer may have a triple rating of 18 000/24 000/32 000 kVA depending on whether it is cooled by the natural circulation of air (18 000 kVA) by forced air cooling with fans (24 000 kVA) or by the forced circulation of oil accompanied by forced air cooling (32 000 kVA). These elaborate cooling systems are nevertheless economical because they enable a much bigger output from a transformer of a given size (Fig. 11-51).

Figure 11-50

Three-phase, type FOA, transformer rated 1300 MVA, 24.5 kV/345 kV, 60 Hz, 65°C temperature rise, impedance: 11.5%. This step-up transformer, installed at a nuclear power generating station, is one of the largest units ever built. The forced-oil circulating pumps can be seen just below the cooling fans. *(Westinghouse)*

Figure 11-51

Three-phase, type OA/FA/FOA transformer rated 36/48/60 MVA, 225 kV/26.4 kV, 60 Hz, impedance 7.4%. The circular tank enables the oil to expand as the temperature rises and reduces the surface of the oil in contact with air. Other details:

weight of core and coils	37.7 t
weight of tank and accessories	28.6 t
weight of oil (44.8 m³)	38.2 t
Total weight	104.5 t

The type of transformer cooling is given by the following typical designations: [*]

    AA - dry-type self-cooled
    AFA - dry-type forced-air cooled
    OA - oil-immersed, self-cooled
    OA/FA - oil-immersed, self-cooled / forced-air cooled
    OA/FA/FOA - oil-immersed self-cooled / forced-air cooled / forced-air, forced-oil-cooled

The temperature rise by resistance of oil-immersed transformers is either $55°C$ or $65°C$. The temperature must be kept low to preserve the quality of the oil. By contrast, the temperature rise by resistance of a dry-type transformer may be as high as $180°C$, depending on the type of insulation used.

[*] Drawn from IEEE Std. 462-1973

QUESTIONS AND PROBLEMS

Practical level

11-1 Name the principal components of a transformer.

11-2 What purpose does the no-load current of a transformer serve?

11-3 Explain how a voltage is induced in the secondary winding of a transformer.

11-4 The secondary winding of a transformer has twice as many turns as the primary. Is the secondary voltage higher or lower than the primary voltage?

11-5 State the voltage and current relationships between the primary and secondary windings of a transformer under load. The primary and secondary windings have N_1 and N_2 turns, respectively.

11-6 Name the losses produced in a transformer.

11-7 Which winding is connected to the load: the primary or secondary?

11-8 What conditions must be met in order to connect two transformers in parallel?

11-9 What is the purpose of taps on a transformer?

11-10 Name three methods used to cool transformers?

11-11 The primary of a transformer is connected to a 600 V, 60 Hz source. If the primary has 1200 turns and the secondary 240, calculate the secondary voltage.

11-12 The windings of a transformer respectively have 300 and 7500 turns. If the low-voltage winding is excited by a 2400 V source, calculate the voltage across the HV winding.

11-13 A 6.9 kV transmission line is connected to a transformer having 1500 turns on the primary and 24 turns on the secondary. If the load across the secondary has an impedance of 5 Ω, calculate:
   a. the secondary voltage;
   b. the primary and secondary currents.

11-14 The primary of a transformer has twice as many turns as the secondary. The primary voltage is 220 V and a 5 Ω load is connected across the secondary. Calculate the power delivered by the transformer, as well as the primary and secondary currents.

11-15 A 3000 kVA transformer has a ratio of 60 kV to 2.4 kV. Calculate the nominal current of each winding.

Intermediate level

11-16 In problem 11-11, calculate the peak value of the flux in the core.

11-17 Explain why the flux in a 60 Hz transformer remains fixed as long as the ac supply voltage is fixed.

11-18 The transformer in Fig. 11-52 is excited by a 120 V, 60 Hz source, and draws a no-load current I_0 of 3 A. The primary and secondary windings respectively possess 200 and 600 turns. If 40 percent of the primary flux is linked by the secondary, calculate:
   a. the voltage indicated by the voltmeter;
   b. the peak value of flux Φ;
   c. the peak value of Φ_m.

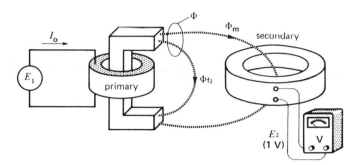

Figure 11-52   See Problem 11-18.

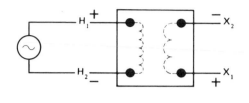

Figure 11-53   See Problem 11-19.

d. Draw the phasor diagram showing E_1, E_2, I_0, Φ_m and Φ_{f1}.

11-19 In Fig. 11-53, when 600 V is applied to terminals H_1, H_2, 80 V is measured across terminals X_1, X_2.

a. What is the voltage between terminals H_1 and X_2?

b. If terminals H_1X_1 are connected together, calculate the voltage across terminals H_2X_2.

c. Does the transformer have additive or subtractive polarity?

11-20a. Referring to Fig. 11-45, what would happen if we reversed terminals H_1 and H_2 of transformer B?

b. Would the operation of the transformer band be affected if terminals H_1H_2 and X_1X_2 of transformer B were reversed? Explain.

11-21 Explain why the secondary voltage of a practical transformer decreases with increasing resistive load.

11-22 What is meant by:

a. transformer impedance;

b. percent impedance of a transformer?

11-23 The transformer in Problem 11-15 has an impedance of 6 percent. Calculate the impedance [Ω] referred to:

a. the 60 kV primary;

b. the 2.4 kV secondary.

11-24 A 2300 V line is connected to terminals 1 and 4 in Fig. 11-44. Calculate:

a. the voltage between terminals X_1 and X_2;

b. the current in each winding, if a 12 kVA load is connected across the secondary.

11-25 A 66.7 MVA transformer has an efficiency of 99.3 percent when it delivers full power to a load having a power factor of 100 percent.

a. Calculate the losses in the transformer under these conditions;

b. Calculate the losses and efficiency when the transformer delivers 66.7 MVA to a load having a power factor of 80 percent.

11-26 If the transformer shown in Fig. 11-47 were placed in a tank of oil, the temperature rise would have to be reduced to 65°. Explain.

Advanced level

11-27 Referring to Fig. 11-54, calculate the peak value of flux in the core if the transformer is supplied by a 50 Hz source.

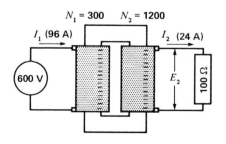

Figure 11-54   See Problem 11-27.

11-28 The impedance of a transformer increases as we reduce the coupling between the primary and secondary windings. Explain.

11-29 The following information is given for the transformer circuit of Fig. 11-34.

R_1 = 18 Ω E_p = 14.4 kV (nominal)
R_2 = 0.005 Ω E_s = 240 V (nominal)
X_{f1} = 40 Ω X_{f2} = 0.01 Ω

If the transformer has a nominal rating of 75 kVA, calculate:

a. the transformer impedance [Ω] referred to the primary side;

b. the percent impedance of the transformer;

c. the impedance [Ω] referred to the secondary side;

d. the percent impedance referred to the secondary side;

e. the total copper losses at full load;

f. the percent resistance and percent reactance of the transformer.

11-30 During a short-circuit test on a 10 MVA, 66 kV/7.2 kV transformer (see Fig. 11-40), the following results were obtained:

E_g = 2640 V I_{sc} = 72 A P_{sc} = 9.85 kW
Calculate:

a. the total resistance and the total leakage reactance referred to the 66 kV primary side;

b. the nominal impedance of the transformer referred to the primary side;

c. the percent impedance of the transformer.

11-31 In Problem 11-30, if the iron losses at rated voltage are 35 kW, calculate the full-load efficiency of the transformer if the power factor of the load is 85 percent.

11-32 If the transformer in Problem 11-31 operates at a peak flux density of 1.6 T, calculate the approximate mass of the core, knowing that it is made of No. 29 gauge, type M-7 laminations (see Fig. 6-16).

11-33a. The windings of a transformer operate at a current density of 3.5 A/mm². If they are made of copper, and operate at a temperature of 75°C, calculate the copper loss per kilogram.

b. If aluminum windings were used, calculate the loss per kilogram under the same conditions.

11-34 If a transformer were actually built according to Fig. 11-54, it would have very poor voltage regulation. Explain why and propose a method of improving it.

12
SPECIAL TRANSFORMERS

Many transformers are specially designed to meet various industrial applications. Nevertheless, they all possess the basic properties of standard transformers discussed in Chapter 11. Furthermore, the following approximations can usually be made when the transformers are under load:

1. the voltage induced in a winding is proportional to the number of turns, and to the flux in the core;
2. the ampere-turns of the primary are equal and opposite to the ampere-turns of the secondary;
3. the exciting current in the primary winding may be neglected;
4. the apparent power input to the transformer is equal to the apparent power output.

12-1 Dual-voltage distribution transformer

Transformers which supply electric power to residential areas generally have two secondary winding, each rated at 120 V. When the two windings are connected in series, the voltage between the lines is 240 V while that between the lines and the middle conductor is 120 V (Fig. 12-1a). The middle conductor, called *neutral*, is always connected to ground. In special cases, the two secondaries can also be connected in parallel by connecting the terminals of the same polarity together (Fig. 12-1b). The resulting output voltage is 120 V, but the power rating remains the same.

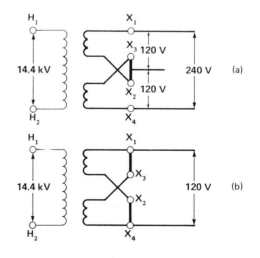

Figure 12-1 Distribution transformer.

209

Figure 12-2 Single-phase pole-mounted distribution transformer rated:
100 kVA, 14.4 kV/240 V/120 V, 60 Hz.

These so-called *distribution transformers* are often mounted on the poles of the electrical utility company (Fig. 12-2). They may supply power to 1, 2, or sometimes as many as 20 customers.

The load on distribution transformers varies greatly throughout the day, depending on customer demand. In residential districts, a peak occurs in the morning and another in the late afternoon. The power peaks never last for more than one or two hours, so that during most of the day the transformers operate far below their normal rating. Because thousands of such units are connected to the public utility system, every effort is made to reduce the no-load losses to a minimum. Consequently, special grain-oriented silicon-steel is used in the core to cut down on power losses.

The nominal rating of distribution transformers ranges from 3 kVA to 500 kVA. The reasons for using such a dual-voltage 120 V/240 V system are covered in Chapter 30.

12-2 Autotransformer

Consider a transformer winding composed of N_1 turns, mounted on a iron core (Fig. 12-3). The winding is connected to a fixed ac source E_1, and the resulting exciting current I_0 creates an ac flux Φ_m in the core. As in any transformer, the flux is fixed so long as E_1 is fixed (Sec. 11-2).

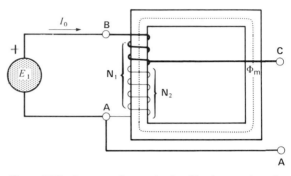

Figure 12-3 Autotransformer having N_1 turns on the primary and N_2 turns on the secondary.

A tap C is taken off the winding, so that there are N_2 turns between terminals A and C. Because the induced voltage between these terminals is proportional to the number of turns, E_2 is given by:

$$E_2 = (N_2/N_1) \times E_1 \qquad (12\text{-}1)$$

If we connect a load to secondary terminals CA, the resulting current I_2 immediately causes a primary current I_1 to flow (Fig. 12-4). Assuming that both the transformer losses and exciting current are negligible, the apparent power drawn by the load must equal the apparent power supplied by the source. Consequently:

$$E_1 I_1 = E_2 I_2 \qquad (12\text{-}2)$$

Referring again to Fig. 12-4, the BC portion of the winding obviously carries current I_1. However, according to Kirchhoff's law, the CA portion carries a current $(I_2 - I_1)$. Furthermore, the mmf due to I_1 must be equal and opposite to the mmf produced by $(I_2 - I_1)$, otherwise the mutual flux Φ_m would change. As a result, we have:

$$I_1 (N_1 - N_2) = (I_2 - I_1) N_2$$

which reduces to:

$$I_1 N_1 = I_2 N_2 \qquad (12\text{-}3)$$

motors, to regulate the voltage of transmission lines and, in general, to transform voltages when the ratio is close to 1. The ratio is rarely greater than 5:1.

Example 12-1:

The autotransformer in Fig. 12-4 has a 66 2/3 per-

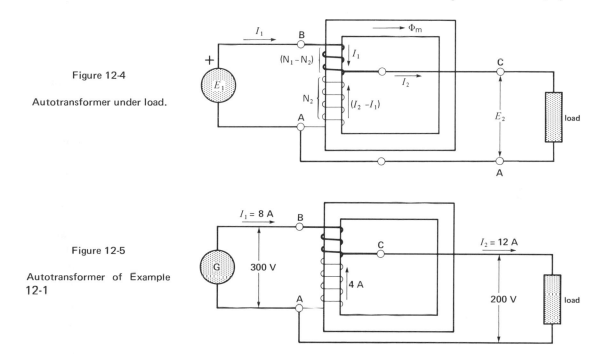

Figure 12-4

Autotransformer under load.

Figure 12-5

Autotransformer of Example 12-1

The foregoing equations are identical to those of a standard transformer having a ratio N_1/N_2. However, in this so-called *autotransformer*, the secondary winding is actually part of the primary winding. In effect, an autotransformer eliminates the need for a separate secondary winding. As a result, autotransformers are always smaller, lighter and cheaper than standard transformers of equal power output. The difference in size becomes particularly important when the ratio of transformation lies betwen 0.5 and 2. On the other hand, the absence of electrical isolation between the high-voltage and low-voltage windings is a serious drawback in some applications.

Autotransformers are used to start induction

cent tap and the supply voltage is 300 V. If a 2.4 kW load is connected across the secondary, calculate:

a. the secondary voltage and current;

b. the currents which flow in the winding;

c. the relative size of the conductors on the windings.

Solution:

a. $E_2 =$ 66 2/3% x 300 = 200 V
   $I_2 =$ 2400/200 = 12 A   (Fig. 12-5)

b. Current supplied by the source is:
   $I_1 =$ 2400/300 = 8 A;
   current in winding BC = 8 A;
   current in winding CA = 12 – 8 = 4 A.

c. The conductors in winding CA can be made half the size of those in winding BC because the current is half as great (see Fig. 12-5).

12-3 Conventional transformer connected as an autotransformer

Any two-winding transformer can be changed into an autotransformer by connecting the primary and secondary windings in series. Depending upon how the connection is made, the secondary voltage may add or substract from the primary voltage. The basic operation and behavior of a transformer is obviously unaffected by a mere change in external connections. Consequently, the following rules apply whenever a standard transformer is connected as an autotransformer:

1. The current in any winding should not exceed its nominal rating,
2. The voltage across any winding should not exceed its nominal rating,
3. If nominal current flows in one winding, nominal current *automatically* flows in the other winding (because the ampere-turns of the two windings must be equal),
4. If nominal voltage appears across one winding, nominal voltage *automatically* appears across the other winding (because the same flux links both windings),
5. If a current I_1 flows into terminal H_1, a current I_2 must flow *out* of terminal X_1 and vice versa.

Example 12-2:
The standard single-phase transformer shown in Fig. 12-6 has a rating of 15 kVA, 600 V/120 V, 60 Hz. We intend to reconnect it as an autotransformer in three different ways to obtain three different voltage ratios:

a. 600 V primary to 480 V secondary;
b. 600 V primary to 720 V secondary;
c. 120 V primary to 480 V secondary.

Calculate the maximum load the transformer can carry in each case.

Solution:
Nominal current of the 600 V winding is:
I_1 = 15 000/600 = 25 A
Nominal current of the 120 V winding is:
I_2 = 15 000/120 = 125 A

a. 1 To obtain 480 V, the secondary voltage (120 V) must subtract from the primary voltage (600 V). Consequently, we connect terminals having the same polarity together, as shown in Fig. 12-7. The corresponding schematic diagram is given in Fig. 12-8.

a. 2 Note that the 120 V winding is connected in series with the load. Because this winding has a nominal current rating of 125 A, the load can draw a maximum power:

$$S_a = 125 \text{ A} \times 480 \text{ V} = 60 \text{ kVA}$$

Figure 12-6

Standard 600 V/120 V transformer.

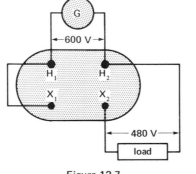

Figure 12-7

Transformer reconnected as an autotransformer.

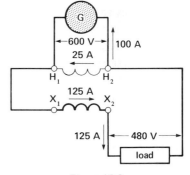

Figure 12-8

Schematic diagram of Fig. 12-7.

The currents flowing in the circuit at full load are shown in Fig. 12-8. Note that:

1. If we assume that the current of 125 A flows *into* terminal X_1, a current of 25 A must flow *out* of terminal H_1. All other currents are then found by applying Kirchhoff's law.

2. The power supplied by the source is equal to that absorbed by the load:

$$S = 100 \text{ A} \times 600 \text{ V} = 60 \text{ kVA}.$$

b. 1 To obtain a ratio of 600 V/720 V, the secondary voltage must add to the primary voltage: 600 + 120 = 720 V. Consequently, terminals of opposite polarity must be connected together, as shown in Fig. 12-9.

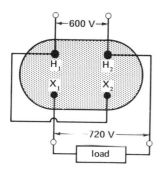

Figure 12-9   Transformer reconnected to give a ratio of 600 V/720 V.

b. 2 The secondary winding is again in series with the load, and the maximum load current is again 125 A. The maximum load is therefore:

$$S_b = 125 \text{ A} \times 720 \text{ V} = 90 \text{ kVA}$$

The previous examples show that when a conventional transformer is connected as an autotransformer, it can supply a load far greater than the nominal capacity of the transformer. However, this depends upon the desired voltage ratio, as the next example shows.

c. 1 To obtain a ratio of 120 V to 480 V, we again connect H_1 and X_1 (as in solution a. 1), but the source is now connected to terminals X_1X_2 (Fig. 12-10).

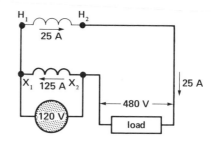

Figure 12-10   Transformer reconnected to give a ratio of 120 V/480 V.

c. 2 This time, it is the 600 V winding that is in series with the load; consequently, the maximum load current is 25 A. The corresponding maximum load is therefore:

$$S_c = 25 \text{ A} \times 480 \text{ V} = 12 \text{ kVA}$$

which is *less* than the nominal rating of the standard transformer.

We want to make one final remark concerning the three autotransformer connections above. The temperature rise of the transformer is the same in each case, whether the load is 60 kVA, 90 kVA, or 12 kVA. The reason is that the losses are the same. In effect, the currents in the windings and the flux in the core are identical for the three connections.

12-4 Potential transformers

Potential transformers are high-precision transformers in which the ratio of primary voltage to secondary voltage is a known constant, which changes very little with load.[*] Furthermore, the secondary voltage is almost exactly in phase with the primary voltage, except for a fraction of a degree. The nominal secondary voltage is usually 115 V, irrespective of the rated primary voltage. In this way standard instruments and relays can be

---

[*]   In the case of potential transformers and current transformers, the "load" is called burden.

used, and maintenance personnel are protected against dangerous voltages. Potential transformers (PT's) serve to measure or monitor the voltage on transmission lines and to isolate the metering equipment from these lines (Fig. 12-11).

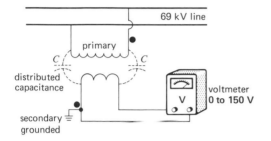

Figure 12-11   Potential transformer installed on a 69 kV line. Note the distributed capacitance between the windings.

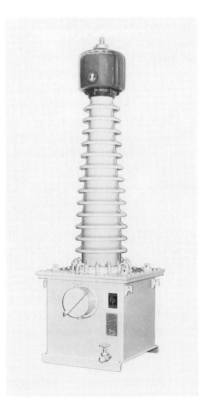

The construction of these transformers is similar to that of conventional transformers. However, their nominal rating is usually less than 500 VA, so that the volume of insulation is often far greater than the volume of copper or steel. For example, the 7000 VA, 80.5 kV transformer shown in Fig. 12-12 has a relatively enormous bushing simply to isolate the HV line from the grounded case. The latter houses the actual transformer. As to the transformer itself, the insulation between the primary and secondary windings must be sufficient to withstand the full line voltage, as well as the very high impulse voltage.

Potential transformers installed on HV lines usually measure the line-to-neutral voltage. This eliminates the need for two HV bushings because one side of the primary is connected to ground.

One terminal of the secondary winding is always connected to ground to eliminate the danger of a fatal shock when touching one of the secondary leads, Although the secondary *appears* to be isolated from the primary, the distributed capacitance between the two windings makes an invisible connection which can produce a very high voltage between the secondary winding and ground. By grounding one of the secondary terminals, the highest voltage between the secondary lines and ground is limited to 115 V.

12-5 Current transformers

Current transformers are high-precision transformers in which the ratio of primary to secondary current is a known constant that changes very little with the burden. The phase angle between the primary and secondary current is very small, usually amounting to much less than one degree. The high-

Figure 12-12

7000 VA, 80.5 kV, 50/60 Hz potential transformer having an accuracy of 0.3% and a BIL of 650 kV. The primary terminal at the top of the bushing is connected to the HV line while the other is connected to ground. The secondary is composed of two 115 V windings each tapped at 66.4 V. Other details: total height: 2565 mm; height of porcelain bushing: 1880 mm; oil: 250 L; weight: 740 kg. *(Ferranti-Packard)*

ly accurate current ratio and the small phase angle are achieved by keeping the exciting current small.

Current transformers are used to measure or monitor the current in a line while isolating the metering and relay equipment. The primary is connected in series with the line, as shown in Fig. 12-13. The nominal secondary current is usually 5 A, irrespective of the primary current rating.

Because current transformers (CT's) are only used for measurement and protection, their power rating is small — generally between 15 and 200 VA. As in the case of conventional transformers, the current ratio is inversely proportional to the number of turns on the primary and secondary windings. A current transformer having a ratio of 150 A/5 A has therefore 30 times more turns on the secondary than on the primary.

For safety reasons, current transformers must

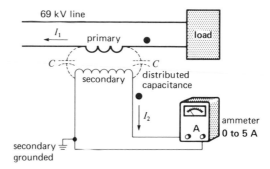

Figure 12-13   Current transformer installed on a 69 kV line.

always be used when measuring currents in HV transmission lines. The insulation between the primary and secondary windings must be sufficient to withstand the full line-to-neutral voltage. The

Figure 12-14   500 VA, 100 A/5 A, 60 Hz current transformer, insulated for a 230 kV line and having an accuracy of 0.6%. *(Westinghouse)*

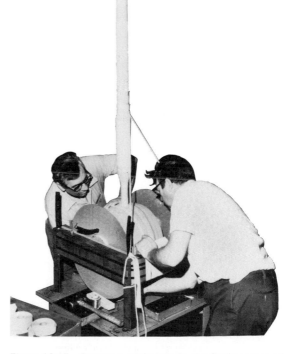

Figure 12-15   Current transformer in the final process of construction. *(Ferranti-Packard)*

maximum voltage the CT can withstand is always shown on the nameplate.

As in the case of potential transformers (and for the same reasons) one of the secondary terminals is always connected to ground.

Figure 12-14 shows a 500 VA, 1000 A/5 A current transformer designed for a 230 kV line. The large bushing serves to lead the line current into and out of the primary winding. The CT is housed in the grounded steel case at the lower end of the bushing. The upper end has two terminals connected in series with the HV line. The construction of a CT is given in Fig. 12-15 and a typical installation is shown in Fig. 12-16.

By way of comparison, the CT shown in Fig. 12-17 is much smaller, mainly because it is insulated for only 36 kV.

Figure 12-16 Current transformer in series with one phase of a 220 kV, 3-phase line inside a substation.

Example 12-3:
The current transformer in Fig. 12-17 has a rating of 50 VA, 400 A/5 A, 36 kV, 60 Hz. It is connected to an ac line, having a line-to-neutral voltage of 24.9 kV, in a manner similar to that shown in Fig. 12-13. The ammeters, relays and connecting wires on the secondary side possess a total impedance (burden) of 1.2 Ω. If the transmission line current is 280 A, calculate:
a. the secondary current;
b. the voltage across the secondary terminals;
c. the voltage drop across the primary.

Solution:
a. 1 The current ratio is:
$$I_1/I_2 = 400/5 = 80$$
a. 2 The turns ratio is:
$$N_1/N_2 = 1/80$$
a. 3 The secondary current is:
$$I_2 = 280/80 = 3.5 \text{ A}$$
b. 1 The voltage across the burden is:
$$E_2 = 3.5 \times 1.2 = 4.2 \text{ V}$$

Figure 12-17 Epoxy-encapsulated current transformer rated 50 VA, 400 A/5 A, 60 Hz and insulated for 36 kV. *(Montel/Sprecher & Schuh)*

b. 2   The secondary voltage is therefore 4.2 V.

c.   The primary voltage is:

$E_1 = 4.2/80 = 0.0525 = 52.5$ mV

This is a negligible drop, compared to the 24.9 kV line voltage.

12-6 Opening the secondary of a CT is dangerous

Every precaution must be taken to *never* open the secondary circuit of a CT while current is flowing in the primary circuit. If the secondary is accidentally opened, the primary current I_1 continues to flow unchanged because it depends only on the electrical load. The line current thus becomes the *exciting* current of the transformer because there is no further bucking effect due to the secondary ampere-turns. Because the line current may be 100 to 200 times greater than the normal exciting current, the flux in the core reaches peaks much higher than normal. The flux is so large that the core is totally saturated for the greater part of every half cycle. Referring to Fig. 12-18, as the pri-

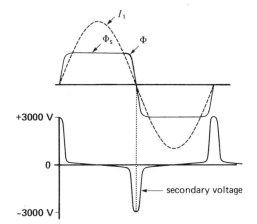

Figure 12-18 Primary current, flux and secondary voltage when a CT is open-circuited.

mary current I_1 rises and falls, flux Φ in the core rises and falls, but it remains at a fixed, saturation level Φ_s for most of the time. During these satu-

rated intervals, the induced voltage is quite low because the flux changes very little. However, during the unsaturated intervals, the flux changes at an extremely high rate, inducing voltage peaks of several thousand volts across the open-circuited secondary. This is a dangerous situation because an unsuspecting operator could easily receive a fatal shock. Furthermore, the high voltage would most certainly damage instruments connected to the secondary side.

In view of the above, if we have to remove a meter in the secondary circuit of a CT, we must first short-circuit the secondary winding *and then* remove the meter. Short-circuiting a current transformer does no harm because the primary current remains unchanged and the secondary current depends exclusively upon the turns ratio. The short-circuit across the winding may be removed after the secondary circuit is again closed.

12-7 Toroidal current transformers

When the line current exceeds 100 A, we can sometimes use a *toroidal* current transformer. It consists of a laminated toroidal (ring-shaped) core which carries the secondary winding. The primary is composed of a single conductor that simply

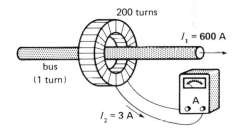

Figure 12-19 Toroidal transformer having a ratio of 1000 A/5 A, connected to measure the current in a line.

passes through the center of the ring (Fig. 12-19). The position of the primary conductor is unimportant as long as it is more or less centered. If the secondary possesses N turns, the ratio of transformation is N. Thus, a toroidal CT having a ratio of 1000 A/5 A has 200 turns on the secondary winding.

Toroidal CT's are simple and inexpensive and are widely used in LV and MV indoor installations. They are also incorporated in circuit-breaker bushings to monitor the line current (Fig. 12-20). If the current exceeds a predetermined limit, the CT causes the circuit-breaker to trip.

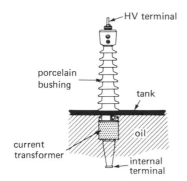

Figure 12-20 Toroidal transformer surrounding a conductor.

Example 12-4:
A potential transformer rated 14 400 V to 115 V and a current transformer of 75/5 A are used to measure the voltage and current in a transmission line. If the voltmeter indicates 111 V and the ammeter reads 3 A, what is the voltage and current in the line?

Solution:
Voltage on the line is:
$$E = 111 \times (14\ 400/115) = 13\ 900 \text{ V}$$
Current in the line is:
$$I = 3 \times (75/5) = 45 \text{ A}$$

12-8 Variable autotransformer

A variable autotransformer is often used when we wish to obtain a variable ac voltage from a fixed ac source. It is composed of a single-layer winding on a toroidal iron core. A movable carbon brush in

contact with the winding serves as a variable tap. The brush can be set in any position between zero and 330°. Manual or motorized control may be used (Figs. 12-21 and 12-23).

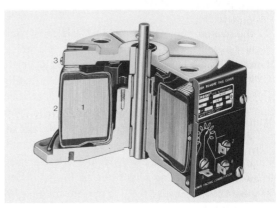

Figure 12-21 Cutaway view of a manually operated 0-140 V, 15 A variable autotransformer showing (1) the laminated toroidal core; (2) the single-layer winding; (3) the movable brush. *(American Superior Electric)*

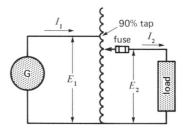

Figure 12-22 Schematic diagram of a variable autotransformer having a fixed 90% tap.

As the brush slides over the bared portion of the windings, the secondary voltage E_2 increases in proportion to the number of turns swept out (Fig. 12-22). The input voltage E_1 is usually connected to a fixed 90 percent tap on the winding. This enables us to vary E_2 from zero to 110 percent of the input voltage.

Variable autotransformers are far more efficient than rheostats are, and they give much better volt-

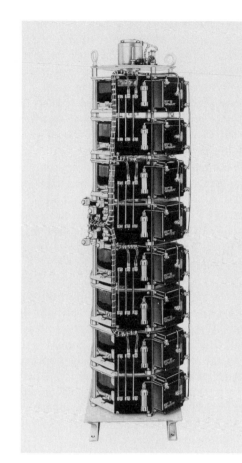

Figure 12-23 Variable autotransformer rated at 200 A, 0-240 V, 50/60 Hz. It is composed of eight 50 A, 120 V units, connected in series-parallel. This motorized unit can vary the output voltage from zero to 240 V in 5 s. Dimensions: 400 mm x 1500 mm. *(American Superior Electric)*

age regulation under variable loads. The secondary line should always be protected by a fuse or circuit-breaker so that the output current I_2 never exceeds the normal current rating of the autotransformer.

12-9 High-impedance transformers

The transformers we have studied so far are all designed to have a relatively low leakage reactance ranging perhaps between 0.03 and 0.1 p.u. (Sec.

11-19). However, some industrial and commercial applications require much higher reactances, sometimes reaching values as high as 0.9 p.u. Such high-impedance transformers are used in the following typical applications:

electric toys	arc welders
fluorescent lamps	electric arc furnaces
neon signs	reactive power regulators
oil-burners	

Let us briefly examine these interesting applications.

1. A toy transformer is often accidentally short-circuited, but, being used by children, it is neither practical nor safe to protect it with a fuse. Consequently, we design the transformer so that its leakage reactance is so high that even a permanent short-circuit will not cause overheating.

   The same remarks apply to some low-voltage bell transformers that distribute signalling power throughout a home. Again, if a short-circuit occurs on the secondary side, the current is automatically limited so as not to burn out the transformer or damage the fragile annunciator wiring.

2. Electric arcs and discharges in gases possess a negative E/I characteristic, meaning that the current increases as the voltage falls. To maintain a steady arc, or a uniform discharge, we must add an impedance in series with the load. The series impedance may be either a resistor or reactor, but we usually prefer the latter because it consumes no active power.

   However, if a transformer is needed to supply the load, it is usually more economical to incorporate the reactance in the transformer itself, by designing it to have a high leakage reactance. A typical example is the neon-sign transformer shown in Fig. 12-24.

   The primary winding P is connected to a 240 V ac source, and two secondary windings S are connected in series across the long neon tube. Owing to the large leakage fluxes Φ_a and

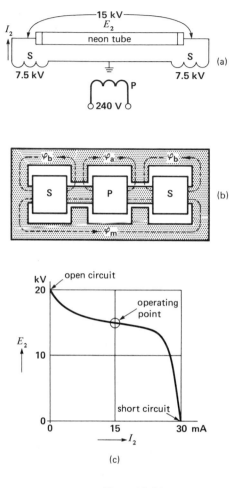

Figure 12-24

a. Schematic diagram of a neon-sign transformer.
b. Construction of the transformer.
c. Typical E-I characteristic of the transformer.

Φ_b, the secondary voltage falls rapidly with increasing current, as seen in Fig. 12-24c. The high open-circuit voltage (20 kV) initiates the discharge, but the normal secondary current is automatically limited to 15 mA. The corresponding voltage across the neon tube falls to 15 kV. The power of these transformers ranges from 50 VA to 1500 VA. The secondary voltages vary from 2 kV to 20 kV, depending main-

ly upon the length of the tube. Referring to Fig. 12-24a, we note that the center of the secondary winding is grounded. This ensures that the secondary line-to-ground voltage is only one-half the voltage across the neon tube.

Fluorescent lamp transformers (called ballasts) have properties similar to neon-sign transformers. Capacitors are usually added to improve the power factor of the total circuit.

Oil-burner transformers possess essentially the same characteristics as neon-sign transformers do. A secondary open-circuit voltage of about 10 kV creates an arc between two closely spaced electrodes situated immediately above the oil jet. The arc continuously ignites the vaporized oil while the burner is in operation.

3. Some electric furnaces generate heat by maintaining an intense arc between carbon electrodes. A relatively low secondary voltage is needed and the large secondary current is limited by the leakage reactance of the transformer. Such transformers have ratings between 100 kVA and 50 MVA. In very big furnaces, the reactance associated with the secondary circuit is usually sufficient to provide the necessary limiting impedance.

4. Arc-welding transformers are also designed to have a high internal impedance so as to stabilize the arc during the welding process. The open-circuit voltage is about 70 V, which facilitates striking the arc when the electrode touches the work. However, as soon as the arc is established, the secondary voltage falls to about 15 V, depending on the length of the arc and the welding current.

5. As a final example of high-impedance transformers, we mention the enormous 3-phase units that absorb reactive power from a 3-phase transmission line. These transformers are intentionally designed to produce leakage flux and, consequently, the primary and secondary windings are very loosely coupled. The three

primary windings are connected to the HV line (typically between 230 kV and 765 kV) while the three secondary windings (typically 6 kV) are connected to an electronic controller (Fig. 12-25). The controller permits more or less secondary current to flow, causing the leakage flux to vary accordingly. A change in the leakage flux produces a corresponding change in the reactive power absorbed by the transformer. The utility of such a static var compensator is explained in Chapter 29.

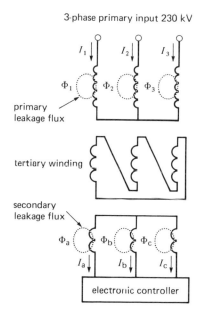

3-phase primary input 230 kV

Figure 12-25   Three-phase static var compensator having high leakage reactance.

12-10 Induction heating transformers

The transformer principle is applied in some high-power induction furnaces to produce high-quality steel and other alloys. The induction principle can be understood by referring to Fig. 12-26. A relatively high-frequency 2000 Hz current is made to flow through a winding composed of a single-conductor ultra-flexible cable. The cable is wrapped around a steel ring, which acts like a short-circuited single-turn secondary. The ac current in the primary produces an enormous current in the secondary (the ring) causing it to heat up. As explained in the caption, the heat causes the ring to expand, enabling it to be slipped over the turbine windings.

Figure 12-26

Special application of the transformer effect. This picture shows one stage in the construction of a turboalternator. It consists of expanding the diameter of a 5 t coil-retaining ring. A coil of asbestos-insulated wire is wound around the ring and connected to a 35 kW, 2000 Hz source (left foreground). The primary creates a 2000 Hz magnetic field, which induces large eddy currents in the ring, bringing its temperature up to 280°C in about 3 h. The resulting expansion enables the ring to be slipped over the coil-ends, where it cools and contracts. This method of induction heating is clean and produces a very uniform temperature rise of the large mass. *(Brown Boveri)*

Coreless induction furnaces operate the same way, but power is applied until the iron actually melts. In Fig. 12-27, for example, a primary coil surrounds a crucible filled with molten iron. The iron acts like a single secondary turn, short-circuited upon itself. Consequently, it carries a very large secondary current. This current provides the energy that keeps the iron in a liquid state, melting other scrap metal as it is added to the pool.

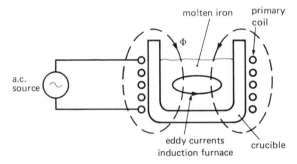

Figure 12-27   Coreless induction furnace.

Such induction furnaces have ratings between 15 kVA and 40 000 kVA. They operate at a frequency of 60 Hz when the power exceeds about 3000 kVA. The power factor is very low (typically 20 percent) because a large magnetizing current is required to drive the flux through the molten iron and through the air. In this regard, we must remember that molten iron has a permeability of one because its temperature is far above the Curie point. Capacitors are installed close to the coil to supply the reactive power it absorbs.

In another type of furnace, known as a *channel furnace*, a transformer having a laminated iron core is made to link with a molten channel of iron, as shown in Fig. 12-28. The primary coil is excited by a 60 Hz source, and the secondary current I_2 flows in the channel and through the molten iron in the crucible. In effect, the channel is equivalent to a single turn shorted upon itself.

The magnetizing current is low because the flux is confined to a highly permeable iron core. On the other hand, the leakage flux is large because the secondary turn is separated from the primary coil

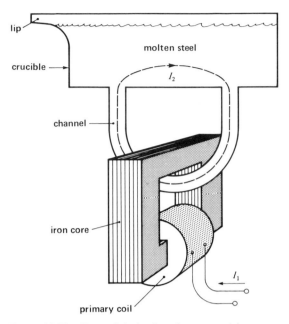

Figure 12-28   Channel induction furnace and its water-cooled transformer.

by a thick layer of ceramic material. Nevertheless, the power factor is higher than that in Fig. 12-27, being typically between 60 and 80 percent. As a result, a smaller capacitor bank is required to furnish the reactive power.

Owing to the very high ambient temperature, the windings of induction furnace transformers are always composed of hollow, water-cooled copper conductors. Induction furnaces are used to melt aluminum, copper, and other metals, as well as iron.

QUESTIONS AND PROBLEMS

Practical level

12-1   What is the difference between an autotransformer and a conventional transformer?

12-2   What is the purpose of a potential transformer? Of a current transformer?

12-3   Why must we never open the secondary of a current transformer?

12-4 Explain why the secondary winding of a CT or PT must be grounded.

12-5 A toroidal current transformer has a ratio of 1500 A/5 A. How many turns does it have?

12-6 A current transformer has a rating of 10 VA, 50 A/5 A, 60 Hz, 2,4 kV. Calculate the nominal voltage drop across the primary winding.

Intermediate level

12-7 A single-phase transformer has a rating of 100 kVA, 7200 V/600 V, 60 Hz. If it is reconnected as an autotransformer having a ratio of 7800 V/7200 V, calculate the load it can carry.

12-8 In Problem 12-7, how should the transformer terminals (H_1, H_2, X_1, X_2) be connected?

12-9 The transformer in Problem 12-7 is reconnected again as an autotransformer having a ratio of 6.6 kV/600 V. What load can it carry and how should the connections be made?

Advanced level

12-10 A current transformer has a rating of 100 VA, 2000 A/5 A, 60 Hz, 138 kV. It has a primary to secondary capacitance of 250 pF. If it is installed on a transmission line where the line-to-neutral voltage is 138 kV, calculate the capacitive leakage current that flows to ground.

12-11 The toroidal current transformer of Fig. 12-19 has a ratio of 1000 A/5 A. The line conductor carries a current of 600 A.

   a. Calculate the voltage across the secondary winding if the ammeter has an impedance of 0.15 Ω.

   b. Calculate the voltage drop the transformer produces on the line conductor.

   c. If the primary conductor is looped four times through the toroidal opening, calculate the new current ratio.

13

THREE-PHASE TRANSFORMERS

To raise or lower the voltage of 3-phase transmission lines, we use transformers, as we do for single-phase lines. The transformers may have three primary and three secondary windings mounted on a single multilegged core. Alternatively, they may be ordinary single-phase transformers connected together to form a 3-phase transformer bank.

13-1 Basic properties of 3-phase transformer banks

When three single-phase transformers are used to transform a 3-phase voltage, the windings can be connected in several different ways. Thus, the primaries may be connected in delta and the secondaries in wye, or vice versa. As a result, the ratio of the 3-phase input voltage to the 3-phase output voltage depends not only upon the turns ratio of the transformers, but also upon how they are connected.

The combination of three single-phase transformers can also produce a *phase-shift* between the 3-phase input voltage and the 3-phase output voltage. The amount of phase-shift depends again

upon the turns ratio of the transformers, and on how the primaries and secondaries are interconnected. In a single-phase system, the phase shift between primary and secondary is either zero or 180°.

The phase-shift feature also enables us to change the *number* of phases. Thus a 3-phase system can be converted into either a 2-phase, a 6-phase, or a 12-phase system. Indeed, if there were a practical application for it, we could even convert a 3-phase system into a 5-phase system by an appropriate choice of single-phase transformers and interconnections.

In making the various connections, it is important to observe transformer polarities. An error in polarity may produce a short-circuit or unbalance the line voltages and currents.

In analyzing the behavior of balanced 3-phase transformer banks, we can make the following assumptions:
1. the exciting currents are negligible;
2. the transformer impedances are negligible;
3. the total apparent input power is equal to the total apparent output power;

4. single-phase transformers connected into a 3-phase system retain all their basic properties, as to current ratio, voltage ratio, flux in the core, and so on.

13-2 Delta-delta connection

The three single-phase transformers P, Q, and R of Fig. 13-1 are connected in *delta-delta*. Terminal H_1 of each transformer is connected to terminal H_2 of the next transformer. Similarly, terminals X_1 and X_2 of successive transformers are connected together. The actual physical layout of the

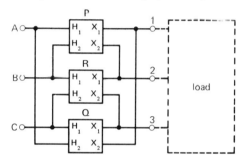

Figure 13-1 Delta-delta connection of three single-phase transformers.

transformers is shown in Fig. 13-1 and the corresponding schematic diagram is given in Fig. 13-2.

The schematic diagram is drawn in such a way to show not only the connections, but also the phasor relationship between the primary and secondary voltages. Thus, each secondary winding is drawn parallel to the corresponding primary winding with which it is coupled. Furthermore, if source G produces voltages E_{AB}, E_{BC}, E_{CA} according to the indicated phasor diagram, the primary windings are oriented the same way, phase by phase.

Because the primary and secondary voltages of a given transformer must be in phase, it follows that E_{12} (secondary voltage of transformer P) must be in phase with E_{AB} (primary of the same transformer). Similarly, E_{23} is in phase with E_{BC}, and E_{31} with E_{CA}. In such a delta-delta connection, the voltages between the incoming and outgoing transmission lines are in phase.

If we connect a balanced load to lines 1-2-3 the resulting secondary currents are equal in magnitude, as are the primary line currents. As in any delta connection, the primary and secondary line currents are $\sqrt{3}$ times greater than the respective currents I_p and I_s flowing in the primary and secondary windings (Fig. 13-2). The rating of the so-called *transformer bank* is 3 times the rating of a single transformer.

Note that although the transformer bank constitutes a 3-phase arrangement, each transformer, considered alone, acts exactly as if it were placed

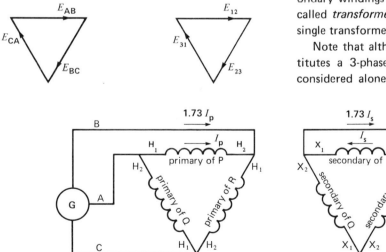

Figure 13-2 Schematic diagram of a delta-delta connection and associated phasor diagram.

in a single-phase circuit. Thus, a current I_p flowing into a primary polarity mark H_1 is associated with a current I_s flowing out of a secondary polarity mark X_1.

Example 13-1:
Three single-phase transformers are connected in delta-delta to step down a line voltage of 138 kV to supply power to a plant operating at 4160 V. The load draws 21 MW at a lagging power factor of 86 percent. Calculate:

a. the apparent power drawn by the load;
b. the apparent power furnished by the HV line;
c. the current in the HV lines;
d. the current in the LV lines;
e. the currents in the primary and secondary windings of each transformer;
f. the load carried by each transformer.

Solution:

a. The apparent power drawn by the load is:
$$S = P/\cos\theta = 21/0.86 \qquad \text{Eq. 9-11}$$
$$= 24.4 \text{ MVA}$$

b. Because the transformer bank itself requires a negligible amount of active and reactive power (the I^2R losses and the reactive power associated with the mutual flux and leakage flux are small) it follows that the apparent power furnished by the HV line is also 24.4 MVA.

c. The current in each HV line is:
$$I_1 = S/(\sqrt{3}E) \qquad \text{Eq. 9-9}$$
$$= (24.4 \times 10^6)/(\sqrt{3} \times 138\,000)$$
$$= 102 \text{ A}.$$

d. The current in the LV lines is in inverse proportion to the line voltages:
$$I_2 = (I_1E_1)/E_2 = (102 \times 138\,000)/4160$$
$$= 3384 \text{ A}$$

e. Referring to Fig. 13-2, we have:
$$I_p = 102/\sqrt{3} = 58.9 \text{ A}$$
$$I_s = 3384/\sqrt{3} = 1954 \text{ A}$$

f. Because the plant load is balanced, each transformer carries one third of the total load, or $24.4/3 = 8.13$ MVA.

The transformer load is also given by the product of the primary voltage times primary current:
$$S = E_pI_p = 138\,000 \times 58.9$$
$$= 8.13 \text{ MVA}$$

Note that we can determine all the line currents and transformer currents even though we do not know how the actual 3-phase load is itself connected. In effect, the factory load (shown as a box in Fig. 13-2) is composed of hundreds of individual loads, some of which are connected in delta, others in wye. Furthermore, some are single-phase loads or 3-phase loads operating at much lower voltages than 4160 V, powered by smaller transformers located inside the plant. The sum total of these loads usually results in a reasonable well-balanced 3-phase load, represented by the box. Fortunately, we don't have to know how all these loads are connected, as far as the transformer bank is concerned.

13-3 Delta-wye connection

The primary windings of the three transformers of Fig. 13-3 are connected the same way as in Fig. 13-1. However, the secondary windings are connected so that all the X_2 terminals are joined together, creating a common neutral N. In such a *delta-wye* connection, the primary voltage across each transformer is equal to the incoming line voltage. However, the outgoing line voltage is $\sqrt{3}$ times the secondary voltage across each transform-

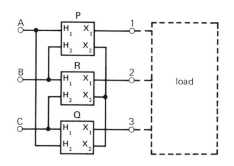

Figure 13-3   Delta-wye connection of three single-phase transformers.

er. The relative values of the corresponding currents in the transformer windings and transmission lines are given in Fig. 13-4. Thus, the line currents in phases A, B, and C are 1.73 times the currents in the primary windings.

A delta-wye connection produces a $30°$ phase shift between the voltages of the incoming and outgoing transmission lines. Thus, outgoing line voltage E_{12} is $30°$ ahead of incoming line voltage

ments of the transformer. In effect, the HV winding has only to be insulated for $1/\sqrt{3}$ or 58 percent of the line voltage. Another advantage is that it can accommodate loads operating at two different voltage levels (see Chapter 30).

Example 13-2:
Three single-phase transformers rated 40 MVA, 13.2 kV/80 kV are connected in delta-wye on a

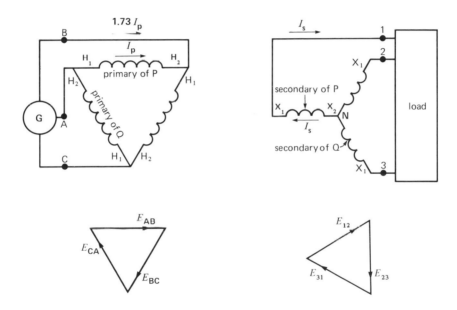

Figure 13-4   Schematic diagram of a delta-wye connection and associated phasor diagram.

E_{AB}, as can be seen from the phasor diagram. If the outgoing line feeds its own group of loads, the difference in phase creates no problem. However, if the outgoing line has to be connected in parallel with a line coming from another source, the $30°$ shift may make such a parallel connection impossible, even if the line voltages are identical.

One of the important advantages of the wye connection is that it reduces the insulation require-

13.2 kV transmission line. If they feed a 90 MVA load, calculate:
a. the secondary line voltage;
b. the currents in the transformer windings;
c. the incoming and outgoing transmission line currents.

Solution:
The easiest way to solve this problem is to consid-

er only one phase at a time (Fig. 13-5).

a. 1 The voltage across each primary winding is obviously 13.2 kV.

a. 2 The secondary voltage across each winding is therefore 80 kV.

a. 3 Secondary voltage between outgoing lines 1, 2, and 3 is:

$$E_s = 80 \sqrt{3} = 138 \text{ kV}$$

13-5 Wye-wye connection

The *wye-wye* connection is only used when the neutral of the primary can be solidly connected to the neutral of the source, usually by way of the ground (Fig. 13-6). When the neutrals are not joined, the line-to-neutral voltages become distorted (non sinusoidal). A wye-wye connection

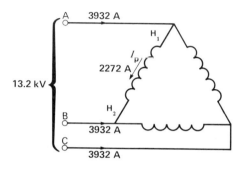

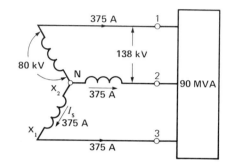

Figure 13-5   See Example 13-2.

b.   Load carried by each transformer = 90/3
            = 30 MVA
     Current in each primary winding:
     I_p = 30 MVA/13.2 kV = 2272 A
     Current in each secondary winding:
     I_s = 30 MVA/80 kV = 375 A

c. 1 Current in each incoming line A, B, C
            = 2272 $\sqrt{3}$ = 3935 A

c. 2 Current in each outgoing line 1, 2, 3
            = 375 A

13-4 Wye-delta connection

The currents and voltages in a *wye-delta* connection are identical to those in the delta-wye connection of Sec. 13-3. The primary and secondary connections are simply interchanged. In other words, the H_2 terminals are connected together to create a neutral, and the $X_1 - X_2$ terminals are connected in delta. Again, there results a 30° phase shift between the voltages of the incoming and outgoing lines.

can be used *without* joining the neutrals, provided that each transformer carries a third winding, called *tertiary* winding. The tertiary windings of the 3 transformers are always connected in delta (Fig. 13-7). They often provide the substation service voltage where the transformers are installed.

Note that there is no phase shift between the incoming and outgoing transmission line voltages of a wye-wye connected transformer.

13-6 Open-delta connection

We can transform the voltage of a 3-phase system by using only 2 transformers, connected in *open-delta*. The open-delta connection is identical to a delta-delta connection, except that one transformer is removed (Fig. 13-8). In effect, one of the advantages of a delta-delta connection is that two transformers can continue to feed the load should one of them become defective.

In medium and high-power installations, the open-delta connection is always temporary, be-

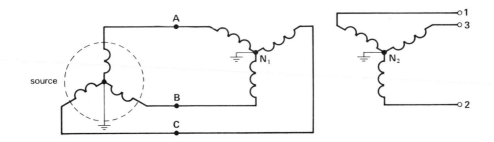

Figure 13-6   Wye-wye connection with neutrals connected.

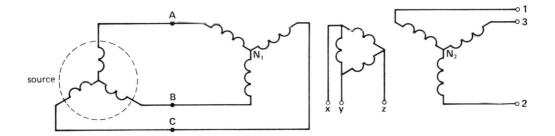

Figure 13-7   Wye-wye connection using a tertiary winding.

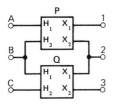

(a)

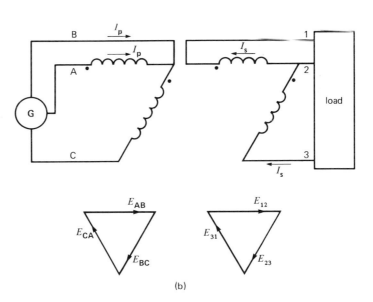

(b)

Figure 13-8   a.   Open delta connection.   b.   Associated schematic and phasor diagram.

cause the load capacity of the transformer bank is only 86.6 percent of the installed transformer capacity.

Example 13-3:

Two single-phase 150 kVA, 7200 V/600 V transformers are connected in open-delta. Calculate the maximum load they can carry.

Solution:

Although each transformer has a rating of 150 kVA, the two together *cannot* carry a load of 300 kVA. The following calculations show why: nominal secondary current of each transformer is:

$$I_s = 150 \text{ kVA}/600 \text{ V} = 250 \text{ A}$$

The current I_s in lines 1, 2, 3 cannot, therefore, exceed 250 A (Fig. 13-8). Consequently, the maximum load is:

$$S = 1.73 \, EI \qquad \qquad \text{Eq. 9-9}$$
$$= 1.73 \times 600 \times 250 = 259\ 500 \text{ VA}$$
$$= 260 \text{ kVA}$$

$$\frac{\text{maximum load}}{\text{installed transformer rating}} = \frac{260 \text{ kVA}}{300 \text{ kVA}}$$

$$= 0.866 \text{ or } 86.6 \text{ percent}$$

13-7 Three-phase transformers

A transformer bank composed of three single-phase transformers may be replaced by one 3-phase transformer (Fig. 13-9). The magnetic core of such a transformer has 3 distinct legs which carry the primary and secondary windings of each phase. The windings are connected internally, either in wye or in delta, with the result that only six terminals have to be brought outside the tank. For a given capacity, a 3-phase transformer is always smaller and cheaper than three single-phase transformers having the same capacity. Nevertheless, single-phase transformers are sometimes preferred, particularly when a replacement unit is essential. For example, suppose a manufacturing plant absorbs 5000 kVA. To guarantee continued service, we install one 3-phase 5000 kVA transformer and

Figure 13-9

Three-phase transformer for an electric arc furnace, rated 36 MVA, 13.8 kV/160 V to 320 V, 60 Hz. The secondary voltage is adjustable from 160 V to 320 V by means of 32 taps on the primary winding (not shown). The three large bus bars in the foreground deliver a current of 65 000 A. Other characteristics: impedance: 3.14%; diameter of each leg of the core: 711 mm; overall height of core: 3500 mm; center line distance between adjacent core legs: 1220 mm. *(Ferranti-Packard)*

keep a second one as a spare. Alternatively, we can install three single-phase transformers each rated 1667 kVA, plus one spare. The 3-phase transformer option is more expensive (total capacity 10 000 kVA) than the single-phase option (total capacity 6667 kVA).

Figure 13-10 shows the different stages of construction of a 3-phase 110 MVA, 222.5 kV/34.5 kV tap-changing transformer.[*] In addition to the three main legs, the magnetic core has two addi-

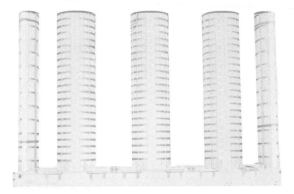

Figure 13-10 a.

Core of a 110 MVA, 222.5 kV/34.5 kV, 60 Hz 3-phase transformer. By staggering laminations of different widths, the core legs can be made almost circular. This reduces the coil diameter to a minimum, resulting in less copper and also lower I^2R losses. The legs are tightly bound to reduce vibration. Mass of the core: 53 560 kg. *(ASEA)*

Figure 13-10 b.

Same transformer with coils in place. The primary windings are connected in wye, and the secondaries in delta. Each primary has 8 taps to change the voltage in steps of ± 2.5%. The motorized tap-changer can be seen in the right upper corner of the transformer. Mass of copper: 15 230 kg. *(ASEA)*

* A tap-changing transformer regulates the secondary voltage by automatically changing from one tap to another on the primary winding. The tap-changer is a motorized device under the control of a sensor that continually monitors the voltage that has to be held constant.

tional lateral legs. They enable the designer to reduce the overall height of the transformer, which simplifies the problem of shipping. In effect, whenever large equipment has to be shipped, the designer is faced with the problem of overhead clearances on highways, rail lines, and so forth.

Figure 13-10 c.

Same transformer ready for shipping. It has been subjected to a 1050 kV impulse test on the HV side and a similar 250 kV test on the LV side. Other details: power rating: 110 MVA/146.7 MVA (OA/FA); total mass including oil: 158.7 t; overall height: 9 m; width: 8.2 m; length: 9.2 m. *(ASFA)*

The 34.5 kV windings (connected in delta) are mounted next to tne core. The HV windings (connected in wye) are mounted on top of the 34.5 kV windings. A space of several centimeters separates the two windings to ensure good isolation and to allow cool oil to flow freely between them. The HV bushings that protrude from the oil-filled tank are connected to a 220 kV line. The medium voltage (MV) bushings are much smaller, and cannot be seen in the photograph.

13-8 Step-up and step-down autotransformer

We can raise or lower the voltage of a 3-phase line by using a 3-phase wye-connected *autotransformer*. The neutral is usually connected to the system neutral, otherwise a tertiary winding must be added to prevent the voltage distortion mentioned previously. The actual physical connections are shown in Fig. 13-11a, and the corresponding schematic diagram is given in Fig. 13-11b. It is obvious that the incoming and outgoing transmission line voltages are in phase.

For a given power output, an autotransformer is smaller and cheaper than a conventional trans-

former (see Sec. 12-2). This is particularly true if the ratio of the incoming line voltage to outgoing line voltage lies between 0.5 and 2.

Figure 13-11c shows a large single-phase autotransformer rated 404 kV/173 kV with a tertiary winding rated 11.9 kV. It is part of a 3-phase bank used to connect a 700 kV transmission line to an existing 300 kV system.

Example 13-4:
The voltage of a 3-phase, 230 kV line has to be stepped up to 345 kV to supply a load of 200 MVA. Three single-phase transformers connected as autotransformers are to be used. Calculate the basic power and voltage rating of each transformer, assuming they are interconnected as shown in Fig. 13-11b.

Solution:
To simplify the calculations, let us consider only one phase (phase A, say).
Line-to-neutral voltage between X_1 and H_2:
$$E_{1N} = 345/\sqrt{3} = 200 \text{ kV}$$
Line-to-neutral voltage between H_1 and H_2:
$$E_{AN} = 230/\sqrt{3} = 133 \text{ kV}$$

Voltage of winding X_1X_2:

E_{1A} = 200 – 133 = 67 kV

Current in each phase of the outgoing line:

I_s = $S/1.73\,E$ Eq. 9-9

   = $(200 \times 10^6)/(1.73 \times 345\,000)$

   = 335 A

Power associated with winding X_1X_2:

S_a = 67 000 × 335 = 22.4 MVA

Winding H_1H_2 must obviously have the same pow-

capacity (67.2 MVA versus 200 MVA). This is in keeping with the fact that the ratio of transformation (345/230 = 1.5) lies between 0.5 and 2.0.

13-9 Phase-shift principle

A 3-phase system enables us to phase-shift voltages and currents very simply. Phase-shifting enables us to create 2-phase, 6-phase and 12-phase systems from an ordinary 3-phase line. Such multi-phase

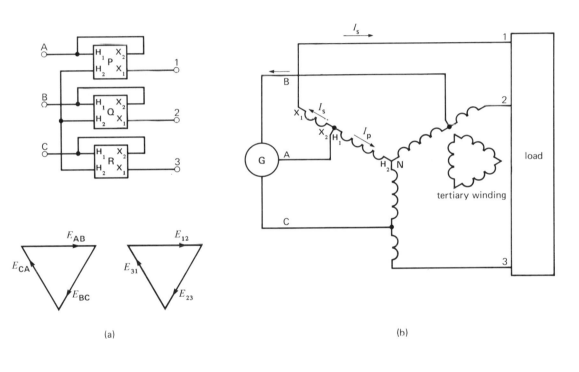

(a)   (b)

Figure 13-11   a. Wye-connected autotransformer.   b. Associated schematic diagram.

er rating. The basic rating of each single-phase transformer is therefore 22.4 MVA.

The basic rating of the 3-phase transformer bank = 22.4 × 3 = 67.2 MVA.

The basic transformer rating (as far as size is concerned) is considerably less than its actual load

systems are used in large electronic converter stations and in special electrical controls. Phase-shifting is also used to control power flow over transmission lines.

To understand the phase-shifting principle, consider a potentiometer connected between phases B

Figure 13-11 c.

Single-phase autotransformer (one of a group of three) connecting a 700 kV, 3-phase, 60 Hz transmission line to an existing 300 kV system. The transformer ratio is 404 kV/173 kV, to give an output of 200/267/333 MVA per transformer, at a temperature rise of 55°C. Cooling is OA/FA/FOA. A tertiary winding rated 35 MVA, 11.9 kV maintains balanced and distorsion-free line-to-neutral voltages, while providing power for the substation. Other properties of this transformer: weight of core and windings: 132 t; tank and accessories: 46 t; oil: 87 t; total weight: 265 t. BIL rating is 1950 kV and 1050 kV on the HV and LV side, respectively. Note the single 700 kV and 300 kV bushings protruding from the tank.

(Hydro-Québec)

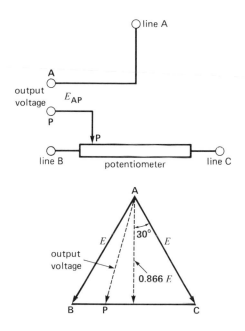

Figure 13-12   Voltage E_{AP} can be phase-shifted with respect to E_{AC} by means of a potentiometer.

and C of a 3-phase line (Fig. 13-12). As we slide contact P from phase B towards phase C, voltage E_{AP} changes both in amplitude and phase. We obtain a 60° phase-shift in moving from one end of the potentiometer to the other. Thus, as we move from B to C, voltage E_{AP} gradually advances in phase with respect to E_{AB}. The magnitude of E_{AP} changes slightly, from E (voltage between the lines) to $0.866\,E$ when the contact is in the middle of the potentiometer.

Such a simple phase-shifter can only be used in circuits where the power output between terminals A and P is in the milliwatt range. If we apply a heavier load, the IR drop in the potentiometer completely changes the voltage and phase-shift from what is was on open-circuit.

To get around this problem, we connect a multi-tap autotransformer between phases B and C (Fig. 13-13). By moving contact P, we obtain the same open-circuit voltages and phase-shifts as before, but they remain essentially unchanged when a load is connected between A and P. Why is this so? The

reason is that the flux in the autotransformer is fixed because E_{BC} is fixed. As a result, the voltage across each turn remains fixed (both in magnitude and phase) whether the autotransformer delivers a current or not.

Figure 13-14 shows 3 tapped autotransformers connected between phases A, B and C. Contacts P_1, P_2, P_3 move in tandem as we switch from one set of taps to the next. This arrangement enables us to create a 3-phase source P_1, P_2, P_3 whose phase angle changes step-wise with respect to source ABC. We can produce a maximum phase-shift of 60° as we move from one end of the auto-transformers to the other. We now discuss some practical applications of the phase-shift principle.

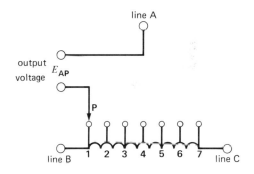

Figure 13-13   Autotransformer used as a phase-shifter.

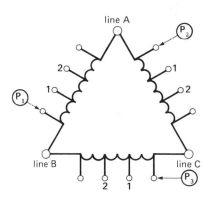

Figure 13-14   Three-phase phase shifter.

13-10 Three-phase to 2-phase transformation

The voltages in a 2-phase system are equal, but displaced from each other by 90°. There are several ways we can create a 2-phase system from a 3-phase source. One of the simplest and cheapest is to use a single-phase autotransformer having taps at 50 percent and 86.6 percent. We connect it between any two phases of a 3-phase line, as shown in Fig. 13-15. If the voltage between phases A, B, C is 100 V, voltages E_{AT} and E_{NC} are both equal to 86.6 V. Furthermore, they are displaced by 90°.

induction motor.

Another way to produce a 2-phase system is to use the so-called Scott connection. It consists of two identical single-phase transformers respectively having a 50% and an 86.6% tap on the primary winding. The transformers are connected as shown in Fig. 13-16. The 3-phase source is connected to lines A, B, C and the 2-phase load is connected to the secondary windings. The ratio of transformation (3-phase line voltage to 2-phase line voltage) is given by a E_{AB}/E_{12}. The Scott connection has the advantage of isolating the 3-phase and 2-phase sys-

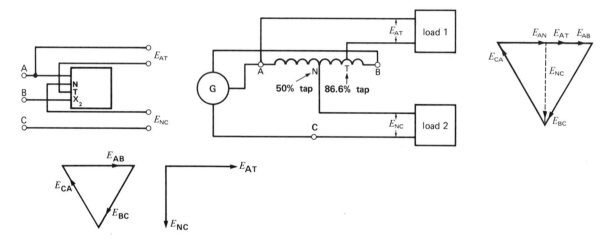

Figure 13-15   Simple method to obtain a 2-phase system from a 3-phase line.

This relationship can be seen by referring to the phasor diagram, and reasoning as follows:

1. Phasors E_{AB}, E_{BC} and E_{CA} are fixed by the source;
2. Phasor E_{AN} is in phase with E_{AB} because the same ac flux links all the turns of the autotransformer;
3. Phasor E_{AT} is in phase with E_{AB} for the same reason;
4. From Kirchhoff's voltage law, $E_{AN} + E_{NC} + E_{CA} = 0$. Consequently, phasor E_{NC} must have the value and direction shown in the figure.

Loads 1 and 2 must be isolated from each other, such as the isolated windings of a 2-phase

tems and providing the desired voltage ratio between them.

Except for servomotor applications, two-phase systems are seldom encountered to-day.

Example 13-5:

A 2-phase, 7.5 kW (10 hp), 240 V, 60 Hz motor has an efficiency of 0.83 and a power factor of 0.80. It is to be fed from a 600 V 3-phase line using a Scott-connected transformer bank. Calculate:

a. the apparent power drawn by the motor;
b. the current in each 2-phase line;
c. the current in each 3-phase line.

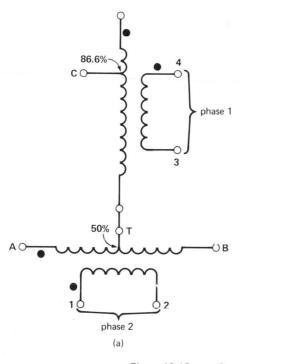

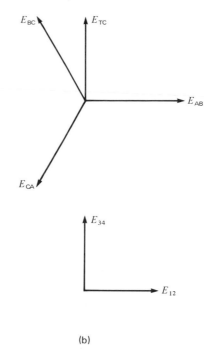

(a)

(b)

Figure 13-16   a. Scott connection.   b. Phasor diagram.

Solution:

a. 1   The active power drawn by the motor is:

$$P_i = P_0/\eta = 7500/0.83$$
$$= 9036 \text{ W}$$

a. 2   Apparent power drawn by the motor:

$$S = P/\cos\theta = 9036/0.8$$
$$= 11\ 295 \text{ VA}$$

b. 1   Apparent power per phase is:

$$S = 11\ 295/2 = 5648 \text{ VA}$$

b. 2   Current in each 2-phase line is:

$$I = S/E = 5648/240$$
$$= 23.5 \text{ A}$$

c. 1   Because the transformer bank itself consumes very little active and reactive power, it follows that the 3-phase line supplies only the active and reactive power absorbed by the motor. The total apparent power drawn from the 3-phase line is therefore 11 295 VA.

c. 2   The 3-phase line current is:

$$I = S/(\sqrt{3}\,E) = 11\ 295/(\sqrt{3} \times 600)$$
$$= 10.9 \text{ A}$$

13-11 Phase-shift transformer

Consider a transmission line connected to the terminals A, B, C of a phase-shift transformer (Fig. 13-17). The transformer twists all the incoming line voltages through an angle α without, however, changing their magnitude. The result is that all the voltages of the outgoing transmission line 1, 2, 3 are shifted with respect to the voltages of the incoming line ABC.

The phase angle is changed in discrete steps by

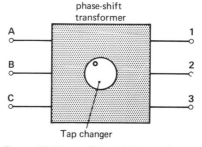

Figure 13-17   a. Phase-shift transformer.

means of a motorized tap-changer. The angle may be leading or lagging, and is usually variable between zero and $\pm 20°$.

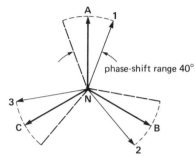

Figure 13-17   b. Phasor diagram showing the range over which the phase angle of the outgoing line can be varied.

The basic power rating of the transformer (which determines its size) depends upon the apparent power carried by the transmission line, and upon the phase-shift. For angles less than $20°$, it is given by the approximate equation:

$$S_T = 0.0175\, S_L\, \alpha_{max} \qquad (13\text{-}1)$$

where

$\qquad S_T =$ basic capacity of the 3-phase transformer bank [VA]

$\qquad S_L =$ apparent power carried by the transmission line [VA]

$\qquad \alpha_{max} =$ maximum transformer phase shift [°]

$\qquad 0.0175 =$ an approximate coefficient [$\approx \pi/180$]

Example 13-6:

A phase-shift transformer is designed to control 150 MVA on a 230 kV 3-phase line. The phase angle is variable between zero and $\pm 15°$.

a. Calculate the approximate basic power rating of the transformer;

b. Calculate the line currents in the incoming and outgoing transmission lines.

Solution:

a. $\quad S_T = 0.0175\, S_L\, \alpha_{max} \qquad$ Eq. 13-1

$\qquad = 0.0175 \times 150 \times 15$

$\qquad = 39$ MVA

b. 1   The line currents are the same in both lines, because the powers and the voltages are the same.

b. 2 $I = S_L/1.73\, E \qquad$ Eq. 9-9

$\qquad = (150 \times 10^6)/(1.73 \times 230\ 000)$

$\qquad = 377$ A

The internal circuit of a tap-changing, phase-shift transformer is fairly complex. However, it rests upon the basic phase-shift principles discussed in Sec. 13-9. The usefulness of such transformers will be covered in Chapter 29.

13-12   Voltage regulation

The voltage regulation of a 3-phase transformer (or transformer bank) may be calculated the same way as for single-phase transformers. In making the calculations, we assume that both the primary and secondary windings are connected in wye, even if they are not (see Sec. 9-11). One phase is then analysed using the respective line-to-neutral voltages and making the nominal power per phase equal to one third of the kVA rating of the 3-phase transformer. The load is similarly treated, and the circuit is solved using the techniques explained in Chapter 11.

Example 13-7:

The 3-phase transformer shown in Fig. 11-50 is rated 1300 MVA, 24.5 kV/345 kV, 60 Hz, impedance 11.5%.

a. Determine the equivalent circuit of this transformer, per phase.

b. Calculate the required generator terminal voltage when the HV line delivers 1200 MVA at 327 kV with a lagging power factor of 0.90.

Solution:

a. 1   First, we note that the primary and secondary winding connections are not specified. We don't need this information. However, we assume that both windings are connected in wye (see Sec. 9-11).

a. 2   Nominal primary voltage is:

$\qquad E_p = 24.5/\sqrt{3} = 14.14$ kV

a. 3 Nominal secondary voltage is:
$$E_s = 345/\sqrt{3} = 199 \text{ kV}$$

a. 4 Ratio of transformation is:
$$a = 199/14.14 = 14.07$$

a. 5 Nominal rating per phase is:
$$S_n = 1300/3 = 433.3 \text{ MVA}$$

a. 6 Nominal impedance per phase (referred to the secondary) is:
$$Z_n = E_n^2/S_n \qquad \text{Eq. 11-17}$$
$$= 199\,000^2/433.3 \times 10^6$$
$$= 91.4 \ \Omega$$

a. 7 Transformer internal impedance (referred to the secondary) is:
$$Z_s = 11.5\% \times 91.4 = 10.5 \ \Omega$$
The impedance is essentially reactive because this is a big transformer.

a. 8 The equivalent circuit is given in Fig. 13-18.

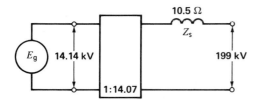

Figure 13-18   See Example 13-7.

b. 1 The line-to-neutral voltage under load is:
$$E_s = 327/\sqrt{3} = 189 \text{ kV}$$

b. 2 The load per phase is:
$$S = 1200/3 = 400 \text{ MVA}$$

b. 3 The current per phase is:
$$I_2 = S/E_s = (400 \times 10^6)/189\,000$$
$$= 2116 \text{ A}$$
This current lags $25.8°$ behind E_s.
$$(\cos 25.8° = 0.9)$$

b. 4 Internal transformer impedance drop is:
$$E_x = I_2 Z_s = 2116 \times 10.5$$
$$= 22.2 \text{ kV}$$

b. 5 Referring to Fig. 13-19 we can calculate the value of E_2:
$$E_2^2 = E_s^2 + E_x^2 - 2E_s E_x \cos (90 + 25.8)$$
$$= 189^2 + 22.2^2$$
$$\qquad - 2 \times 189 \times 22.2 \cos 115.8°$$
$$= 39\,866$$
$$= 199.7 \text{ kV}$$

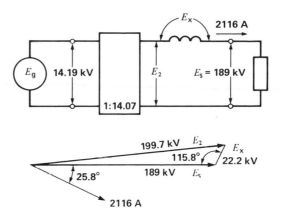

Figure 13-19   See Example 13-7.

b. 6 The corresponding generator line-to-neutral voltage is:
$$E_g = E_2/a = 199.7/14.07 = 14.19 \text{ kV}$$

b. 7 The alternator terminal voltage is therefore:
$$E = \sqrt{3}\,E_g = \sqrt{3} \times 14.19$$
$$= 24.6 \text{ kV}$$

13-13 Polarity marking of 3-phase transformers

The HV terminals of a 3-phase transformer are marked H_1, H_2, H_3 and the LV terminals are marked X_1, X_2, X_3. The following rules have been standardized:

1. If the primary windings and secondary windings are connected wye-wye or delta-delta, the voltages between similarly-marked terminals are in phase. Thus:

$E_{H1, H2}$ is in phase with $E_{X1, X2}$;
$E_{H2, H1}$ is in phase with $E_{X2, X1}$;
$E_{H1, H3}$ is in phase with $E_{X1, X3}$
    . . . and so on . . .

2. If the primary and secondary windings are connected in wye-delta or delta-wye, there results a $30°$ phase shift between the primary and secondary line voltages. The internal connections are made so that the voltages on the HV side *always* lead the voltages of similarly-marked terminals on the LV side. Thus:

$E_{H1, H2}$ leads $E_{X1, X2}$ by $30°$;
$E_{H2, H1}$ leads $E_{X2, X1}$ by $30°$;

$E_{H3, H2}$ leads $E_{X3, X2}$ by $30°$
. . . and so on . . .

3. These rules are not affected by the phase sequence of the line voltage applied to the primary side.

QUESTIONS AND PROBLEMS

Practical level

13-1 Assuming that the transformer terminals have polarity marks H_1, H_2, X_1, X_2, make schematic drawings of the following connections:
   a. delta-wye;   b. open-delta.

13-2 Three single-phase transformers rated at 250 kVA, 7200 V/600 V, 60 Hz, are connected in wye-delta on a 12 470 V 3-phase line. If the load is 450 kVA, calculate the currents:
   a. in the incoming and outgoing transmission lines;
   b. in the primary and secondary windings.

13-3 The transformer in Fig. 13-9 has a rating of 36 MVA, 13.8 kV/320 V. Calculate the nominal currents in the primary and secondary lines.

13-4 Calculate the nominal currents in the primary and secondary windings of the transformer shown in Fig. 11-50, knowing that the windings are connected in delta-wye.

Intermediate level

13-5 The transformer shown in Fig. 11-51 operates in the forced-air mode during the morning peaks.
   a. Calculate the currents in the secondary lines if the primary line voltage is 225 kV and the primary line current is 150 A.
   b. Is the transformer overloaded?

13-6 The transformers in Problem 13-2 are used to raise the voltage of a 3-phase 600 V line to 7.2 kV.
   a. How must they be connected?
   b. Calculate the line currents for a 600 kVA load.
   c. Calculate the corresponding primary and secondary currents.

13-7 In order to meet an emergency, three single-phase transformers rated 100 kVA, 13.2 kV/2.4 kV are connected in wye-delta on a 3-phase 18 kV line.
   a. What is the maximum load that can be connected to the transformer bank?
   b. What is the outgoing line voltage?

13-8 Two transformers rated 250 kVA, 2.4 kV/600 V are connected in open-delta to supply a load of 400 kVA.
   a. Are the transformers overloaded?
   b. What is the maximum load the bank can carry on a continuous basis?

13-9 Referring to Figs. 13-3, and 13-4, the line voltage between phases A-B-C is 6.9 kV and the voltage between lines 1, 2 and 3 are balanced and equal to 600 V. Then, in a similar installation the secondary windings of transformer P are by mistake connected in reverse.
   a. Determine the voltages measured between lines 1-2, 2-3 and 3-1.
   b. Draw the new phasor diagram.

PART V

ROTATING MACHINERY

14

DIRECT
CURRENT
GENERATORS

Direct current generators are not as common as they used to be, because direct current, when required, is mainly produced by electronic rectifiers. These devices can convert the current of an ac system into direct current without using any moving parts. Nevertheless, an understanding of dc generators is important because it represents a logical introduction to the behavior of dc motors. Indeed,

many dc motors in industry actually operate as generators for brief periods.

Commercial dc generators and motors are built the same way; consequently, any dc generator can operate as a motor and vice versa. Direct current machines possess four main components: the field, the armature, the commutator and the brushes (Fig. 14-1).

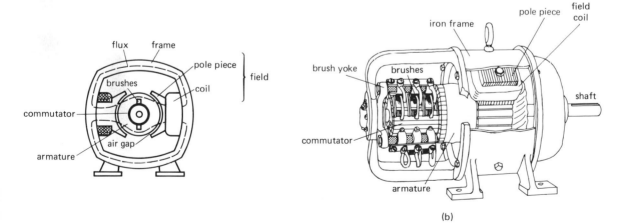

(b)

Figure 14-1 a. Cross section of 2-pole generator.   b. Cutaway view of a 4-pole dc generator.

14-1 Field

The field produces the magnetic flux in the machine. It is basically a stationary electromagnet composed of a set of salient poles bolted to the inside of a circular frame. Field coils, mounted on the poles, carry the dc exciting current. The frame is usually made of solid cast steel, whereas the pole pieces are composed of stacked iron laminations. In some generators, called magnetos, the flux is created by permanent magnets.

The number of poles depends mainly upon the physical size of the machine; the bigger it is, the more poles it will have. By using a multi-pole design, we can reduce the dimensions and cost of large machines, and also improve their performance.

The field coils of a multipole machine are connected together so that adjacent poles have opposite magnetic polarities (Fig. 14-2). The coils are composed of several hundred turns of wire carrying a relatively small current. The coils are well insulated from the pole pieces to prevent short-

magnetic materials having excellent permeability, most of the mmf is available to force the flux across the air gap. Consequently, by reducing its length, we can diminish the size of the field coils.

14-2 Armature

The armature is keyed to a shaft and revolves between the field poles. It is composed of slotted, iron laminations that are stacked to form a solid cylindrical core. The laminations are individually coated with an insulating film so that they do not come in electrical contact with each other. The slots are lined up to provide space to lodge the armature conductors.

The armature conductors carry the current which is delivered by the machine. They are insulated from the iron core by several layers of paper or mica and are firmly held in place by fiber slot sticks. If the armature current is small, we use round wire, but for currents exceeding 20 A, we prefer rectangular conductors because they make better use of the available slot space. The outward

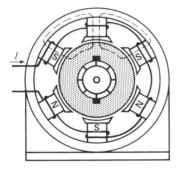

Figure 14-2 Adjacent poles have opposite polarities.

circuits.

The mmf developed by the coils produces a magnetic field that passes through the pole pieces, the frame, the armature and the air gap. The air gap is the short space between the armature and the pole pieces. It ranges from 1.5 to 5 mm as the generator output increases from 1 to 100 kW.

Because the armature and field are composed of

Figure 14-3 a. Armature showing the commutator, stacked laminations, slots and shaft. *(General Electric Company, USA)*

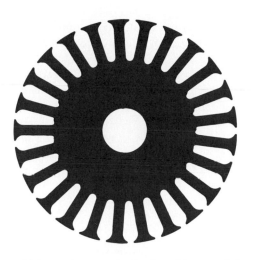

Figure 14-3 b. Armature lamination with tapered slots.

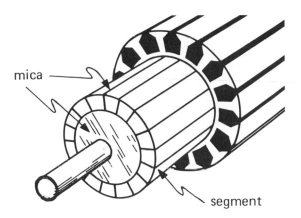

Figure 14-4 Commutator of a dc machine.

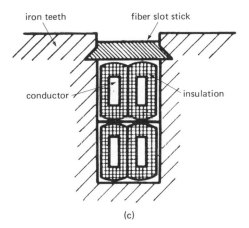

(c)

Figure 14-3 c. Cross section of a slot containing 4 armature conductors.

appearance and general construction of an armature is shown in Fig. 14-3.

14-3 Commutator and brushes

The commutator is composed of an assembly of tapered copper segments insulated from each other by mica sheets, and mounted on the shaft of the machine (Fig. 14-4). The armature conductors are connected to the commutator in a manner we explain below.

Great care is taken in building the commutator because any eccentricity will cause the brushes to bounce, producing unacceptable sparking. The sparks burn the brushes and overheat and carbonize the commutator.

A two-pole generator has two fixed brushes diametrically opposite to each other (Fig. 14-5a). They slide on the commutator and ensure good electrical contact between the revolving armature and the stationary external load.

Multipole machines possess as many brush sets as they have poles. The brush sets are composed of one or more brushes, depending upon the current

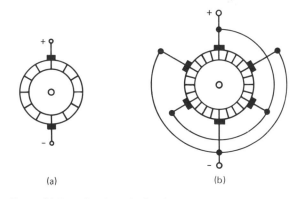

(a)   (b)

Figure 14-5 a. Brushes of a 2-pole generator.

b. Brushes and connections of a 6-pole generator.

that has to be carried. In Fig. 14-1b, for example, each brush set has three brushes. The brush sets are spaced at equal intervals around the commutator. They are supported by a movable brush yoke that permits the entire brush assembly to be shifted and then locked in the desired position. As we go around the commutator, the successive brush sets have positive and negative polarities. Brushes having the same polarity are connected together and the leads are brought out to one positive and one negative terminal (Fig. 14-5b).

The brushes are made of carbon because it has good electrical conductivity and is soft enough to not score the commutator. To improve the conductivity, we sometimes add a small amount of copper to the carbon. The brush pressure may be set by means of adjustable springs (Fig. 14-6). If

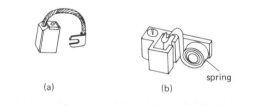

(a)   (b)   spring

(c)

Figure 14-6   a. Carbon brush.
b. Brush holder and spring.
c. Brush assembly mounted on rocker arm.
(General Electric Company, USA)

the pressure is too great, the friction produces excessive heating of the commutator and brushes; on the other hand, if it is too weak, the imperfect contact may produce sparking. The pressure is usually about 15 kPa ($\approx$ 2 lb/in^2) and the current density approximately 10 A/cm^2 ($\approx$ 65 A/in^2). Thus, a typical brush having a cross section of 3 cm x 1 cm ($\approx$ 1.2 in x 0.4 in) exerts a pressure of 4.5 N ($\approx$ 1 lb) and carries a current of 30 A.

Figure 14-7 shows the construction of a 4-pole dc generator.

14-4   Lap winding

The armature coils may be connected to the commutator in several different ways; one of the most common is known as the *lap winding*. In this winding, the coils are effectively connected in series so that the voltages induced under the N and S poles add up.

Figure 14-8 shows a lap winding composed of four coils, A, B, C, D, placed in four armature slots and connected to the four segments of a commutator. Coil A, for example, is connected to segments a and b. For reasons of mechanical symmetry, one side of the coil is placed in the upper part of the slot 1, whereas the other coil-side is placed in the bottom of slot 3. Each slot contains the conductors of two different coils. Thus, slot 1 contains the conductors belonging to coil A and to coil C. The schematic diagram of Fig. 14-8b shows that the coils are effectively connected in series and form a closed loop around the armature. The position of the coils corresponds to that in Fig. 14-8a. Thus, coil A is moving across the space between the pole tips, while coil B is sweeping across the center of the poles.

We usually employ much more than four coils on an armature.* Figure 14-9 shows a winding composed of 12 coils, placed in 12 slots, and connected to a commutator having 12 segments. Some of the coils are labelled with two numbers, corresponding to the slots into which the coil-sides are placed (Fig. 14-10). Thus, coil A is placed in slots 1 and 7.

*   See Section 2-15.

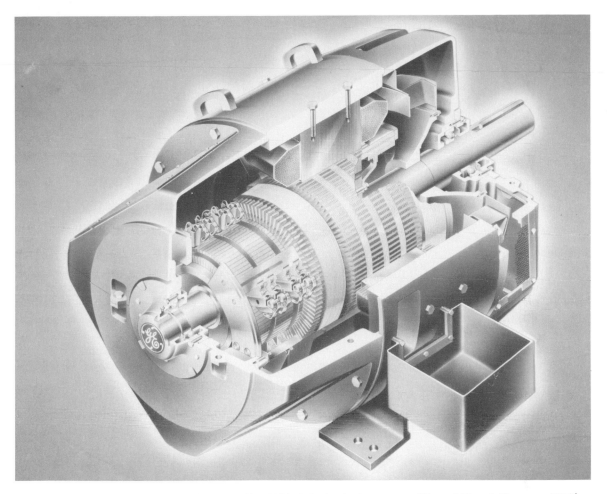

Figure 14-7   Sectional view of a 100 kW, 250 V, 1750 r/min 4-pole dc generator. *(General Electric Company, USA)*

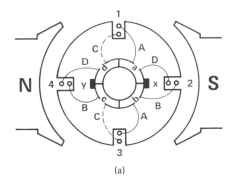

(a)

Figure 14-8   a.   Lap winding composed of 4 coils.

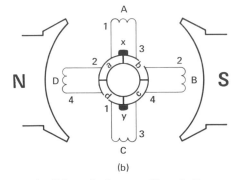

(b)

b.   Schematic diagram of lap winding.

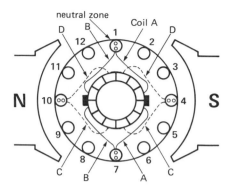

Figure 14-9 Lap winding composed of 12 coils.

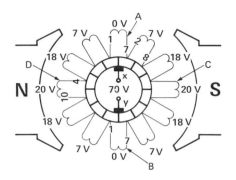

Figure 14-10 Schematic diagram of lap winding

14-5 Induced voltage

When the armature rotates, the voltage E induced in each conductor depends upon the flux density which it cuts, expressed by

$$E = Blv \qquad \text{Eq. 2-2}$$

Because the density in the air gap varies from point to point, the induced voltage per conductor depends upon its instantaneous position. It follows that the voltage induced in each *coil* depends upon *its* instantaneous position. Consider, for example, the voltages induced in the armature when it occupies the position shown in Fig. 14-9. The conductors in slots 1 and 7 are exactly between the poles, where the flux density is zero. The voltage induced in the two coils lodged in slots 1 and 7 is therefore zero. On the other hand, the conductors in slots 4 and 10 are directly under the center of the poles, where the flux density is greatest. The voltage induced in the two coils lodged in *these* slots in therefore maximum. Finally, owing to magnetic symmetry, the voltage induced in the coils lodged in slots 3 and 9 is the same as that induced in the coils lodged in slots 5 and 11.

Figure 14-10 is a schematic diagram showing the instantaneous voltages induced in each of the 12 coils of the armature. They are 0-7-18 and 20 V, respectively. Note that the brushes straddle the coils in which the voltage is momentarily zero.

Taking polarities into account, we can see that the voltage between the two brushes is (7 + 18 + 20 + 18 + 7) = 70 V, and brush x is positive with respect to brush y. This voltage remains essentially constant as the armature rotates, because the number of coils between the brushes is always the same, irrespective of armature position.

If we were to shift the brush yoke by 90°, the voltage between the brushes would become (+ 18 + 7 + 0 – 7 – 18) = 0 V. Furthermore, in this position, the brushes would continually short-circuit coils that generate 20 V. Large short-circuit currents would flow in the shorted coils and brushes, and sparking would result.

14-6 Neutral zones

Neutral zones are those places on the surface of the armature where the flux density is zero. When the generator operates at no load, the neutral zones are located exactly between the poles. No voltage is induced in a coil that cuts through the neutral zone. We always try to set the brushes so that they are in contact with coils that are momentarily in a neutral zone. If we shift the brushes to another position, they short-circuit coils in which the induced voltage is not zero. Heavy short-circuit currents result, producing sparking underneath the brushes.

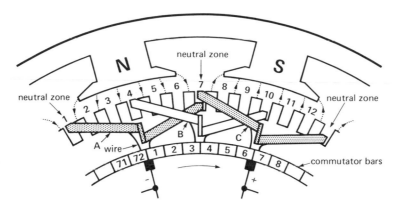

Figure 14-11 Partial view of a 12-pole generator.

14-7 Multipole generators

Multipole generators have the same type of lap winding on the armature as two-pole machines have. The coil width (or coil pitch) must be such that each coil-side cuts the flux coming respectively from a N and a S pole. Figure 14-11 shows a partial view of a 12-pole machine that has 72 slots on the armature, 72 segments on the commutator, and 72 coils.* Only three coils A, B, C are shown so as not to confuse the diagram. Coil A has its coil-sides in slots 1 and 7, while those of coil B are in slots 4 and 10. Furthermore, coil A is connected to commutator segments 72 and 1, while coil B is connected to segments 3 and 4.

In the position shown, the coil-sides of coil A are in the neutral zone between the poles. Consequently, no voltage is induced in coil A. On the other hand, the coil-sides of B are directly under the N and S poles. The voltage in coil B is maximum at this moment. Consequently, the voltage between adjacent commutator segments 3 and 4 is maximum.

The voltage in coil C is also zero because its coil-sides are sweeping across the neutral zone. Note that the positive and negative brushes both straddle coils having zero voltage.

* In practice, a 12-pole machine has more coils and commutator bars than given in this example.

The voltage between the brushes is equal to the sum of the voltages induced in the five coils connected to segments 1-2, 2-3, 3-4, 4-5, and 5-6. The machine has six positive and six negative brush sets (Fig. 14-12). The positive brushes are connected together and the negative brushes are connected together to yield one positive and one negative terminal.

Figure 14-12 shows schematically how all the coils are connected together and how they are joined to the commutator bars. Coils A, B, and C mentioned previously are identified on the diagram. It is obvious that the voltage between any two brush sets is equal to the voltages induced by an adjacent pair of N, S poles.

Example 14-1:
The generator in Fig. 14-12 generates 240 V between adjacent brushes and delivers a current of 2400 A to the load. Calculate:
a. the current delivered per brush-set;
b. the current flowing in each coil;
c. the average voltage induced per coil.

Solution:
a. 1 A current of 2400 A flows out of the (+) terminal and back into the (–) terminal of the generator.
a. 2 The current per brush set is
$$I = 2400/6 = 400 \text{ A}$$

b.   Each positive brush set gathers current from the windings to the right and left of the brush. Consequently, the current per coil is 400/2 = 200 A.

c.   There are six coils between adjacent brush sets. The average voltage per coil is
$$E_{avge} = 240/6 = 40 \text{ V}$$

14-8 Value of the induced voltage

The voltage induced in a dc generator is given by:

$$\boxed{E_0 = Zn\Phi/60} \qquad (14\text{-}1)$$

where

Z = total number of conductors on the armature

n = speed rotation [r/min]

Φ = flux per pole [Wb]

This important equation shows that for a given generator, the voltage is directly proportional to the flux per pole and to the speed of rotation.

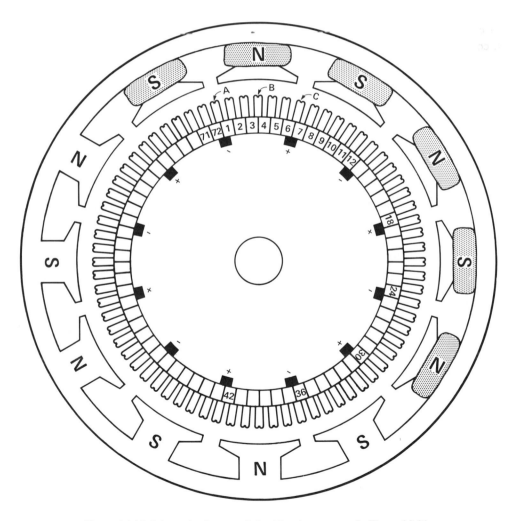

Figure 14-12 Schematic diagram of the 12-pole generator in Figure 14-11.

Example 14-2:

The armature of a six-pole, 600 r/min generator, has 90 slots. If each coil has 4 turns and the flux per pole is 0.04 Wb, calculate the value of the induced voltage.

Solution:

Each turn has two active conductors, and 90 coils are required to fill the 90 slots. The total number of armature conductors is:

$Z = 90 \times 4 \times 2 = 720$

also

$n = 600$

Consequently,

$E_0 = Zn\Phi/60 = 720 \times 600 \times 0.04/60$
$\quad = 288$ V

The voltage between the brushes at no load is therefore 288 V.

14-9 Separately excited generator

When the dc field current in a generator is supplied by an independent source (such as a storage cell or another generator, called *exciter*), the generator is said to be separately excited. Thus, in Fig. 14-13, the dc source connected to terminals a and b causes exciting current I_x to flow. If the armature is driven by a turbine or other machine, a voltage E_0 appears between brush terminals x and y.

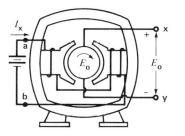

Figure 14-13 Separately excited dc generator.

14-10 No-load operation

When a separately excited dc generator runs at no

load (armature circuit open), a change in either the exciting current or the speed of rotation causes a corresponding change in the induced voltage.

Induced voltage vs exciting current. As we raise the exciting current I_x, the mmf of the field increases, which increases the flux Φ per pole. By plotting Φ as a function of I_x, we obtain the *saturation curve* of Fig. 14-14a. When the exciting current is relatively small, the flux is small, and the iron in the machine is unsaturated. Very little mmf

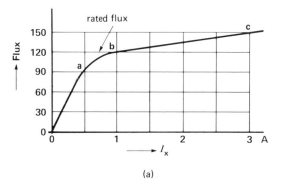

Figure 14-14 a. Flux per pole versus exciting current.

is needed to force the flux through the iron, with the result that the mmf developed by the field coils is almost entirely available to force flux through the air-gap. Because the permeability of air is constant, the flux increases in direct proportion to the exciting current, as shown by the portion **oa** of the saturation curve. However, as we continue to raise the exciting current, the iron in the field and armature begins to saturate. A large mmf is now required to produce a small increase in flux, as shown by portion **bc** of the curve. The machine is now said to be saturated. Saturation of the iron begins to be important when we reach the so-called "knee" **ab** of the saturation curve.

How does the saturation curve relate to the induced voltage E_0? If we drive the generator at constant speed, E_0 is directly proportional to the flux Φ. Consequently, by plotting E_0 as a function of I_x, we obtain a curve whose shape is identical to the saturation curve of Fig. 14-14a. The result is shown in Fig. 14-14b; it is called the no-load saturation curve of the generator.

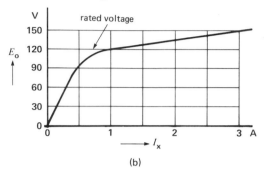

(b)

Figure 14-14   b. Induced voltage versus exciting current.

The nominal voltage of a generator is usually a little above the knee of the curve. Thus, in Fig. 14-14b, the nominal voltage is about 125 V. By varying the exciting current, we can vary the induced voltage as we please. Furthermore, by reversing the current, the flux reverses and so, too, will the polarity of the induced voltage.

Induced voltage vs speed. For a given exciting current, the induced voltage increases in direct proportion to the speed, a result that follows from Eq. 14-1.

If we reverse the direction of rotation, the polarity of the induced voltage also reverses. However, if we reverse both the exciting current *and* the direction of rotation, the polarity of the induced voltage remains the same.

14-11   Shunt generator

A shunt-excited generator is a machine whose field coils are directly connected to the armature terminals, so that the generator is self-excited (Fig. 14-15). The principal advantage of this connection is that it eliminates the need for an external source of excitation.

How is self-excitation achieved? When a shunt generator is started up, a small voltage is induced owing to the remanent flux in the poles. This voltage produces a small exciting current I_x in the shunt field. The resulting small mmf acts in the

same direction as the remanent flux, causing the flux per pole to increase. The increased flux in-

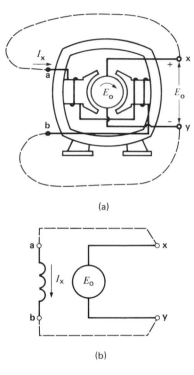

(a)

(b)

Figure 14-15   a. Self-excited shunt generator.
b. Schematic diagram.

creases E_0 which increases I_x ... and this progressive build up continues until E_0 reaches a maximum value determined by the field resistance and degree of saturation.

14-12   Controlling the voltage

It is easy to control the induced voltage of a shunt-excited generator. We simply vary the exciting current by means of a rheostat connected in series with the field (Fig. 14-16).

To understand how the output voltage varies, suppose that E_0 is 120 V when the movable brush **p** is in the center of the rheostat. If we move the brush towards extremity m, the resistance between points a and b diminishes, which causes the exciting current to increase. This increases the flux and, consequently, the induced voltage E_0. On the

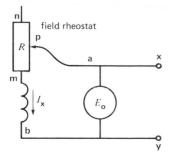

Figure 14-16   Controlling the generator voltage with a field rheostat.

other hand, if we move the brush towards extremity n, the resistance increases, the exciting current diminishes, the flux diminishes, and E_0 will fall.

We can easily determine the value of E_0 if we know the saturation curve of the generator and the total resistance R_t of the rheostat and field. We simply draw a straight line corresponding to the resistance R_t, superimposing it on the saturation curve. The point where the line intersects the curve, corresponds to the induced voltage. For example, if the field has a resistance of 50 Ω, and the rheostat is set to zero, then $R_t = 50 \Omega$. The straight line corresponding to R_t must pass through the point $E = 50$ V, $I = 1$ A. This line intersects the saturation curve at a point corresponding to a voltage of 150 V (Fig. 14-17). This

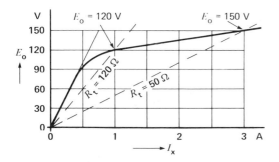

Figure 14-17   The no-load voltage depends upon the rheostat setting.

is the maximum voltage which the shunt generator can produce. By changing the position of the rheostat, the total resistance of the field circuit increases. Thus, when R_t is 120 Ω, we obtain a line which cuts the curve at a voltage E_0 of 120 V.

14-13   Equivalent circuit

The armature winding contains a set of identical coils, all of which possess a certain resistance. The total armature resistance R_0 is that which exists between the armature terminals when the machine is stationary. The resistance is usually very small, often less than one hundredth of an ohm. Its value depends mainly upon the power and voltage of the generator. To simplify the generator circuit, we can represent R_0 as if it were in series with one of the brushes. The equivalent circuit of a generator is thus composed of a resistance R_0 in series with a voltage E_0 (Fig. 14-18). The latter is the voltage in-

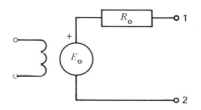

Figure 14-18 Equivalent circuit of a dc generator.

duced in the revolving conductors. Terminals 1, 2 are the external terminals of the machine.

14-14   Generator under load

Let us assume that a separately excited generator runs at constant speed and fixed excitation. Induced voltage E_0 is therefore fixed. When the machine operates at no-load, terminal voltage E_{12} is equal to the induced voltage E_0 because the voltage drop in the armature resistance is zero. However, if we connect a load across the armature (Fig. 14-19), the resulting load current I produces a voltage drop across resistance R_0. Terminal voltage

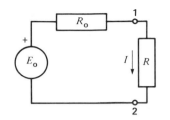

Figure 14-19 Generator under load.

E_{12} is now less than the induced voltage E_0. As we increase the load, the terminal voltage diminishes progressively, as shown in Fig. 14-20. The graph of terminal voltage as a function of load current is called the *load curve* of the generator.

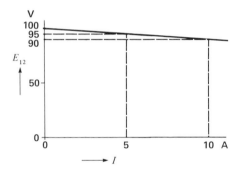

Figure 14-20   Generator load characteristic.

The terminal voltage of a self-excited shunt generator falls off more sharply with increasing load than that of a separately excited generator. The reason is that E_0 in a separately excited machine remains essentially constant. This is not so for a self-excited generator, because its exciting current falls as the terminal voltage drops. For a self-excited generator, the voltage drop from no load to full load is about 15 percent of the no-load voltage, whereas for a separately excited generator it is usually less than 10 percent.

In addition to armature resistance, another phenomenon known as *armature reaction* also causes the terminal voltage to fall. In effect, when current flows in the armature, it creates a mmf that tends to reduce the field flux. The flux reduction produces a corresponding drop in the induced voltage. This phenomenon will be studied in greater detail in Chapter 15.

14-15 Compound generator

In some applications, we can tolerate a reasonable drop in terminal voltage as the load increases, but this is unacceptable in lighting circuits. For example, the distribution system of a ship supplies power to both dc machinery and incandescent lamps. The current delivered by the generator fluctuates continually, in response to the varying loads. These current variations produce corresponding changes in the generator terminal volt-

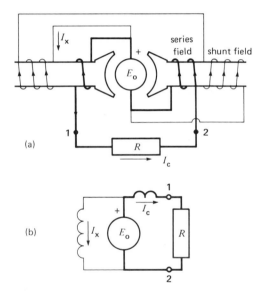

Figure 14-21   a. Compound generator under load.
b. Schematic diagram.

age, causing the lights to flicker. To eliminate such voltage fluctuations, we use *compound generators*.

A compound generator (Fig. 14-21a) is similar to a shunt generator, except that it has additional exciting coils connected in series with the armature. These *series field* coils are composed of a few turns of heavy wire, big enough to carry the armature current. The total resistance of the series coils is therefore small. Figure 14-21b is a schematic diagram of the shunt and series field connections. When the generator runs at no load, the current in the series coils is zero. The shunt coils, however, carry exciting current I_x which produces the field flux, as in a standard self-excited shunt generator. As we load the generator, the terminal voltage tends to drop, but load current I_c now flows through the series field coils. The mmf developed by these coils acts in the same direction as the mmf of the shunt field. Consequently, the field flux under load rises above its original no-load value, which raises the value of E_0. If the series coils are properly designed, the terminal voltage remains practically constant with increasing load. In effect, the rise in the induced voltage compensates for the armature IR drop.

In some cases, we have to compensate not only for the armature voltage drop, but also for the IR drop in the line between the generator and load. The generator manufacturer then places a few ad-ditional turns on the series windings so that the terminal voltage increases as the load current rises. Such machines are called *over-compound generators*. If the compounding is too strong, a low resistance can be placed in parallel with the series field. This reduces the current in the field and has the same effect as reducing the number of turns. For example, if the *diverter* resistance is equal to that of the series field, the current in the latter is reduced by half.

14-16 Differential compound generator

If the mmf of the series field acts opposite to the shunt field, the terminal voltage falls drastically with increasing load. We can make such a *differential compound generator* by simply reversing the series field of a standard compound generator. Differential compound generators were formerly used in dc arc welders, because they tended to limit the short-circuit current and to stabilize the arc during the welding process.

Figure 14-22

This direct-current Thomson generator was first installed in 1889, to light the streets of Montreal. It delivered a current of 250 A at a voltage of 110 V. Other properties of this pioneering machine:

Speed	1300 r/min
Total weight	2390 kg
Armature diameter	292 mm
Stator internal diameter	330 mm
Number of commutator bars	76
Armature conductor size	# 4
Shunt field conductor size	#14

A modern generator having the same power and speed weighs 7 times less and occupies only 1/3 the floor space.

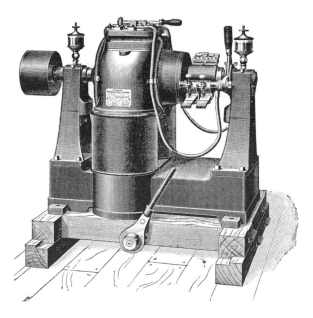

14-17   Load characteristics

The load characteristics of some shunt and compound generators are given in Fig. 14-23. The voltage of an over-compound generator increases by 10 percent when full load is applied, whereas that of a flat-compound generator remains constant. On the other hand, the full-load voltage of a shunt generator is 15 percent below its no-load value, while that of a differential-compound generator is 30 percent lower.

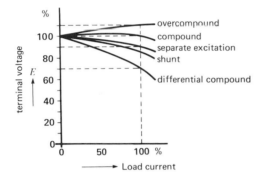

Figure 14-23 Typical load characteristics of dc generators.

14-18   Mechanical forces under load

The load current delivered by a generator flows through all the armature conductors. If we could look inside the machine, we would discover that current always flows in the same direction in those conductors that are momentarily under a N pole.

The same is true for conductors that are momentarily under a S pole. However, the currents under the N pole flow opposite to those under a S pole.

Because all the conductors lie in a magnetic field, they all experience a force, according to Lorentz's law. If we take into account the direction of current flow and the direction of flux, we find that the forces on the conductors all act in the same sense. In effect, they produce a torque which acts opposite to the direction in which the generator is driven. To keep the machine going, we must exert a torque on the shaft to overcome this opposing electromagnetic torque. The resulting mechanical power is converted into electrical power, which is delivered to the generator load.

The preceding explanation becomes clearer if we refer to Fig. 14-24, which shows a portion of a 12-pole dc generator. The generator is being driven counterclockwise by a motor or engine of some kind. The armature conductors under the S pole carry currents that flow *into* the page, away from the reader. Conversely, the armature currents under the N pole flow *out* of the page, towards the reader. The force on every conductor acts therefore towards the right, producing a net clockwise torque. This braking torque acts opposite to the direction of rotation.

14-19   Generator specifications

The nameplate of a generator indicates the power, voltage, speed, and so forth, of the machine. These specifications, or *nominal characteristics*, are the

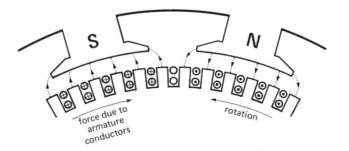

Figure 14-24 A generator under load exerts a braking torque.

values guaranteed by the manufacturer. Consider, for example, the information punched on the nameplate of a 100 kW generator:

Power	100 kW	Speed	1200 r/min
Voltage	250 V	Type	Compound
Exciting current	20 A	Class	B
Temperature rise	50°C		

These specifications tell us that the machine can deliver, continuously, a power of 100 kW at a voltage of 250 V, without exceeding a temperature rise of 50°C. It can therefore supply a load current of 100 000/250 = 400 A. It possesses a series winding, and the current in the shunt field is 20 A. In practice, the voltage is adjusted to a value close to 250 V, and we may draw any amount of power, as long as it does not exceed 100 kW. The class B designation refers to the class of insulation used in the machine (see Sec. 4-7).

QUESTIONS AND PROBLEMS

Practical level

14-1 Sketch the main components of a dc generator.

14-2 Why are the brushes of a dc machine always placed at the neutral points?

14-3 Describe the construction of a commutator.

14-4 How is the induced voltage of a separately excited dc generator affected if:
   a. the speed increases;
   b. the exciting current is reduced?

14-5 How do we adjust the voltage of a shunt generator?

14-6 The terminal voltage of a shunt generator decreases with increasing load. Explain.

14-7 Explain why the output voltage of a compound generator increases as the load increases.

14-8 Explain the difference between shunt, compound and differential compound generators:
   a. as to construction;
   b. as to electrical properties.

Intermediate level

14-9 A separately excited dc generator turning at 1400 r/min produces an induced voltage of 127 V. The armature resistance is 2 Ω and the machine delivers a current of 12 A. Calculate:
   a. the terminal voltage;
   b. the heat dissipated in the armature;
   c. the braking torque exerted by the armature.

14-10 A separately excited dc generator produces a no-load voltage of 115 V. What happens if:
   a. the speed is increased by 20 percent?
   b. the direction of rotation is reversed?
   c. the exciting current is increased by 10 percent?
   d. the polarity of the field is reversed?

14-11 Each pole of a 100 kW, 250 V compound generator has a shunt field of 2000 turns and a series field of 7 turns. If the total shunt-field resistance is 100 Ω, calculate the mmf when the machine operates:
   a. at no-load;
   b. at full-load.

14-12 Figure 14-14b shows the no-load saturation curve of a separately excited dc generator when it revolves at 1500 r/min. Calculate the exciting current needed to generate 120 V at 1330 r/min.

14-13 Referring to Fig. 14-8, the induced voltage in coil D is momentarily 18 V, in the position shown. Calculate the voltages induced in coils A, B, and C at the same instant.

14-14 Referring to Fig. 14-10, calculate the voltage induced in coil A, when the armature has rotated by 90°; by 120°.

14-15 Brush **x** is positive with respect to brush **y** in Fig. 14-10. Show the polarity of each of the 12 coils. Does the polarity reverse when a coil turns through $180°$?

14-16 The generator of Fig. 14-11 revolves at 960 r/min and the flux per pole is 20 mWb. Calculate the no-load armature voltage if each armature coil has 6 turns.

14-17 a. How many brush sets are needed for the generator in Fig. 14-11?

b. If the machine delivers a load current of 1800 A, calculate the current flowing in each armature coil.

Advanced level

14-18 Draw a simple diagram of the armature circuit (similar to that of Fig. 14-10) for the generator shown in Fig. 14-11.

14-19 Referring to Fig. 14-8, determine the polarity of E_{xy} when the armature turns counterclockwise.

14-20 a. In Fig. 14-11, determine the polarity of E_{34} between commutator segments 3 and 4, knowing that the armature is turning clockwise.

b. At the same instant, what is the polarity of segment 35 with respect to segment 34?

14-21 The armature shown in Fig. 15-4 has 81 slots and the commutator has 243 segments. It will be wound to give a 6-pole lap winding having 1 turn per coil. If the flux per field pole is 30 mWb, calculate:

a. the induced voltage at a speed of 1200 r/min;

b. the average flux density per pole;

c. the time needed to reverse the current in each armature coil, knowing that the brushes are 15 mm wide and that the diameter of the commutator is 450 mm.

14-22 If we want to induce a voltage in a straight conductor by using the earth's magnetic field, how should the conductor be oriented, and in what direction should it be displaced?

Calculate the voltage induced in such a conductor if it is 1 m long and moves at 120 km/h through a field of 50 μT.

15

DIRECT CURRENT MOTORS

Direct current motors transform electrical energy into mechanical energy. They drive machines and devices such as hoists, fans, pumps, calenders, punch-presses, and cars. These devices may have a definite torque-speed characteristic (such as a pump or fan) or a highly variable one (such as a hoist or car). The torque-speed characteristic of the motor must be adapted to the type of load it has to drive, and this requirement has given rise to three basic types of motors:

> 1. shunt motors,
> 2. series motors,
> 3. compound motors.

Direct current motors are seldom used in ordinary industrial applications because all electric utility systems furnish alternating current. However, for special applications such as in steel mills, mines and electric trains, it is advantageous to transform the alternating current into direct current in order to use dc motors. The reason is that the torque-speed characteristics of dc motors can be varied over a wide range while retaining high efficiency.

15-1 Counter-emf

Direct current motors are built the same way as generators are; consequently, a dc machine can operate either as a motor or as a generator. To illustrate the point, consider a dc generator in which the armature, initially at rest, may be connected to a dc source E_s by means of a switch (Fig. 15-1). The armature has a resistance R, and the field is created by a set of permanent magnets.

As soon as the switch is closed, a large current flows in the armature because its resistance is very low. The individual armature conductors are immediately subjected to a force because they are

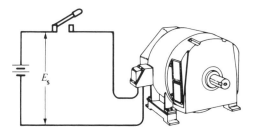

Figure 15-1   Starting a dc motor.

immersed in the magnetic field created by the permanent magnets. These forces add up to produce a powerful torque, causing the armature to rotate.

On the other hand, as soon as the armature begins to turn, a second phenomenon takes place: the generator effect. We know that a voltage E_0 is induced in the armature conductors as soon as they cut a magnetic field. This is always true, *no matter what causes the rotation*. The value and polarity of the induced voltage are the same as those obtained when the machine operates as a generator. The induced voltage E_0 is therefore proportional to the speed of rotation n of the motor and to the flux Φ per pole, as already given by Eq. 14-1:

$$E_0 = Zn\Phi/60 \qquad \text{Eq. 14-1}$$

As in the case of a generator, Z is a constant which depends upon the number of turns on the armature and the type of winding. For lap windings, Z is equal to the number of armature conductors.

The induced voltage E_0 is called *counter-electromotive force* (cemf) because its polarity usually acts "against" the source voltage E_s. It acts against the voltage in the sense that the net voltage acting in the series circuit of Fig. 15-2 is equal to $(E_s - E_0)$ volts and not $(E_s + E_0)$ volts.

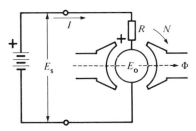

Figure 15-2   Counter electromotive force (cemf) in a dc motor.

15-2 Acceleration of the motor

The net voltage acting in the armature circuit in Fig. 15-2 is $(E_s - E_0)$ volts. The resulting armature current I is limited only by the armature resistance

R, and so

$$I = (E_s - E_0)/R \qquad (15-1)$$

When the motor is at rest, the induced voltage $E_0 = 0$, and the above equation becomes

$$I = E_s/R$$

This means that at the moment we close the switch, the current is very high because the armature resistance is always low. The starting current may be 20 to 30 times greater than the nominal full-load current of the motor. The large forces acting on the armature conductors produce a powerful starting torque and a consequent rapid acceleration of the armature.

As the speed increases, the counter-emf E_0 increases, with the result that the value of $(E_s - E_0)$ diminishes. It follows from Eq. 15-1 that the armature current I drops progressively as the speed increases.

Although the armature current decreases, the motor continues to accelerate until it reaches a definite, maximum speed. At no-load, this speed produces a counter-emf E_0 slightly less than the source voltage E_s. In effect, if E_0 were equal to E_s, the net voltage $(E_s - E_0)$ would become zero and so, too, would the current I. The driving forces would cease to act on the armature conductors, and the mechanical drag produced by the fan and the bearings would immediately cause the motor to slow down. As the speed decreases the net voltage $(E_s - E_0)$ increases, and so does the current I. The speed will cease to fall as soon as the torque developed by the armature current is equal to the dragging torque. Thus, when a motor runs at no-load, the counter-emf must be slightly less than E_s so as to enable a small current to flow, sufficient to produce the required torque.

Example 15-1:
The armature of a permanent-magnet dc generator has a resistance of 1 Ω and generates a voltage of 50 V when the speed is 500 r/min. If the armature is connected to a source of 150 V, calculate:
a. the starting current;
b. the counter-emf when the motor runs at 1000 r/min? at 1460 r/min?

c. the armature current at 1000 r/min?
   at 1460 r/min?

Solution:

a.   At the moment of start-up, the armature is stationary, so E_0 = 0 V (Fig. 15-3a). The initial starting current is limited only by the armature resistance:

   I = 150 V/1 Ω = 150 A

simple equations that enable us to calculate them:

1. According to Eq. 14-1, the cemf induced in a lap-wound armature is

$$E_0 = Zn\Phi/60 \qquad \text{Eq. 14-1}$$

The electrical power P_a supplied to the armature is equal to the supply voltage E_s multiplied by the armature current I:

$$P_a = E_s I \qquad (15\text{-}2)$$

However, E_s is equal to the sum of E_0 plus the

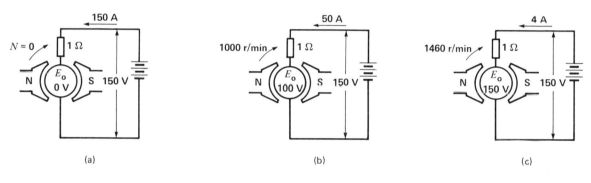

Figure 15-3   See Example 15-1.

b.   Because the generator voltage is 50 V at 500 r/min, the cemf of the motor will be 100 V at 1000 r/min   —   and 146 V at 1460 r/min.

c. 1   The net voltage in the armature circuit at 1000 r/min is:

   $E_s - E_0$ = 150 – 100 = 50 V

   The corresponding armature current is:

   I = $(E_s - E_0)/R$ = 50/1 = 50 A (Fig. 15-3b)

c.2   If the motor speed rises to 1460 r/min, the cemf will be 146 V, almost equal to the source voltage. Under these conditions, the armature current is only 4 A and the motor torque is much smaller than before (Fig. 15-3c).

15-3 Mechanical power and torque

The power and torque of a dc motor are two of its most important properties. We now derive two

IR drop in the armature:

$$E_s = E_0 + IR \qquad (15\text{-}3)$$

It follows that

$$P_a = E_s I \qquad (15\text{-}4)$$
$$= (E_0 + IR)\,I$$
$$= E_0 I + I^2 R \qquad (15\text{-}4)$$

The I^2R term represents heat dissipated in the armature, but the extremely important $E_0 I$ term is the electrical power that is converted into mechanical power. The mechanical power of the motor is therefore exactly equal to the product of the cemf multiplied by the armature current:

$$\boxed{P = E_0 I} \qquad (15\text{-}5)$$

where

P = mechanical power of the motor [W]
E_0 = induced voltage in the armature (cemf) [V]
I = total current supplied to the armature [A]

2. Turning our attention to torque T, we know that mechanical power P is given by the expression

$$P = nT/9.55 \qquad \text{Eq. 3-5}$$

where n is the speed of rotation.

Combining Eqs. 3-5, 14-1 and 15-5, we obtain

$$nT/9.55 = E_0I$$
$$= Zn\Phi I/60$$

Whence $T = Z\Phi I/6.28$

The torque developed by a lap-wound motor is therefore given by the expression

$$\boxed{T = Z\Phi I/6.28} \qquad (15\text{-}6)$$

where

T = torque [N·m]

Z = number of conductors on the armature

Φ = effective flux per pole [Wb] *

I = armature current [A]

6.28 = constant, to take care of units [exact value = 2π]

* The effective flux is given by $\Phi = 60\,E_0/Zn$.

Equation 15-6 shows that we can double the torque of a motor either by doubling the current in the armature or by doubling the flux created by the poles.

Example 15.2:

The following details are given on a 225 kW ($\approx$ 300 hp), 250 V, 1200 r/min dc motor (see Fig. 15-4 and 15-5):

armature coils	243
turns per coil	1
type of winding	lap
armature slots	81
commutator segments	243
field poles	6
diameter of armature	559 mm
axial length of armature	235 mm

Calculate:

a. the armature current;

b. the number of conductors per slot;

c. the flux per pole.

Solution:

a. 1 We can assume that the induced voltage E_0 is nearly equal to the applied voltage (250 V).

Figure 15-4   Bare armature and commutator of a dc motor rated 225 kW, 250 V, 1200 r/min. The armature core has a diameter of 559 mm and an axial length of 235 mm. It is composed of 400 stacked laminations 0.56 mm thick. The armature has 81 slots and the commutator has 243 bars. *(H. Roberge)*

a. 2 The armature current is:
$$I = P/E_0 = 225\,000/250$$
$$= 900\ A$$

b. 1 Each coil is made up of 2 conductors, so altogether there are 243 x 2 = 486 conductors on the armature.

Conductors per slot = 486/81 = 6

Coil sides per slot = 6

c. 1 Motor torque is:
$$T = 9.55\,P/n \qquad \text{Eq. 3-5}$$
$$= 9.55 \times 225\,000/1200$$
$$= 1790\ \text{N·m}$$

c. 2 The flux per pole is:
$$\Phi = 6.28\,T/ZI \qquad \text{Eq. 15-6}$$
$$= (6.28 \times 1790)/(486 \times 900)$$
$$= 25.7\ \text{mWb}$$

15-4 Speed of rotation

When a dc motor runs under *normal* conditions, the IR drop due to armature resistance is always small compared to the supply voltage E_s. This means that the counter-emf E_0 is very nearly equal to E_s.

On the other hand, we have already seen that E_0 may be expressed by the equation
$$E_0 = Zn\Phi/60 \qquad \text{Eq. 14-1}$$
Replacing E_0 by E_s, we obtain
$$E_s = Zn\Phi/60$$

That is, $$\boxed{n = \frac{60\,E_s}{Z\Phi}}\ \text{(approximately)} \qquad (15\text{-}7)$$

where

n = speed of rotation [r/min]

E_s = armature voltage [V]

Z = total number of armature conductors

This important equation shows that motor speed is directly proportional to the armature supply voltage, and inversely proportional to the flux per pole.

Figure 15-5 Armature of Fig. 15-4 in the process of being wound: (a) Coil forming machine gives the coils the desired shape; (b) One of the 81 coils ready to be placed in the slots; (c) Connecting the coil ends to the commutator bars; (d) Commutator connections ready for brazing. *(H. Roberge)*

15-5 Armature speed control

If the flux per pole Φ is kept constant (permanent magnet field or field with fixed excitation), the speed depends only upon the armature voltage E_s. By raising or lowering E_s, the motor speed will rise and fall in proportion.

In practice, we can vary E_s by connecting the motor armature M to a variable-voltage dc generator G (Fig. 15-6). The field excitation of the motor is kept constant, but the generator excitation can be varied from zero to maximum, and even reversed. The generator output voltage E_s can

flows into the positive terminal.

Let us now reduce E_s by reducing the generator excitation Φ_G. As soon as E_s becomes less than E_0, current I reverses. As a result, (a) the motor torque reverses and (b) the armature of the motor *delivers* power to generator G. In effect, the motor suddenly becomes a generator. The electric power which the motor now delivers to G is derived from the kinetic energy of the revolving armature and its connected load. As a result, the motor is suddenly forced to slow down.

What happens to the power received by generator G? In a Ward-Leonard system, the generator is

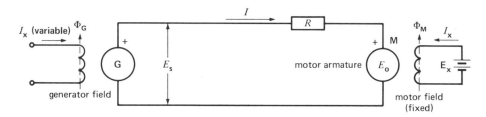

Figure 15-6   Ward-Leonard speed control system.

therefore be varied from zero to maximum, with either positive or negative polarity. This enables the motor speed to be varied from zero to maximum, in either direction. This method of speed control, known as the Ward-Leonard system, is employed in steel mills, high-rise elevators, mines, and paper mills.

In modern installations, the generator is often replaced by a high-power electronic converter which changes the ac power of the electrical utility to dc, by electronic means. Such converters are discussed in Chapter 24.

The Ward-Leonard system is more than just a simple way of applying a variable dc voltage to the armature of a dc motor. It actually forces the motor to develop the torque and speed required by the load.

For example, if E_s is slightly higher than E_0, current flows in the direction given in Fig. 15-6, and the motor develops a positive torque. The armature of the motor absorbs power because I

always driven by an ac motor connected to the ac mains supply. When G receives electric power, it operates as a motor, driving its own ac motor as a generator![*] The result is that ac power is fed back into the line that normally feeds the ac motor. The fact that power can be recovered this way makes the Ward-Leonard system very efficient, and constitutes another of its advantages.

Example 15-3:
A 2000 kW, 500 V, variable-speed motor is driven by a 2500 kW generator, using a Ward-Leonard control system shown in Fig. 15-6. The total resistance of the motor and generator armature circuit is 10 mΩ. The motor turns at a nominal speed of 300 r/min, when E_0 is 500 V.
Calculate:
a. the motor torque and speed when
$\qquad E_s =$ 400 V and $E_0 =$ 380 V;

[*]   This generator effect is explained in Sec. 18-16.

b. the motor torque and speed when
$E_s = 350$ V and $E_0 = 380$ V.

Solution:

a. 1 Armature current:
$$I = (E_s - E_0)/R = (400 - 380)/0.01$$
$$= 2000 \text{ A}$$

a. 2 Power to motor armature:
$$P = E_0 I = 380 \times 2000 = 760 \text{ kW}$$

a. 3 Motor speed:
$$n = (380 \text{ V}/500 \text{ V}) \times 300 = 228 \text{ r/min}$$

a. 4 Motor torque:
$$T = 9.55 \, P/n \qquad \text{Eq. 3-5}$$
$$= (9.55 \times 760 \, 000)/228$$
$$= 31.8 \text{ kN}$$

b. 1 Because $E_0 = 380$ V, the motor speed is still 228 r/min.

b. 2 Armature current:
$$I = (380 - 350)/0.01 = 3000 \text{ A}$$
The current flows in reverse; consequently, the motor torque also reverses.

b. 3 Power delivered by the motor to the generator and the 10 mΩ resistance:
$$P = E_0 I = 380 \times 3000 = 1140 \text{ kW}$$

b. 4 Braking torque developed by the motor:
$$T = 9.55 \, P/n - (9.55 \times 1 \, 140 \, 000)/228$$
$$= 47.8 \text{ kN}$$
The speed of the motor and its connected load will rapidly drop under the influence of this electromechanical braking torque.

Another way to change the voltage at the armature terminals is to place a rheostat in series with the armature (Fig. 15-7). The current in the rheostat produces a voltage drop which subtracts from the fixed source voltage E_s, yielding a smaller supply voltage across the armature. This method enables us to *reduce* the speed below its nominal speed. It is only recommended for small motors because a lot of power and heat is wasted in the rheostat, and the overall efficiency is low. Furthermore, the speed regulation is poor, even for a fixed setting of the rheostat. In effect, the *IR* drop across the rheostat increases as the armature current increases. This produces a substantial drop in

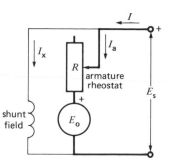

Figure 15-7   Armature speed control using a rheostat.

speed with increasing mechanical load.

15-6 Field speed control

Let us now keep the armature voltage E_s constant so that the numerator in Eq. 15-7,

$$n = \frac{60 \, E_s}{Z\Phi}$$

is constant. Consequently, the motor speed now changes in inverse proportion to the flux Φ: if we increase the flux, the speed will drop and vice versa.

This second method of speed control is frequently used because it is simple and inexpensive. To control the flux (and hence, the speed), we connect a rheostat R_f in series with the field (Fig. 15-8a).

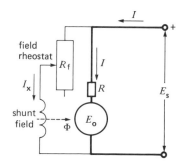

Figure 15-8   a. Schematic diagram of a shunt motor including the field rheostat.

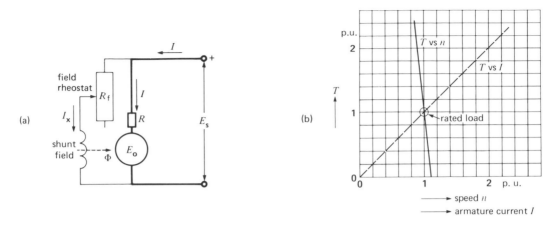

Figure 15-8   a. Schematic diagram of a shunt motor including the field rheostat.
b. Torque-speed and torque-current characteristic.

To understand this method of speed control, let us suppose that the motor in Fig. 15-8a is initially running at constant speed. The counter-emf E_0 is slightly less than the armature supply E_s, owing to the IR drop in the armature. If we suddenly raise the resistance of the rheostat, both the exciting current I_x and the flux Φ will diminish. This immediately reduces the counter-emf E_0, causing the armature current I to jump to a much higher value. The current changes dramatically because it depends upon the very small *difference* between E_s and E_0. Despite the weaker field, the motor develops a greater torque than before. It will accelerate until E_0 is again almost equal to E_s. Clearly, to develop the same E_0 with a weaker flux, the motor must turn faster. We can therefore raise the motor speed above its nominal value by introducing a resistance in series with the field. For shunt-wound motors, this method of speed control enables high-speed/low-speed ratios as high as 3 to 1. Broader speed ranges tend to produce instability and poor commutation.

Under certain abnormal conditions, the flux may drop to dangerously low values. For example, if the exciting current of a shunt motor is interrupted accidentally, the only flux remaining is that due to remanent magnetism in the poles.* This flux is so small that the motor has to spin at a dan-

gerously high speed to induce the required cemf. Safety devices are introduced to prevent such runaway conditions.

15-7 Shunt motor under load

Consider a dc motor running at no-load. If a mechanical load is suddenly applied to the shaft, the small no-load current does not produce enough torque to carry the load and the motor begins to slow down. This causes the cemf to diminish, resulting in a higher current and a corresponding higher torque. When the torque developed by the motor is exactly equal to the torque imposed by the mechanical load, then, and only then, will the speed remain constant (see Sec. 3-11). To sum up, as the mechanical load increases, the armature current rises and the speed drops.

The speed of a shunt motor stays relatively constant from no-load to full-load. In small motors, it only drops by 10 to 15 percent when full load is

---

*   The term *residual* magnetism is also used. However, the IEEE Standard Dictionary of Electrical and Electronic Terms states " . . . If there are no air gaps . . . in the magnetic circuit, the remanent induction will equal the residual induction; if there are air gaps . . . the remanent induction will be less than the residual induction".

applied. In big machines, the drop is even less, owing, in part, to the very low armature resistance. By adjusting the field rheostat, the speed can, of course, be kept absolutely constant as the load changes.

Typical torque-speed and torque-current characteristics of a shunt motor are shown in Fig. 15-8b.

Example 15-4:

A shunt motor rotating at 1500 r/min is fed by a 120 V source. (Fig. 15-9a). The line current is 51 A and the shunt field resistance is 120 Ω. If the armature resistance is 0.1 Ω, calculate:

a. the current in the armature;

b. the counter-emf;

c. the mechanical power developed by the motor.

c. 3 Power dissipated in the armature:
$$P = I^2R = 50^2 \times 0.1 = 250 \text{ W}$$

c. 4 Mechanical power developed by the armature:
$$P = 6000 - 250 = 5750 \text{ W (equivalent to } 5750/746 \approx 7.71 \text{ hp)}$$

The actual mechanical output is slightly less than 5750 W (7.71 hp) because some of the mechanical power is dissipated in bearing friction losses, in windage losses and in armature iron losses.

15-8 Series motor

A series motor is identical in construction to a shunt motor except that the field is connected in

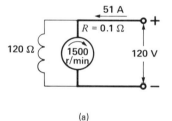

(a)

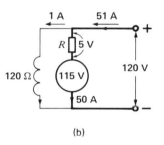

(b)

Figure 15-9   See Example 15-4.

Solution:

a. 1   Field current (Fig. 15-9b):
$$I_x = 120 \text{ V}/120 \ \Omega = 1 \text{ A}$$

a. 2   Armature current:
$$I = 51 - 1 = 50 \text{ A}$$

b. 1   Armature voltage:
$$E = 120 \text{ V}$$
Voltage drop due to armature resistance:
$$IR = 50 \times 0.1 = 5 \text{ V}$$

b. 2   cemf generated by the armature:
$$E_0 = 120 - 5 = 115 \text{ V}$$

c. 1   Power supplied to the motor:
$$P = EI = 120 \times 51 = 6120 \text{ W}$$

c. 2   Power absorbed by the armature:
$$P = EI = 120 \times 50 = 6000 \text{ W}$$

series with the armature and carries the full armature current (Fig. 15-10a). The series field is composed of a few turns of wire having a cross section sufficiently large to carry the current.

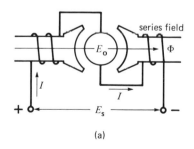

(a)

Figure 15-10   a. Series motor connection diagram.

Although the construction is similar, the properties of a series motor are completely different from those of a shunt motor. In a shunt motor, the flux Φ per pole is constant at all loads because the shunt field is connected to the line. In the series motor, the flux per pole depends upon the armature current, and hence upon the load. When the current is large, the flux is large and vice versa. Despite these differences, the same basic principles and equations apply to both machines.

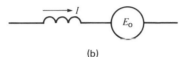

(b)

Figure 15-10 b. Schematic diagram of a series motor.

When a series motor operates at normal load, the flux per pole is the same as that of a shunt motor of identical power and speed. However, when the series motor starts up, the armature current is higher than normal, with the result that the flux per pole is also greater than normal. It follows that the starting torque of a series motor is considerably greater than that of a shunt motor.

On the other hand, if the load is less than normal, the armature current and the flux per pole are smaller than normal. The weaker field causes the speed to rise, in the same way as it would for a shunt motor with a weak shunt field. For example, if the load current of a series motor drops to half its normal value, the flux diminishes almost by half and so the speed doubles. Obviously, if the load is small, the speed may rise to dangerously high values. For this reason, we never permit a series motor to operate at no-load. It tends to run away, and the resulting centrifugal forces may tear the windings out of the armature and destroy the machine.

15-9 Series motor speed control

When a series motor carries a load, its speed may be increased by placing a low resistance in parallel with the field. The field current is then smaller than before, which produces a drop in flux and an increase in speed.

Conversely, the speed may be lowered by connecting an external resistor in series with the armature and the field. The total IR drop across the re-

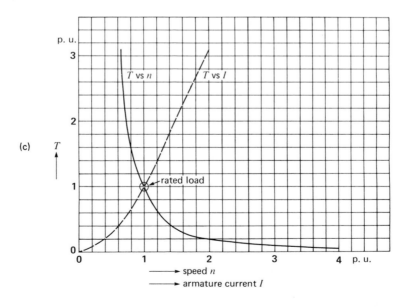

Figure 15-10   c. Typical torque-speed and torque-current characteristic of a series motor.

sistor and field reduces the armature supply voltage, and so the speed must fall.

Typical torque-speed and torque-current characteristics are shown in Fig. 15-10c.

15-10 Applications of the series motor

Series motors are used on equipment requiring a high starting torque. They are also used to drive devices which must run at high speed at light loads. The series motor is particularly well adapted for traction purposes, such as in electric trains (Fig. 15-11). Acceleration is rapid because the torque is high at low speeds. Furthermore, the series motor automatically slows down as the train goes up a grade, yet turns at top speed on flat ground. The power of a series motor tends to be constant because high torque is accompanied by low speed and vice versa. Series motors are also used in electric cranes and hoists: light loads are lifted quickly and heavy loads more slowly.

15-11 Compound motor

A compound dc motor carries both a series field and a shunt field. In a *cumulative compound* motor, the mmf of the two fields add. The shunt field is always stronger than the series field. If the series field is connected so that it opposes the shunt field, we obtain a *differential compound* motor.

Figure 15-11   Direct current series motors are often used to drive electric trains. *(General Electric)*

Figure 15-12 shows the connection and schematic diagrams of a compound motor. When the motor runs at no load, the current I in the series winding is low and the mmf of the series field is negligible. However, the shunt field is fully excited by current I_x and so the motor behaves like a shunt machine: it does not tend to run away at no load.

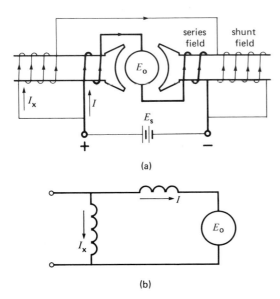

(a)

(b)

Figure 15-12　a. Connection diagram of a dc compound motor.
　　　　　　　b. Schematic diagram.

As the load increases, the mmf of the series field increases, but the shunt field strength remains constant. The total mmf (and the resulting flux per pole) is therefore greater under load than at no-load. The motor speed falls with increasing load and the speed drop from no-load to full-load is generally between 10 percent and 30 percent.

In differential-compound motors, the series winding opposes the shunt winding; consequently, the total mmf decreases with increasing load. The speed rises as the load increases, and this usually leads to instability. The differential compound motor has very few applications.

Figure 15-13 shows the typical torque-speed

curves of shunt, compound and series motors having the same power and speed ratings. Figure 15-14 shows a typical application of dc motors in a steel mill (see p. 274).

15-12　Reversing the direction of rotation

To reverse the direction of rotation of a dc motor, we must reverse the current in either (a) the armature, or (b) both the shunt and series fields.

15-13　Starting a shunt motor

If we apply full voltage to a stationary shunt motor, the starting current in the armature will be very high, and we run the risk of:

a. burning out the armature;
b. damaging the commutator and brushes, owing to heavy sparking;
c. overloading the feeder;
d. snapping off the shaft due to mechanical shock;
e. damaging the driven equipment because of the sudden mechanical hammerblow.

All dc motors must therefore be provided with means to limit the starting current to reasonable values, usually between 1.5 and twice full-load current. One solution is to simply connect a rheostat in series with the armature. The resistance is gradu-

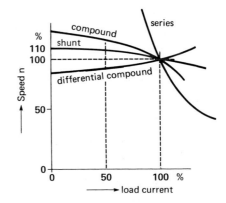

Figure 15-13　Typical speed versus load characteristics of various dc motors.

ally reduced as the motor accelerates and is eventually eliminated entirely, when the machine has attained full speed.

Today, we often employ electronic methods to limit the starting current and to provide speed control (Chapter 24).

15-14 Face-plate starter

Figure 15-15 shows the schematic diagram of a manual face-plate starter for a shunt motor. Bare copper contacts are connected to resistors $R_1, R_2,$ R_3 and R_4. Conducting arm **1** sweeps across the contacts when it is pulled to the right by means of insulated handle **2**. In the position shown, the arm touches dead copper contact **M** and the motor circuit is open. As we draw the handle to the right,

tact, and so forth, until the arm finally touches the last contact. The arm is magnetically held in this position by a small electromagnet **4**, which is in series with the shunt field.

If the supply voltage is suddenly interrupted, or if the field excitation should accidentally be cut, the electromagnet releases the arm, allowing it to return to its dead position, under the pull of spring **3**. This safety feature prevents the motor from starting by itself when the supply voltage is re-established.

15-15 Stopping a motor

One is inclined to believe that stopping a dc motor is a simple, almost trivial, operation. Unfortunately, this is not always true. When a large dc motor

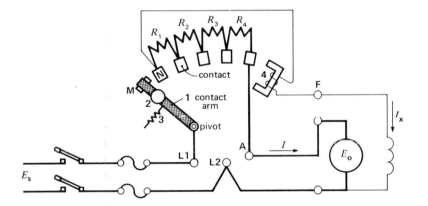

Figure 15-15   Manual face-plate starter for a shunt motor.

the conducting arm first touches fixed contact **N**.

The supply voltage E_s immediately causes full field current I_x to flow, but the armature current I is limited by the four resistors in the starter box. The motor begins to turn and, as the cemf builds up, the armature current gradually falls. When the motor speed ceases to rise any more, we pull the arm to the next contact, thereby removing resistor R_1 from the armature circuit. The current immediately jumps to a higher value and the motor quickly accelerates to the next higher speed. When the speed is again stable, we move to the next con-

is coupled to a heavy inertia load, it may take an hour or more for the system to come to a halt. For many reasons, such a lengthy deceleration time is often unacceptable and, under these circumstances, we must apply a braking torque to ensure a rapid stop. One way to brake the motor is by simple mechanical friction, in the same way we stop a car. A more elegant method consists of circulating a reverse current in the armature, so as to brake the motor electrically. Two methods are employed to create such an electromechanical brake: (a) dynamic braking and (b) plugging.

15-16 Dynamic braking

Consider a shunt motor whose field is directly connected to a source E_s, and whose armature is connected to the same source by means of a switch. The direction of the armature current I_1 and the polarity of the cemf E_0 are shown in Fig. 15-16a. Neglecting the armature IR drop, E_0 is equal to E_s.

sequently, the braking torque becomes smaller and smaller, finally becoming zero when the armature ceases to turn. The speed drops quickly at first and then more slowly, as the armature comes to a halt. To illustrate the usefulness of dynamic braking, Fig. 15-18 compares the speed-time curves for a motor equipped with dynamic braking and one which simply coasts to a stop. The speed decreases

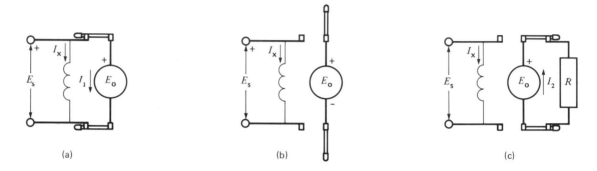

Figure 15-16   a. Armature connected to dc source E_s. b. Armature on open circuit generating a voltage E_0. c. Dynamic braking.

If we suddenly open the switch (Fig. 15-16b), the motor continues to turn, but its speed will gradually drop owing to friction and windage losses. On the other hand, because the shunt field is still excited, induced voltage E_0 continues to exist, falling at the same rate as the speed. In essence, the motor is now a generator whose armature is on open-circuit.

Let us close the switch so that the armature is suddenly connected to an external resistor R (Fig. 15-16c). Voltage E_0 will immediately produce an armature current I_2. However, this current flows in the *opposite* direction to the original current I_1. It follows that a *reverse* torque is developed whose magnitude depends upon I_2. The reverse torque brings the machine to a rapid, but very smooth stop. In practice, resistor R is chosen so that the initial braking current is about twice the rated motor current. The initial braking torque is then twice the normal torque of the motor.

As the motor slows down, the gradual decrease in E_0 produces a corresponding decrease in I_2. Con-

exponentially, like the voltage across a discharging capacitor. Consequently, the speed decreases by half in equal intervals of time.

15-17 Plugging

We can stop the motor even more rapidly by using a method called *plugging*. It consists of suddenly reversing the current in the armature by reversing the terminals of the source (Fig. 15-17).

Under normal motor conditions, armature current I_1 is given by

$$I_1 = (E_s - E_0)/R_0$$

where R_0 is the armature resistance. If we suddenly reverse the terminals of the source, the net voltage acting on the armature circuit becomes $(E_0 + E_s)$. The so-called "counter" emf of the armature is no longer counter to anything, but actually *adds* to the supply voltage E_s. This net voltage would produce an enormous reverse current, perhaps 50 times greater than full-load armature current. This current would initiate an arc all around the commutator, destroying segments, brushes and supports, even before the line circuit-breakers could open.

To prevent such a catastrophe, we must limit the reverse current by introducing a resistor R in series with the reversing circuit (Fig. 15-17b). Like in dynamic braking, the resistor is designed to limit the initial braking current I_2 to about twice full-load current. With this arrangement, a reverse torque is developed even when the armature has come to a stop. In effect, at zero speed, $E_0 = 0$, but $I_2 = E_s/R$, which is about one-half its initial value. As soon as the motor stops, we immediately

open the armature circuit, otherwise it will begin to run in reverse. Circuit interruption is usually effected by an automatic null-speed device, mounted on the motor shaft.

The curves of Fig. 15-18 enable us to compare plugging and dynamic braking for the same initial braking current. Note that plugging stops the motor completely after an interval $2\,T_0$. On the other hand, if dynamic braking is used, the speed is still 25 percent of its original value at this time. Nevertheless, the comparative simplicity of dynamic braking renders it more popular in most applications.

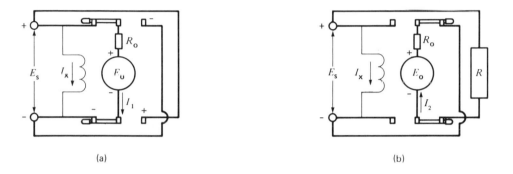

(a)   (b)

Figure 15-17   a. Armature connected to dc source E_s.   b. Plugging.

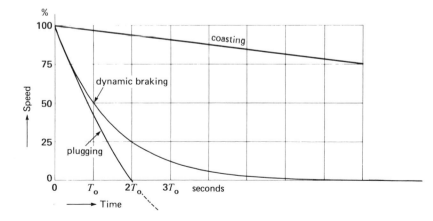

Figure 15-18   Speed versus time for various braking methods.

Figure 15-14 Hot strip finishing mill, composed of 6 stands each driven by a 2500 kW dc motor. The wide steel strip is delivered to the runout table (left foreground) driven by 161 dc motors rated 3 kW. *(General Electric)*

15-18 Dynamic braking and mechanical time constant

We mentioned that the speed decreases exponentially with time when a dc motor is stopped by dynamic braking. We can therefore speak of a mechanical time constant T in much the same way we speak of the electrical time constant of a discharging capacitor.

In essence, T is the time it takes for the speed of the motor to fall to 36.8 percent of its initial value. However, it is much easier to draw the speed-time curves by defining a time constant T_0 which is the time for the speed to decrease to 50 percent of its original value. There is a direct mathematical relationship between the conventional time constant T and the "half" time constant T_0:

$$T_0 = 0.693\,T \qquad (15\text{-}8)$$

We can prove that this mechanical time constant is given by:

$$T_0 = \frac{Jn_1^2}{131.5\,P_1} \qquad (15\text{-}9)$$

where

$T_0 =$ time for the motor speed to fall to one-half its previous value [s]

$J \;=$ moment of inertia of the rotating parts, referred to the motor shaft [kg·m^2]

$n_1 =$ initial speed of the motor when braking starts [r/min]

$P_1 =$ initial power delivered by the motor to the braking resistor [W]

$131.5 =$ a constant [exact value = $(30/\pi)^2/\log_e 2$]

$0.693 =$ a constant [exact value = $\log_e 2$]

This equation is based upon the assumption that the braking effect is entirely produced by the energy dissipated in the braking resistor. If, in addition, the motor is subjected to an extra braking torque due to its load the braking time will be shorter than that indicated by Eq. 15-9.

Example 15-5:

A 225 kW ($\approx$ 300 hp), 250 V, 1200 r/min dc motor has windage, friction, and iron losses of 8 kW. It drives a large flywheel and the total moment of inertia of the flywheel and armature is 177 kg·m². The motor is connected to a 210 V dc source, and its speed is 1280 r/min just before the armature is switched across a braking resistor of 0.2 Ω. Calculate:

a. the mechanical time constant T_0 of the braking system;

b. the time for the motor speed to drop to 20 r/min;

c. the time for the motor speed to drop to 20 r/min, if the only braking force is that due to the windage, friction and iron losses.

Solution:

a. 1 We note that the armature voltage is 210 V and the speed is 1280 r/min. The higher-than-normal speed is due to the setting of the field rheostat.

a. 2 When the armature is switched to the braking resistor, the induced voltage is still very close to 210 V. The initial power delivered to the resistor is:

$$P_1 = E^2/R = 210^2/0.2 = 220\,500 \text{ W}$$

a.3 $T_0 = Jn_1^2/(131.5\,P_1)$ Eq. 15-9
$$= (177 \times 1280^2)/(131.5 \times 220\,500)$$
$$= 10 \text{ s}$$

b. 1 The motor speed drops by 50 percent every 10 s. The speed versus time curve follows the sequence given below:

speed 1280 640 320 160 80 40 20 r/min
time   0   10 20 30 40 50 60 s

c. 1 The initial windage, friction and iron losses

are about 8 kW. These losses do not vary in exactly the same way as do the losses in a braking resistor. However, the behavior is comparable, which enables us to make a rough estimate of the braking time. We have:

$n_1 = 1280,$ $P_1 = 8000$
$T_0 = Jn_1^2/(131.5\,P_1)$
$$= (177 \times 1280^2)/(131.5 \times 8000)$$
$$= 276 \text{ s}$$

c. 2 The stopping time increases in proportion to the time constant. Consequently, the time to reach 20 r/min is approximately:

$t = (276/10) \times 60 = 1656 \text{ s}$
$\approx 28 \text{ min}$

Theoretically, a motor which is dynamically braked never comes to a complete stop. In practice, however, we can assume that the machine stops after an interval equal to 5 T_0 seconds.

On the other hand, if the motor is plugged, the stopping time has a definite value given by

$$\boxed{t_s = 2\,T_0} \qquad (15\text{-}10)$$

where
t_s = stopping time [s]
T_0 = time as given in Eq. 15-9 [s]

Example 15-6:

The motor in Example 15-5 is plugged, and the braking resistor is increased to 0.4 Ω, so that the initial braking current is the same as before. Calculate:

a. the initial braking current and braking power;

b. the stopping time.

Solution:

a. 1 The net voltage acting across the resistor is:
$E = E_0 + E_s = 210 + 210 = 420 \text{ V}$

a. 2 The initial braking current is:
$I_1 = E/R = 420/0.4 = 1050 \text{ A}$

a. 3 The initial braking power is:
$P_1 = E_0 I_1 = 210 \times 1050 = 220.5 \text{ kW}$

a. 4 T_0 therefore has the same value as before:
10 s

b. $t_s = 2\,T_0 = 20 \text{ s}$

15-19 Armature reaction

Up till now, we have assumed that the only mmf acting in a dc machine is that due to the field. However, the current flowing in the armature conductors also creates a magnetomotive force that distorts and weakens the flux coming from the poles. This distortion and field weakening takes place in both motors and generators. The magnetic action of the armature mmf is called *armature reaction*.

15-20 Direction of current flow in the armature

To understand armature reaction, we must first establish the direction of current flow in the armature conductors immediately under the field poles.

Consider, for example, the conductors situated under the N pole of a *motor* which is turning *clockwise* (Fig. 15-19). The cw rotation is obvious-

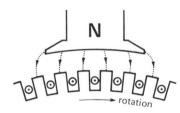

Figure 15-19   Armature currents under a N pole (motor).

ly due to forces which act towards the right on each conductor. Based upon the principles explained in Sec. 2-26, the currents must flow out of the page, towards the reader. For the same reason,

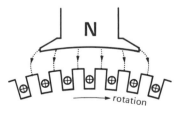

Figure 15-20   Armature currents under a N pole (generator).

the currents in the conductors under the S pole must flow in the opposite direction, into the page.

Now, consider the armature currents under the N pole of a *generator* which is also turning clockwise (Fig. 15-20). Clearly, the force on the conductors opposes the motion because we have to exert a torque to drive the machine. The conductor force tends to rotate the armature counterclockwise and so the current under the N pole must flow into the page, away from the reader.

15-21 Flux distortion due to armature reaction

When a motor runs at no-load, the small current flowing in the armature does not appreciably affect the flux Φ_1 coming from the poles (Fig. 15-21). But when the armature carries its normal current, it produces a strong magnetomotive force which, if it acted alone, would create a flux Φ_2 (Fig. 15-22). By superimposing Φ_1 and Φ_2, we obtain the resulting flux Φ_3 (Fig. 15-23). In our example, the flux density increases under the left of the pole while it decreases under the right half. This unequal distribution produces two important effects. First, the neutral zone shifts towards the left (against the direction of rotation). Second, owing to the higher flux density in pole tip A, saturation sets in. Consequently, the increase of flux under the left-hand side of the pole is less than the decrease under the right-hand side. Flux Φ_3 at full-load is therefore slightly less than flux Φ_1 at no-load. For large machines, the decrease in flux may be as much as 10 percent and it causes the speed to increase with load. Such a condition tends to be unstable; to eliminate the problem, we sometimes add a series field of one or two turns to strengthen the shunt field. Such motors are said to have a *stabilized-shunt winding*.

Owing to the shift in the neutral zone, we must move the brushes to ensure good commutation. For a motor, the brushes are shifted to the new neutral zone by moving them against the direction of rotation. For generators, the brushes are shifted with the direction of rotation.

As soon as the brushes are moved, the commu-

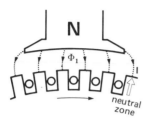

Figure 15-21   Flux distribution in a motor running at no-load.

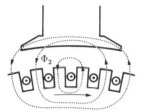

Figure 15-22   Flux created by the full-load armature current.

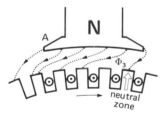

Figure 15-23   Resulting flux distribution in a motor running at full load.

tation improves. However, if the load fluctuates, the armature mmf rises and falls and the neutral zone oscillates between the no-load and full-load positions. We must therefore move the brushes continually to obtain sparkless commutation. This procedure is not practical when the load current varies frequently and suddenly. For small motors, the brushes are set in an intermediate position, to produce acceptable commutation at all loads.

15-22   Commutating poles

To counter the effect of armature reaction, we usually place a set of *commutating* poles between the main poles of a dc machine (Fig. 15-24).* These narrow poles develop a magnetomotive force equal and opposite to the mmf of the armature. The commutating-pole windings are connected in series with the armature so that the respective magnetomotive forces rise and fall together as the load current varies. By nullifying the armature mmf in this way, we no longer have to shift the brushes.

Figure 15-25 shows how the commutating poles of a two-pole motor are connected. Clearly, the mmf of the commutating poles acts opposite to the mmf of the armature and therefore neutralizes

* Commutating poles are sometimes called interpoles.

Figure 15-24   The narrow commutating poles are placed between the main poles of this 6-pole motor.

its effect. However, the neutralization is restricted to the narrow zone where commutation takes place. The distorted flux distribution under the pole face unfortunately remains the same.

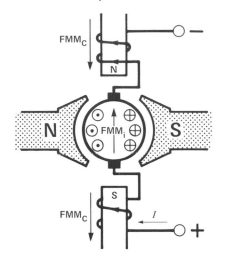

Figure 15-25 Commutating pole connections for a dc motor turning clockwise.

15-23 Compensating winding

Some dc motors in the 100 kW to 10 MW, (≈ 134 hp to 13 400 hp) range employed in steel mills, perform a series of rapid, heavy-duty operations. They accelerate, decelerate, stop, reverse, all in a matter of seconds. The corresponding armature current increases, decreases, reverses in stepwise fashion, producing very sudden changes in armature reaction.

For such motors, the commutating poles and series stabilizing windings do not adequately neutralize the armature mmf. Torque and speed control is difficult under such transient conditions and flash-overs may occur across the commutator. To eliminate this problem, additional *compensating windings* are connected in series with the armature. They are distributed in slots, cut into the pole faces of the main field poles (Fig. 15-26). Like commutating poles, these windings produce a mmf equal and opposite to the mmf of the armature. However, because the windings are distrib-

uted across the pole faces, the armature mmf is bucked from point to point, which eliminates the field distortion shown in Fig. 15-23. With compensating windings, the field distribution remains essentially undisturbed from no-load to full-load, retaining the general shape shown in Fig. 15-21.

The addition of compensating windings has a profound effect on the design and performance of a dc motor;

1. A shorter air gap can be used because we no longer have to worry about the demagnetizing effect of the armature. A shorter gap means that the shunt field strength can be reduced, and hence the size of the coils.

2. Fewer turns can be placed on the commutating poles, because most of the required mmf is now developed by the compensating winding.

3. The inductance of the armature circuit is reduced by a factor of 4 or 5; consequently, the armature current can change more quickly and the motor gives a much better response. This is particularly true in big machines.

4. A motor equipped with compensating windings can briefly develop 3 to 4 times its rated torque. The torque of an uncompensated motor tends to reach a limit when the armature current is large. The reason is that the effective flux in the air gap falls off rapidly with increasing current.

We conclude that compensating windings are essential in large motors subjected to severe duty.

QUESTIONS AND PROBLEMS

Practical level

15-1 Name three types of dc motors and make a sketch of the connections.

15-2 Explain what is meant by the generator effect in a motor.

15-3 What determines the magnitude and polarity of the counter emf in a dc motor?

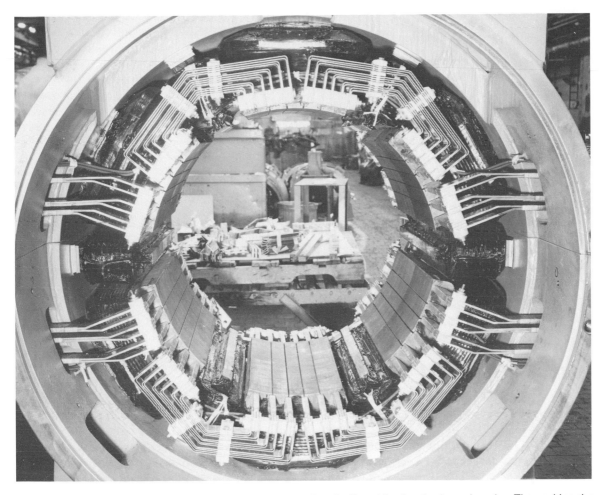

Figure 15-26 Six-pole dc motor having a compensating winding distributed in slots in the main poles. The machine also has 6 commutating poles. *(General Electric Company, USA)*

15-4 The counter-emf of a motor is always slightly less than the applied armature voltage. Explain.

15-5 Name some of the methods used to vary the speed of a dc motor.

15-6 Explain why the armature current of a shunt motor decreases as the motor accelerates.

15-7 Why is a starting resistor needed?

15-8 Show how we can reverse the direction of rotation of a compound motor.

15-9 A 230 V shunt motor has a nominal armature current of 60 A. If the armature resistance is 0.15 Ω, calculate:

a. the counter-emf [V] ;

b. the power supplied to the armature [W] ;

c. the mechanical power developed by the motor, [kW] and [hp] .

15-10 a. In Problem 15-9, calculate the initial starting current if the motor is directly connected across the 230 V line;

b. Calculate the value of the starting resistor needed, to limit the initial current to 115 A.

Intermediate level

15-11 The compound motor of Fig. 15-12 has 1200 turns on the shunt winding and 25 turns on the series winding, per pole. The shunt field has a total resistance of 115 Ω, and the nominal armature current is 23 A. If the motor is connected to a 230 V line, calculate:
   a. the mmf per pole at full load;
   b. the mmf at no-load.

15-12 A separately excited dc motor turns at 1200 r/min when the armature is connected to a 115 V source. Calculate the armature voltage we must apply so that the motor runs at 1500 r/min? 100 r/min?

15-13 The following details are known about a 250 hp, 230 V, 435 r/min dc shunt motor:
   nominal full-load current - 862 A
   insulation class - H
   weight - 3400 kg
   external diameter of the frame - 915 mm
   length of frame - 1260 mm
   a. Calculate the total losses and efficiency at full load.
   b. Calculate the approximate exciting current if the shunt field produces 20% of the total losses.
   c. Calculate the value of the armature resistance, as well as the counter-emf, knowing that 50% of the total losses at full load are due to armature resistance.
   d. If we wish to attain a speed of 1100 r/min, what should be the approximate exciting current?

15-14 We wish to stop a 120 hp, 240 V, 400 r/min motor by using the dynamic braking circuit shown in Fig. 15-16. If the nominal armature current is 400 A, calculate:
   a. the value of the braking resistor R if we want to limit the maximum braking current to 125% of its nominal value;
   b. the braking power [kW], when the motor turns at 200 r/min? 50 r/min? 0 r/min?

15-15 a. The motor in Problem 15-14 is now stopped by using the plugging circuit of Fig. 15-17. Calculate the new braking resistor R so that the maximum braking current is 500 A.
   b. Calculate the braking power [kW] when the motor turns at 200 r/min? 50 r/min? 0 r/min?
   c. Compare the braking power developed, at 200 r/min to the instantaneous power dissipated in resistor R.

Advanced level

15-16 The armature of a 225 kW, 1200 r/min motor has a diameter of 559 mm and an axial length of 235 mm. Calculate:
   a. the approximate moment of inertia, knowing that iron has a density of 7900 kg/m^3;
   b. the kinetic energy of the armature alone when it turns at 1200 r/min;
   c. the total kinetic energy of the revolving parts at a speed of 600 r/min, if the J of the windings and commutator is equal to the J calculated in (a).

15-17 If we reduce the exciting current of a practical shunt motor by 50%, the speed increases, but it never doubles. Explain.

15-18 The speed of a series motor drops with rising temperature, while that of a shunt motor increases. Explain.

15-19 In Problems 15-14 and 15-15 calculate the respective braking torques [ft·lb] developed when the motor turns at 400 r/min? at 0 r/min?

15-20 A wide paper sheet coming off a mill is rolled up on a cylinder. The diameter of the roll increases gradually from 300 mm to 1000 mm. The paper is delivered at a uniform rate and at constant tension. The driving motor develops 50 kW at 600 r/min when the roll is begun. Calculate the power and speed required when the roll is finished.

16

EFFICIENCY AND HEATING OF ELECTRICAL MACHINES

Whenever a machine transforms energy from one form to another, there is always a certain loss. The loss takes place in the machine itself, causing (a) an increase in temperature and (b) a reduction in efficiency.

In this chapter, we analyse the losses in dc machines, but the same losses are also found in most machines operating on alternating current. The study of power losses is important because it gives us a clue as to how they may be reduced.

Electrical machines may be divided into two groups: those which have revolving parts (motors, generators, etc.) and those which do not (transformers, reactors, etc.). Electrical and mechanical losses are produced in rotating machines, while only electrical losses are produced in stationary machines.

16-1 Mechanical losses

Mechanical losses are due to bearing friction, brush friction and windage. The friction losses depend upon the speed of the machine, and upon the design of the bearings, brushes, commutator, and

slip rings. Windage losses depend on the speed and design of the cooling fan, and on the turbulence produced by the revolving parts. In the absence of prior information, we usually conduct tests on the machine itself to determine the value of these mechanical losses.

16-2 Electrical losses

Electrical losses are composed of:
(1) conductor I^2R losses (sometimes called *copper* losses);
(2) brush losses;
(3) iron losses.

1. Conductor losses. In dc motors and generators, copper losses occur in the armature, the series field, the shunt field, the commutating poles, and the compensating winding. These I^2R losses show up as heat, causing the conductor temperatures to rise above ambient temperature.

Instead of using the I^2R equation, we sometimes prefer to express the losses in terms of the number of watts per kilogram of conductor material. The losses are then given by:

$$P_c = 1000 \, J^2 \rho / \varsigma \qquad (16\text{-}1)$$

where

P_c = conductor loss [W/kg]
J = current density [A/mm²]
ρ = resistivity of conductor [nΩ·m]
ς = density of conductor [kg/m³]
1000 = constant, to take care of units

According to this equation, the loss per unit mass is proportional to the square of the current density. For copper conductors, we use densities between 1.5 A/mm² and 6 A/mm². The corresponding losses vary from 5 W/kg to 90 W/kg (Fig. 16-1). The higher densities require an efficient cooling system to prevent an excessive temperature rise.

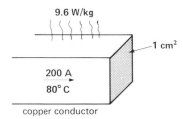

Figure 16-1   Copper losses may be expressed in W/kg.

2. Brush losses. The I^2R losses in the brushes are negligible because the current density is only about 0.1 A/mm² which is far less than that used in copper. However, the voltage contact drop between the brushes and commutator may produce significant losses. The drop varies from 0.8 V to 1.3 V, depending on the type of brush, the applied pressure and the brush current (Fig. 16-2).

3. Iron losses. Iron losses are produced in the armature of a dc machine. They are due to hysterisis and eddy currents, as previously explained in Sec. 6-7 and 6-10. Iron losses depend upon the flux density, the speed of rotation, the quality of the steel and the size of the armature. They typically range from 0.5 W/kg to 20 W/kg. The higher values occur in the armature teeth, where the flux density may be as high as 1.7 T. The losses in the arma-

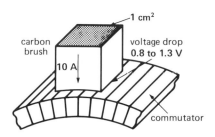

Figure 16-2   Brush contact voltage drop.

ture core are usually much lower. The losses can be further reduced by annealing the steel (Fig. 16-3). Some iron losses are also produced in the pole faces. They are due to flux pulsations created as successive armature teeth and slots sweep across the pole face.

Strange as it may seem, iron losses impose a mechanical drag on the armature, producing the same effect as mechanical friction.

Example 16-1:
A dc machine turning at 875 r/min carries an armature winding whose total weight is 40 kg. The current density is 5 A/mm² and the operating temperature is 80°C. The total iron losses amount to 1100 W. Calculate:
a. the copper losses;
b. the mechanical drag [N·m] due to the iron losses.

Solution:
a. 1   The resistivity of copper at 80°C is:
$$\rho = \rho_0 \, (1 + \alpha t) \qquad \text{Eq. 5-2}$$
$$= 15.88 \, (1 + 0.004 \, 27 \times 80) \quad \text{Table 5B}$$
$$= 21.3 \, \text{nΩ·m}$$
a. 2   The density of copper is: Table 5B
8890 kg/m³

Figure 16-3   This **150 kW** electric oven is used to anneal punched steel laminations. This industrial process, carried out in a controlled atmosphere of 800°C, significantly reduces the iron losses. The laminations are seen as they leave the oven. *(General Electric)*

a. 3 $P_c = 1000\,J^2\rho/\zeta$ Eq. 16-1
$= 1000 \times 5^2 \times 21.3/8890$
$- 60\ \text{W/kg}$

a. 4   Total copper loss is:
$P = 60 \times 40 = 2400\ \text{W}$

b.   The braking torque due to iron losses can be calculated from:
$P = nT/9.55$ Eq. 2-5
$1100 = 875\ T/9.55$
$T = 12\ \text{N·m}\ (\approx 8.85\ \text{ft·lb})$

16-3 Losses as a function of load

A dc motor running at no-load develops no useful power. However, it must absorb some power to continue to rotate. This no-load power overcomes the friction, windage, and iron losses, and provides for the copper losses in the shunt field. The I^2R losses in the armature, series field and commutating field are negligible because the no-load current is seldom more than 10% of the nominal full-load current.

As we load the machine (mechanically, if it is a motor, electrically if it is a generator), the current increases in the armature circuit. Consequently, the I^2R losses in the armature circuit will rise. On the other hand, the no-load losses, mentioned above, remain essentially constant as the load increases, unless the speed of the machine changes appreciably. It follows that the total losses increase with load. Because they are converted into heat, the temperature of the machine rises progressively as the load increases.

However, the temperature must not exceed the limiting temperature, corresponding to the insulation used in the machine. Consequently, there is a limit to the power which the machine may deliver. This temperature-limited power enables us to establish the *nominal* or *rated power* of the machine. A machine loaded beyond its nominal rating will usually overheat. The insulation deteriorates more rapidly, which inevitably shortens the service life.

If a machine runs *intermittently*, it can carry heavy overloads without overheating, provided

Figure 16-4

Totally enclosed, water-cooled, 450 kW, 3600 r/min motor for use in a hostile environment. Warm air inside the machine is blown upwards and through a water-cooled heat exchanger, situated immediately above the Westinghouse nameplate. After releasing its heat to a set of water-cooled pipes, the cool air re-enters the machine by way of two rectangular pipes leading into the end bells. The cooling air therefore moves in a closed circuit, and the surrounding contaminated atmosphere never reaches the motor windings. The circular capped pipes located diagonally on the heat exchanger serve as cooling-water inlet and outlet respectively. *(Westinghouse)*

that the operating time is short. Thus, a motor having a nominal rating of 10 kW can easily carry a load of 15 kW, if it only operates a few minutes per hour. However, for higher loads, the capacity is limited by other factors, usually electrical. For instance, it is physically impossible for a 10 kW machine to yield an output of 100 kW.

Rotating machines are usually cooled by an internal fan mounted on the motor shaft. It draws in cool air from the surroundings, blows it over the windings, and expels it again through suitable vents. In hostile environments, special cooling methods are sometimes used, as illustrated in Fig. 16-4.

16-4 Efficiency curve

The efficiency of a machine is the ratio of the useful output power P_0 to the input power P_i (Sec. 3-7). Furthermore, input power is equal to useful power plus the losses ρ. We therefore have:

$$\eta = \frac{P_0}{P_i} \times 100 = \frac{P_0}{P_0 + \rho} \times 100 \qquad (16\text{-}2)$$

where

η = efficiency [%]
P_0 = output power [W]
P_i = input power [W]
ρ = losses [W]

The next example shows how to calculate the efficiency of a dc machine.

Example 16-2:

A dc compound motor having a rating of 10 kW, 1150 r/min, 230 V, 50 A, has the following losses *at full load*:

	bearing friction loss	=	40 W
	brush friction loss	=	50 W
	windage loss	=	200 W
(1)	total mechanical losses	=	290 W
(2)	iron losses	=	420 W
(3)	copper loss in the shunt field	=	120 W
	copper loss:		
	a. in the armature	=	500 W
	b. in the series field	=	25 W
	c. in the commutating winding	=	70 W
(4)	total copper loss in the armature circuit	=	595 W

Calculate the losses and efficiency at no-load and at 25, 50, 75, 100 and 150 percent of the nominal rating of the machine. Draw a graph showing efficiency as a function of mechanical load (neglect the losses due to brush contact drop).

Solution:

No-load. The no-load losses are equal to the sum of the mechanical losses (1), the iron losses (2), and the shunt field losses (3):

no-load losses = 290 + 420 + 120 = 830 W

The losses remain more or less constant as the load varies. The copper losses in the armature circuit are negligible at no-load.

The efficiency is zero because no useful power is developed by the motor.

25% load. When the motor is loaded to 25% of its nominal rating, the armature current is approximately 25% (or 1/4) of its full-load value. Because the copper losses vary as the square of the current, we have:

copper losses in the armature circuit
$= (1/4)^2 \times 595 = 37$ W
no-load losses = 830 W
total losses = 37 + 830 = 867 W

Useful power developed by the motor at 25% load:

$P_0 = 10$ kW $\times (1/4) = 2500$ W (≈ 3.35 hp)
power supplied to the motor:

$P_i = 2500 + 867 = 3367$ W

$\eta = (P_0/P_i) \times 100$ Eq. 16-2

$= (2500/3367) \times 100 = 74\%$

In the same way, we can find the losses at 50, 75, 100 and 150 percent of the nominal load:

At 50% load, the losses are $(1/2)^2 \times 595 + 830$
$= 980$ W

At 75% load, the losses are $(3/4)^2 \times 595 + 830$
$= 1165$ W

At 100 %load, the losses are $595 + 830$
$= 1425$ W

At 150% load, the losses are $(3/2)^2 \times 595 + 830$
$= 2170$ W

The efficiency calculations for the various loads are listed in Table 16A and the results are shown graphically in Fig. 16-5. It is important to remember that at light loads the efficiency of any motor is poor. Consequently, we always try to select a motor having a power rating roughly equal to the load it has to drive.

TABLE 16A   LOSSES AND EFFICIENCY OF A DC MOTOR

Load [%]	Total Losses [W]	Output Power P_0 [W]	Input Power P_i [W]	Efficiency [%]
0	830	0	830	0
25	867	2 500	3 367	74
50	980	5 000	5 980	83.6
75	1 165	7 500	8 665	86.5
100	1 425	10 000	11 425	87.5
150	2 170	15 000	17 170	87.4

We can prove that the efficiency of any dc machine reaches a maximum at that load where the armature circuit copper losses are equal to the no-load losses. In our example, this corresponds to a total loss of 1660 W, an output of 11 800 W (15.8 hp) and an efficiency of 87.67 percent. The reader may wish to check these results.

16-6 Limiting ambient temperature and hot-spot temperature rise

We saw in Sec. 4-7 that insulators are classified according to the maximum temperature they can tolerate. Thus, class B insulation has a reasonable life expectancy, provided its temperature does not

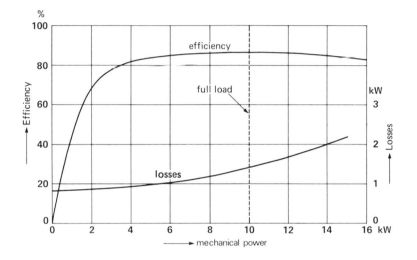

Figure 16-5   Losses and efficiency as a function of mechanical power. See Example 16-2.

16-5 Temperature rise

The temperature rise of a machine is the difference between the temperature of its warmest accessible part and the ambient temperature. It may be measured by simply taking the difference of two thermometer readings. However, owing to the practical difficulty of placing a thermometer close to the really warmest spot inside the machine, this method is seldom used. We usually rely upon more sophisticated methods, described in the following sections.

Temperature rise has a direct bearing on the power rating of a machine or device. Consequently, it is a very important quantity.

exceed $130°C$.

Standards organizations have also established a maximum *ambient temperature*, which is usually $40°C$. This limiting temperature was established for the following reasons:

1. It enables electrical manufacturers to foresee the worst ambient temperature conditions which their machines are likely to encounter. Consequently,

2. it enables them to standardize the size of their machines and to give performance quarantees.

The temperature of a machine varies from point to point, but there are places where the temperature is warmer than anywhere else. This so-called *hottest-spot temperature* must not exceed the tem-

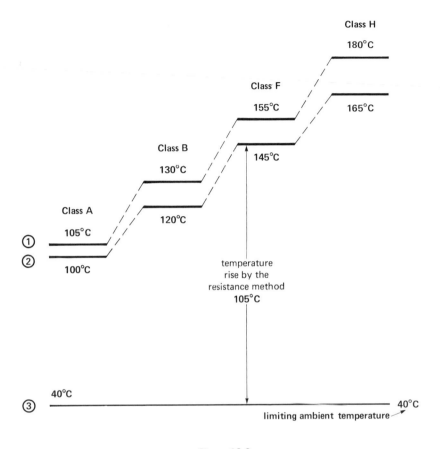

Figure 16-6

Typical temperature limits of some dc and ac industrial machines, according to the insulation class:
(1) shows the maximum permissible temperature of the insulation to obtain a reasonable service life;
(2) shows the maximum permissible temperature using the resistance method;
(3) shows the limiting ambient temperature.

perature limit for the particular class of insulation used.

Figure 16-6 shows the hot-spot temperature limits for class A, B, F, and H insulation (curve 1). They are equal to the temperature limits previously mentioned in Sec. 4-7. The limiting ambient temperature of 40°C is also shown (curve 3). The temperature difference between these two curves gives the *maximum permissible temperature rise* for each insulation class. This limiting temperature rise enables the manufacturer to establish the

physical size of the motor, relay, and so forth, he intends to put on the market. Thus, for class B insulation, the limiting temperature rise is (130 – 40) = 90°C.

To show how the temperature rise affects the size, suppose a manufacturer has designed and built a 10 kW motor using class B insulation. To test the motor, he places it in a constant ambient temperature of 40°C and loads it up until it delivers 10 kW of mechanical power. Special temperature detectors, located at strategic points throughout the machine, record the temperature of the windings. After the temperatures have stabilized (which may take several hours), the hottest temperature is noted, and this is called the hot-spot temperature. If the hot-spot temperature so recorded exceeds 130°C, the manufacturer is not permitted to sell his product. The reason is that the temperature rise exceeds the maximum permissible rise of 90°C for class B insulation.

On the other hand, if the hottest-spot temperature is only 100°C, the temperature rise is only (100 – 40) = 60°C. The manufacturer immediately perceives that he can make a more economical design and still remain within the permissible temperature rise limits. For instance, he can reduce the conductor size until the hot-spot temperature rise is very close to 90°C. Obviously, this reduces the weight and cost of the windings. But the manufacturer also realizes that the reduced conductor size now enables him to reduce the size of the slots. This, in turn, reduces the amount of iron. By thus redesigning the motor, the manufacturer ultimately ends up with a product which operates within the permissible temperature rise limits and has the smallest possible physical size.

In practice, it is not convenient to carry out performance tests in a controlled ambient temperature of 40°C. The motor is usually loaded to its rated capacity in much lower (and more comfortable) ambient temperatures. The hottest-spot temperature is recorded as before. If the temperature rise under these conditions is equal to or less than 90°C (for class B insulation), the manufacturer is allowed to sell his product. There is, however, one further condition: In carrying out the test, the ambient temperature must lie between 10°C and 40°C. These upper and lower ambient temperature limits are set by standards.

Example 16-3:
A 75 kW motor, insulated class F, operates at full load in an ambient temperature of 32°C. If the hot-spot temperature is 125°C, does the motor meet the temperature standards?

Solution:
The hot-spot temperature rise is (125 – 32) = 93°C. According to Fig. 16-6, the permissible hot-spot temperature rise is (155 – 40) = 115°C. The motor easily meets the temperature standards. The manufacturer could reduce the size of the motor and, thereby, produce a more competitive product.

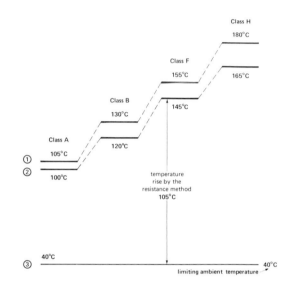

Figure 16-6

16-7 Temperature rise by the resistance method

The hot-spot temperature rise is rather difficult to measure because it has to be taken at the very inside of the winding. This can be done by embedding a small temperature detector, such as a thermocouple or thermistor. However, this direct method of measuring hot-spot temperature is costly, and is only justified for large machines.

To simplify matters, accepted standards permit a second method of determining temperature rise. It is based upon the *average* winding temperature rather than the hot-spot temperature. The limiting average winding temperatures for the various insulation classes are shown in curve 2, Fig. 16-6. For example, in the case of class B insulation, an average winding temperature of 120°C is assumed to correspond to a hot-spot temperature of 130°C. Similarly, an average temperature rise of (120 – 40) = 80°C is assumed to correspond to a hot-spot temperature rise of (130 – 40) = 90°C.

The average temperature of a winding is found by the so-called *resistance method*. It consists of measuring the winding resistance at a known temperature, and measuring it again when the machine is hot. Knowing the temperature coefficient of the conductor, we can easily calculate the temperature (Sec. 5-9). For example, if the winding is made of copper, we can use the following equation to determine its average temperature:

$$t_2 = \frac{R_2}{R_1}(234 + t_1) - 234 \qquad (16\text{-}3)$$

where

t_2 = average temperature of the winding when hot [°C]

234 = a constant equal to $1/\alpha = 1/0.004\ 27$ [see Table 5B]

R_2 = hot resistance of the winding [Ω]

R_1 = cold resistance of the winding [Ω]

t_1 = temperature of the winding when cold [°C]

Knowing the winding temperature by the resistance method, we can immediately calculate the corresponding temperature rise by subtracting the ambient temperature. If this temperature rise falls within the permissible limit (80°C for class B insulation), the product is acceptable, from a standards point of view. Note that when performance tests are carried out using the resistance method, the ambient temperature must again lie between 10°C and 40°C.

Example 16-4:
A 1000 hp motor, which has been idle for several days, possesses a shunt-field resistance of 22 Ω. The ambient temperature is 19°C. The motor then operates at full load and, when temperatures have stabilized, the field resistance is found to be 30 Ω. The corresponding ambient temperature is 24°C. If the motor is insulated class B, calculate:

a. the average temperature of the winding, when hot;

b. the temperature rise by the resistance method;

c. whether the motor meets the temperature standards.

Solution:

a. $t_2 = (R_2/R_1)(234 + t_1) - 234$

   $= (30/22)(234 + 19) - 234$

   $= 111°C$

b. Temperature rise $= 111 - 24 = 87°C$

c. The temperature rise permitted for class B insulation is (120 – 40) = 80°C. Consequently, the motor does not meet the standards. Either its rating will have to be reduced, or the cooling system improved, before it can be put on the market. Alternatively, it may be rewound using class F insulation. As a very last resort, its size could be increased.

A final word of caution: temperature rise standards depend not only on the class of insulation, but also on the type of apparatus (motor, transformer, relay, etc.), its construction (drip-proof, totally enclosed, etc.) and its field of application (commercial, industrial, naval, etc.). Consequently, we should always consult the pertinent standards before conducting a heat-run test on a specific machine or device.

16-8 Relationship between the speed and size of a machine

Although temperature limits establish the nominal power rating of a machine, its basic physical size depends upon the power and speed of rotation.

Consider a 100 kW, 250 V, 2000 r/min generator shown in Fig. 16-7. Suppose we have to build another generator having the same power and voltage, but running at half the speed.

To generate the same voltage at half the speed, we either have to double the number of conductors on the armature, or double the flux from the poles. Consequently, we must either increase the size of the armature, or increase the size of the

Figure 16-7   100 kW, 2000 r/min, mass: 300 kg.

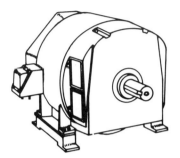

Figure 16-8   100 kW, 1000 r/min, mass: 500 kg.

poles. In practice, we increase both. We conclude that for a given power output, a low-speed machine is always bigger than a high-speed machine (Fig. 16-8). This is true for both ac and dc machines.

Basically, the size of a machine depends upon its torque. Thus, a 100 kW, 2000 r/min motor has about the same physical size as a 10 kW motor running at 200 r/min.

Low-speed motors are therefore more costly than high-speed motors of equal power. Consequently, for low-speed drives, it is often cheaper to use a high-speed motor with a gear box, than to use a low-speed motor directly.

QUESTIONS AND PROBLEMS

Practical level

16-1   Name the losses in a dc motor.
16-2   What causes iron losses and how can they be reduced?

16-3   Explain why the temperature of a machine increases as the load increases.
16-4   What determines the power rating of a machine?
16-5   If we cover up the vents in a motor, its output power must be reduced. Explain.
16-6   May we load a motor above its rated power in a cold area?   Why?

Intermediate level

16-7   A dc motor operating on a 240 V line produces a mechanical output of 160 hp. Knowing that the losses are 12 kW, calculate the input power and the line current.
16-8   A 115 V dc generator delivers 120 A. If it has an efficiency of 81%, calculate the mechanical power needed to drive it [hp].
16-9   Calculate the full load current of a 250 hp, 230 V dc motor having an efficiency of 92%.
16-10   A machine having class B insulation, attains a temperature of 208°C (by resistance) in an ambient temperature of 180°C.
   a. What is the temperature rise?
   b. Is the machine running too hot and, if so, by how much?
16-11   The efficiency of a machine is always low when it operates at 10% of its nominal power rating. Explain.
16-12   Calculate the efficiency of the motor in Example 16-2 when it delivers an output of 40 hp.
16-13   An electric motor driving a skip hoist withdraws 1.5 metric tons of minerals from a trench 20 m deep, every 30 seconds. If the hoist has an overall efficiency of 94%, calculate the power output of the motor, in horsepower and in kilowatts.
16-14   Thermocouples are used to measure the internal temperature of a 1200 kW ac motor insulated class F. If the motor runs normally, what is the maximum temperature these detectors should indicate in an ambient temperature of 40°C?   30°C?   14°C?
16-15   A 60 hp ac motor with class F insulation

has a winding resistance of 12 Ω at 23°C. When it runs at rated load in an ambient temperature of 31°C, the winding resistance is found to be 17.4 Ω.

a. Calculate the hot winding temperature;

b. Calculate the temperature rise of the motor;

c. Could the manufacturer increase the nameplate rating of the motor? Explain.

16-16 The motor in Problem 16-15 is overloaded and delivers 75 hp in an ambient of 12°C.

a. Knowing that the windings can operate at 145°C (measured by resistance) without reducing the expected life of the machine, calculate the permissible temperature rise under these overload conditions;

b. What is the temperature rise stamped on the nameplate?

16-17 A No. 10 round copper wire 210 m long carries a current of 12 A. Knowing that the temperature of the conductor is 105°C, calculate:

a. the current density [A/mm^2];

b. the copper losses [W/kg].

Advanced level

16-18 An aluminum conductor operates at a current density of 2 A/mm^2.

a. If the conductor temperature is 120°C,

calculate the losses [W/kg].

b. Express the current density in circular mils per ampere.

16-19 The temperature rise of a motor is roughly proportional to its losses. On the other hand, its efficiency is reasonably constant in the range between 50% and 150% of its nominal rating (see, for example, Fig. 16-5). Based on these facts, if a 20 kW motor has a full-load temperature rise of 80°C, what power can it deliver at a temperature rise of 105°C?

16-20 The armature in Fig. 15-4 is designed for a 6-pole dc motor running at 1200 r/min. It possesses 81 slots and the peak flux density in the teeth is 1.4 T. Each tooth is 10 mm wide and 35 mm deep. If the laminations are type M-36 No. 24 gauge, calculate:

a. the frequency of the ac flux in each tooth;

b. the iron losses in the teeth (see Fig. 6-15).

16-21 An 11 kW ac motor having class B insulation would normally have a life span of 20 000 h provided the winding temperature by resistance does not exceed 120°C. By how many hours is the lifetime reduced if the motor runs for 3 h at a temperature (by resistance) of 200°C?   (See Sec. 4-6).

16-22 The armature of a 300 kW dc motor is made of No. 24 gauge type M-36 laminations. The peak flux density is 1.2 T. Calculate the percent reduction in iron losses if No. 29 gauge type M-14 laminations were used.

17
THREE-PHASE INDUCTION MOTORS

Three-phase induction motors are the motors most frequently encountered in industry. They are simple, rugged, low-priced, and easy to maintain. They run at essentially constant speed from zero to full load. However, the speed is frequency dependent and, consequently, these motors are not easily adapted to speed control. We usually prefer direct-current motors when large speed variations are required. Nevertheless, variable frequency electronic sources are now available to drive commercial induction motors (Chapter 25).

17-1 Principal components

A 3-phase induction motor has two main parts: a stationary stator and a revolving rotor. The rotor is separated from the stator by a small air-gap which ranges from 0.4 mm to 4 mm, depending on the power of the motor.

The *stator* consists of a steel frame which encloses a hollow, cylindrical core made up of stacked laminations. A number of evenly spaced slots, punched out of the internal circumference of the laminations, provide the space for the stator

winding (Fig. 17-1).

The *rotor* is also composed of punched laminations. These are carefully stacked to create a series of rotor slots to provide space for the rotor winding. We use two types of rotor windings: (a) conventional three-phase windings made of insulated wire and (b) squirrel-cage windings. The type of winding gives rise to two main classes of motors: *squirrel-cage induction motors* and *wound-rotor induction motors*.

A **squirrel-cage rotor** (Fig. 17-2) is composed of bare copper bars, slightly longer than the rotor, which are pushed into the slots. The opposite ends are welded to two copper end-rings, so that all the bars are short-circuited together. The entire construction (bars and end-rings) ressembles a squirrel-cage, from which the name is derived. In small and medium-size motors, the bars and end-rings are made of die-cast aluminum, molded to form an integral block (Fig. 17-3).

A **wound rotor** has a three-phase winding, similar to the one on the stator. The winding is uniformly distributed in the slots and is usually con-

Figure 17-1 Stator of a 2 kW, 1725 r/min, 60 Hz, 3-phase induction motor. *(Brook Crompton Parkinson Ltd)*

Figure 17-2 Squirrel-cage rotor of a 2 kW, 1725 r/min, 60 Hz, 3-phase induction motor. *(Brook Crompton Parkinson Ltd)*

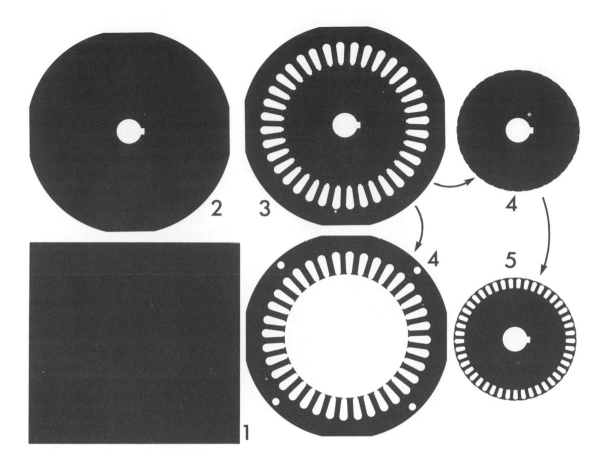

Figure 17-3   a. Progressive steps in the manufacture of stator and rotor laminations. Sheet steel is sheared to size (1), blanked (2), punched (3), blanked (4) and punched (5). *(Lab-Volt)*

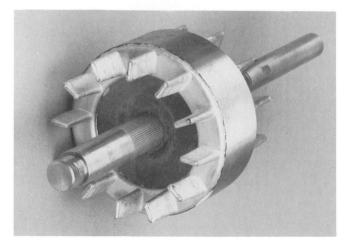

Figure 17-3 b

Die-cast aluminum squirrel-cage rotor with integral cooling fan. *(Lab-Volt)*

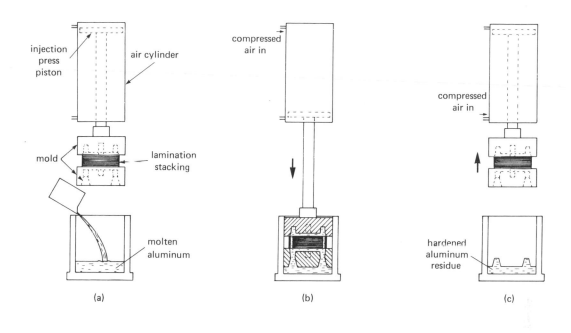

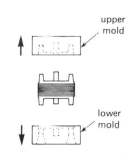

Figure 17-3 c. Progressive steps in the injection molding of a squirrel-cage rotor.

a. Molten aluminum is poured into a cylindrical cavity. The laminated rotor stacking is firmly held between two molds.

b. Compressed air rams the mold assembly into the cavity. Molten aluminum is forced upwards through the rotor bar holes and into the upper mold.

c. Compressed air withdraws the mold assembly, now completely filled with hot (but hardened) aluminum.

d. The upper and lower molds are pulled away, revealing the die-cast rotor. The cross section view shows that the upper and lower end rings are joined by the rotor bars. *(Lab-Volt)*

Figure 17-4   a.   Exploded view of a 5 hp, 1730 r/min wound-rotor induction motor.

Figure 17-4   b.   Closeup of the slip ring end of the rotor. *(Brook Crompton Parkinson Ltd)*

nected in wye. The terminals are connected to three slip-rings which turn with the rotor (Fig. 17-4). The revolving slip-rings and associated stationary brushes enable us to connect external resistors in series with the rotor winding. The external resistors are mainly used during the start-up period; under normal running conditions, the three brushes are short-circuited.

17-2 Principle of the induction motor

The operation of a 3-phase induction motor is based upon the application of Faraday's Law and the Lorentz force on a conductor (Secs. 2-8, 2-9, and 2-25). The behavior can readily be understood by means of the following example.

Consider a series of conductors of length l whose extremities are short-circuited by two bars A and B (Fig. 17-5). A permanent magnet, placed

the magnetic field of the permanent magnet, it experiences a mechanical force (Lorentz force);
4. the force always acts in a direction to drag the conductor along with the magnetic field (Sec. 2-26).

If the conducting "ladder" is free to move, it will accelerate towards the right. However, as it picks up speed, the conductors will be cut less rapidly, with the result that the induced voltage E and the current I will diminish. Consequently, the force acting on the conductors will also decrease. If the ladder were to move at the same speed as the magnetic field, the induced voltage E, the current I, and the force would all be zero.

In an induction motor, the ladder is closed upon itself to form a squirrel-cage, and the moving magnet is replaced by a rotating field. The field is produced by the 3-phase currents which flow in the stator windings, as we shall now explain.

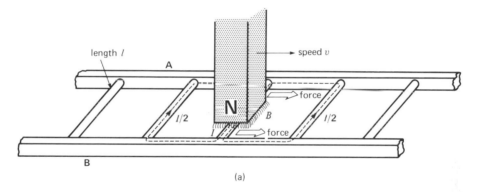

Figure 17-5 a. Moving magnet cutting across a conducting ladder.

above this conducting "ladder," moves rapidly to the right at a speed v, so that its magnetic field B sweeps across the conductors. The following sequence of events then takes place:
1. a voltage $E = Blv$ is induced in each conductor while it is being cut by the flux (Faraday's Law);
2. the induced voltage immediately produces a current I, which flows down the conductor, through the end-bars, and back through the other conductors;
3. because the current-carrying conductor lies in

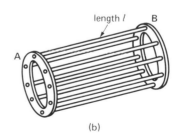

Figure 17-5 b. Ladder bent upon itself to form a squirrel-cage.

17-3 The rotating field

Consider a simple stator having six salient poles, connected as shown in Fig. 17-6. Three identical windings AN, BN, CN, each composed of two coils in series, are arranged to be mechanically spaced at 120° to each other. The respective coil ends aa, bb, cc are joined but we have neglected to show the connections so as not to encumber the drawing. The three sets of coils are connected in wye, and have a common neutral N. Owing to the perfectly symmetrical arrangement, the line-to-neutral impedances are identical. In other words, with regard to terminals A, B, C, the windings constitute a balanced 3-phase system.

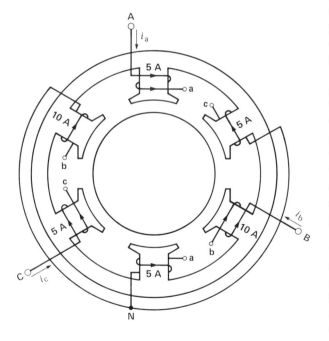

Figure 17-6   Elementary stator having terminals A, B, C, connected to a 3-phase source (not shown).

If we apply a 3-phase source to terminals A, B, C, alternating currents I_a, I_b and I_c will flow in the windings. The currents will have the same magnitude, but will be displaced in time by an angle of 120°. These currents produce magnetomotive forces which, in turn, create a magnetic flux. It is this flux we are interested in.

In order to follow the sequence of events, we assume that positive currents always flow from line to neutral. Conversely, negative currents flow from neutral to line. Furthermore, to enable us to work with numbers, suppose that each coil has 5 turns and that the peak current per phase is 10 A. Thus, when $I_a = +7$ A, the two coils of phase A will together produce a mmf of 70 ampere-turns and a corresponding value of flux. Because the current is positive, the flux is directed vertically upwards, according to the right-hand rule.

As time goes by, we can determine the instantaneous value and direction of the current in each winding, and thereby establish the successive flux patterns. Thus, referring to Fig. 17-7 at instant 1, current I_a has a value of + 10 A, whereas I_b and I_c both have a value of – 5 A. The mmf of phase A is 10 A x 10 turns = 100 ampere-turns, while the mmf of phases B and C are each 50 ampere-turns. The direction of the mmf depends upon the instantaneous current flows and, using the right-hand rule, we find that the resulting magnetic field has the shape shown in Fig. 17-8a. Note that as far as the rotor is concerned, the six poles together produce a magnetic field having essentially one north and one south pole. The combined magnetic field points upwards.

At instant 2, one sixth cycle later, current I_c attains a peak of – 10 A, while I_a and I_b both have a value of + 5 A (Fig. 17-8b). We discover that the new field has the same shape as before, except that it has moved clockwise by an angle of 60°. In other words, the flux makes one sixth of a turn between instants 1 and 2.

Proceeding in this way for each of the successive instants 3, 4, 5, 6, and 7, separated by intervals of 1/6 cycle, we find that the magnetic field makes one complete turn during one cycle (see Figs. 17-8a to 17-8d).

The rotational speed of the field depends, therefore, upon the frequency of the source. If the frequency is 60 Hz, the resulting field makes one

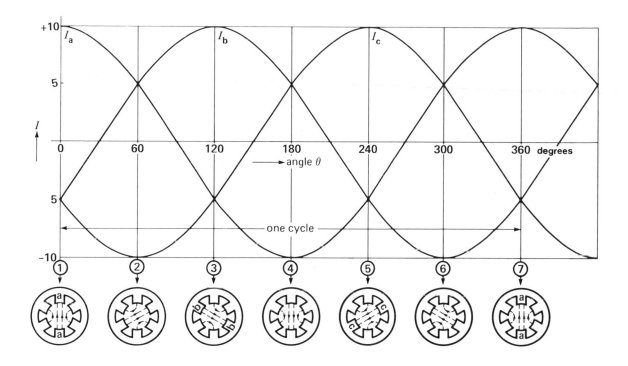

Figure 17-7   Instantaneous values of currents in Fig. 17-6.

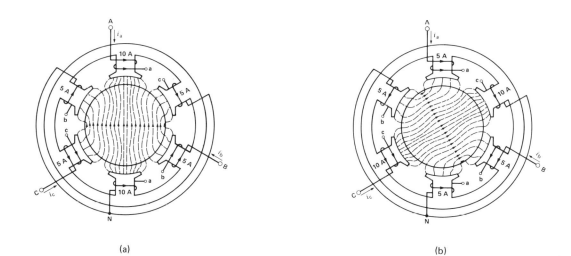

(a)   (b)

Figure 17-8   The field rotates by 60° between instant 1 and instant 2.
            a.   flux pattern at instant 1   b.   flux pattern at instant 2.

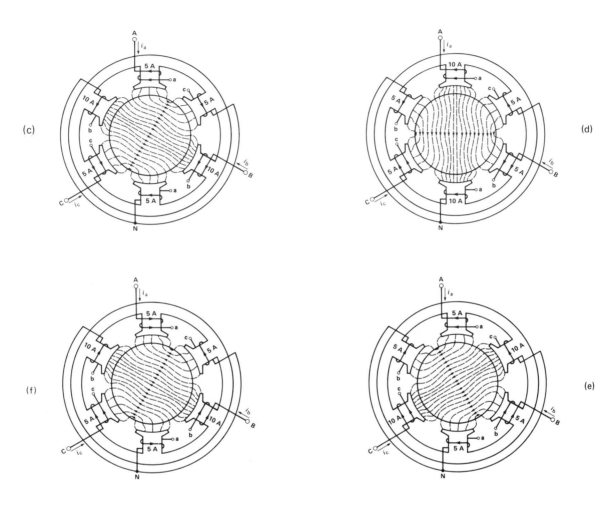

Figure 17-8   The field rotates through a further 240° between instants 2 and 6.
c. flux pattern at instant 3   e. flux pattern at instant 5
d. flux pattern at instant 4   f. flux pattern at instant 6.

turn in 1/60 s, or 3600 revolutions per minute. On the other hand, if the frequency were 5 Hz, the field would make one turn in 1/5 s, giving a speed of only 300 r/min. Because the speed of the rotating field is necessarily synchronized with the frequency of the source, it is called *synchronous speed*.

17-4 Direction of rotation

The positive crests of the currents in Fig. 17-7 fol-low each other in the order A-B-C. This phase sequence produces a field that rotates clockwise. If we interchange any two of the lines connected to the stator, the new phase sequence will be A - C - B. Following the same line of reasoning developed in Sec. 17-3, we find that the field now revolves at synchronous speed in the opposite, or counter-clockwise direction. Interchanging any two lines of a 3-phase motor will therefore reverse its direction of rotation.

17-5 Number of poles - synchronous speed

Although early machines were built with salient poles, the stators of modern motors are smooth. Thus, the two-pole stator of Fig. 17-6 is built as shown in Fig. 17-9a. In effect, the two stator coils of phase A are stretched to cover the full 180° span of one pole. The coils of the other two phases are identical and, as in Fig. 17-6, all the coils are displaced at 120° to each other. The individual coils are called phase groups, or simply groups. The resulting field is similar and consists of 2 poles.

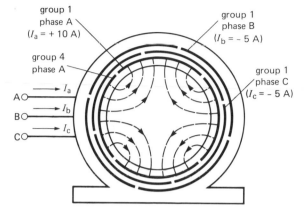

Figure 17-9 b. Four-pole, full-pitch lap wound stator and resulting magnetic field when I_a = + 10 A.

windings are connected to a 3-phase source, a revolving field having 4 poles is created (Fig. 17-9b). This field rotates at only half the speed of the 2-pole field shown in Fig. 17-9a.

We can increase the number of poles as much as we please provided there are enough slots. Thus, Fig. 17-9c shows a 3-phase 8-pole stator. Each phase consists of 8 groups and the groups of all the phases together produce an 8-pole rotating field, (Fig. 17-9c). When connected to a 60 Hz source,

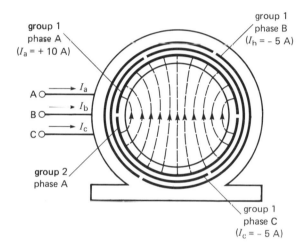

Figure 17-9 a. Two-pole, full-pitch lap wound stator and resulting magnetic field when the current in phase A = + 10 A.

To construct a 4-pole stator, the coils are distributed as shown in Fig. 17-9b. The 4 groups of phase A are identical but they now span only 90° of the stator circumference. The groups are connected in series and in such a way that adjacent groups produce magnetomotive forces acting in opposite directions. In other words, if a current were to flow only in the stator winding of phase A, it would create 4 alternate N-S poles. The windings of the other two phase are identical but are displaced from each other (and from phase A) by a mechanical angle of 60°. When the wye-connected

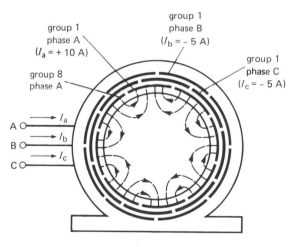

Figure 17-9 c. Eight-pole pole, full-pitch lap wound stator and resulting magnetic field when I_a = + 10 A.

the poles turn, like the spokes of a wheel, at a synchronous speed of 900 r/min.

Is there any way we can tell what the synchronous speed should be? Without going into all the details of current flow in the three phases, let us restrict our attention to phase A. In Fig. 17-9c, each phase group covers a mechanical angle of 360/8 = 45°. Suppose the current in phase A is at its maximum positive value. The magnetic flux is then centered with respect to phase A, and the N-S poles are located as shown in the figure. A few

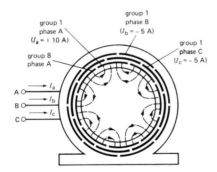

group 1
phase A
($I_a = \cup 10$ A)

group 1
phase B
($I_b = -5$ A)

group 1
phase C
($I_c = -5$ A)

group 8
phase A

A $\circ$ $\xrightarrow{I_a}$
B $\circ$ $\xrightarrow{I_b}$
C $\circ$ $\xrightarrow{I_c}$

Figure 17-9 c. Determining the synchronous speed.

moments later, the current in phase A will reverse and when it reaches its maximum negative value, the flux pattern will be the same as before, except that all the N poles will become S poles and vice versa. In effect, the entire magnetic field shifts by an angle of 45° — and this gives us the key to finding the speed of rotation. The flux moves 45° in one half cycle, and so it takes 8 half cycles (= 4 cycles) to move 360°. On a 60 Hz system the time to make one turn is therefore 4 x 1/60 = 1/15 s. The flux turns at the rate of 15 r/s or 900 r/min.

The speed of a rotating field depends therefore upon the frequency of the source and the number of poles on the stator. Using the same reasoning as above, we can prove that the speed is always given by:

$$n_s = \frac{120 f}{p} \qquad (17\text{-}1)$$

where

n_s = synchronous speed [r/min]
f = frequency of the source [Hz]
p = number of poles per phase

This equation shows that the synchronous speed increases with frequency and decreases with the number of poles.

Example 17-1:

Calculate the synchronous speed of a three-phase induction motor having 20 poles per phase when it is connected to a 50 Hz source.

Solution:

$$n_s = 120 f/p = 120 \times 50/20 = 300 \text{ r/min}$$

17-6 Starting characteristics of a squirrel-cage motor

Let us connect the stator of an induction motor to a 3-phase source, with the rotor locked. The revolving field created by the stator cuts across the rotor bars and induces a voltage in all of them.

This is an ac voltage because each conductor is cut, in rapid succession, by a N pole followed by a S pole. The frequency depends upon the number of N and S poles that sweep across a conductor per second; when the rotor is at rest, it is always equal to the frequency of the source.

Because the rotors bars are shorted by the end-rings, the induced voltage causes a large current to flow - usually several hundred amperes per bar in machines of medium power.

The current-carrying conductors are in the path of the flux created by the stator, consequently, they all experience a strong mechanical force. These forces tend to drag the entire rotor along with the revolving field.

In summary:

1. a revolving magnetic field is set up when a 3-phase voltage is applied to the stator of an induction motor;

2. the revolving field induces a voltage in the rotor bars;

3. the induced voltage creates large circulating cur-

rents which flow in the rotor bars and end-rings;

4. the current-carrying rotor bars are immersed in the magnetic field created by the stator; they are therefore subjected to a strong mechanical force;

5. the sum of the mechanical forces on all the rotor bars produces a torque which tends to drag the rotor along in the direction of the revolving field.

17-7 Acceleration of the rotor - slip

As soon as we release the rotor, it rapidly accelerates in the direction of the rotating field. As it picks up speed, the relative velocity of the field with respect to the rotor diminishes progressively. This causes both the value and the frequency of the induced voltage to decrease because the rotor bars are cut more slowly. The rotor current, very large at first, decreases rapidly as the motor picks up speed.

The speed will continue to increase, but it will never catch up with the revolving field. In effect, if the rotor *did* turn at the same speed as the field (synchronous speed), the flux would no longer cut the rotor bars and the induced voltage and current would fall to zero. Under these conditions, the force acting on the rotor bars would also become zero and the friction and windage would immediately cause the rotor to slow down.

The rotor speed must always be slightly less than synchronous speed so as to produce a current in the rotor bars sufficiently large to overcome the braking torque. At no-load the difference in speed between the rotor and field (called slip), is very small: usually less than 0.1% of synchronous speed.

17-8 Motor under load

If we apply a mechanical load to the shaft, the motor will begin to slow down and the revolving field will cut the rotor bars at a higher and higher rate. The induced voltage and the resulting current in the bars will increase progressively, producing a

greater and greater motor torque. The question is, for how long can this go on? Will the speed continue to drop until the motor comes to a halt?

No; the motor and the mechanical load will reach a state of equilibrium, making peace with each other so to speak, when the motor torque is exactly equal to the load torque. When this state is reached, the speed will cease to drop any more and the motor will turn at a constant rate. It is very important to understand that a motor only turns at constant speed when its torque is *exactly* equal to the torque exerted by the mechanical load. The moment this state of equilibrium is upset, the motor speed will start to change (Sec. 3-11).

A motor is essentially a willing workhorse, doing its best to develop a torque always equal to that imposed by the load. If the load torque exceeds the maximum the motor can produce, it will simply come to a stop.

Under normal loads, induction motors run very close to synchronous speed. Thus, at full load, the slip for large motors (1000 kW and more) rarely exceeds 0.5% of synchronous speed, and for small machines (10 kW and less), it seldom exceeds 3%. This is why induction motors are considered to be constant-speed machines. However, because they never actually turn at synchronous speed, they are sometimes called asynchronous machines.

17-9 Slip

The slip s of an induction motor is the difference between the synchronous speed and the rotor speed expressed as a percent of synchronous speed.

$$s = \frac{n_s - n}{n_s} \qquad (17\text{-}2)$$

where

s = slip
n_s = synchronous speed [r/min]
n = rotor speed [r/min]

The slip is practically zero at no-load, and it is equal to 1 (or 100%) when the rotor is locked.

Example 17-2:

A 6-pole induction motor is excited by a 3-phase, 60 Hz source. If the full-load speed is 1140 r/min, calculate the slip.

Solution:

The synchronous speed of the motor is:

$$n_s = 120\,f/p = 120 \times 60/6 \qquad \text{Eq. 17-1}$$
$$= 1200 \text{ r/min}$$

The difference between the synchronous speed and rotor speed is the slip speed:

$$n_s - n = 1200 - 1140 = 60 \text{ r/min}$$
$$s = (n_s - n)/n_s = 60/1200 \qquad \text{Eq. 17-2}$$
$$= 0.05 \text{ or } 5\%$$

17-10 Voltage and frequency induced in the rotor

The voltage and frequency induced in the rotor both depend upon the slip. They are given by the following equations:

$$f_2 = sf_0 \qquad (17\text{-}3)$$

$$E_2 = sE_{0c} \qquad (17\text{-}4)$$

where

f_2 = rotor frequency [Hz]
f_0 = frequency of the source connected to the stator [Hz]
s = slip
E_2 = voltage induced in the rotor at slip s [V]
E_{0c} = open-circuit voltage induced in the rotor when at rest [V]

It should be noted that Eq. 17-3 always holds true, but Eq. 17-4 is valid only if the revolving flux (in webers) remains constant.

Example 17-3:

A 6-pole wound-rotor induction motor is excited by a 3-phase 60 Hz source. Calculate the frequency of the rotor current under the following conditions:

a. at standstill;

b. motor turning at 500 r/min in the same direction as the revolving field;

c. motor turning at 500 r/min in the opposite direction to the revolving field;

d. motor turning at 2000 r/min in the same direction as the revolving field.

Solution:

From Example 17-2, the synchronous speed of the motor is 1200 r/min.

a. 1 At standstill, the motor speed $n = 0$. Consequently, the slip is:

$$s = (n_s - n)/n_s = (1200 - 0)/1200 = 1$$

a. 2 The frequency of the induced voltage (and of the induced current) is:

$$f_2 = sf_0 = 1 \times 60 = 60 \text{ Hz}$$

b. 1 When the motor turns in the same direction as the field, the motor speed n is positive. The slip is:

$$s = (n_s - n)/n_s = (1200 - 500)/1200$$
$$= 7/12 = 0.583$$

b. 2 The frequency of the induced voltage (and of the rotor current) is:

$$f_2 = sf_0 = 0.583 \times 60 = 35 \text{ Hz}$$

c. 1 When the motor turns in the opposite direction to the field, the motor speed is *negative*: $n_s = -500$. The slip is:

$$s = (n_s - n)/n_s = \{1200 - (-500)\}/1200$$
$$= (1200 + 500)/1200 = 17/12$$
$$= 1.417$$

c. 2 The frequency of the induced voltage and rotor current is:

$$f_2 = sf_0 = 1.417 \times 60 = 85 \text{ Hz}$$

d. 1 The motor speed is positive: $n = +2000$.

d. 2 The slip is:

$$s = (n_s - n)/n_s = (1200 - 2000)/1200$$
$$= -2/3 = -0.667$$

d. 3 The frequency of the induced voltage and rotor current is:

$$f_2 = sf_0 = -0.667 \times 60 = -40 \text{ Hz}$$

A negative frequency means that the phase sequence of the voltages induced in the rotor windings is reversed. Thus, if the phase sequence of the voltages is A-B-C when the frequency is positive, the phase sequence is A-C-B when the frequency is negative. As far as a frequency meter is concerned, a "negative" frequency gives the same reading as a "positive" frequency does. Consequently, we can say that the frequency is simply 40 Hz.

17-11 Characteristics of squirrel-cage induction motors

Table 17A lists the typical properties of squirrel-cage induction motors in the power range between 1 kW and 20 000 kW. The following explanations will help us understand the values and the percentage given in the table.

1. Motor at no load. When the motor runs at no load, the stator current lies between 0.5 and 0.3 p.u. (of full-load current). The no-load current is composed of a magnetizing component which creates the revolving flux and a small active component that supplies the windage and friction losses in the rotor plus the iron losses in the stator.

Considerable reactive power is needed to create the revolving field and, to keep it within acceptable limits, we use as short an air-gap as mechani-cal tolerances will permit. The power factor at no-load is therefore low; it ranges from 0.2 (or 20%) for small machines to 0.05 for large machines. The efficiency is zero because the output power is zero.

2. Motor under load. Both the exciting current and reactive power under load remain about the same as at no load. However, the active power (kW) absorbed by the motor increases in proportion to the mechanical load. It follows that the power factor of the motor improves as the mechanical load increases. At full load, it ranges from 0.70 for small machines to 0.90 for large machines. The efficiency at full load is particularly high; it can attain 98% for very large machines.

3. Locked rotor characteristics. The locked-rotor current is 5 to 6 times the full-load current, making the I^2R losses 25 to 36 times higher than normal. The rotor must therefore never remain locked for more than a few moments.

Although the mechanical power is zero, the motor develops a strong torque. The power factor is low because considerable reactive power is needed to produce the leakage flux in the rotor and stator windings. These leakage fluxes are much

TABLE 17A   TYPICAL CHARACTERISTICS OF SQUIRREL - CAGE INDUCTION MOTORS *(per unit values)*

Loading	Current		Torque		Slip		Efficiency		Power factor	
motor size →	small*	big*	small	big	small	big	small	big	small	big
Full load	1	1	1	1	0.03	0.004	0.7 to 0.9	0.96 to 0.98	0.8 to 0.85	0.87 to 0.9
No-load	0.5	0.3	0	0	≈ 0	≈ 0	0	0	0.2	0.05
Locked rotor	5 to 6	4 to 6	1.5 to 3	0.5 to 1	1	1	0	0	0.4	0.1

*   **small** means under 11 kW (15 hp)   **big** means over 1120 kW (1500 hp) and up to 25 000 hp.

larger than in a transformer because the stator and the rotor windings are not as tightly coupled (see Sec. 11-3).

17-12   Calculation of approximate current

The full-load current of a three-phase induction motor may be calculated by means of the following approximate equation:

$$I = 600\,P_\text{h}/E \qquad (17\text{-}5)$$

where
I = full load current [A]
P_h = output power [horsepower]

Recalling that the starting current is 5 to 6 times full-load current and the no-load current is between 0.5 and 0.3 p.u., we can readily estimate the critical currents for any induction motor.

Example 17-4:
a. Calculate the approximate full-load current, locked-rotor current and no-load current of a 3-phase induction motor having a rating of 500 hp, 2300 V.
b. What is the apparent power drawn under locked rotor conditions?
c. Give the nominal rating of this motor, expressed in kilowatts.

Solution:
a. The full-load current is:
I = $600\,P_\text{h}/E$ Eq. 17-5
     = 600 x 500/2300
     = 130 A (approx.)
No-load current:
I = 0.3 x 130 = 39 A (approx.)

Starting current:
I = 6 x 130 = 780 A (approx.)
b. The apparent power under locked-rotor conditions is:
S = $\sqrt{3}\,EI$ = 1.73 x 2300 x 780   Eq. 9-9
     = 3100 kVA

c. When the power of a motor is expressed in kilowatts, it always relates to the mechanical output, and *not* to the electrical input. The nominal rating of this motor expressed in SI units is therefore:
P = 500/1.34
     = 373 kW (see Power conversion chart)

17-13   Active power flow

Voltages, currents, and phasor diagrams enable us to understand the detailed behavior of an induction motor. However, it is easier to see how electrical energy is converted into mechanical energy by following the active power as it flows through the machine. Thus, referring to Fig. 17-10, active power P_e flows from the line into the three-phase stator. Owing to the stator copper losses, a portion P_js is dissipated as heat in the windings. Another portion P_f is dissipated as heat in the stator core, owing to the iron losses. The remaining active power P_r is carried across the air-gap and transferred to the rotor by electro-magnetic induction.

Owing to the I^2R losses in the rotor, a third portion P_jr is dissipated as heat and the remainder is finally available in the form of mechanical power P_m. By subtracting a small fourth portion P_v, representing windage and bearing-friction losses, we finally obtain P_L, the mechanical power available at the shaft and supplied to the load.

The power flow diagram of Fig. 17-10 enables us to identify and to calculate three important properties of the induction motor: 1) *its efficiency*, 2) *its power*, and 3) *its torque*.

1. Efficiency. By definition, the efficiency of a motor is the ratio of the output power to the input power:

$$\text{efficiency } (\eta) = P_\text{L}/P_\text{e} \qquad (17\text{-}6)$$

2. I^2R losses in the rotor. It can be shown[*] that

* See page 312.

the rotor I^2R losses P_{jr} are related to the rotor input power P_r by the equation:

$$P_{jr} = sP_r \qquad (17\text{-}7)$$

where

P_{jr} = rotor losses [W]
s = slip
P_r = power transmitted to the rotor [W]

Equation 17-7 shows that the rotor I^2R losses become more and more important as the slip increases. A rotor turning at half synchronous speed ($s = 0.5$) dissipates in the form of heat 50% of the active power it receives. When the rotor is locked ($s = 1$), all the power is dissipated as heat.

where

T_m = torque developed by the motor [N·m]
P_r = power transmitted to the rotor [W]
n_s = synchronous speed [r/min]
9.55 = multiplier to take care of units [exact value: $60/2\pi$]

The actual torque T_L available at the shaft is slightly less than T_m, owing to the torque required to overcome the windage and friction losses. However, in most calculations we can neglect this small difference.

Equation 17-9 shows that the torque is directly proportional to the active power transmitted to the rotor. Thus, to develop a high locked rotor torque, the rotor must absorb a large amount of

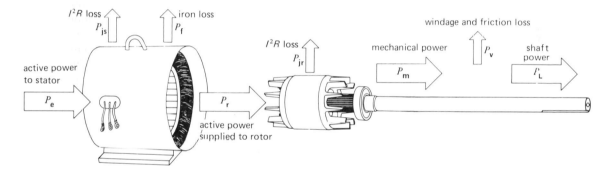

Figure 17-10 Active power flow in a 3-phase induction motor.

3. Mechanical power. The mechanical power P_m developed by the motor is obviously:

$$P_m = P_r - P_{jr} \qquad \text{Eq. 17-7}$$
$$= P_r - sP_r$$

whence

$$P_m = (1 - s)P_r \qquad (17\text{-}8)$$

4. Motor torque. The torque T_m developed by the motor is given by:

$$T_m = 9.55\,P_r/n_s \qquad (17\text{-}9)$$

active power. The latter is dissipated in the form of heat, consequently, the temperature of the rotor rises very rapidly.

Example 17-5:

A 3-phase induction motor having a synchronous speed of 1200 r/min draws 80 kW from a 3-phase feeder. The copper losses and iron losses in the stator amount to 5 kW. If the motor runs at 1152 r/min, calculate:

a. the active power transmitted to the rotor;
b. the rotor I^2R losses;
c. the mechanical power developed;
d. the mechanical power delivered to the load,

knowing that the windage and friction losses are equal to 2 kW;

e. the efficiency of the motor.

Solution:

a.   Active power to the rotor is:

$$P_r = P_e - P_{js} - P_f$$
$$= 80 - 5 = 75 \text{ kW}$$

b. 1   The slip is:

$$s = (n_s - n)/n_s = (1200 - 1152)/1200$$
$$= 48/1200 = 0.04$$

b. 2   Rotor I^2R losses are:

$$P_{jr} = sP_r = 0.04 \times 75 = 3 \text{ kW}$$

c.   The mechanical power developed is:

$$P_m = P_r - I^2R \text{ losses in rotor}$$
$$= 75 - 3 = 72 \text{ kW}$$

d.   The mechanical power P_L delivered to the load is slightly less than P_m owing to the friction and windage losses.

$$P_L = P_m - P_v = 72 - 2 = 70 \text{ kW}$$

e. $$\eta = P_L/P_e = 70/80$$
$$= 0.875 \text{ or } 87.5\%$$

rotor is:

$$P_r = P_e - P_{js} - P_f$$
$$= 40 - 5 - 1 = 34 \text{ kW}$$
$$T_m = 9.55 \, P_r/n_s \qquad \text{Eq. 17-9}$$
$$= 9.55 \times 34\,000/900$$
$$= 361 \text{ N·m } (\approx 266 \text{ ft·lb})$$

Note that the solution to this problem is independent of the actual speed of rotation. The motor could be at a standstill or running at full speed, but as long as the power P_r transmitted to the rotor is equal to 34 kW, the motor develops a torque of 361 N·m.

Example 17-7:

A 3-phase induction motor having a nominal rating of 100 hp ($\approx$ 75 kW) is excited by a 600 V source (Fig. 17-11a). The two-wattmeter method shows a total power consumption of 70 kW, and an ammeter indicates a line current of 78 A. Precise measurements give a rotor speed of 1763 r/min. In addition, the following characteristics are known about the motor:

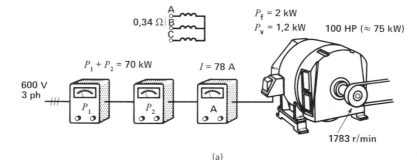

(a)

Figure 17-11   a.   See Example 17-7.

Example 17-6:

A 3-phase squirrel-cage induction motor, connected to a 60 Hz line, possesses a synchronous speed of 900 r/min. The motor absorbs 40 kW, and the stator copper and iron losses amount to 5 kW and 1 kW, respectively. Calculate the torque developed by the motor.

Solution:

The power transmitted across the air gap to the

stator iron losses $P_f = 2$ kW
windage and friction losses $P_r = 1.2$ kW
resistance between two stator terminals $= 0.34 \, \Omega$

Calculate:

a. power supplied to the rotor;

b. rotor I^2R losses;

c. mechanical power supplied to the load, in horsepower;

d. efficiency;

e. torque developed at 1763 r/min.

Solution:

a. 1   Power supplied to the stator is:
$P_e = 70$ kW;

a. 2   Stator resistance per phase (assume a wye connection) is:
$R = 0.34/2 = 0.17 \ \Omega$;

a. 3   Stator I^2R losses are:
$P_{js} = 3 \ I^2R = 3 \times 0.17 \times (78)^2 = 3.1$ kW;

a. 4   Iron losses $P_f = 2$ kW;

a. 5   Power supplied to the rotor $P_r = P_e - P_{js} - P_f$
$= (70 - 3.1 - 2) = 64.9$ kW;

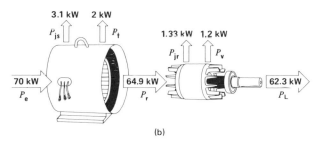

(b)

Figure 17-11   b. Power flow in Example 17-7.

b. 1   The slip is:
$s = (n_s - n)/n_s = (1800 - 1763)/1800$
$= 0.0205$;

b. 2   Rotor I^2R losses:
$P_{jr} = sP_r = 0.0205 \times 64.9 = 1.33$ kW;

c. 1   Mechanical power developed is:
$P_m = P_r - P_{jr} = 64.9 - 1.33 = 63.5$ kW

c. 2   Mechanical power P_L to the load:
$P_L = 63.5 - P_v = 63.5 - 1.2 = 62.3$ kW
$= 62.3 \times 1.34 = 83.5$ hp

The previous calculations are summarized in Fig. 17-11b.

d.   Efficiency of the motor is:
$\eta = P_L/P_e = 62.3/70 = 0.89$ or 89%

e.   Torque at 1763 r/min:
$T = 9.55 \ P_r/n_s = 9.55 \times 64 \ 900/1800$
$= 344$ N·m

17-14   Torque versus speed curve

The torque a motor develops depends upon its speed, but the relationship between the two cannot be expressed by a simple equation. Consequently, we prefer to show the relationship in the form of a curve. Figure 17-12 shows the torque-speed curve of a conventional 3-phase induction motor whose nominal, full-load torque is T. The starting torque is $1.5 \ T$ and the maximum torque (called breakdown torque) is $2.5 \ T$.

At full load, the motor runs at a speed n. If the mechanical load increases, the speed will drop until the motor torque is again equal to the load torque. As soon as the two torques are in balance, the motor will turn at a constant but lower, speed. However, if the load torque exceeds $2.5 \ T$, the motor will suddenly stop.

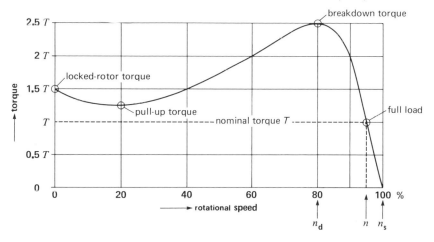

Figure 17-12   Typical torque-speed curve of a 3-phase squirrel-cage induction motor.

Small motors (10 kW and less) develop their maximum torque at a speed n_d of about 80% of synchronous speed. Big motors (1000 kW and more) attain their breakdown torque at about 98% of synchronous speed.

17-15 Effect of rotor resistance

If we increase the rotor resistance, the torque-speed curve will change. The breakdown torque is not affected, but it occurs at a lower speed. The starting torque, full-load speed, and other motor characteristics are also affected, as we see in the following example.

Figure 17-13a shows the torque-speed curve of a 10 kW (13.4 hp), 50 Hz, 380 V motor having a synchronous speed of 1000 r/min and a full-load torque of 100 N·m ($\approx$ 73.7 ft·lb). The full-load current is 20 A and the locked-rotor current, 100 A. The rotor has a nominal resistance R.

If we increase the resistance by a factor of 2.5, the starting torque doubles and the locked-rotor current decreases from 100 A to 90 A (Fig. 17-13b). The motor develops its breakdown torque at a speed n_d of 500 r/min, compared to the original breakdown speed of 800 r/min.

If we again double the resistance so that it becomes 5 R, the locked-rotor torque attains a maximum value of 250 N·m for a corresponding current of 70 A (Fig. 17-13c). A further increase in rotor resistance decreases both the locked-rotor torque and locked-rotor current. For example, if the rotor resistance is increased 25 times (25 R), the locked-rotor current drops to 20 A, but the motor develops the same starting torque (100 N·m), as it did when the locked-rotor current was 100 A (Fig. 17-13d).

To sum up, a high rotor resistance is desirable because it produces a high starting torque and a relatively low starting current (Fig. 17-13c). Unfortunately, it also produces a rapid fall-off in speed with increasing load. Furthermore, because the slip at rated torque is high, the rotor I^2R losses

are high. The efficiency is low and the motor tends to get hot.

Under running conditions, it is preferable to use a low rotor resistance (Fig. 17-13a). The speed decreases much less with increasing load, and the slip at rated torque is small. Consequently, the efficiency is high and the motor tends to run cool.

We can obtain both a high starting resistance and a low running resistance by designing the rotor bars in a special way (Sec. 18-3.2). However, if the rotor resistance has to be varied over a wide range, it may be necessary to use a wound-rotor induction motor. Such a motor enables us to vary the rotor resistance at will by means of an external rheostat.

17-16 Wound-rotor motor

We explained the basic difference between a squirrel-cage motor and a wound-rotor motor in Sec. 17-1. Although the wound-rotor motor costs more than a squirrel-cage motor does, it offers the following advantages:

1. The locked-rotor current can be drastically reduced by inserting three external resistors in series with the rotor. Nevertheless, the locked-rotor torque is still as high as that of a squirrel-cage motor;
2. The speed can be varied by varying the external rotor resistors;
3. The motor is ideally suited to accelerate high-inertia loads, which require a long time to come up to speed.

Figure 17-14 is a schematic diagram of the circuit used to start a wound-rotor motor. The rotor windings are connected to three wye-connected external resistors by means of a set of slip-rings and brushes. Under locked-rotor (LR) conditions, the variable resistors are set to their highest value. As the motor speeds up, we gradually reduce the resistance until full-load speed is reached, whereupon the brushes are short-circuited. By properly selecting the resistance values, we can produce a high accelerating torque with a stator current that never exceeds twice full-load current.

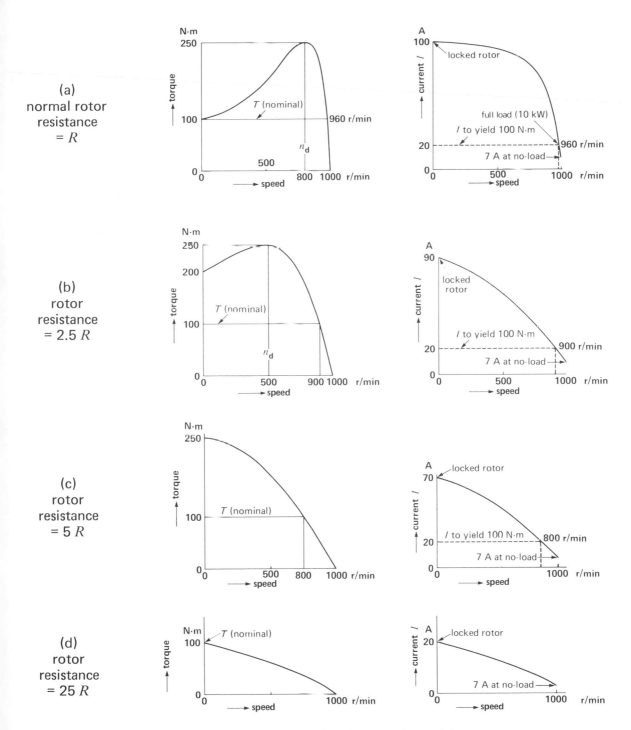

Figure 17-13   Rotor resistance affects the motor characteristics.

DERIVATION OF SOME FUNDAMENTAL INDUCTION MOTOR EQUATIONS

We have explained the torque, power, and slip of an induction motor in terms of the interaction between the induced rotor current and the magnetic field created by the stator. The behavior of the induction motor is, however, dictated by an even more fundamental principle which may be described as follows.

Consider a permanent magnet that can be rotated around a rotor. The rotor is made of a material that produces losses when subjected to a varying magnetic field. Specifically, these losses may be due to eddy currents, hysteresis, or a combination of the two.

Let us rotate the magnet clockwise at a constant speed n_s [r/min] while keeping the rotor locked. The resulting rotor losses P_{jr} [W] are entirely converted into heat. Obviously, the heat is produced at the expense of the mechanical power P_{mec} required to turn the magnet. Furthermore, to rotate the magnet, we must exert a torque T [N·m]. We can therefore write:

$$P_{jr} = P_{mec} = n_s T/9.55 \qquad \text{Eq. 3-5}$$
$$\therefore \ T = 9.55\,P_{jr}/n_s$$

Now, from Newton's third law we know that to every action there is an equal and opposite reaction. Consequently, the torque we exert on the magnet is exactly equal to the torque which the magnet exerts on the rotor. Furthermore, according to the same law, the torque tends to rotate the rotor in the same direction as the magnet. The locked rotor torque developed by the rotor is therefore given by:

$$T = 9.55\,P_{jr}/n_s$$

We conclude that the locked rotor torque is proportional to the power dissipated in the rotor, and inversely proportional to the speed of the revolving field.

Suppose, now, that the rotor is allowed to rotate while driving a load. Let it turn at a constant speed n, while the magnet continues to turn at the original constant speed n_s. Losses P_{jr} will still occur in the rotor, but they will usually be different from their locked-rotor value. Finally, suppose the rotor develops a torque T_m. Referring to the figure, and based upon Sec. 3-11, we can write the following equations:

$T_m = T_L$, otherwise the rotor speed n will change;
$T_1 = T_2$, otherwise the magnet speed n_s will change;
$T_2 = T_m$, because of Newton's third law of motion.
Consequently, $T_1 = T_2 = T_L = T_m$

The power P_r [W] transmitted to the rotor is obviously equal to the power required to drive the magnet. Thus:

$$P_r = n_s T_1/9.55 = n_s T_2/9.55 = n_s T_m/9.55$$

we can therefore write:

$$T_m = 9.55\,P_r/n_s \qquad \text{Eq. 17-9}$$

The mechanical power P_m delivered to the load is:

$$P_m = n T_L/9.55 = n T_m/9.55$$

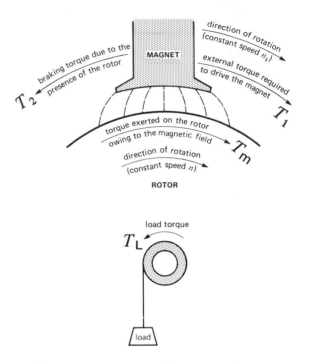

If the new rotor losses are P_{jr}, then according to the law of **conservation** of energy, the power P_r transmitted to the rotor must be:

$$P_r = P_m + P_{jr}$$

substituting, we have:

$$n_s T_m/9.55 = n T_m/9.55 + P_{jr}$$
$$(n_s - n)\,T_m = 9.55\,P_{jr}$$
$$(n_s - n)\,9.55\,P_r/n_s = 9.55\,P_{jr}$$

$$P_{jr} = \frac{(n_s - n)}{n_s}\,P_r$$

$$\therefore \qquad P_{jr} = sP_r \qquad \text{Eq. 17-7}$$

In an induction motor the revolving magnet is replaced by a rotating field, but the effect on the rotor is identical. In effect, the rotor has no way of "knowing" whether the field is due to a revolving magnet or to a 3-phase stator. Consequently, the equations developed here are basic. The reader will note that the relationships are independent of the mechanism producing the rotor losses, so long as they are due to a revolving field. Equations 17-7, 17-8 and 17-9 are therefore fundamental: they apply no matter how the rotor is constructed or of what material it is made.

To start large motors, we often use liquid rheostats because they are easy to control and have a large thermal capacity. A liquid rheostat is composed of three electrodes immersed in a suitable electrolyte. To vary its resistance, we simply vary the level of the electrolyte surrounding the electrodes. The large thermal capacity of the electrolyte limits the temperature rise. For example, a liquid rheostat is typically used with a 1260 kW wound-rotor motor to bring a large synchronous motor up to speed.

17-17 Three-phase windings

In 1883, a 27-year old Yougoslav scientist named Nikila Tesla invented the three-phase induction motor. His first model had a pole-type stator winding similar to the one shown in Fig. 17-6. Since then, the design of induction motors has evolved considerably; modern machines are built with so-called *lap windings* distributed in slots around the stator.

A lap winding consists of a set of phase groups

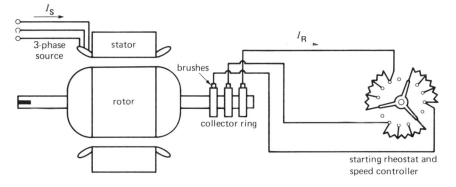

Figure 17-14   External resistors connected to the 3 slip rings of a wound-rotor induction motor.

We can also regulate the speed of a wound-rotor motor by varying the resistance of the rheostat. As we increase the resistance, the speed will drop. This method of speed control has the disadvantage that a lot of heat is dissipated in the resistors; the efficiency is therefore low. Furthermore, for a given rheostat setting, the speed varies considerably if the mechanical load varies.

The power of a wound-rotor motor depends upon its speed. Thus, for the same temperature rise, a motor which can develop 100 kW at 1800 r/min will deliver only about 40 kW at 900 r/min.

If we have to vary the speed of a very large motor, the continual power loss in the external resistors becomes unacceptable. Under these circumstances, we use electronic converters which, instead of wasting the rotor energy, pump it back into the 3-phase system (see Chapter 25).

evenly distributed around the stator circumference. The number of groups is given by the equation:

$$\text{groups} = \text{poles} \times \text{phases}$$

Thus, a 4-pole 3-phase stator must have 4 x 3 = 12 phase groups. Because a group must have at least one coil, it follows that the minimum number of coils is equal to the number of groups. A 4-pole, 3-phase stator must therefore have at least 12 coils. Furthermore, in a lap winding, the stator has as many slots as it has coils. Consequently, a 4-pole, 3-phase stator must have at least 12 slots. However, motor designers have discovered that it is preferable to use 2, 3, or more coils per group rather than only one. The number of coils and slots increases in proportion. For example, a 4-pole, 3-phase stator having 5 coils per group must have a total of (4 x 3 x 5) = 60 coils, lodged in 60 slots. The coils in each group are connected in series and staggered at one-slot intervals (Fig. 17-15).

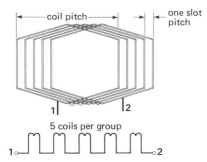

Figure 17-15 The 5 coils are connected in series to create one phase group.

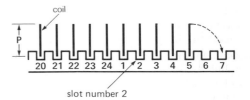

Figure 17-16 a. Coils held upright in stator slots.

Figure 17-16 b. Coils laid down to make a typical lap winding.

Such a *distributed* winding is obviously more costly to build than a concentrated winding having only 1 coil per group. However, it improves the starting torque and reduces the noise under normal running conditions. The coils are identical and may possess one or more turns. The width of each coil is called the *coil pitch*.

When the stator windings are excited from a 3-phase source, a multipole revolving field is produced. The width of one pole is called the *pole pitch*. It is equal to the internal circumference of the stator divided by the number of poles. For example, a 12-pole stator having a circumference of 600 mm has a pole pitch of 600/12 or 50 mm. In practice, the coil pitch is between 80% and 100% of the pole pitch. The coil pitch is usually less than the pole pitch in order to save copper and to improve the flux distribution in the air gap. The shorter coil width results in lower cost and weight, while the more sinusoidal flux distribution improves the torque during start-up, and often results in a quieter machine. In the case of two-pole machines, the shorter pitch also makes the coils much easier to insert in the slots.

To get an overall picture of a lap winding, let us consider a 24-slot stator laid out flat as shown in Fig. 17-16a. The 24 coils are held upright, with one coil-side standing in each slot. If the windings are now laid down so that all the other coil sides fall into the slots, we obtain the classical appearance of a 3-phase lap winding (Fig. 17-16b). The coils are connected together to create 3 identical windings, one for each phase. Each winding consists of a number of groups equal to the number of poles. Finally, the groups of each phase are symmetrically distributed around the circumference of the stator.

Example 17-8:
The stator of a 3-phase, 10-pole induction motor possesses 120 slots. If a lap winding is used, calculate:

a. the total number of coils;
b. the number of coils per phase;
c. the number of coils per group;
d. the pole pitch;
e. the coil pitch (expressed as a percent of the pole pitch), if the coil width extends from slot 1 to slot 11.

Solution:

a. A 120-slot stator requires 120 coils.
b. Coils per phase = $120 \div 3 = 40$.
c. 1 Number of groups per phase = number of poles = 10.
c. 2 Coils per group = $40 \div 10 = 4$.
d. The pole pitch corresponds to:
   pole pitch = slots/poles = 120/10
   = 12 slots.
   The pole pitch extends therefore from slot 1 (say) to slot 13.
e. The coil pitch covers 10 slots (slot 1 to slot

11). The percent coil pitch = 10/12 = 83.3%.

The next example shows in greater detail how the coils are interconnected in a typical 3-phase stator winding.

Example 17-9:

A stator having 24 slots has to be wound with a 3 phase, 4-pole winding. Determine:

a. the connections between the coils;

b. the connections between the phases.

Solution:

The winding has 24 coils. Assume that they are standing upright, with one coil-side in each slot. We shall first determine the coil distribution for phase A and then proceed with the connections. Similar connections are made for phases B and C. Here is the line of reasoning to follow:

a. The revolving field creates 4 poles; the motor therefore has 4 groups per phase, or 12 phase groups in all. Each rectangle in Fig. 17-17a rep-

resents one group. Because the stator contains 24 coils, each group consists of 24/12 = 2 consecutive coils. In effect, we have a distributed winding composed of 2 coils per phase group.

b. The groups of each phase must be uniformly spaced around the stator. The group distribution for phase A is shown in Fig. 17-17b. Each rectangle represents 2 upright coils connected in series, producing the two terminals shown. Note that the mechanical distance between two successive groups always corresponds to an *electrical* phase angle of 180°.

c. Successive groups must have opposite polarities. Consequently, the 4 groups of phase A are connected in series to produce successive N-S-N-S poles (Fig. 17-17c). The groups could be connected in parallel, or in series-parallel, so long as the N-S-N-S sequence is respected. Phase A now has two terminals, a "starting" terminal A_1 and a "finishing" terminal A_2.

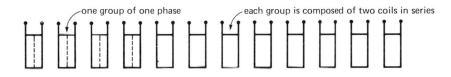

Figure 17-17   a. Coils connected in series two-by-two to make 12 groups.

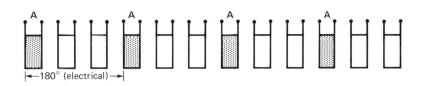

Figure 17-17   b. The 4 groups of phase A are evenly spaced from each other.

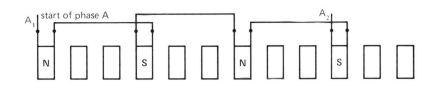

Figure 17-17   c. The groups of phase A are connected in series to create alternate N-S poles.

d. The phase groups of phases B and C are spaced the same way around the stator. However, the "starting" terminals B_1 and C_1 are respectively located at $120°$ and $240°$ (electrical) with respect to the starting terminal of phase A (Fig. 17-17d).

e. The groups in phases B and C are connected in series in the same way as those of phase A are (Fig. 17-17e). This yields six terminals: A_1A_2, B_1B_2 and C_1C_2. They may be connected either in wye or in delta. The 3 wires corresponding to the 3 phases are brought out to the terminal box of the machine (Fig. 17-17f). In practice,

the connections are made, not while the coils are upright (as shown) but only after they are laid down in the slots.

f. Because the pole pitch corresponds to a breadth of 24/4 = 6 slots, the coil pitch may be shortened to 5 slots (slot 1 to slot 6). Thus, the first coil of phase A is lodged in the first and sixth slots (Fig. (17-18). All the other coils and connections follow suit, according to Fig. 17-17e.

Figures 17-19 and 17-20 show the coil and stator of a 450 kW (600 hp) induction motor. Figure 17-21 illustrates the procedure used in winding a smaller 37.5 kW (50 hp) stator.

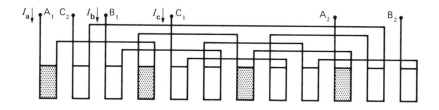

Figure 17-17   d. Start of phases B and C begins $120°$ and $240°$ after phase A.

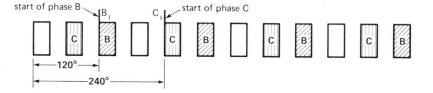

Figure 17-17   e. When all phase groups are connected, only leads remain.

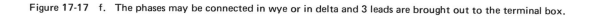

Figure 17-17   f. The phases may be connected in wye or in delta and 3 leads are brought out to the terminal box.

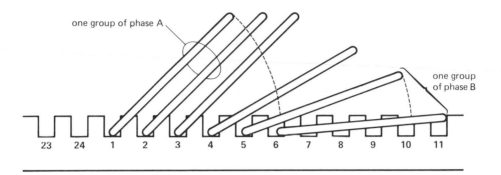

one group of phase A

one group
of phase B

23 24 1 2 3 4 5 6 7 8 9 10 11

Figure 17-18   The pole pitch is from slot 1 slot 7; the coil pitch from slot 1 to slot 6.

Figure 17-19

Stator of a 3-phase, 450 kW, 1180 r/min, 575 V, 60 Hz induction motor. The lap winding is composed of 108 preformed coils having a pitch from slots 1 to 15. One coil side falls into the bottom of a slot and the other at the top. Rotor diameter: 500 mm; axial length; 460 mm. (H. Roberge)

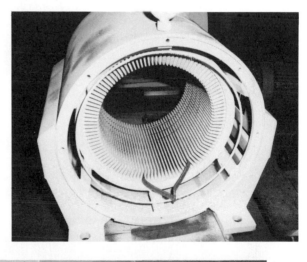

Figure 17-20

Closeup view of the preformed coil in Fig. 17-19.

Figure 17-21   Stator winding of a 3-phase, 50 hp, 575 V, 60 Hz, 1765 r/min induction motor. The stator possesses 48 slots carrying 48 coils.

    a. Coil made of 27 turns # 15 copper wire, covered with a high-temperature polyimide insulation, ready to be placed into 2 slots.

    b. One coil-side is threaded into slot 1 (say) and the other side goes into slot 12. The coil-pitch is therefore from 1 to 12.

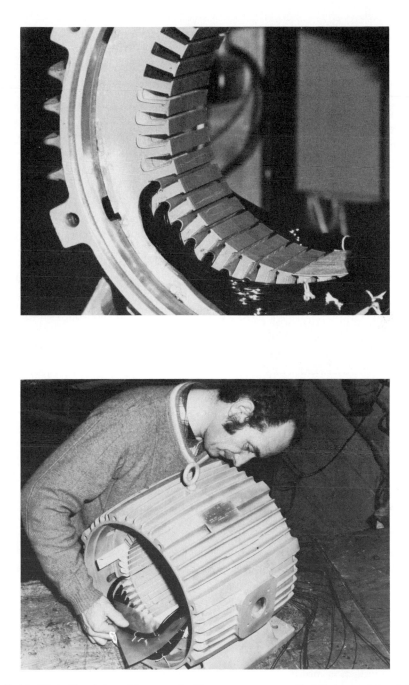

c. Each coil-side fills half a slot, and is covered with a paper spacer so that it does not touch the second coil-side placed in the same slot. The photograph shows 3 empty and uninsulated slots and 4 empty slots insulated with a composition paper liner. The remaining 10 slots each carry one coil-side.

d. A varnished cambric cloth, cut in the shape of a triangle, provides extra insulation between adjacent phase groups. *(Electro Mecanik)*

17-18 Sector motor

Consider a standard 3-phase, 4-pole, wye-connected motor having a synchronous speed of 1800 r/min. Let us cut the stator in half, so that half the winding is removed and only two complete N and S poles are left. Next, let us connect the three phases in wye, without making any other changes to the existing coil connections. Finally, we mount the original rotor above this *sector stator*, leaving a small air gap (Fig. 17-22).

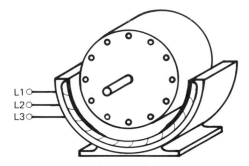

Figure 17-22 Two-pole sector induction motor.

If we connect the stator terminals to a 3-phase, 60 Hz source, the rotor again turns at close to 1800 r/min. To prevent saturation, the voltage should be reduced to half its original value because the stator winding now has only one-half the original number of turns. Under these conditions, this remarkable truncated, sector motor still develops about 30 percent of its original rated power.

The sector motor produces a "revolving" field that moves at the same peripheral speed as the flux in the original 3-phase motor. However, instead of making a complete turn, the field simply travels continuously from one end of the stator to the other.

17-19 Linear induction motor

It is obvious that the sector stator could be laid out flat, without affecting the shape or speed of the magnetic field. Such a flat stator produces a field that moves at constant speed, in a straight line. Using the same reasoning as in Sec. 17-5, we can prove that the flux travels at a linear synchronous speed given by:

$$v_s = 2\,wf \qquad (17\text{-}10)$$

where

v_s = linear synchronous speed [m/s]
w = width of one pole-pitch [m]
f = frequency [Hz]

Note that the speed does not depend upon the number of poles but only on the pole-pitch. Thus, it is possible for a 2-pole linear stator to create a field moving at the same speed as that of a 6-pole linear stator (say), provided they have the same pole-pitch.

If a flat squirrel-cage winding is brought near the flat stator, the travelling field drags the squirrel cage along with it (Sec. 17-2). In practice, we generally use a simple aluminum or copper plate as a "rotor" (Fig. 17-23). Furthermore, to increase the power and to reduce the reluctance of the magnetic path, two flat stators are usually mounted, face-to-face, on opposite sides of the aluminum

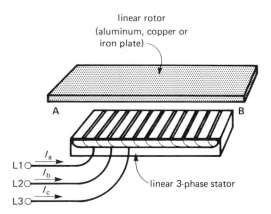

Figure 17-23 Components of a 3-phase linear induction motor.

plate. The combination is called a *linear induction motor*. The direction of the motor can be reversed by interchanging any two stator leads.

In many practical applications, the rotor is stationary, while the stator moves. For example, in some high-speed trains, the rotor is composed of a thick aluminum plate fixed to the ground extending over the full length of the track. The linear stator is bolted to the undercarriage of the train and straddles the plate. Train speed is varied by changing the frequency applied to the stator (Fig. 17-27).

Example 17-10:

The stator of a linear induction motor is excited from a 75 Hz electronic source. If the distance between consecutive phase groups of phase A is 300 mm, calculate the linear speed of the magnetic field.

Solution:

The pole pitch is 300 mm. Consequently,

$$v_s = 2\ wf \qquad \text{Eq. 17-10}$$
$$= 2 \times 0.3 \times 75$$
$$= 45 \text{ m/s or } 162 \text{ km/h}$$

17-20 Traveling waves

We are sometimes left with the impression that when the flux reaches the end of a linear stator, there must be a delay before it "returns" to restart once more at the beginning. This is not the case. The linear motor produces a traveling wave of flux which moves continuously and smoothly from one end of the stator to the other. Figure 17-24 shows how the flux moves from left to right in a 2-pole linear motor. The flux cuts off sharply at extremities A, B of the stator. However, as fast as a N pole "disappears" at the right, it builds up again at the left.

17-21 Properties of a linear induction motor

The properties of a linear induction motor are almost identical to those of a standard rotating machine. Consequently, the equations for slip, thrust, power, etc., are also very similar.

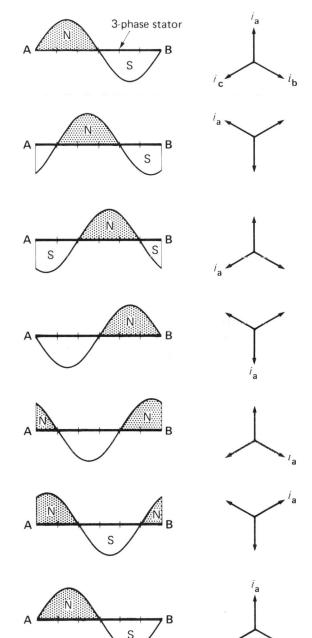

Figure 17-24 Shape of the magnetic field created by a 2-pole, 3-phase linear stator, over one complete cycle. The successive frames are separated by an interval of time equal to 1/6 cycle or 60°.

1. Slip. Slip is defined by:

$$s = (v_s - v)/v_s \qquad (17\text{-}11)$$

where
s = slip
v_s = synchronous linear speed [m/s]
v = speed of rotor (or stator) [m/s]

2. Active power flow. Referring to Fig. 17-10, active power flows through a linear motor in the same way it does through a rotating motor, except that the stator and rotor are flat. Consequently, Eqs. 17-6, 17-7 and 17-8 apply to both types of machines:

$$\eta = P_L/P_c \qquad \text{Eq. 17-6}$$
$$P_{jr} = sP_r \qquad \text{Eq. 17-7}$$
$$P_m = (1 - s)\,P_r \qquad \text{Eq. 17-8}$$

3. Thrust. The thrust or force developed by a linear induction motor is given by:

$$F = P_r/v_s \qquad (17\text{-}12)$$

where
F = thrust [N]
P_r = power transmitted to the rotor [W]
v_s = linear synchronous speed [m/s]

Example 17-11:
An overhead crane in a factory is driven horizontally by means of two linear induction motors whose "rotors" are the two steel I-beams upon which the crane rolls. The 3-phase, 4-pole linear stators (mounted on opposite sides of the crane and facing the respective webs of the I-beams) have a pole pitch of 8 cm and are driven by a variable-frequency electronic source. During a test on one of the motors, the following results were obtained:

Stator frequency:	15 Hz
Power to stator:	5 kW
Copper loss and iron loss in stator:	1 kW
Crane speed:	1.8 m/s

Calculate:
a. synchronous speed and slip;

b. power to the rotor;
c. I^2R loss in rotor;
d. mechanical power and thrust.

Solution:
a. 1   Linear synchronous speed is:
$$v_s = 2\,wf \qquad \text{Eq. 17-10}$$
$$= 2 \times 0.08 \times 15$$
$$= 2.4 \text{ m/s}$$
a. 2 $s = (v_s - v)/v_s \qquad \text{Eq. 17-11}$
$$= (2.4 - 1.8)/2.4$$
$$= 0.25$$

b.   Power to rotor:
$$P_r = P_e - P_{js} - P_f \qquad \text{(see Fig. 17-10)}$$
$$= 5 - 1$$
$$= 4 \text{ kW}$$
c. I^2R loss in rotor:
$$P_{jr} = sP_r \qquad \text{Eq. 17-7}$$
$$= 0.25 \times 4$$
$$= 1 \text{ kW}$$
d. 1   Mechanical power:
$$P_m = P_r - P_{jr} \qquad \text{(Fig. 17-10)}$$
$$= 4 - 1$$
$$= 3 \text{ kW}$$
d. 2   Thrust:
$$F = P_r/v_s \qquad \text{Eq. 17-12}$$
$$= 4000/2.4$$
$$= 1666 \text{ N} = 1.67 \text{ kN} \ (\approx 375 \text{ lb})$$

17-22 Magnetic levitation

In Sec. 17-2, we saw that a moving permanent magnet, sweeping across a conducting ladder, tends to drag the ladder along with the magnet. We will now show that this horizontal tractive force is also accompanied by a *vertical* force which tends to push the magnet away from the ladder.

Referring to Fig. 17-25, suppose that the center of the N pole of the magnet is sweeping across the top of conductor 2. The voltage induced in this conductor is maximum because the flux density is greatest at the center of the pole. If the magnet moves very slowly, the resulting induced current reaches *its* maximum value at virtually the same time. This current, returning by conductors 1 and 3, creates magnetic poles NNN and SSS as shown in Fig. 17-25. According to the laws of attraction

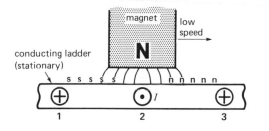

Figure 17-25   Currents and magnetic poles at low speed.

and repulsion, the front half of the magnet is repelled upwards while the rear half is attracted downwards. Because the distribution of the NNN and SSS poles is symmetrical with respect to the center of the magnet, the vertical forces of attraction and repulsion are equal, and the resulting force is nil. Consequently, there is only a horizontal tractive force.

But now, suppose that the magnet moves very rapidly. Owing to coil inductance, the current in conductor 2 reaches its maximum value a fraction of a second after the voltage has attained *its* maximum. Consequently, by the time the current in conductor 2 is maximum, the center of the magnet is already some distance ahead of the conductor (Fig. 17-26). The current returning by conduc-

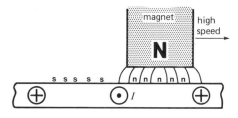

Figure 17-26   Currents and magnetic poles at high speed.

tors 1 and 3 again creates NNN and SSS poles; however, the N pole of the magnet is now directly above a NNN pole, with the result that a strong

vertical force tends to push the magnet upwards.* This effect is called the principle of *magnetic levitation*.

Magnetic levitation is used in some ultra-high-speed trains that glide on a magnetic cushion rather than on wheels. A powerful electromagnet, fixed underneath the train, moves across a conducting rail inducing currents in the rail, in the same way as in our ladder. The force of levitation is always accompanied by a small horizontal braking force which must, of course, be overcome by the motor that propels the train. See Figs. 17-27 and 17-28.

QUESTIONS AND PROBLEMS

Practical level

17-1 Name the principal components of an induction motor.

17-2 Explain how a revolving field is set up in 3-phase induction motor.

17-3 If we double the number of poles on the stator of an induction motor, will its synchronous speed also double?

17-4 The rotor of an induction should never be locked while full voltage is being applied to the stator. Explain.

17-5 Why does the rotor of an induction motor turn slower than the revolving field?

17-6 What happens to the rotor speed and current when the mechanical load on an induction motor increases?

---

* The current is always delayed (even at low speeds) by an interval of time Δt, which depends upon the L/R time constant of the rotor. This delay is so brief that, at low speeds, the current reaches its maximum at virtually the same time and place as the voltage does. On the other hand, at high speeds, the same delay Δt produces a significant shift *in space* between the points where the voltage and current reach their respective maximum values.

Figure 17-27

This 17 t electric train is driven by a linear motor. The motor consists of a stationary rotor and a flat stator fixed to the undercarriage of the train. The rotor is the vertical aluminum plate mounted in the center of the track. The 3-tonne stator is energized by a 4.7 MVA electronic dc to ac inverter whose frequency can be varied from zero to 105 Hz. The motor develops a maximum thrust of 35 kN ($\approx$ 7800 lb) and the top speed is 200 km/h. Direct current power at 4 kV is fed into the inverter by means of a brush assembly in contact with 6 stationary dc bus bars mounted on the left-hand side of the track.

Electromagnetic levitation is obtained by means of a superconducting electromagnet. The magnet is 1300 mm long, 600 mm wide and 400 mm deep and weighs 500 kg. The coils of the magnet are maintained at a temperature of 4 K by the forced circulation of liquid helium. The current density is 80 A/mm^2 and the resulting flux density is 3 T. The vertical force of repulsion attains a maximum of 60 kN and the vertical gap between the magnet and the reacting metallic track varies from 100 mm to 300 mm depending on the dc current. *(Siemens)*

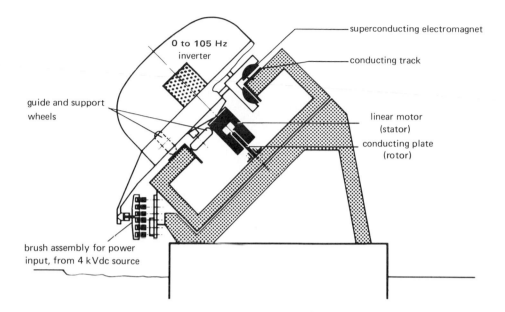

Figure 17-28 Cross section view of the main components of the high-speed train shown in Fig. 17-27. *(Siemens)*

17-7 Would you recommend using a 50 hp induction motor to drive a 10 hp load? Explain.

17-8 Give two advantages of a wound-rotor motor over a squirrel-cage motor.

17-9 Both the voltage and frequency induced in the rotor of an induction motor decrease as the rotor speeds up. Explain.

17-10 A 3-phase, 20-pole induction motor is connected to a 600 V, 60 Hz source.
   a. What is the synchronous speed?
   b. If the voltage is reduced to 300 V, will the synchronous speed change?
   c. How many groups are there, per phase?

17-11 Describe the principle of operation of a linear induction motor.

17-12 Calculate the approximate values of starting current, full-load current and no-load current of a 150 horsepower, 575 V, 3-phase induction motor.

17-13 Make a drawing of the magnetic field created by a 3-phase, 12-pole induction motor.

17-14 How can we change the direction of rotation of a 3-phase induction motor?

Intermediate level

17-15 a. Calculate the synchronous speed of a 3-phase, 12-pole induction motor excited by a 60 Hz source.
   b. What is the nominal speed if the slip at full load is 6 percent?

17-16 A 3-phase 6-pole induction motor is connected to a 60 Hz supply. The voltage induced in the rotor bars is 4 V when the rotor is locked. Calculate the voltage and frequency induced:
   a. at 300 r/min;
   b. at 1000 r/min;
   c. at 1500 r/min.

17-17 a. Calculate the approximate values of full-load current, starting current and no-load current of a 75 kW, 4000 V, 3-phase, 900 r/min induction motor.
   b. Calculate the nominal full-load speed and torque knowing that the slip is 2 percent.

17-18 A 3-phase, 75 hp, 440 V induction motor has a full-load efficiency of 91% and a power factor of 83%. Calculate the nominal current per phase.

17-19 An open-circuit voltage of 240 V appears across the slip-rings of a wound rotor induction motor when the rotor is locked. The stator has 6 poles and is excited by a 60 Hz source. If the rotor is driven by a variable-speed dc motor, calculate the open-circuit voltage and frequency across the slip-rings if the dc motor turns:

a. at 600 r/min, in the same direction as the rotating field;

b. at 900 r/min, in the same direction as the rotating field;

c. at 3600 r/min, opposite to the rotating field.

17-20 a. Referring to Fig. 17-7, calculate the instantaneous values of I_a, I_b and I_c for an angle of 150°.

b. Determine the actual direction of current flow in the three phases at this intant and calculate the mmf developed by the windings.

c. Does the resulting mmf point in a direction intermediate between the mmf's corresponding to instants 3 and 4?

17-21 A 3-phase lap-wound stator possessing 72 slots, produces a synchronous speed of 900 r/min when connected to a 60 Hz source. Calculate the number of coils per phase group as well as the probable coil pitch. Draw the complete coil connection diagram, following steps (a) to (f) in Fig. 17-17.

17-22 The 3-phase 4-pole stator of Fig. 17-21 has an internal diameter of 250 mm and a stacking (axial length) of 200 mm. If the maximum flux density per pole is 0.7 T, calculate:

a. the peripheral speed [m/s] of the revolving field when the stator is connected to a 60 Hz source;

b. the peak voltage induced in the rotor bars;

c. the pole pitch.

17-23 A large 3-phase, 4000 V, 60 Hz squirrel-cage induction motor draws a current of 385 A and a total active power of 2344 kW when operating at full load. The corresponding speed is accurately measured and is found to be 709.2 r/min. The stator is connected in wye and the resistance between two stator terminals is 0.10 Ω. The total iron losses are 23.4 kW and the windage and friction losses are 12 kW. Calculate:

a. the power factor at full load;

b. the active power supplied to the rotor;

c. the total I^2R losses in the rotor;

d. the load mechanical power [kW], torque [kN·m] and efficiency [%].

17-24 If we increase the rotor resistance of an induction motor, what effect does this have (increase or decrease) upon:

a. starting torque   d. efficiency

b. starting current   e. power factor

c. full load speed   f. motor temperature

17-25 Explain the principle of magnetic levitation.

Advance level;

17-26 In Fig. 17-5, the permanent magnet has a width of 100 mm and moves at 30 m/s. The flux density in the air-gap is 0.5 T and the effective resistance per rotor bar is 1 mΩ. Calculate the current I and the tractive force.

17-27 If the conducting ladder in Fig. 17-5 is pulled along with a force of 20 N, what is the braking force exerted on the magnet?

17-28 A 3-phase, 5000 hp, 6000 V, 60 Hz 12-pole wound rotor induction motor turns at 594 r/min. What are the approximate rotor I^2R losses?

17-29 The motor in problem 17-28 has the following characteristics:

1. dc resistance between stator terminals
   = 0.112 Ω at 17°C

2. dc resistance between rotor slip-rings
   = 0.0073 Ω at 17°C

3. open-circuit voltage induced between slip-rings with rotor locked
   = 1600 V

4. line-to-line stator voltage
    = 6000 V
5. no-load current, per phase
    = 100 A
6. active power supplied to motor at no-load
    = 91 kW
7. windage and friction losses
    = 51 kW
8. iron losses in the stator
    = 39 kW
9. locked rotor current at 6000 V
    = 1800 A
10. active power to stator with rotor locked
    = 2207 kW

Calculate:

a. rotor and stator resistance per phase at 75°C (assume a wye connection);
b. voltage and frequency induced in the rotor when it turns at 200 r/min and at 594 r/min;
c. reactive power used to create the revolving field, at no-load;
d. I^2R losses in the stator when the motor runs at no-load (winding temperature 75°C);
e. active power supplied to the rotor at no-load;

17-30 Referring to the motor described in Problem 17-29, calculate, under full-voltage LR (locked rotor) conditions:

a. reactive power absorbed by the motor;
b. I^2R losses in the stator;
c. active power supplied to the rotor;
d. mechanical power output;
e. torque developed by the rotor.

17-31 We wish to control the speed of the motor given in Problem 17-29 by inserting resistors in series with the rotor (see Fig. 17-14). If the motor has to develop a torque of 20 kN·m at a speed of 450 r/min, calculate:

a. voltage between the slip rings;
b. rotor resistance, per phase and the total power dissipated;
c. approximate rotor current, per phase.

17-32 The train shown in Fig. 17-27 moves at 200 km/h when the stator frequency is 105 Hz. By supposing a negligible slip, calculate the length of the pole-pitch of the linear motor [mm].

17-33 A 3-phase, 300 kW, 2300 V, 60 Hz, 1780 r/min induction motor is used to drive a compressor. It has a full-load efficiency and power factor of 92% and 86%, respectively. If the terminal voltage rises to 2760 V while the motor operates at full load, determine the effect (increase or decrease) upon:

a. mechanical power delivered by the motor;
b. motor torque;
c. rotational speed;
d. full-load current;
e. power factor and efficiency;
f. starting torque;
g. starting current;
h. breakdown torque;
i. motor temperature rise;
j. flux per pole;
k. exciting current;
l. iron losses.

17-34 A 3-phase, 60 Hz linear induction motor has to reach a top no-load speed of 12 m/s and it must develop a standstill thrust of 10 kN. Calculate the required pole-pitch and the minimum I^2R loss in the rotor, at standstill.

18

SELECTION AND APPLICATION OF THREE-PHASE INDUCTION MOTORS

When purchasing a three-phase induction motor for a particular application, we often discover that several types can fill the need. Consequently, we have to make a choice. The selection is generally simplified because the manufacturer of the lathe, fan, pump, and so forth indicates the type of motor which is best suited to drive the load. Nevertheless, it is useful to know something about the basic construction and characteristics of the various types of three-phase induction motors that are available on the market.

In this chapter, we also cover some special applications of induction machines, such as asynchronous generators and frequency converters. These interesting devices will enable the reader to gain an even better understanding of induction motors in general.

18-1 Standardization and classification of induction motors[1]

The frames of all industrial motors motors under 500 hp have standardized dimensions. Thus, a 25 hp, 1725 r/min, 60 Hz motor of one manufac-

turer can be replaced by that of any other manufacturer, without having to change the mounting holes, the shaft height, or the type of coupling. The standardization covers not only frame sizes, but also establishes limiting values for electrical, mechanical and thermal characteristics. Thus, motors must satisfy minimum requirements as to starting torque, locked-rotor current, overload capacity, and temperature rise.

18-2 Classification according to environment and cooling methods

Motors are grouped into several distinct categories, depending upon the environment in which they have to operate. We shall limit our discussion to five important classes.

---

[1] Standards in the United States are governed by National Electrical Manufacturers (NEMA) publication MG-1 titled **Motors and Generators**.
Standards in Canada are similarly governed by Canadian Standards Association (CSA) publication C 154. The two standards are essentially identical.

1. Drip-proof motors. The frame protects the windings against liquid drops and solid particles which fall at any angle from 0 to 15 degrees downward from the vertical. The motors are cooled by means of a fan directly coupled to the rotor. Cool air, drawn into the motor through vents in the frame, is blown over the windings and then expelled. The maximum allowable temperature rise (measured by the change in winding resistance) may be 60°C, 80°C, 105°C, or 125°C according to whether the windings possess class A, B, F, or H insulation. Drip-proof motors can be used in most locations (see Fig. 18-1).

outside of the case. They are designed for very wet and dusty locations. Most of these motors are rated below 10 kW because it is difficult to get rid of the heat. The motor losses are dissipated by natural convection and radiation from the frame. The permissible temperature rise is 65°C, 85°C, 110°C or 130°C according to whether the windings possess class A, B, F or H insulation (see Fig. 18-2).

4. Totally enclosed, fan-cooled motors. Medium - and high-power motors that are totally enclosed are usually cooled by an external blast of air. An external fan, directly coupled to the shaft, blows

Figure 18-1   Energy efficient drip-proof, 3-phase squirrel-cage induction motor rated 230 V/460 V, 3 hp, 1750 r/min, 60 Hz. *(Gould)*

Figure 18-2   Totally enclosed induction motor for a centrifugal pump. *(Brook Crompton-Parkinson Ltd)*

2. Splash-proof motor. The frame protects the windings against liquid drops and solid particles which fall at any angle from 0 to 100 degrees downward from the vertical. Cooling is similar to that in drip-proof motors and the maximum temperature rise is also the same. These motors are mainly used in wet locations.

3. Totally enclosed, nonventilated motors. These machines have closed frames which prevent the free exchange of air between the inside and the

air over the ribbed motor frame. A concentric outer shield prevents physical contact with the fan and serves to channel the air-stream. The permissible temperature rise is the same as for drip-proof motors (see Fig. 18-3).

5. Explosion-proof motors. These motors are used in highly inflamable or explosive surroundings such as are found in coal mines, oil refineries and grain elevators. They are totally enclosed (but not air-tight) and the frames are designed to withstand

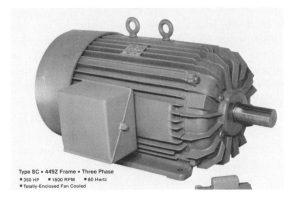

Figure 18-3 Totally-enclosed fan-cooled induction motor rated 350 hp, 1760 r/min, 440 V, 3-phase, 60 Hz. *(Gould)*

Figure 18-4 Totally enclosed, fan-cooled, explosion-proof motor. Note the particularly rugged construction of this type of motor. *(Brook Crompton-parkinson Ltd)*

the enormous pressure which may build up inside the motor due to an internal explosion. Furthermore, the flanges on the end-bells are made extra-long in order to cool any escaping gases generated by such an explosion. The permissible temperature rise is the same as for totally enclosed motors (see Fig. 18-4).

18-3 Classification according to electrical and mechanical properties

In addition to special enclosures, 3-phase squirrel-cage motors may have various electrical and mechanical characteristics, as listed below.

1. Motors with standard locked-rotor torque (NEMA Design B). Most induction motors belong to this group. The locked-rotor torque depends upon the size of the motor, and ranges from 130% to 70% of full-load torque, as the power increases from 20 hp to 200 hp (15 kW to 150 kW). The corresponding locked-rotor current should not exceed 6.4 times the rated full-load current. These general-purpose motors are employed to drive fans, centrifugal pumps, machine-tools, and so forth.

2. High starting-torque motors (NEMA Design C). These motors are employed when starting condi-

tions are difficult. Pumps and piston-type compressors that have to start under load are two typical applications. In general, these motors have a special double-cage rotor. In the range from 20 hp to 200 hp, the locked-rotor torque is 200% of full-load torque and the locked-rotor current should not exceed 6.4 times the rated full-load current.

The excellent performance of a double-cage rotor (Fig. 18-5) is based upon the fact that:

a. the frequency of the rotor current diminishes as the motor speeds up;
b. a conductor that lies close to the rotor surface (cage 1) has a lower inductive reactance than one buried deep inside the core (cage 2).

When the motor is connected to the line with the rotor at standstill, the frequency of the rotor current is equal to line frequency. Owing to the high inductive reactance of squirrel-cage 2, the rotor current flows mainly in the small bars of cage 1. The effective motor resistance is therefore high, being almost equal to that of cage 1. As the motor speeds up, the rotor frequency falls, with the result that the inductive reactance of both squirrel-cage windings diminishes. At normal speed, the rotor frequency is so low that the reactance of both windings is negligible. The rotor current is then limited only by the *resistance* of cage 1 and cage 2, respectively. The conductors of

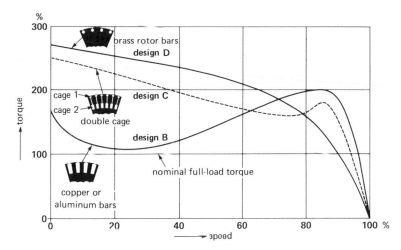

Figure 18-5   Typical torque-speed curves of NEMA design B, C and D motors. Each curve corresponds to the minimum NEMA values of locked-rotor torque, pull-up torque and breakdown torque of a 3-phase 1800 r/min, 10 hp, 60 Hz squirrel-cage induction motor. The cross section of the respective rotors indicates the type of rotor bars used.

cage 2 are much larger than those of cage 1, with the result that the effective rotor resistance at normal speed is much lower than at standstill. For this reason, the double-cage rotor develops both a high starting torque and a low slip at full load.

Despite their high torque, Design C motors are not recommended for prolonged starting of high-inertia loads. The reason is that most of the rotor I^2R losses during startup are concentrated in cage 1. Owing to its small size, it tends to overheat and the bars may melt.

3. High-slip motors (NEMA Design D). The nominal speed of high-slip motors is between 85% and 95% of synchronous speed. These motors develop even higher starting-torques than Design C motors do, and the locked-rotor currents are somewhat lower. These motors can drive high inertia loads (such as centrifugal dryers) which take a relatively long time to reach full speed. The high-resistance squirrel cage is made of brass, and the motors are usually designed for intermittent operation to prevent overheating. The large drop in speed with increasing load is also ideal to drive impact machine-tools such as shears, punch presses, and so forth that are equipped with a flywheel. The flywheel stores mechanical energy during the idling period

and releases it during the sharp impulse load.

The graphs of Fig. 18-5 enable us to compare the torque-speed characteristics of these various motors. The rotor construction is also shown, and it can be seen that the distinguishing properties are obtained by changing the rotor design. For example, as we increase the rotor resistance (by using brass instead of copper or aluminum), the locked-rotor torque increases, but the nominal speed drops.

18-4   Motor size

Table 18A gives the approximate dimensions and cost of commercial induction motors in the range between 1 hp and 1000 hp. If we compare the 1 hp and 1000 hp machines, we discover that a thousandfold increase in power requires only a 150-fold increase in weight. This dramatic increase in power with increasing size is an inherent characteristic of all electrical machines. As a rule-of-thumb, we can say that for similar machines operating at the same speed, a doubling of the weight produces about 2.5 times as much power. Thus, if a 10 hp motor weighs 70 kg, a 25 hp motor of similar design and speed will weigh about 140 kg.

TABLE 18A   APPROXIMATE DIMENSIONS AND COST OF INDUCTION MOTORS

Drip-proof squirrel-cage motors, Design B, 3-phase, 60 Hz, n_s = 1800 r/min

Power P		Mass	Volume	External diameter	External length	Cost (1980)	Cost/P (1980)
hp	kW	kg	dm^3	mm	mm	$	$/hp
1	0.75	16	5.5	180	220	120	120
10	7.5	60	20	270	350	320	32
100	75	350	130	500	650	1 900	19
1000	750	2500	650	750	1500	27 000	270

18-5 Choice of motor speed

The choice of motor speed is rather limited because the synchronous speed of induction motors changes by quantum jumps, depending upon the frequency and number of poles. For example, it is impossible to build a conventional 60 Hz induction motor having an acceptable efficiency and running at a speed, say, of 2000 r/min.

The speed of a motor is obviously determined by the speed of the machine it has to drive. However, for low-speed machines, it is often preferable to use a high-speed motor and a gear box instead of directly coupling a low-speed motor to the load. There are several advantages to the gear-box approach:

1. For a given output power, the size and cost of a high-speed motor is less than that of a low-speed motor, and its efficiency and power factor are higher;

2. The locked-rotor torque of high-speed motors is always greater (as a percent of full-load torque) than that of low-speed motors of similar type and power.

By way of example, Table 18B compares the properties of two commercial 3-phase 60 Hz induction motors having the same power but different synchronous speeds. The difference in price alone would justify the use of a high-speed motor and a gear box to drive a load operating at, say, 900 r/min.

When equipment has to operate at very low speeds (100 r/min or less), a gear box is mandatory. The gears are often an integral part of the motor, making for a very compact unit (Fig. 18-8).

By the same token, a gear box is mandatory when equipment has to run above 3600 r/min. For example, a large gear unit is needed to drive a 5000 r/min centrifugal compressor using a 3560 r/min induction motor.

18-6 Two-speed motors

We can design the stator of a squirrel-cage induction motor so that the motor can operate at two different speeds. The speed is changed by simply changing the external stator connections. The synchronous speeds obtained are usually in the ratio 2 to 1: 3600/1800 r/min; 1200/600 r/min; etc.

Consider, for example, the stator winding of a

TABLE 18B		COMPARISON BETWEEN TWO MOTORS OF DIFFERENT SPEEDS					
Power		Synchronous speed	Power factor	Efficiency	Locked rotor torque	Mass	Cost (1980)
hp	kW	r/min	%	%	%	kg	$
10	7.5	3600	89	87	150	50	340
10	7.5	900	82	83	125	115	960

4-pole 60 Hz motor (Fig. 18-6a). If we apply power to terminals 1 and 2, two N poles and two S poles are created and, according to Eq. 17-1, the synchronous speed is:

$$n_s = 120 f/p$$

$$= 120 \times 60/4 = 1800 \text{ r/min}$$

On the other hand, if we short leads 1 and 2 and apply power to terminals 2 and 4, currents will flow in the direction shown in Fig. 18-6b. It is clear that this connection produces four N poles.

Because every N pole must be accompanied by a S pole, it follows that 4 south poles will appear between the 4 north poles. The south poles created in this ingenious way are called *consequent* poles. The new connection produces 8 poles in all and the synchronous speed is 900 r/min. Thus, we can double the number of poles by simply changing the stator connections. It is upon this principle that two-speed motors are based.

Figure 18-8 shows the stator connections for a

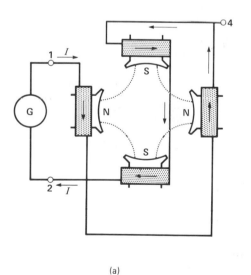

(a)

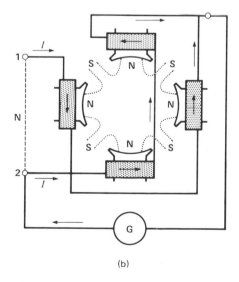

(b)

Figure 18-6   Method of doubling the number of poles by changing external connections.

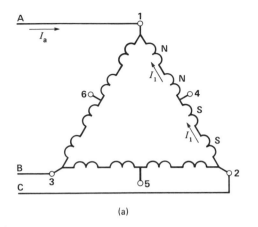

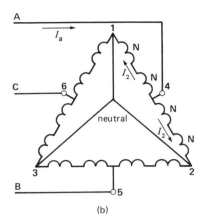

Figure 18-7   a. High-speed connection of a 3-phase stator, yielding 4 poles.   b. Low-speed connection of same motor, yielding 8 poles.

two-speed, 3-phase motor. Six leads, numbered 1 to 6, are brought out from the stator winding. For the high-speed connection power is applied to terminals 1-2-3 and terminals 4-5-6 are open. The resulting delta connection produces 4 groups per phase having alternate N and S polarities (Fig. 18-7a).

The low-speed connection is made by shorting terminals 1-2-3 and applying power to terminals 4-5-6. The resulting double-wye connection again produces 4 groups per phase but now they all possess the same polarity (Fig. 18-7b). The coil-pitch of two-speed motors is always one half the high-speed pole-pitch. A short coil-pitch facilitates the creation of the consequent poles at the lower speed.

Two-speed motors have a relatively lower efficiency and power factor than single-speed motors do. They can be designed to develop (at both speeds) either constant power, constant torque or variable torque, depending upon the load that has to be driven.

18-7 Plugging an induction motor

In some industrial applications, the induction motor operates in ways we have not studied so far. For example, if we wish to bring the motor to a

rapid stop, we can simply interchange two stator leads, so that the revolving field suddenly turns in the opposite direction to the rotor. During this so-called *plugging* period, the motor acts as a *brake*.

Figure 18-8   Gear motor rated at 2.25 kW, 1740 r/min, 60 Hz. The output torque and speed are respectively 172 N·m and 125 r/min. *(Reliance Electric)*

It absorbs kinetic energy from the still-revolving load, causing its speed to fall. The associated power P_m is entirely dissipated as heat in the rotor. Unfortunately, the rotor also continues to receive

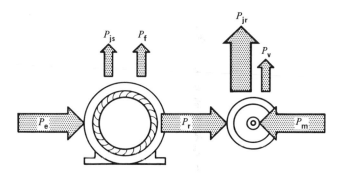

P_e — active power supplied to stator

P_{js} — stator copper losses

P_f — stator iron losses

P_r — active power transmitted to the rotor

P_m — mechanical power supplied to the rotor

P_{jr} — I^2R losses in the rotor

P_v — windage and friction losses

Figure 18-9   When a 3-phase induction motor is plugged, the rotor I^2R losses are very high.

power P_r from the stator, which is also dissipated as heat (Fig. 18-9). Consequently, plugging produces I^2R losses in the rotor which even exceed those when the rotor is locked. Motors should not be plugged too frequently because high rotor temperatures may melt the rotor bars or overheat the stator winding.

18-8 Effect of inertia

High-inertia loads put a heavy strain on induction motors because they prolong the starting period. The starting current in both the stator and rotor is high during this interval so that overheating becomes a major problem. A prolonged starting period may even put a strain on the transmission line feeding the plant where the motor is installed. The line voltage may fall below normal for several seconds, thus affecting other connected loads. To relieve the problem, induction motors are often started on reduced voltage. This limits the power drawn by the motor, and consequently reduces the voltage drop and the heating rate. Unfortunately, it also increases the starting time, but this is usually not too important.

Plugging induction motors, to bring them to a stop, creates even worse problems. The reason is that the plugging time may be long, and severe overheating, particularly of the rotor, may result. In this regard, two important rules affecting starting and plugging operations of *unloaded* motors are worth remembering:

Rule 1 - The heat dissipated in the rotor during the starting period (from zero to nominal speed) is equal to the final kinetic energy stored in all the revolving parts;

Rule 2 - The heat dissipated in the rotor during the plugging period (nominal-speed to zero) is *three times* the original kinetic energy of all the revolving parts.

These rules always hold true, irrespective of the stator voltage or the torque-speed curve of the motor.

Example 18-1:
A 100 kW, 60 Hz, 1175 r/min motor is coupled to a flywheel by means of a gear box. The kinetic energy of all the revolving parts is 300 kJ when the motor runs at nominal speed. The motor is plugged to a stop and allowed to run up to 1175 r/min in the reverse direction. Calculate the energy dissipated in the rotor if the flywheel is the only load acting on the motor.

Solution:
During the plugging period, the motor speed drops from 1175 r/min to zero. The heat dissipated in the rotor is 3 x 300 kJ = 900 kJ. The motor then accelerates to nominal speed in the reverse direction. The energy dissipated in the rotor during this period is 300 kJ. The total heat dissipated in the rotor from start to finish is therefore 900 + 300 = 1200 kJ.

When very high inertia loads have to be accelerated or brought to a stop, wound-rotor motors are recommended because most of the thermal energy absorbed in the rotor circuit is dissipated by the external resistors. Furthermore, we can maintain a consistently high torque by gradually varying the rotor resistance during the acceleration or deceleration periods.

18-9 Braking with direct current

An induction motor and its high-inertia load can be brought to a quick stop by circulating dc current in the stator winding. Any two terminals can be connected to the dc source.

The direct current produces stationary N, S poles in the stator. The number of poles created is equal to the number of poles which the motor develops normally. Thus, a 3-phase, 4-pole induction motor produces 4 dc poles, no matter how the motor terminals are connected to the dc source.

When the rotor sweeps past the stationary field, an ac voltage is induced in the rotor bars. The voltage produces an ac current and the resulting rotor I^2R losses are dissipated at the expense of the kinetic energy stored in the revolving parts. The motor finally comes to rest when all the kinetic energy has been dissipated as heat in the rotor.

The advantage of dc braking is that it produces far less heat than does plugging. In effect, the energy dissipated in the rotor is only equal to the original kinetic energy stored in the revolving masses, and not three times that energy. The energy dissipated in the rotor is independent of the magnitude of the dc current. However, a smaller dc current increases the braking time, with the result that the temperature rise of the rotor is somewhat reduced. The dc current can be two or three times the rated current of the motor. Even larger values can be used, provided that the stator does not become too hot. The braking torque is proportional to the square of the dc braking current.

Example 18-2:
A 50 hp, 1760 r/min, 440 V, 3-phase induction motor drives a load having a total moment of inertia of 25 kg·m². The dc resistance between two terminals is 0.32 Ω, and the rated motor current is 62 A. We want to stop the motor by connecting a 24 V battery across the stator. Calculate:
a. the dc current in the stator;
b. the energy dissipated in the rotor;
c. the average braking torque if the stopping time is 4 min.

Solution:
a. The dc current is:
$$I = E/R = 24/0.32 = 75 \text{ A}$$
This current is slightly higher than the rated current of the motor. However, the stator will not overheat, because the braking time is relatively short.
b. The kinetic energy in the rotor and load at 1760 r/min is:
$$E_k = 5.48 \times 10^{-3} Jn^2 \qquad \text{Eq. 3-8}$$
$$= 5.48 \times 10^{-3} \times 25 \times 1760^2$$
$$= 424 \text{ kJ}$$
c. The average braking torque can be calculated from:
$$\Delta n = 9.55\, T\Delta t/J \qquad \text{Eq. 3-14}$$
$$1760 = 9.55\, T \times (4 \times 60)/25$$
$$T = 19.2 \text{ N·m}$$

18-10 Abnormal conditions

Abnormal motor operation may be due to internal causes (short-circuit in the stator, overheating of the bearings, etc.), or to external conditions. External problems may be caused by:
1. mechanical overload;
2. supply voltage changes;
3. single phasing;
4. frequency changes.

According to national standards, a motor shall operate successfully on any voltage within ± 10% of the nominal voltage, and for any frequency within ± 5% of the nominal frequency. If the voltage and frequency *both* vary, the sum of the two percentage changes must not exceed 10%. Finally, all motors are designed to operate satisfactorily at altitudes up to 1000 m above sea level. At higher altitudes, the temperature may exceed the permissible level owing to the poor cooling afforded by the thinner air.

18-11 Mechanical overload

Although standard induction motors can develop twice their rated power for short periods, they should not be allowed to run continuously beyond their rated capacity. Overloads cause overheating, which deteriorates the insulation and reduces its useful life. In practice, the higher motor current causes the thermal overload relays to trip, bringing the motor to a stop before its temperature gets too high.

Some drip-proof motors are designed to carry a continuous overload of 15 percent. This overload capacity is shown on the nameplate by the so-called *service factor* 1.15. The allowable temperature rise is then 10°C higher than that permitted for drip-proof motors operating at normal load.

During emergencies, a drip-proof motor can be made to carry overloads of as much as 150%, as long as strong external ventilation is provided. This is not recommended for extended periods, because even if the external frame is cool, the winding temperature may be very high.

18-12 Line voltage changes

The most important consequence of a line voltage change is its effect upon the torque-speed curve of the motor. In effect, the torque at any speed is proportional to the *square* of the applied voltage. Thus, if the stator voltage decreases by 10%, the torque will drop by 20%. A line voltage drop is often produced during start-up, due to the heavy starting current drawn from the line.

On the other hand, if the line voltage is too high, the flux per pole will be too high. This increases both the iron losses and the magnetizing current, with the result that the temperature increases slightly and the power factor is somewhat

reduced.

A slight *unbalance* of the 3-phase voltages produces a serious unbalance of the three line currents. This condition increases the stator and rotor losses, yielding a higher temperature. A voltage unbalance of as little as 3.5% can cause the temperature to increase by 15°C. The utility company should be notified whenever the phase-to-phase line voltages differ by more than 1%.

18-13 Single-phasing

If one line of a 3-phase line is accidentally opened, or if a fuse blows while the motor is running, the machine will continue to run as a *single-phase motor*. The current drawn from the remaining two lines will almost double, and the motor will begin to overheat. The thermal relays protecting the motor will eventually trip the circuit-breaker, thereby disconnecting the motor from the line.

The torque-speed curve is seriously affected when a 3-phase motor operates on single phase. The breakdown torque decreases to about 40% of its original value, and the motor develops no starting torque at all. Consequently, a fully loaded 3-phase motor may simply stop if one of its lines is suddenly opened. The resulting locked-rotor current is about 90% of the normal 3-phase L.R. current. It is therefore large enough to trip the circuit breaker or to blow the fuses.

Figure 18-10 shows the typical torque-speed curves of a 3-phase motor when it runs normally and when it is single-phasing. Note that the curves follow each other closely until the torque approaches 50% of the 3-phase breakdown torque.

18-14 Frequency variation

Important frequency changes never take place on a large distribution system, except during a major disturbance. However, the frequency may vary significantly on isolated, low-power systems where electrical energy is generated by diesel engines or gas turbines. The emergency supply in a hospital, the electrical system on a ship, the generators in a

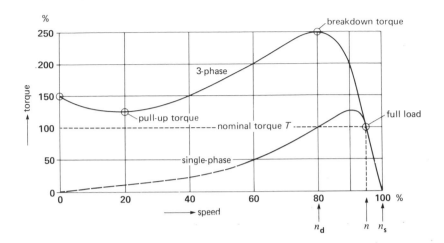

Figure 18-10   Typical torque-speed curves when a 3-phase squirrel-cage motor operates normally and when it operates on single-phase.

lumber camp, etc., are examples of this type of supply.

The most important consequence of a frequency change is the resulting change in motor speed: if the frequency drops by 20%, the motor speed drops by 20%.

Machine tools and other motor-driven equipment imported from countries where the frequency is 50 Hz, may cause problems when they are connected to a 60 Hz system. Everything runs 20% faster than normal, and this may not be acceptable in some applications. In such cases, we either have to gear the motor speed down, or supply an expensive auxiliary 50 Hz source.

A 50 Hz motor operates well on a 60 Hz line, but its terminal voltage should be raised to 6/5 (or 120%) of the nameplate rating. The new breakdown torque is then equal to the original breakdown torque and the starting torque is only slightly reduced. Power factor, efficiency and temperature rise remain satisfactory.

A 60 Hz motor can also operate on a 50 Hz line, but its terminal voltage should be reduced to 5/6 (or 83%) of its nameplate value. The breakdown torque and starting torque are then about the same as before, and the power factor, efficiency and temperature rise remain satisfactory.

18-15 Complete torque-speed characteristic under load

Most asynchronous machines operate as motors supplying a torque that varies from zero to a nominal full-load torque T_n. It so happens that between these limits, the torque-speed curve is essentially a straight line (Fig. 18-11). The slope of the line depends mainly upon the rotor resistance: the higher the resistance, the sharper the slope. This linear relationship between torque and speed, enables us to establish a very simple equation between the various parameters of a motor. In effect, when we know the characteristics of a motor under a given load condition, we can calculate its speed, torque, power, etc., under any other load condition.

We can show that these quantities are related by the equation:

$$s_x = s_n \left[\frac{T_x}{T_n} \right] \left[\frac{R_x}{R_n} \right] \left[\frac{E_n}{E_x} \right]^2 \qquad (18\text{-}1)$$

where subscripts n, x refer to the initial or nominal load condition (n) and new load condition (x), respectively,

and s = slip

T = torque [N·m]

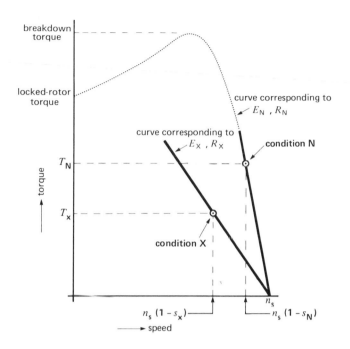

Figure 18-11   The torque-speed curve is essentially a straight line between the no-load and full-load operating points.

R – rotor resistance $[\Omega]$
E = stator voltage $[V]$

In applying this equation, the only restriction is that the new torque T_x must not be greater than $T_n \, (E_x/E_n)^2$. Under these conditions, Eq. 18-1 yields an accuracy of better than 5%, which is sufficient for most practical problems.

Example 18-3:

A 3-phase, 208 V induction motor having a synchronous speed of 1200 r/min turns at a speed of 1140 r/min when connected to a 215 V line. Calculate the speed if the voltage increases to 240 V.

Solution:

a. 1   The slip at 215 V is:

$$s = (n_s - n)/n_s = (1200 - 1140)/1200$$
$$= 0.05$$

a. 2   In applying Eq. 18-1, everything remains the same except the voltage and slip. Consequently:

$$s_x = s_n \, (E_n/E_x)^2 = 0.05 \, (215/240)^2$$
$$= 0.04$$

a. 3   The slip speed is therefore $0.04 \times 1200 = 48$ r/min

a. 4   The new speed is:
$$n_x = 1200 - 48 = 1152 \text{ r/min}$$

Example 18 4:

A 3-phase induction motor driving a fan has an initial speed of 873 r/min when connected to a fixed 460 V, 60 Hz line. The initial rotor temperature is $23°C$. The speed drops to 864 r/min after the machine has reached its final temperature. Calculate:

a. the increase in rotor resistance;

b. the approximate hot temperature of the rotor, knowing it is made of copper.

Solution:

a. 1   The initial and final slips are:
$$s_n = (900 - 873)/900 = 0.03$$
$$s_x = (900 - 864)/900 = 0.04$$

a. 2   The voltage, frequency, etc., are fixed; consequently, the speed change is entirely due to the change in rotor resistance.

$$s_x = s_n (R_x/R_n)$$
$$0.04 = 0.03 (R_x/R_n)$$
$$R_x = 1.33 R_n$$

b.   The hot temperature is:

$$t_2 = \frac{R_2}{R_1} (234 + t_1) - 234 \qquad \text{Eq. 16-3}$$

$$= 1.33 (234 + 23) - 234$$

$$= 108°C$$

Example 18-5:

A 3-phase wound-rotor induction motor has a rating of 110 kW ($\approx$ 150 hp), 1760 r/min, 2.3 kV, 60 Hz. Three external resistors of 2 Ω are connected in wye across the rotor. The motor develops a torque of 300 N·m at a speed of 1000 r/min.

a. Calculate the speed at a torque of 400 N·m;
b. Calculate the value of the external resistors so that the motor develops 10 kW at 200 r/min.

Solution:

a. 1   The given conditions are:

$$T_n = 300 \text{ N·m}$$
$$s_n = (1800 - 1000)/1800 = 0.444$$

All other conditions being fixed, we have:

$$s_x = s_n (T_x/T_n) = 0.44 (400/300)$$
$$= 0.592$$

a. 2   The slip speed $= 0.592 \times 1800 = 1067$ r/min

Consequently, the motor turns at a speed:

$$n = 1800 - 1067$$
$$= 733 \text{ r/min}$$

Note that the speed drops considerably with increasing load.

b. 1   The torque corresponding to 10 kW at 200 r/min is:

$$T_x = 9.55 P/n = 9.55 \times 10\,000/1200$$
$$= 478 \text{ N·m}$$

b. 2   The slip is:

$$s_x = (1800 - 200)/1800 = 0.89$$

All other conditions being fixed, we have:

$$s_x = s_n (T_x/T_n) (R_x/R_n)$$
$$0.89 = 0.44 (478/300) (R_x/2)$$

whence $R_x = 1.25 \ \Omega$

b. 3   Three 1.25 Ω wye-connected resistors in the rotor circuit will enable the motor to develop 10 kW at 200 r/min.

18-16 Induction motor operating as a generator

Consider an electric train powered by an induction motor which is directly coupled to the wheels. As the train climbs the side of a hill, the motor runs at slightly less than synchronous speed, developing a torque sufficient to overcome both friction and the force of gravity. At the top of the hill, on level ground, the force of gravity no longer comes into play and the motor has only to overcome the friction of the rails and the moving air. The motor runs essentially at no-load and very close to synchronous speed.

What happens when the train begins to move downhill? The force of gravity causes the train to accelerate and its speed (even without a motor) could easily exceed that on level ground. Because the motor is coupled to the wheels, it begins to rotate *above* synchronous speed. However, as soon as this takes place, the motor develops a counter torque which *opposes* the increase in speed. This torque has the same effect as a brake. However, instead of being dissipated as heat, the mechanical braking power at the wheels is returned to the line in the form of electrical energy. An induction motor which turns faster than synchronous speed acts therefore as a generator. It converts the mechanical energy it receives into electrical energy, and this energy is released by the stator. Such a machine is called an *asynchronous generator*.

Although induction motors are rarely used to drive trains (Fig. 18-12), there are several industrial applications which may cause a motor to run above synchronous speed. In cranes, for example, during the lowering cycle, the motor receives power from the mechanical "load" and returns it to the line.

We can make an asynchronous generator by connecting an ordinary squirrel-cage motor to a 3-phase line and coupling it to a gasoline engine

Figure 18-12

This electric train makes the round trip between Zermatt (1604 m) and Gornergrat (3089 m), in Switzerland. The drive is provided by four 3-phase wound-rotor induction motors, rated 78 kW, 1470 r/min, 700 V, 50 Hz. Two aerial conductors constitute phases A and B, and the rails provide phase C. A toothed gear-wheel 573 mm in diameter engages a stationary rack on the roadbed to drive the train up and down the steep slopes. The speed can be varied from zero to 14.4 km/h by means of variable resistors in the rotor circuit. The rated thrust is 78 kN. *(Brown Boveri)*

(Fig. 18-13). As soon as the engine speed exceeds the synchronous speed, the motor becomes a source, delivering active power to the electrical system to which it is connected. However, to create its magnetic field, the motor has to absorb reactive power. This power can only come from the supply line with the result that the reactive power Q flows in the opposite direction to the active power P (Fig. 18-13).

The active power delivered is directly proportional to the slip above synchronous speed. Thus, a higher engine speed produces a greater output. However, the rated output is reached at very small slips, usually less than 3%.

The reactive power may be supplied by a group of capacitors connected to the terminals of the motor. With this arrangement, we can supply a 3-phase load without using an external source (Fig. 18-14). The frequency generated is slightly less than that corresponding to the speed of rotation. Thus, a 4-pole motor driven at a speed of 2400 r/min produces a frequency slightly less than $f = pn/120 - 4 \times 2400/120 = 80$ Hz.

The terminal voltage increases with the capacitance, but its magnitude is limited by saturation in the iron. If the capacitance is insufficient, the generator voltage will not build up. The capacitor bank must be large enough to supply the reactive

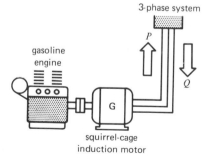
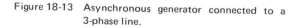

Figure 18-13 Asynchronous generator connected to a 3-phase line.

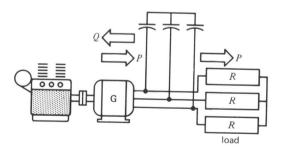

Figure 18-14 Capacitors can provide the reactive power for an asynchronous generator.

power the machine normally absorbs when operating as a motor.

Example 18-6;
We wish to use a 40 hp, 1760 r/min, 440 V, 3-phase squirrel-cage induction motor as an asynchronous generator. The rated current of the motor is 41 A, and the full-load power factor is 84%.
a. Calculate the capacitance required per phase if the capacitors are connected in delta.
b. At what speed should the driving engine run to generate a frequency of 60 Hz?

Solution:
a. 1 The apparent power drawn by the machine when it operates as a motor:
$$S = 1.73\,EI = 1.73 \times 440 \times 41$$
$$= 31.2 \text{ kVA}$$
a. 2 The corresponding active power absorbed is:
$$P = S \cos \theta = 31.2 \times 0.84$$
$$= 26.2 \text{ kW}$$
a. 3 The corresponding reactive power absorbed is:
$$Q = \sqrt{S^2 - P^2} = \sqrt{31.2^2 - 26.2^2}$$
$$= 17 \text{ kvar}$$
a. 4 When the machine operates as an asynchronous generator, the capacitor bank must supply $17 \div 3 = 5.7$ kvar per phase. The voltage per phase is 440 V because the capacitors are connected in delta. Consequently, the capacitive current per phase is:
$$I_c = Q/E = 5700/440$$
$$= 13 \text{ A}$$
a. 5 The capacitive reactance per phase is:
$$X_c = E/I = 440/13$$
$$= 34 \,\Omega$$
a. 6 The capacitance per phase is:
$$C = 1/2\pi f X_c = 1/(2\pi \times 60 \times 34)$$
$$= 78 \,\mu\text{F}$$
Figure 18-15 shows how the generating system is connected. Note that if the load itself absorbs reactive power, the capacitor bank must be increased to provide it.
b. 1 The driving engine must turn at slightly

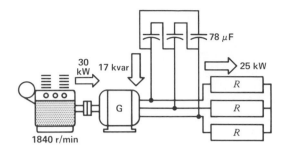

Figure 18-15   See Example 18-7.

more than synchronous speed. Typically, the slip should be equal to the normal slip when the machine operates as a motor. Consequently,
$$\text{slip} = 1800 - 1760$$
$$= 40 \text{ r/min}$$
b. 2 The engine should therefore run at an approximate speed of:
$$n = 1800 + 40$$
$$= 1840 \text{ r/min}$$

18-17 Frequency converter

A conventional wound-rotor motor may be used as a *frequency converter* to generate a frequency different from that of the utility company. The stator of the wound-rotor machine is connected to the utility line, and the rotor is driven at an appropriate speed by a motor M (Fig. 18-16). The wound-rotor machine then behaves as a rotating

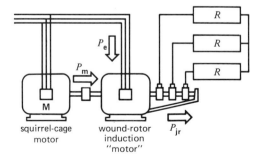

Figure 18-16 Wound-rotor motor used as a frequency converter.

transformer in which the stator is the primary and the rotor the secondary. The rotor supplies power to the three-phase load at a voltage E_2 and frequency f_2 that depends upon the slip. Thus, according to Eqs. 17-3 and 17-4, we have:

$$f_2 = sf_0 \qquad\qquad \text{Eq. 17-3}$$
$$E_2 = sE_{0c} \qquad\qquad \text{Eq. 17-4}$$

In general, the desired frequency is two or three times that of the utility company.

The operation of a frequency converter is identical to that of an induction motor, except that the power P_{jr}, usually dissipated as heat in the rotor, is now available to supply power to the load. The converter acts as a generator, and the active power flow is as shown in Fig. 18-17.

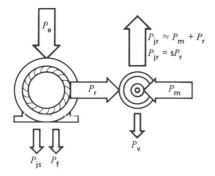

Figure 18-17   Power flow in a frequency converter.

Example 18-7:

A 3-phase wound-rotor induction motor has a rating of 150 hp ($\approx$ 110 kW), 1760 r/min, 2.3 kV, 60 Hz. The open-circuit rotor voltage between the slip rings is 500 V. The rotor is driven by variable-speed dc motor. Calculate:

a. the turns ratio of the stator to rotor windings;
b. the rotor voltage and frequency when the rotor is driven at 720 r/min in the same direction as the revolving field;
c. the rotor voltage and frequency when the rotor is driven at 720 r/min opposite to the revolving field;
d. the rotor voltage and frequency when the rotor is driven at 2880 r/min.

Solution:

a.   The turns ratio is:
$$a = E_1/E_{0c} = 2300/500$$
$$= 4.6$$

b. 1   The slip is:
$$s = (n_s - n)/n_s = (1800 - 720)/1800$$
$$= 0.6$$

b. 2   The rotor voltage is:
$$E_2 = sE_{0c} = 0.6 \times 500$$
$$= 300 \text{ V}$$

b. 3   The rotor frequency is:
$$f_2 = sf_0 = 0.6 \times 60$$
$$= 36 \text{ Hz}$$

c. 1   The motor speed is considered to be negative (–) when it turns opposite to the revolving field. The slip is:
$$s = (n_s - n)/n_s = (1800 - (-720))/1800$$
$$= (1800 + 720)/1800$$
$$= 1.4$$

c. 2   The rotor voltage and frequency are:
$$E_2 = sE_{0c} = 1.4 \times 500$$
$$= 700 \text{ V}$$
$$f_2 = sf_0 \ 1.4 \times 60$$
$$= 84 \text{ Hz}$$

d. 1   At 2880 r/min, the motor runs above synchronous speed:
$$s = (n_s - n)/n_s = (1800 - 2880)/1800$$
$$= -0.6$$

d. 2   The rotor voltage is:
$$E_2 = sE_{0c} = -0.6 \times 500 = -300 \text{ V}$$
The negative sign indicates that the voltage is 180° out of phase with the voltage calculated in b. 2. The negative sign also tells us that power is flowing from the rotor to the stator, rather than the other way around.

d. 3   The rotor frequency is:
$$f_2 = sf_0 = -0.6 \times 60$$
$$= -36 \text{ Hz}$$
The negative sign indicates the phase sequence of the rotor voltages is the reverse of that in part b. 3.

Example 18-8:

We wish to use a 30 kW, 900 r/min, 60 Hz wound-rotor motor to generate 60 kW at a frequency of

180 Hz. If the supply-line frequency is 60 Hz, calculate:

a. the driving speed of the frequency converter;
b. the active power input to the stator of the frequency converter;
c. the power of the induction motor M driving the converter.
d. Will the converter overheat under these conditions?

Solution:

a. 1
$$f_2 = sf_0 \qquad \text{Eq. 17-3}$$
$$180 = s \times 60$$
from which $\quad s = 3$

a. 2 On the other hand,
$$s = (n_s - n)/n_s \qquad \text{Eq. 17-2}$$
$$3 = (900 - n)/900$$
from which $\quad n = -1800$ r/min

The converter must therefore be driven at a speed of 1800 r/min. The negative sign indicates that the rotor must run opposite to the revolving field created by the stator.

b. 1 The rotor delivers an output of 60 kW. This corresponds to P_{jr}, but instead of being "dissipated" in the rotor, P_{jr} is useful power delivered to a load (Fig. 18-18). The power P_r transferred from the stator to the rotor is:
$$P_r = P_{jr}/s = 60/3 = 20 \text{ kW} \qquad \text{Eq. 17-7}$$

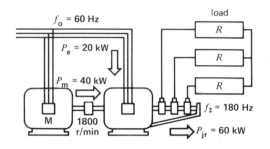

Figure 18-18   See Example 18-9.

b. 2 The power input to the stator of the frequency converter is equal to 20 kW plus the small copper losses and iron losses in the stator.

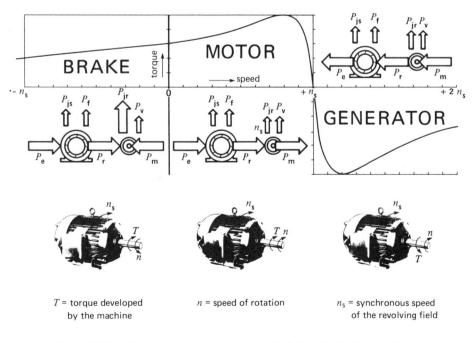

T = torque developed by the machine

n = speed of rotation

n_s = synchronous speed of the revolving field

Figure 18-19   Complete torque-speed curve of a 3-phase induction machine.

c. The remaining power input to the rotor (60 – 20 = 40 kW) is derived from the mechanical input to the shaft. By referring to Fig. 18-17 and Fig. 18-18, we can see how the active power flows into (and out of) the converter.

In summary, the rotor receives 20 kW of electrical power from the stator and 40 kW of mechanical power from the driving motor M. The rotor converts this power into 60 kW of electrical power at a frequency of 180 Hz. Induction motor M must have a rating of 40 kW, 60 Hz, 1800 r/min.

d. The stator of the converter will not overheat because the 20 kW it absorbs is much less than its nominal rating of 30 kW. The rotor will not overheat either, even though it delivers 60 kW. The increased power arises from the fact that the voltage induced in the rotor is three times higher than at standstill. In effect, the relative speed of the rotor with respect to the revolving field is three times greater than under locked-rotor conditions. The iron losses in the rotor will be high because the frequency is 180 Hz; however, because the rotor turns at twice normal speed, the cooling is more effective, and the rotor will probably not overheat. The stator frequency is 60 Hz, consequently the stator iron losses are normal.

18-18 Complete torque-speed characteristic of an induction machine

We have seen that a 3-phase induction machine may function as a motor, as a generator or as a brake. These three modes of operation merge into each other, as can be seen from the torque-speed curve of Fig. 18-19. This curve, together with the power flow diagrams illustrate the overall properties of a 3-phase induction machine.

QUESTIONS AND PROBLEMS

Practical level

18-1 What is the difference between a drip-proof motor and an explosion-proof motor?

18-2 What is the approximate life expectancy of a motor?

18-3 Explain why a NEMA design D motor is unsatisfactory for driving a pump.

18-4 Identify the motor components shown in Fig. 18-3a.

18-5 Show the flow of active power in a 3-phase induction motor when it operates:
a. as a motor;   b. as a brake.

18-6 Will a 3-phase motor continue to rotate if one of the lines becomes open? Will the motor start on such a line?

18-7 What type of ac motor would you recommend for the following application:
a. a saw in a lumber mill?
b. a variable speed pump?

18-8 Give some of the advantages of standardization as it relates to induction motors.

18-9 We can bring an induction motor to a quick stop either by plugging it or by exciting the stator from a dc source. Which method produces the least amount of heat in the motor? Explain.

18-10 A standard squirrel-cage induction motor rated at 50 hp, 440 V, 60 Hz, 1150 r/min is connected to a 208 V, 3-phase line. By how much are the breakdown torque and locked-rotor torque reduced?

Intermediate level

18-11 A 3-phase squirrel-cage induction motor having a rated voltage of 575 V, is connected to a 208 V line. Explain how the following

parameters are affected:
  a. locked-rotor current;
  b. locked-rotor torque;
  c. no-load current;
  d. no-load speed;
  e. full-load current;
  f. full-load power factor;
  g. full-load efficiency.

18-12 a. Referring to Fig. 18-6, if we eliminated the gear box and used a motor directly coupled to the load, what would its power output be [hp]?
  b. How many poles would the motor have?

18-13 Draw the typical torque-speed curve of a NEMA Design C squirrel-cage induction motor, rated at 30 hp, 900 r/min (see Fig. 18-5). Give the values of the LR, pull-up and breakdown torques and the corresponding speeds [ft·lb and r/min].

18-14 A 300 hp, 2300 V, 3-phase, 60 Hz squirrel-cage induction motor runs at a full-load speed of 590 r/min.
  a. Calculate the approximate value of the rotor I^2R losses. If the line voltage then drops to 1944 V, calculate:
  b. the new speed, knowing that the load torque remains the same;
  c. the new power output;
  d. the new I^2R losses in the rotor.

18-15 We wish to make an asynchronous generator using a standard squirrel-cage induction motor rated at 40 hp, 208 V, 870 r/min, 60 Hz (Fig. 18-14). The generator is driven at 2100 r/min by a gasoline engine, and the load consists of three 5 Ω resistors connected in wye. The generator voltage builds up when three 100 μF capacitors are connected in wye across the terminals. If the line voltage is 520 V, calculate:
  a. the approximate frequency generated;
  b. the active power supplied to the load;
  c. the reactive power supplied by the capacitor bank;
  d. the stator current;
  e. the following gasoline engines are avail-

able: 30 hp, 100 hp, 150 hp. Which is best suited for the purpose?
  f. discuss the copper and iron losses when the machine operates as above, compared to when it functions as a standard motor at 60 Hz.

18-16 A 30 000 hp, 13.2 kV, 3-phase, 60 Hz air-to-water cooled induction motor drives a turbocompressor in a large oxygen-manufacturing plant. The motor runs at an exact full-load speed of 1792.8 r/min and by means of a gearbox, it drives the compressor at a speed of 4930 r/min. The motor has an efficiency of 98.1% and a power factor of 0.90. The LR torque and current are respectively 0.7 p.u. and 4.9 p.u.
Calculate:
  a. the full-load current;
  b. the total losses at full load;
  c. the exact rotor I^2R losses if the windage and friction losses amount to 62 kW;
  d. the LR current and torque;
  e. the torque developed at the compressor shaft [ft·lb].

18-17 The motor in Problem 18-16 is cooled by circulating 350 gallons (U.S.) of water through the heat exchanger per minute. Calculate the increase in temperature as the water flows through the heat exchanger.

18-18 The motor and compressor in Problem 18-16 are started on reduced voltage and the average starting torque is 0.25 p.u. The compressor has a moment of inertia of 130 000 lb·ft^2 referred to the motor shaft. The squirrel cage rotor alone has a J of 18 000 lb·ft^2.
  a. How long will it take to bring the motor and compressor up to speed, at no-load;
  b. what is the energy dissipated in the rotor during the starting period [Btu]?

18-19 A 3-phase induction motor rated at 10 kW, 1450 r/min, 380 V, 50 Hz has to be connected to a 60 Hz line.
  a. What line voltage should be used, and what is the approximate speed of the motor?

b. What is the power [hp] the motor can deliver without overheating?

Advanced level

18-20 A 1 hp squirrel-cage Design B induction motor accelerates an inertia load of 1.4 kg·m^2, from zero to 1800 r/min. Could this motor be replaced by a class D motor and if so:

a. which motor has the shortest acceleration time from zero to 1200 r/min?

b. which of the two rotors will be the hottest, after reaching the top no-load speed?

18-21 A three-phase wound-rotor induction motor having a rating of 150 hp, 1760 r/min, 2.3 kV, 60 Hz, drives a belt conveyor. The rotor is connected in wye and the nominal open circuit voltage between the slip-rings is 530 V. Calculate:

a. the rotor winding resistance per phase;

b. the resistance which must be placed in series with the rotor (per phase) so that the motor will deliver 40 hp at a speed of 600 r/min, knowing that the line voltage is 2.4 kV.

18-22 A 150 hp, 1165 r/min, 440 V, 60 Hz, 3-phase induction motor is running at no-load, close to its synchronous speed of 1200 r/min. The stator leads are suddenly reversed, and the stopping time is clocked at 1.3 s. Assuming that the torque exerted during the plugging interval is equal to the starting torque (1.2 p.u.), calculate:

a. the magnitude of the plugging torque;

b. the moment of inertia of the rotor.

18-23 In Problem 18-22 calculate the energy dissipated in the rotor during the plugging interval.

18-24 A 3-phase, 8-pole induction motor has a rating of 40 hp, 575 V, 60 Hz. It drives a steel flywheel having a diameter of 31.5 inches and a thickness of 7 7/8 in. The torque-speed curve corresponds to a design D motor given in Fig. 18-5. Calculate:

a. the mass of the flywheel and its moment of inertia [lb·ft^2] ;

b. the rated speed of the motor and the corresponding torque [ft·lb] ;

c. the locked-rotor torque [ft·lb] .

d. draw the torque-speed curve and give the torques at 0, 180, 360, 540, 720 and 810 r/min.

18-25 a. In Problem 18-24, calculate the average torque between zero and 180 r/min.

b. Using Eq. 3-14, calculate the time required to accelerate the flywheel from zero to 180 r/min, assuming no other load on the motor.

c. Using Eq. 3-8, calculate the kinetic energy in the flywheel at 180 r/min.

d. Calculate the time required to accelerate the flywheel from zero to 540 r/min, knowing that this time, the load exerts a fixed counter torque of 300 N·m, in addition to the flywheel load.

18-26 The train in Fig. 18-12 has a mass of 78 500 lb and can carry 240 passengers. Calculate:

a. the speed of rotation of the gear wheels when the train moves at 9 miles per hour.

b. the gear ratio between the motor and the gearwheel.

c. the approximate transmission line current when the motors are operating at full load;

d. the total mass if the average weight of a passenger is 60 kg;

e. the energy required to climb from Zermatt to Gornergrat [MJ] ;

f. the minimum time required to make the trip [min] ;

g. assuming that 80% of the electrical energy is converted into mechanical energy when the train is going uphill, and that 80% of the mechanical energy is reconverted to electrical energy when going downhill, calculate the total electrical energy consumed during a round trip [kW·h] .

19

THREE-PHASE
ALTERNATORS

Three-phase alternators are the primary source of all the electrical energy we consume. These machines are the largest energy converters found in the world. They convert mechanical energy into electrical energy, in powers ranging up to 1500 MW. In this chapter, we shall study the construction and characteristics of these large, modern alternators. They are based upon the elementary principles covered in Sec. 9-4, and the reader may wish to review this material before proceeding further.

19-1 Commercial alternators

Commercial 3-phase alternators have either a stationary or a rotating dc magnetic field. A *stationary-field alternator* has the same outward appearance as a dc generator. The salient poles create the dc field which is cut by a revolving armature. The armature possesses a 3-phase winding whose terminals are connected to three slip-rings mounted on the shaft. A set of brushes, sliding on the slip-rings, enables us to connect the armature to an external 3-phase load. The armature is driven by a gasoline engine, or some other source of motive power. As it rotates, a 3-phase voltage is induced, whose value depends upon the speed of rotation and upon the dc exciting current in the stationary poles. Stationary-field alternators are ideal when the output power is less than 5 kVA. However, for greater outputs, it is cheaper, safer and more practical to employ a stationary armature (or stator) and a revolving dc field.

The field is excited by a dc generator, usually mounted on the same shaft (Fig. 19-1). The stationary 3-phase stator winding is directly connected to the load, without going through large, unreliable slip-rings and brushes. A stationary stator also makes it easier to insulate the windings because they are not subjected to centrifugal forces.

19-2 Number of poles

The number of poles on an alternator depends upon the speed of rotation and the frequency we wish to produce. Consider, for example, a stator conductor that is successively swept by the north

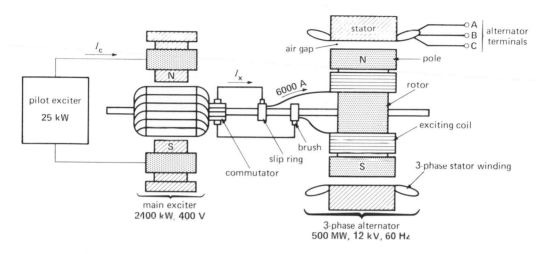

Figure 19-1 Schematic diagram of a typical 500 MW alternator and its 2400 kW dc exciter. The dc exciting current I_x (6000 A) flows through the commutator and two slip rings. The dc control current I_c from the pilot exciter permits variable field control of the main exciter, which, in turn, controls I_x.

and south poles of the rotor. If a *positive* voltage is induced when a north pole sweeps across the conductor, a similar *negative* voltage is induced when the south pole sweeps by. Every time a *pair* of poles crosses the conductor, the induced voltage goes through a *complete cycle*. The same is true for every other conductor on the stator; we can therefore deduce that the alternator frequency is given by:

$$f = \frac{pn}{120} \qquad (19\text{-}1)$$

where

f = frequency of the induced voltage [Hz]
p = number of poles on the rotor
n = speed of the rotor [r/min]

Example 19-1:

A hydraulic turbine turning at 200 r/min is connected to an alternator. If the induced voltage has a frequency of 60 Hz, how many poles does the rotor have?

Solution:
From Eq. 19-1, we have:

p = 120 f/n
  = 120 × 60/200
  = 36 poles or 18 pair of poles

19-3 Stator

From an electrical standpoint, the stator of an alternator is *identical* to that of a three-phase induction motor (Sec. 17-17). It is composed of a cylindrical laminated core containing a set of slots that carry the 3-phase lap winding (Figs. 19-2, 19-3). The winding is always connected in wye and the neutral is connected to ground. We prefer a wye connection to a delta connection because:

1. The voltage per phase is only $1/\sqrt{3}$ or 58% of the voltage between the lines. This means that the highest effective voltage between a stator conductor and the grounded stator core is only 58% of the line voltage. We can therefore reduce the amount of insulation in the slots which, in turn, enables us to increase the cross section of the conductors. A larger conductor permits us to increase the current and hence, the power output of the machine.

2. When an alternator is under load, the voltage

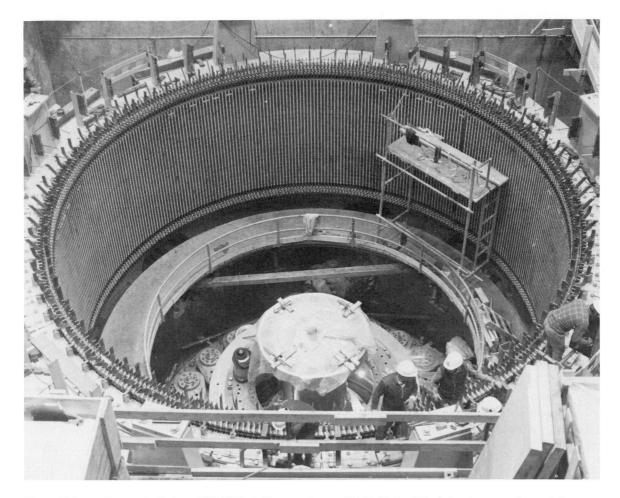

Figure 19-2 a. Stator of a 3-phase, 500 MVA, 0.95 power factor, 15 kV, 60 Hz, 200 r/min, alternator. Internal diameter: 9250 mm; effective axial length of iron stacking: 2350 mm; 378 slots. *(Marine Industrie)*

per phase becomes distorted, and the waveform is no longer sinusoidal. The distortion is mainly due to an undesired *third harmonic* voltage whose frequency is three times that of the fundamental frequency. With a wye connection, the distorting line-to-neutral harmonics do not appear between the lines because they effectively cancel each other. Consequently, the line voltages remain sinusoidal under all load conditions. Unfortunately, when a delta connec-

tion is used, the harmonic voltages do not cancel, but add up. Because the delta is closed, they produce a large third-harmonic circulating current, which increases the I^2R losses.

The nominal line voltage of an alternator depends upon its kVA rating. In general, the greater the power, the higher the voltage. However, the nominal voltage seldom exceeds 25 kV because the increased slot insulation takes up valuable space at the expense of the copper conductors.

Figure 19-2   b. The copper bars connecting successive stator poles are designed to carry a current of 3200 A. The total
output is 19 250 A per phase. *(Marine Industrie)*

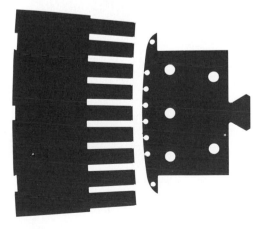

Figure 19-2 c.

The stator is built up from toothed segments of high-quality
silicon-iron steel laminations (0.5 mm thick), covered with an insulating varnish. The slots are 22.3 mm wide and 169 mm deep.
The salient poles of the rotor are composed of much thicker
(2 mm) iron laminations. These laminations are not insulated because the dc flux they carry does not vary. The width of the
poles from tip-to-tip is 600 mm and the air gap length is 33 mm.
The 8 round holes in the face of the salient pole carry the bars of
a squirrel-cage winding.

Figure 19-3 Stator of a 3-phase, 722 MVA, 3600 r/min, 19 kV, 60 Hz turboalternator during the construction phase. The windings are water-cooled. The stator will eventually be completely enclosed in a metal housing (see background). The housing contains hydrogen under pressure to further improve the cooling. *(Brown Boveri)*

19-4 Rotor

Alternators are built with two types of rotors: salient-pole rotors and smooth, cylindrical rotors. The first are usually coupled to low-speed hydraulic turbines while the second are driven by high-speed steam-turbines.

1. Salient pole rotors. Most hydraulic turbines turn at low speeds (between 50 and 300 r/min) in order to extract the maximum power from a water fall. Because the rotor is directly coupled to the water-wheel, and because a frequency of 60 Hz is required, we must place a large number of poles on the rotor. Low-speed rotors always possess a large diameter to provide the necessary space for the poles. The salient poles are mounted on a large circular steel frame which is fixed to a revolving vertical shaft (Fig. 19-4). To ensure good cooling, the field coils are made of bare copper bars, insulated from each other by strips of mica (Fig. 19-5). The coils are connected in series so that adjacent poles have opposite polarities.

In addition to the dc winding, we often add a squirrel-cage winding, embedded in the pole-faces (Fig. 19-6). Under normal conditions, this winding does not carry any current because the rotor turns at synchronous speed. However, when the load on the alternator changes suddenly, the rotor speed begins to oscillate, producing momentary speed variations above and below synchronous speed. This induces a voltage in the squirrel-cage winding, causing a large current to flow. The current reacts with the magnetic field of the stator, producing forces which dampen the oscillation of the rotor. For this reason, the squirrel-cage winding is sometimes called a *damper winding*.

The damper winding also tends to maintain balanced three-phase voltages under unbalanced load conditions.

2. Cylindrical rotors. Steam turbines are smaller and more efficient when they turn at high speed. The same is true of alternators. However, to generate the required frequency, we can use no less than 2 poles, and this fixes the highest possible speed. On a 60 Hz system, it is 3600 r/min. The next lower speed is 1800 r/min, corresponding to a 4-pole machine. Consequently, these so-called *turbo-alternators* possess either 2 or 4 poles. The rotor is a long, solid steel cylinder which contains a series of longitudinal slots milled out of the cylindrical mass (Fig. 19-7). Concentric field coils, firmly wedged into the slots and retained by high-strength end-rings,* serve to create the poles.

The high speed of rotation produces strong centrifugal forces which impose an upper limit on the diameter of the rotor.† On the other hand, to build powerful turbo-alternators, we have to use massive rotors. It follows that high-power, high-speed rotors have to be very long.

19-5 Exciters

The main exciter is usually a dc generator that feeds the exciting current to the rotor by way of brushes and slip-rings. Under normal conditions, the exciter voltage lies between 125 V and 600 V. It may be regulated manually or automatically, by varying the current I_c, produced by a so-called *pilot exciter* (Fig. 19-1).

The power rating of the main exciter depends upon the capacity of the alternator. Typically, a 25 kW exciter is needed to excite a 1000 kVA alternator (2.5% of its rating) whereas a 2500 kW exciter suffices for an alternator of 500 MW (only 0.5% of its rating).

Under normal conditions, the excitation is varied automatically. It follows the load changes so as to maintain a constant ac line voltage or to control the reactive power delivered to the electric utility system. A serious disturbance on the system

---

* See Fig. 12-26.

† In the case of a rotor turning at 3600 r/min, the elastic limit of the steel requires the manufacturer to limit the diameter to a maximum of 1.2 m. We can double the diameter when the speed is 1800 r/min, but because of the transportation problems, we seldom go beyond 1.8 m.

Figure 19-4   This 36-pole rotor is being lowered into the stator shown in Fig. 19-2. The 2400 A dc exciting current is supplied by a 330 V, electronic rectifier. Other details are: mass: 600 t; moment of inertia: 4140 t·m²; air gap: 33 mm. *(Marine Industrie)*

Figure 19-5   This rotor winding for a 250 MVA salient-pole alternator is made of 18 turns of bare copper bars having a width of 89 mm and a thickness of 9 mm.

Figure 19-6   Salient-pole of a 250 MVA alternator, showing 12 slots to carry the squirrel-cage winding.

Figure 19-7   a. Rotor of a 3-phase turboalternator rated 1530 MVA, 1500 r/min, 27 kV, 50 Hz. The 40 slots are being milled out of the solid steel mass. They will carry the dc winding. Effective axial magnetic length: 7490 mm; diameter: 1800 mm.

Figure 19-7   b. Rotor with its 4-pole dc winding. Total mass: 204 t; moment of inertia: 85 t·m²; air gap: 120 mm. The dc exciting current of 11.2 kA is supplied by a 600 V dc brushless exciter, bolted to the end of the main shaft. *(Allis-Chalmers Power Systems Inc., West Allis, Wisconsin)*

may produce a sudden voltage drop across the terminals of the alternator. The exciter must then react very quickly to keep the ac voltage constant. For example, the exciter voltage may have to rise to twice its normal value in 300 to 400 milliseconds. This represents a very quick response, considering that the power of the exciter may be several thousand kilowatts.

19-6 Brushless excitation

Owing to brush wear and carbon dust, we constantly have to maintain brushes, slip-rings and commutators on conventional dc excitation systems. To eliminate the problem, *brushless excitation systems* have been developed in which a 3-phase ac exciter and a group of rectifiers supply the direct current to the main alternator (Fig. 19-8). The ac exciter and rectifiers are mounted on the main shaft and turn together with the alterna-

regulates the main exciter output I_x, as in the case of a conventional dc exciter. The main exciter is actually a stationary-field alternator. Its ouput frequency is generally two to three times the main alternator frequency (60 Hz). Figure 19-9 shows the rotating portion of a typical brushless exciter.

19-7 Factors affecting alternator size

The prodigious amount of energy generated by electrical utility companies has made them very sensitive about the efficiency of their machines. For example, if the efficiency of a 1000 MW generating station improves by only 1%, it represents savings of several thousand dollars per day. In this regard, the *size* of the alternator is particularly important, because we know that its efficiency automatically improves as the power increases. For example, if an alternator of 1 kilowatt has an efficiency of 50%, a larger, but similar model having a

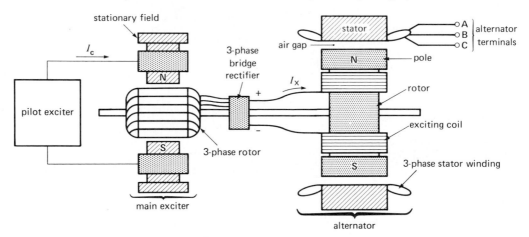

Figure 19-8   Typical brushless exciter system.

tor. In comparing the excitation system of Fig. 19-8 with that of Fig. 19-1, we can see they are identical, except that the 3-phase rectifier replaces the commutator. In other words, the commutator (or mechanical rectifier) is replaced by an electronic rectifier. The result is that the brushes and slip rings are no longer needed.

A dc control current, from the pilot exciter,

capacity of 10 MW *inevitably* has an efficiency of about 90%. This improvement in efficiency with size is why alternators of 1000 MW and up, possess efficiencies of the order of 99%.

A further advantage of large machines is that the power output per kilogram increases as the power increases. Again referring to our example, if an alternator of 1 kW weighs 20 kg (50 W/kg), a

Figure 19-9

This brushless exciter provides the dc current for the rotor shown in Fig. 19-7. The exciter consists of a 7000 kVA alternator and two sets of diodes. Each set corresponding respectively to the positive and negative terminals is housed in the circular rings mounted on the shaft, and visible in the center of the photograph. The ac exciter is seen to the right. The two round conductors protruding from the center of the shaft (foreground) lead the exciting current to the 1530 MVA alternator. *(Allis-Chalmers Power Systems Inc., West Allis, Wisconsin)*

10 MW machine of similar construction will weigh only 20 000 kg, yielding 500 W/kg. From a power standpoint, large machines weigh relatively less than small machines do; consequently, they are cheaper.

Everything, therefore, seems to favor the large machines. However, as they grow in size, we run into serious cooling problems. In effect, large machines inherently produce high power losses per unit surface area (W/m^2); consequently, they tend to overheat. To prevent an unacceptable temperature rise, we must therefore design efficient cooling systems that become ever more elaborate as the power increases. For example, a circulating cold-air system is adequate to cool turboalternators whose rating is below 50 MW, but between 50 MW and 300 MW, we have to resort to hydrogen cooling (Sec. 4-4). Very big alternators in the 1000 MW range have to be equipped with hollow, water-cooled conductors. Ultimately, a point is reached where the increased cost of cooling exceeds the savings made elsewhere, and this fixes the upper limit to size.

To sum up, the evolution of big alternators has mainly been dictated by the evolution of sophisticated cooling techniques (Fig. 19-10, 11). Other technological breakthroughs, such as better materials, new windings, and so forth have also played a major part in modifying the design of early machines (Fig. 19-12).

As regards speed, low-speed alternators are al-

Figure 19-10   Partial view of a 3-phase, salient-pole alternator rated 87 MVA, 428 r/min, 50 Hz. Both the rotor and stator are water-cooled. The high resistivity of pure water and the use of insulating plastic tubing enables the water to be brought into direct contact with the live parts of the machine. *(Brown Boveri)*

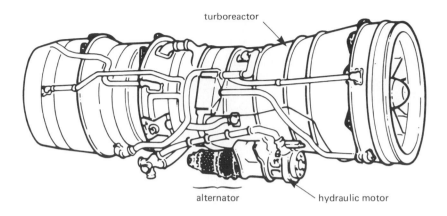

turboreactor

alternator   hydraulic motor

Figure 19-11   The electrical energy needed on board the Concord aircraft is supplied by four alternators rated 60 kVA, 200/115 V, 12 000 r/min, 400 Hz. Each alternator is driven by a hydraulic motor which absorbs a small portion of the enormous power developed by the turboreactor engines. The hydraulic fluid streaming from the hydraulic motor is used to cool the alternator and then recycled. The alternator itself weighs only 54.5 kg. *(Air France)*

Figure 19-12   This rotating-field alternator was first installed in North America in 1888. It was used in a 1000-lamp street lighting system. The alternator was driven by an 1100 r/min steam engine and had a rated output of 2000 V, 30 A at a frequency of 110 Hz. It weighed 2320 kg, which represents 26 W/kg. A modern alternator of equal speed and power produces about 140 W/kg and occupies only one-third the floor space.

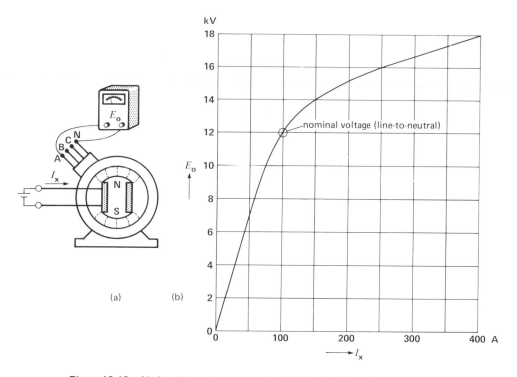

Figure 19-13   No-load saturation curve of a 36 MW, 21 kV, 3-phase alternator.

ways bigger than high-speed machines of equal power. Bigness simplifies the cooling problem; a good air-cooling system, completed perhaps with a heat exchanger, usually suffices. For example, the large, slow-speed 500 MVA, 200 r/min alternators installed in a typical hydropower plant, are air-cooled whereas the much smaller high-speed 500 MVA, 1800 r/min units installed in a steam plant, are hydrogen-cooled.

19-8 No-load operation

Figure 19-13 shows a two-pole alternator operating at no-load. It is driven at constant speed by a turbine (not shown). The leads from the 3-phase wye-connected stator are brought out to terminals A, B, C, N and a variable exciting current I_x produces the flux in the air gap.

Let us gradually increase the exciting current while observing the ac voltage E_0 between terminal

A, say, and the neutral N. For small values of I_x, the voltage increases in proportion to the exciting current. However, as the iron begins to saturate, the voltage rises much less for the same increase in I_x. If we draw the curve of E_0 versus I_x, we obtain the *no-load saturation curve* of the alternator. It is similar to that of a dc generator (Sec. 14-10).

Figure 19-13 is the actual no-load saturation curve of a 36 MW alternator having a nominal voltage of 12 kV (line to neutral). Up to about 9 kV, the voltage increases in proportion to the current, but then the iron begins to saturate. Thus, an exciting current of 100 A produces an output of 12 kV, but if the current is doubled, the voltage rises only to 15 kV.

19-9 Synchronous reactance - equivalent circuit of an alternator

Consider an alternator having terminals A, B, C,

feeding a balanced 3-phase load (Fig. 19-14). The alternator is driven by a turbine (not shown), and is excited by a dc current I_x. We can assume that the machine and its load are connected in wye, yielding the circuit of Fig. 19-15. Although neutrals N_1 and N_2 are not connected, they are at the same potential because the load is balanced. Consequently, we *could* connect them together (dotted line) without affecting the behavior of the voltages or currents in the circuit. The reason for making this connection will soon become clear.

The revolving field carries an exciting current which produces a flux Φ. The flux induces, in the stator, three equal voltages E_0, that are $120°$ out of phase (Fig. 19-16).

Each phase of the stator winding possesses a resistance R and a certain inductance L. Because this is an alternating-current machine, the inductance manifests itself as a reactance X_s, given by:

$$X_s = 2\pi f L$$

where

X_s = synchronous reactance, per phase [Ω]

f = alternator frequency [Hz]

L = apparent inductance of the stator winding, per phase [H]

The so-called *synchronous reactance* of an alternator is an internal impedance. Just like its internal resistance R, the impedance is there, but it can neither be seen, nor touched. Nevertheless, the synchronous reactance behaves like an ac inductor, in series with E_0 and R. The value of X_s is 10 to 100 times greater than R; consequently, we can neglect the resistance, unless we are interested in efficiency or heating effects.

The schematic diagram of Fig. 19-16 is quite elaborate, but we can simplify it by showing only one phase of the stator. In effect, the two other phases are identical, except that their respective voltages (and currents) are out of phase by $120°$. Furthermore, if we neglect the resistance of the windings, we obtain the very simple circuit of Fig. 19-17.

In this circuit, the exciting current I_x produces

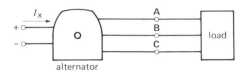

Figure 19-14   Alternator connected to a load.

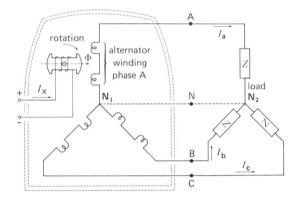

Figure 19-15   Electric circuit representing the installation of Fig. 19-14.

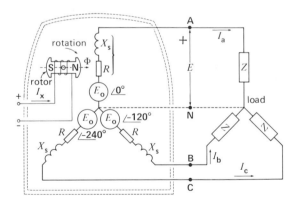

Figure 19-16   Voltages and impedances in a 3-phase alternator and its connected load.

the flux Φ which induces the internal voltage E_0. Voltage E at the terminals of the alternator depends upon E_0 and the load Z. Note that E_0 and E

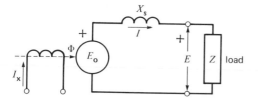

Figure 19-17   Equivalent circuit of a 3-phase alternator, showing only one phase.

are line-to-neutral voltages and I is the line current. An alternator can therefore be represented by an equivalent circuit composed of an induced voltage E_0 in series with a reactance X_s.

We can determine the approximate value of X_s by exciting the alternator so that it generates its rated open-circuit voltage E_n (line-to-neutral). The three armature terminals are then shorted, and the exciting current I_x is brought up to the same value it had before. The short-circuit current I_{sc} is measured and X_s is calculated by using the expression

$$\boxed{X_s = E_n/I_{sc}} \quad * \qquad (19\text{-}2)$$

where

X_s = synchronous reactance, per phase [Ω]

E_n = rated open-circuit line-to-neutral voltage [V]

I_{sc} = short-circuit current, per phase, using the same exciting current I_x that was required to produce E_n [A]

Example 19-2:

A 3-phase alternator generates an open-circuit line voltage of 6920 V when the dc exciting current is 50 A. The ac terminals are then short-circuited, and the three line currents are found to be 800 A.

a. Calculate the synchronous reactance per phase;

b. Calculate the terminal voltage if three 12 Ω resistors are connected in wye across the alternator.

---

* This value of X_s corresponds to the direct-axis synchronous reactance. It is widely used to describe synchronous machine behavior.

Solution:

a. 1   The induced line-to-neutral voltage is:
$$E_0 = E_L/\sqrt{3} = 6920/1.73 \qquad \text{Eq. 9-4}$$
$$= 4000 \text{ V}$$

a. 2   When the terminals are shorted, the only impedance limiting the current flow is that due to the synchronous reactance. Consequently:
$$X_s = E_0/I = 4000/800$$
$$= 5 \ \Omega$$
The synchronous reactance per phase is therefore 5 Ω.

b. 1   The equivalent circuit per phase is shown in Fig. 19-18a.

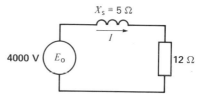

Figure 19-18   a.   See Example 19-2.

b. 2   The impedance of the circuit is:
$$Z = \sqrt{R^2 + X_s^2} = \sqrt{12^2 + 5^2} \qquad \text{Eq. 7-15}$$
$$= 13 \ \Omega$$

b. 3 $I = E_0/Z = 4000/13$
$$= 308 \text{ A}$$

b. 4   The voltage across the resistor is:
$$E = IR = 308 \times 12 = 3696 \text{ V}$$

b. 5   The line voltage under load is:
$$E_L = \sqrt{3} \, E = 1.73 \times 3696$$
$$= 6394 \text{ V}$$
The schematic diagram of Fig. 19-18b helps us visualize what is happening in the actual circuit.

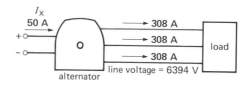

Figure 19-18   b. See Example 19-2.

19-10 Nominal impedance, per unit X_s and short-circuit ratio

The nominal impedance of an alternator is the ratio of the rated line-to-neutral voltage divided by the rated line current. We can readily prove that its value is given by:

$$Z_n = \frac{E_n^2}{S_n} \qquad (19\text{-}3)$$

where

Z_n = nominal impedance (line-to-neutral) of the alternator $[\Omega]$

E_n = rated line voltage $[V]$

S_n = rated power of the alternator $[VA]$

The nominal impedance is used as a base of comparison for other impedances that the alternator possesses. Thus, the synchronous reactance may be expressed as a per-unit value of Z_n. In general, X_s lies between $0.8\,Z_n$ and $2\,Z_n$, depending upon the design of the machine.

Example 19-3:

A 30 MVA, 15 kV, 60 Hz alternator has a synchronous reactance of 1.2 p.u. and a resistance of 0.02 p.u. Calculate:

a. the nominal impedance of the alternator;

b. the value of the synchronous reactance;

c. the winding resistance, per phase;

d. the total full-load copper losses.

Solution:

a. 1   The nominal line-to-neutral impedance is:

$Z_n = E_n^2/S_n$ Eq. 19-3

$\quad = 15\,000^2/(30 \times 10^6)$

$\quad = 7.5\ \Omega$

b. 1   The synchronous reactance is:

$X_s = 1.2\,Z_n = 1.2 \times 7.5$

$\quad = 9\ \Omega$

c. 1   The resistance per phase is:

$R = 0.02\,Z_n = 0.02 \times 7.5$

$\quad = 0.15\ \Omega$

Note that all impedance values are from line to neutral.

d. 1   The copper losses for all 3 phases are:

$P = 0.02\,S_n = 0.02 \times 30 = 0.6$ MW

$\quad = 600$ kW

The synchronous reactance does not remain constant, but varies with the degree of saturation. When the iron is heavily saturated, the value of X_s may be only half its unsaturated value. Despite this rather broad variation, we usually take the unsaturated value for X_s because it yields sufficient accuracy in most cases.

Instead of expressing the synchronous reactance as a per-unit value of Z_n, the term *short-circuit ratio* is sometimes used. It is the ratio of the field current I_{x1} needed to generate rated open-circuit armature voltage to the field current I_{x2} needed to produce rated current I_n, on a sustained short-circuit. The short-circuit ratio (I_{x1}/I_{x2}) is a number whose value is exactly equal to the reciprocal of the per-unit value of X_s as defined in Eq. 19-2. Thus, if the per-unit value of X_s is 1.2, the short-circuit ratio is 1/1.2 or 0.833.

19-11 Alternator under load

The behavior of an alternator depends upon the type of load it has to supply. Although there are many types of loads, they can all be reduced to 2 basic categories:

1. isolated loads, supplied by one alternator;

2. the infinite bus.

We shall begin our study with isolated loads leaving the discussion of the infinite bus to Sec. 19-13.

Consider a 3-phase alternator that supplies power to a load having a lagging power factor. Figure 19-19 represents the equivalent circuit for one phase, and line current I lags behind voltage E by an angle θ. The current produces a voltage drop E_x across the synchronous reactance, given by:

$$E_x = IX_s$$

This internal drop is $90°$ ahead of the current because X_s is an inductive reactance. The induced voltage E_0 is equal to the phasor sum of E and E_x. As we would expect, it is larger than the terminal

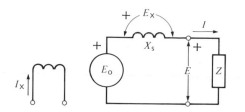

Figure 19-19   Equivalent circuit of an alternator under load.

voltage E. Voltage E_0 is generated by the dc exciting current I_x. The complete phasor diagram is given in Fig. 19-20.

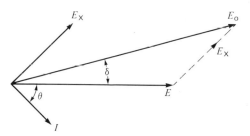

Figure 19-20   Phasor diagram for a lagging power factor load.

In some cases, the load is somewhat capacitive, so that current I leads the terminal voltage by an angle θ. What effect does this have on the phasor diagram? The answer is given in Fig. 19-21. The voltage E_x across the synchronous reactance is still 90° ahead of the current. Furthermore, E_0 is again equal to the phasor sum of E and E_x. However, the terminal voltage is now greater than the induced voltage, which is a very surprising result. In effect, the inductive reactance X_s enters into partial reso-

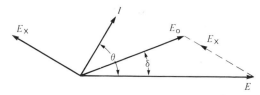

Figure 19-21   Phasor diagram for a leading power factor load.

nance with the capacitive reactance of the load. Although it may appear we are getting something for nothing, the higher terminal voltage does not yield any more power.

If the load is entirely capacitive, a very high terminal voltage can be produced with a small exciting current. However, in later chapters, we shall see that under-excitation is undesirable.

Example 19-4:
A 36 MVA, 21 kV, 3-phase alternator has a synchronous reactance of 9 Ω, and a nominal current of 1 kA. The no-load saturation curve giving the relationship between E_0 and I_x is given in Fig. 19-13. If the excitation is adjusted so that the terminal voltage remains fixed at 21 kV, calculate the exciting current required and draw the phasor diagram for the following conditions:
a. no-load;
b. resistive load of 36 MW;
c. capacitive load of 12 Mvar.

Solution:
We shall immediately simplify the circuit to show only one phase. The line-to-neutral terminal voltage for all cases is fixed at:
$E = 21/1.73 = 12$ kV
a. 1   At no-load, there is no voltage drop in the synchronous reactance; consequently,
$E_0 = E = 12$ kV
a. 2   The exciting current is:
$I_x = 100$ (see Fig. 19-13)
a. 3   The phasor diagram is given in Fig. 19-22a.

(a)

Figure 19-22   a. Phasor diagram at no-load.

b. 1   The power per phase is:
$P = 36/3 = 12$ MW
b. 2   The full-load line current is:
$I = P/E = 12 \times 10^6/12\,000 = 1000$ A
This current is in phase with the terminal voltage.

b. 3   The voltage drop across X_s is:
$$E_x = IX_s = 1000 \times 9 = 9 \text{ kV}$$
This voltage is $90°$ ahead of I.

b. 4   The voltage generated by I_x is equal to the phasor sum of E and E_x. Its value is given by:
$$E_0 = \sqrt{E^2 + E_x{}^2} = \sqrt{12^2 + 9^2}$$
$$= 15 \text{ kV}$$

b. 5   The corresponding exciting current is:
$$I_x = 200 \text{ A} \qquad \text{(see Fig. 19-13)}$$

b. 6   The phasor diagram is given in Fig. 19-22b.

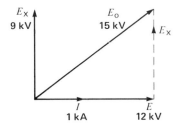

Figure 19-22   b. Phasor diagram with a unity power factor load.

c. 1   With a capacitive load of 12 Mvar the reactive power per phase is:
$$Q = 12/3 = 4 \text{ Mvar}$$

c. 2   The line current is:
$$I = Q/E = 4 \times 10^6/12\,000 = 333 \text{ A}$$

c. 3   The voltage drop across X_s is:
$$E_x = IX_s = 333 \times 9 = 3 \text{ kV}$$
As before, E_x, leads I by $90°$ (Fig. 19-22c).

c. 4   The voltage generated by I_x is equal to the phasor sum of E and E_x.
$$E_0 = E + E_x = 12 + (-3) = 9 \text{ kV}$$

c. 5   The corresponding exciting current is:
$$I_x = 70 \text{ A} \qquad \text{(see Fig. 19-13)}$$

c. 6   The phasor diagram for this capacitive load is given in Fig. 19-22c.

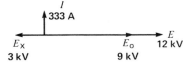

Figure 19-22   c. Phasor diagram with a capacitive load.

19-12   Synchronization of an alternator

We sometimes have to couple two or more alternators in parallel to supply a common load. For example, as the power requirements of a large utility system build up during the day, alternators are successively connected to the system to provide the extra power. Later, when the power demand falls, selected alternators are disconnected from the system until power again builds up the following day. Alternators are therefore regularly being connected and disconnected from a large power grid, in response to customer demand.

Before connecting an alternator to a system (or in parallel with another alternator), it must be *synchronized*. An alternator is said to be synchronized when it meets all of the following conditions:

1. the alternator frequency is equal to the system frequency;
2. the alternator voltage is equal to the system voltage;
3. the alternator voltage is in phase with the system voltage.

To synchronize an alternator, we proceed as follows:
1. adjust the speed regulator of the turbine so that the alternator frequency is close to the system frequency;
2. adjust the excitation so that the alternator voltage E_0 is equal to the system voltage E;
3. observe the phase angle between E_0 and E by means of a *synchroscope* (Fig. 19-23). This instrument has a pointer that continually indicates the phase angle between the two voltages, covering the entire range from zero to 360 degrees. Although the degrees are not shown, the dial has a zero marker to indicate when the voltages are in phase. In practice, when we synchronize an alternator, the pointer rotates slowly as it follows the phase angle between the alternator and system voltages. If the alternator frequency is slightly higher than the system frequency, the pointer rotates clockwise, indicating that the alternator has a tendency to lead

Figure 19-23   Synchroscope. *(Lab-Volt)*

the system frequency. Conversely, if the alternator frequency is slightly low, the pointer rotates counterclockwise. The turbine speed-regulator is fine-tuned accordingly, so that the pointer barely creeps across the dial. A final check is made to see that the alternator voltage is still equal to the system voltage. Then, the moment the pointer crosses zero . . .

4. the line circuit breaker is closed, connecting the alternator to the system.

In modern generating stations, synchronization is usually done automatically.

19-13   Alternator on an infinite bus

We seldom have to couple only two alternators in parallel except, perhaps, in isolated areas (Fig. 19-24). It is much more common to connect an alternator to a large power system that already has dozens of alternators connected to it. Such a system is called an *infinite bus*.

An infinite bus is a system so powerful that it imposes its own voltage and frequency upon any apparatus connected to its terminals. Once connected to a large system (infinite bus), an alternator becomes part of a network comprising hundreds of other alternators that deliver power to thousands of loads. It is impossible, therefore, to specify the nature of the load (large or small, resistive or capacitive) connected to the terminals of this particular alternator. What, then, determines the power the machine delivers? To answer this question, we must remember that the value and frequency of the terminal voltage across the alternator is fixed. Consequently, we can control only two machine parameters:

1. the exciting current I_x;
2. the mechanical power of the turbine.

Let us see how a change in these parameters affects the performance of the machine.

Figure 19-24

This floating oil derrick provides its own energy needs. Four diesel-driven alternators rated 1200 kVA, 440 V, 900 r/min, 60 Hz supply all the electrical energy. Although ac power is generated and distributed, all the motors on board are thyristor-controlled dc motors. *(Siemens)*

19-14   Infinite bus -
effect of varying the exciting current

Immediately after we synchronize an alternator with an infinite bus, the induced voltage E_0 is equal to, and in phase with, the terminal voltage E of the system (Fig. 19-25a). There is no difference of potential across the synchronous reactance and, consequently, the load current I is zero. Although the alternator is connected to the system, it delivers no power; it is said to "float" on the line.

If we now increase the exciting current, the voltage E_0 will increase and the synchronous reactance X_s will experience a difference of potential $E_x = E_0 - E$. A current $I = (E_0 - E)/X_s$ will circulate in the circuit and, because the synchronous reactance is inductive, the current lags $90°$ behind E_x (Fig. 19-25b). The current is, therefore, $90°$ behind E, which means that the alternator "sees" the system as if it were an inductor. Consequently, when we over-excite an alternator, it supplies reactive power to the infinite bus. The reactive power increases as we raise the dc exciting current. Contrary to what we might expect, it is impossible to change the active power of an alternator by varying its excitation.

Let us now decrease the exciting current so that E_0 becomes smaller than E. Phasor $E_x = E_0 - E$ now points to the left, and the resulting current $I = E_x/X_s$ again lags $90°$ behind E_x (Fig. 19-25c). However, I is now $90°$ ahead of E, which means that the alternator "sees" the system as if it were a capacitor. Consequently, when we under-excite an alternator, it draws reactive power from the system. This reactive power produces part of the magnetic field required by the machine; the remainder is supplied by exciting current I_x.

19-15   Infinite bus -
effect of varying the mechanical torque

Let us return to the situation with the alternator floating on the line, E_0 and E being equal and in phase. If we open the steam valve of the turbine driving the alternator, the immediate result is an increase in mechanical power. The rotor will accelerate and, consequently, E_0 will attain its maximum value a little sooner than before. Phasor E_0 will slip ahead of phasor E, leading it by a phase angle δ. Although both voltages have the same value, the phase angle produces a difference of potential $E_x = E_0 - E$ across the synchronous reactance (Fig. 19-26b). A current I will flow (again lagging $90°$ behind E_x), but it is now almost in phase with E. It follows that the alternator pumps active power into the system. The rotor will continue to accelerate, the angle δ will continue to diverge, and the electrical power delivered to the system will gradually build up. However, as soon as the electrical power delivered to the system is equal to the mechanical power supplied by the turbine, the rotor will cease to accelerate. The alternator will again run at synchronous speed, and the so-called *torque angle* δ between E_0 and E remains constant.

It is important to understand that a difference of potential is created when two equal voltages are out of phase. Thus, in Fig. 19-26, a potential difference of 4 kV exists between E_0 and E, although both voltages have a value of 12 kV.

19-16   Physical interpretation
of alternator behaviour

The phasor diagram of Fig. 19-26b shows that when the phase angle between E_0 and E increases, the active power delivered by the alternator also increases. To understand the physical meaning of the diagram, let us examine the currents, fluxes, and position of the poles inside the machine.

Whenever 3-phase currents flow in the stator of an alternator, they produce a rotating magnetic field, identical to that in an induction motor. In an alternator, this field rotates at the same speed and in the same direction as the rotor. Furthermore, it has the same number of poles. The respective fields produced by the rotor and stator are, therefore, stationary with respect to each other. Depending on the relative position of the poles, either

Figure 19-25 a. Alternator "floating" on an infinite bus.

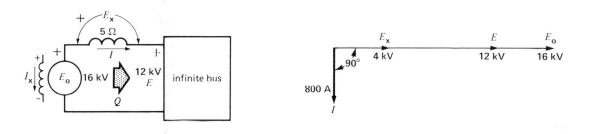

Figure 19-25 b. Over-excited alternator on an infinite bus.

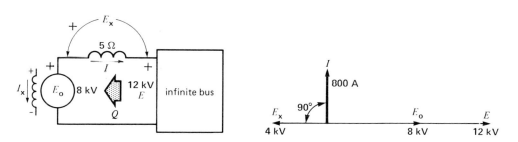

Figure 19-25 c. Under-excited alternator on an infinite bus.

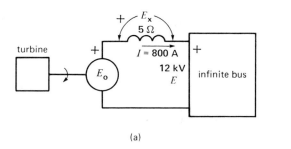
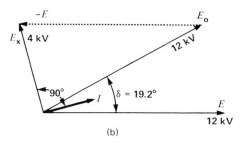

(a)   (b)

Figure 19-26 a. Turbine driving the alternator.   b. Phasor diagram showing the torque angle δ.

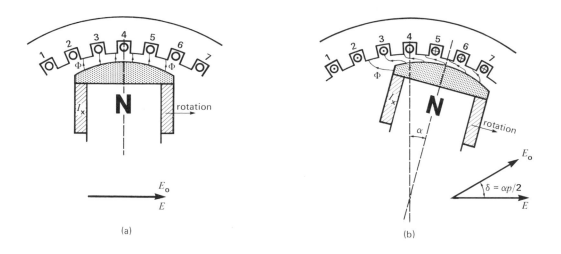

Figure 19-27   Relationship between the mechanical displacement angle α and the torque angle δ.

weak or powerful forces of attraction and repulsion are set up between them. When the alternator floats on the line, the stator current is zero and no forces are developed. The only flux is that created by the rotor, and it induces voltages E_0 (Fig. 19-27a).

If a mechanical torque is applied to the alternator (by admitting more steam to the turbine, for example), the rotor accelerates and gradually advances by a mechanical angle α, compared to its original position (Fig. 19-27b). Stator currents immediately begin to flow, owing to the electrical phase angle δ between induced voltage E_0 and terminal voltage E. The stator currents create a revolving field and a corresponding set of N and S poles. Forces of attraction and repulsion are developed between the stator poles and rotor poles, and these magnetic forces produce a torque that tends to bring the rotor back to its former position. When the electromagnetic torque is equal to the mechanical torque, the mechanical angle will no longer increase, but will remain at a constant value α.

There is a direct relationship between the mechanical angle α and the phase angle δ, given by:

$$\boxed{\delta = p\alpha/2} \qquad (19\text{-}4)$$

where

δ = torque angle between the terminal voltage and the excitation voltage [electrical degrees]

p = number of poles on the alternator

α = mechanical angle between the centers of the stator and rotor poles [mechanical degrees]

Example 19-5:

The rotor poles of an 8-pole alternator shift by 10 mechanical degrees, between no-load and full-load.

a. Calculate the torque angle between E_0 and the terminal voltage at full load;

b. Which voltage is leading?

Solution:

a. $\delta = p\alpha/2 = 8 \times 10/2$ Eq. 19-4
   $= 40°$

b. When an alternator delivers active power E_0 *always* leads E.

19-17 Active power delivered

We can prove that the active power delivered by an alternator is given by the equation:

$$P = \frac{E_0 E}{X_s} \sin \delta \quad {}^*$$ (19-5)

where

P = active power, per phase [W]
E_0 = induced voltage, per phase [V]
E = terminal voltage, per phase [V]
X_s = synchronous reactance per phase [Ω]
δ = torque angle between E_0 and E [°]

* This equation is derived below.

To understand the meaning of this equation, suppose an alternator is connected to an infinite bus having a voltage E. Furthermore, assume that the dc excitation is kept constant, so that E_0 is constant. The term $E_0 E/X_s$ is then fixed, and the power will vary directly with $\sin \delta$, the sine of the torque angle. The active power which the alternator delivers to the bus will increase with the torque angle. Thus, as we admit more steam, δ will increase and so, too, will the power. However, there is an upper limit to the active power the alternator can deliver. This limit is reached when δ is 90°. The peak power is then $P = E_0 E/X_s$. If we try to exceed this limit (such as by admitting more steam to the turbine), the alternator will accelerate and lose synchronism with the infinite bus. The rotor

---

POWER TRANSFER BETWEEN TWO SOURCES

The circuit of Fig. 19-27A is used to represent several important power devices and systems. For example, it is encountered in the study of alternators, synchronous motors and transmission lines. In such circuits, we are often interested in the active power transmitted from source E_1 to source E_2, or vice versa. Applying Kirchhoff's voltage law to this circuit, we obtain the equation:

$$E_1 = E_2 + jIX$$

If we assume that I lags behind E_2 by an arbitrary angle θ_1, and that E_1 leads E_2 by an angle δ, we obtain the phasor diagram shown. Phasor IX leads I by 90°. The active power absorbed by E_2 is:

$$P = E_2 I \cos \theta$$ (19-a)

From the sine law for triangles, we have:

$$\begin{aligned} IX/\sin \delta &= E_1/\sin \psi \\ &= E_1/\sin (90 + \theta) \\ &= E_1/\cos \theta \end{aligned}$$

Consequently, $I \cos \theta = E_1 \sin \delta/X$ (19-b)
Substituting in Eq. 19-a, we find:

$$P = \frac{E_1 E_2}{X} \sin \delta$$ (19-c)

where

P = active power transmitted [W]
E_1 = voltage of source 1 [V]
E_2 = voltage of source 2 [V]
δ = phase angle between E_1 and E_2 [°]
X = reactance connecting the sources [Ω]

The active power received by E_2 is equal to that delivered by E_1, because the reactance consumes no active power. Its magnitude is determined by the phase angle between E_1 and E_2: the angle θ between E_2 and I does not have to be known.

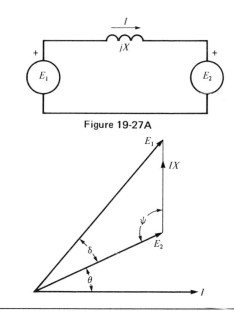

Figure 19-27A

will turn faster than the rotating field of the stator, and large, pulsating currents will begin to flow in the stator. In practice, this condition is never reached because the circuit-breakers trip as soon as synchronism is lost. We then have to resynchronize the alternator before it can again pick up the load.

Example 19-6:

A 36 MVA, 21 kV, 1800 r/min alternator has a synchronous reactance of 5 Ω per phase. If the exciting voltage is 12 kV (line-to-neutral), and the system voltage is 17.3 kV (line-to-line), calculate:

a. the power which the machine delivers when the phase angle δ is 30°;

b. the peak power which the machine can deliver before it falls out of step.

Solution:

a. 1   We have E_0 = 12 kV;
$$E = 17.3 \text{ kV}/\sqrt{3} = 10 \text{ kV};$$
$$\delta = 30°$$

a. 2   The power delivered is:
$$\begin{aligned} P &= (E_0 E/X_s) \sin \delta \\ &= (10 \times 12/5) \times 0.5 \\ &= 12 \text{ MW} \end{aligned}$$
The total power delivered by all three phases is 36 MW.

b.   The maximum power, per phase, is attained when δ = 90°:
$$\begin{aligned} P &= (E_0 E/X_s) \sin 90 \\ &= (10 \times 12/5) \times 1 \\ &= 24 \text{ MW} \end{aligned}$$
The peak power of the alternator is, therefore, 72 MW.

19-18   Transient reactance

An alternator connected to a system is subject to unpredictable load changes that sometimes occur very quickly. In such cases, the simple equivalent circuit shown in Fig. 19-17 does not reflect the behavior of the alternator. This circuit is only valid under steady-state conditions, or when the load changes gradually.

For sudden load current changes, the synchro-

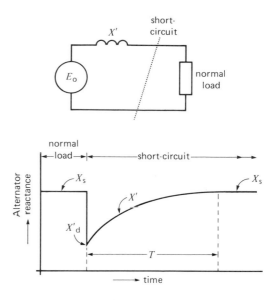

Figure 19-28   Variation of alternator reactance following a short-circuit.

nous reactance X_s must be replaced by another reactance X' whose value varies as a function of time. Figure 19-28 shows how X' varies when the normal alternator load is suddenly short-circuited. Prior to the short-circuit, the reactance is simply X_s. However, at the instant of short-circuit, the reactance immediately falls to a much lower value X'_d. It then increases gradually until it is again equal to X_s after a time interval T. The duration of the interval depends upon the size of the alternator. For machines below 100 kVA it only lasts a fraction of a second, but for machines in the 1000 MVA range, it may last as long as 10 seconds.

The minimum reactance X'_d is called the *transient reactance* of the alternator. It may be as low as 15% of the synchronous reactance. Consequently, the initial short-circuit current is much higher than that corresponding to the synchronous reactance X_s. This has a direct bearing on the size of the circuit breakers at the alternator output. In effect, because they are designed to interrupt a circuit in two or three *cycles*, it follows they have to interrupt a very high current.

On the other hand, the low reactance accompanying rapid load changes, simplifies the voltage regulation problem. First, the internal voltage drop due to X' is smaller than it would be if the synchronous reactance X_s were acting. Second, X' stays at a value far below X_s for a sufficiently long time to enable us to quickly raise the exciting current I_x. Raising the excitation, increases E_0, which helps maintain a reasonably stable terminal voltage.

Example 19-7:
A 250 MVA, 25 kV, 3-phase turboalternator has a synchronous reactance of 1.6 p.u. and a transient reactance X'_d of 0.23 p.u. It delivers its rated output at a power factor of 100%. A short-circuit suddenly occurs on the line, close to the generating station. Calculate:
a. the induced voltage E_0 prior to the short-circuit;
b. the initial value of the short-circuit current;
c. the final value of the short-circuit current if the circuit-breakers fail to open.

Solution:
a. 1   The nominal impedance of the alternator is:
$$Z_n = E_n^2/S_n = 25\ 000^2/(250 \times 10^6)$$
$$= 2.5\ \Omega$$
a. 2 $X_s = 1.6 \times 2.5$
$$= 4\ \Omega$$
a. 3   The terminal voltage per phase is:
$$E = 25/\sqrt{3} = 14.4\ \text{kV}$$
a. 4   The load current per phase is:
$$I = S/1.73\ E$$
$$= 250 \times 10^6/(1.73 \times 25\ 000)$$
$$= 5780\ \text{A}$$
a. 5   The internal voltage drop is:
$$E_x = IX_s = 5780 \times 4$$
$$= 23.1\ \text{kV}$$
a. 6   The current is in phase with E; consequently, according to the phasor diagram (Fig. 19-29), E_0 is:
$$E_0 = \sqrt{E^2 + E_x^2} = \sqrt{14.4^2 + 23.1^2}$$
$$= 27.2\ \text{kV}$$
b. 1   The transient reactance is:
$$X'_d = 0.23 \times 2.5$$
$$= 0.575\ \Omega$$

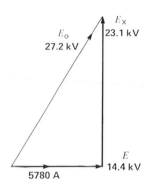

Figure 19-29   See Example 19-7.

b. 2   The initial short circuit current is:
$$I_{sc} = E_0/X'_d = 27.2/0.575$$
$$= 47.3\ \text{kA}$$
which is 8.2 times rated current.

c. 1   If the short-circuit is sustained and the excitation is unchanged, the current will eventually level off at a steady-state value of:
$$I = E_0/X_s = 27.2/4$$
$$= 6.8\ \text{kA}$$
which is only 1.2 times rated current.

Figure 19-30 shows the alternator current before and during the short-circuit. We assume a time interval T of 5 seconds. Note that in practice the circuit breakers would certainly trip within 0.1 s after the short-circuit occurs. Consequently, they have to interrupt a current of about 47 kA.

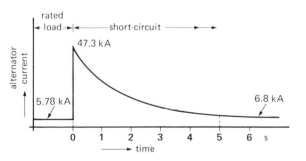

Figure 19-30   See Example 19-7.

19-19 Control of active power

When a single alternator supplies power to a system, its speed and frequency are kept constant by an extremely sensitive governor. This device can detect speed changes as small as 0.01%. An automatic control system, sensitive to such small speed changes, immediately modifies the valve (or gate) opening of the turbine so as to maintain a constant speed and frequency.

On a big utility network, the power delivered by each alternator depends upon a program established in advance between the various generating stations. The operators communicate with each other to modify the power delivered by each station so that the generation and transmission of energy is as efficient as possible. In more elaborate systems, the entire network is under the control of a computer.

In addition, individual overspeed detectors are always ready to respond to a large speed change, particularly if an alternator, for one reason or another, should suddenly become disconnected from the system. Because the steam valves (or water gates) are still wide open, the alternator will rapidly accelerate and may attain a speed 50% above normal in 4 to 5 seconds. The centrifugal forces at synchronous speed are already close to the limit the materials can withstand, so any excess speed can quickly create a very dangerous situation. Consequently, steam valves and water gates must immediately be closed off, during such emergencies.

QUESTIONS AND PROBLEMS

Practical level

19-1 What are the advantages of a stationary armature in large alternators? Why is the stator always connected in wye?

19-2 State the main differences between turboalternators and salient-pole alternators. For a given power output, which of these machines is the larger?

19-3 In analyzing a hydropower site, it is found that the turbines should turn close to 350 r/min. If the directly-coupled alternator must generate a frequency of 60 Hz, calculate:
a. the number of poles on the rotor;
b. the exact turbine speed.

19-4 An alternator generates a no-load line voltage of 13.2 kV. If a load having a lagging power factor of 0.8 is connected to the machine, must the excitation be increased or decreased in order to maintain constant line voltage?

19-5 What conditions must be met before an alternator can be connected to a 3-phase system?

19-6 Calculate the number of poles on the alternator in Fig. 19-12 using the information given.

19-7 Calculate the number of poles on the aircraft alternator shown in Fig. 19-11.

19-8 An alternator turning at 1200 r/min generates a no-load voltage of 9 kV, 60 Hz. How will the terminal voltage be affected if the following loads are connected to its terminals?
a. resistive load;
b. inductive load;
c. capacitive load.

19-9 In Problem 19-8, calculate the no-load voltage and frequency if the speed is:
a. 1000 r/min   b. 5 r/min.

Intermediate level

19-10 What is meant by the synchronous reactance of an alternator? Draw the equivalent circuit of an alternator, and explain the meaning of all the parameters.

19-11 State the advantages of brushless excitation systems over conventional systems. Using a schematic diagram, show how the rotor in Fig. 19-7 is excited.

19-12 Referring to Fig. 19-13, calculate the exciting current needed to generate a no-load line voltage of: a. 24.2 kV;   b. 12.1 kV.

19-13 An alternator possesses a synchronous reactance of 6 Ω and an excitation voltage E_0 of 3 kV per phase (ref. Fig. 19-19). Calculate the line-to-neutral voltage E for a resistive load of 8 Ω and draw the phasor diagram.

19-14 a. In Problem 19-13, draw the curve of E versus I for the following resistive loads: infinity, 24, 12, 6, 3, 0 ohms.

    b. Calculate the active power P per phase, in each case.

    c. Draw the curve of E versus P. For what value of load resistance is the power output a maximum?

19-15 Referring to Fig. 19-2, calculate the length of one pole pitch measured along the internal circumference of the stator.

19-16 The 3-phase alternator shown in Fig. 19-16 has the following characteristics: E_0 = 2400 V; X_s = 144 Ω; R = 17 Ω; load impedance Z = 175 Ω (resistive). Calculate:

    a. the synchronous impedance Z_s, per phase;

    b. the total resistance of the circuit, per phase;

    c. the total reactance of the circuit, per phase;

    d. the line current;

    e. the line-to-neutral voltage across the load;

    f. the line voltage across the load;

    g. the power of the turbine driving the alternator;

    h. the phase angle between E_0 and the voltage across the load.

19-17 A 3-phase alternator rated 3000 kVA, 20 kV, 900 r/min, 60 Hz delivers power to a 2400 kVA, 16 kV load having a lagging power factor of 0.8.

    a. If the synchronous reactance is 100 Ω, calculate the value of E_0, per phase;

    b. Calculate the exciting current required, knowing that the saturation curve of Fig. 19-13 applies.

19-18 The alternator in Fig. 19-2 has a synchronous reactance of 0.4 Ω, per phase. It is connected to an infinite bus having a line voltage of 14 kV, and the excitation voltage is

adjusted to 1.14 p.u. Calculate:

    a. the torque angle δ;

    b. the mechanical displacement angle when the alternator delivers 420 MW;

    c. the linear pole shift (measured along the inside stator circumference) corresponding to this angle [in].

19-19 A test taken on the 500 MVA alternator of Fig. 19-2 yields the following results:

    1. open-circuit line voltage is 15 kV for a dc exciting current of 1400 A;

    2. when the armature is shorted, the corresponding ac line current is 21 000 A.

Calculate:

    a. the nominal impedance of the alternator, per phase;

    b. the value of the synchronous reactance;

    c. the per unit value of X_s;

    d. the short-circuit ratio.

Advanced level

19-20 The alternator in Fig. 19-2 has an efficiency of 98.4% when it delivers an output of 500 MW. Knowing that the dc exciting current is 2400 A at a dc voltage of 300 V, calculate:

    a. the total losses in the machine;

    b. the copper losses in the rotor;

    c. the torque developed by the turbine;

    d. the average difference in temperature between the cool incoming air and warm outgoing air, if the air flow is 280 m^3/s.

19-21 Referring to Fig. 19-4, each coil on the rotor has 21.5 turns, and carries a dc current of 500 A. Knowing that the air gap length is 1.3 inches, calculate the flux density in the air gap at no-load. Neglect the mmf required for the iron portions of the magnetic circuit.

19-22 Referring to Fig. 19-17, the following information is given about an alternator: E_0 = 12 kV; E = 14 kV; X_s = 2 Ω; E_0 leads E by 30°.

    a. Calculate the total active power output of the alternator;

b. draw the phasor diagram for one phase;

c. calculate the power factor of the load.

19-23 The turboalternator shown in Fig. 19-3 has a synchronous reactance of 1.3 p.u. The excitation voltage E_0 is adjusted to 1.2 p.u. and the machine is connected to an infinite bus of 19 kV. If the torque angle δ is $20°$, calculate:

a. the active power output;

b. the line current;

c. draw the phasor diagram, for one phase.

19-24 In Problem 19-23, calculate the active power output of the alternator if the steam valves arc closed. Does the alternator receive or deliver reactive power and how much?

19-25 The alternator in Problem 19-20 is driven by a hydraulic turbine whose moment of inertia is 54×10^6 lb·ft^2.

a. If the line circuit breakers suddenly trip, calculate the speed of the generating unit (turbine and alternator) 1 second later, assuming that the wicket gates remain wide open.

b. By how many mechanical degrees do the poles advance (with respect to their normal position) during the 1 second interval? By how many electrical degrees?

20

SYNCHRONOUS MOTORS

The alternators described in the previous chapter may operate either as generators or as motors. When operating as motors (by connecting them to a 3-phase source), they are called *synchronous motors*. As the name implies, synchronous motors run in synchronism with the revolving field. The speed of rotation is therefore tied to the frequency of the source. Because the frequency is fixed, the motor speed stays constant, irrespective of the load or voltage of the 3-phase line. However, synchronous motors are used not so much because they run at constant speed, but because they possess other unique electrical properties. We shall study these properties in this chapter.

Most synchronous motors are rated between 150 kW (200 hp) and 15 MW (20 000 hp) and turn at speeds ranging from 150 to 1800 r/min. Consequently, these machines are mainly used in heavy industry (Fig. 20-1). At the other end of the power spectrum, we find tiny single-phase synchronous motors used in control devices and electric clocks. They are discussed in Chapter 21.

20-1 Construction

Synchronous motors are identical in construction to salient-pole alternators. The *stator* is composed of a slotted magnetic core which carries a 3-phase lap winding. The winding is identical to that of a 3-phase induction motor.

The *rotor* has a set of salient poles that are excited by dc current (Fig. 20-2). The exciting coils are connected in series to two slip rings, and the dc current is fed into the winding from an external exciter. Slots are also punched out along the circumference of the salient poles. They carry a squirrel-cage winding similar to that in a 3-phase induction motor. This so-called *damper winding* serves to start the motor.

Modern synchronous motors often employ brushless excitation, similar to that used in alternators. Referring to Fig. 20-3, a 3-phase exciter and a rectifier are mounted at the end of the motor shaft. The dc output from the rectifier is fed directly into the salient-pole windings, without going

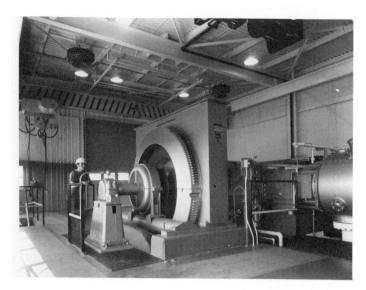

Figure 20-1

Three-phase, unity power factor synchronous motor rated 3000 hp, (2200 kW), 327 r/min, 400 V, 60 Hz driving a compressor used in a pumping station on the Trans-Canada pipeline. Brushless excitation is provided by a 21 kW, 250 V alternator/rectifier which is mounted on the shaft between the bearing pedestal and the main rotor. *(General Electric)*

through brushes and slip rings. The output of the exciter can be varied by controlling the small exciting current I_c that flows in the stationary field winding. Figure 20-4 shows how the exciter, rectifier and salient poles are mounted in a 3000 kW synchronous motor.

The rotor and stator always have the same number of poles. As in the case of an induction motor, the number of poles determines the synchronous speed of the motor:

$$n_s = 120 \frac{f}{p} \qquad (20\text{-}1)$$

where

n_s = motor speed [r/min]
f = frequency of the source [Hz]
p = number of poles

Figure 20-2

Rotor of a 50 Hz to 16 2/3 Hz frequency converter used to power an electric railway. The 4-pole rotor at the left is associated with a single-phase alternator rated 7000 kVA, 16 2/3 Hz, PF 85%. The rotor on the right is for a 6900 kVA, 50 Hz, 90% PF synchronous motor which drives the single-phase alternator. Both rotors are equipped with squirrel-cage windings. *(Brown Boveri)*

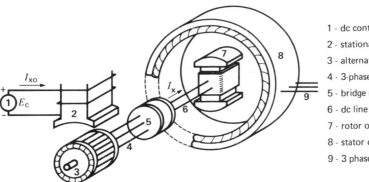

1 - dc control source
2 - stationary exciter poles
3 - alternator (3-phase exciter)
4 - 3-phase connection
5 - bridge rectifier
6 - dc line
7 - rotor of synchronous motor
8 - stator of synchronous motor
9 - 3 phase input to stator

Figure 20-3   Diagram showing the main components of a brushless exciter for a synchronous motor. It is similar to that of an alternator.

Figure 20-4 a.

Synchronous motor rated 4000 hp (3000 kW), 200 r/min, 6.9 kV, 60 Hz, 80% power factor designed to drive an ore crusher. The brushless exciter (alternator/rectifier) is mounted on the overhung shaft and is rated 50 kW, 250 V. *(General Electric)*

Figure 20-4 b.

Closeup of the 50 kW exciter, showing the armature winding and 5 of the 6 diodes used to rectify the ac current. *(General Electric)*

Example 20-1:

Calculate the number of salient poles on the rotor of the synchronous motor shown in Fig. 20-4a.

Solution:

The motor operates at 60 Hz and runs at 200 r/min; consequently:

$$n_s = 120 \, f/p$$
$$200 = (120 \times 60)/p$$
$$p = 36 \text{ poles}$$

The rotor possesses 18 north poles and 18 south poles.

20-2 Starting a synchronous motor

A synchronous motor cannot start by itself; consequently, the rotor is usually equipped with a squirrel-cage winding so that it can start up as an induction motor. When the stator is connected to the 3-phase line, the motor accelerates until it reaches a speed slightly below synchronous speed. The dc excitation is suppressed during this period.

While the rotor accelerates, the rotating flux created by the stator sweeps across the slower moving salient poles. Because the rotor coils possess a relatively large number of turns, a high voltage is induced in the rotor winding when it turns at low speeds. The voltage decreases as the rotor accelerates, and eventually becomes negligible when the rotor approaches synchronous speed. To limit the voltage, and to improve the starting torque, we either short-circuit the dc winding or connect it to an auxiliary resistor during the staing period.

If the capacity of the supply line is limited, we sometimes have to apply reduced voltage to the stator. As in the case of induction motors, we use either autotransformers or series reactors, to limit the starting current (see Chapter 22). Very large synchronous motors (20 MW and more) are sometimes brought up to speed by an auxiliary motor, called a *pony motor*. Finally, in some big installations, the motor may be brought up to speed by a variable-frequency electronic source. Such sources are covered in Chapter 25.

20-3 Pull-in torque

As soon as the motor approaches synchronous speed, we excite the rotor with dc current. If the resulting magnetic poles on the rotor face poles of *opposite* polarity on the stator at the moment the rotor is excited, a strong magnetic attraction is set up between them (Fig. 20-5). The mutual attraction locks the rotor and stator together, and so the rotor is literally yanked into step with the revolving field.

The *pull-in torque* of a synchronous motor is powerful, but the dc current must be applied at the right moment to produce the best effect. For example, if excitation is applied when the N-S poles of the rotor are opposite the N-S poles of the stator, the resulting magnetic repulsion produces a violent mechanical shock. The motor will immediately slow down and the circuit breakers will trip. In practice, starters for synchronous motors are designed to detect the precise moment when excitation should be applied. The motor then pulls automatically and smoothly into step with the revolving field.

Once the motor turns at synchronous speed, no voltage is induced in the squirrel-cage winding. Consequently, the behavior of a synchronous motor is entirely different from that of an induction motor. Basically, a synchronous motor rotates because of the magnetic attraction between the poles of the rotor and the opposite poles of the stator.

To reverse the direction of rotation, we simply interchange any two lines connected to the stator.

20-4 Motor under load - general description

When a synchronous motor runs at no-load, the rotor poles are directly opposite the stator poles and their axes coincide (Fig. 20-5). However, if we apply a mechanical load, the rotor poles fall slightly behind the stator poles, while continuing to turn at synchronous speed. The mechanical angle α between the poles increases progressively as we increase the load (Fig. 20-6). Nevertheless, the

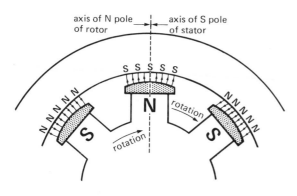

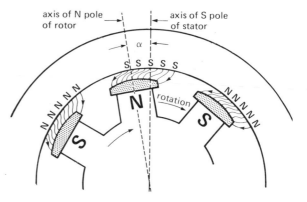

Figure 20-5   The poles of the rotor are attracted to the opposite poles on the stator. At no-load, the axes of the poles coincide.

Figure 20-6   The rotor poles are displaced with respect to the stator poles when the motor delivers mechanical power.

magnetic attraction keeps the rotor locked to the revolving field, and the motor develops an ever more powerful torque.

But there is a limit. If the mechanical load exceeds the *pull-out torque* of the motor, the rotor suddenly pulls away from the stator and the motor comes to a halt. A motor that pulls out of step creates a major disturbance on the line, and the circuit breakers immediately trip. This protects the motor because both the squirrel-cage and stator windings overheat rapidly when the machine ceases to run at synchronous speed. The pull-out torque depends upon the magnetic strength of both the rotor and stator poles. The strength of the rotor depends upon the dc excitation, while that of the stator depends upon the ac line voltage. The pull-out torque is usually 1.5 to 2.5 times the nominal full-load torque.

The mechanical angle α between the rotor and stator poles has a direct bearing on the stator current. As the angle increases, the current increases. This is to be expected because a larger angle corresponds to a bigger mechanical load, and the increased power can only come from the ac source.

20-5   Motor under load - simple calculations

We can get a better understanding of the operation of a synchronous motor by referring to the equivalent circuit shown in Fig. 20-7a. It applies to one phase of a wye-connected motor. It is identical to the equivalent circuit of an alternator, because both machines are built the same way. Thus, the flux Φ created by the rotor induces a voltage E_0 in the stator. This flux depends on the dc exciting current I_x. Consequently, E_0 varies with the excitation.

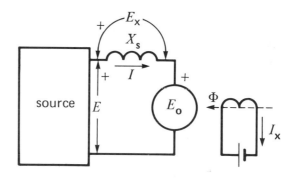

Figure 20-7   a. Equivalent circuit of a synchronous motor, showing one phase.

As already mentioned, the rotor and stator poles are lined up at no-load. Under these conditions, induced voltage E_0 is in phase with line volt-

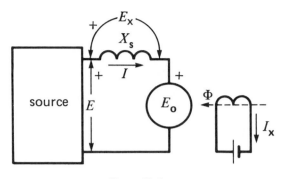

Figure 20-7 a.

age E (Fig. 20-7b). If, in addition, we adjust the excitation so that $E_0 = E$, the motor "floats" on the line, and the line current I is zero.

Figure 20-7 b. No-load conditions. E and E_0 are equal and in phase.

What happens if we now apply a mechanical load? The motor will begin to slow down, causing the rotor poles to fall behind the stator poles by an angle α. Owing to this mechanical lag, E_0 reaches its maximum value a little later than before. Thus, referring to Fig. 20-7c, E_0 is now δ electrical degrees behind E. In effect, the mechanical displacement α produces an electrical phase shift δ between E_0 and E.

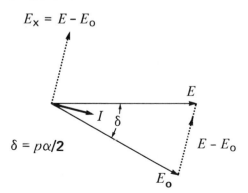

Figure 20-7 c. Motor under load. E_0 has the same value as in Fig. 20-7b, but it lags behind E_s.

The phase shift gives rise to a difference of potential $E_x = E - E_0$. This voltage appears across the synchronous reactance X_s. Consequently, a current I must flow in the circuit, given by:

$$I = (E - E_0)/X_s = E_x/X_s$$

Furthermore, the current lags $90°$ behind E_x because X_s is inductive. The phasor diagram under load is shown in Fig. 20-7c. Because I is nearly in phase with E, the motor absorbs active power. This power is entirely transformed into mechanical power, except for the relatively small copper and iron losses in the stator.

Example 20-2:
A synchronous motor connected to a 3980 V, 3-phase line generates an excitation voltage E_0 of 1790 V (per phase) when the dc exciting current is 25 A. The synchronous reactance is 22 Ω and the torque angle between E_0 and E is $30°$. Calculate:
a. the value of E_x;
b. the ac line current;
c. the power factor of the motor.

Solution:
a. 1 Although we can calculate E_x by using trigonometry, we can also determine its value by drawing a phasor diagram to scale.
a. 2 The line-to-neutral voltage is:
$$E = E_L/1.73 = 3980/1.73 \qquad \text{Eq. 9-4}$$
$$= 2300 \text{ V}$$
a. 3 E_0 lags $30°$ behind E. If E and E_0 are drawn to scale, we find that the length of E_x corresponds to $E_x = 1170$ V (Fig. 20-8).
b. 1 The line current is:
$$I = E_x/X_s = 1170/22$$
$$= 53 \text{ A}$$
c. 1 The current lags $90°$ behind E_x; consequently, we draw it perpendicular to E_x, as shown in Fig. 20-8.
c. 2 The angle between E and I is measured by a protractor. It is found to be:
$$\theta = 40°$$

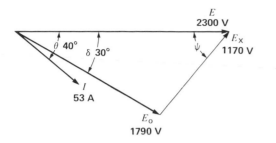

Figure 20-8   See Example 20-2.

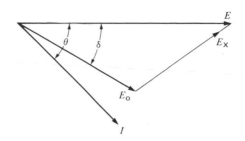

Figure 20-9   Basic phasor diagram of a synchronous motor.

c. 3   The power factor of the motor is:

$$\cos \theta = \cos 40° = 0.766 \qquad \text{Eq. 8-6}$$

The solution can be found quickly and accurately by trigonometry. Thus; referring to Fig. 20-8:

$$\begin{aligned} E_x{}^2 &= E^2 + E_0{}^2 - 2\,E\,E_0 \cos \delta \\ &= 2300^2 + 1790^2 \\ &\quad - 2 \times 2300 \times 1790 \times \cos 30 \\ &= 1.363 \times 10^6 \end{aligned}$$

$$\therefore\ E_x = 1167 \text{ V}$$

The phase angle ψ between E and E_x can be found from the sine law:

$$\begin{aligned} E_0/\sin \psi &= E_x/\sin \delta \\ 1790/\sin \psi &= 1167/\sin 30 \\ \sin \psi &= 0.7669 \\ \therefore\ \psi &= 50° \end{aligned}$$

Because I lags 90° behind E_x it follows that

$$\begin{aligned} \theta &= 90 - \psi = 90 - 50 \\ &= 40° \end{aligned}$$

20-6 Power and torque

When a synchronous motor operates under load, it draws active power from the line. The power is given by the same equation we previously encountered for the synchronous alternator in Chapter 19:

$$P = (E_0 E/X_s) \sin \delta \qquad \text{19-5}$$

As in the case of an alternator, the active power absorbed by the motor depends upon the line voltage E, the excitation voltage E_0 and the phase angle δ between them. If we neglect the relatively small I^2R losses in the stator, all the power is transmitted across the air gap to the rotor. This is analogous to the power P_r transmitted across the air gap of an induction motor (See 17-13). However, in a synchronous motor, the rotor I^2R losses are entirely supplied by the dc source. Consequently, all the power transmitted across the air gap is available in the form of mechanical power. Referring to Fig. 20-9, the mechanical power developed by a synchronous motor is therefore expressed by the equation:

$$P - \frac{E_0 E}{X_s} \sin \delta \qquad (20\text{-}2)$$

where

P = mechanical power of the motor, per phase [W]

E_0 = line-to-neutral voltage induced by I_x [V]

E = line-to-neutral voltage of the source [V]

X_s = synchronous reactance per phase [Ω]

δ = torque angle between E_0 and E [electrical degrees]

This equation shows that the mechanical power increases with the torque angle, and its maximum value is reached when δ is 90°. The poles of the rotor are then mid-way between the N and S poles of the stator. The peak power P_{max} (per-phase) is given by:

$$P_{max} = \frac{E_0 E}{X_s} \qquad (20\text{-}3)$$

As far as torque is concerned, it is directly proportional to the mechanical power because the rotor speed is fixed. It is derived from Eq. 3-5:

$$T = \frac{9.55\,P}{n_s} \qquad (20\text{-}4)$$

where

T = torque, per phase [N·m]
P = mechanical power, per phase [W]
n_s = synchronous speed [r/min]
9.55 = a constant [exact value = $60/2\pi$]

The maximum torque the motor can develop is called the pull-out torque, mentioned previously. It occurs when $\delta = 90°$ (Fig. 20-10).*

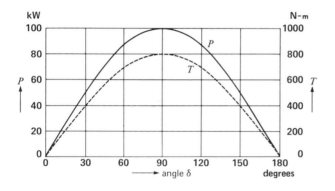

Example 20-3:
A 150 kW, 1200 r/min, 460 V, 3-phase synchronous motor has a synchronous reactance of 0.8 Ω, per phase. If the excitation voltage E_0 is fixed at 300 V, per phase, determine:
a. the power versus δ curve;
b. the torque versus δ curve;
c. the pull out torque of the motor.

---

* The remarks in this section apply to motors having smooth rotors. Most synchronous motors have salient poles; in this case the pull-out torque occurs at an angle of about 70°. Furthermore, the so-called reluctance torque created by the salient poles increases the indicated pull-out torque by about 8%.

Solution:
a. 1   The line voltage, per phase, is:
$$E = E_L/1.73 = 460/1.73 \qquad \text{Eq. 9-4}$$
$$= 266\ \text{V}$$

a. 2 $P = (E_0E/X_s)\sin\delta \qquad \text{Eq. 20-2}$
$\quad\quad = (266 \times 300/0.8)\sin\delta$
$\quad\quad = 99\ 750\sin\delta\ \text{[W]}$
$\quad\quad = 100\sin\delta\ \text{[kW]}$

a. 3   By selecting different values for δ, we can calculate the corresponding values of P:

δ [°]	P [kW]	δ [°]	P [kW]
0	0	120	86.6
30	50	150	50
60	86.6	180	0
90	100		

Figure 20-10

Power and torque per phase as a function of the torque angle δ. Synchronous motor rated 150 kW (200 hp), 1200 r/min, 3-phase, 60 Hz. See Example 20-3.

These values are plotted in the curve of Fig. 20-10.

b. 1   The torque curve can be found by applying Eq. 20-4:
$$T = 9.55\,P/n_s = 9.55\,P/1200$$
$$= P/125$$

c. 1   The pull-out torque coincides with the maximum power output:
$$T_{max} = P_{max}/125 = 100\ 000/125$$
$$= 800\ \text{N·m}$$

The actual pull-out torque is 3 times as great (2400 N·m) because this is a 3-phase machine. Similarly, the power and torque values given in Fig. 20-10 must also be multiplied by 3. Consequently, this 150 kW motor can develop a maximum output of 300 kW, or 400 hp.

20-7 Mechanical and electrical angles

As in the case of alternators, there is a precise relationship between the mechanical angle α (Fig. 20-6), the torque angle δ, and the number of poles p. It is given by:

$$\delta = p\alpha/2 \qquad (20\text{-}4)$$

Example 20-4:

A 3-phase, 6000 kW, 4 kV, 180 r/min, 60 Hz motor has a synchronous reactance of 1.2 Ω. At full load, the rotor poles are displaced by a mechanical angle of 1° from their no-load position. If the line-to-neutral excitation $E_0 = 2.4$ kV, calculate the mechanical power developed.

Solution:

a. 1 $\delta = p\alpha/2 = (40 \times 1)/2 = 20°$

a. 2 $E = E_L/1.73 = 4 \text{ kV}/1.73$
           $= 2.3$ kV

   also, $\sin 20° = 0.342$
       $P = (E_0 E/X_s) \sin \delta$ Eq. 20-2
         $= (2.4 \times 2.3/1.2) \times 0.342$
         $= 1.57$ MW (per phase)

a. 3   Total power $= 3 \times 1.57$
                  $= 4.71$ MW (≈ 6300 hp)

20-8 Excitation and reactive power

Consider a wye-connected synchronous motor connected to a 3-phase source whose line-to-neutral voltage E is fixed (Fig. 20-11). Referring to phase A, line current I_a produces an ac magnetomotive force U_a in the stator. The rotor produces a dc magnetomotive force U_r. However, as far as the stator is concerned, the dc mmf appears as an ac mmf because the rotor is constantly turning. Furthermore, U_r has the same frequency as U_a because the rotor turns at synchronous speed. The total flux Φ_a linking phase A, is therefore due to the combined action of U_a and U_r.

Pursuing our reasoning, flux Φ_a induces a voltage E_a in the stator (Fig. 20-11). If we neglect the very small IR drop, we have $E_a = E$. Because E is fixed, it follows that Φ_a is also fixed, as in the case of a transformer (see Sec. 11-2).

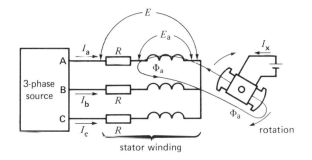

Figure 20-11 The flux linking phase A is constant as long as E is constant. The fluxes linking phases B and C are identical but they do not reach their peak values at the same time.

The mmf needed to create the constant flux Φ_a may be produced either by the stator or the rotor, or by both. If the rotor exciting current I_x is zero, all the flux has to be produced by the stator. The stator must then absorb considerable reactive power from the 3-phase line (see Sec. 8-8). On the other hand, if we excite the rotor with a dc current I_x, the rotor mmf helps produce part of the flux Φ_a. Consequently, less reactive power is drawn from the line. If we gradually raise the excitation, the rotor will eventually produce all the required flux by itself. The stator then draws no more reactive power, with the result that the power factor of the motor becomes unity (1.0).

What happens if we excite the motor above this critical level? The stator, instead of absorbing reactive power, actually *delivers* reactive power to the 3-phase line. The motor then behaves like a source of reactive power, as if it were a capacitor. Because of this important property, synchronous motors are sometimes used to correct the power factor of a plant at the same time as they deliver mechanical power.

Most synchronous motors operate at unity power factor. However if they also have to deliver reactive power, they are usually designed to operate at a power factor of 0.8 (leading). A motor designed for a power factor of 0.8 can deliver reactive power equal to 75 percent of its rated mechanical load. Thus, the 3000 kW motor shown in Fig. 20-4 can supply 75% x 3000 = 2250 kvar to the line at the same time as it develops its rated mechanical output. Motors designed to operate at leading power factors are bigger and more costly than unity power factor motors are. The reason is that the exciting current and the stator current are higher.

motor is therefore:

$$S = \sqrt{P^2 + Q^2} = \sqrt{800^2 + 600^2} \qquad \text{Eq. 8-4}$$
$$= 1000 \text{ kVA}$$

Finally, if we increase the excitation to 200 A, the machine delivers 600 kvar while still absorbing 800 kW. The apparent power is again 1000 kVA, at a leading power factor of 0.8.

Thus, to sum up, a synchronous motor *absorbs* reactive power when it is *under-excited* and *delivers* reactive power when it is *over-excited*.

20-9 V-curves

If we plot the apparent power of a synchronous

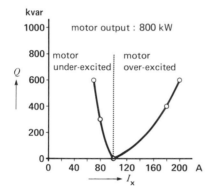

Figure 20-12   Reactive power as a function of dc exciting current for a 1000 hp, (800 kW), 60 Hz synchronous motor, running at full load.

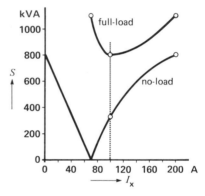

Figure 20-13   No-load and full-load V-curves of a 1000 hp synchronous motor.

Figure 20-12 shows how the reactive power of a fully loaded 800 kW motor varies with the exciting current I_x. When the current is adjusted to 100 A, the rotor produces all the flux. Consequently, the reactive power drawn from the ac line is zero and the motor only absorbs the 800 kW needed to drive the mechanical load. On the other hand, if we reduce I_x to 70 A, the motor has to absorb 600 kvar from the line to produce the same flux as before. At the same time, it continues to draw 800 kW from the line to produce the mechanical output. The apparent power S absorbed by the

motor as a function of the dc exciting current, we obtain a V-shaped curve. Thus, by redrawing Fig. 20-12 to show apparent power instead of reactive power, we obtain the full-load V-curve shown in Fig. 20-13. The no-load V-curve is also shown, to illustrate the large reactive power that can be absorbed or delivered by simply changing the excitation.

Example 20-5:
A 4000 hp (3000 kW), 6600 V, 60 Hz, 200 r/min synchronous motor operates at full load at a

leading power factor of 0.8. If the synchronous reactance is 11 Ω, calculate:

a. the apparent power of the motor, per phase;
b. the ac line current;
c. the value of E_x;
d. draw the phasor diagram to scale;
e. measure the length of phasor E_0 and determine
   its value;
f. determine the torque angle δ, using a protractor.

Solution:

We shall immediately reduce the values to correspond to one phase of a wye-connected motor.

a. 1   The active power per phase is:
$$P = 3000/3 = 1000 \text{ kW}$$

a. 2   The apparent power per phase is:
$$S = P/\cos\theta = 1000/0.8 \qquad \text{Eq. 8-5}$$
$$= 1250 \text{ kVA}$$

b. 1   The line-to-neutral voltage is:
$$E = E_L/1.73 = 6600/1.73$$
$$= 3815 \text{ V}$$

b. 2   The line current is:
$$I = S/E = 1250 \times 1000/3815$$
$$= 328 \text{ A}$$

c. 1   The voltage across X_s is:
$$E_x = IX_s = 328 \times 11$$
$$= 3608 \text{ V}$$

d. 1   To draw the phasor diagram, we refer to the
       equivalent circuit (Fig. 20-7a) and start with

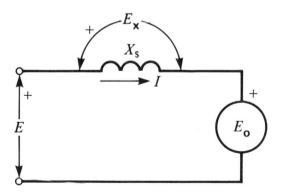

Figure 20-7 a.

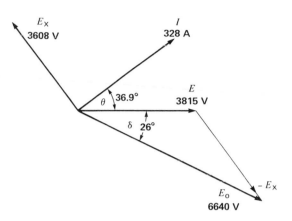

Figure 20-14   See Example 20-5.

phasor E (3815 V), laying it out horizontally
to an appropriate scale, say 1 mm = 100 V
(Fig. 20-14).

d. 2   The phase angle θ between E and I is given
by:
$$\cos\theta = 0.8$$
$$\therefore \quad \theta = 36.9°$$
Consequently, phasor I (328 A) is drawn
36.9° ahead of E.

d. 3   Phasor E_x (3608 V) is drawn so that it leads
I by 90°.

d. 4   Phasor E_0 is found from the fact that:
$$E_0 = E - E_x \quad \text{(Fig. 20-7a)}$$

e. 1   Upon scaling off the length of phasor E_0, we
find it corresponds to a value of 6640 V.

f. 1   Torque angle δ is measured to be about 26°.
The reader familiar with trigonometry can solve
this problem without having to draw the phasors
to scale. Nevertheless, the scaling method is often
useful because it serves as a double check.

20-10 Synchronous capacitor

A synchronous capacitor is essentially a synchronous motor running at no-load. Its only purpose is
to absorb or deliver reactive power on a 3-phase
system, in order to stabilize the voltage (see Chap.
29). The machine acts as an enormous 3-phase capacitor (or inductor) whose reactive power can be
varied by changing the dc excitation.

Most synchronous capacitors are rated between 20 Mvar and 200 Mvar and many are hydrogen-cooled (Fig. 20-15). They are started up like synchronous motors. However, if the system cannot furnish the required starting power, we use a pony motor to bring them up to synchronous speed. For example, in one installation, a 160 Mvar synchronous capacitor is started up by means of a 1270 kW wound-rotor motor.

Example 20-6:
A synchronous capacitor is rated at 160 Mvar, 16 kV, 1200 r/min, 60 Hz. It has a synchronous reactance of 0.8 p.u. and is connected to a 16 kV line. Calculate the value of E_0 so that the machine:
a. absorbs 160 Mvar;
b. delivers 120 Mvar.

Figure 20-15 a.

Three-phase, 16 kV, 60 Hz, 900 r/min synchronous capacitor rated – 200 Mvar (supplying reactive power) to + 300 Mvar (absorbing reactive power). It is used to regulate the voltage of a 735 kV transmission line. Other characteristics: mass of rotor: 143 t; rotor diameter: 2670 mm; axial length of stator iron: 3200 mm; air gap length: 39.7 mm.

Figure 20-15 b.

Synchronous capacitor enclosed in its steel housing containing hydrogen under pressure (300 kPa, or about 44 lb/in²). *(Hydro-Québec)*

Solution:

a. 1 The nominal impedance of the machine is:

$Z_n = E_n^2/S_n$ Eq. 19-3

     $= 16\ 000^2/(160 \times 10^6)$

     $= 1.6\ \Omega$

a. 2 $X_s = 0.8 \times 1.6$

     $= 1.28\ \Omega$

a. 3 The rated current is:

$I_n = S_n/(1.73\ E_n)$

     $= 160 \times 10^6/(1.73 \times 16\ 000)$

     $= 5780\ A$

a. 4 The drop across the synchronous reactance is:

$E_x = IX_s = 5780 \times 1.28$

     $= 7400\ V$

a. 5 $E = E_L/1.73 = 16\ 000/1.73$

     $= 9250\ V$

a. 6 The current lags $90°$ behind E; consequently:

$E_0 = E - E_x = 9250 - 7400$

     $= 1850\ V$ (see Fig. 20-16a)

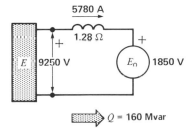

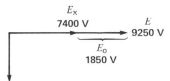

Figure 20-16   a. Under-excited synchronous capacitor absorbs reactive power (Example 20-6).

b. 1 The load current at 120 Mvar is:

$I = Q/(1.73\ E_n)$

     $= 120 \times 10^6/(1.73 \times 16\ 000)$

     $= 4335\ A$

b. 2 Voltage drop across X_s is:

$E_x = IX_s = 4335 \times 1.28$

     $= 5550\ V$

b. 3 The current leads E by $90°$; consequently:

$E_0 = E + E_x = 9250 + 5550$

     $= 14\ 800\ V$ (see Fig. 20-16b)

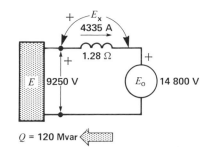

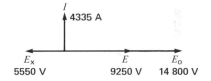

Figure 20-16   b. Over-excited synchronous capacitor delivers reactive power (Example 20-6).

20-11 Stopping synchronous motors

Owing to the inertia of the rotor and load, large synchronous motors may take several minutes to stop after they are disconnected from the line. To reduce the stopping time, we use the following braking methods:

1. maintain full dc excitation with the armature in short-circuit;
2. maintain full dc excitation with the armature connected to 3 external resistors;
3. apply mechanical braking.

In methods (1) and (2), the motor slows down because it functions as an alternator, dissipating its energy in the resistive elements of the circuit. Mechanical braking is usually applied after the motor has reached half speed or less, to prevent undue wear of the brake shoes.

Example 20-7:

A 1500 kW, 4600 V, 600 r/min, 60 Hz synchronous motor possesses a synchronous reactance of 16 Ω and a stator resistance of 0.4 Ω, per phase. The excitation voltage E_0 is 2400 V and the moment of inertia of the motor and its load is 275 kg·m². We wish to stop the motor by short-circuiting the armature while keeping the dc rotor current fixed. Calculate:

a. the power dissipated in the armature at 600 r/min;

b. the power dissipated in the armature at 150 r/min;

c. the kinetic energy at 600 r/min;

d. the kinetic energy at 150 r/min;

e. the time required for the speed to fall from 600 r/min to 150 r/min.

Solution:

a. 1 Referring to Fig. 20-17a, the impedance per phase is:

$$Z = \sqrt{R^2 + X_L{}^2} \qquad \text{Eq. 7-15}$$
$$= \sqrt{0.4^2 + 16^2}$$
$$= 16\ \Omega$$

a. 2 The current per phase is:

$$I = E_0/Z = 2400/16$$
$$= 150\ \text{A}$$

a. 3 The power dissipated in the 3 phases at 600 r/min is:

$$P = 3I^2R = 3 \times 150^2 \times 0.4$$
$$= 27\ \text{kW}$$

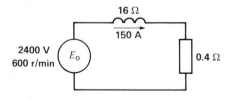

Figure 20-17 a. Motor turning at 600 r/min (Example 20-7).

b. 1 Because the exciting current is fixed, the induced voltage E_0 is proportional to the speed. Consequently, at 150 r/min:

$$E_0 = 2400 \times (150/600) = 600\ \text{V}$$

b. 2 The frequency is also proportional to the speed, so that:

$$f = 60 \times (150/600) = 15\ \text{Hz}$$

b. 3 The synchronous reactance is proportional to the frequency (Sec. 19-9); consequently,

$$X_s = 16 \times (150/600) = 4\ \Omega$$

b. 4 Referring to Fig. 20-17b, the new impedance per phase at 150 r/min is:

$$Z = \sqrt{0.4^2 + 4^2} = 4\ \Omega$$

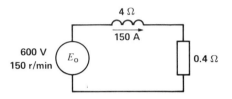

Figure 20-17 b. Motor turning at 150 r/min (Example 20-7).

b. 5 The current per phase is:

$$I = E_0/Z = 600/4 = 150\ \text{A}$$

b. 6 The power dissipated in the 3 phases is the same as before:

$$P = 27\ \text{kW}$$

c. The kinetic energy at 600 r/min is:

$$E_{k1} = 5.48 \times 10^{-3}\,Jn^2 \qquad \text{Eq. 3-8}$$
$$= 5.48 \times 10^{-3} \times 275 \times 600^2$$
$$= 542.5\ \text{kJ}$$

d. The kinetic energy at 150 r/min is:

$$E_{k2} = 5.48 \times 10^{-3} \times 275 \times 150^2$$
$$= 33.9\ \text{kJ}$$

e. 1 The loss in kinetic energy in going from 600 r/min to 150 r/min is:

$$W = E_{k1} - E_{k2} = 542.5 - 33.9$$
$$= 508.6\ \text{kJ}$$

e. 2 This energy is lost as heat in the armature resistance. The time is given by:

$$P = W/t \qquad \text{Eq. 3-4}$$
$$27 = 508.6/t$$

Whence $t = 18.8\ \text{s}$

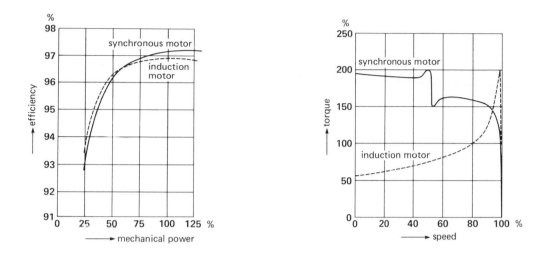

Figure 20-18 Comparison between the efficiency and starting torque of a squirrel-cage induction motor and a synchronous motor, both rated at 4000 hp, 1800 r/min, 6.9 kV, 60 Hz.

20-12 Uses of a synchronous motor - comparison with induction motor

We have already seen that induction motors are excellent for speeds above 600 r/min. At lower speeds, they become heavy, costly, and have relatively low power factor and low efficiency.

Synchronous motors are particularly attractive for low-speed drives because the power factor can always be adjusted to 1.0 and the efficiency is high. Although more complex to build, their weight and cost are often less than those of induction motors of equal power and speed. This is particularly true for speeds below 300 r/min.

A synchronous motor can improve the power factor of a plant while carrying its rated load. Furthermore, its starting torque can be made considerably greater than that of an induction motor. The reason is that we can raise the resistance of the squirrel-cage winding without affecting the speed or efficiency at synchronous speed. Figure 20-18 compares the properties of a squirrel-cage induction motor and a synchronous motor having the same nominal rating. The biggest difference is in the starting torque.

High-power electronic converters generating very low frequencies enable us to run synchronous motors at ultra-low speeds. Thus, huge motors in the 10 MW range drive crushers, rotary kilns, and variable-speed ball mills. Electronic control of synchronous motors is covered in Chapter 24.

QUESTIONS AND PROBLEMS

Practical level

20-1 Compare the construction of an alternator, a synchronous motor and a squirrel-cage induction motor.

20-2 Explain how a synchronous motor starts up. When should the dc excitation be applied?

20-3 Why does the speed of a synchronous motor remain constant even under variable load?

20-4 Name some of the advantages of a synchronous motor compared to a squirrel-cage induction motor.

20-5 What is meant by a synchronous capacitor and what is it used for?

20-6 a. What is meant by an "under-excited" synchronous motor?

b. If we over-excite a synchronous motor, does its mechanical power output increase?

20-7 A synchronous motor draws 2000 kVA at a power factor of 90% leading. Calculate the approximate power developed by the motor [hp].

20-8 A synchronous motor driving a pump operates at a power factor of 100%. What happens if the dc excitation is increased?

20-9 A 3-phase, 225 r/min synchronous motor connected to a 4 kV, 60 Hz line draws a current of 320 A and absorbs 2000 kW. Calculate:

a. the apparent power supplied to the motor;

b. the power factor;

c. the reactive power absorbed;

d. the number of poles on the rotor.

20-10 A synchronous motor draws 150 A from a 3-phase line. If the exciting current is raised, the current drops to 140 A. Was the motor over- or under-excited before the excitation was changed?

Intermediate level

20-11 a. Calculate the approximate full-load current of the 3000 hp motor in Fig. 20-1, if it has an efficiency of 97%

b. What is the value of the field resistance?

20-12 Referring to Fig. 20-2, at what speed must the rotor turn to generate the indicated frequencies?

20-13 A 3-phase synchronous motor rated 800 hp, 2.4 kV, 60 Hz operates at unity power factor. The line voltage suddenly drops to 1.8 kV, but the exciting current remains unchanged. Explain how the following quantities are affected:

a. motor speed and mechanical power output;

b. torque angle δ;

c. position of the rotor poles;

d. power factor;

e. stator current.

20-14 A synchronous motor has the following parameters, per phase (ref. Fig. 20-7a):
$E = 2.4$ kV; $E_0 = 3$ kV; $X_s = 2$ Ω; $I = 900$ A. Draw the phasor diagram to scale and determine:

a. torque angle δ;

b. active power, per phase;

c. power factor of the motor;

d. reactive power absorbed (or delivered), per phase.

20-15 a. In Problem 20-14, calculate the line current and the new torque angle δ if the mechanical load is suddenly removed.

b. Calculate the new reactive power absorbed (or delivered) by the motor, per phase.

20-16 A 500 hp synchronous motor drives a compressor and its excitation is adjusted so that the power factor is unity. If the excitation is increased without making any other change, what is the effect upon:

a. the active power absorbed by the motor?

b. the line current?

c. the reactive power absorbed (or delivered) by the motor?

d. the torque angle?

Advanced level

20-17 The 4000 hp, 6.9 kV motor shown in Fig. 20-4 possesses a synchronous reactance of 10 Ω, per phase. The stator is connected in wye, and the motor operates at full load (4000 hp) with a leading power factor of 0.89. If the efficiency is 97%, calculate:

a. the apparent power;

b. the line current;

c. the value of E_0, per phase;

d. the mechanical displacement of the poles from their no-load position;

e. the total reactive power supplied to the electrical system;

f. the approximate maximum power the

motor can develop, without pulling out of step [hp].

20-18 In Problem 20-17, we wish to adjust the power factor to unity. Calculate:

a. the exciting voltage E_0 required, per phase;

b. the new torque angle.

20-19 A 3-phase, unity power factor synchronous motor rated 400 hp, 2300 V, 450 r/min, 80 A, 60 Hz drives a compressor. The stator has a synchronous reactance of 0.88 p.u., and the excitation E_0 is adjusted to 1.2 p.u. Calculate:

a. the value of X_s and of E_0, per phase;

b. the pull-out torque [ft·lb];

c. the line current when the motor is about to pull out of synchronism.

20-20 The synchronous capacitor in Fig. 20-15 possesses a synchronous reactance of 0.6 Ω, per phase. The resistance per phase is 0.007 Ω. If the machine coasts to a stop, it will run for about 3 h. In order to shorten the stopping time, the stator is connected to three large 0.6 Ω braking resistors connected in wye. The dc excitation is fixed at 250 A so that the initial line voltage across the resistors is one tenth of its rated value, or 1600 V, at 900 r/min. Calculate:

a. the total braking power and braking torque at 900 r/min;

b. the braking power and braking torque at 450 r/min;

c. the average braking torque between 900 r/min and 450 r/min;

d. the time for the speed to fall from 900 r/min to 450 r/min, knowing that the moment of inertia of the rotor is 1.7 x 10^6 lb·ft^2.

21

SINGLE-PHASE MOTORS

Single-phase motors are the most familiar of all electric motors because they are used in home appliances and portable machine tools. In general, they are employed when 3-phase power is not available.

There are many kinds of single-phase motors on the market, each designed to meet a specific application. However, we shall limit our study to a few basic types, with particular emphasis on the widely used split-phase induction motor.

21-1 Construction of a single-phase induction motor

Single-phase induction motors are very similar to 3-phase induction motors. They are composed of a squirrel-cage rotor (identical to that in a 3-phase motor) and a stator (Fig. 21-1). The stator carries a so-called *main winding* which creates a set of N-S poles. It also carries a smaller *auxiliary winding* that only operates during the brief period when the motor starts up. The auxiliary winding has the same number of poles as the main winding has.

Figure 21-2 shows the progressive steps in wind-

ing a 4-pole, 36-slot stator. Starting with the bare stator, the main winding is first laid in the slots (Fig. 21-2b). The auxiliary winding is then inserted so that is straddles the main winding (Fig. 21-2c).

Each pole of the main winding consists of a

Figure 21-1   Cutaway view of a 5 hp, 1725 r/min, 60 Hz single-phase capacitor-start motor. *(Gould)*

392

Figure 21-2   a. Bare, laminated stator of a 1/4 hp (187 W), single-phase motor. The 36 slots are insulated with a paper liner. The squirrel-cage rotor is identical to that of a 3-phase motor.

Figure 21-2   b. Four poles of the main winding are inserted in the slots.

Figure 21-2   c. Four poles of the auxiliary winding straddle the main winding. *(Lab-Volt)*

group of concentric coils, connected in series (Fig. 21-3a). Adjacent poles are connected to produce alternate N, S polarities. The empty slot in the center of each pole, and the partially filled slots on either side of it, are used to lodge the auxiliary winding. The latter has 2 concentric coils per pole (Fig. 21-3b).

Figure 21-4 shows a 2-pole stator; the large main winding and the smaller auxiliary winding are displaced at right angles to each other. The reason for this arrangement will be explained shortly.

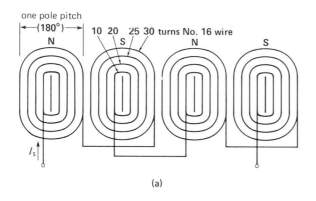

(a)

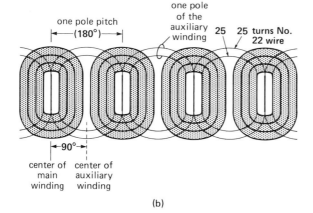

(b)

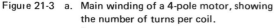

Figure 21-3 a. Main winding of a 4-pole motor, showing the number of turns per coil.
b. Position of the auxiliary winding with respect to the main winding.

Figure 21-4 Main and auxiliary windings in a 2-pole single-phase motor. The stationary contact in series with the auxiliary winding opens when the centrifugal switch, mounted on the shaft, reaches 75 percent of synchronous speed.

21-2 Synchronous speed

As in the case of 3-phase motors, the synchronous speed of all single-phase induction motors is given by Eq. 17-1:

$$n_s = \frac{120\,f}{p}$$ Eq. 17-1

where
n_s = synchronous speed [r/min]
f = frequency of the source [Hz]
p = number of poles

The rotor always turns at slightly less than synchronous speed, and the full-load slip is typically 3% to 5% for fractional horsepower motors.

Example 21-1:
Calculate the speed of the single-phase motor shown in Fig. 21-1, if the slip is 3.4%. The line frequency is 60 Hz.

Solution:

a. 1 The motor has 4 poles; consequently,

$$n_s = 120\,f/p = (120 \times 60)/4$$
$$= 1800 \text{ r/min}$$

a. 2 The full-load speed is given by:

$$s = (n_s - n)/n_s \qquad \text{Eq. 17-2}$$
$$0.034 = (1800 - n)/1800$$
$$n = 1739 \text{ r/min}$$

21-3 Torque-speed characteristic

Figure 21-5 is a schematic diagram of the rotor and main winding of a 2-pole single-phase induction motor. When an ac voltage is applied to the stator, the resulting current I_s produces an ac flux Φ_s. The flux alternates back and forth but, unlike the flux in a 3-phase stator, no revolving field is produced. The flux induces an ac voltage in the stationary rotor which, in turn, creates large ac rotor currents. In effect, the rotor behaves like the short-circuited secondary of a transformer; consequently, the motor has no tendency to start by itself.

However, if we spin the rotor in one direction or the other, it continues to rotate in the direction of spin. As a matter of fact, the rotor quickly accelerates until it reaches a speed slightly below synchronous speed. The acceleration indicates that the motor develops a positive torque as soon as it begins to turn. Figure 21-6 shows the typical torque-speed curve when the main winding is excited. Although the starting torque is zero, the motor develops a powerful torque as it approaches synchronous speed.

21-4 Principle of operation

The principle of operation of a single-phase induction motor is quite complex, and may be explained as follows by the so-called cross-field theory.*

---

* The double revolving field theory (not discussed here) is also used to explain the behavior of the single-phase motor.

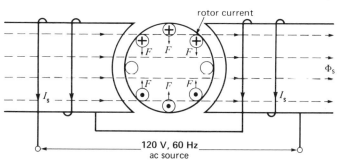

Figure 21-5   Currents in the rotor bars when the rotor is locked.

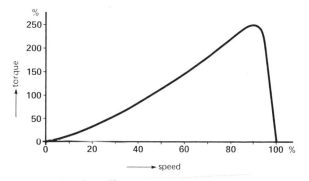

Figure 21-6   Typical torque-speed curve of a single-phase motor.

As soon as the rotor begins to turn, a revolving field is set up. This field is produced by the combined action of the stator and rotor magnetomotive forces. Referring to Fig. 21-7, a "speed emf" E is induced in the rotor conductors, as they cut the stator flux Φ_s. This voltage increases as the rotor speed increases. It causes currents I_r to flow in the rotor bars facing the stator poles. These currents produce an ac flux Φ_r which acts at right angles to the stator flux Φ_s. Equally important is the fact that Φ_r does not reach its maximum value at the same time as Φ_s does. In effect, Φ_r lags almost $90°$ behind Φ_s, owing to the inductance of the rotor.

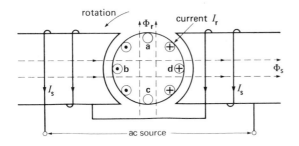

Figure 21-7   Currents induced in the rotor bars due to rotation.

The combined action of Φ_s and Φ_r produces a revolving magnetic field, similar to that in a 3-phase motor. The value of Φ_r increases with increasing speed, becoming almost equal to Φ_s at synchronous speed. This explains, in part, why the torque increases as the motor speeds up.

We can understand how the revolving field is produced by referring to Fig. 21-8. It gives a snapshot of the currents and fluxes created respectively by the rotor and stator, at successive intervals of time. We assume that I_r lags $90°$ behind I_s; consequently, Φ_r is zero when Φ_s is maximum, and vice versa. We also assume that the motor is running far below synchronous speed, and so Φ_r is much smaller than Φ_s. By following the successive pictures in Fig. 21-8, it is obvious that the combination of Φ_s and Φ_r produces an elliptical, revolving

field. It rotates counterclockwise in the same direction as the rotor. The field rotates at synchronous speed, irrespective of the actual speed of the rotor. As the motor approaches synchronous speed, Φ_r becomes almost equal to Φ_s, and a nearly perfect revolving field is produced.

21-5 Locked-rotor torque

To produce a starting torque in a single-phase motor, we must somehow create a revolving field. This is done by adding an auxiliary winding, as shown in Fig. 21-9. When the main and auxiliary windings are connected to an ac source, the main winding produces a flux Φ_s, while the auxiliary winding produces a flux Φ_a. If the fluxes are out of phase, so that Φ_a either lags or leads Φ_s, a rotating field is set up.

The reader will immediately see that the auxiliary winding produces a strong flux Φ_a during the starting period when the rotor flux Φ_r is weak. As a result, Φ_a "strengthens" Φ_r, thereby producing a powerful torque both at standstill and at low speeds. The locked-rotor torque is given by:

$$T = kI_aI_s \sin \alpha \qquad (21\text{-}1)$$

where

T = locked-rotor torque [N·m]
I_a = locked-rotor current in the auxiliary winding [A]
I_s = locked-rotor current in the main winding [A]
α = phase angle between I_s and I_a [°]
k = a constant, depending on the design of the motor

To obtain the desired phase shift between I_s and I_a (and hence between Φ_s and Φ_a), we add an impedance Z in series with the auxiliary winding. The impedance may be resistive, inductive, or capacitive, depending upon the desired starting torque. The choice of impedance gives rise to various types of so-called *split-phase* motors.

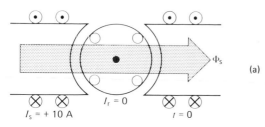

(a)

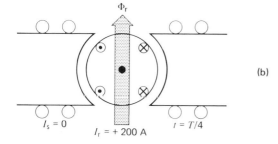

(b)

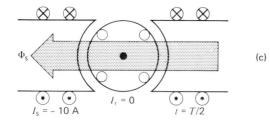

(c)

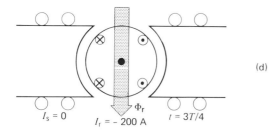

(d)

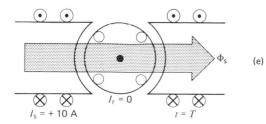

(e)

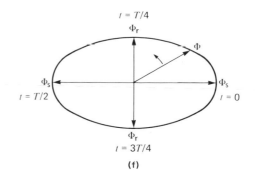

(f)

Figure 21-8

Instantaneous currents and flux in a single-phase motor with the main winding excited. The duration of 1 cycle is T seconds, and conditions are shown at successive quarter-cycle intervals.

a. Stator current I_s is maximum, rotor current I_r is zero.
b. Stator current is zero, rotor current is maximum; however, Φ_r is smaller than Φ_s.
c. Stator current is maximum, but negative.
d. Rotor current is maximum, but negative.
e. After one complete cycle ($t = T$), the conditions repeat.
f. Resulting flux Φ in the air gap rotates ccw at synchronous speed. Its amplitude varies from a maximum of Φ_s to a minimum Φ_r.

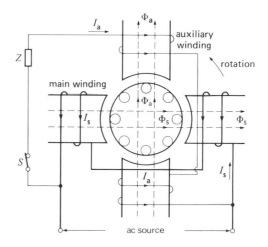

Figure 21-9 Currents and fluxes at standstill when the main and auxiliary windings are energized. An elliptical revolving field is produced.

A special switch is also connected in series with the auxiliary winding. It disconnects the winding when the motor reaches about 75% of synchronous speed. Speed-sensitive centrifugal switches, mounted on the shaft, are often used for this purpose (Fig. 21-10).

resistance is low. As a result, the starting current I_s lags considerably behind the applied voltage E (Fig. 21-11).

In a resistance split-phase motor (often simply called split-phase motor), the auxiliary winding has a relatively small number of turns of fine wire. Its

(a)   (b)

Figure 21-10   Centrifugal switch

a. In the closed, or stopped position. The stationary contact is closed.
b. In the open, or running position. Owing to centrifugal force, the rectangular weights have swung out against the restraining tension of the springs. This has caused the plastic collar to move to the left along the shaft, thus opening the stationary contact.

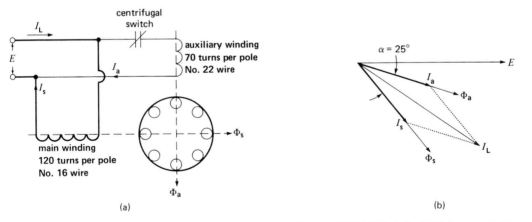

(a)   (b)

Figure 21-11   a. Resistance split-phase motor (1/4 hp, 115 V, 1725 r/min, 60 Hz) at standstill.
b. Corresponding phasor diagram.

21-6 Resistance split-phase motor

The main winding of a single-phase motor is always made of relatively large wire, to reduce the I^2R losses. The winding also has a relatively large number of turns. Consequently, under locked rotor conditions, the inductive reactance is high and the

resistance is higher and its reactance lower than that of the main winding, with the result that locked rotor current I_a is more nearly in phase with E. The resulting phase angle α between I_a and I_s produces the starting torque.

The line current I_L is equal to the phasor sum of I_s and I_a. It is usually 6 to 7 times the nominal current of the motor.

Owing to the small wire used on the auxiliary winding, the current density is high and the winding heats up very quickly. If the starting period lasts for more than 5 seconds, the winding begins to smoke and may burn out, unless the motor is protected by a built-in-thermal relay. This type of split-phase motor is well suited for infrequent starting of low-inertia loads.

Example 21-2:

A resistance split-phase motor is rated at 1/4 hp (187 W), 1725 r/min, 115 V, 60 Hz. When the rotor is locked, a test at reduced voltage on the main and auxiliary windings yields the following results.

	main winding	auxiliary winding
applied voltage	23 V	23 V
current	4 A	1.5
active power	60 W	30 W

Calculate:

a. the phase angle between I_a and I_s;
b. the locked-rotor current drawn from the line at 115 V.

Solution:

We first calculate the phase angle between I_s and E of the main winding.

a. 1 The apparent power is:
$$S = EI = 23 \times 4 = 92 \text{ VA}$$

a. 2 The power factor is:
$$\cos \phi_s = P/S = 60/92 = 0.65$$
$$\therefore \quad \phi_s = 49.2°$$
I_s lags 49.2° behind the voltage.

We now calculate the phase angle between I_a and E of the auxiliary winding.

a. 3 The apparent power is:
$$S = EI = 23 \times 1.5 = 34.5 \text{ VA}$$

a. 4 The power factor is:
$$\cos \phi_a = P/S = 30/34.5 = 0.87$$
$$\therefore \quad \phi_a = 29.6°$$
I_a lags 29.6° behind the voltage.

a. 5 The phase angle between I_s and I_a is:
$$\alpha = \phi_s - \phi_a = 49.2 - 29.6$$
$$= 19.6°$$

b. To determine the total line current, we first calculate the total value of P and Q drawn by both windings, and then deduce the total apparent power S.

b. 1 $P = P_s + P_a$
$$= 60 + 30 = 90 \text{ W}$$

b. 2 $Q = Q_s + Q_a$
$$= \sqrt{92^2 - 60^2} + \sqrt{34.5^2 - 30^2}$$
$$= 86.8 \text{ var}$$

b. 3 $S = \sqrt{90^2 + 86.8^2} = 125 \text{ VA}$

b. 4 $I_1 = S/E = 125/23 = 5.44 \text{ A (at 23 V)}$

b. 5 The current drawn at 115 V is:
$$I_L = 5.44 \times (115/23) = 27.2 \text{ A}$$

Owing to their low cost, split-phase induction motors are the most popular single-phase motors on the market. They are used where a moderate starting torque is required and where the starting periods are infrequent. They drive fans, pumps, washing machines, oil burners, small machine tools, and other devices too numerous to mention. The power rating usually lies between 60 W and 250 W (1/12 hp to 1/3 hp).

21-7 Capacitor-start motor

The capacitor-start motor is identical to a split-phase motor, except that the auxiliary winding has about as many turns as the main winding has. Furthermore, a capacitor is connected in series with the auxiliary winding (Fig. 21-12).

The capacitor is chosen so that I_a leads I_s by about 80°, which is considerably more than the 25° found in a split-phase motor. Consequently, for equal starting torques, the current in the auxiliary winding is only about half that in a split-phase motor. It follows that the auxiliary winding of a capacitor motor heats up less quickly and the locked-rotor line current I_L is also smaller. The current is typically 4 to 5 times the nominal full-load current.

Owing to the high starting torque and the rela-

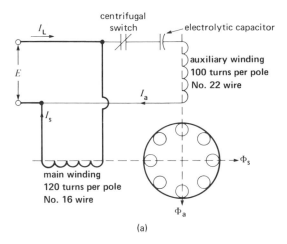

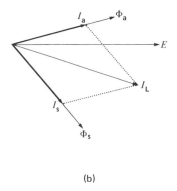

Figure 21-12   a. Capacitor-start motor.   b. Corresponding phasor diagram.

tively low value of I_a, the capacitor-start motor is well suited to applications involving either frequent or prolonged starting periods. Although the *starting* characteristics of this motor are better than those of a split-phase motor, both machines possess the same load characteristics because the main windings are identical.

The wide use of capacitor-start motors is a direct result of the availability of small, reliable, low-cost electrolytic capacitors. Prior to the development of these capacitors, repulsion-induction motors had to be installed whenever a high starting torque was required. Repulsion-induction motors possess a special commutator and brushes, that require considerable maintenance. Most motor manufacturer have stopped making them.

Capacitor-start motors are used when a high starting torque is required. They are built in sizes ranging from 120 W to 7.5 kW ($\approx$ 1/6 hp to 10 hp). Typical loads are compressors, large fans, pumps, and high-inertia loads.

21-8 Characteristics of single-phase induction motors

The efficiency and power factor of fractional horsepower single-phase motors are usually low. Thus, at full load, a 186 W motor (1/4 hp) has an efficiency and power factor of about 60%. The low power factor is mainly due to the large magnetizing current, which ranges between 70% and 90% of full-load current. Consequently, even at no-load, these motors reach temperatures close to full-load temperature.

Table 21A gives the properties of a capacitor-start motor having a rating of 250 W (1/3 hp), 1760 r/min, 115 V, 60 Hz. Figure 21-13 also shows the torque-speed curve for the same machine. Note that during the acceleration phase (0 to 1370 r/min), the main and auxiliary windings together produce a very high starting torque. When the rotor reaches 1370 r/min, the centrifugal switch snaps open, causing the motor to operate along the torque-speed curve of the main winding. The torque suddenly drops from 9.5 N·m to 2.8 N·m, but the motor continues to accelerate until it reaches 1760 r/min, the rated full-load speed.

21-9 Vibration of single-phase motors

If we touch the stator of a single-phase motor, we note that it vibrates rapidly, whether it operates at full-load or no-load. These vibrations do not exist in 2-phase or 3-phase motors and, consequently, single-phase motors are more noisy.

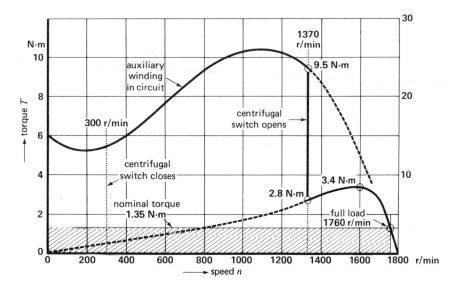

Figure 21-13   Torque-speed curves of a capacitor-start motor, rated 1/3 hp (250 W), 1760 r/min, 115 V, 60 Hz, class A insulation.

TABLE 21A

CHARACTERISTICS OF A CAPACITOR-START MOTOR
RATING: 250 W, 1760 r/min, 115 V, 60 Hz, INSULATION CLASS 105°C

full load			**no load**		
voltage	—	115 V	voltage	—	115 V
power	—	250 W	current	—	4.0 A
current	—	5.3 A	losses	—	105 W
P.F.	—	64 %	**locked rotor**		
efficiency	—	63.9 %	voltage	—	115 V
speed	—	1760 r/min	current I_s	—	23 A
torque	—	1.35 N·m	current I_a	—	19 A
break down			current I_L	—	29 A
torque	—	3.4 N·m	torque	—	6 N·m
speed	—	1600 r/min	capacitor	—	320 μF
current	—	13 A			

What causes this vibration? It is due to the fact that a single-phase motor always receives *pulsating* electric power whereas it delivers *constant* mechanical power. Consider the 250 W motor having the properties given in Table 21A. The full-load current is 5.3 A and it lags 50° behind the line voltage. If we draw the waveshapes of voltage and current, we can easily plot the instantaneous power supplied to the motor (Fig. 21-14). We find that i_s oscillates between + 1000 W and – 218 W. When the power is positive, the motor receives energy from the line. Conversely, when it is negative, the motor returns energy to the line. However, whether the instantaneous electric power is positive, negative, or zero, the mechanical power delivered is always 250 W.

Obviously, the motor will slow down during the brief periods when the electric power it receives is less than 250 W. On the other hand, it will accelerate whenever the electric power exceeds the mechanical output plus the losses. The acceleration intervals coincide with the positive peaks of the power curve. Similarly, the deceleration intervals coincide with the negative peaks. Consequently, the acceleration/deceleration intervals occur twice per cycle, or 120 times per second, on a 60 Hz system. As a result, both the stator and rotor vibrate at twice the line frequency.

The stator vibrations are transmitted to the baseplate which, in turn, generates additional vibration and noise. To eliminate the problem, the motor is cradled in a resilient mounting (Fig. 21-15). It consists of two soft rubber rings placed between each end bell and a supporting metal bracket. Because the rotor also vibrates, we sometimes place a rubber isolator between the shaft and the mechanical load.

Two-phase and 3-phase motors do not vibrate because the total instantaneous power they receive from *all* the phases is constant (see Sec. 9-3).

21-10 Capacitor-run motor

The capacitor-run motor is essentially a 2-phase

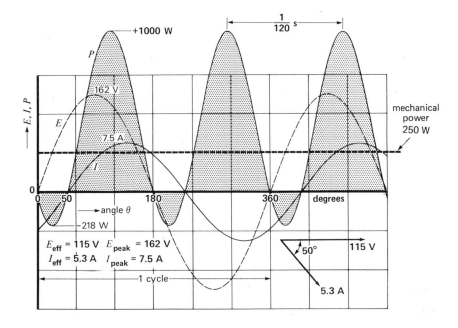

Figure 21-14   The instantaneous power absorbed by a single-phase motor varies between + 1000 W and – 218 W. The power output is constant at 250 W; consequently, vibrations are produced.

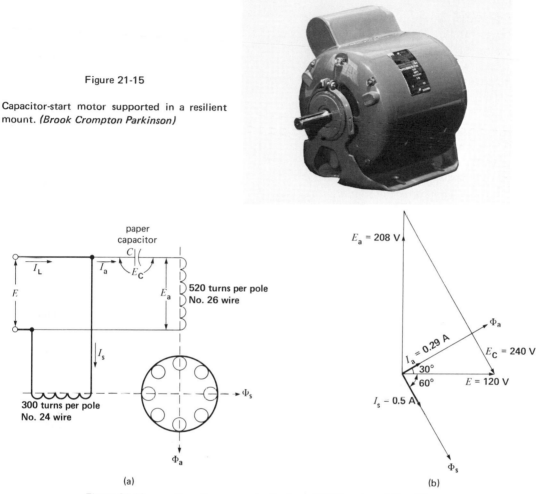

Figure 21-15

Capacitor-start motor supported in a resilient mount. *(Brook Crompton Parkinson)*

(a)

(b)

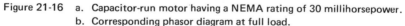

Figure 21-16   a. Capacitor-run motor having a NEMA rating of 30 millihorsepower.
b. Corresponding phasor diagram at full load.

motor that receives its power from a single-phase source. It has two windings, one of which is directly connected to the source. The other winding is also connected to the source, but in series with a paper capacitor (Fig. 21-16). The capacitor-fed winding has a large number of turns of relatively small wire, compared to the directly connected winding.

This particularly quiet motor is used to drive fixed loads in hospitals, studios, and other places where silence is important. It has a high power factor and efficiency, and no centrifugal switch is required. However, the starting torque is low.

The motor acts as a true 2-phase motor only when it operates at full load. Under these conditions, fluxes Φ_a and Φ_s created by the two windings are equal and out of phase by $90°$. The motor is then vibration-free. Capacitor-run motors are usually rated below 500 W.

21-11 Shaded-pole motor

The shaded-pole motor is very popular for ratings below 0.05 hp ($\approx$ 40 W) because of its extremely simple construction (Fig. 21-17). It is basically a small squirrel-cage motor in which the auxiliary winding is composed of a copper ring surrounding a portion of each pole.

The main winding is a simple coil connected to the ac source. The coil produces a total flux Φ that may be considered to be made up of three components Φ_1, Φ_2 and Φ_3, all in phase. Flux Φ_1 links the short-circuited ring on the left-hand pole, inducing a rather large current. This current produces a flux Φ_a that lags behind Φ_1. Consequently, Φ_a also lags behind Φ_2 and Φ_3. The combined action of $(\Phi_2 + \Phi_3)$ and Φ_a produces a weak revolving field, which starts the motor. The direction of rotation is from the "unshaded" to the "shaded" (ring side) of the pole. A similar torque is set up by the pole on the right. Flux Φ_2 induces a current in the ring, and the resulting flux Φ_b lags behind Φ_2. As before, the combined action of $(\Phi_1 + \Phi_3)$ and Φ_b produces a weak revolving field.

Although the starting torque, efficiency, and power factor are very low, the simple construction and absence of a centrifugal switch give this motor a marked advantage in low-power applications. The direction of rotation cannot be changed, because it is fixed by the position of the copper rings. Table 21B gives the typical properties of a shaded-pole motor having a rated output of 6 W.

Example 21-2:
Calculate the full-load efficiency and slip of the shaded-pole motor whose properties are listed in Table 21B.

Solution:
a. $\eta = (P_0/P_i) \times 100 = (6/21) \times 100$ Eq. 3-6
     $= 28.6\%$
b. The slip is:
     $s = (n_s - n)/n_s = (3600 - 2900)/3600$
       $= 0.194 = 19.4\%$

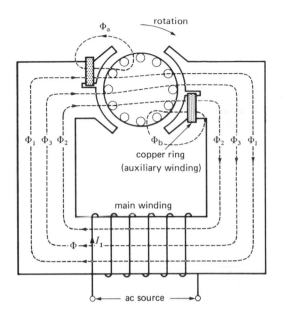

Figure 21-17   a. Fluxes in a shaded-pole motor.

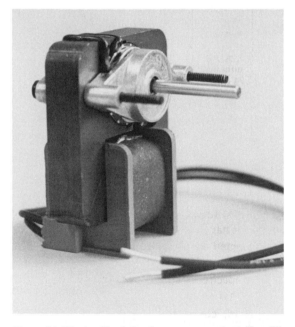

Figure 21-17   b. Shaded-pole motor rated at 5 milli-horsepower, 115 V, 60 Hz, 2900 r/min. *(Gould)*

TABLE 21B

Properties of a shaded-pole motor, rated 6 W, 115 V, 60 Hz.

No load

current	0.26 A
input power	15 W
speed	3550 r/min

Locked rotor

current	0.35 A
input power	24 W
torque	10 mN·m

Full load

current	0.33 A
Input power	21 W
speed	2900 r/min
torque	19 mN·m
mechanical power	6 W
breakdown speed	2600 r/min
breakdown torque	21 mN·m

21-12 Series motor

The single-phase series motor is very similar to a dc series motor (Sec. 15-8). The basic construction of a small ac series motor is shown in Fig. 21-18. The entire magnetic circuit is laminated to reduce eddy current losses. Such a motor can operate on either ac or dc, and the resulting-torque-speed curve is about the same, in each case. This is why it is sometimes called a *universal* motor.

When the motor is connected to an ac source, the ac current flows through the armature and the series field. The field produces an ac flux Φ that reacts with the current flowing in the armature to produce a torque. Because the armature current and the flux reverse simultaneously, the torque always acts in the same direction. No revolving field is produced in this type of machine; the principle of operation is the same as that of a dc series motor and it possesses the same basic characteristics.

The main advantage of fractional horsepower series motors is their high speed (and corresponding small size) and high starting torque. They can therefore be used to drive high-speed vacuum cleaners and small portable tools, such as electric saws and drills. No-load speeds as high as 5000 to 15 000 r/min are possible, but, as in any series motor, the speed drops rapidly with increasing load.

Series motor are built in many different sizes, starting from small toy motors to very large traction motors used in some electric locomotives.

21-13 Hysteresis motor

To understand the operating principle of a hysteresis motor, let us first consider Fig. 21-19. It shows a stationary rotor surrounded by a pair of N, S

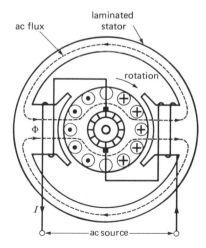

Figure 21-18   Alternating current series motor.

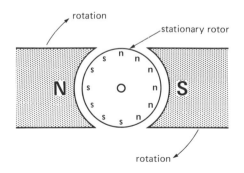

Figure 21-19   Permanent magnet rotor and a mechanical revolving field.

poles that can be rotated mechanically in a clockwise direction. The rotor is composed of a ceramic permanent magnet material whose resistivity approaches that of an insulator (Sec. 6-20). Consequently, it is impossible to set up eddy currents in such a rotor.

As the mechanical field rotates, it magnetizes the rotor; consequently, poles of opposite polarity are continuously being produced under the moving N, S poles. In effect, the revolving field is continuously reorienting the magnetic domains in the rotor. Clearly, the domains go through a complete cycle (or hysteresis loop) each time the field makes one complete revolution. Hysteresis losses are therefore produced in the rotor, proportional to the area of the hysteresis loop (Sec. 6-6). These losses are dissipated as heat in the rotor.

Let us assume that the hysteresis loss per revolution is E_h joules and the field rotates at n r/min. The energy dissipated in the rotor per minute is:

$$W = nE_h$$

The corresponding power (dissipated as heat) is:

$$P_h = W/t \qquad \text{Eq. 3-4}$$
$$= nE_h/60 \ [\text{W}]$$

However, the power dissipated in the rotor can only come from the mechanical power used to drive the poles. This power is given by:

$$P = nT/9.55 \qquad \text{Eq. 3-5}$$

Because $P = P_h$, we have:

$$nT/9.55 = nE_h/60$$

whence
$$\boxed{T = E_h/6.28} \qquad (21\text{-}2)$$

where

T = torque exerted on the rotor [N·m]

E_h = hysteretic energy dissipated in the rotor, per turn [J/r]

6.28 = constant [exact value = 2π]

The torque needed to drive the magnets is therefore constant, irrespective of the speed of rotation. In other words, whether the poles just barely creep around the rotor or whether they move at high speed, the torque exerted on the rotor is always

the same. It is this basic property that distinguishes hysteresis motors from all other motors.

In practice, the revolving field is produced by a 3-phase stator, or by a single-phase stator having an auxiliary winding. When a hysteretic rotor is placed inside such a stator, it immediately accelerates up to synchronous speed. The accelerating torque is essentially constant as shown by Fig. 21-20. This is quite different from a squirrel-cage induction motor, whose torque falls toward zero as it approaches synchronous speed.

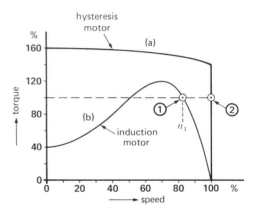

Figure 21-20   Typical torque-speed curves of two capacitor-run motors:
a. hysteresis motor; b. induction motor.

Thanks to the fixed frequency of large distribution systems, the hysteresis motor is employed in electric clocks, and other precise timing devices (Fig. 21-21). It is also used to drive tape-decks, turn-tables and other precision audio equipment. In such devices, the constant speed is, of course, the feature we are looking for. However, the hysteresis motor is particularly well suited to drive such devices because of their high inertia. Inertia prevents many synchronous motors (such as reluctance motors) from coming up to speed because to reach synchronism, they have to *suddenly* lock with the revolving field. No such abrupt transition occurs in the hysteresis motor because it develops a powerful torque right up to synchronous speed.

In audio equipment, these salient features are further enhanced by designing the motor to function as a vibration-free capacitor-run motor (Sec. 21-10).

Figure 21-21   Single-phase hysteresis clock motor having 32 poles and a ferrite rotor.

Example 21-3:

A small 60 Hz hysteresis clock motor possesses 32 poles. In making one complete turn with respect to the revolving field, the hysteresis loss in the rotor amounts to 0.8 J. Calculate:

a. the pull-in and pull-out torques,

b. the maximum power output before the motor stalls,

c. the rotor losses when the motor is stalled,

d. the rotor losses when the motor runs at synchronous speed.

Solution:

a.   The pull-in and pull-out torques are about equal in a hysteresis motor:

$$T = E_h/6.28 = 0.8/6.28 \qquad \text{Eq. 21-2}$$
$$= 0.127 \text{ N·m}$$

b. 1   The synchronous speed is:

$$n_s = 120 f/p = 120 \times 60/32$$
$$= 225 \text{ r/min}$$

b. 2   The maximum power is:

$$P = nT/9.55 = (225 \times 0.127)/9.55$$
$$= 3 \text{ W (or 3/746} = 1/250 \text{ hp)}$$

c. 1   When the motor stalls, the rotating field

moves at 225 r/min with respect to the rotor. The energy loss per minute is therefore:

$$W = 225 \times 0.8 = 180 \text{ J}$$

c. 2   The power dissipated in the rotor is:

$$P = W/t = 180/60 = 3 \text{ W}$$

d.   There is no energy loss in the rotor when the motor runs at synchronous speed because the magnetic domains no longer reverse.

21-14   Synchronous reluctance motor

We can build a synchronous motor by milling out a standard squirrel-cage rotor so as to create a number of salient poles. The number of poles must be equal to the number of poles on the stator. Figure 21-22 shows a rotor milled out to create four salient poles.

Such a *reluctance motor* starts up as a standard squirrel cage motor but, when it approaches synchronous speed, the salient poles lock with the revolving field, and so the motor runs at synchronous speed. Both the pull-in and pull-out torques are weak, compared to those of a hysteresis motor of equal size. Furthermore, reluctance motors cannot accelerate high-inertia loads to synchronous speed. The reason can be seen by referring to Fig. 21-20. Suppose the motor has reached a speed n_1 corresponding to full-load torque (operating point 1).

The stator poles are slipping past the rotor poles at a rate that corresponds to the slip. If the rotor is to lock with the revolving field, it must do so in the time it takes for one stator pole to sweep past a rotor pole. If pull-in is not achieved during this interval (Δt), it will never be achieved. The problem is that in going from speed n_1 to synchronous speed, the kinetic energy must increase by an amount:

$$\Delta E_k = 5.48 \times 10^{-3} J (n_s^2 - n_1^2) \text{ See Eq. 3-8 (21-3)}$$

Consequently, to reach synchronous speed, the motor must develop an accelerating power P_a of at least:

$$P = \Delta E_k/\Delta t \qquad (21-4)$$

Furthermore, the motor must continue to supply the power P_L demanded by the load. If the sum of $P_a + P_L$ exceeds the capacity of the motor, it will never pull into step. In essence, a reluctance motor can only synchronize when the slip is small and the moment of inertia is low.

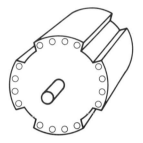

Figure 21-22   Rotor of a synchronous reluctance motor.

Despite this drawback, the reluctance motor is cheaper than any other type of synchronous motor. It is particularly well adapted to variable-frequency electronic speed control. Inertia is then no problem because the speed of the revolving field always tracks with the speed of the rotor. Three-phase reluctance motors of several hundred horsepower have been built, using this approach.

21-15 Selsyn drive

In some remote control systems, we may have to move the position of a small rheostat that is one or two meters away. This problem is easily solved by using a flexible shaft. But if the rheostat is 100 m away, the flexible-shaft solution becomes impractical. We then employ an electrical shaft to tie the knob and rheostat together. How does such a shaft work?

Consider two conventional wound-rotor induction motors whose 3-phase stators are connected in parallel (Fig. 21-23). Two phases of the respective rotors are also connected in parallel, and energized from a *single-phase* source. The remarkable feature about this arrangement is that the rotor on one machine automatically tracks the rotor on the other. Thus, if we turn one rotor cw through 17 degrees, the other rotor will move cw through 17°. Obviously, such a system would enable us to control a rheostat from a remote location. Two tiny wound-rotor motors are required, one (the transmitter) coupled to the control knob, and the other (the receiver) to the rheostat. A 5-conductor cable linking the transmitter and receiver constitutes the flexible electrical shaft.

The behavior of this so-called *selsyn* or *servo*

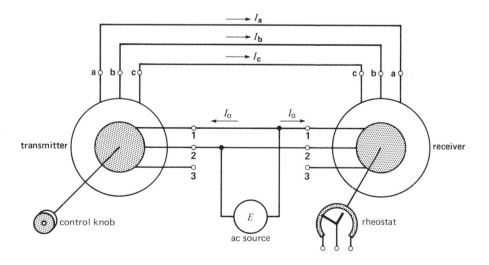

Figure 21-23   Components and connections of a servo system.

control system is explained as follows. Assume the transmitter and receiver are identical, and that the rotors are in identical positions. When the rotors are excited, they behave like the primaries of two transformers, inducing voltages in the stator windings. The voltages induced in the three stator windings of one selsyn are always unequal because the windings are displaced by $120°$. Indeed, the voltage induced in one winding may be zero, depending upon the position of the rotor.

Nevertheless, no matter what the respective stator voltages may be, they are identical in both machines when the rotors occupy the same position. The stator voltages balance each other and, consequently, no current flows in the lines connecting the stators. The rotors, however, carry a small exciting current I_0.

Now if we turn the rotor of one selsyn, its three stator voltages will change. They will no longer balance the stator voltages of the other selsyn; consequently, currents will flow in the lines connecting the two devices. These currents produce a torque on both rotors tending to line them up. Since the rotor of the receiver is free to move, it will line up with the transmitter. As soon as they are aligned, the stator voltages are again in balance, and the torque-producing currents disappear.

Selsyns are often employed to indicate the positive of an antenna, valve, gun turret, and so on, with the result that the torque requirements are small. Such transmitters and receivers are built with watch-like precision to ensure they will track with as little an error as possible.

If high torques have to be transmitted, standard 3-phase wound-rotor induction motors can be used.

QUESTIONS AND PROBLEMS

Practical level

21-1 A 6-pole single-phase motor is connected to a 60 Hz source. What is its synchronous speed?

21-2 What is the purpose of the auxiliary winding in a single-phase induction motor? How can we change the rotation of such a motor?

21-3 State the main difference between a split-phase motor and a capacitor-start motor. What are their relative advantages?

21-4 Explain briefly how a shaded-pole motor operates.

21-5 List some of the properties and advantages of a series ac motor.

21-6 Why are some single-phase motors equipped with a resilient mounting? Is such a mounting necessary on 3-phase motors?

21-7 What is the main advantage of a capacitor-run motor?

21-8 Which of the motors discussed in this chapter is best suited to drive the following loads:
   a. a small portable drill;
   b. a 3/4 hp air compressor;
   c. a vacuum cleaner;
   d. a 1/100 hp blower;
   e. a 1/3 hp centrifugal pump;
   f. a 1/4 hp fan for use in a hospital ward;
   g. an electric timer;
   h. a hi-fi turn-table.

Intermediate level

21-9 Referring to Fig. 21-11, the effective impedance of the main and auxiliary windings under locked rotor conditions are given as follows:

	Effective resistance	Effective reactance
Main winding	4Ω	7.5Ω
Auxiliary winding	7.5Ω	4Ω

If the line voltage is 119 V, calculate:
   a. the magnitude of I_a and I_s;
   b. the phase angle between I_a and I_s;
   c. the line current I_L;
   d. the power factor under locked-rotor conditions.

21-10 The palm of the human hand can just barely tolerate a temperature of $130°$F. If the no-load temperature of the frame of a 1/4 hp motor is 64°C in an ambient temperature of 76°F:

a. can a person keep his hand on the frame?

b. is the motor running too hot?

21-11 Referring to Fig. 21-13, if the load torque is constant at 4 N·m, explain the resulting behavior of the motor.

21-12 a. A single-phase motor vibrates at a frequency of 100 Hz. What is the frequency of the power line?

b. A capacitor-run motor does not have to be set in a resilient mounting. Why?

c. A 4-pole, 60 Hz single-phase hysteresis motor develops a torque of 6 in·lb when running at 1600 r/min. Calculate the hysteresis loss per revolution [J].

21-13 Referring to the 6 W shaded-pole motor in Table 21B, calculate:

a. the rated power output in millihorsepower;

b. the full-load power factor;

c. the slip at the breakdown torque;

d. the per unit no-load current and locked-rotor current.

21-14 Referring again to Fig. 21-13, calculate:

a. the locked-rotor torque [ft·lb];

b. the p.u. value of the LR torque;

c. the starting torque when only the main winding is excited;

d. the per-unit breakdown torque;

e. how are the torque-speed curves affected if the line voltage falls from 115 V to 100 V?

Advanced level

21-15 In Table 21A, calculate:

a. the voltage across the capacitor at locked-rotor conditions;

b. the corresponding phase angle between I_s and I_a.

21-16 Referring to Fig. 21-16, if the motor operates at full load, calculate:

a. the line current I_L;

b. the power factor of the motor;

c. the active power absorbed by each winding;

d. the efficiency of the motor.

21-17 The motor described in Table 21A, has a LR power factor of 0.9 lagging. It is installed in a workshop situated 600 ft from a home, where the main service entrance is located. The line is composed of a 2-conductor cable made of No. 12 gauge copper. The ambient temperature is 25°C and the service entrance voltage is 122 V. Calculate:

a. the resistance of the transmission line;

b. the starting current and the voltage at the motor terminals;

c. the starting torque [N·m].

PART VI

ELECTRICAL AND ELECTRONIC CONTROLS

22

INDUSTRIAL MOTOR CONTROL

Industrial control, in its broadest sense, encompasses all the methods used to control the performance of an electrical system. When applied to machinery, it involves the starting, acceleration, reversal, deceleration and stopping of a motor and its load. In this chapter, we shall study the electrical (but not electronic) control of 3-phase alternating current motors. Our study is limited to elementary circuits because industrial circuits are usually too intricate to explain in a few words. However, the basic principles covered here apply to any system of control, no matter how complex it may appear to be.

22-1 Control devices

Every control circuit is composed of a number of basic components connected together to achieve the desired performance. The size of the components varies with the power of the motor, but the principle of operation remains the same. Using only a dozen basic components, we can design control systems that are very complex. The basic components are:

1. Disconnecting switches
2. Manual circuit breakers
3. Cam switches
4. Pushbuttons
5. Relays
6. Magnetic contactors
7. Thermal relays and fuses
8. Pilot lights
9. Limit switches and other special switches
10. Resistors, reactors, transformers and capacitors

Table 22A illustrates these devices, and states their main purpose and application. The symbols for these and other devices are given in Table 22B.

22-2 Normally open and normally closed contacts

Control circuit diagrams always show components in a state of rest, that is, when they are not energized (electrically) or activated (mechanically). In this state, some electrical contacts are open while others are closed. They are respectively called *normally open contacts* (NO) *and normally closed contacts* (NC) and are designated by the following

TABLE 22A	BASIC COMPONENTS FOR CONTROL CIRCUITS

Disconnecting switches

A disconnecting switch isolates the motor from the power source. It consists of 3 knife-switches and 3 line fuses enclosed in a metallic box. The knife-switches open and close simultaneously by means of an external handle. An interlocking mechanism prevents the hinged cover from opening when the switch is closed. Disconnecting switches are designed to carry the nominal full-load current indefinitely, and to withstand short-circuit currents for brief intervals.

Figure 22-1 Three-phase, fused disconnecting switch rated 600 V, 30 A. *(Square D)*

Manual circuit breakers

A manual circuit breaker opens and closes a circuit, like a toggle switch. It trips (opens) automatically when the current exceeds a predetermined limit. After tripping, it can be reset manually. Circuit breakers are often used instead of disconnecting switches because no fuses have to be replaced.

Figure 22-2 Three-phase circuit breaker, 600 V, 100 A. *(Square D)*

Cam switches

A cam switch has a group of fixed contacts and an equal number of moveable contacts. The contacts can be made to open and close in a preset sequence by rotating a handle or knob. Cam switches are used to control the motion and position of hoists, callenders, machine tools, etc.

Figure 22-3 Three-phase surface-mounted cam switch, 230 V, 2 kW. *(Klockner-Moeller)*

TABLE 22A	BASIC COMPONENTS FOR CONTROL CIRCUITS

Pushbuttons

A pushbutton is a switch activated by finger pressure. Two or more contacts open or close when the button is depressed. Pushbuttons are usually spring loaded so as to return to their normal position when pressure is removed.

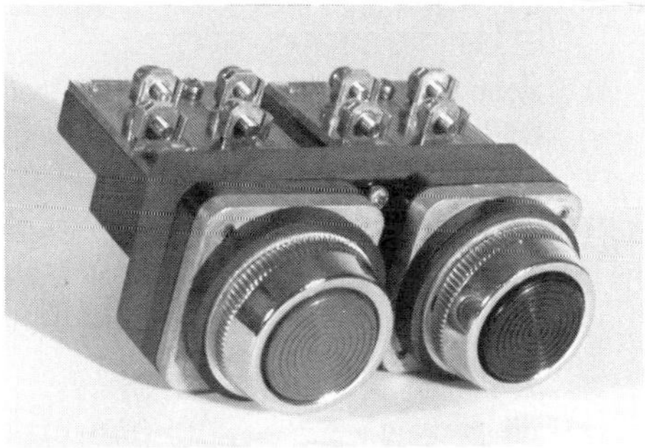

Figure 22-4   Mechanically-interlocked pushbuttons with NO and NC contacts; rated to interrupt an ac current of 6 A one million times. *(Siemens)*

Control relays

A control relay is an electromagnetic switch that opens and closes a set of contacts when a so-called relay coil is energized. The coil produces a strong magnetic field which attracts a moveable armature bearing the contacts. Control relays are mainly used in low-power circuits. They include time-delay relays whose contacts open or close after a definite time interval.

Figure 22-5   Single-phase relays: 25 A, 115/230 V and 5 A, 115 V. *(Potter and Brumfield)*

Thermal relays

A thermal relay (or overload relay) is a temperature-sensitive device whose contacts open or close when the motor current exceeds a preset limit. The current flows through a small, calibrated heating element which raises the temperature of the relay. Thermal relays are inherent time-delay devices because the temperature cannot follow the instantaneous changes in current.

Figure 22-6   Three-phase thermal relay with variable current setting, 6 A to 10 A. *(Klockner-Moeller)*

TABLE 22A	BASIC COMPONENTS FOR CONTROL CIRCUITS

Magnetic contactors

A magnetic contactor is basically a large control relay designed to open and close a power circuit. It possesses a relay coil which activates a set of contacts. Magnetic contactors are used to control motors ranging from 0.5 hp to several hundred horsepower. The size, dimensions and performance of contactors are standardized.

Figure 22-7 a

Figure 22-7 a. Three-phase magnetic contactor rated 50 hp, 575 V, 60 Hz. Width: 158 mm; height: 155 mm; depth: 107 mm; weight: 3.5 kg. *(Siemens)*

Figure 22-7 b. Three-phase magnetic contactor with mercury contacts, 480 V, 60 A. The contactor is particularly quiet on opening and closing. *(Davis Controls Ltd)*

Figure 22-7 b

Pilot lights

A pilot light indicates the on/off state of a remote component in a control system.

Figure 22-8 Pilot light, 120 V, 3 W mounted in a start-stop pushbutton station. *(Siemens)*

TABLE 22A   BASIC COMPONENTS FOR CONTROL CIRCUITS

Limit switches and special switches

A limit switch is a low-power snap-action device which opens or closes a contact, depending upon the position of a mechanical part. Other limit switches are sensitive to pressure, temperature, liquid level, direction of rotation, and so on.

Figure 22-9 a. Limit switch with one NC contact; rated for ten million operations; position accuracy: 0.5 mm. *(Square D)*

Figure 22-9 b. Liquid level switch. *(Square D)*

(a)   (b)

symbols:

   normally open contact ─┤├─

   normally closed contact ─┤╱├─

22-3 Relay coil exciting current

When a magnetic contactor is in the open position, the magnetic circuit has a very long air gap, compared to when the contactor is closed. Consequently, the inductance (and inductive reactance) of the relay coil is much lower when the contacts are open than when they are closed. Because the coil is excited by a fixed ac voltage, the magnetizing current is much higher in the open than in the closed contactor position. In other words, a considerable inrush current is drawn by the relay coil at the moment it is excited. This places a heavier than expected duty on auxiliary contacts that energize the coil.

Example 22-1:
A 3-phase NEMA size 5 magnetic contactor rated at 270 A, 460 V possesses a 120 V, 60 Hz relay coil. The coil absorbs an apparent power of 2970 VA and 212 VA, respectively, in the open and closed contactor position. Calculate:
a. the inrush exciting current;
b. the normal, sealed exciting current;
c. the ratio of the control power to the power handled by the contactor.

Solution:
a. 1 The inrush current is:
$$I = S/E = 2970/120 = 24.75 \text{ A}$$
b. 1 The normal current when the contactor is sealed is:
$$I = S/E = 212/120 = 1.77 \text{ A}$$
c. 1 The ratio of controlled power to controlling power is:
$$S_1/S_2 = 460 \times 270 \times \sqrt{3}/212$$
$$= 1015$$

22-4 Control diagrams

A control system can be represented by four types of circuit diagrams, listed as follows, in order of increasing detail and completeness:

- block diagram
- one-line diagram*
- wiring diagram
- schematic diagram

A *block diagram* is composed of a set of rectangles, each representing a control device, together with a brief description of its function. The rectangles are connected by arrows which indicate the direction of power flow (Fig. 22-10).

A *schematic diagram* shows all the electrical connections between components, without regard to their physical location or terminal arrangement. This type of diagram is indispensable when troubleshooting a circuit or analyzing its mode of operation (Fig. 22-13).

The reader should note that the four diagrams in Figs. 22-10 to 22-13 all relate to the same control circuit. The symbols used to designate the various components are given in Table 22B.

22-5 Starting methods

Three-phase squirrel-cage motors are started either by connecting them directly across the line or by

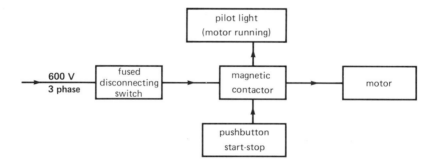

Figure 22-10   Block diagram of a combination starter.

A *one-line diagram* is similar to a block diagram except that the components are shown by their symbols, rather than by rectangles. The symbols give us an idea of the nature of the components; consequently, one-line diagrams yield more information. The lines connecting the various components represent two or more conductors (Fig. 22-11).

A *wiring diagram* shows the connections between the components, taking into account the physical location of the terminals and even the color of wire. These diagrams are employed when installing equipment or when troubleshooting a circuit (Fig. 22-12).

applying reduced voltage to the stator. The starting method depends upon the power capacity of the supply line and the type of load.

Across-the-line starting is simple and inexpensive. The main disadvantage is the high starting current, which is 5 to 6 times the rated full-load current. It can produce a significant voltage drop, which may affect other customers connected to the same line. Voltage-sensitive devices such as incandescent lamps, television sets, and high-precision machine tools respond badly to such voltage dips.

Mechanical shock is another problem that should not be overlooked. Equipment can be seriously damaged if full-voltage starting produces a hammer-blow impulse. Conveyor belts are another example where sudden starting may not be acceptable.

---

*   Also called single-line diagram.

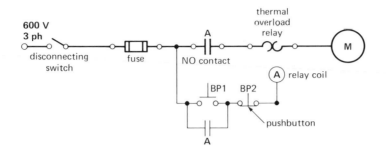

Figure 22-11   One-line diagram of a combination starter.

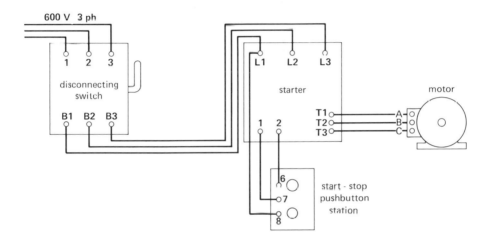

Figure 22-12   Wiring diagram of a combination starter.

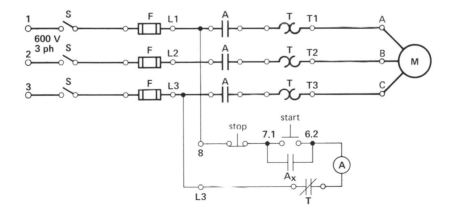

Figure 22-13   Schematic diagram of a combination starter.

TABLE 22B   GRAPHIC SYMBOLS FOR ELECTRICAL DIAGRAMS

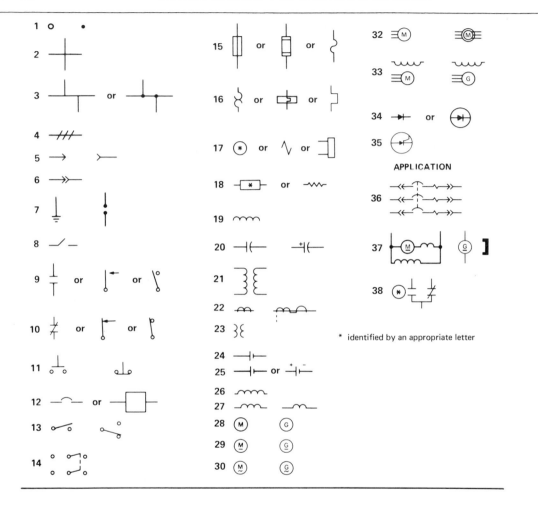

* identified by an appropriate letter

1. terminal; connection   2. conductors crossing   3. conductors connected   4. three conductors   5. plug; receptacle
6. separable connector   7. ground connection; arrester   8. disconnecting switch   9. normally open contact (NO)
10. normally closed contact (NC)   11. pushbutton NO; NC   12. circuit-breaker   13. single pole switch; three-way switch
14. double pole double throw switch   15. fuse   16. thermal overload element   17. relay coil   18. resistor
19. winding, inductor or reactor   20. capacitor; electrolytic capacitor   21. transformer   22. current transformer; bushing type
23. potential transformer   24 dc source (general)   25. cell   26. shunt winding   27. series winding; commutating pole or compensating winding
28. motor; generator (general symbols)   29. dc motor; dc generator (general symbols)   30. ac motor; ac generator (general symbols)
32. 3-phase squirrel-cage induction motor; 3-phase wound-rotor motor   33. synchronous motor; 3-phase alternator   34. diode
35. thyristor or SCR   36. 3-pole circuit breaker with magnetic overload device, drawout type
37. dc shunt motor with commutating winding; permanent magnet dc generator   38. magnetic relay with one NO and one NC contact.

For a complete list of graphic symbols and references see *"IEEE Standard and American National Standard Graphic Symbols for Electrical and Electronics Diagrams"* (ANSI Y32.2/IEEE No. 315) published by the Institute of Electrical and Electronic Engineers, Inc., New York, N.Y. 10017. Essentially the same symbols are used in Canada and several other countries.

Figure 22-14   Manual starters for single-phase motors rated 1 hp (0.75 kW); left: surface mounted; center: flush mounted; right: waterproof enclosure. *(Siemens)*

In large industrial installations, we can sometimes tolerate across-the-line starting even for motors rated up to 10 000 hp. Obviously, the fuses and circuit-breakers must be designed to carry the starting current during the acceleration period.

A disconnect switch or manual circuit-breaker is always placed between the supply line and the starter. The switch and starter are sometimes mounted in the same enclosure to make what is called a *combination starter*. The line fuses (when used) are rated at about 3.5 times full-load current; consequently, they do not protect the motor against continuous overloads. Their primary function is to protect the motor and supply line against catastrophic currents resulting from a short-circuit in the motor or starter, or a failure to start up. The fuse rating, in amperes, must comply with the requirements of the National Electrical Code.

22-6 Manual across-the-line starters

Manual 3-phase starters (Fig. 22-14) are composed of a circuit breaker and either two or three thermal relays, all mounted in an appropriate enclosure.

They are used for small motors (10 hp or less) at voltages ranging from 120 V to 600 V. The interchangeable thermal relays trip the circuit-breaker whenever the current in one of the phases exceeds the rated value.

Figure 22-15   a. Three-phase across-the-line magnetic starter, 30 hp, 600 V, 60 Hz. *(Klockner-Moeller)*

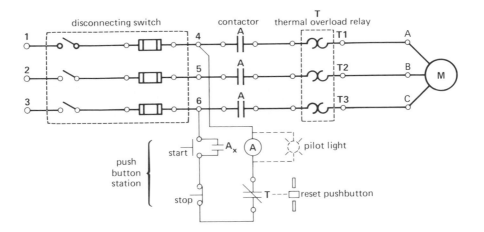

Figure 22-15   b. Schematic diagram of a 3-phase across-the-line magnetic starter.

22-7 Magnetic across-the-line starters

Magnetic across-the-line starters are employed whenever we have to control a motor from a remote location. They are also used whenever the power rating exceeds 10 kW.

Figure 22-15 shows a typical magnetic starter and its associated connection diagram. The starter has three main components: a magnetic contactor, a thermal relay, and a control station.

1. The *magnetic contactor* A possesses three heavy contacts A and one small auxiliary contact A_x. Contacts A must be big enough to carry the starting current and the nominal full-load current, without overheating. The relay coil is represented by the symbol Ⓐ . Contacts A and A_x remain closed as long as the coil is energized.

2. The *thermal relay* T protects the motor against sustained overloads.* The relay comprises three individual heating elements, respectively connected in series with the three phases. A small, normally closed contact T forms part of the relay assembly. It opens when the relay gets too hot, and stays open until the relay is manually reset.

* The thermal relay is often designated by the letters OL (overload).

The current rating of the relay is chosen to protect the motor against sustained overloads. Contact T opens after a period of time which depends upon the magnitude of the overload current. Thus, Figure 22-16 shows the tripping time as a function of the rated relay current. At rated current (multiple 1), the relay never trips, but at twice rated current, it trips after an interval of 40 s. The thermal relay is equipped with a reset button enabling us to reclose contact T following an overload. It is preferable to wait a few minutes before pushing the button, to allow the relay to cool down.

3. The *control station*, composed of start-stop pushbuttons, may be located either close to, or far away from the starter. The pilot light is optional.

To start the motor, we first close the disconnecting switch and then depress the *start* button momentarily. Coil A is immediately energized, contacts A and A_x close, and full voltage appears across the motor. When the pushbutton is released, it returns to its normal position, but the relay coil remains excited because auxiliary contact A_x is now closed. Contact A_x is said to be a *self-sealing* contact.

To stop the motor, we simply push the *stop*

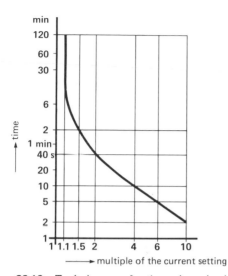

Figure 22-16   Typical curve of a thermal overload relay, showing tripping time versus line current. The tripping time is measured from cold-start conditions. If the motor has been operating at full load for one hour or more, the tripping time is reduced by about 30 percent.

Figure 22-17   Three-phase across-the-line combination starter, 150 hp, 575 V, 60 Hz. The protruding knob controls the disconnecting switch; the pushbutton station is set in the transparent polycarbonate cover. *(Klockner-Moeller)*

Figure 22-18   Three-phase across-the-line combination starter rated 100 hp, 575 V, 60 Hz. The isolating circuit breaker is controlled by an external handle. The magnetic contactor is mounted in the bottom left-hand corner of the water proof enclosure. The small 600 V/120 V transformer in the lower right-hand corner supplies low-voltage power for the control circuit. *(Square D)*

button, which opens the circuit to the coil. In case of an overload, the opening of contact T produces the same effect.

It sometimes happens that a thermal relay will occasionally stop a motor for no apparent reason. This condition can occur when the ambient temperature around the starter is too high. We can remedy the situation by changing the location of the starter, or by replacing the relay by another one having a higher current rating. Care must be taken before making such a change because if the ambient temperature around the motor is also high, the occasional tripping may actually be a warning.

Figure 22-17 shows a typical combination starter. Figure 22-18 shows another combination starter equipped with a small step-down transformer to excite the control circuit. Such transformers are always used on high-voltage starters (above 600 V) because they eliminate the insulation problem, permit the use of standard components, and reduce the hazard to operating personnel.

22-8 Inching and jogging

In some mechanical systems, we have to adjust the position of a motorized part very precisely. To accomplish this, we energize the motor in short spurts so that it barely starts before it again comes to a halt. A double-contact pushbutton J is added to the usual start/stop circuit, as shown in Fig. 22-19. This arrangement permits conventional start-stop control as well as so-called *jogging*, or *inching*.

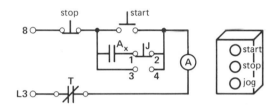

Figure 22-19   Control circuit and pushbutton station for start-stop-jog operation. Terminals 8, L3 correspond to terminals 8, L3 in Fig. 22-13.

Jogging imposes severe duty on the power contacts because they continually make *and break* currents that are 6 times greater than normal. It is estimated that each impulse corresponds to 30 normal start/stop operations. Thus, a contactor that can normally start and stop a motor 3 million times, can only jog the motor 100 000 times before the contacts have to be replaced.[*] Furthermore, jogging should not be repeated too quickly because the intense heat of the breaking arc may cause the contacts to weld together. When jogging is required, the contactor is usually selected to be one NEMA size bigger than for normal duty.

22-9 Reversing the direction of rotation

We can reverse the direction of rotation of a motor by interchanging any two lines. This can be done

---

[*] A magnetic contactor has an estimated *mechanical* life of about 20 million open/close cycles, but the electrical contacts should be replaced after 3 million such normal interruptions.

by using two magnetic contactors A and B and a manual 3-position cam switch (Fig. 22-20). In the forward direction, the cam switch engages contact 1, which energizes relay coil A, causing contactor A to close.[*]

To reverse the rotation, we move the switch to position 2. However, in doing so, we have to move past the off position (0). Consequently, it is impossible to energize coils A and B simultaneously. Occasionally, however, a mechanical defect may prevent a contactor from dropping out, even after its relay coil is deenergized. This is a serious situation, because when the *other* contactor closes, a dead short-circuit results across the line. The short-circuit current can easily be 50 to 500 times greater than normal, and both contactors may be severely damaged. To eliminate this danger, the contactors are mounted side by side and mechanically interlocked, thus making it physically impossible for both to be closed at the same time. The interlock is a simple steel bar, pivoted at the center, whose extremities are tied to the moveable contact assembly of each contactor.

During an emergency, push-button U, equipped with a large red bull's-eye, can be used to stop the motor (Fig. 22-20c). In practice, operators find it easier to hit a large button than to turn a cam switch to the off position.

22-10 Plugging

We have already seen that an induction motor can be brought to a rapid stop by reversing two of the lines (Sec. 18-7). However, to prevent the motor from turning in reverse, a zero-speed switch must open the line as soon as the machine has come to rest. The circuit of Fig. 22-21a shows the basic elements needed to provide for this type of plugging. The circuit operates as follows:

---

[*] The contacts and relay coils may be designated by any appropriate letters. Thus, the letters F and R are often used to designate "forward" and "reverse" operating components. In this text, we have adopted the letters A and B mainly for reasons of continuity from one circuit to the next.

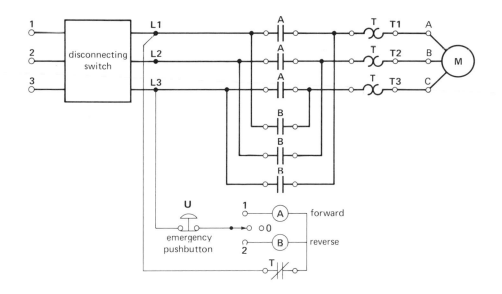

Figure 22-20   a. Simplified schematic diagram of a reversible magnetic starter.

Figure 22-20   b. Three-position cam switch in Fig. 22-20a. *(Siemens)*

Figure 22-20   c. Emergency stop push-button in Fig. 22-20a. *(Square D)*

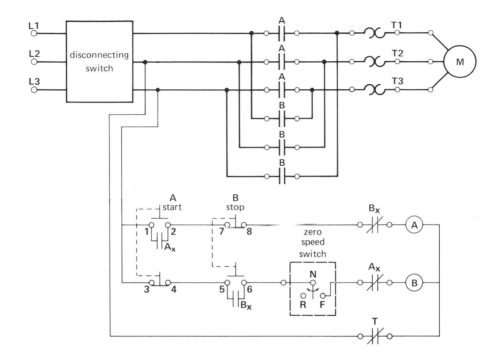

Figure 22-21   a. Simplified schematic diagram of a starter with plugging control.

1. Start/stop push-buttons A and B each have a normally open and a normally closed contact. The contacts are interlocked mechanically (dotted lines). The interlock provides additional safety because it prevents the two contactors from closing simultaneously. In effect, if we analyze the circuit, we can see that it is impossible to energize coil A before deenergizing coil B, and vice versa.

2. Contact F-N of the zero-speed switch is normally open, but it closes as soon as the motor turns in the forward direction. This "prepares" the plugging circuit for the eventual operation of coil B.

3. Auxiliary contacts A_x and B_x are normally closed and constitute an electrical interlock. It offers additional safety because coil A cannot become energized until contactor B has actually dropped out, and vice versa.

Several types of zero-speed switches are on the market and Fig. 22-21b shows one that operates

on the principle of an induction motor. It consists of a permanent magnet rotor and a bronze ring, or "drag-cup," that is free to pivot between stationary contacts F and R. As soon as the rotor turns clockwise, it drags the ring along in the same direction, closing contacts F-N. When the motor stops turning, the drag-cup returns to the off position.

22-11 Reduced-voltage starting

Some industrial loads have to be started very gradually. Examples are coil winders, printing presses, and other machines that process fragile products. In other industrial applications, we cannot connect a motor directly to the line because the starting current is too high. In all these cases, we have to reduce the voltage applied to the motor either by connecting resistors (or reactors) in series with the line or by employing an autotransformer. In reducing the voltage, we recall that:

1. The locked-rotor current is proportional to the

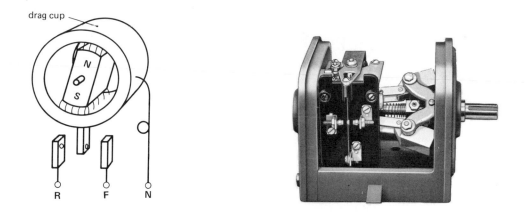

Figure 22-21   b. Typical zero speed switches for use in Fig. 22-21a; left; drag cup type; right: centrifugal switch type. *(Hubbell)*

voltage: reducing the voltage by half reduces the current by half;

2. the locked-rotor torque is proportional to the square of the voltage: reducing the voltage by half reduces the torque by a factor of four.

22-12 Primary resistance starting

Resistance starting consists of placing three resistors in series with the motor during the start-up period (Fig. 22-22). When the motor runs close to synchronous speed, a magnetic contactor B short-circuits the resistors. This method gives a very smooth start with complete absence of mechanical shock. The voltage drop across the resistors is high at first, but gradually diminishes as the current falls. Consequently, the voltage across the motor increases with speed, and so the electrical and mechanical shock is negligible when full voltage is finally applied (closure of contactor B). The resistors are shorted after a delay that depends upon the setting of time-delay relay RT.

The schematic control diagram (Fig. 22-22b) reveals the following circuit elements:

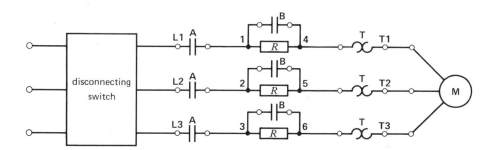

Figure 22-22   a. Simplified schematic diagram of the power section of a reduced-voltage primary resistor startor.

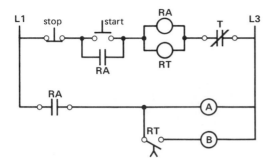

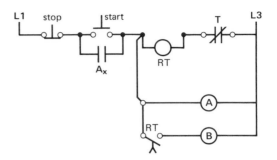

Figure 22-22 b. Control section of Fig. 22-22 a.

Figure 22-23 Control circuit of Fig. 22-22b, after eliminating auxiliary relay RA, and adding auxiliary contact A$_x$.

RA - small auxiliary relay possessing two NO contacts

RT - time-delay relay that closes the circuit of coil B after a preset interval of time

A, B - magnetic contactors and their associated contacts and relay coils

The purpose of relay RA is to carry the exciting currents of relay coils A and B. The magnetic contactors are assumed to be particularly large, and the exciting currents could damage the pushbutton contacts if they were directly used to control the coils (Fig. 22-23). Consequently, it is better to add an auxiliary relay having heavy contacts. Other circuit components are straightforward, and the reader should have no difficulty in analyzing the operation of the circuit.

Figures 22-24a and 22-24b respectively show the torque-speed curve and current-speed curve when full voltage is applied to a typical 3-phase, 1800 r/min induction motor. Corresponding curves are also shown when resistors are inserted in series with the line. The resistors are adjusted so that the locked-rotor voltage is 0.65 p.u. The locked-rotor torque is then considerably less than full-load torque, so the motor must be started at light load. When the speed reaches about 1700 r/min, we short-circuit the resistors. The current jumps from about 1.8 p.u. to 2.5 p.u. - a very moderate jump.

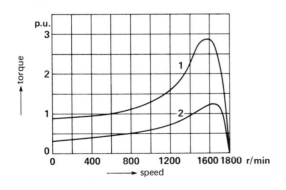

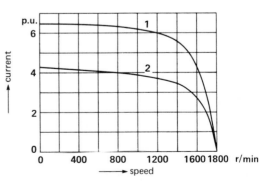

Figure 22-24 a. Typical torque-speed curves of a 3-phase squirrel-cage induction motor: (1) full voltage starting; (2) primary resistance starting with voltage reduced to 0.65 p.u.

Figure 22-24 b. Typical current-speed curves of a 3-phase squirrel-cage induction motor: (1) full voltage starting; (2) primary resistance starting with voltage reduced to 0.65 p.u.

Example 22-2:

A 150 kW (200 hp), 460 V, 3-phase, 3520 r/min, 60 Hz induction motor has a locked-rotor torque of 600 N·m and a locked-rotor current of 1400 A. Three resistors are connected in series with the line so as to reduce the voltage across the motor to 0.65 p.u. Calculate:

a. the apparent power absorbed by the motor under full-voltage, locked-rotor conditions;
b. the apparent power absorbed by the motor when the resistors are in the circuit;
c. the apparent power drawn from the line, with the resistors in the circuit;
d. the locked-rotor torque developed by the motor.

Solution:

a. 1   At full voltage, the apparent power is:
$$S = \sqrt{3}\,EI \qquad \text{Eq. 9-9}$$
$$= 1.73 \times 460 \times 1400$$
$$= 1114 \text{ kVA}$$

b. 1   The voltage across the motor at 0.65 p.u. is:
$$E = 0.65 \times 460 = 299 \text{ V}$$

b. 2   The current drawn by the motor decreases in proportion to the voltage:
$$I = 0.65 \times 1400 = 910 \text{ A}$$

b. 3   The apparent power drawn by the motor is:
$$S_m = 1.73\,EI = 1.73 \times 299 \times 910$$
$$= 471 \text{ kVA}$$

c. 1   The apparent power drawn from the line is:
$$S_L = 1.73\,EI = 1.73 \times 460 \times 910$$
$$= 724 \text{ kVA}$$

d. 1   The torque varies as the square of the voltage:
$$T = 0.65^2 \times 600 = 0.42 \times 600$$
$$= 252 \text{ N·m } (\approx 186 \text{ ft·lb})$$

The results of these calculations are summarized in Fig. 22-25.

Example 22-3:

In Example 22-2, if the locked-rotor power factor of the motor alone is 0.35, calculate the required value and power of the series resistors.

Solution:

We will solve this problem by considering active and reactive powers and without using phasor diagrams.

a. 1   The apparent power drawn by the motor is:
$$S_m = 471 \text{ kVA} \quad \text{(from Example 22-2)}$$

a. 2   The apparent power drawn by the line is:
$$S_L = 724 \text{ kVA} \quad \text{(from Example 22-2)}$$

a. 3   The active power drawn by the motor is:
$$P_m = S_m \cos\theta = 471 \times 0.35$$
$$= 165 \text{ kW}$$

a. 4   The reactive power absorbed by the motor is:
$$Q_m = \sqrt{S_m^2 - P_m^2} = \sqrt{471^2 - 165^2}$$
$$= 441 \text{ kvar}$$

a. 5   The resistors can only absorb active power in the circuit. Consequently, the reactive power supplied by the line *must* be equal to that absorbed by the motor:
$$Q_L = 441 \text{ kvar}$$

a. 6   The active power supplied by the line is:
$$P_L = \sqrt{S_L^2 - Q_L^2} = \sqrt{724^2 - 441^2}$$
$$= 574 \text{ kW}$$

a. 7   The active power absorbed by the three resistors is:
$$P_R = P_L - P_m = 574 - 165$$
$$= 409 \text{ kW}$$

a. 8   The active power per resistor is:
$$P = P_R/3 = 409/3 = 136 \text{ kW}$$

a. 9   The current in each resistor is:
$$I = 910 \text{ A} \quad \text{(from Example 22-2)}$$

a. 10 The value of each resistor is:
$$P = I^2 R$$

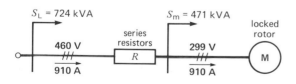

Figure 22-25   See Example 22-2.

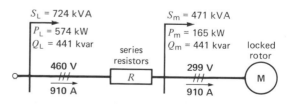

Figure 22-26   See Example 22-3.

$$136\ 000 = 91^2 R$$
$$\therefore \quad R = 0.164\ \Omega$$

This is an interesting example of the usefulness of active and reactive power in solving a relatively difficult problem. The results are summarized in Fig. 22-26.

22-13 Autotransformer starting

For a given torque, autotransformer starting draws a lower line current than does resistance starting. The disadvantage is that autotransformers cost more, and the transition from reduced-voltage to full-voltage is not quite as smooth.

Autotransformers usually have taps to give output voltages of 0.8, 0.65, and 0.5 p.u. The corresponding starting torques are respectively 0.64, 0.42, and 0.25 of the full-voltage starting torque. Furthermore, the starting currents on the *line* side are also reduced to 0.64, 0.42, and 0.25 of the full-voltage locked-rotor current.

Figure 22-27 shows a simple starter using two autotransformers connected in open delta. They have a 65% tap. The time-delay relay possesses three contacts RT. The contact in parallel with the start button closes as soon as coil RT is energized. The other two contacts operate after a delay that depends upon the relay setting. Contactors A and B are mechanically interlocked to prevent them from closing simultaneously.

Contactor A closes as soon as the start button is depressed. This excites the autotransformer and reduced voltage appears across the motor terminals. A few seconds later, the two RT contacts in series with coils A and B respectively open and close. Contactor A drops out, followed almost immediately by the closure of contactor B. This action applies full voltage to the motor, and simultaneously disconnects the autotransformer from the line.

In transferring from contactor A to contactor B, the motor is momentarily disconnected from the line. This creates a problem because when contactor B closes, a large transient current is drawn from the line. This transient surge is hard on the contacts and also produces a mechanical shock. For this reason, we sometimes employ more elaborate circuits in which the motor is never completely disconnected from the line. Figure 22-28a and 22-28b compare the torque and line current when autotransformer and resistance starting is used. The locked-rotor voltage in each case is 0.65 p.u. The reader will note that the locked-rotor torques are identical, but the locked-rotor line current is much lower using an autotransformer (2.7 versus 4.2 p.u.).

However, once the motor reaches 90% of synchronous speed, resistance starting yields a higher torque because the terminal voltage is then higher than the 65% value that existed at the moment of start-up. On the other hand, the line current at all speeds is smaller when using an autotransformer.

Because the autotransformers and resistors operate for very short periods, they can be wound with much smaller wire than continuously rated devices. This enables us to drastically reduce the size, weight and cost of these components.

Figure 22-27   a.   Reduced   voltage   autotransformer starter, 100 hp, 575 V, 60 Hz. *(Square D)*

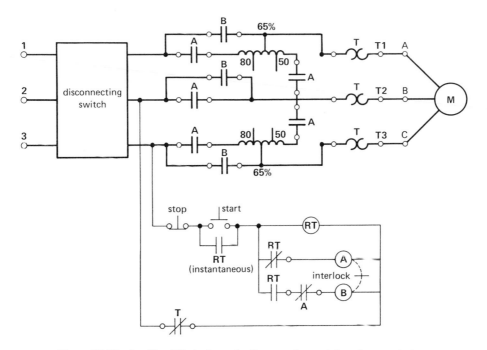

Figure 22-27   b. Simplified schematic diagram of an autotransformer starter.

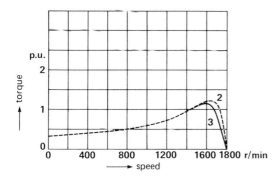

Figure 22-28   a. Typical reduced voltage (0.65 p.u.) torque-speed curves of a 3-phase squirrel-cage induction motor: (2) primary resistance starting; (3) autotransformer starting.

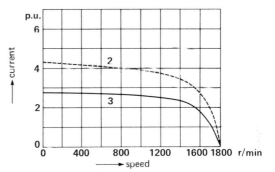

Figure 22-28   b. Typical reduced voltage (0.65 p.u.) current-speed curves of a 3-phase squirrel-cage induction motor: (2) primary resistance starting; (3) autotransformer starting.

Example 22-4:

A 200 hp (150 kW), 460 V, 3-phase, 3520 r/min, 60 Hz induction motor has a locked rotor torque of 600 N·m and a locked-rotor current of 1400 A. Two autotransformers, connected in open delta,

and having a 65% tap, are employed to provide reduced-voltage starting. Calculate:

a. the apparent power absorbed by the motor;

b. the apparent power supplied by the 460 V line;

c. the current supplied by the 460 V line;

d. the locked-rotor torque.

Solution:

a. 1   The voltage across the motor is:
$$E = 0.65 \times 460 = 299 \text{ V}$$

a. 2   The current drawn by the motor is:
$$I = 0.65 \times 1400 = 910 \text{ A}$$

a. 3   The apparent power drawn by the motor is:
$$S_m = 1.73\,EI = 1.73 \times 299 \times 910$$
$$= 471 \text{ kVA}$$

b. 1   The apparent power supplied by the line is equal to that absorbed by the motor because the active and reactive power consumed by the autotransformers is negligible (Sec. 13-1). Consequently,
$$S_L = 471 \text{ kVA}$$

c. 1   The current drawn from the line is:
$$I = S_L/(1.73\,E) \qquad \text{Eq. 9-9}$$
$$= 471\,000/(1.73 \times 460)$$
$$= 592 \text{ A}$$

Note that this current is considerably smaller than the line current with resistance starting.

d.   The locked-rotor torque varies as the square of the motor voltage:
$$T = 0.65^2 \times 600$$
$$= 252 \text{ N·m}$$

The results of these calculations are summarized in Fig. 22-29.

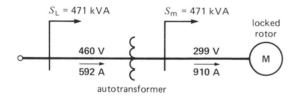

$S_L = 471$ kVA $S_m = 471$ kVA

locked rotor

460 V   299 V

592 A   910 A   M

autotransformer

Figure 22-29   See Example 22-4.

22-14 Other starting methods

Several other methods are employed to limit the current and torque when starting induction motors. Some only require a change in the stator winding connections. Thus, with *part-winding* starting, two identical sections of one phase are first connected in series during the starting period and then in parallel during the running period. Part-winding starting is only used on big motors.

In *wye-delta* starting, all six stator leads are brought out to the terminal box. The windings are connected in wye during start-up, and in delta during normal running conditions. This starting method gives the same results as an autotransformer starter having a 58% tap. The reason is that the voltage across each wye-connected winding is only $1/\sqrt{3}$ (= 0.58) of its rated value.

To start wound-rotor motors, we progressively short-circuit the external rotor resistors in one, two, or more steps. The number of steps depends upon the size of the machine and the nature of the load.

22-15 Cam switches

Some industrial operations have to be under the continuous control of an operator. In hoists, for example, an operator has to vary the lifting and lowering rate, and the load has to be carefully set down at the proper place. Such a control sequence can be done with cam switches.

Figure 22-30 shows a 3-position cam switch designed for the *forward*, *reverse* and *stop* operation of a 3-phase induction motor. For each position of the knob, some contacts are closed while others are open. This information is given in a table, usually glued to the side of the switch. A cross **X** designates a closed contact, while a blank space is an open contact. In the "forward" position, for example, contacts 2, 4, and 5 are closed and contacts 1 and 3 are open. When the knob is turned to the stop position, all contacts are open. Figure 22-30b shows the shape of the cam that controls the opening and closing of contact 1.

The schematic diagram (Fig. 22-31) shows how to connect the cam switch to a 3-phase motor. The state of the contacts (open or closed) is shown directly on the diagram for each position of the knob. The 3-phase line and motor are connected to the appropriate cam-switch terminals. Note that several jumpers J are also required to complete the connections. The reader should analyze the circuit

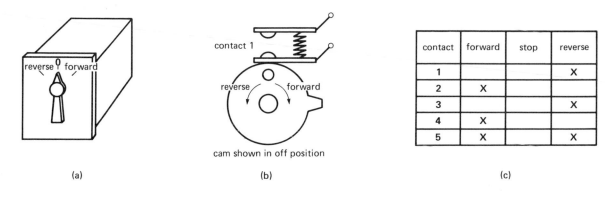

contact	forward	stop	reverse
1			X
2	X		
3			X
4	X		
5	X		X

(a)   (b)   (c)

cam shown in off position

Figure 22-30   Cam switch: a. external appearance; b. detail of the cam controlling contact 1, in the "stop" position; c. table listing the on-off state of the 5 contacts.

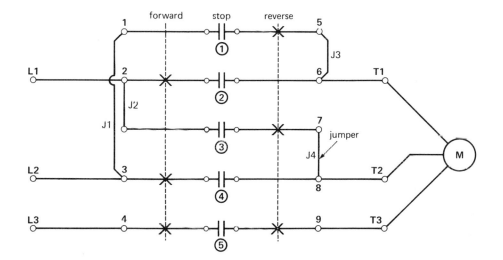

Figure 22-31   Schematic diagram of a cam switch permitting forward-reverse and stop operation of a 3-phase motor.

connections and current flow for each position of the switch.

Some cam switches are designed to carry several hundred amperes, but we often prefer to use magnetic contactors to handle large currents. In such cases, a small cam switch is employed to control the relay coils of the contactors. Very elaborate control schemes can be designed with multicontact cam switches.

QUESTIONS AND PROBLEMS

Practical level

22-1 Name four types of circuit diagrams and describe the purpose of each.

22-2 Without referring to the text, describe the operation of the starter shown in Fig. 22-15b, and state the use of each component.

22-3 Give the symbols for a NO and a NC contact, and for a thermal relay.

22-4 Identify all the components shown in Fig. 22-20a using the list given in Table 22A. Where are contact T and coil A situated physically?

22-5 If the start and stop pushbuttons in Fig. 22-21a are pushed simultaneously, what will happen?

22-6 Referring to Fig. 22-13, if the A_x contact in parallel with the start pushbutton were removed, what effect would it have on the operation of the starter?

22-7 If a dead short-circuit occurs in motor M of Fig. 22-13, which device will open the circuit?

22-8 A small short-circuit between the turns of the stator winding in Fig. 22-13 produces a 50% increase in the line current of one phase. Which device will shut down the motor?

22-9 Under what circumstances is reduced-voltage starting required?

Intermediate level

22-10 A thermal relay having a tripping curve given in Fig. 22-16 has to protect a 40 hp, 575 V, 3-phase, 720 r/min induction motor having a nominal current rating of 40 A. If the relay is set to 40 A, how long will it take to trip if the motor current is (a) 60 A; (b) 240 A?

22-11 a. If the control circuit of Fig. 22-19 is used in place of that shown in Fig. 22-13, show that the motor will start and continue to run if we momentarily press the "start" button.

b. Show that if we press the jog button, the motor only runs for as long as the button is depressed.

22-12 A magnetic contactor can make 3 million "normal" circuit interruptions before its contacts need to be replaced. If an operator jogs the motor so that it starts and stops once per minute, after how many working days (approx) will the contacts have to be replaced, assuming the operator works an 8-hour day?

22-13 a. Referring to Fig. 22-21a, and assuming the motor is initially at rest, explain the operation of the circuit when the "start" button is momentarily depressed.

b. If the motor is running normally, what happens if we momentarily press the "stop" button?

22-14 a. Explain the sequence of events that takes place when the start button in Fig. 22-22 is momentarily depressed, knowing that relay RT is adjusted for a delay of 10 s.

b. With the motor running, explain what happens when the stop button is depressed.

22-15 Referring to Fig. 22-27b, describe the sequence of events that takes place when the start button is momentarily depressed, knowing that relay RT is set for a delay of 5 s. Draw the actual circuit connections, in sequence, until the motor reaches its final speed.

22-16 A 100 hp, 460 V, 3-phase induction motor possesses the characteristics given by curve 1, Figs. 22-24a and 22-24b. The full-load current is 120 A, and the thermal relays are set to this value. If the relay tripping curve is given by Fig. 22-16, calculate the approximate tripping time if the load current suddenly rises to 240 A. (Assume the motor had been running for several hours at full load).

Advanced level

22-17 a. The curves in Fig. 22-24 relate to a 100 hp, 460 V, 1765 r/min, 3-phase, 60 Hz induction motor, whose full-load current is 120 A. Calculate the breakdown torque for curves 1 and 2 [ft·lb].

b. Calculate the torque developed when the resistors are in the circuit and the line current is 480 A [ft·lb].

23

FUNDAMENTAL ELEMENTS OF POWER ELECTRONICS

Electronic systems and controls have gained wide acceptance in power technology; consequently, it has become almost indispensable to know something about power electronics. Naturally, we cannot cover all aspects of this broad subject in a single chapter. Nevertheless, we can explain in simple terms the behavior of a large number of electronic power circuits, including those most commonly used today.

As far as devices are concerned, we shall limit the discussion to diodes and thyristors. They are found in all electronic systems involving the conversion of ac power to dc power and vice versa. Other devices such as triacs, gate turn-off thyristors, and switching power transistors are, of course, important, but their action on a circuit is basically no different from that of a thyristor and its associated switching circuitry. All these devices are basically ultra-high-speed switches, so much so, that much of power electronics can be explained by the opening and closing of circuits at precise instants of time. However, we should not conclude that circuits containing these components and devices are simple - they are not - but the circuits can

be understood without having an extensive background in semiconductor theory.

23-1 Voltage across some circuit elements

The concept of potential levels is very useful in analyzing and understanding electronic power circuits. Consequently, the reader should review Sec. 2-7 before undertaking the study of this chapter.

Let us first look at the voltage levels that appear across some active and passive circuit elements commonly found in electronic circuits. Specifically, we examine sources, switches, resistors, coils and capacitors.

1. **Sources.** By definition, ac and dc voltage sources impose "rigid" potential levels; nothing that happens in a circuit can modify these levels. On the other hand, ac and dc *current* sources deliver a constant current, and the voltage levels must respond accordingly.

2. **Potential across a switch.** When a switch is open (Fig. 23-1), the voltage across its terminals depends exclusively upon the external elements that make up the circuit. On the other hand,

when the switch is closed, the potential level of both terminals must be the same. Thus, if we happen to know the level of terminal 2, then the level of terminal 1 is also known. This simple rule also applies to thyristors and diodes, because they behave like switches.

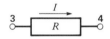

Figure 23-1   Potential across a switch.

3. **Potential across a resistor.** If no current flows in a resistor, its terminals must be at the same potential, because the IR drop is zero (Fig. 23-2). Consequently, if we happen to know the potential level of one of the terminals, the level of the other is also known. On the other hand, if the resistor carries a current I, the IR drop

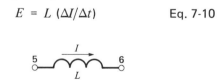

Figure 23-2   Potential across a resistor.

produces a corresponding potential difference between the terminals. For example, if current flows in the direction shown in Fig. 23-2, the potential of terminal 3 is above that of terminal 4, by an amount equal to IR.

4. **Potential across a coil or inductance.** The terminals of a coil are at the same potential only during those moments when the current is not changing. If the current varies, the potential difference is given by

$$E = L (\Delta I/\Delta t) \qquad \text{Eq. 7-10}$$

Figure 23-3   Potential across an inductor.

Thus, if the current in Fig. 23-3 is *increasing* while flowing in the direction shown, the potential of terminal 5 is *above* that of terminal 6,

by an amount equal to $L \Delta I/\Delta t$.

5. **Potential across a capacitor.** The terminals of a capacitor are at the same potential only when the capacitor is completely discharged. Furthermore, the potential difference between the terminals remains constant whenever current I is zero (Fig. 23-4).

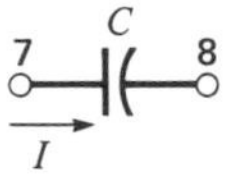

Figure 23-4   Potential across a capacitor.

6. **Initial potential level.** A final rule regarding potential levels is worth remembering. Unless we know otherwise, we assume the following initial conditions:
   a. all currents in the circuit are zero;
   b. all capacitors are discharged.

These hypothetical starting points enable us to analyse the behavior of any circuit from the moment power is applied.

23-2 The diode

A diode is an electronic device possessing two terminals, respectively called anode (A) and cathode (K) (Fig. 23-5). Although it has no moving parts, a diode acts like a high-speed switch whose contacts open and close according to the following rules:

Rule 1. When no voltage is applied across a diode, it acts like an open switch. The circuit is therefore open between terminals A and K (Fig. 23-5a).

Rule 2. If we apply an *inverse voltage* E_2 across the diode, so that the anode is *negative* with respect to the cathode, the diode continues to act as an open switch (Fig. 23-5b).

Rule 3. If a momentary *forward voltage* E_1 is applied across the terminals, so that anode A is slightly *positive* with respect to the cathode, the terminals become short-circuited. The diode acts like a closed switch and a current immediately begins to flow from anode to cathode (Fig. 23-5c). Only a few tenths of a volt are needed to trigger the diode so that its "contacts" close.

While the diode conducts, a small voltage drop appears across its terminals. However, it is always less than 1.5 V, so we can neglect it in most electronic circuits. It is precisely because of this small drop that we can assume the diode is essentially a closed switch when it conducts.

Rule 4. As long as current flows, the diode acts like a closed switch. However, if it stops flowing for even as little as 10 μs, the diode immediately returns to its original open state (Fig. 23-5d). Conduction will only resume when the anode again becomes slightly positive with respect to the cathode (Rule 3).

In conclusion, a diode behaves like a normally open switch whose contacts close as soon as the anode *voltage* becomes slightly positive with re-

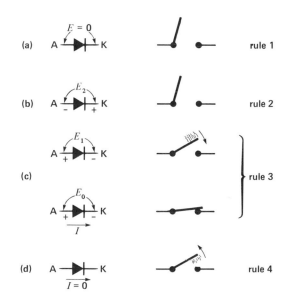

(a) A ▷|K E = 0   rule 1

(b) A ▷|K E_2 rule 2

(c) A ▷|K E_1
    A ▷|K E_0 I   } rule 3

(d) A ▷|K I = 0   rule 4

Figure 23-5   Basic rules governing diode behavior.

spect to the cathode. Its contacts only reopen when the *current* (not the voltage) has fallen to zero. This simple rule is crucially important to an understanding of circuits involving diodes and thyristors.

Symbol for a diode. The symbol for a diode bears an arrow that indicates the direction of conventional current flow when the diode conducts.

23-3   Principal characteristics of a diode

Peak inverse voltage. A diode can withstand only so much inverse voltage before it breaks down. The *peak inverse voltage* (PIV) ranges from 50 V to 2000 V, depending on the construction. If we exceed the rated PIV, the diode begins to conduct in reverse and, in many cases, it is immediately destroyed.

Maximum average current. There is also a limit to the current a diode can carry. The maximum current may range from a few hundred milliamperes to over 2000 A, depending on the construction and size of the diode.

Maximum temperature. A diode must never operate beyond its rated temperature. Most silicon diodes can operate satisfactorily provided the internal temperature lies between $-50°$C and $+200°$C. The temperature of a diode can change very quickly owing to its small dimensions and mass. To improve heat transfer, diodes are usually mounted on thick metallic supports, called *heat sinks*. Furthermore, in large installations, the diodes may be cooled by fans, by oil, or by a continuous flow of deionized water. Table 23A gives the specifications of some typical diodes. Figure 23-6 shows a range of low to very high power diodes.

Diodes have many applications and, to illustrate their properties, we now analyze a few simple circuits.

23-4   Battery charger with series resistor

The circuit of Fig. 23-7a represents a battery charger. Transformer T, connected to a 120 V ac supply, furnishes a sinusoidal secondary voltage having

TABLE 23A			PROPERTIES OF SOME TYPICAL DIODES					
relative	I_0	E_0	I_{cr}	E_2	I_2	T_J	d	l
power	A	V	A	V	mA	$^\circ$C	mm	mm
low	1	0.8	30	1000	0.05	175	3.8	4.6
medium	12	0.6	240	1000	0.6	200	11	32
high	100	0.6	1 600	1000	4.5	200	25	54
very high	1000	1.1	10 000	2000	50	200	47	26

I_0 - average dc current

E_0 - voltage drop corresponding to I_0

I_{cr} - peak value of surge current for 1 cycle

E_2 - peak inverse voltage

I_2 - reverse leakage current corresponding to E_2

T_J - maximum junction temperature (inside the diode)

d - diameter

l - length

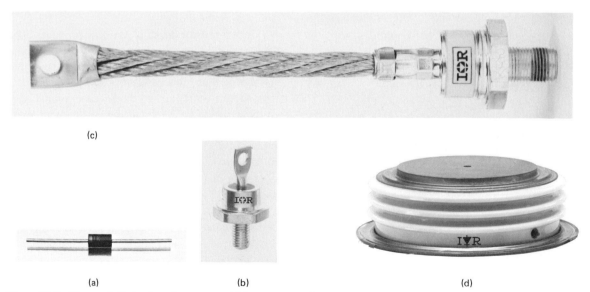

(c)

(a)　　　　　　　　(b)　　　　　　　　(d)

Figure 23-6　Range of diodes from low to very high power capacity.

a. Average current: 4 A; PIV: 400 V; body length: 10 mm; diameter: 5.6 mm.

b. Average current: 15 A; PIV: 500 V; stud type; length less thread: 25 mm; diameter: 17 mm.

c. Average current: 500 A; PIV: 2000 V; length less thread: 244 mm; diameter: 40 mm.

d. Average current: 2600 A; PIV: 2500 V; Hockey Puk; distance between pole faces: 35 mm; overall diameter: 98 mm.
(Photos courtesy of International Rectifier)

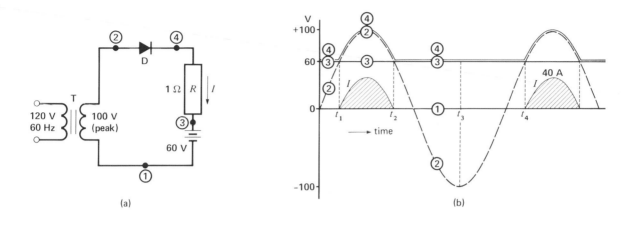

Figure 23-7   a. Simple battery charger circuit.   b. Corresponding voltage and current waveforms.

a peak of 100 V. A 60 V battery, a 1 Ω resistor, and a diode D are connected in series across the secondary.

To explain the operation of the circuit, let us choose point 1 as the reference terminal. The potential of this terminal is therefore a straight horizontal line. The potential of terminal 2 swings above and below point 1, according to whether 2 is positive or negative with respect to 1. The level of terminal 3 is always 60 V above terminal 1, because the battery voltage is constant. The potential levels are shown in Fig. 23-7b.

Circuit analysis

a. Prior to $t = 0$, we assume all currents are zero. The potential of point 4 is therefore at the same level as point 3.

b. During the interval from 0 to t_1, anode 2 is negative with respect to cathode 4; consequently, the diode cannot conduct (Rule 2);

c. At instant t_1, terminal 2 becomes positive with respect to 4, and the diode begins to conduct (Rule 3). From this moment on, the diode acts like a closed switch. Consequently, point 4 follows point 2.

d. From t_1 to t_2, the current in the circuit is given by:

$$I = E_{43}/1\ \Omega$$

The current reaches a peak of 40 A when $E_{43} = +40$ V. As long as current flows, the diode behaves like a closed switch;

e. At instant t_2, the current is zero, and the diode immediately "opens" the circuit (Rule 4). From this moment on, point 4 follows point 3.

f. From t_2 to t_4, point 2 is negative with respect to point 3. Because point 4 follows point 3 (no IR drop in the resistor), the PIV across the diode reaches a maximum of 160 V, at instant t_3;

g. Finally, from t_4 on, the cycle (a to f) repeats itself.

This circuit produces a pulsating current that always flows into the positive terminal of the battery. Consequently, the latter receives energy and progressively charges up.

23-5 Battery charger with series inductor

The current flow in the battery charger of Fig. 23-7 is limited by resistor R. Unfortunately, this produces large I^2R losses and a corresponding poor efficiency. We can get around this problem by replacing the resistor by an inductor, as shown in Fig. 23-8a. Let us analyze the operation of this circuit.

a. As in the previous example, the diode begins to

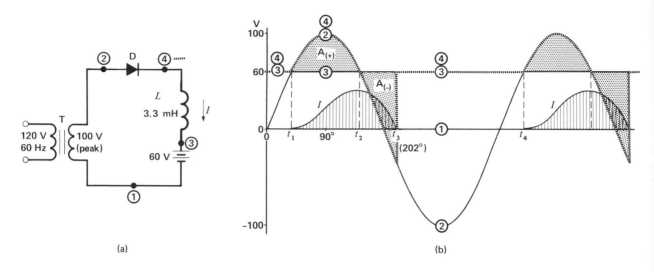

Figure 23-8   a. Battery charger using a series inductor.   b. Corresponding voltage and current waveforms.

conduct at instant t_1 when anode 2 becomes positive with respect to cathode 4. From this moment on, point 4 follows point 2, and voltage E_{43} appears across the inductor (Fig. 23-8b). The latter begins to "accumulate" volt-seconds and the current increases gradually, until it reaches a maximum given by

$$I_{max} = A+/L \qquad \text{Eq. 7-11}$$

where $A+$ = dotted area between t_1 and t_2 [V·s]
   L = inductance [H]

Note that the current reaches its peak at instant t_2 whereas it was zero at this moment, when a resistor was used. This is consistent with the fact that current through an inductor lags behind the voltage.

b. The current decreases between t_2 and t_3, becoming zero at t_3 when $A-$ is equal to $A+$.

c. As soon as the current is zero, the diode "opens" the circuit, whereupon point 4 jumps to the level of point 3. It stays at this level until instant t_4, where the whole cycle repeats itself.

This is an interesting example of the use of an inductor to store and release electrical energy. During the interval from t_1 to t_2, the inductor stores energy and, from t_2 to t_3, it returns it again to the circuit (See Sec. 7-12).

Example 23-1:
The coil in Fig. 23-8 has an inductance of 3.3 mH. Calculate the peak current if the supply frequency is 60 Hz.

Solution:

a. To calculate the peak current, we must find the value of area $A+$. This can be done by integral calculus, but we shall employ a much simpler graphical method. Thus, referring to Fig. 23-8c, we have redrawn the voltage levels using graph paper. The voltage cycle is divided into 24 equal parts, each representing an interval

$$\Delta t = (1/24) \times (1/60) = 1/1440 \text{ s}$$

Similarly, the ordinates are scaled off in 10 V intervals. Consequently, each small square represents an area of $(1/1440) \times 10 = 6.944$ mV·s.

b. By counting squares, we find that $A+$ contains 19 squares; consequently, its area corresponds to:

$$A+ = 19 \times 6.944 = 132 \text{ mV·s} = 0.132 \text{ V·s}$$

c. The peak current is:

$$I_{max} = A+/L = 0.132/0.0033 = 40 \text{ A}$$

The peak current is the same with an inductor

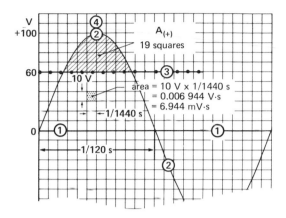

Figure 23-8   c. See Example 23-1.

of 3.3 mH as it was for a resistor of 1 Ω. However, the big advantage of the inductor is that it has essentially no losses. The conversion of ac to dc power is therefore much more efficient.

23-6   Single-phase bridge rectifier

The circuit of Fig. 23-9a enables us to rectify both the positive and negative half cycles of an ac source, to supply a dc load R. The four diodes together make up what is called a *single-phase bridge rectifier*.

The circuit operates as follows. When source voltage E_{12} is positive, terminal 1 is positive with respect to terminal 2 and current i_a flows through R by way of diodes A1 and A2. Consequently, point 3 follows point 1 and point 4 follows point 2 during this conduction interval. Conduction ceases when i_a falls to zero at instant t_1 (Fig. 23-9b). The polarity then reverses and E_{21} becomes positive, meaning that terminal 2 is positive with respect to terminal 1. Current i_b now flows through R in the same direction as before, but this time by way of diodes B1 and B2. Consequently, point 3 now follows point 2 while point 4 follows point 1. Voltage E_{34} across the load is therefore composed of a series of half-cycle sine waves that are always positive. The voltage pulsates between zero and a maximum value E_m equal to the peak voltage of the source. The average value of this rectified voltage is given by:

$$E_d = 0.90\,E \qquad (23\text{-}1)$$

where

E_d =   dc voltage of a single-phase bridge rectifier [V]

E =   effective value of the ac line voltage [V]

0.90 =   constant [exact value = $2\sqrt{2}/\pi$]

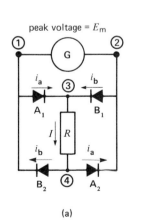

(a)

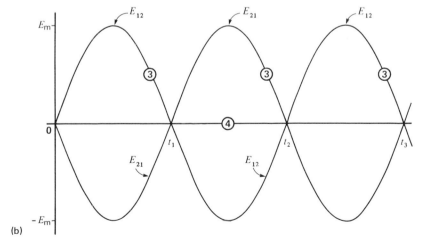

(b)

Figure 23-9   a. Single-phase bridge rectifier.   b. Voltage levels.

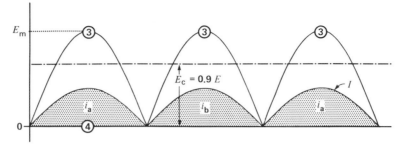

Figure 23-9   c. Voltage and current waveforms.

Referring to Fig. 23-9b, the potential levels of terminals 1 and 2 are treated as reference potentials on alternate half cycles. Thus, terminal 2 is the reference during the first half cycle, while terminal 1 is the reference during the second half cycle, and so on. Figure 23-9c shows the rectified voltage and current of the load.

Load voltage E_{34} contains an ac component whose fundamental frequency is twice the line frequency. Furthermore, the so-called peak-to-peak *ripple* is equal to E_m.

In the case of a resistive load, the current has the same waveshape as the voltage; its average, or dc value, is given by:

$$I_d = E_d/R$$

Example 23-2:
The ac source in Fig. 23-9a has an effective voltage of 120 V, 60 Hz. The load draws a dc current of 20 A. Calculate:
a. the dc voltage across the load;
b. the average dc current in each diode;

Solution:
a. The dc voltage across the load is given by Eq. 23-1:

$$E_d = 0.90\,E$$
$$= 0.90 \times 120$$
$$= 108\ V$$

b. The dc current in the load is 20 A, but the diodes only carry the current on alternate half cycles. Consequently, the average dc current in each diode is:

$$I = I_d/2 = 20/2 = 10\ A$$

23-7 Filters

The rectifier circuits we have studied so far produce pulsating voltages and currents. In some types of loads, we cannot tolerate such pulsations, and *filters* must be used to smooth out the valleys and peaks. The basic purpose of a dc filter is to produce a smooth power flow into a load. Consequently, a filter must absorb energy whenever the dc voltage or current tends to increase, and it must release energy whenever the voltage or current tends to fall. In this way, the filter tends to maintain a constant voltage and current in the load.

The most common filters are inductors and capacitors. Inductors store energy in their magnetic field. They tend to maintain a constant current; consequently, they are placed *in series* with the load (Fig. 23-10a). Capacitors store energy in their electric field. They tend to maintain a constant voltage; consequently, they are placed *in parallel* with the load (Fig. 23-10b).

The filtering action improves as we increase the amount of energy stored in the filter. In the case of an inductor, we obtain good current smoothing (peak-to-peak ripple less than 5% of the dc current) provided that:

$$\boxed{W_L > P/f} \qquad (23\text{-}2)$$

where
W_L = dc energy stored in the smoothing inductor [J]
P = dc power absorbed by the load [W]
f = frequency of the source [Hz]

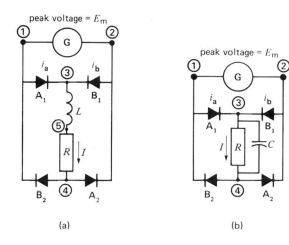

peak voltage = E_m

(a)

(b)

Figure 23-10   a. Rectifier with inductive filter.

b. Rectifier with capacitive filter.

The current in Fig. 23-10a is much more constant than in Fig. 23-9a. The voltage between terminals 3 and 4 still pulsates badly, but it is very smooth across the load (Fig. 23-11). The dc voltage across the load is again given by Eq. 23-1.

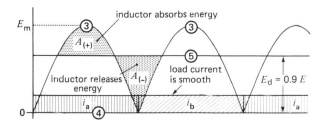

Figure 23-11   Current and voltage waveforms with inductive filter.

Bridge rectifiers provide dc current for relays, electromagnets, motors, and many other magnetic devices. In most cases, the self-inductance of the coils is sufficient to provide good filtering. Thus, although the voltage across the coils may pulsate very strongly, the dc current through them is smooth. Consequently, the magnetic field pulsates very little.

Example 23-3:

We wish to build a 135 V, 20 A dc power supply using a single-phase bridge rectifier and an inductive filter. The peak-to-peak current ripple should be less than 5%. If a 60 Hz ac source is available, calculate:

a. the effective value of the ac voltage;

b. the energy stored in the inductor;

c. the inductance of the inductor;

d. the peak-to-peak current ripple.

Solution:

a. The ac voltage is given by Eq. 23-1:

$$E_d = 0.9 E$$
$$135 = 0.9 E$$
$$\therefore E = 150 \text{ V}$$

b. The dc power output of the rectifier is:

$$P = E_d I_d$$
$$= 135 \times 20$$
$$= 2700 \text{ W}$$

The energy to be stored in the inductor (or choke) is given by:;

$$W_L > P/f \qquad \text{Eq. 23-2}$$
$$W_L > 2700/60$$
$$> 45 \text{ J}$$

Consequently, to keep the current ripple within 5%, the inductor must store at least 45 J in its magnetic field.

c. The inductance of the choke can be calculated from:

$$W_L = \frac{1}{2} L I_d^2 \qquad \text{Eq. 7-12}$$

$$45 = \frac{1}{2} L (20)^2$$

$$\therefore L = 0.225 \text{ H}$$

d. The peak-to-peak ripple is about 5% of the dc current:

$$I_{\text{peak-to-peak}} = 0.05 \times 20 = 1 \text{ A}$$

The dc output current therefore pulsates between 19.5 A and 20.5 A.

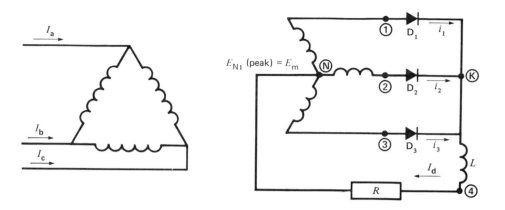

Figure 23-12   Three-phase, 3-pulse rectifier with inductive filter.

23-8 Three-phase 3-pulse rectifier

The simplest 3-phase rectifier is composed of three diodes connected in series with the secondaries of a 3-phase, delta-wye transformer (Fig. 23-12). A very large inductance L is connected in series with the load, so that current I_d remains ripple-free. Although the load is represented by a resistance R, in reality it is always a useful energy-consuming device, and not a heat-dissipating resistor. Thus, the load may be a dc motor, a large magnet or an electroplating bath. This simple rectifier has some serious drawbacks, but it provides a good introduction to 3-phase rectifiers in general. We now analyse its behavior.

1. **Voltage across the load.** Choosing the transformer neutral as a reference point, let us assume that the secondary voltages follow the levels 1, 2, 3, shown in Fig. 23-13. These potential levels are fixed by the ac source and they reach a peak value E_m.
   Before we energize the circuit, points K, 4, N are at the same level because I is zero. However, the moment we apply power ($t = 0$) the potential of point 1 suddenly becomes positive with respect to K. This immediately initiates conduction in diode D1 (Sec. 23-2, Rule 3). Current i_1 increases rapidly, attaining a final value I which depends upon load R. During this inter-

val, K is at the same level as point 1 because the diode is conducting.

As points K and 1 move together in time, they eventually reach a critical moment, corresponding to an angle θ_0 of 60° (Fig. 23-13). The moment is critical because immediately after, terminal 2 becomes positive with respect to K and 1. According to Rule 3, this initiates conduction in diode D2, so that *it* begins to carry current I. At the same time that conduction starts in D2, it ceases in diode D1. Consequently, beyond 60°, point K follows the level of point 2.

The sudden switchover from one diode to another is called *commutation*. When the switchover takes place automatically (as it does in our example), it is called *natural* commutation. Commutation from one diode to another does not really take place instantaneously, as we have indicated. Owing to transformer leakage reactance, the current gradually increases in in diode D2 as it decreases in diode D1. This gradual transition continues until all the load current is carried by diode D2. However, the commutation period is very short (typically less than 2 ms) and, for our purposes, we can assume it occurs instantaneously.

The next critical moment occurs at 180°, because terminal 3 then becomes positive with re-

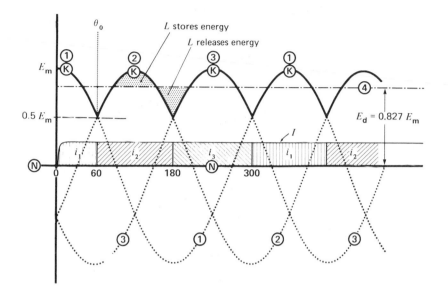

Figure 23-13   Voltage and current waveforms in a 3-phase, 3-pulse rectifier.

spect to K. Commutation again takes place as the load current switches from diode D2 to diode D3. Point K therefore follows the positive peaks of waves 1, 2 and 3, and each diode carries the full load current for equal intervals of time (120°). The diode currents have a rectangular waveshape composed of positive current intervals of 120° followed by zero current intervals of 240°.

Voltage E_{KN} across the load and inductor pulsates between $0.5\,E_m$ and E_m. The ripple voltage is therefore smaller than that produced by a single-phase bridge rectifier (Fig. 23-11). Furthermore, the fundamental ripple frequency is three times the supply frequency, which makes it easier to achieve good filtering. The dc voltage across the load is given by:

$$\boxed{E_d = 0.675\,E} \qquad (23\text{-}4)$$

where

E_d = average or dc voltage of a 3-pulse rectifier [V]

E = effective ac line voltage [V]

0.675 = a constant [exact value = $3/(\pi\sqrt{2})$]

If we reverse the diodes in Fig. 23-12, the rectifier operates the same way, except that the load current reverses. Voltage E_{KN} becomes negative, and point K follows the negative peaks of waves 1, 2 and 3.

2. **Line currents.** Currents i_1, i_2, i_3, which flow in the diodes, also flow in the secondary windings of the transformer. These currents have a chopped rectangular waveshape which is quite different from the sinusoidal currents we are familiar with. Furthermore, the currents flow for only 1/3 of the time in a given winding. Owing to this intermittent flow, the nominal rating of the transformer cannot be harnessed to deliver an equivalent amount of dc power. For example, to deliver 100 kW of dc power in Fig. 23-12, we would have to install a transformer having a rating of 135 kVA. The utilization factor of the transformer is said to be 100/135 = 0.74 or 74 percent.

The chopped secondary currents are reflected into the primary windings, with the result that the transmission line currents feeding the transformer also change very sharply. The sudden jumps in currents I_a, I_b and I_c produce rapid

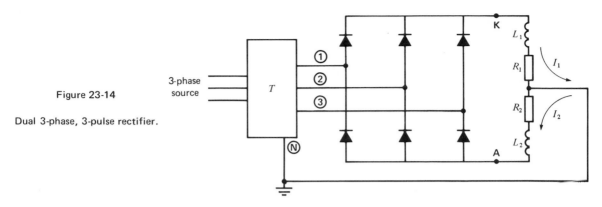

Figure 23-14

Dual 3-phase, 3-pulse rectifier.

fluctuations in the magnetic field surrounding the transmission line. These fluctuations can induce substantial voltages and noise in nearby telephone lines.

Owing to these drawbacks, we try to design rectifiers so that the line currents change more gradually and so that the transformer windings carry current for more than one-third of the time. This is done by using 3-phase, 6-pulse rectifiers.

23-9 Three-phase, 6-pulse rectifier*

Consider the circuit of Fig. 23-14 in which a transformer T (identical to the one shown in Fig. 23-12), supplies power to 6 diodes and their associated loads R_1 and R_2. The upper set of diodes together with inductor L_1 and load R_1 are identical to the 3-phase, 3-pulse rectifier we have just studied. Load current I_1 flows in the neutral line, as shown. The lower set of diodes, together with R_2 and L_2, also constitute a 3-phase, 3-pulse rectifier. The corresponding load current I_2 flows in the neutral as shown. The two 3-phase rectifiers operate quite independently of each other, K following the positive peaks of points 1, 2, 3, while A follows the negative peaks. All diodes conduct during 120° intervals.

If we make $R_1 = R_2$, then $I_1 = I_2$ and the current in the neutral becomes zero. Consequently, we can remove the neutral conductor, yielding the circuit

* Also called 3-phase bridge rectifier.

of Fig. 23-15. The two loads and the two inductors are simply combined into one, shown as R and L, respectively. The 6 diodes constitute what is called a 3-phase, 6-pulse rectifier. It is called 6-pulse because the currents flowing in the 6 diodes all start at different times. However, each diode still conducts for only 120°.

The line currents supplied by the transformer are given by Kirchhoff's law:

$$I_a = i_1 - i_4 \qquad I_b = i_2 - i_5 \qquad I_c = i_3 - i_6$$

They consist of identical ac pulses that are out of phase by 120° (Fig. 23-16). They flow for 2/3 of the time in the secondary windings, with the result that the utilization factor of the transformer rises to 95%. Consequently, a 100 kW dc load requires a transformer having a rating of only 100/0.95 = 105 kVA.

The average dc output voltage is twice that of a 3-phase, 3-pulse rectifier, and its value is given by:

$$\boxed{E_d = 1.35\,E} \qquad (23\text{-}5)$$

where

E_d = average or dc voltage of a 6-pulse rectifier [V]

E = effective line voltage [V]

1.35 = a constant [exact value = $3\sqrt{2}/\pi$]

It is much easier to visualize the shape of the output voltage E_{KA}, by using A as a reference point. Thus, in Fig. 23-17, we show the line volt-

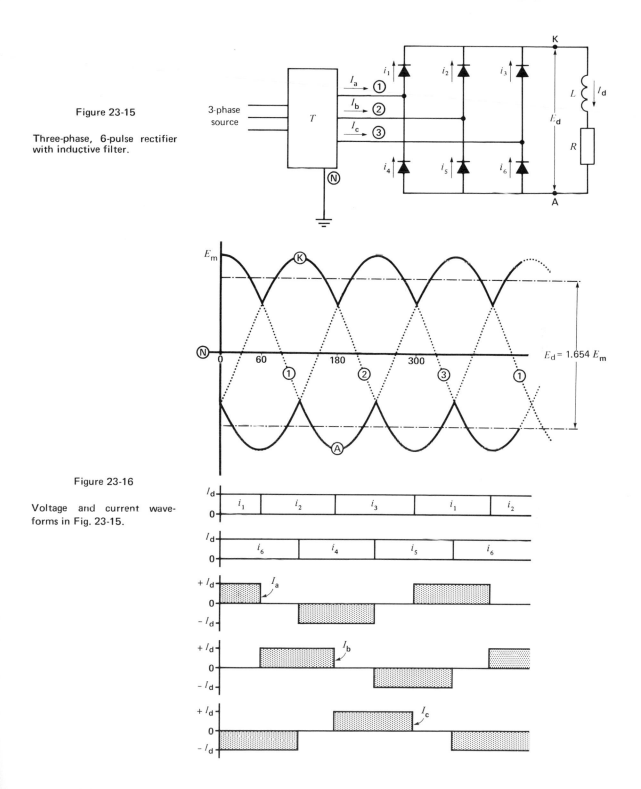

Figure 23-15

Three-phase, 6-pulse rectifier with inductive filter.

Figure 23-16

Voltage and current waveforms in Fig. 23-15.

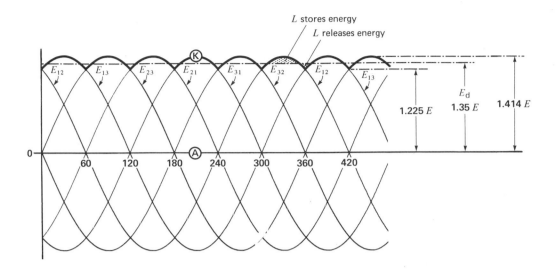

Figure 23-17   Another way of showing E_{KA} using line voltage potentials.

ages rather than the line-to-neutral voltages used in Fig. 23-16. The level of K follows the tops of the successive sine waves. The output voltage fluctuates between $1.414\,E$ and $1.225\,E$, where E is the effective value of the line voltage.* The average value of E_{KA} is $1.35\,E$, as given by Eq. 23-5.

The peak-to-peak ripple is only $(1.414 - 1.225)$ $E = 0.189\,E$ and the fundamental ripple frequency is six times the line frequency. Consequently, the ripple is much easier to filter. In effect, we obtain good inductor filtering (peak-to-peak current ripple less than 5%) provided that:

$$\boxed{W_L > 0.03\,P/f} \qquad (23\text{-}6)$$

where
W_L = dc energy stored in the inductor [J]
P = dc power absorbed by the load [W]
f = frequency of the source [Hz]

Figure 23-17 shows that the inductor stores energy whenever the rectifier voltage exceeds the average value E_d. This energy is then released during the brief interval when the rectifier voltage

* $\sqrt{2}\,E = 1.414\,E$; $\sqrt{2}\,E \cos 30 = 1.225\,E$.

is less than E_d.

The PIV across each diode is equal to the peak value of the line voltage, or $\sqrt{2}\,E$.

The 3-phase, 6-pulse rectifier is a big improvement over the 3-phase, 3-pulse rectifier. It constitutes the basic building block of most large rectifier installations.

Another way of looking at the 3-phase bridge rectifier is to consider the diodes to be in a box (Fig. 23-18). The box is fed by three ac lines and it has two output terminals K and A. The diodes act like automatic switches that successively connect these terminals to the ac lines. The connections can be made in six distinct ways, as shown in Fig. 23-18. It follows that the output voltage E_{KA} is composed of segments of the ac line voltages. This is why we draw line voltages in Fig. 23-17 instead of line-to-neutral voltages.

Each dotted connection in Fig. 23-18 represents a diode that is conducting. The successive 60-degree intervals correspond to those in Fig. 23-17. Thus, in the interval from $300°$ to $360°$, currents i_1 and i_5 are flowing. Consequently, diodes D1 and D5 are conducting. It follows from Fig. 23-15 that K is connected to line 1 while A is connected to line

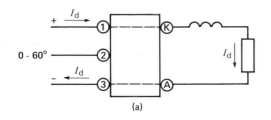

0 - 60°

(a)

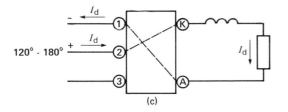

60° - 120°

(b)

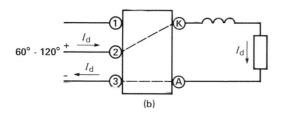

120° - 180°

(c)

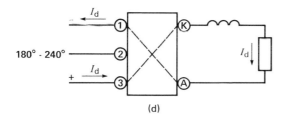

180° - 240°

(d)

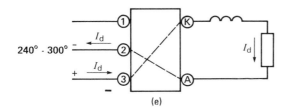

240° - 300°

(e)

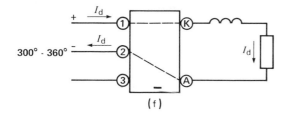

300° - 360°

(f)

2. These connections correspond exactly to those shown in Fig. 23-18f.

Because the diode voltage drop is small, we can assume that each dotted line represents a loss-free connection. The dc power absorbed by the load must therefore be equal to the active power drawn from the 3-phase source.

Example 23-4:

A 3-phase bridge rectifier has to supply power to a 360 kW, 240 V dc load. If a 600 V, 3-phase, 60 Hz feeder is available, calculate:

a. power and voltage rating of the 3-phase transformer;

b. dc current per diode;

c. PIV across each diode;

d. peak-to-peak ripple in the output voltage and its frequency.

Solution:

a. 1 Transformer rating $= \dfrac{P_{dc}}{\text{utilization factor}}$

$= 360/0.95 = 378 \text{ kVA}$

a. 2 Secondary line voltage is:

$E = E_d/1.35 = 240/1.35 = 177 \text{ V}$

Thus, a 3-phase, 378 kVA transformer having a line voltage ratio of 600 V/177 V would be satisfactory. The primary and secondary windings may be connected either in wye or in delta.

b. 1 dc load current $= 360 \text{ kW}/240 \text{ V} = 1500 \text{ A}$

b. 2 dc current per diode $= 1500/3 = 500 \text{ A}$

b. 3 peak current in each diode $= 1500 \text{ A}$

c. PIV across each diode

$= \sqrt{2} E = 1.414 \times 177 = 250 \text{ V}$

d. 1 The output voltage E_{KA} fluctuates between $1.225 E$ and $1.414 E$ (Fig. 23-17). In other words, the voltage fluctuates between

$E_{min} = 1.225 \times 177 = 217 \text{ V}$

and

Figure 23-18 Successive diode connections between the 3-phase input and dc output terminals of a 3-phase, 6-pulse rectifier.

$$E_{max} = 1.414 \times 1.77 = 250 \text{ V}$$

The peak-to-peak ripple is therefore:

$$E_{peak\text{-}to\text{-}peak} = 250 - 217 = 33 \text{ V}$$

d. 2 Fundamental ripple frequency

$$= 6 \times 60 = 360 \text{ Hz}$$

Example 23-5:

a. Calculate the inductance of the smoothing choke required in Example 23-4, if the peak-to-peak ripple in the current is not to exceed 5%.

b. Does the presence of the choke modify the peak-to-peak ripple in the output voltage E_{KA}?

Solution:

a. Using Eq. 23-6, we have:

$$W_L > 0.03 \, P/f$$
$$> 0.03 \times 360\,000/60$$
$$> 180 \text{ J}$$

Consequently, the inductor must store 180 J in its magnetic field. The inductance is found from:

$$W_L = \frac{1}{2} LI^2{}_d$$

$$180 = \frac{1}{2} L \, (1500)^2$$

$$\therefore L = 1.6 \times 10^{-4}$$

$$= 0.16 \text{ mH}$$

b. No, the choke does not affect the voltage ripple.

23-10 The thyristor*

A thyristor may be considered to be a diode whose conduction can be controlled. Like a diode, a thyristor possesses an anode and a cathode, plus a third terminal called gate (Fig. 23-19). If the gate is connected to the cathode, the thyristor will not conduct, even if the anode is positive†. The thyristor is said to be *blocked* (Fig. 23-20a). To initiate conduction, two conditions have to be met:

a. the anode must be positive;

b. a current I_g must flow into the gate for a few microseconds. In practice, the current is injected

Figure 23-19   Symbol of a thyristor or SCR.

by applying a short, positive voltage pulse E_g to the gate (Fig. 23-20b).

(a)   (b)

Figure 23-20   a. A thyristor does not conduct when the gate and cathode are connected.

b. A thyristor conducts when the anode is positive and a current pulse is injected into the gate.

As soon as conduction starts, the gate loses all further control. Conduction will only stop when anode current I falls to zero, after which the gate again exerts control.

Basically, a thyristor behaves the same way a diode does except that the gate enables us to initiate conduction precisely when we want it to. This seemingly slight advantage is actually of profound

---

* Thyristor is a generic term that applies to all four-layer controllable semiconductor devices. However, we use it in this book to refer specifically to the reverse-blocking triode thyristor, commonly called SCR (semiconductor controlled rectifier). This is in response to a seemingly general trend in the literature to use the terms SCR and thyristor interchangeably.

† To simplify the wording, we adopt the following terms:
1. When the anode is positive with respect to the cathode, we simply say the anode is positive;
2. When the gate is positive with respect to the cathode, we simply say the gate is positive.

TABLE 23B			PROPERTIES OF SOME TYPICAL THYRISTORS						
relative	I_1	I_{cr}	E_2	E_P	I_G	E_G	T_J	d	l
power	A	A	V	V	mA	V	$°$C	mm	mm
medium	8	60	500	– 10	50	2.5	105	11	33
high	110	1 500	1200	– 5	50	1.25	125	27	62
very high	1200	10 000	1200	– 20	50	1.5	125	58	27

I_1 - maximum effective current during conduction d - diameter

I_{cr} - peak value of surge current for 1 cycle l - length

E_2 - peak inverse anode voltage

E_P - peak inverse gate voltage

E_G - positive gate voltage to initiate conduction

I_G - gate current corresponding to E_G

T_J - maximum junction temperature

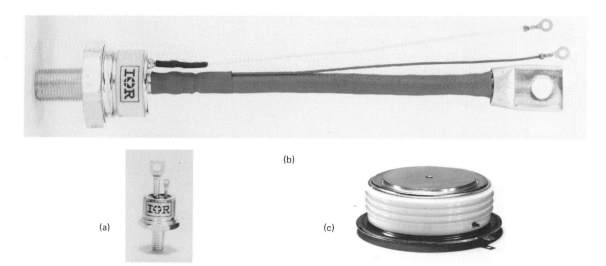

Figure 23-21   Range of SCRs from medium to very high power capacity.

a. Average current: 50 A; voltage: 400 V; length less thread: 31 mm; diameter: 17 mm.

b. Average current: 285 A; voltage: 1200 V; length less thread: 244 mm; diameter: 37 mm.

c. Average current: 1000 A; voltage: 1200 V; distance between pole faces: 27 mm; overall diameter: 73 mm. *(Photos courtesy of International Rectifier)*

importance. It enables us not only to convert ac power into dc power, but also to do the reverse: convert dc power into ac power. Thanks to the development of reliable SCRs we are witnessing a fundamental change in the control of large blocks of power. Table 23B lists some of the properties of typical thyristors. See also Fig. 23-21.

23-11 Principles of gate firing

Consider Fig. 23-22a in which a thyristor and a resistor are connected in series across an ac source. A number of short positive pulses E_g is applied to the gate, of sufficient amplitude to initiate conduction. These pulses may be generated by a manual switch or by a special electronic control circuit.

Referring to Fig. 23-22b, the gate pulses occur at angles θ_1, θ_2, θ_3, θ_4 and θ_5. Table 23C explains how the circuit reacts to these pulses.

To sum up, we can control the current in an ac circuit by delaying the gate pulses with respect to the start of each positive half-cycle. If the pulses occur at the beginning of each half-cycle, conduction lasts for 180°, and the thyristor behaves like an ordinary diode. On the other hand, if the pulses are delayed, say by 150°, current only flows during the remaining 30° of each half-cycle.

23-12 Power gain of a thyristor

When a voltage pulse is applied to the gate, a certain gate current flows. Because the pulses last only a few microseconds, the average power supplied to the gate is very small, compared to the average power supplied to the load. The ratio of the two powers, called power gain, may exceed one million. Thus, a gate input of only 1 W may control a load of 1000 kW.

An SCR does not, of course, have the magical property of turning one watt into a million watts. The large power actually comes from an appropriate power source, and the SCR only serves to control the power flow. Thus, in the same way that a small power input to the accelerator of an automobile produces a tremendous increase in motive

power, so does a small input to the gate of an SCR produce a tremendous increase in electrical power.

23-13 Current interruption and forced commutation

A thyristor ceases to conduct and the gate regains control only after the anode current falls to zero. The current may cease flowing quite naturally (as it did in Fig. 23-22) or we can force it to zero artificially. Such *forced commutation* is required on some circuits where the anode current has to be interrupted at a specific instant.

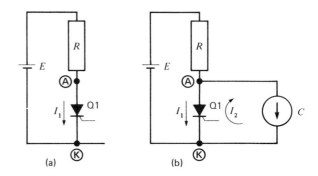

Figure 23-23   a. Thyristor connected to a dc source.

b. Forced commutation.

Consider Fig. 23-23a in which a thyristor and a resistor are connected in series across a dc source E. If we apply a single positive pulse to the gate, the resulting dc load current I_1 will flow indefinitely. However, we can stop conduction in the SCR in one of 3 ways:

1. momentarily reduce the supply voltage E to zero;
2. open the load circuit by means of a switch;
3. force the anode current to zero for a brief period.

The first two solutions are trivial, so let us examine the third method. In Fig. 23-23b, a current source C delivering a current I_2 is connected in parallel with thyristor Q1. As we increase I_2, the net current $(I_1 - I_2)$ flowing in the thyristor decreases. However, the thyristor continues to conduct, with the result that the current flowing in the resistor is

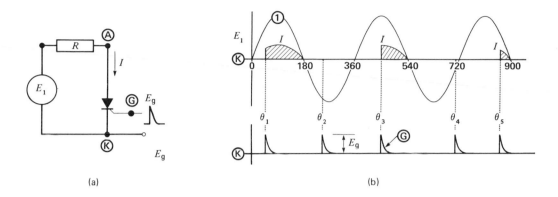

Figure 23-22   a. Thyristor connected to an ac source.   b. Thyristor behavior depends on the timing of the gate pulses
(see description below).

TABLE 23C	DESCRIPTION OF THYRISTOR BEHAVIOR (see Fig. 23-22)
angle or time interval	**Explanation of circuit operation**
zero to θ_1	Although the anode is positive, conduction is impossible because the gate voltage is zero. The thyristor behaves like an open switch.
angle θ_1	Conduction starts because both the anode and gate are positive.
θ_1 to 180°	Conduction continues even though the gate voltage has fallen to zero. Gate pulses have no further effect once the thyristor conducts. The anode to cathode voltage drop is less than 1.5 V; consequently, we can consider that the anode and cathode are shorted. The thyristor behaves like a closed switch.
angle 180°	The thyristor current is zero, conduction ceases and the gate regains control.
180° to 360°	Conduction is impossible because the anode is negative. Although the gate is triggered at angle θ_2, it produces no effect. The thyristor experiences an inverse voltage during this half cycle.
360° to 540°	Conduction starts at θ_3 and ceases again as soon as the current is zero. The gate pulse is delayed more than during the first positive half-cycle. Consequently, the anode current flows for a shorter time.
720° to 900°	Conduction starts at angle θ_5, but the resulting anode current is very small because of the long delay in firing the gate.

unchanged. But, if we increase I_2 until it is equal to I_1, the thyristor ceases to conduct, and the gate regains control. In practice, I_2 is a brief current pulse, usually supplied by triggering a second thyristor. For example, in Fig. 23-24, a capacitor, initially charged as shown, is suddenly discharged by triggering the gate of thyristor Q2. Discharge cur-

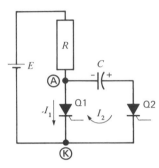

Figure 23-24   A discharging capacitor can force-commutate a thyristor.

rent I_2 immediately cancels the current in Q1; consequently, it stops conducting. Current now flows through resistor R, capacitor C, and thyristor Q2. The capacitor quickly charges up and when its voltage is close to E, the charging current becomes so small that Q2 also stops conducting. Thus, current ceases to flow in the load shortly after Q2 is triggered.

This type of forced commutation, using a commutating capacitor, is employed in some converters* that generate their own frequency.

23-14 Basic thyristor power circuits

Thyristors are used in many different ways. However, in power electronics, seven basic circuits cover about 90 percent of all industrial applications. These circuits, and some of their applications, are listed in Table 23D.

---

* A converter is any device that converts power of one frequency into power of another frequency. A converter may be a rectifier, an inverter, a cycloconverter or even a rotating machine.

To explain the principle of operation of these basic circuits, we shall use single-phase sources. In practice, 3-phase sources are mainly used, but single-phase examples are less complex, and they enable us to focus attention on the essential principles involved.

23-15 Controlled rectifier supplying a passive load (Circuit 1)

Figure 23-25a shows a resistive load and a thyristor connected in series across a single-phase source. The source produces a sinusoidal voltage having a peak value E_m. The gate pulses are synchronized with the line frequency and, in our example, they are delayed by an angle of $90°$. Conduction is therefore initiated every time the ac voltage reaches its maximum positive value. Based upon explanations given in Sec. 23-10, it is obvious that current will flow for $90°$.

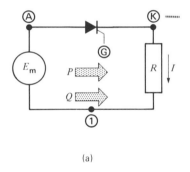

(a)

Figure 23-25   a. SCR supplying a passive load.

In Fig. 23-25b, the current "lags" behind the voltage because it only flows during the final 90 degrees. This lag produces the same effect as an inductive load. Consequently, the ac source has to supply reactive power Q in addition to the active power P (see Sec. 8-12). The power factor decreases as we delay the triggering pulse. On the other hand, if the SCR is triggered at zero degrees, no reactive power is absorbed by the rectifier.

Circuit No.	Thyristor circuit	Typical applications
TABLE 23D	SOME BASIC THYRISTOR POWER CIRCUITS	
1	Controlled rectifier supplying a passive load	Electroplating, dc arc welding, electrolysis
2	Controlled rectifier supplying an active load	Battery charger, dc motor control, dc transmission line
3	Naturally commutated inverter supplying an active ac load	DC motor control, wound-rotor motor speed control, dc transmission line
4	AC electronic contactor	Spot welding, lighting control, ac motor speed control
5	Cycloconverter	Low-speed synchronous motor control, electroslag refining of metals
6	Self-commutated inverter	Induction motor control, portable ac sources
7	DC chopper	Electric traction

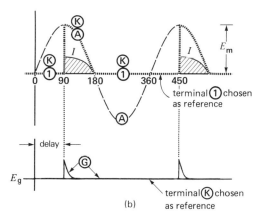

(b)

Figure 23-25   b. Voltage and current waveforms.

23-16 Controlled rectifier supplying an active load (Circuit 2)

Figure 23-26 shows an ac source E_m and a dc load E_d connected by an SCR in series with an inductor. The load (represented by a battery), receives energy because current I enters the positive terminal. Smoothing inductor L limits the peak current to a value within the SCR rating. Gate pulses E_g initiate conduction at angle θ_1 (Fig. 23-26b).

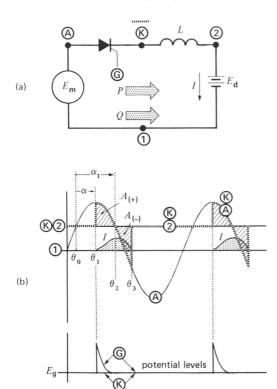

Figure 23-26   a. SCR supplying an active load.

b. Voltage and current waveforms.

If the SCR were replaced by a diode, conduction would begin at angle θ_0 because this is the instant when the anode becomes positive. However, in our example, conduction only begins after a delay of α degrees. As soon as conduction starts,

point K jumps to the level of point A, and voltage E_{A2} appears across the inductor. The latter begins storing volt-seconds, and current I increases accordingly. The volt-seconds reach a maximum at θ_2, where area $A+$ is maximum. The corresponding peak current is given by:

$$I_{max} = A+/L \qquad \text{Eq. 7-11}$$

The current then gradually decreases and becomes zero at angle θ_3 where $A-$ is equal to $A+$. As soon as conduction stops, point K jumps to the level of point 2 and stays there until the next gate pulse.

As in circuit 1 (Fig. 23-25), the load current lags behind voltage E_m; consequently, the source again has to supply reactive power Q as well as active power P.

If we reduce the firing angle α, area $A+$ increases, and so does current I. We can therefore vary the active power supplied to the load from zero ($\alpha = \alpha_1$) to a maximum ($\alpha = 0$).

From a practical point of view, the circuit could be used as a variable battery charger. Another application is to control the speed and torque of a dc motor. In this case, E_d represents the counter emf of the armature, and L the armature inductance.

23-17 Naturally commutated inverter (Circuit 3)

An *inverter*, by definition, changes dc power into ac power. It performs the reverse operation of a rectifier, which converts ac power into dc power. There are two main types of inverters:
1. self-commutated inverters (or oscillators) in which the commutation elements are included within the power inverter;
2. externally commutated inverters (or naturally commutated inverters) in which the means of commutation is not included within the power inverter.

In this section, we examine the operating principle of a naturally commutated inverter. The circuit of such an inverter is identical to that of a controlled rectifier, except that the thyristor terminals are reversed (Fig. 23-27). Because current can only flow from anode to cathode, the dc source E_d *delivers* power whenever the thyristor conducts.

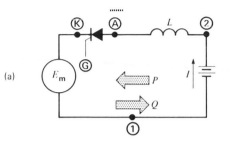

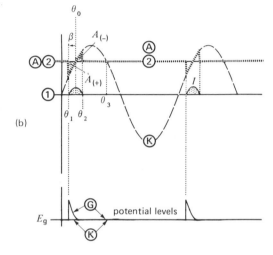

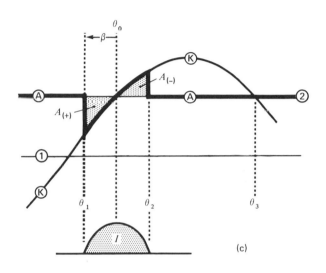

"advanced" ahead of θ_0 by β degrees. The enlarged view of the conduction zone (Fig. 23-27c), shows that the resulting current reaches a peak at θ_0 where area $A+$ is maximum. It then falls to zero and conduction stops at θ_2, when $A- = A+$. The current peaks lead the voltage peaks but because current flows into the positive terminal, the source still has to supply reactive power Q to the inverter

Figure 23-27   a. Naturally-commutated inverter.

b. Voltage and current waveforms.

Figure 23-27   c. Enlarged view of the current-carrying interval.

On the other hand, this power must be absorbed by the ac terminals because no losses occur in the inductor or the thyristor. Consequently, the circuit of Fig. 23-27 is potentially able to convert dc power into ac power.

To achieve this power conversion, the thyristor has to be triggered within a precisely defined range. First, the current must flow into the ac terminals only during those intervals when K is positive with respect to 1. Only then does the ac system act as a load (Sec. 2-2). Second, to initiate conduction, A must be positive with respect to K. The triggering pulse must therefore be applied either prior to θ_0 or after θ_3 (Figs. 23-27b and 23-27c). For reasons that will soon become obvious, the gate must be triggered *prior* to θ_0. In effect, the firing angle is

(Sec. 8-5, rule 3). Consequently, P and Q flow in opposite directions in an inverter.

To increase the active power flow, we simply advance the triggering angle β. However, we cannot carry this process too far. In order for conduction to cease, $A-$ must equal $A+$. However, the maximum area that $A-$ can have is that bounded by the crest of the sine wave and the level of point 2 (Fig. 23-27c). As we advance the firing angle, $A+$ becomes bigger and bigger; but if it should exceed the maximum available value of $A-$, conduction will never stop. The dc current will then build up with each cycle, until the circuit breakers trip. For the same reason, conduction must never be initiated after angle θ_3.

The current flowing into the ac terminals is far

from sinusoidal and a stiff ac system is needed to maintain a sinusoidal voltage. In practice, we add appropriate filters to ensure that the current flowing into the ac line is reasonable sinusoidal. We should also remember that naturally commutated inverters always involve 3-phase systems and not the "simple" single-phase circuit of Fig. 23-27.

23-18   AC electronic contactor (Circuit 4)

An ac electronic contactor is composed of two thyristors connected in anti-parallel (back-to-back), so that current can flow in both directions. Thus, in Fig. 23-28, a symmetrical ac current flows

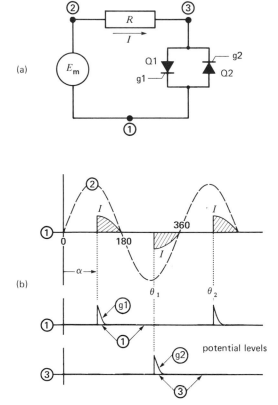

(a)

(b)

Figure 23-28   a. Electronic contactor.

b. Waveforms with a resistive load.

through load resistor R. Gate pulses g1 and g2 are synchronized with the line frequency and, depending on the firing angle α, a greater or lesser ac current will flow in the load.

An important advantage of this contactor is that the ac current can be precisely controlled. Furthermore, in contrast to a magnetic contactor, an electronic contactor is absolutely silent and its "contacts" never wear out.

Figure 23-28 c.

Single-phase water-cooled contactor composed of two Hockey Puk thyristors. Continuous current rating: 1200 A (RMS) at 2000 V; cooling water requirements: 4.5 L/min at 35°C max. For intermittent (10% duty) spot welding applications, this unit can handle 2140 A for 20 cycles. Width: 175 mm; length: 278 mm; depth: 114 mm. *(Photo courtesy of International Rectifier)*

23-19   Cycloconverter (Circuit 5)

A cycloconverter produces low-frequency ac power directly from a higher-frequency ac source. Referring to Fig. 23-29, three groups of thyristors, mounted back-to-back, are connected to a 3-phase source. They supply single-phase power to a resistive load R.

To understand the operation of the circuit, sup-

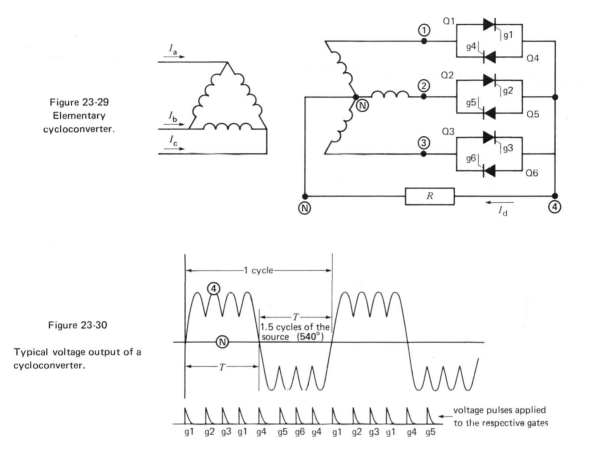

Figure 23-29
Elementary
cycloconverter.

Figure 23-30

Typical voltage output of a
cycloconverter.

pose all thyristors are initially blocked (noncon-ducting). Then, for an interval T, the gates of thy-ristors Q1, Q2 and Q3 are triggered by 4 successive pulses g1, g2, g3, g1, so that the thyristors func-tion as ordinary diodes. As a result, terminal 4 is positive with respect to N (Fig. 23-30). During the next interval T, thyristors Q4, Q5, Q6, are fired by 4 similar pulses g4, g5, g6, g4. This makes terminal 4 *negative* with respect to N. The process is then repeated for the Q1 to Q3 thyristors, and so on, with the result that a low-frequency ac voltage ap-pears across the load. The duration of 1 cycle is $2\,T$ seconds. Compared to a sine wave, the low-frequency waveshape is rather poor. However, this is of secondary importance because means are avail-able to improve it.

Referring to Fig. 23-30 and assuming a 60 Hz source, we can show that each half-cycle corre-sponds to 540°, on a 60 Hz base. The duration of T is therefore $(540/360) \times (1/60) = 0.025$ s, which corresponds to a frequency of 20 Hz.

Obviously, by repeating the firing sequence g1, g2, g3, g1, . . . , we can keep terminal 4 positive for as long as we wish, followed by an equally long negative period, where g4, g5, g6, g4 . . . are fired. In this way, we can generate frequencies as low as we please. The high end of the frequency spectrum is limited to about 40 percent of the supply fre-quency. The reader should also note that the cy-cloconverter can supply a single-phase load from a 3-phase system, without unbalancing the 3-phase lines.

23-20 Self-commutated inverter (Circuit 6)

A self-commutated inverter changes dc power into ac power. There are many types of self-commutated inverters, but all are based on forced commutation. The output frequency may be as high as 20 kHz, depending on the switching capability of the thyristors. The load may be passive, such as a resistor or inductor, or active, such as an ac motor.

As an example of a simple self-commutated inverter, consider the circuit of Fig. 23-31. It con-

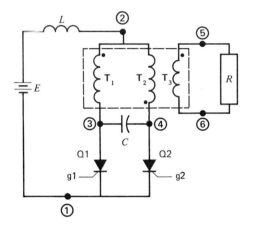

Figure 23-31   Typical self-commutated inverter.

sists of the following components:
1. dc source E;
2. thyristors Q1 and Q2;
3. smoothing inductor L to maintain a constant source current;
4. commutating capacitor C;
5. transformer having two primaries T_1 and T_2 and one secondary T_3;
6. load R;
7. external control system (not shown) that triggers gates g1 and g2, to obtain the desired output frequency.

To understand how the inverter operates, suppose that initially Q1 is conducting and Q2 is blocked. Current I_1 is flowing in primary T_1 and capacitor C is fully charged with polarities as shown in Fig. 23-32a. This condition is stable until

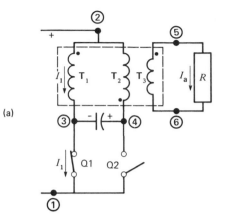

(a)

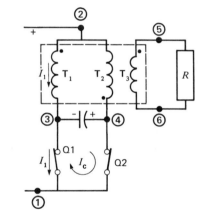

(b)

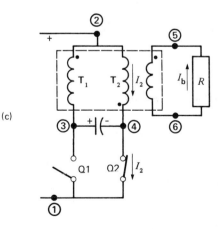

(c)

Figure 23-32   Commutating sequence in a self-commutated inverter.

we apply a pulse to gate g2, initiating conduction in Q2. As soon as Q2 conducts, the level of point 4 drops to point 1. Simultaneously, the capacitor discharges, producing a commutating current I_c that flows opposite to I_1 (Fig. 23-32b). This current rises very fast and as soon as $I_c = I_1$, the current in Q1 becomes zero whereupon it acts like an open switch (Fig. 23-32c).

With Q1 blocked, the capacitor charges up rapidly via primary T_1; consequently, terminal 3 quickly becomes positive with respect to terminal 4. As charging current I_1 decreases in winding T_1, current I_2 builds up in winding T_2. When conditions stabilize, T_2 carries the same current as T_1 carried before.

The next pulse initiates conduction in Q1, causing the level of point 3 to drop to point 1. The capacitor again discharges, thus blocking Q2 and bringing us back to the starting point (Fig. 23-32a). In effect, the alternate gate pulses transfer conduction from one thyristor to the other, owing to the presence of commutating capacitor C.

With regard to the transformer, when Q1 conducts, primary current I_1 induces a secondary current I_a. Based on the polarity marks, I_a must flow in the direction shown. Similarly, when Q2 conducts, primary current I_2 induces a secondary current I_b. The result is that secondary currents I_a and I_b are equal, but flow in opposite directions. Consequently, load R carries an alternating current.

To change the inverter frequency, we simply vary the frequency of the gate pulses. In this way, we can generate frequencies ranging from 4 Hz to 5000 Hz, depending on the design of the transformer and the properties of the SCRs.

23-21 DC chopper (Circuit 7)

In some electronic circuits, power has to be transferred from a high-voltage dc source E_s to a lower-voltage dc load E_0. One solution is to connect the source and load by a resistor, as shown in Fig. 23-33. However, the large I^2R loss renders this solution inefficient, unless the two voltages are nearly equal. Another solution is to connect an induc-

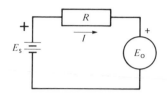

Figure 23-33   Inefficient power transfer.

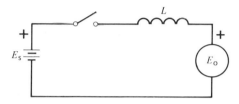

Figure 23-34   Energy transfer using an inductor.

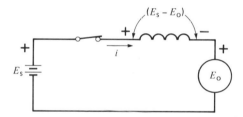

Figure 23-35   Energy is stored in the inductor.

tor and a switch between the source and load (Fig. 23-34). In order to follow the transfer of energy, assume the switch closes for a time T_a. During this interval, the voltage across the inductor is $E_s - E_0$, as shown in Fig. 23-35. The inductor accumulates volt-seconds, and the resulting current i increases at a constant rate. At time T_a (when the switch is about to open), the current is:

$$I_a = A + /L = (E_s - E_0) T_a/L \qquad (23-7)$$

The corresponding energy stored in the inductor is:

$$W = \frac{1}{2} LI_a^2 \qquad \text{Eq. 7-12}$$

When the switch opens, the current collapses and all the stored energy is dissipated in the arc across the switch. At the same time, a very high

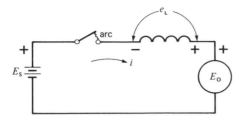

Figure 23-36   Energy is dissipated in the arc.

voltage e_L is induced across the inductor. The polarity of this voltage is *opposite* to what it was when the current was increasing (Fig. 23-36).

Although some energy is transferred from E_s to E_0 while the switch is closed, there is a great loss of energy every time the switch opens. The efficiency is therefore poor.

We can prevent this energy loss by adding a diode to the circuit as shown in Fig. 23-37. When the switch opens, current i again begins to fall, inducing a voltage e_L. However, e_L cannot jump to the high value it had before because as soon as it exceeds E_0, the diode begins to conduct. Assuming the diode voltage drop is negligible, it follows that

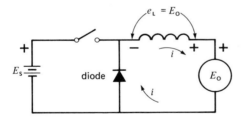

Figure 23-37   Energy transferred without loss.

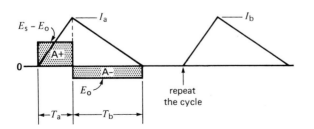

Figure 23-38 E and I in the inductor of Fig. 23-37.

$e_L = E_0$. Because E_0 is constant, current i falls at a uniform rate, and eventually becomes zero after a time $T_b{}^*$ (Fig. 23-38).

When the current is zero, the inductor will have supplied all its energy to load E_0. Simultaneously, the diode will cease to conduct. We can therefore reclose the switch for another interval T_a and repeat the cycle indefinitely. Consequently, this circuit enables us to transfer energy from a high-voltage dc source to a lower voltage dc load without incurring any losses. In effect, the inductor absorbs energy at a relatively high voltage $(E_s - E_0)$ and delivers it at a lower voltage E_0.

Instead of letting the load current swing between zero and I_a, we can open and close the switch rapidly so that the current increases and decreases by small increments. Referring to Fig. 23-39a, the switch is closed for an interval T_a and open during an interval T_b. When the switch is open, the load current falls from an initial value I_a to a final value I_b. During this interval, current flows in the inductor, the load and the so-called free-wheeling diode. No current is supplied by the source during this interval.

When the current has fallen to a value I_b, the switch recloses. The current in the diode immediately stops flowing, and the source now supplies current I_b. The current builds up and when it reaches the value I_a, the switch suddenly reopens. The free-wheeling diode again comes into play and the cycle repeats. The current supplied to the load oscillates therefore between I_a and I_b (Fig. 23-39b). Its average or dc value is obviously given by:

$$I_0 = (I_a + I_b)/2 \qquad (23\text{-}9)$$

On the other hand, the current supplied by the source is composed of a series of sharp pulses, as

---

* We can calculate T_b because the volt seconds accumulated during the "charging" period T_a must equal the volt seconds released during the "discharge" interval T_b. Referring to Fig. 23-38, we have:

V.s during charge period = V.s during discharge period

$$A+ = A-$$
$$(E_s - E_0)\,T_a = E_0 T_b$$

Consequently, $T_b = (E_s - E_0)\,T_a/E_0 \qquad (23\text{-}8)$

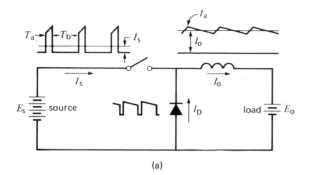

(a)

Figure 23-39   a. Currents in a chopper circuit.

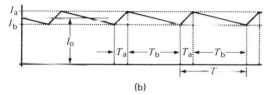

(b)

Figure 23-39   b. Current in the load.

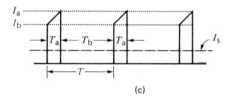

(c)

Figure 23-39   c. Current drawn from the source.

shown in Fig. 23-39c. What is the average value of these pulses? It is easily found by noting that the average current during each *pulse* is obviously $(I_a + I_b)/2 = I_0$. Consequently, the average current I_s during one *cycle* (time T) is:

$$I_s = I_0 (T_a/T)$$

that is

$$\boxed{I_s = I_0 f T_a} \qquad (23\text{-}10)$$

where

I_s = dc current drawn from the source [A]
I_0 = dc current absorbed by the load [A]
T_a = time during which the switch is closed [s]
T = duration of one cycle [s]
f = switching frequency (= $1/T$) [Hz]

The circuit of Fig. 23-39a clearly shows the cur-

rent waveshapes in the source, the load, and the diode. Although the waveshapes are choppy and discontinuous, they still obey Kirchhoff's current law, instant by instant.

Example 23-6:
The switch in Fig. 23-39a opens and closes at a frequency of 20 Hz and remains closed for 4 ms per cycle. A dc ammeter connected in series with the load E_0 indicates a current of 50 A.
a. If a dc ammeter is connected in series with the source, what current will it indicate?
b. What is the average current per pulse?

Solution:
a. Using Eq. 23-10, we have:

$$\begin{aligned} I_s &= I_0 f T_a \\ &= 50 \times 20 \times 0.004 \\ &= 4\ \text{A} \end{aligned}$$

b. The average current per pulse is 50 A. The source may have to be specially designed to supply such a high current pulse. In most cases, a capacitor is connected across the terminals of the source. It can readily furnish the high current pulses as it discharges.

Turning our attention to the power aspects, the dc power drawn from the source must equal the dc power absorbed by the load because there is no power loss in either the switch, the inductor or the free-wheeling diode. We can therefore write:

$$E_s I_s = E_0 I_0$$

If we substitute Eq. 23-10 in the above equation, we obtain the useful relationship:

$$\boxed{E_0 = E_s f T_a} \qquad (23\text{-}11)$$

where

E_0 = dc voltage across the load [V]
E_s = dc voltage of the source [V]
f = switching frequency [Hz]
T_a = "on" time of the switch [s]

Equations 23-10 and 23-11 indicate that we can

control the load voltage and current by varying the switching frequency f or the "on" time T_a, or both.

In practice, the mechanical switch is replaced by a thyristor that is turned on and off at a frequency f. The frequency may range from 30 Hz to 2000 Hz. Forced commutation is used to turn the thyristor off at the end of each cycle. The combination of the thyristor, inductor and diode is called a dc *chopper*. It is widely used in electric trains and trolleybuses and wherever dc-to-dc energy conversion is needed.

Example 23-7:

We wish to charge a 120 V battery from a 600 V dc source using a dc chopper. The average battery current should be 20 A, with a peak-to-peak ripple of 2 A. If the chopper "on" time is fixed at 1 ms, calculate:

a. the dc current drawn from the source;
b. the dc current in the diode;
c. the chopper frequency;
d. the value of the inductor.

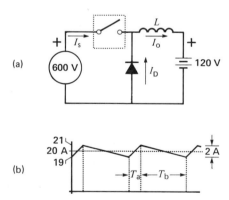

(a)

(b)

Figure 23-40   a. Circuit of Example 23-7.

b. Current in the load.

Solution:

The circuit diagram is shown in Fig. 23-40a and the desired battery current is given in Fig. 23-40b. It fluctuates between 19A and 21A, thus yielding an average of 20 A with a peak-to-peak ripple of 2 A.

a. 1   The power supplied to the battery is:
$$P = 120 \times 20 = 2400 \text{ W}$$

a. 2   The power supplied by the source is therefore 2400 W.

a. 3   The dc current from the source is:
$$I_s = 2400/600 = 4 \text{ A}$$

b.   To calculate the average current in the diode, we refer to Fig. 23-40a. Current I_0 is 20 A and I_s was found to be 4 A. By applying Kirchhoff's current law to the diode-inductor junction, it is obvious that the diode current is:
$$I_D = I_0 - I_s$$
$$= 20 - 4$$
$$= 16 \text{ A}$$

c. 1   We have $E_0 = 120$ V, E_s 600 V, and $T_a = 1$ ms
$$E_0 = E_s f T_a \qquad \text{Eq. 23-12}$$
$$120 = 600 f (0.001)$$
$$f = 200$$

The chopper frequency is therefore 200 Hz.
$$T = 1/f = 1/200 = 5 \text{ ms}$$

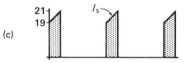

(c)

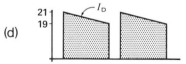

(d)

Figure 23-40   c. Current drawn from the source.

d. Current in the freewheeling diode.

Consequently, we also find that the "off" time T_b is:
$$T_b = T - T_a = 5 - 1 = 4 \text{ ms}$$

The waveshapes of I_s and I_D are shown in Figs. 23-40c and 23-40d, respectively. Note the sharp pulses delivered by the source.

d. 1   During interval T_a, the voltage across the inductor is (600 – 120) = 480 V.

d. 2   The volt-seconds accumulated by the inductor during this interval is 480 x 1 ms = 480 mV·s. The change in current during the interval = (21 – 19) = 2 A.

$$I = A/L \qquad \text{Eq. 1-24}$$
$$2 = 0.48/L$$
$$L = 0.24 \text{ H}$$

Consequently, the inductor has to have an inductance of 0.24 H.

So far, we have assumed the chopper feeds power to an active load E_0. However, it may also be used to connect a high-voltage dc source E_s to a low-voltage load resistor R_0 (Fig. 23-41). Equations 23-10 and 23-11 still apply, but we now have the additional relationship $E_0 = I_0 R_0$. Furthermore, the apparent dc resistance R_s across the terminals of the source is given by

$$R_s = E_s/I_s$$

We can therefore write

$$R_0 = E_0/I_0$$
$$= E_s f T_a \times (f T_a / I_s)$$
$$= (E_s/I_s)\, f^2 T_a^2$$
$$= R_s f^2 T_a^2$$

Consequently,

$$\boxed{R_s = R_0/f^2 T_a^2} \qquad (23\text{-}12)$$

where

R_s = apparent dc resistance across the source [Ω]

R_0 = actual load resistance [Ω]

f = chopper frequency [Hz]

T_a = "on" time [s]

The chopper, therefore, has the ability to change the apparent resistance of a fixed resistor by simply varying either the chopper frequency or the "on" time.

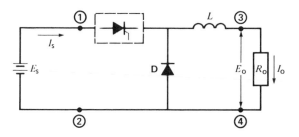

Figure 23-41   A chopper can make a fixed resistor R_0 appear as a variable resistance between terminals 1-2.

Example 23-8:

The chopper in Fig. 23-41 operates at a frequency of 30 Hz and the "on" time is 200 μs. Calculate the apparent resistance across the source, knowing that R_0 = 36 mΩ.

Solution:

Applying Equation 23-12, we have:

$$R_s = R_0/f^2 T_a^2$$
$$= 0.036/30^2 \times (200 \times 10^{-6})^2$$
$$= 1000 \ \Omega$$

This example shows that the actual value of a resistor can be increased many times by using a chopper.

23-22   3-phase, 6-pulse converter

The 3-phase, 6-pulse thyristor converter is the most widely used rectifier/inverter unit in power electronics. Owing to its practical importance, we shall explain how it operates in some detail. As in 3-phase converters, the waveforms become rather complex, although not particularly difficult to understand. Even the simplest circuits yield chopped voltages and currents that pile on top of each other, and it taxes the mind to keep track of everything that is going on. Consequently, in the ensuing text we shall keep the waveforms as simple as possible, so as to highlight the basic principle of operation.

Three-phase, 6-pulse converters have 6 thyristors connected to the secondaries of a 3-phase

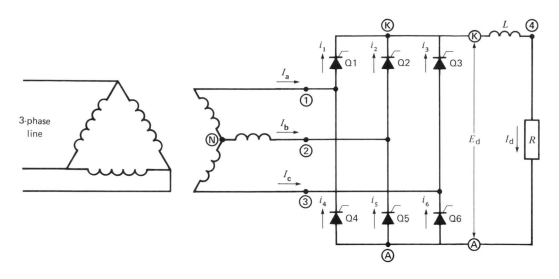

Figure 23-42   Three-phase, 6-pulse thyristor converter.

transformer (Fig. 23-42). The arrangement is identical to the rectifier circuit of Fig. 23-15, except that the diodes are replaced by thyristors. Because we can initiate conduction whenever we please, the thyristors enable us to vary the dc output voltage when the converter operates in the rectifier mode. The converter can also function as an inverter, provided that a dc source is used in place of the load resistor R.

23-23 Basic principle of operation

We can gain a basic understanding of how the converter works by referring to Fig. 23-43. In this figure, the SCRs are assumed to be enclosed in a box, where they successively switch the output terminals K, A to the ac lines 1, 2, 3. The load is represented by a resistor in series with an inductor L. The inductor is assumed to have a very large inductance, so that the load current I_d remains constant. In Fig. 23-43a, the two thyristors between terminals K-1 and A-2 are conducting. A moment later, the thyristors between K-2 and A-1 conduct (Fig. 23-43b). The other thyristors are similarly switched, in sequence. When these steps have been completed, the entire switching cycle repeats. The reader will note that the dc current I_d also flows in

the ac lines. However, the current in each line reverses periodically, and so it is a true ac current of amplitude I_d. It is also evident that the current in a particular line is zero for brief intervals.

The sequence we have just described is similar to that of the diode bridge rectifier of Fig. 23-18. There is, however, an important difference. The thyristors can be made to conduct at precise moments on the ac voltage cycle. Thus, conduction can be initiated when the instantaneous voltage between the ac lines is either high or low. If the voltage is low, the dc output voltage will obviously be low. Conversely, if the thyristors conduct when the ac line voltage is momentarily near its peak, the dc output voltage will be high. In effect, the output voltage E_{KA} is composed of short segments of the ac line voltage. The average value of E_{KA} is the dc output voltage E_d.

In examining Fig. 23-43, it can be seen that the line current always flows out of a line that is momentarily positive. This must be so because the line delivers active power to the load.

Now that we know how the thyristor converter behaves as a rectifier, the question arises: how can it be made to operate as an inverter? Three basic conditions have to be met.

First, we must have a source of dc current I_d. It

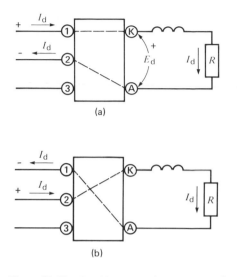

(a)

(b)

Figure 23-43   Rectifier mode (see Fig. 23-42)

a. Q1 and Q5 conducting.

b. Q2 and Q4 conducting.

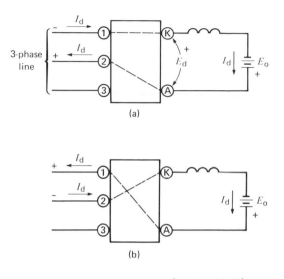

(a)

(b)

Figure 23-44   Inverter mode (see Fig. 23-42)

a. Q1 and Q5 conducting.

b. Q2 and Q4 conducting.

can be provided by a voltage source E_0 in series with a large inductance (Fig. 23-44a).

Second, the converter must be connected to a 3-phase line that can maintain an undistorted sinusoidal voltage, even when the line current is nonsinusoidal. The voltage may be taken from a power utility, or generated by a local alternator.

Third, the thyristors must be switched so that current I_d flows into an ac line that is momentarily positive. The gate firing must therefore be precisely synchronized with the line frequency.

The inverter operation can best be understood by referring to Fig. 23-44. The SCRs enclosed in the box are arranged the same way as in Fig. 23-43. In other words, the converters in the two figures are absolutely identical. Looking first at the dc side, the dc current I_d must flow in the same direction as before because SCRs cannot conduct in reverse. On the other hand, because we want the dc source to deliver power, I_d must flow out of the positive terminal, as shown. In other words, the positive side of E_0 must be connected to terminal A. On the ac side, the 3-phase line is simply connected to terminals 1, 2, 3.

We are now ready to fire the thyristors. However, the firing must be properly timed so that the ac line receives power. This is consistently done in Fig. 23-44 because current I_d always flows into an ac terminal that is momentarily positive. Note that the line polarities in the inverter mode are consistently opposite to those in the rectifier mode.

The reader can see that the line current alternates as before, and it has a peak value equal to I_d. Indeed, the waveshape of the ac line currents is the same in Figs. 23-43 and 23-44; it is only the instantaneous voltages that differ.

If the dc supply voltage E_0 is low, the thyristors must be fired when the instantaneous ac voltage is low. Conversely, if the dc voltage is high, the thyristors must be triggered when the ac line voltage is at or near its peak.

We wish to make one final important observation. The voltage that appears between terminals K and A is composed of segments of the ac line voltages. Consequently, E_{KA} is a fluctuating voltage whose average value is E_d. This average voltage must be equal to E_0 because the dc voltage drop across the inductor is negligible.

23-24 Three-phase, 6-pulse converter

Consider the circuit of Fig. 23-45 in which a 3-phase, 6-pulse converter supplies power to a load. The load is composed of a dc voltage E_0 and a resistor R, in series with a smoothing inductor. The converter is fed from a 3-phase transformer. The gates of thyristors Q1 to Q6 are triggered in succession, at 60-degree intervals. We assume that the converter has been in operation for some time, so that conditions are stable. Initially, thyristors Q5 and Q6 are conducting, carrying load current I_d (Fig. 23-46). Then, at the $0°$ point (θ_0), thyristor Q1 is triggered by gate pulse g1. Commutation occurs and Q1 starts conducting, taking over from Q5.

At $60°$, thyristor Q2 is fired and the resulting commutation transfers the load current from Q6 to Q2. This switching process continues indefinitely and, as in Fig. 23-17, point K follows the peaks of the successive waves. The thyristors are labelled according to the order in which they are fired. Two SCR's conduct at a time; the conduction pairs are therefore Q1 - Q2, Q2 - Q3, Q3 - Q4, and so on. Thus, by referring to Fig. 23-45, we can tell at a glance which thyristors are conducting at any given time.

The converter acts as a rectifier and the average or dc voltage between K and A is $E_d = 1.35\,E$. Because there is no appreciable dc voltage drop in an inductor, the dc voltage between points 4 and A is also $1.35\,E$. Consequently, the dc current I_d is given by:

$$I_d = (E_d - E_0)/R \qquad (23\text{-}13)$$

The triggering time has to be very precise to obtain the rectified voltage shown in Fig. 23-46. Thus, if g1 fires slightly ahead of θ_0, conduction cannot start because anode 1 is then negative. On the other hand, if g1, fires after θ_0, Q5 will continue to conduct until g1 *is* fired.

23-25 Delayed triggering - rectifier mode

Let us now delay all triggering pulses by an angle α of $15°$ (Fig. 23-47a). Current I_d, instead of switching over to Q1 at θ_0, will continue to flow in Q5 until gate pulse g1 triggers Q1. Commutation occurs, and the potential of point K jumps from line 3 to line 1. A similar switching action takes place for the other thyristors. The resulting choppy waveshape between terminals K and A is shown in Fig. 23-47a.

Note that the triggering delay does not shorten the conduction period: each thyristor still conducts for a full $120°$. Furthermore, the current remains constant and ripple-free, owing to the presence of the big inductor. The level of point K follows the tops of the individual sine waves, but the average voltage E_d, between K and A, is obviously smaller than before. We can prove that it is given by:

$$\boxed{E_d = 1.35\,E\cos\alpha} \qquad (23\text{-}14)$$

where

E_d = dc voltage produced by the 3-phase, 6-pulse converter [V]

E = effective value of the ac line voltage [V]

α = firing angle [°]

According to Eq. 23-14, E_d becomes smaller and smaller as α increases. However, if E_d becomes equal to or less than E_0, the load current ceases to flow. Ordinarily, the current would reverse when E_d is smaller than E_0. However, this is impossible, because the SCRs can only conduct in the forward direction.

Figures 23-47b and 23-47c show the waveform between K and A for $\alpha = 45°$ and $75°$, respectively. The ac component in E_{KA} is now very large, compared to the dc component. Indeed, at $\alpha = 75°$, the dc component E_d is so small that load voltage E_0 has to be nearly zero for a significant current to flow.

Example 23-8:

The 3-phase converter of Fig. 23-45 is connected to a 3-phase 480 V, 60 Hz source. The load consists of a 500 V dc source having an internal resistance of 2 Ω. Calculate the power supplied to the load for triggering delays of: a. $15°$; b. $75°$.

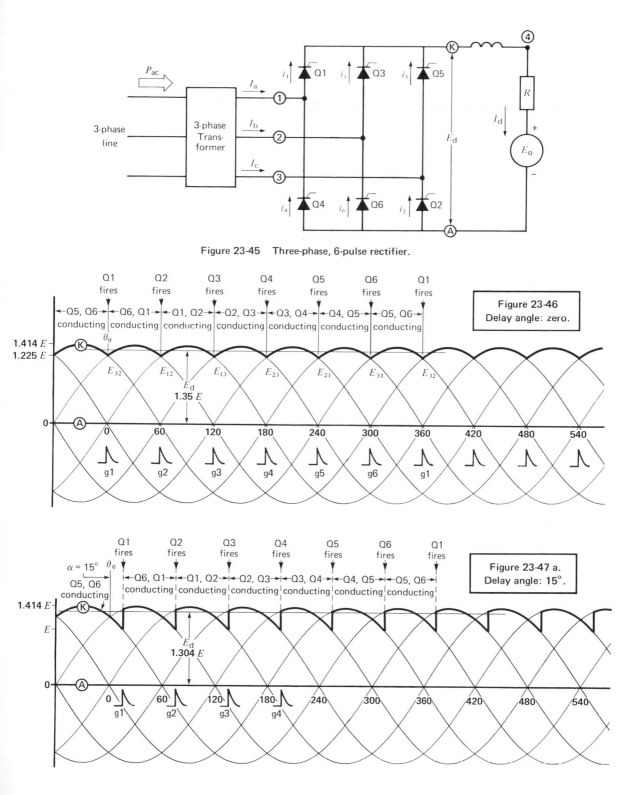

Figure 23-45   Three-phase, 6-pulse rectifier.

Figure 23-46
Delay angle: zero.

Figure 23-47 a.
Delay angle: 15°.

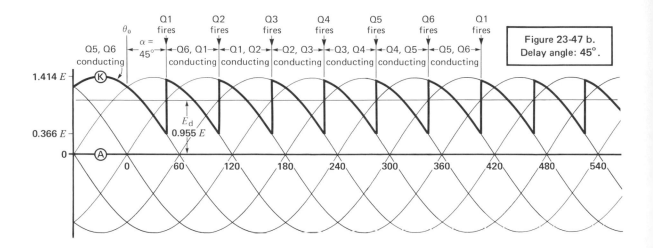

Figure 23-47 b.
Delay angle: 45°.

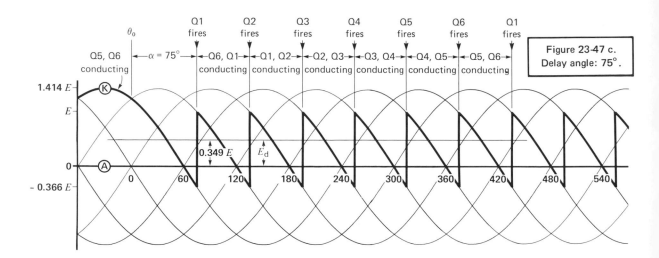

Figure 23-47 c.
Delay angle: 75°.

Solution:

a. 1 The dc output voltage of the converter is:

$$E_d = 1.35\,E\cos\alpha$$
$$= 1.35 \times 480 \times \cos 15°$$
$$= 626\text{ V}$$

Because the dc voltage drop across the inductor is negligible, the IR drop across the 2 Ω internal resistance is:

$$E = E_d - E_0$$
$$= 626 - 500 = 126\text{ V}$$

a. 2 The dc load current is therefore:

$$I_d = E/R = 126/2 = 63\text{ A}$$

a. 3 The power supplied to the load is:

$$P = E_d I_d$$
$$= 626 \times 63 = 39.4\text{ kW}$$

b. 1 With a phase angle delay of 75°, the converter voltage is:

$$E_d = 1.35\,E\cos\alpha$$
$$= 1.35 \times 480 \times \cos 75°$$
$$= 167.7\text{ V}$$

Because E_d is less than E_0, the current tends to flow in reverse. This is impossible, and consequently the current is simply zero, and so, too, is the power.

23-26 Delayed triggering - inverter mode

If triggering is delayed by more than $90°$, E_d becomes negative. This does not produce a negative current because, as we said, SCRs conduct in only one direction. Consequently, the load current is simply zero. However, we can *force* a current to flow by connecting a dc voltage of proper magnitude and polarity across the converter terminals. This external voltage E_0 must be slightly greater than E_d in order for current to flow (Fig. 23-48). The load current is given by:

$$I = (E_0 - E_d)/R$$

the inverter mode, the firing angle lies between $90°$ and $180°$, and a dc source of proper polarity *must* be provided.

Figure 23-49 shows the waveshapes at firing angles of $105°$, $135°$ and $165°$. The dc voltage E_d generated by the inverter is still given by Eq. 23-14. It reaches a maximum value of $E_d = 1.35\,E$ at $180°$.

23-27 Triggering range

The triggering range of a given thyristor is usually kept between $15°$ and $165°$. The thyristor acts as a

Figure 23-48   Three-phase, 6-pulse converter in the inverter mode.

Because current flows out of the positive terminal of E_0, the "load" is actually a source, delivering a power output $P = E_d I_d$. Part of this power is dissipated as heat in the circuit resistance and the remainder is absorbed by the secondaries of the 3-phase transformer. If we subtract the small transformer losses and the virtually negligible SCR losses, we are left with a net active power P_{ac} that is returned to the 3-phase line.

The original rectifier has now become an *inverter*, converting dc power into ac power. The transition from rectifier to inverter is smooth, and requires no change in the converter connections. In the rectifier mode, the firing angle lies between $0°$ and $90°$, and the load may be active or passive. In

rectifier between $15°$ and $90°$ and as an inverter between $90°$ and $165°$. The dc voltage developed reaches its maximum value at $15°$ and $165°$; it is zero at $90°$.

The triggering angle is seldom less than $15°$ in the rectifier mode. The reason is that sudden line voltage changes might cause a thyristor to misfire, thus producing a discontinuity in the dc output current.

In the inverter mode, we seldom permit the firing angle to exceed $165°$. If we go beyond this point, the inverter may lose its ability to commutate and the currents build up very quickly until the circuit breakers trip. In some cases, the firing angle is not allowed to exceed $150°$, to ensure an

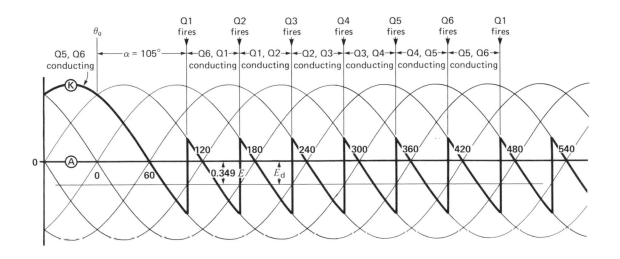

Figure 23-49   a. Triggering sequence and waveforms with a delay angle of 105°.

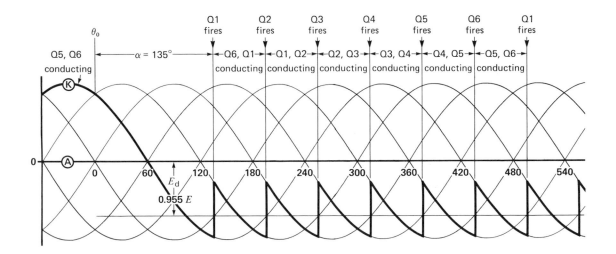

Figure 23-49   b. Triggering sequence and waveforms with a delay angle of 135°.

adequate safety margin.

Figure 23-50 shows the allowed and forbidden gate firing zones for a particular thyristor in a 3-phase, 6-pulse converter. Specifically, it refers to Q1 in Fig. 23-45. The other thyristors have similar firing zones, but they occur at different times.

23-28 Equivalent circuit of a converter

We may think of a converter as being a static ac/dc motor-generator set whose dc output voltage E_d changes both in magnitude and polarity depending upon the gate pulse delay. However, the dc "gener-

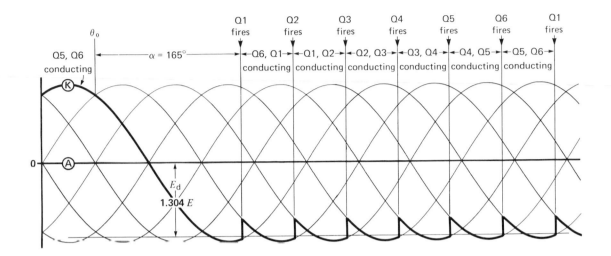

Figure 23-49   c. Triggering sequence and waveforms with a delay angle of 165°.

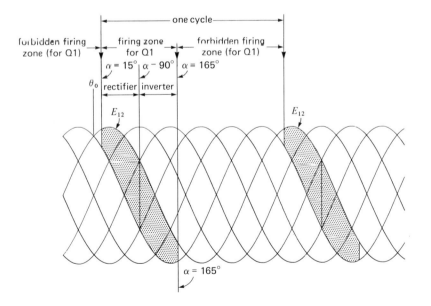

Figure 23-50   Permitted gate firing zones for thyristor Q1.

ator'' has some special properties:

1. it can carry current in only one direction;
2. it generates an increasingly large ac ripple voltage as the dc voltage decreases.

The analogy may be represented by the circuit of Fig. 23-51, in which:

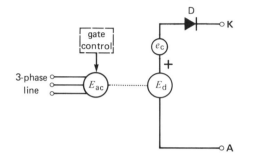

Figure 23-51   Equivalent circuit of a converter.

- E_{ac} represents the 3-phase line voltage
- E_d is the dc voltage generated by the converter
- e_c is the ac voltage generated by the converter
- D is a diode to remind us that current can flow in only one direction
- The dotted line between E_{ac} and E_d indicates that active power can flow between the ac and dc systems

When the converter is operating as a rectifier, the equivalent circuit is shown in Fig. 23-52. When operating as in inverter, the circuit is given by Fig. 23-53. The ac voltage generated by the converter appears across inductor L. Its inductance is sufficiently large to ensure an almost ripple-free dc current.

23-29 Currents in a 3-phase, 6-pulse converter

Figure 23-54 shows the voltage and current waveshapes when the converter functions as a rectifier at a firing angle of 45°. The current in each thyristor flows for 120°, and the peak value is equal to the dc current I_d. This is obviously true for any firing angle between zero and 180°. Consequently, the currents in a thyristor converter are identical to those in a plain 3-phase diode rectifier (Fig.

23-16). The only difference is that they flow later in the cycle.

The waveshapes of the ac line currents are easily found because they are equal to the difference between the respective thyristor currents. Thus, in Fig. 23-48, line current $I_a = i_1 - i_4$. These currents

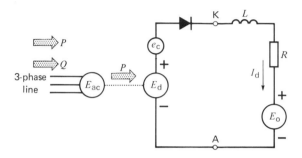

Figure 23-52   Equivalent circuit of a 3-phase converter in the rectifier mode.

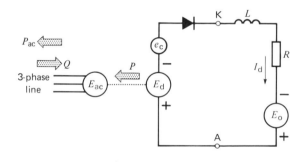

Figure 23-53   Equivalent circuit of a 3-phase converter in the inverter mode.

also have a peak value I_d, and they flow in positive and negative pulses of 120°.

The heating effect of the ac line currents is important because they usually flow in the windings of a converter transformer. The I^2R loss depends upon the effective value I of the current. Using the method explained in Sec. 7-5, we calculate its value as follows:

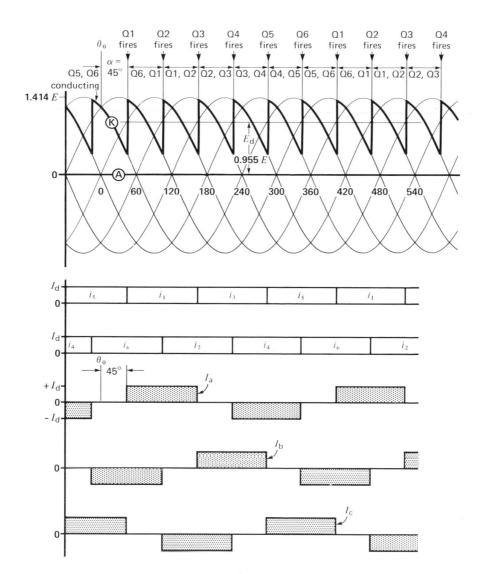

Figure 23-54   Voltage and current waveforms in the converter of Figure 23-45 with a delay angle of 45°.

$$I^2 \times 180° = I^2_d \times 120°$$

$$\boxed{I = 0.816\, I_d} \qquad (23\text{-}15)$$

where

I = effective ac line current [A]

I_d = dc current of the converter [A]

0.816 = a constant [exact value = $\sqrt{2/3}$]

The effective value of the ac line current is therefore directly related to the dc output current, and is unaffected by the firing angle. Clearly, the same is true when the converter operates as an inverter.

23-30 Power factor

Returning briefly to the 3-phase, 3-pulse rectifier (Figs. 23-12 and 23-13), we note that the currents in lines 1, 2, and 3 are symmetrical with respect to the line-to-neutral voltages. Thus, rectangular current i_2 is exactly in the middle of the positive E_{2N} wave. In essence (and in actual fact), i_2 is in phase with E_{2N}. This is also true for the other currents, as regards their respective voltages. This condition is reflected back into the primary of the transformer, and from there, to the primary feeder. Because the currents are in phase with the voltages, the power factor is 100%. The same remarks apply to a 3-phase, 6-pulse rectifier (Fig. 23-16).

Referring now to Fig. 23-54, where triggering has been delayed by 45°, we note that the thyristor currents have all been shifted by 45° to the right. Consequently, the currents lag the respective voltages by 45°; the power factor is no longer unity but only 0.707 (cos 45° = 0.707). This means that a converter absorbs reactive power from the ac system to which it is connected. This is true whether the converter operates as a rectifier or inverter. The reactive power is given by:

$$\boxed{Q = P \tan \alpha} \qquad (23\text{-}16)$$

where

Q = reactive power absorbed by the converter [var]

P = dc power of the converter (positive for a rectifier, negative for an inverter) [W]

α = triggering angle [°]

In practice, the reactive power absorbed is about 40% of the active power.

Example 23-9:
Calculate the reactive power absorbed by the converter of Example 23-8, for a triggering angle of 15°.

Solution:
The total dc power absorbed by the load is:

$P = E_d I_d$ = 39.4 kW

$Q = P \tan \alpha$

   = 39.4 tan 15

   = 10.6 kvar

Example 23-10:
A 16 kV dc source having an internal resistance of 1 Ω supplies 900 A to a 12 kV, 3-phase, 60 Hz inverter (Fig. 23-55). Calculate:

a. the dc current carried by each SCR;

b. the dc voltage generated by the inverter;

c. the required firing angle α;

d. the effective value of the ac line currents;

e. the reactive power absorbed by the inverter;

f. the apparent power supplied by the ac line.

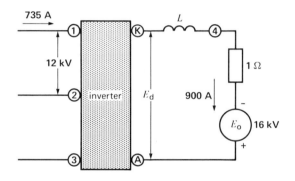

Figure 23-55   See Example 23-9.

Solution:

a. Each SCR carries the load current for one third of the time. The dc current is therefore:

$I = I_d/3$ = 900/3

   = 300 A

b. the voltage E_d generated by the inverter is equal to E_0 less the IR drop. Thus,

$E_d = E_0 - I_d R$

   = 16 000 – 900 x 1

   = 15 100 V

c. Knowing the effective ac line voltage is 12 000 V, the firing angle can be found from Eq. 23-14:

$E_d = 1.35\ E \cos \alpha$

15 100 = 1.35 x 12 000 cos α

cos α = 0.932

$\alpha = 21.2°$

This is the firing angle that would be required if the converter operated as a rectifier. However, because it is in the inverter mode, the actual firing angle is:

$\alpha = 180 - 21.2 = 158.8°$

d. The effective value of the ac line current is:

$I = 0.816 I_d$ Eq. 23-15
$\quad = 0.816 \times 900$
$\quad = 734$ A

e. The dc power absorbed by the inverter is:

$P = E_d I_d$
$\quad = 15\ 100 \times 900$
$\quad = 13.6$ MW

P is actually negative because the inverter *absorbs* dc power; hence, $P = -13.6$ MW.

The reactive power absorbed by the inverter is:

$Q = P \tan \alpha$ Eq. 23-16
$\quad = -13.6 \tan 158.8$
$\quad = 5.27$ Mvar*

f. The apparent power furnished by the transformer is:

$S = EI\sqrt{3}$ Eq. 9-9
$\quad = 12\ 000 \times 734\sqrt{3}$
$\quad = 15.3$ MVA†

QUESTIONS AND PROBLEMS

Practical level

23-1 State the basic properties of a diode.

23-2 State the basic properties of a thyristor.

23-3 What is the approximate voltage drop across a diode or SCR when it conducts?

23-4 What is the approximate maximum operating temperature of a thyristor?

---

* In practice, the actual reactive power is higher than the calculated value, owing to commutation overlap.

† The apparent power calculated in this way is approximate because the current is rectangular rather than sinusoidal. However, 15.3 MVA is a close approximation to the real value.

23-5 Explain the meaning of the following terms:

anode	harmonic
cathode	commutation
gate	natural commutation
choke	converter
filter	inverter
chopper	cycloconverter
peak inverse voltage	bridge rectifier
rectifier	forced commutation
self-commutated inverter	

23-6 The 3-phase transformer shown in Fig. 23-12 produces a secondary line voltage of 2.4 kV. The dc load current I_d is 600 A. Calculate:
  a. the dc voltage across the load;
  b. the average current carried by each diode;
  c. the maximum current carried by each diode.

23-7 The 3-phase transformer shown in Fig. 23-15 produces a secondary line voltage of 2.4 kV. If the dc load current is 600 A, calculate:
  a. the dc voltage across the load;
  b. the average current carried by each diode.

23-8 An ac source having an effective voltage of 600 V, 60 Hz is connected to a single-phase bridge rectifier as shown in Fig. 23-10a. The load resistor has a value of 30 Ω. Calculate:
  a. the dc voltage E_{34};
  b. the dc voltage E_{54};
  c. the dc load current I;
  d. the average current carried by each diode;
  e. the active power supplied by the ac source.

23-9 The chopper shown in Fig. 23-41 is connected to a 3000 V dc source. The chopper frequency is 50 Hz and the on-time is 1 ms. Calculate:
  a. the voltage across resistor R_0;
  b. the value of I_s if $R_0 = 2\ \Omega$.

23-10 In Problem 23-11, if we double the on-time, calculate the new power absorbed by the load.

Intermediate level

23-11 a. In Problem 23-7 calculate the power dissipated by the six diodes if the average voltage drop during the conduction period is 0.6 V.

b. What is the efficiency of the rectifier alone?

23-12 a. The current in Fig. 23-2 has a value of − 6 A. What is the polarity of E_{34}?

b. The current in Fig. 23-3 has a value of + 6 A and E_{65} is negative. Is the current increasing or decreasing?

23-13 The single-phase bridge rectifier shown in Fig. 23-10a is connected to a 120 V, 60 Hz source. If the load resistance is 3 Ω, calculate:

a. the dc load current;

b. the PIV across the diodes;

c. the energy that must be stored in the choke so that the peak-to-peak ripple is about 5% of the dc current;

d. the inductance of the choke;

e. the peak-to-peak ripple across the choke.

23-14 The line voltage is 240 V, 60 Hz on the secondary side of the converter transformer in Fig. 23-15. The dc load draws a current of 750 A. Calculate:

a. the dc voltage produced by the rectifier;

b. the active power supplied by the 3-phase source;

c. the peak current in each diode;

d. the duration of current flow in each diode [ms];

e. the effective value of the secondary line current;

f. the reactive power absorbed by the converter;

g. the peak-to-peak ripple across the inductor.

23-15 The chopper shown in Fig. 23-41 is connected to a 2000 V dc source, and the load resistor R_0 has a value of 0.15 Ω. The on-time is fixed at 100 μs and the dc voltage across the resistor is 60 V. Calculate:

a. the power supplied to the load;

b. the power drawn from the source;

c. the dc current drawn from the source;

d. the peak value of I_s;

e. the chopper frequency;

f. the apparent resistance across the dc source;

g. draw the waveshapes of I_s, I_0 and I_D.

23-16 The 3-phase, 6-pulse converter shown in Fig. 23-42 is directly connected to a 3-phase, 208 V line. Calculate:

a. the dc output voltage for a firing angle of 90°;

b. the firing angle needed to generate 60 V (rectifier mode);

c. the firing angle needed to generate 60 V (inverter mode).

23-17 The converter shown in Fig. 23-42 is connected to a transformer that produces a secondary line voltage of 40 kV, 60 Hz. The resistance R is negligible and the load draws a dc current of 450 A. If the delay angle is 75°, calculate:

a. the dc output voltage;

b. the active power drawn from the ac line;

c. the effective value of the secondary line current;

d. the reactive power absorbed by the converter.

23-18 In Problem 23-17, calculate:

a. the peak positive value of E_{KA};

b. the peak negative value of E_{KA};

c. the peak-to-peak ripple across the inductor.

Advanced level

23-19 The rectifier shown in Fig. 23-15 produces a dc output of 1000 A at 250 V. Inductor L reduces the current ripple, but an additional purpose is to limit the rapid build-up of dc current should load R become shorted. This enables the circuit breakers to trip before the dc current becomes too large. Assuming the initial current is 1000 A, calculate the minimum value of L so that the short-circuit current does not exceed 3000 A after 50 ms.

23-20 A diode having a PIV rating of 600 V is used in a battery charger similar to the one shown in Fig. 23-7a. The battery voltage is 120 V and R = 10 Ω.

    a. Calculate the maximum effective secondary voltage of the transformer so that the diode will not break down.

    b. For how many electrical degrees will the diode conduct if the RMS secondary voltage is 300 V?

    c. What is the peak current in the diode?

23-21 The cycloconverter in Fig. 23-29 is connected to a 60 Hz source. Calculate the time interval between the firing of gates g1 and g4 if we wish to generate an output frequency of 12 Hz? Draw the waveshape of the voltage across the load resistance.

23-22 a. Referring to Fig. 23-11, and recognizing that area $A-$ is almost triangular, calculate the approximate value of $A+$ if the effective voltage produced by the source is 2000 V, 60 Hz [V·s].

    b. If the peak-to-peak current ripple must not exceed 7 A, calculate the inductance of choke L.

23-23 The chopper shown in Fig. 23-37 transfers power from a 400 V source to a 100 V load.

The inductor has an inductance of 5 H. If the chopper is "on" for 2 s and "off" for 10 s, calculate:

    a. the maximum current in the inductor;

    b. the energy transferred to the load, per cycle [J] ;

    c. the average power delivered to the load [W] ;

    d. draw the waveshape of i as a function of time and compare it with Fig. 23-38.

23-24 In Problem 23-23, if the chopper is on for 2 s and off for 2 s, what is the value of the current:

    a. after 2 s?   b. after 4 s?

    c. after 6 s?   d. after 8 s?

    e. Will anything prevent the current from building up indefinitely?

23-25 A 3-phase, 6-pulse converter shown in Fig. 23-48 is to be used as an inverter. The dc side is connected to a 120 V battery and R = 10 mΩ. The ac side is connected to a 3-phase, 120 V, 60 Hz line. If the battery delivers a current of 500 A, calculate:

    a. the firing angle required;

    b. the active power delivered to the ac line;

    c. the reactive power absorbed by the converter.

24

ELECTRONIC CONTROL OF DIRECT CURRENT MOTORS

High-speed reliable and inexpensive semiconductor devices have produced a dramatic change in the control of dc motors. In this chapter, we examine some of the basic principles of such electronic controls. The circuits involve rectifiers and inverters already covered in Chapter 23. The reader should therefore review this chapter before proceeding further.

In describing the various methods of control, we shall only study the behavior of power circuits. Consequently, the many ingenious ways of shaping and controlling triggering pulses are not covered here. The reason is that they constitute, by themselves, a complex subject involving sophisticated electronics, logic circuits, integrated circuits, and even microprocessors. Nevertheless, the omission of this important subject, covered in other texts, does not detract from the thrust of this chapter, which is to explain the fundamentals of electronic dc drives.

24-1 Fundamentals of electric drives

Some industrial drives require an electric motor to function at various torques and speeds, both in forward and reverse. In addition to operating as a motor, the machine often has to function for brief periods as a generator or brake. In electric locomotives, for example, the motor may run clockwise or counterclockwise and the torque may act either with, or against, the direction of rotation. In other words, the speed and torque may be positive or negative.

In describing industrial drives, the various operating modes can best be shown in graphical form. The positive and negative speeds are plotted on a horizontal axis, and the positive and negative torques on a vertical axis (Fig. 24-1). This gives rise to four operating quadrants, labelled respectively quadrants 1, 2, 3, and 4.

If a machine operates in quadrant 1, both the torque and speed are positive, meaning that they act in the same direction (clockwise, say). Consequently, a machine operating in this quadrant functions as a motor. As such, it delivers mechanical power to the load. The same remarks apply to quadrant 3, except that both the torque and speed are reversed.

A machine that operates in quadrant 2 develops a positive torque but its speed is negative. In other words, the torque acts clockwise while the machine turns counterclockwise. In this quadrant, the machine *absorbs* mechanical power from the load; consequently, it functions basically as a generator. The mechanical power is converted into electric power and the latter is usually fed back into the line. However, the electric power may be dissipated in an external resistor, such as in dynamic braking.

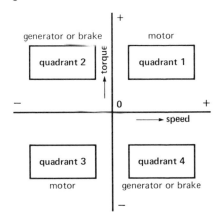

Figure 24-1   Electric drives can operate in four distinct quadrants.

Depending on the way it is connected, a machine may also function as a *brake* when operating in quadrant 2. The mechanical power absorbed is again converted to electric power, but the latter is immediately and unavoidably converted into heat. In effect, when a machine functions as a brake, it absorbs electric power from the supply line at the same time as it absorbs mechanical power from the shaft. Both power inputs are dissipated as heat - often inside the machine itself. For example, whenever a machine is plugged, it operates as a brake. In large power drives, we seldom favor the brake mode of operation because it is very inefficient. Consequently, the circuit is usually arranged so that the machine functions as a generator when operating in quadrant 2.

Quadrant 4 is identical to quadrant 2, except that the torque and speed are reversed; consequently, the same remarks apply.

24-2 Typical torque-speed curves

The torque-speed curve of a 3-phase induction motor is an excellent example of the motor-generator-brake behavior of an electrical machine. Referring to the solid curve in Fig. 24-2, the machine acts as a motor in quadrant 1, as a brake in quad-

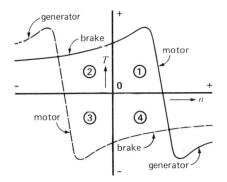

Figure 24-2   Typical torque-speed curve of a squirrel-cage induction motor.

rant 2 and as a generator in quadrant 4 (Sec. 18-18). If the stator leads are reversed, another torque-speed curve is obtained. This dotted curve shows that the machine now operates as a motor in quadrant 3, as a generator in quadrant 2, and as a brake in quadrant 4. Note that the machine can function either as a generator or brake in quadrants 2 and 4. On the other hand, it always runs as a motor in quadrants 1 and 3.

To give another example, Figure 24-3 shows the complete torque-speed curve of a dc shunt motor. The motor-generator-brake modes are again apparent. If the armature leads are reversed, we obtain the dotted curve.

In designing electric drives, we try to vary the speed and torque in a smooth, continuous way to satisfy the load requirements. This is usually done by shifting the entire torque-speed characteristic back and forth along the horizontal axis. For example, the torque-speed curve of Fig. 24-3 may be shifted back and forth by varying the armature voltage. Similarly, we can shift the curve of an induction motor by varying the voltage and frequency. The shift can be produced electrically, but more recently it is accomplished by means of power electronics.

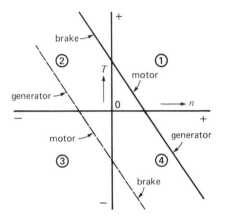

Figure 24-3   Typical torque-speed curve of a dc motor.

Power electronics can even change the *shape* of the torque-speed curve. Thus, by using feedback, a shunt motor can be made to duplicate the properties of a series motor, and a squirrel-cage induction motor can imitate the behavior of a dc compound motor. Consequently, the distinction between motors has become somewhat blurred owing to the tremendous versatility afforded by electronic control. The truth of this statement will be realized as the reader discovers some of the novel control methods used today.

24-3   First quadrant speed control

We begin our study with a variable speed drive for a dc shunt motor. We assume its operation is restricted to quadrant 1. The field excitation is fixed,

and the speed is varied by changing the armature voltage. A 3-phase, 6-pulse converter is connected between the armature and a 3-phase source (Fig. 24-4). The field is separately excited by a single-phase bridge rectifier. External inductor L ensures a relatively smooth armature current. If the armature inductance L_a is large enough, the external inductor can be dispensed with. The armature is initially at rest and the disconnecting switch S is open.

A gate triggering processor receives external inputs such as actual speed, actual current, actual torque, etc. These inputs are picked off the power circuit by means of suitable transducers. In addition, the processor can be set for any desired motor speed and torque. The actual values are compared with the desired values, and the processor automatically generates gate pulses to bring them as close together as possible. Limit setting are also incorporated so that the motor never operates beyond acceptable values of current, voltage and speed.

Gate pulses are initially delayed by an angle $\alpha = 90°$ so that converter output voltage E_d is zero. Switch S is then closed and α is gradually reduced so that E_d begins to build up. Armature current I_d starts flowing and the motor gradually accelerates. During the starting period, the current is monitored automatically. Furthermore, the gate triggering processor is preset so that the pulses can never produce a current in excess of 1.6 p.u., say.

Three features deserve our attention as regards the start-up period:
1. no armature resistors are needed; consequently, there are no I^2R losses;
2. the power loss in the thyristors is negligible; consequently, all the active power drawn from the ac source is available to drive the load;
3. even if an inexperienced operator tried to start the motor too quickly, the current-limit setting would override the manual command. In effect, the armature current can never exceed the allowable preset value.

When the motor reaches full speed, the firing angle is usually between $15°$ and $20°$. Converter voltage E_d is slightly greater than induced voltage E_0 by an amount equal to the armature circuit IR_a drop. The converter voltage is given by:

$$E_d = 1.35\, E \cos \alpha \qquad \text{Eq. 23-14}$$

To reduce the speed, we increase α so that E_d becomes less than E_0. In a Ward-Leonard system,

because the SCR losses are small. However, the ripple voltage generated by the converter is greater than under full-load conditions because α is greater (Sec. 23-28). Consequently, the armature current is not as smooth as before, which tends to increase the armature copper losses. An equally serious

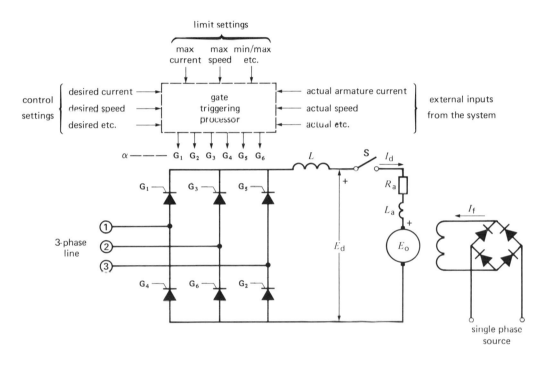

Figure 24-4   Armature torque and speed control using a thyristor converter.

this would immediately cause the armature current to reverse (see Sec. 15-5). Unfortunately, the current cannot reverse in Fig. 24-4 because the SCRs conduct in only one direction. As a result, when we increase α, the current is simply cut off and the motor *coasts* to the lower speed. During this interval, E_0 gradually falls, and when it eventually becomes less than the new setting of E_d, the armature current again starts to flow. The torque quickly builds up and the motor will continue to run at the lower speed.

The efficiency at the lower speed is still high

problem is the large reactive power absorbed by the converter as the firing angle is increased.

To stop the motor, we delay the pulses by $90°$ so that $E_d = 0$ V. The motor will coast to a stop at a rate that depends on the mechanical load and the inertia of the revolving parts.

Example 24-1:
A 750 hp, 250 V, 1200 r/min dc motor is connected to a 208 V, 3-phase, 60 Hz line using a 3-phase bridge converter (Fig. 24-5a). The full-load armature current is 2500 A and the armature resis-

tance is 4 mΩ. Calculate:

a. the required firing angle α under rated full-load conditions;

b. the firing angle so the motor will run at 600 r/min at no-load;

c. the firing angle required so that the motor develops its rated torque at 400 r/min.

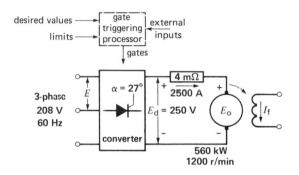

Figure 24-5   a. See Example 24-1.

Solution:

a. 1   At full load, the converter must develop a dc output of 250 V:

$$E_d = 1.35\,E \cos \alpha \qquad \text{Eq. 23-14}$$
$$250 = 1.35 \times 208 \cos \alpha$$
$$\cos \alpha = 0.89$$
$$\therefore \quad \alpha = 27°$$

b. 0   Full-load armature IR drop:

$$= 2500\,\text{A} \times 0.004\,\Omega = 10\,\text{V}$$
$$E_0 = 250 - 10 = 240\,\text{V}$$

b. 1   At 600 r/min:

$$E_0 = 240 \times (600/1200) = 120\,\text{V}$$

The IR drop at no load is negligible; consequently:

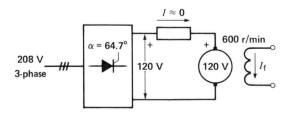

Figure 24-5   b. Motor running at no-load.

$$E_d = E_0 = 120\,\text{V}$$

b. 2   To determine the firing angle at no-load, we have:

$$E_d = 1.35\,E \cos \alpha \qquad \text{Eq. 23-14}$$
$$120 = 1.35 \times 208 \cos \alpha$$
$$\cos \alpha = 0.427$$
$$\therefore \quad \alpha = 64.7° \quad \text{(see Fig. 24-5b)}$$

c. 1   To develop rated torque, the armature current must be 2500 A

cemf at 400 r/min:

$$E_0 = (400/1200) \times 240 = 80\,\text{V}$$

Armature IR drop $= 2500 \times 0.004 = 10\,\text{V}$
Armature terminal voltage is:

$$E_d = 80 + 10 = 90\,\text{V}$$

c. 2   To determine the firing angle, we have:

$$E_d = 1.35\,E \cos \alpha \qquad \text{Eq. 23-14}$$
$$90 = 1.35 \times 208 \cos \alpha$$
$$\alpha = 71° \quad \text{(see Fig. 24-5c)}$$

Example 24-2:

Referring to Example 24-1, calculate the reactive power absorbed by the converter when the motor develops full torque at 400 r/min.

Solution:

The load condition is given in Fig. 24-5c. The dc power absorbed by the motor is:

$$P = E_d I_d = 90 \times 2500 = 225\,\text{kW}$$

Because the converter losses are negligible, the active power supplied by the ac source is also 225 kW.

The reactive power drawn from the ac source is given by:

$$Q = P \tan \alpha \qquad \text{Eq. 23-16}$$

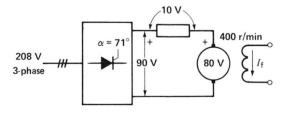

Figure 24-5   c. Rated torque at 600 r/min.

Q = 225 tan 71°

= 653 kvar (compare with the active power of 225 kW)

This example shows that an enormous amount of reactive power is required as the firing angle is increased. It even exceeds the reactive power needed at full load. Capacitors should be installed on the ac side of the converter to reduce the burden on the transmission line. Alternatively, a variable-tap transformer can be placed between the 3-phase source and the converter. By reducing the ac voltage at the lower speeds, the reactive power can be reduced dramatically. The reason is that the firing angle can then be kept between 15° and 20°.

24-4 Two-quadrant control, field reversal

We cannot always tolerate a situation where a motor simply coasts to a lower speed. To obtain a quicker response, we have to modify the circuit so that the motor acts temporarily as a generator. By controlling the generator output, we can make the speed fall as fast as we please. We often resort to dynamic braking using an external resistor. However, the converter can also be made to operate as an inverter, feeding power back into the 3-phase line. Such *regenerative braking* is often preferred because the kinetic energy is not lost. Furthermore, the generator output can be precisely controlled to obtain the desired rate of change in speed.

To make the converter act as an inverter, the polarity of E_d must be reversed. This means we have to reverse the polarity of E_0. Finally, E_d must be adjusted to be slightly less than E_0 to obtain the desired braking current (Fig. 24-6).

These changes are not as simple as they may at first appear. The polarity of E_d can be changed almost instantaneously by delaying the gate pulses by more than 90°. However, to change the polarity of E_0, we must reverse either the field or the armature, and this requires additional equipment. The reversal also takes a significant length of time. Furthermore, after the generator phase is over, we must again reverse the armature or field so that the machine runs as a motor. Bearing these conditions

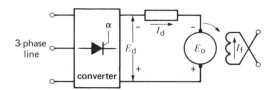

Figure 24-6   Motor control by field reversal.

in mind, we list the steps to be taken when *field reversal* is employed.

Step 1: Delay the gate pulses by nearly 180° so that E_d becomes quite large and negative. This operation takes a few milliseconds, after which current I_d is zero.

Step 2: Reverse the field current as quickly as possible so as to reverse the polarity of E_0. The total reversing time may last from 1 to 2 seconds, owing to the high inductance of the shunt field. The armature current is still zero during this interval.

Step 3: Reduce α so that E_d becomes slightly less than E_0, enabling the desired armature current to flow. The motor now acts as a generator, feeding power back into the ac line. Its speed drops rapidly towards the lower setting.

What do we do when the lower speed is reached? We quickly rearrange matters so that the dc machine again runs as a motor. This involves the following steps:

Step 4: Delay the gate pulses by nearly 180° so that E_d becomes quite large and negative. This operation takes a few milliseconds, after which current I_d is again zero.

Step 5: Reverse the field current as quickly as possible so as to make E_0 positive. The reversing time again lasts from 1 to 2 seconds. During this interval, the armature current is zero.

Step 6: Reduce α so that E_d becomes positive and slightly *greater* than E_0, enabling the desired armature current to flow. The machine now acts as a motor, and the converter is back to the rectifier mode.

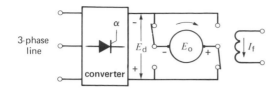

Figure 24-7   Motor control by armature reversal.

24-5 Two-quadrant control - armature reversal

In some industrial drives, the long delay associated with field reversal is unacceptable. In such cases, we reverse the armature instead of the field. This requires a high-speed reversing switch designed to carry the full armature current. The control system is arranged so that switching occurs only when the armature current is zero. Although this reduces contact wear and arcing, the switch still has to be fairly large to carry a current, say, of several thousand amperes.

Owing to its low inductance, the armature can be reversed in about 150 ms, which is about 10 times faster than reversing the field. Figure 24-7 is a simplified circuit showing a dc shunt motor connected to the converter by means of a reversing

contactor. To reduce speed, the same steps are followed as in the case of field reversal, except that the armature is reversed instead of the field.

24-6 Two-quadrant control - two converters

When speed control has to be even faster, we use two identical converters connected in reverse parallel. Both are connected to the armature, but only one operates at a given time, acting either as a rectifier or inverter (Fig. 24-8). The other converter is on "standby," ready to take over whenever power to the armature has to be reversed. Consequently, there is no need to reverse the armature or field. The time to switch from one converter to the other is typically 10 ms. Reliability is considerably improved, and maintenance is reduced. Balanced against these advantages are higher cost and increased complexity of the triggering source.

Because one converter is always ready to take over from the other, the respective converter voltages are close to the existing armature voltage, both in value and polarity. Thus, in Fig. 24-9, converter 1 acts as a rectifier, supplying power to the motor at a voltage slightly higher than the cemf E_0. During this period, gate pulses are withheld

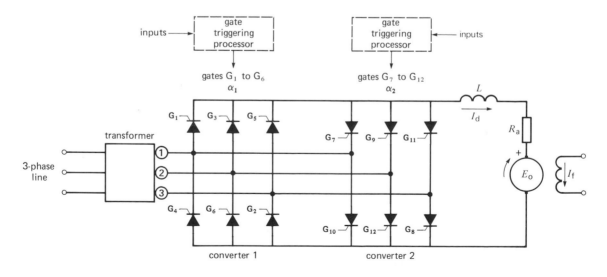

Figure 24-8   Two-quadrant control using two converters without circulating currents.

from converter 2 so that it is inactive. Nevertheless, the control circuit continues to generate pulses having a delay α_2 so that E_{d2} *would* be equal to E_{d1} if the pulses were allowed to reach the gates (G7 to G12).

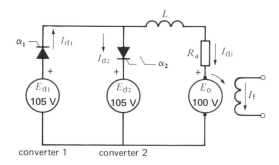

Figure 24-9 a. Converter 1 in operation, converter 2 blocked.

To reduce the speed, gate pulses α_1 are delayed and, as soon as the armature current has fallen to zero, the control circuit withholds the pulses to converter 1 and simultaneously unblocks the pulses to converter 2. Converter 1 becomes inactive and the delay angle α_2 is then reduced so that

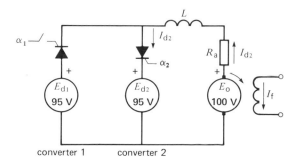

Figure 24-9 b. Converter 2 in operation, converter 1 blocked.

E_{d2} becomes slightly less than E_0, thus permitting reverse current I_{d2} to flow (Fig. 24-9b).

This current reverses the torque, and the motor speed decreases rapidly. During the deceleration

phase, α_2 is varied automatically so that E_{d2} follows the rapidly decreasing value of E_0. In some cases, α_2 is varied to maintain a constant braking current. During this period, the control circuit continues to generate gate pulses for converter 1 and the delay angle α_1 tracks α_2 so that E_{d1} *would* be equal to E_{d2} if the pulses were allowed to reach the gates (G1 to G6).

If the motor only operates in quadrants 1 and 4, it never reverses. Consequently, converters 1 and 2 always act respectively as rectifier and inverter.

Figure 24-10 Precise 4-quadrant electronic speed control is provided in a modern steel mill. *(Siemens)* ·

24-7 Two-quadrant control - two converters with circulating current

Some industrial drives require precise speed and torque control right down to zero speed. This means that the converter voltage may at times be close to zero. Unfortunately, the converter current is discontinuous under these circumstances. In other words, the current in each thyristor no longer flows for 120°. Thus, at low speeds, the torque and speed tend to be erratic, and precise control is difficult to achieve.

To get around this problem, we use two con-

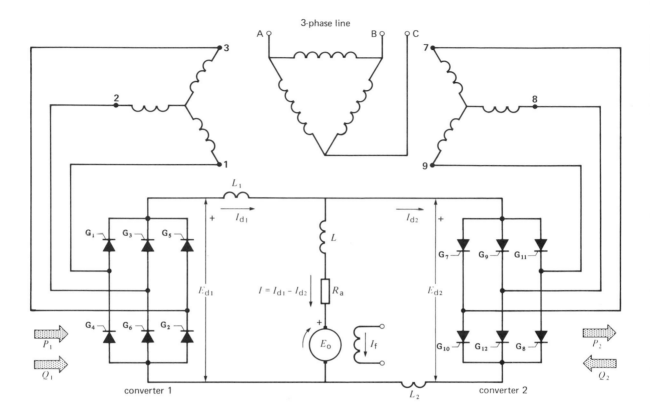

Figure 24-11   Two-quadrant control using two converters with circulating currents.

verters that function *simultaneously*. They are connected back-to-back across the armature. When one functions as a rectifier, the other functions as an inverter, and vice versa. The armature current I is the difference between currents I_{d1} and I_{d2} flowing in the two converters (Fig. 24-11). With this arrangement, the currents in both converters flow for $120°$, even when $I = 0$. Obviously, with two converters continuously in operation, there is no delay at all in switching from one to the other. The armature current can be reversed almost instantaneously; consequently, this represents the most sophisticated control system available. It is also the most expensive. The reason is that when converters operate simultaneously, each must be provided with a large series inductor (L_1, L_2) to limit the ac circulating currents. Furthermore, the

converters must be fed from independent sources, such as the isolated secondary windings of a 3-phase transformer. A typical circuit composed of a delta-connected primary and two wye-connected secondaries is shown in Fig. 24-11. Other transformer circuits are sometimes used to optimize performance, to reduce cost, to enhance reliability or to limit short-circuit currents.

Example 24-3:

The dc motor in Fig. 24-11 has an armature voltage of 450 V while drawing a current of 1500 A. Converter 1 delivers a current I_{d1} of 1800 A, and converter 2 absorbs 300 A. If the ac line voltage for each converter is 360 V, calculate:

a. the dc power associated with converters 1 and
   2

b. the active power drawn from the incoming 3-phase line;

c. the firing angles for converters 1 and 2;

d. the reactive power drawn from the incoming 3-phase line.

Solution:

a. The dc power delivered by converter 1 is:

$$P_1 = E_{d1}I_{d1}$$
$$= 450 \times 1800$$
$$= 810 \text{ kW}$$

The power absorbed by converter 2 (operating as an inverter) is:

$$P_2 = E_{d2}I_{d2}$$
$$= 450 \times 300$$
$$= 135 \text{ kW}$$

Note that the converter voltages are essentially identical because the voltage drops in L_1 and L_2 are negligible. This means that the respective triggering of converters 1 and 2 cannot be voltage-controlled (by E_0, for example). In effect, the triggering is current-controlled and is made to depend on the desired converter currents I_{d1} and I_{d2}.

b. The active power drawn from the incoming ac line is:

$$P = P_1 - P_2$$
$$= 810 - 135$$
$$= 675 \text{ kW}$$

Secondary winding 1, 2, 3 delivers 810 kW while secondary winding 7, 8, 9 receives 135 kW. It follows that the net active power drawn from the line (neglecting losses) is 675 kW.

c. The firing angle for converter 1 can be found from Eq. 23-14:

$$E_{d1} = 1.35 \, E \cos \alpha_1$$
$$450 = 1.35 \times 360 \cos \alpha_1$$
$$\cos \alpha_1 = 0.926$$
$$\therefore \quad \alpha_1 = 22.2°$$

The firing angle for converter 2 is found to have the same value. However, because it operates as an inverter, the angle is:

$$\alpha_2 = 180 - \alpha_1$$
$$= 180 - 22.2$$
$$= 157.8°$$

Note that it would take a very small change in either α_1 or α_2 to make a very big change in I_{d1} and I_{d2}. This is why α_1 and α_2 are current-controlled.

d. The reactive power drawn by converter 1 is:

$$Q_1 = P_1 \tan \alpha_1$$
$$= 810 \tan 22.2$$
$$= 330 \text{ kvar}$$

The reactive power drawn by converter 2 is:

$$Q_2 = P_2 \tan \alpha_2$$
$$= -135 \tan 157.8$$
$$= 55 \text{ kvar}$$

Consequently, the reactive power drawn from the incoming 3-phase line is:

$$Q = Q_1 + Q_2$$
$$= 330 + 55$$
$$= 385 \text{ kvar}$$

It is interesting to note that whereas the active powers substract, $(P = P_1 - P_2)$, the reactive powers add: $(Q = Q_1 + Q_2)$. The reason is that a naturally commutated converter always absorbs reactive power, whether it functions as a rectifier or inverter.

24-8 Two-quadrant control with positive torque

So far, we have discussed various ways to obtain torque-speed control when the torque reverses. However, many industrial drives involve torques that always act in one direction, even when the speed reverses. Hoists and elevators fall into this category because gravity always acts downwards whether the load moves up or down. Operation is therefore in quadrants 1 and 2.

Consider a hoist driven by a shunt motor having constant field excitation. The armature is connected to the output of a 3-phase, 6-pulse converter. When the load is being raised, the motor absorbs power from the converter. Consequently, it acts as a rectifier (Fig. 24-12). The lifting speed de-

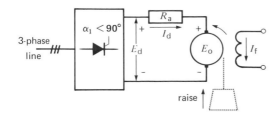

Figure 24-12   Hoist raising a load.

pends directly upon converter voltage E_d. The armature current depends upon the weight of the load.

When the weight is being lowered, the motor reverses, which changes the polarity of E_o. However, the descending weight delivers power to the motor, and so it becomes a generator. We can feed the electric power into the ac line by making the converter act as an inverter. The gate pulses are simply delayed by more than 90°, and E_d is adjusted to obtain the desired current flow (Fig. 24-13).

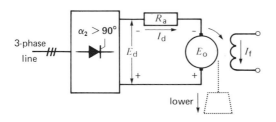

Figure 24-13   Hoist lowering a load.

Hoisting and lowering can therefore be done in a stepless manner, and no field or armature reversal is required. However, the empty hook may not descend by itself. The downward motion must then be produced by the motor, which means that either the field or armature has to be reversed.

24-9   Four-quadrant control

We can readily achieve 4-quadrant control of a dc machine by using a *single* converter, combined with either field or armature reversal. However, a great deal of switching may be required. Four-quadrant control is possible without field or arma-

ture reversal by using two converters operating back-to-back. They may function either alternately or simultaneously, as previously described.

The following example illustrates 4-quadrant control of an industrial drive.

Example 24-4:
An industrial drive has to develop the torque-speed characteristic given in Fig. 24-14. A dc shunt motor is used, powered by two converters operating back-to-back. The converters function alternately (only one at a time). Determine the state of each converter over the 26-second operating period, and indicate the polarity at the terminals of the dc machine. The speed and torque are considered positive when acting clockwise.

Solution:
The analysis of such a drive is simplified by subdividing the torque-speed curve into the respective 4 quadrants. In doing so, we look for those moments when either the torque or speed pass through zero. These moments always coincide with the transition from one quadrant to another. Referring to Fig. 24-14, the speed or torque passes through zero at 2, 8, 15, 21, and 25 s.

We draw vertical lines through these points (Fig. 24-15). We then examine whether the torque and speed are positive or negative during each interval. Knowing the respective signs, we can immediately state in which quadrant the motor is operating. For example, during the interval from 2 s to 8 s, both the torque and speed are positive. Consequently, the machine is operating in quadrant 1. On the other hand, in the interval from 21 s to 25 s, the speed is negative and the torque positive, indicating operation in quadrant 2.

Knowing the quadrant, we know whether the machine functions as a motor or generator. Finally, assuming that a positive (clockwise) speed corresponds to a "positive" armature voltage (Fig. 24-16a), we can deduce the required direction of current flow. This tells us which converter is in operation, and whether it acts as a rectifier or inverter.

Thus, taking the interval from 21 to 25 seconds,

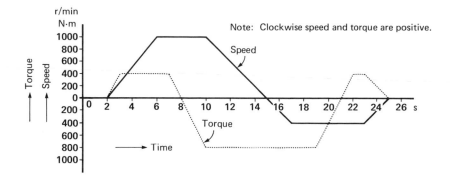

Figure 24-14   Torque-speed characteristic of an industrial drive.

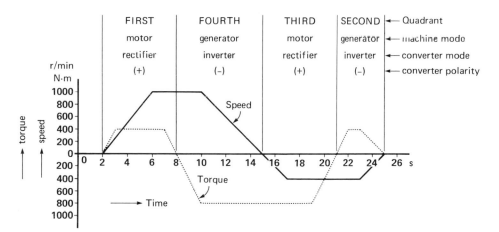

Figure 24-15   See Example 24-4.

it is clear that the machine acts as a generator. Consequently, one of the two converters *must* function as an inverter. But which one? To answer the question, we first look at the polarity of the armature. Because the speed is negative, the armature polarity is negative, as shown in Fig. 24-16b.

Current flows out of the positive terminal because the machine acts as a generator. Only converter 1 can carry this direction of current flow, and so it is the one in operation.

A similar line of reasoning enables us to determine the operating mode of each converter for the

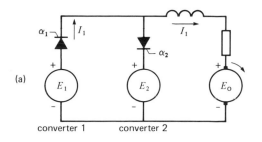

Figure 24-16   a. Polarity when the speed is positive.   b. Interval from 21 s to 25 s.

other intervals. The results are tabulated below; we encourage the reader to verify them.

Time interval	Operating mode	
	converter 1	converter 2
2 - 8 s	rectifier	off
8 - 15 s	off	inverter
15 - 21 s	off	rectifier
21 - 25 s	inverter	off

24-10 DC traction

Electric trains and buses have for years been designed to run on direct current, principally because of the special properties of the dc series motor. Many are now being modified to make use of the advantages offered by thyristors. Existing trolley lines still operate on dc and, in most cases, dc series motors are still used. To modify such systems, high-power *electronic choppers* are installed on board the vehicle (see Sec. 23-21). Such choppers can drive motors rated at several hundred horsepower, with outstanding results. To appreciate the improvement that has taken place, let us review some of the features of the older systems.

A train equipped with, say, two dc motors, is started with both motors in series with an external resistor. As the speed picks up, the resistor is shorted out. The motors are then paralleled and connected in series with another resistor. Finally, the last resistor is shorted out, as the train reaches its nominal torque and speed. The switching sequence produces small jolts, which, of course, are repeated during the electric braking process. Although a jolt affects passenger comfort, it also produces slippage on the tracks, with consequent loss of traction. The dc chopper overcomes these problems because it permits smooth and continuous control of torque and speed. We now study some simple chopper circuits used in conjunction with series motors.

Figure 24-17 shows the armature and field of a series motor connected to the output of a chopper. Supply voltage E_s is picked off from two overhead trolley wires. The inductor-capacitor combination L_1C_1 acts as a dc filter, preventing the sharp current pulses I_s from reaching the trolley line. The capacitor can readily furnish these high current pulses. The presence of the inductor has a smoothing effect so that current I drawn from the line has a relatively small ripple.

As far as the motor is concerned, the total inductance of the armature and series field is large enough to store and release the energy needed during the chopper cycle. Consequently, no external inductor is required. When the motor starts up, a low chopper frequency is used, typically 50 Hz. The corresponding "on" time T_a is typically 500 μs. In many systems, T_a is kept constant while the switching frequency varies. The top frequency (about 2000 Hz) is limited by the switching and turn-off time of the thyristors.

Figure 24-17

Direct-current series motor driven by a chopper. The chopper is not a switch as shown, but a force-commutated SCR.

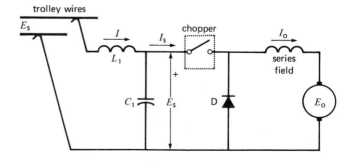

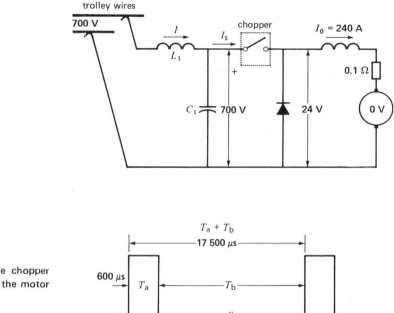

Figure 24-18 a.
See Example 24-5.

Figure 24-18 b.

Current pulses drawn by the chopper from the 700 V source when the motor si stalled.

Other choppers function at constant frequency, but with a variable "on" time T_a. In still more sophisticated controls, both the frequency and T_a are varied. In such cases, T_a may range from 20 μs to 800 μs. Nevertheless, the basic chopper operation remains the same, no matter how the on-off switching times are varied. Thus, the chopper output voltage E_0 is related to the input voltage E_s by the equation:

$$E_0 = E_s f T_a \qquad \text{Eq. 23-11}$$

where
f is the chopper frequency and T_a the "on" time.

Example 24-5:
A trolleybus is driven by a 150 hp, 1500 r/min, 600 V series motor. The nominal full-load current is 200 A and the total resistance of the armature and field is 0.1 Ω. The bus is fed from a 700 V dc line.

A chopper controls the torque and speed. The chopper frequency varies from 50 Hz to 1600 Hz, but the "on" time T_a is fixed at 600 μs:

a. calculate the chopper frequency and the current drawn from the line when the motor is at standstill and drawing a current of 240 A;
b. calculate the chopper frequency when the motor delivers its rated output.

Solution:

a. 1   Referring to Fig. 24-18a, the armature IR drop is 240 A x 0.1 Ω = 24 V, and the cemf is zero because the motor is at standstill.
　　Consequently, E_0 = 24 V, and E_s = 700 V. We can find the frequency from:

$$E_0 = E_s f T_a \qquad \text{Eq. 23-11}$$
$$24 = 700 f \times 600 \times 10^{-6}$$
$$f = 57.14 \text{ Hz}$$
$$T_a + T_b = 1/f = 1/57.14$$
$$= 17\,500 \ \mu s \quad \text{(Fig. 24-18b)}$$

a. 2   dc current drawn from line is:
$$I = I_s = P/E_s = 24 \times 240/700$$
$$= 8.23 \text{ A}$$

(Note the very low current drawn from the line during start-up)

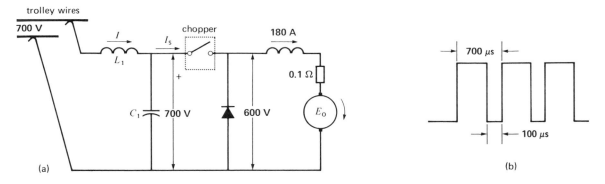

Figure 24-19   a. Conditions when the motor is running at rated torque and speed.

b. Corresponding current pulses drawn by the chopper from the 700 V source.

b. 1   At rated output, the voltage across the motor terminals is 600 V (Fig. 24-19a).

$$E_0 = E_s f T_a \qquad \text{Eq. 23-11}$$
$$600 = 700 f \times 600 \times 10^{-6}$$
$$\therefore f = 1429 \text{ Hz}$$
$$T_a + T_b = 1/f = 1/1429$$
$$= 700 \ \mu s \quad \text{(Fig. 24-19b)}$$

Line current $I = I_s = P/E_s$
$$= 600 \times 200/700$$
$$= 171 \text{ A}$$

Example 24-6:

Referring to Example 24-5 and Fig. 24-18a, calculate the peak value of currents I_s and I when the motor is at standstill.

Solution:

a. 1   Although the average value of I_s is 8.23 A, its peak value is 240 A. The current flows in a series of brief, sharp pulses. On the other hand, the armature current is steady at 240 A.

a. 2   The average value of line current I is 8.23 A, but since it flows continuously, the peak value is also 8.23 A. In practice, the current will have a ripple because inductor L does not have infinite inductance. Consequently, the peak value of I will be somewhat greater than the average value.

24-11 Current-fed dc motor

Some electronic drives involve direct current motors that do not look at all like dc machines. The reason is that the usual rotating commutator is replaced by a stationary electronic converter. We now discuss the theory behind these so-called "commutatorless" dc machines.

Consider a 2-pole dc motor having 3 independent armature coils, A, B, and C spaced at 120° to each other (Fig. 24-20). The two ends of each coil are connected to diametrically opposite segments of a 6-segment commutator. Two narrow brushes are connected to a *constant-current* source that successively feeds current into the coils as the armature rotates. A permanent magnet N, S creates the magnetic field.

With the armature in the position shown, current flows in coil A and the resulting torque causes the armature to turn counterclockwise. As soon as contact is broken with this coil, it is immediately established in the next coil. Consequently, conductors facing the N pole always carry currents that flow into the page, while those facing the S pole carry currents that flow out of the page (towards the reader). The motor torque is therefore continuous and may be expressed by:

$$\boxed{T = kIB} \qquad (24\text{-}2)$$

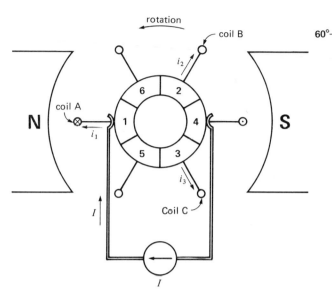

Figure 24-20   Special current-fed dc motor.

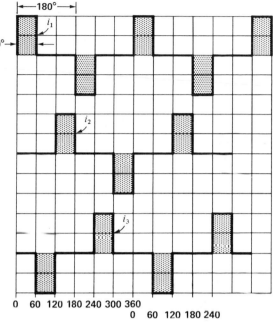

Figure 24-21   The dc current changes to ac in the coils.

where

T = motor torque [N·m]

I = current in the conductors [A]

B = average flux density surrounding the current-carrying conductors [T]

k = a constant, dependent upon the number of turns per coil, and the size of the armature

If the current and flux density are fixed, the resulting torque is also fixed, independent of motor speed.

The commutator segments are 60° wide; consequently, the current in each coil flows in 60° pulses. Furthermore, the current reverses every time the coil makes half a turn (Fig. 24-21). The alternating nature of the current is of crucial importance. If the current did *not* alternate, the torque developed by each coil would act first in one, then the opposite direction, as the armature rotates. The net torque would be zero, and so the motor would not develop any power.

Figure 24-21 shows that the ac currents in the 3

coils are out of phase by 120°. Consequently, the armature behaves as if it were excited by a 3-phase source. The only difference is that the current waveshapes are rectangular instead of sinusoidal. Basically, the commutator acts as a mechanical converter, changing the dc current from the dc source into ac current in the coils. The frequency is given by:

$$f = pn/120 \qquad (24\text{-}3)$$

where p is the number of poles and n the speed [r/min]. The frequency in the coils is automatically related to the speed because the faster the machine rotates, the faster the commutator switches from one coil to the next. In effect, the commutator generates a frequency which at all times is appropriate to the speed.

As the coils rotate, they cut across the magnetic field created by the N, S poles. An ac voltage is therefore induced in each coil, and its frequency is also given by Eq. 24-3. Furthermore, the voltages are mutually displaced at 120° owing to the way the coils are mounted on the armature. The in-

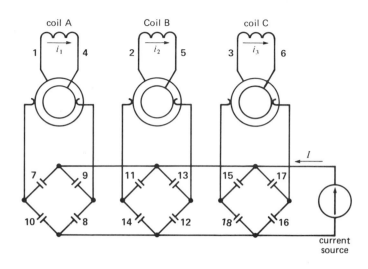

Figure 24-22

The commutator can be replaced by an array of mechanical switches and a set of slip rings.

duced ac voltages appear as a dc voltage between the brushes. The reason is that the brushes are always in contact with coils that are moving in the same direction through the magnetic field; consequently, the polarity is always the same (see Sec. 2-14).

If the brushes were connected to a dc *voltage source* E, the armature would accelerate until the induced voltage E_0 is about equal to E (Sec. 15-2). What determines the speed when the armature is fed from a *current source*, as it is in our case? The speed will increase until the load torque is equal to the torque developed by the motor. Thus, while the speed of a voltage-fed armature depends upon equilibrium between induced voltage and applied voltage, the speed of a *current-fed* armature depends upon equilibrium between motor torque and load torque. The torque of a mechanical load always rises with increasing speed. Consequently, for a given motor torque, a state of torque equilibrium is always reached, provided the speed is high enough. Care must be taken so that current-fed motors do not run away when the load torque is removed.

24-12 Commutator replaced by reversing switches

Recognizing that each coil in Fig. 24-20 carries an alternating current, we can eliminate the commu-

tator by connecting each coil to a pair of slip rings and bringing the leads out to a set of mechanical reversing switches (Fig. 24-22). Each switch has 4 normally open contacts. Considering coil A, for example, switch contacts 7 and 8 are closed during the 60° interval when coil side 1 faces the N pole (Fig. 24-23). The contacts are then open for 120° until coil side 4 faces the N pole, whereupon contacts 9 and 10 close for 60°. Consequently, by synchronizing the switch with the *position* of coil

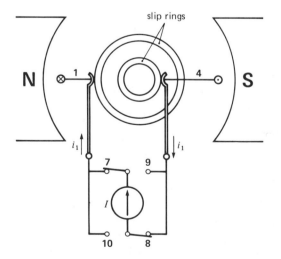

Figure 24-23   Circuit showing how current is controlled in coil A.

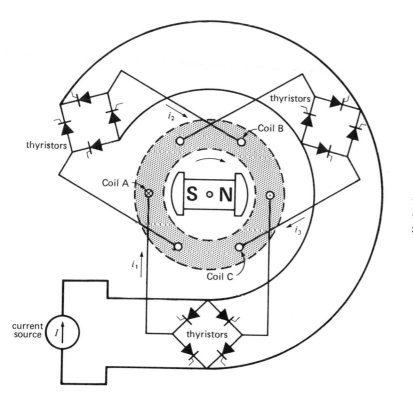

Figure 24-24

The armature is now the stator, and the switches have been replaced by SCRs.

A, we obtain the same result as if we used a commutator.

Coils B and C operate the same way, but they are energized at different times. Figure 24-22 shows how the array of 12 contacts and 6 slip rings are connected to the current source. The reversing switches really act as a 3-phase mechanical inverter, changing dc power into ac power. The slip rings merely provide electrical contact between the revolving armature and the stationary switches and power supply.

Clearly, the switching arrangement of Fig. 24-22 is more complex than the original commutator. However, we can simplify matters by making the armature stationary and letting the permanent magnets rotate. By thus literally turning the machine inside out, we can eliminate 6 slip rings. Then, as a final step, we can replace each contact by a thyristor (Fig. 24-24). The 12 thyristors are triggered by gate signals that depend upon the in-

stantaneous *position* of the revolving rotor.

The dc motor in Fig. 24-24 looks so different from the one in Fig. 24-20 that we would never suspect they have the same properties. And yet they do.

For example:

1. If we increase the dc current I or the field strength of poles N, S, the torque increases, and consequently, the speed will increase.

2. If we shift the brushes against the direction of rotation in Fig. 24-20, current will start flowing in each coil a little earlier than before. Consequently, the ac current in each coil will *lead* the ac voltage induced across its terminals. We can produce exactly the same effect by firing the thyristors a little earlier in Fig. 24-24. Under these circumstances, the machine furnishes reactive power to the three thyristor bridges, at the same time as it absorbs active power from them.

3. If we shift the brushes by 180°, the current in each *coil* flows in the opposite direction to that shown in Fig. 24-20. However, the induced voltage in each coil remains unchanged because it depends only on the speed and direction of rotation. Consequently, the machine becomes a generator, feeding dc power back into the current source.

The same result occurs if we fire the thyristors 180° later in Fig. 24-24. The thyristors then behave as inverters feeding power back to the dc current source.

It is now clear that the machines in Figs. 24-20 and 24-24 behave the same way. The only difference between them is that one is equipped with a rotating mechanical commutator, while the other has a stationary electronic commutator composed of 12 thyristors. By firing the thyristors earlier or later, we produce the same effect as shifting the brushes.

24-13 Synchronous motor as a commutatorless dc machine

The revolving-field motor in Fig. 24-24 is built like a 3-phase synchronous motor. However, because of the way it receives its ac power, it behaves like a "commutatorless" dc machine. This has a profound effect upon its performance.

First, the "synchronous motor" can never pull out of step because the stator frequency is not fixed, but changes automatically with speed. The reason is that the frequency produced by the SCRs depends upon the speed of the rotor. For the same reason, the machine has no tendency to oscillate or hunt, under sudden load changes.

Second, the phase angle between the ac current in a winding and the ac voltage across it can be modified by altering the timing of the gate pulses. This enables the synchronous motor to operate at leading, lagging, or unity power factor.

Third, because the phase angle between the respective voltages and currents can be fully controlled, the machine can even function as a generator, feeding power back to the dc current source.

The thyristor bridges then operate as inverters.

Currents i_1, i_2, i_3 in Fig. 24-24 flow only during 60 degree intervals, as they did in the original dc machine. In practice, the conduction period can be doubled to 120°, by connecting the coils in wye and exciting them by a 3-phase, 6-pulse converter (Fig. 24-25). This reduces the number of thyristors by half. Furthermore, it improves the current-carrying capacity of the windings because the duration of current flow is doubled. Gate triggering is again dependent upon the position of the rotor. The phase angle between line voltage E_s and line current I is modified by firing the gates earlier or later. In the circuit of Fig. 24-25, the power factor of the motor has to be slightly leading to provide the reactive power absorbed by the converter.

As a matter of interest, the converter and motor of Fig. 24-25 could be replaced by the dc motor shown in Fig. 24-26. The armature coils are connected in wye and the 3 leads are soldered to a 3-segment commutator. The respective voltages and currents are identical in the two figures.

24-14 Standard synchronous motor and commutatorless dc machine

The machine shown in Fig. 24-25 can be made to function as a conventional synchronous motor by applying a *fixed* frequency to the SCR gates. Under these conditions, the input to the gate triggering processor no longer depends on rotor position or rotor speed.

Obviously then, the behavior of the machine as a commutatorless dc motor or synchronous motor depends upon the way the gates are fired. If the triggering frequency is constant, the machine acts as a synchronous motor. On the other hand, if the triggering frequency depends on the speed of the rotor, it behaves like a commutatorless dc motor.[*]

---

[*] Readers familiar with feedback theory will recognize that the basic distinction between the two machines is that one functions on open loop while the other operates on closed loop.

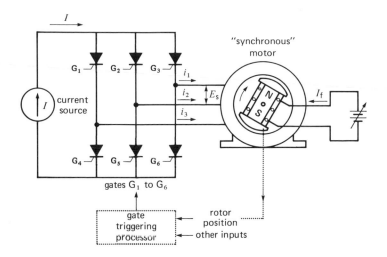

Figure 24-25   Commutatorless dc motor being driven by a converter.

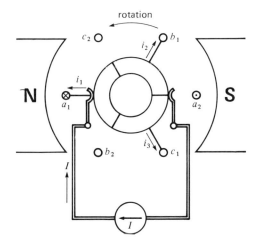

Figure 24-26   This elementary dc motor is equivalent to the entire circuit of Fig. 24-25.

24-15 Synchronous motor drive using current-fed dc link

Figure 24-27 shows a typical commutatorless dc motor circuit. It consists of two converters connected between a 3-phase source and the synchronous motor. Converter 1 acts as a controlled rectifier, feeding dc power to converter 2. The latter behaves as a naturally commutated inverter whose ac voltage and frequency are established by the motor.

Gate triggering of converter 1 is done at line frequency (60 Hz) while that of converter 2 is done at motor frequency.

With regard to controls, information picked off from various points is assimilated in the gate triggering processors which then send out appropriate gate firing pulses to converters 1 and 2. Thus, the processors receive information as to desired speed of rotation, actual speed, instantaneous rotor position, stator voltage, stator current, field current, etc. They interpret whether these inputs represent

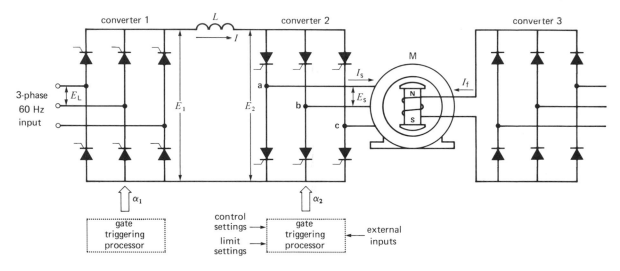

Figure 24-27 Commutatorless dc motor driven by a converter with a dc link. The output frequency can be considerably greater than 60 Hz, thus permitting high speeds.

A smoothing inductor L maintains a ripple-free current in the so-called *dc link* between the two converters. Current I is controlled by converter 1, which acts as a current source. A smaller bridge rectifier (converter 3) supplies the field excitation for the rotor.

Converter 2 is naturally commutated by voltage E_s induced across the terminals of the synchronous motor. This voltage is created by the revolving magnetic flux in the air gap. The flux depends upon the stator currents and the exciting current I_f. The flux is usually kept fixed; consequently, the induced voltage E_s is proportional to the motor speed.

normal or abnormal conditions, and emit appropriate gate pulses to correct the situation or to meet a specific command.

The gate pulses of converter 2 are controlled by the position of the rotor. This is accomplished by a set of transducers mounted on the stator next to the air gap. Other methods employ position transducers mounted on the end of the shaft. Owing to this method of gate control, the synchronous motor acts as a commutatorless dc machine. The motor speed may be increased by raising either the dc link current I or the field current I_f.

Stator voltage E_s produces a dc emf given by:

$$E_2 = 1.35\, E_s \cos \alpha_2 \qquad \text{Eq. 23-14}$$

where

E_2 = dc voltage generated by converter 2 [V]
E_s = effective line-to-line stator voltage [V]
α_2 = firing angle of converter 2 [°]

Similarly, the voltage produced by converter 1 is given by:

$$E_1 = 1.35\,E_L \cos \alpha_1$$

Link voltages E_1 and E_2 are almost equal, differing only by the negligible IR drop in the inductor. Firing angle α_1 is automatically controlled so that the magnitude of the link current is just sufficient to develop the required torque.

The stator line current I_s flows in 120° rectangular pulses, as shown in Fig. 24-28. These step-like currents produce a revolving MMF that moves in jerks around the armature. This produces torque pulsations, but they are almost completely damped out (except at low speeds) owing to the inertia of the rotor. The shaft therefore turns smoothly when running at rated speed.

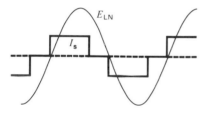

Figure 24-28   Typical voltage and current waveshapes in Fig. 24-27.

The motor line-to-neutral voltage E_{LN} and line voltage E_s are essentially sinusoidal. The field current, line current, and triggering are adjusted so that line current I_s leads the line-to-neutral voltage (Fig. 24-28). The reason is that the synchronous motor must operate at leading power factor to supply the reactive power absorbed by converter 2.

Converter 1 is designed so that under full-load conditions, firing angle α_1 is about 15° to minimize the reactive power drawn from the ac line.

Regenerative braking is accomplished by shifting the gate firing pulses so that converter 2 acts as a rectifier while converter 1 operates as an inverter. The polarity of E_2 and E_1 reverses, but

the link current continues to flow in the same direction. Power is therefore pumped back into the ac line and the motor slows down. During this period, the machine functions as a commutatorless dc generator.

Starting the motor creates a problem, because the stator voltage E_s is zero at standstill. Consequently, no voltage is available to produce the natural commutation of converter 2. To get around this difficulty, the converters are fired in such a way that short current pulses flow successively in phases ab, bc, and ca. The successive pulses create N, S poles in the stator that are always just ahead of their opposite poles on the rotor. Like a dog chasing its tail, the rotor accelerates and, when it reaches about 10 percent of rated speed, converter 2 takes over and commutation takes place normally. This pulse mode of operation can also be used to continually brake the motor as it approaches zero speed.

The speed control of synchronous motors using a current-fed dc link is applied to motors ranging from 1 kW to several megawatts. Permanent-magnet synchronous motors for the textile industry and brushless synchronous motors for nuclear reactor circulating pumps are two examples. Pumped-storage hydropower plants also use this method to bring the huge synchronous machines up to speed so that they may be smoothly synchronized with the line.

24-16 Synchronous motor and cycloconverter

We have seen that cycloconverters can directly convert ac power from a higher frequency to a lower frequency (Sec. 23-19). These converters are sometimes used to drive slow-speed commutatorless dc motors rated up to several megawatts. If a 60 Hz source is used, the cycloconverter output frequency is typically variable from zero to 10 Hz. Such a low frequency permits excellent control of the wave-shape of the output voltage by computer-controlled firing of the thyristor gates. As in drives with a dc link, no forced commutation is needed, with the result that the complexity of the elec-

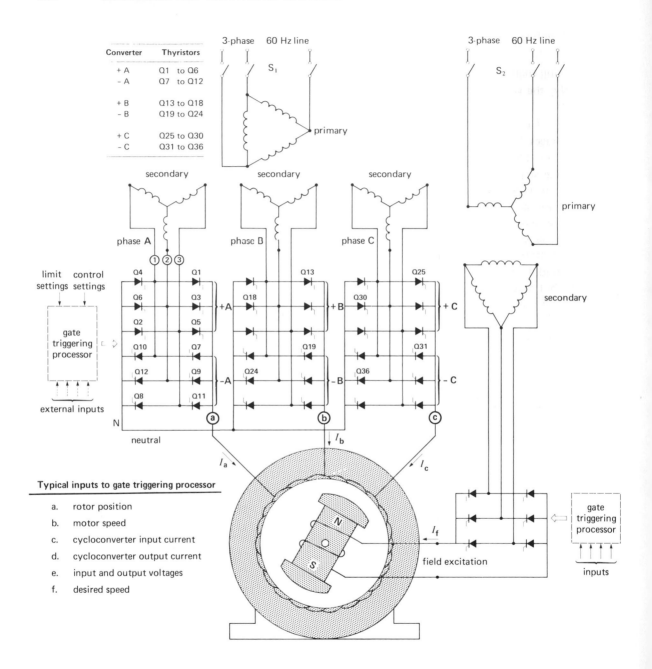

Figure 24-29 Cycloconverter driving a commutatorless dc machine. The output voltage associated with phase A is a slowly changing sine wave having a frequency of 6 Hz, which is 10 times less than the supply frequency. Thyristors Q1 to Q12 are triggered so as to track the desired sine wave as closely as possible. This produces the saw-tooth output voltage shown in Fig. 24-30. The power factor at the input to the motor is assumed to be unity. The corresponding power factor at the input to the cycloconverter is less than unity, owing to the delay angles.

tronics surrounding each SCR is considerably reduced.

Figure 24-29 shows three cycloconverters connected to the wye-connected stator of a 3-phase synchronous motor. Each cycloconverter produces a single-phase output, based upon the principle explained in Sec. 23-19. Referring to phase A, the associated cycloconverter is composed of two 3-phase bridges, +A and –A, each fed by the same 3-phase 60 Hz line.

Bridge +A generates the positive half cycle for line 1, while bridge –A generates the negative half. The two bridges are prevented from operating at the same time so as to eliminate circulating currents between them. The resulting low-frequency wave is composed of segments of the original 60 Hz voltage. By appropriate gate firing, the low frequency voltage can be made to approach a sine wave quite closely (Fig. 24-30). However, to reduce the reactive power absorbed from the 60 Hz line, the output voltage is usually designed to have a trapezoidal, flat-topped, wave-shape.

Both the cycloconverter and the 3-phase controlled rectifier supplying field current I_f function as current sources. The magnitudes of the stator currents and of I_f are controlled so as to keep a constant flux in the air gap. Furthermore, the gate pulses are timed so that the motor operates at unity power factor. However, even at unity power factor, the cycloconverter absorbs reactive power

from the 60 Hz line. The input power factor is typically 85 percent when the motor runs at rated power and speed.

Figure 24-31 shows a large slow-speed synchronous motor that is driven by a cycloconverter. The speed can be continuously varied from zero to 15 r/min. The low speed permits direct drive of the ball-mill without using a gear reducer. The motor is stopped by altering the gate firing so as to feed power back into the ac line.

Similar high-power low-speed systems are being introduced as propeller drives on board ships.

24-17 Cycloconverter voltage and frequency control

Returning to Fig. 24-30, we can see that the low-frequency output voltage is composed of selected segments of the ac line voltage. The segments are determined by the gate firing of the SCRs. The triggering is identical to that of a conventional 6-pulse rectifier, except that the firing angle is continually being varied so as to obtain an output voltage that approaches the desired sine wave. During the positive half cycle, thyristors Q1 to Q6 are triggered in sequence, followed by thyristors Q7 to Q12 for the negative half cycle. In our example, the low-frequency output voltage has the same peak amplitude as the 3-phase line voltage; consequently, it has the same effective value. The fre-

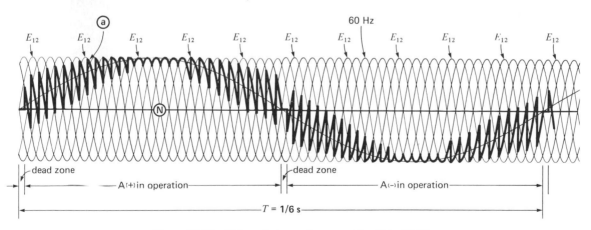

Figure 24-30   Voltage between lines a and N of Fig. 24-29.

Figure 24-31   a. Stator of a 3-phase synchronous motor rated 6400 kW (8576 hp), 15 r/min, 5.5 Hz, 80°C used to drive
a ball-mill in a cement factory. The stator is connected to a 50 Hz cycloconverter whose output frequency is variable from zero to 5.5 Hz. Internal diameter of stator: 8000 mm; active length of stacking:
950 mm; slots: 456. *(Brown Boveri)*

Figure 24-31 b.

The 44 rotor poles are directly mounted on the ball-mill so as to eliminate the need of a gear box. The two slip rings on the right-hand side of the poles bring the dc current into the windings.

Figure 24-31 c.
End view of the ball-mill showing the enclosed stator frame in the background. The mill contains 470 tons of steel balls and 80 tons of crushable material. The motor is cooled by blowing 40 000 m³ of fresh air over the windings, per hour. *(Brown Boveri)*

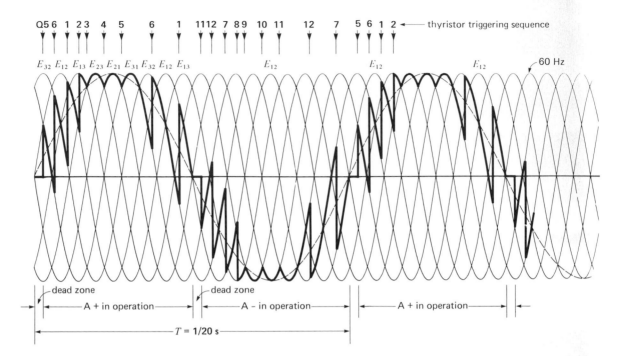

Figure 24-32   Waveshape of the output voltage E_{aN} in Fig. 24-29, at a frequency of 20 Hz. The effective output voltage has the same value as the effective input voltage between the 3-phase lines.

quency is 1/10 of the line frequency, or 6 Hz on a 60 Hz system.

We can gain a better understanding of the triggering process by referring to Fig. 24-32. In this case, the output frequency is 20 Hz on a 60 Hz system. The 60 Hz line voltages are indicated, as well as the firing sequence for the various SCRs. Although the resulting waveshape is very jagged, it does follow the general shape of the desired sine wave (shown dotted). The gate triggering times are quite irregular (not evenly spaced) to obtain the desired output voltage. That is why the firing program has to be under computer control.

If this 20 Hz voltage is applied to the motor of Fig. 24-29, the resulting current will be a reasonably good sine wave. In effect, the inductance of the windings smooths out the ragged edges that would otherwise be produced by the saw-tooth voltage wave.

To reduce the speed, both the frequency and

voltage have to be reduced in the same proportion. Thus, in Fig. 24-33, the frequency is now 10 Hz and the amplitude of the output voltage is also reduced by one half. The gate pulses are altered accordingly and, as we can see, a very jagged voltage is produced. Nevertheless, the current flowing in the windings will still be quite sinusoidal. A low output voltage requires a large firing angle delay, which in turn, produces a very low power factor on the 60 Hz line.

Although we have only discussed the behavior of phase A, the same remarks apply to phases B and C in Fig. 24-27. The gate firing is timed so that the low-frequency line curents I_a, I_b, I_c are mutually out of phase by 120°.

The cycloconverter drive is excellent when high starting torque and relatively low speeds are needed. However, it is not suitable for frequencies exceeding one half the system frequency. In such cases we use drives with a dc link, as previously discussed.

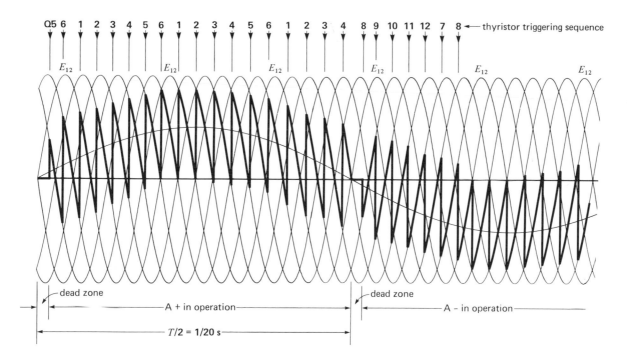

Figure 24-33 Waveshape of E_{aN} in Fig. 24-29 at a frequency of 10 Hz. The SCRs are triggered later in the cycle so that the effective value of the output voltage is only half that between the 3-phase lines. As a result, the flux in the air gap is the same in this figure as it is in Fig. 24-32.

QUESTIONS AND PROBLEMS

Practical level

24-1 State in which quadrants a machine operates: a. as a brake; b. as a motor; c. as a generator.

24-2 A machine is turning clockwise in quadrant 3. Does it develop a clockwise or counter-clockwise torque?

24-3 A 2-pole dc motor runs at 5460 r/min. What is the frequency of the voltage induced in the coils?

24-4 Referring to Fig. 24-4, the converter is connected to a 3-phase 480 V, 60 Hz line and the delay angle is set at 15°. Switch S is closed and the armature current is 270 A. Calculate:
a. the dc voltage across the armature;
b. the power supplied to the motor;
c. the average current in each diode;
d. the power output [hp] if the armature circuit has a resistance of 0.07 Ω.

24-5 Explain why the field or armature has to be reversed in order that the converter in Fig. 24-4 may feed power back into the ac line.

24-6 Compare the basic behavior of the power circuit of Fig. 24-8 with that of Fig. 24-11.

24-7 What is meant by the term commutatorless dc machine? Describe its construction and principle of operation.

Intermediate level

24-8 a. It is impossible for a machine to instantaneously change from a point in quadrant 1 to point in quadrant 2. Why?
b. Can it move instantaneously from quadrant 1 to quadrant 4?

24-9 A 4-pole, shunt-wound dc motor has an armature circuit resistance of 4 Ω. It is connected to a 240 V dc source and the no-load speed is 1800 r/min; the corresponding armature current is negligible. Assuming constant field excitation and assuming that armature reaction effects can be neglected, calculate:

a. the armature current at 900 r/min;

b. the mechanical power output [hp] at 1200 r/min;

c. the torque [N·m] at 300 r/min;

d. the starting torque [ft·lb].

e. Draw the torque-speed curve that passes through quadrants 1, 2 and 4 (see Fig. 24-3).

24-10 a. In Problem 24-9, draw the torque-speed curve if 60 V is applied to the armature, while maintaining the same field excitation;

b. what is the frequency of the current in the armature coils at a speed of 300 r/min?

24-11 The motor shown in Fig. 24-4 has a shunt field rated at 180 V, 2 A.

a. Calculate the effective value of the 60 Hz ac voltage that should be applied to the single-phase bridge circuit;

b. what is the peak-to-peak voltage ripple across the field terminals?

c. Does the field current contain a substantial ripple? Explain.

d. Draw the waveshape of the current in the ac line.

e. What is the effective value of the ac line current?

24-12 A 10 hp, 240 V, 1800 r/min permanent magnet dc motor has an armature resistance of 0.4 Ω and a rated armature current of 35 A. It is energized by the converter shown in Fig. 24-4. If the ac line voltage is 208 V, 60 Hz and the motor operates at full-load, calculate:

a. the delay angle required so that the motor operates at its rated voltage;

b. the reactive power absorbed by the converter;

c. the effective value of the line currents.

d. The induced voltage E_0 at 900 r/min.

24-13 The motor in Problem 24-12 is started at reduced voltage, and the starting current is limited to 60 A. Calculate:

a. the delay angle required;

b. the reactive power absorbed by the inverter;

c. does inductor L absorb reactive power from the ac line?

24-14 Referring to Fig. 24-8, an ac ammeter inserted in series with line 1 gives a reading of 280 A. Furthermore, a 3-phase power factor meter indicates a lagging power factor of 0.83. Calculate:

a. the value of the dc load current I_d;

b. the approximate delay angle if converter 1 is operating alone as an inverter.

24-15 The hoist motor shown in Fig. 24-12 is lifting a mass of 5000 lb at a constant speed of 400 ft/min.

a. Neglecting gear losses, calculate the value of E_0 if the armature current I_d is 150 A.

b. Knowing that $R_a = 0.1\ \Omega$, calculate the value of converter voltage E_d.

24-16 In Problem 24-15 (and referring to Fig. 24-12), if the same mass is lowered at a constant speed of 100 ft/min, calculate:

a. the armature current and its direction;

b. the value of E_0 and its polarity;

c. the value of E_d and its polarity.

d. In which direction does the active power flow in the ac line?

24-17 In Problem 24-15, if the mass is simply held still in midair, calculate:

a. the value of E_0;

b. the armature current I_d;

c. the value and polarity of E_d.

24-18 If the 3-phase line voltage is 240 V, 60 Hz, calculate the delay angle required:

a. in Problem 24-16;

b. in Problem 24-17.

24-19 a. Referring to Fig. 24-18a, calculate the average current and also the peak current carried by the freewheeling diode;
   b. what is the PIV across the diode?

24-20 An electronic chopper is placed between a 600 V dc trolley wire and the armature of a series motor. The switching frequency is 800 Hz and each power pulse lasts for 400 μs. If the dc current in the trolley wire is 80 A, calculate:
   a. the armature voltage;
   b. the armature current;
   c. draw the waveshape of the current in the freewheeling diode, assuming the armature inductance is high.

Advanced level

24-21 Referring to Fig. 24-11, the cemf E_0 has the polarity shown when the armature turns clockwise. Furthermore, when the armature current actually flows in the direction shown and the machine is turning clockwise, it operates in quadrant 1. State whether converters 1 and 2 are acting as rectifiers or inverters when the machine operates:
   a. in quadrant 2;
   b. in quadrant 3;
   c. in quadrant 4.
   d. Make a sketch of the actual direction of current flow and the actual polarity of E_0 in each case.

24-22 A 50 hp dc motor is required to drive a centrifuge at a speed ranging between 18 000 and 30 000 r/min. Owing to commutation problems associated with a standard commutator at these speeds, it is decided to use a commutatorless dc motor driven by two converters with a dc link (Fig. 24-27). A 2-pole motor is selected having a nominal rating of 50 hp, 30 000 r/min, 460 V, 60 A, 90 percent power factor leading. When the motor delivers its rated output, the delay angle for converter 2 is 155°. If the available 60 Hz line voltage is 575 V, calculate:
   a. the triggering frequency of converter 2;
   b. the dc link voltage;
   c. the delay angle of converter 1;
   d. the dc link current if the motor draws an input power of 41.5 kW;
   e. the fundamental ripple frequency in E_1;
   f. the fundamental ripple frequency in E_2;
   g. the power factor of the 60 Hz line;
   h. the effective value of I_s;
   i. show the flow of active and reactive power in the converters.

24-23 The following specifications are given for the motor shown in Fig. 24-20:

armature diameter:	100 mm
armature axial length:	50 mm
turns per coil:	200
rotational speed:	840 r/min
flux density in air gap:	0.5 T
armature current:	5 A

Using this information, calculate:
   a. the voltage induced in each coil;
   b. the dc voltage between the brushes;
   c. the frequency of the voltage in each coil;
   d. the frequency of the current in each coil;
   e. the power developed by the motor;
   f. the torque exerted by the motor.

24-24 a. Referring to Fig. 24-17, what would be the effect if capacitor C_1 were removed?
   b. In Fig. 24-18a, calculate the approximate value of capacitor C_1 [μF] so that the voltage across it does not drop by more than 50 V during the time of a current pulse.

25

ELECTRONIC CONTROL OF ALTERNATING CURRENT MOTORS

We saw in Chapter 24 that the electronic control of dc motors enables us to obtain high efficiency at all torques and speeds. Full 4-quadrant control is possible to meet precise high-speed industrial standards. The same remarks apply to the electronic control of ac motors. Thus, we find that squirrel-cage and wound-rotor induction motors, as well as synchronous motors, lend themselves well to electronic control. Whereas dc machines are controlled by varying the voltage and current, ac machines are often controlled by varying the voltage and frequency. Now we may ask, if dc machines do such an outstanding job, why do we also use ac machines? There are several reasons:

1. AC machines have no commutators; consequently, they require less maintenance;
2. AC machines cost less (and often weigh less) than dc machines;
3. AC machines are more rugged and work better in hostile environments.

In this chapter, we cover 3-phase motor controls, in keeping with the power emphasis of the book. However, the reader should first review the basic principles covered in Chapters 23 and 24.

25-1 Types of ac drives

Although there are many kinds of electronic ac drives, the majority can be grouped under the following broad classes:
1. Static frequency changers
2. Variable voltage controllers
3. Rectifier - inverter systems with natural commutation
4. Rectifier - inverter systems with self-commutation

Static frequency changers convert the incoming line frequency directly into the desired load frequency. Cycloconverters fall into this category, and they are used to drive both synchronous and squirrel-cage induction motors.

Variable voltage controllers enable speed and torque control by varying the ac voltage. They are used in squirrel-cage and wound-rotor induction motors. This method of speed control is the least expensive of all, and provides a satisfactory solution for small and medium-power machines used on fans, centrifugal pumps and electric hoists.

Rectifier - inverter systems rectify the incoming line frequency to dc, and the dc is reconverted to ac by an inverter. The inverter may be self-commutated, generating its own frequency, or it may be naturally commutated by the very motor it drives. Such rectifier-inverter systems with a dc link are used to control squirrel-cage and wound-rotor inductor motors and, in some cases, synchronous motors.

To better understand the basic principles of electronic control, we will first show how variable frequency affects the behavior of a squirrel-cage induction motor.

that the synchronous speed depends on the frequency. The question now arises, how is the torque-speed curve affected when both the voltage and frequency are varied? In practice, they are varied in the same proportion so as to maintain a constant flux in the air gap. Thus, when the frequency is doubled, the stator voltage is doubled. Under these conditions, the shape of the torque-speed curve remains the same, but its position along the speed axis shifts with the frequency.

Figure 25-1 shows the torque-speed curve of an 15 hp, (11 kW) 3-phase, 460 V, 60 Hz squirrel-cage induction motor. The full-load speed and

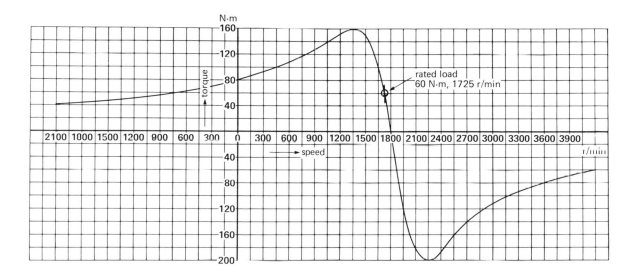

Figure 25-1   Torque-speed curve of a 15 hp, 460 V, 60 Hz, 3-phase squirrel-cage induction motor.

25-2 Shape of the torque-speed curve

The torque-speed curve of a 3-phase squirrel-cage induction motor depends upon the voltage and frequency applied to the stator. We already know that if the frequency is fixed, the torque varies as the square of the applied voltage. We also know

torque are respectively 1725 r/min and 60 N·m; the breakdown torque is 160 N·m and the locked rotor torque is 80 N·m.

If we reduce both the voltage and frequency to one fourth their original value (115 V and 15 Hz), the new torque-speed curve is shifted towards the left. The curve crosses the axis at a synchronous

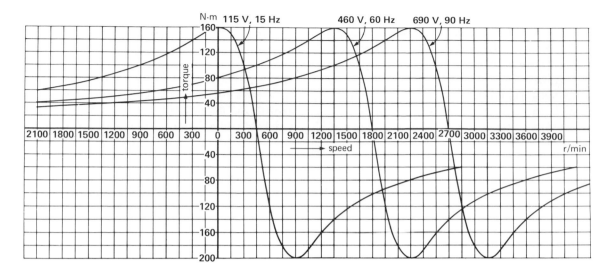

Figure 25-2 Torque-speed curve at three different frequencies.

speed of 1800/4 = 450 r/min (Fig. 25-2). Similarly, if we raise the voltage and frequency by 50 percent (690 V, 90 Hz), the curve is shifted to the right and the new synchronous speed is 2700 r/min.

Even if we bring the frequency down to zero (dc), the torque-speed curve retains essentially the same shape. Current can be circulated in any two lines of the stator leaving the third line open. The motor develops a symmetrical braking torque that increases with increasing speed, reaching a maximum in both directions, as shown in Fig. 25-3. In this figure, the dc current in the windings is adjusted to produce the rated breakdown torque.

Because the shape of the torque-speed curve is the same at all frequencies, it follows that the torque developed by an induction motor is the same whenever the slip speed (in r/min) is the same.

Example 25-1:

A standard 3-phase, 10 hp, 575 V, 1750 r/min, 60 Hz squirrel-cage induction motor develops a torque of 110 N·m at a speed of 1440 r/min. If the motor is excited at a frequency of 25 Hz, calculate:

a. the required stator voltage to maintain the same flux in the machine;

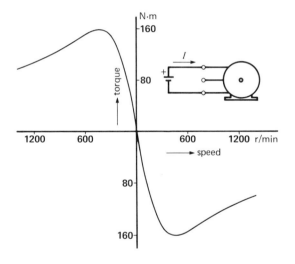

Figure 25-3 Stator excited by dc current.

b. the new speed at a torque of 110 N·m.

Solution:

a.   The voltage must be reduced in proportion to the frequency.

$$E = (25/60) \times 575 = 240 \text{ V}$$

b. 1 The synchronous speed of the 4-pole, 60 Hz motor is obviously 1800 r/min. Consequent-

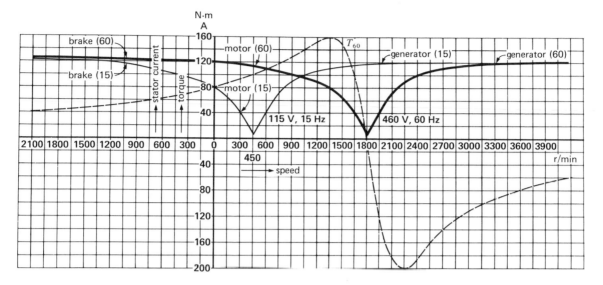

Figure 25-4   Current-speed curve at 60 Hz and 15 Hz. Also T-n curve at 460 V, 60 Hz.

ly, the slip speed at a torque of 110 N·m is:

$$n_1 = n_s - n$$
$$= 1800 - 1440 = 360 \text{ r/min}$$

b. 2   The slip speed is the same for the same torque, irrespective of the frequency. The synchronous speed at 25 Hz is:

$$n_s = (25/60) \times 1800 = 750 \text{ r/min}$$

b. 3   The new speed at 110 N·m is:

$$n = 750 - 360 = 390 \text{ r/min}$$

25-3   Current-speed curves

The current-speed characteristic of an induction motor is a V-shaped curve having a minimum value at synchronous speed. The minimum is equal to the magnetizing current needed to create the flux in the machine. Because the flux is intentionally kept constant, the magnetizing current is the same at all synchronous speeds.

Figure 25-4 shows the current-speed curve of the 15 hp, 460 V, 60 Hz squirrel-cage induction motor mentioned previously. We have plotted the effective values of current; consequently, the current is always positive. The locked rotor current is 120 A and the corresponding torque is 80 N·m.

If the stator voltage and frequency are varied in the same proportion, the current-speed curve re-

tains the same shape, but shifts along the horizontal axis. In effect, the torque-speed and current-speed curves move in unison as the frequency is varied. This produces a very important result.

Suppose, for example, that the voltage and frequency are reduced by 75 percent to 115 V, 15 Hz. The locked rotor current decreases to 80 A, but the corresponding torque *increases* to 160 N·m, equal to the full breakdown torque. Thus, by reducing the frequency, we obtain a larger torque with a smaller current (Fig. 25-5). This is one of the big advantages of frequency control. In effect,

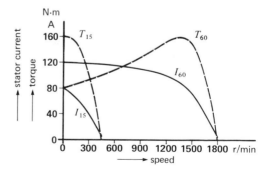

Figure 25-5   The starting torque increases and the current decreases with decreasing frequency.

we can gradually accelerate a motor and its load by progressively increasing the voltage and frequency. During the start-up period, the voltage and frequency can be varied automatically so that the motor develops close to its breakdown torque all the way from zero to rated speed. This ensures a rapid start at practically constant current.

In conclusion, an induction motor has excellent characteristics under variable frequency conditions. For a given frequency, the speed changes very little with increasing load. In many ways, the torque-speed characteristic resembles that of a dc shunt motor with variable armature-voltage control.

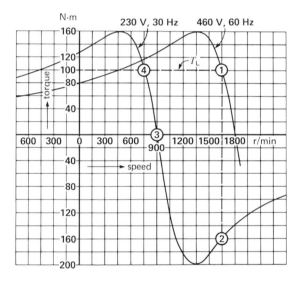

Figure 25-6   Effect of suddenly changing the stator frequency.

Example 25-2:

Using the information revealed by the 60 Hz torque-speed and current-speed curves of Fig. 25-4, calculate the voltage and frequency required so that the machine will run at 3200 r/min while developing a torque of 100 N·m. What is the corresponding stator current?

Solution:

a. 1   We first have to find the slip speed corresponding to a torque of 100 N·m. When the motor operates at 60 Hz and a torque of 100 N·m, the speed is 1650 r/min. Consequently, the slip speed is:

$$n_1 = n_s - n = 1800 - 1650 = 150 \text{ r/min}$$

The slip speed is the same when the motor develops 100 N·m at 3200 r/min. Consequently, the synchronous speed must be:

$$n_s = 3200 + 150 = 3350 \text{ r/min}$$

a. 2   The corresponding frequency is therefore:

$$f = (3350/1800) \times 60 = 111.7 \text{ Hz}$$

a. 3   The corresponding stator voltage is:

$$E = (3350/1800) \times 460 = 856 \text{ V}$$

The given current-speed curve shows that the stator current is 40 A when the torque is 100 N·m. Because the curve shifts along with the torque-speed curve, the current is again 40 A at 3200 r/min and 100 N·m.

25-4   Regenerative braking

A further advantage of frequency control is that it permits regenerative braking. Referring to Fig. 25-6, suppose the motor is connected to a 460 V, 60 Hz line. It is running at 1650 r/min driving a load of constant torque $T_L = 100$ N·m (operating point 1). If we suddenly reduce the frequency and voltage by 50 percent, the motor will immediately operate along the 30 Hz, 230 V torque-speed curve. Because the speed cannot change instantaneously (owing to inertia), we suddenly find ourselves at operating point 2 on the new torque-speed curve. The motor torque is negative; consequently, the speed will drop very quickly, following the 50 percent curve until we reach torque T_L (operating point 4). The interesting feature is that in moving along the curve from point 2 to point 3, energy is returned to the ac line, because the motor acts as an asynchronous generator during this interval.

The ability to develop a high torque from zero to full speed, together with the economy of regenerative braking, is the main reason why frequency-controlled induction motor drives are becoming

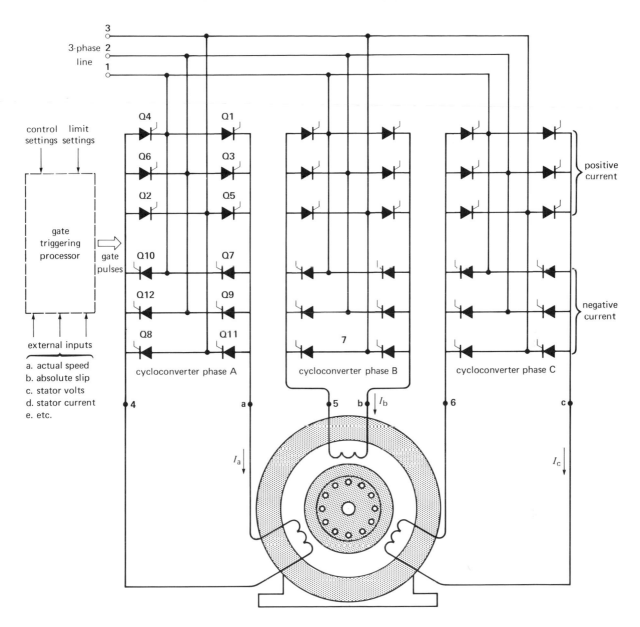

Figure 25-7   Squirrel-cage induction motor fed from a 3-phase cycloconverter.

so popular.

Large induction motors (100 hp and up) have characteristic curves much flatter than those shown in the previous figures. For example, the nominal breakdown torque is reached at slips of only 3 to 5 percent, compared to the 16.7 percent shown.

25-5 Squirrel-cage induction motor with cycloconverter

Figure 25-7 shows a 3-phase squirrel-cage induction motor connected to the output of a 3-phase cycloconverter. The circuit arrangement is similar to that of Fig. 24-29, except that the windings are fed from a common 3-phase source. Consequently, the windings cannot be connected in wye or delta, but must be isolated from each other. Motor speed is varied by applying appropriate gate pulses to the thyristors, to vary the output voltage and frequency. For example, the speed of a 2-pole induction motor can be varied from zero to 2400 r/min, on a 60 Hz line by varying the output frequency of the cycloconverter from zero to 40 Hz.

Excellent torque-speed characteristics in all 4 quadrants can be obtained, which, of course, includes regenerative braking. Standard 60 Hz motors can be used. The stator voltage is adjusted in proportion to the frequency to maintain a constant flux in the machine. Consequently, the torque-speed curves are similar to those shown in Fig. 25-2.

The stator current and voltage are jagged sine waves owing to the constant switching between output and input. Consequently, cycloconverter-fed motors run about $10°C$ hotter than normal, and adequate cooling must be provided. A separate ventilating fan may be needed at low speeds.

The cycloconverter can furnish the reactive power absorbed by the induction motor. However, a lot of reactive power is drawn from the 60 Hz line; the power factor is therefore poor. Cycloconverter drives are feasible on small and medium power induction motors operating at top speeds of about 2000 r/min on a 60 Hz line.

25-6 Squirrel-cage motor and variable voltage controller

We can vary the speed of a 3-phase squirrel-cage induction motor by simply varying the stator voltage. Consider, for example, a motor driving a blower or centrifugal pump. The stator is con-

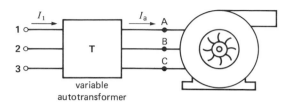

Figure 25-8   Variable-speed blower motor.

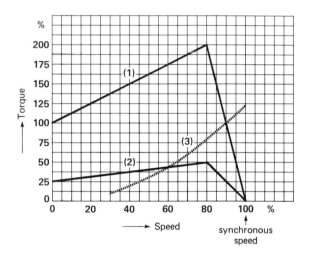

Figure 25-9   Torque-speed curve of motor at rated voltage (1) and 50% rated voltage (2). Curve (3) is the torque-speed characteristic of the fan.

nected to a variable 3-phase autotransformer (Fig. 25-8). At rated voltage, the torque-speed characteristic of the motor is given by curve 1 of Fig. 25-9. To simplify the drawing, the curve is shown as two straight lines. If we apply half the rated voltage, we obtain curve 2. Because torque is proportional to the square of the applied voltage, the torques in curve 2 are only 1/4 of the corresponding torques in curve 1.

The load torque of a blower varies nearly as the square of the speed. This typical characteristic, shown by curve 3, is superimposed on the motor torque-speed curves. Thus, at rated voltage, the intersection of curves 1 and 3 shows that the blower runs at 90 percent of synchronous speed. On the

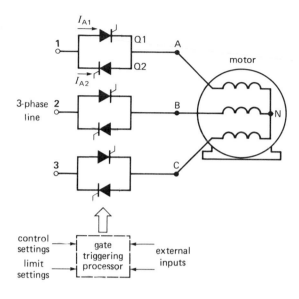

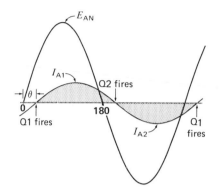

Figure 25-11   Waveshapes at rated voltage.

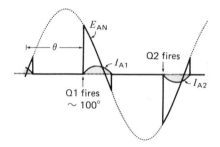

Figure 25-10   Variable-voltage speed control of a squirrel-cage induction motor.

Figure 25-12   Waveshapes at 50% rated voltage.

other hand, at half rated voltage, the blower rotates at only 60 percent of synchronous speed. By reducing the voltage still more, we can make the speed as low as we please.

We can replace the variable autotransformer by 3 sets of SCRs connected back-to-back, as shown in Fig. 25-10. To produce rated voltage across the motor, the respective thyristors are fired with a delay θ equal to the phase angle lag that *would* exist if the motor were directly connected to the line. Figure 25-11 shows the resulting current and line-to-neutral voltage for phase A. The thyristors in phases B and C are triggered the same way, except for an additional delay of $120°$ and $240°$, respectively.

To reduce the motor voltage, we delay the firing angle θ. For example, to obtain 50 percent rated voltage, all the pulses are delayed by about $100°$. The resulting voltage and current for phase A are given in Fig. 25-12. The voltage is distorted, and there is a considerable lag between the current and the voltage. The distorsion increases the losses in the motor compared to the autotransformer method. Furthermore, the power factor is consid-

erably lower, because of the large phase angle lag θ. Nevertheless, to a first approximation, the basic torque-speed characteristics given in Fig. 25-9 still apply.

Owing to the considerable I^2R losses and low power factor, this type of electronic speed control is only feasible for motors rated below 20 hp. Small hoists are well suited to this type of control, because they operate intermittently. Consequently, they can cool off during the idle and light-load periods.

Special thyristors have been developed which can both turn on and turn off a current by applying appropriate signals to the gates. Such *gate turn-off* thyristors (GTOs) can improve the power factor by forcing the current to flow during the voltage peaks. Thus, in Fig. 25-13, the current is

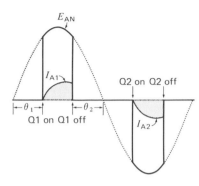

Figure 25-13   The input power factor can be raised to unity by controlling the turn-on and turn-off time.

essentially in phase with the voltage, because firing angle θ_1 is equal to extinction angle θ_2. We can even produce a leading power factor by decreasing θ_1 and simultaneously increasing θ_2.

25-7 Self-commutated inverters for squirrel-cage motors

In Section 24-15, we saw that a synchronous motor can be driven by a naturally commutated inverter. This is possible because the synchronous motor can provide the reactive power needed by the inverter. Unfortunately, if we replace the synchronous motor by an induction motor, the frequency conversion system breaks down, because an induction motor cannot deliver reactive power. Worse still, it actually absorbs reactive power.

Nevertheless, we can drive an induction motor using a *self-commutated* inverter. Such an inverter operates quite differently from a naturally-commutated inverter. First, it is able to generate its own frequency, determined by the frequency of the pulses applied to the gates. Second, it can either absorb or deliver reactive power. The reactive power generated or absorbed depends upon the switching action of the power thyristors.

Self-commutated inverters convert dc power into ac power. The power thyristors are arranged in a conventional 3-phase bridge circuit. However, each thyristor is surrounded by an array of capacitors, inductors, diodes and auxiliary thyristors.

The purpose of these auxiliary components is to force some power thyristors to conduct when normally they would not, and to force other power thyristors to stop conducting before their "natural" time. It is precisely this special switching action that enables these converters to generate reactive power.

Owing to the complexity and variety of the switching circuits used, we show the self-commutated inverter as a simple 5-terminal device having a dc input and a 3-phase output (Fig. 25-14). This simple representation also helps us understand the basic features of all self-commutated inverters:

1. The inverter power loss is negligible; consequently, the dc input power is equal to the active ac output power.
2. The reactive power generated by the inverter is not produced by the commutating capacitors included in its circuitry. The reactive power is due to the switching action alone.
3. The reactive power output requires no dc power input.
4. The thyristors switch the dc input terminals to the ac output terminals in a controlled sequence, with negligible voltage drop. It follows that:
   a. in a *voltage-fed* inverter, the ac line voltages are successively equal to the dc input voltage;
   b. in a *current-fed* inverter, the ac line currents are successively equal to the dc current.

To control the speed of a squirrel-cage motor, we use a rectifier-inverter system with a dc link. The rectifier is connected to the 3-phase supply line and the inverter is connected to the stator (Fig. 25-14a). Two types of dc links are used - constant current and constant voltage. The constant current link supplies a constant current to the inverter, which is then fed sequentially into the three phases of the motor (Fig. 25-14). Similarly, the constant-voltage link furnishes a constant voltage to the inverter which is switched from one phase to the next of the induction motor (Fig. 25-15). The switching sequence may be simple or complex, depending on the design of the converter. In the following sections, we describe some of the more common switching schemes used.

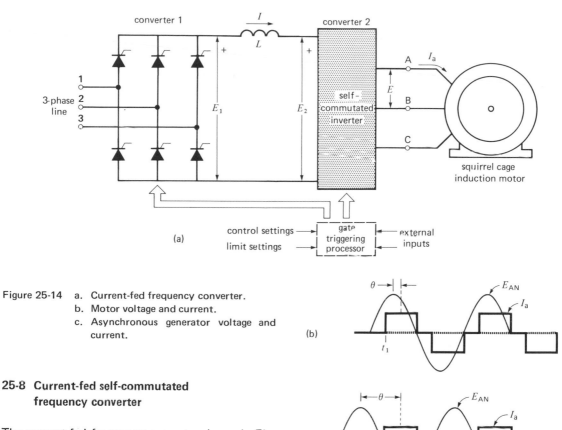

Figure 25-14 a. Current-fed frequency converter.
          b. Motor voltage and current.
          c. Asynchronous generator voltage and
              current.

25-8 Current-fed self-commutated frequency converter

The current-fed frequency converter shown in Fig. 25-14a is used to control the speed of *individual* squirrel-cage motors. The current in each phase is a rectangular pulse which flows for 120°. On the other hand, the voltage between the lines is reasonably sinusoidal. The reason is that the 3-phase rectangular current pulses together produce a revolving magnetic field that is almost sinusoidal. Figure 25-14b shows the line current I_a in one phase, and the associated line-to-neutral voltage E_{AN}. Phase angle θ corresponds to the operating power factor of the motor. It depends upon the properties of the motor itself and not upon the switching action of the inverter. In effect, although instant t_1 coincides with the firing of the thyristor connected to phase A, the *timing* of the pulse is not determined by the zero crossing point of voltage E_{AN}. The voltage finds its own place, so to speak, depending upon the particular speed, torque, and direction of rotation the motor happens to have.

We can obtain regenerative braking (generator action) by changing the firing angle of converter 2. This reverses the polarity of E_2. By simultaneously retarding the triggering of the thyristors in converter 1, we also reverse the polarity of E_1. Consequently, converter 1 acts as an inverter, feeding power back into the ac line. The new relationship between stator voltage and stator current is shown in Fig. 25-14c. Note that the inverter continues to supply reactive power to the motor during this regenerative braking period.

The direction of rotation is easily changed by altering the phase sequence of the pulses that trigger the gates of converter 2. Consequently, this static frequency converter can operate in all four

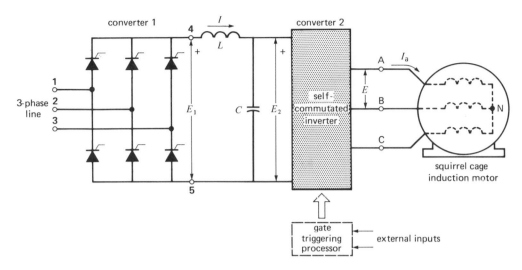

Figure 25-15   a.   Voltage-fed frequency converter.

quadrants with high efficiency. The torque-speed curve can be moved around to any position, as already shown in Fig. 25-2. High inertia loads can be quickly brought up to speed by designing the control system so that the full breakdown torque is developed as the motor accelerates.

In practice, the output frequency of an inverter may be varied over a range of 20:1, with top frequencies of the order of 1000 Hz. The ac voltage is changed in proportion to the frequency so as to maintain a constant stator flux. Consequently, the dc link voltage E_1 must be reduced as the speed decreases. This is done by increasing the firing angle of the thyristors in converter 1. Unfortunately, this increases the reactive power drawn from the 3-phase line, and capacitors may have to be installed to correct the low power factor.

Various control systems have been developed to improve the dynamic performance of the motor. The inputs to the gate triggering processor are selected so that the system approaches the behavior of a dc drive. Some of these control schemes are very elaborate, involving the monitoring of speed, slip, and even the instantaneous position of the revolving flux inside the motor. However, these sophisticated controls do not change the basic operation of the current-fed frequency converter.

25-9 Voltage-fed self-commutated frequency converter

In some industrial applications, such as in textile mills, the speeds of several motors have to move up and down together. These motors must be connected to a common bus in order to function at the same voltage. The current-fed frequency converter is not feasible in this case because it tends to supply a constant current to the total ac load, irrespective of the mechanical loading of individual machines. Under these circumstances, we use a *voltage-fed* frequency converter. It consists of a rectifier and a self-commutated inverter connected by a dc link (Fig. 25-15a).

A 3-phase bridge rectifier produces a dc voltage E_1. An LC filter ensures a "hard" dc voltage at the input to the inverter. The inverter successively switches this input voltage, in positive and negative pulses of 120° duration across the lines of the 3-phase load (Fig. 25-15b).

The inverter output voltage E is varied in proportion to the frequency so as to maintain a constant flux in the motor (or motors). Because the peak ac voltage is equal to the dc voltage E_2, it follows that rectifier voltage E_1 must be varied as the frequency varies. The speed of the motor can be

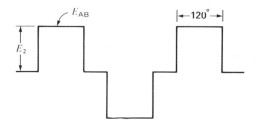

Figure 25-15   b. Motor line-to-line voltage.

controlled from zero to maximum while developing full torque.

Power reversal is possible but, owing to the special arrangement of the auxiliary components in converter 2, the link current I reverses when the motor acts as a generator. Voltage E_2 does not change polarity as it does in a current-fed inverter. Because converter 1 cannot accept reverse current flow, a third converter (not shown) has to be installed in reverse parallel with converter 1 to permit regenerative braking. The third converter functions as an inverter, and while it operates, converter 1 is blocked.

25-10   Chopper speed control
of a wound-rotor induction motor

We have already seen that the speed of a wound-rotor induction motor can be controlled by placing

three variable resistors in the rotor circuit (Sec. 17-16). Another way to control the speed is to connect a 3-phase bridge rectifier across the rotor terminals and feed the rectified power into a *single* variable resistor. The resulting torque-speed characteristic is identical to that obtained with a 3-phase rheostat. Unfortunately, the single rheostat still has to be varied mechanically in order to change the speed. We can make an all-electronic control by adding a chopper and a fixed resistor to the secondary circuit, as shown in Fig. 25-16. In this circuit, capacitor C supplies the high current pulses drawn by the chopper. The purpose of inductor L and free-wheeling diode D has already been explained in Sec. 23-21. By varying the chopper on-time T_a, the apparent resistance across the bridge rectifier may be made either high or low. The relationship is given by:

$$R_d = R_0/f^2T_a^2 \qquad\qquad \text{Eq. 23-12}$$

where R_d is the apparent external rotor resistance and the other terms have the meaning previously explained in Sec. 23-21.

Example 25-3:

The wound-rotor motor shown in Fig. 25-16 is rated at 30 kW (40 hp), 1170 r/min, 460 V, 60 Hz. The open-circuit rotor line voltage is 400 V and the load resistor R_0 is 0.5 Ω. If the chopper frequency is 200 Hz, calculate time T_a so that the motor develops a torque of 200 N·m at 900 r/min.

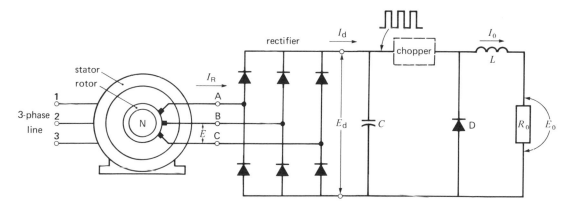

Figure 25-16   Speed control of a wound rotor induction motor using a chopper.

Solution:

This problem can be solved by applying the principles covered in Chapters 17 and 23.

a. 1   The rated synchronous speed is clearly 1200 r/min. The slip at 900 r/min is:

$$s = (n_s - n)/n_s \qquad \text{Eq. 17-2}$$
$$= (1200 - 900)/120$$
$$= 0.25$$

a. 2   The rotor line voltage at 900 r/min is:

$$E = sE_{0c} \qquad \text{Eq. 17-4}$$
$$= 0.25 \times 400$$
$$= 100 \text{ V}$$

a. 3   The dc voltage developed by the bridge rectifier is:

$$E_d = 1.35\,E \qquad \text{Eq. 23-5}$$
$$= 1.35 \times 100$$
$$= 135 \text{ V}$$

a. 4   Knowing the torque, we can calculate the power P_r delivered to the rotor:

$$T = 9.55\,P_r/n_s \qquad \text{Eq. 17-9}$$
$$200 = 9.55\,P_r/1200$$
$$P_r = 25\,130 \text{ W}$$

a. 5   Part of P_r is dissipated as heat in the rotor:

$$P_{jr} = sP_r \qquad \text{Eq. 17-7}$$
$$= 0.25 \times 25\,130$$
$$= 6282 \text{ W}$$

The power is actually dissipated in resistor R_0, but it is obviously equal to the rectifier output $E_d I_d$. Thus,

$$E_d I_d = P_{jr}$$
$$135\,I_d = 6282$$
$$I_d = 46.5 \text{ A}$$

a. 6   The apparent resistance at the input to the chopper circuit is therefore:

$$R_d = E_d/I_d$$
$$= 135/46.5$$
$$= 2.9 \ \Omega$$

Applying Eq. 23-12, we have;

$$R_0 = R_d f^2 T_a^2$$
$$0.5 = 2.9 \times 200^2 \times T_a^2$$
$$T_a = 2.08 \text{ ms}$$

The chopper on-time is therefore 2.08 ms.

This may be compared with the time T of one chopper cycle:

$$T = 1/200 = 5 \text{ ms}$$

Example 25-4:

In Example 25-3, calculate the magnitude of the current pulses drawn from the capacitor.

Solution:

a. 1   The current in R_0 is:

$$I_0^2 R_0 = P_{jr}$$
$$I_0^2 \times 0.5 = 6282$$
$$I_0 = 112 \text{ A}$$

The capacitor therefore delivers current pulses having an amplitude of 112 A and pulse width of 2.08 ms at a rate of 200 pulses per second. On the other hand, the rectifier continuously charges the capacitor with a current I_d of 46.5 A.

25-11   Recovering power in a wound rotor induction motor

Instead of dissipating the rotor power in a resistor, we could use it to charge a large dc battery (Fig. 25-17). Assuming the battery voltage E_2 can be varied from zero to some arbitrary maximum value, let us analyze the behavior of the circuit.

Voltage E across the rotor terminals is given by:

$$E = sE_{0c} \qquad \text{Eq. 17-4}$$

where s is the slip and E_{0c} is the open-circuit rotor voltage at standstill (Sec. 17-10).

On the other hand, rectified output voltage E_d is given by:

$$E_d = 1.35\,E \qquad \text{Eq. 23-5}$$

Because the IR drop in the smoothing inductor is negligible, $E_d = E_2$. Combining equations 17-4 and 23-5, we obtain:

$$\boxed{s = \frac{E_2}{1.35\,E_{0c}}} \qquad (25\text{-}1)$$

Equation 25-1 shows that the slip depends exclusively upon the battery voltage E_2, and is indepen-

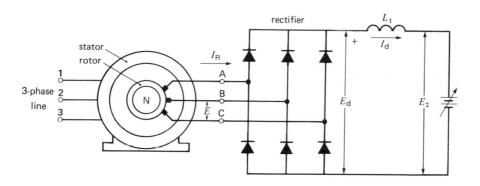

Figure 25-17   Speed control using a variable-voltage battery.

dent of the motor load. Consequently, we can vary the speed from essentially synchronous speed ($s = 0$) to zero ($s = 1$) by varying the battery voltage from zero to $1.35\,E_{oc}$.

In practice, instead of using a battery to absorb the rotor power, we use a 3-phase inverter which returns the power to the ac source. The naturally commutated inverter is connected to the same line that supplies power to the stator (Fig. 25-18). A transformer T is usually added, so that effective voltage E_T lies between 80 and 90 percent of E_2. The firing angle is then reasonably close to $180°$

thus reducing the reactive power absorbed by the inverter. As usual, the voltages are related by the equation:

$$E_2 = 1.35\,E_T \cos \alpha \qquad \text{Eq. 23-14}$$

where

E_2 = dc voltage developed by the inverter [V]

E_T = secondary line voltage of transformer T [V]

α = firing angle [°]

This method of speed control is very efficient because the rotor power is not dissipated in a

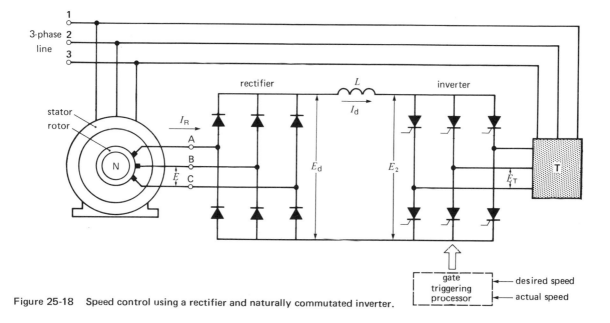

Figure 25-18   Speed control using a rectifier and naturally commutated inverter.

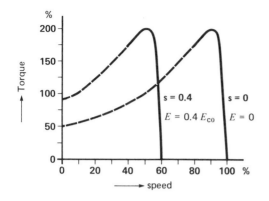

Figure 25-19 a. Torque-speed characteristics of a wound-rotor motor for two settings of voltage E_2.

group of resistors, but is returned to the line. Another advantage is that the slip is practically constant from no-load to full-load. Its value depends only on the setting of E_2. Referring to Fig. 25-19a, the torque-speed curve for a given voltage setting is identical to the torque-speed curve with the slip rings shorted, except that it is shifted to the left. The slip increases slightly with increasing torque, owing to the rotor resistance.

The rotor current I_R is rectangular and flows during 120° intervals (Fig. 25-19b). It is symme-

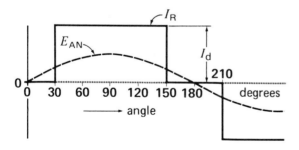

Figure 25-19 b. Rotor voltage and current.

trical with respect to the respective line-to-neutral rotor voltage E_{AN}. Consequently, the power factor of the "load" across the rotor is always unity.

This method of control is economical because the converters only have to carry the *slip* power of the rotor, which is considerably less than the input power to the stator. For example, if the lowest motor speed is 80% of synchronous speed, the power handled by the converters (at rated torque) is only 20% of the input power to the stator. It follows that the converters are much smaller than if they were placed in the stator circuit where the full stator power would have to be controlled.

Example 25-5:
A 3-phase, 3000 hp, 4000 V, 60 Hz, 8-pole wound-rotor induction motor drives a variable-speed centrifugal pump. When the motor is connected to a 4160 V line, the open-circuit rotor line voltage is 1800 V. A 3-phase 4000 V/480 V transformer is connected between the inverter and the line (Fig. 25-20). If the motor has to develop 800 kW at a speed of 700 r/min, calculate:
a. the power output of the rotor;
b. rotor voltage and current;
c. link voltage E_d and link current I_d;
d. firing angle of the inverter;
e. current in the primary and secondary lines of transformer T.

Solution:
a. 1 Synchronous speed is:

$$n_s = 120\,f/p = 120 \times 60/8 \qquad \text{Eq. 19-1}$$
$$= 900 \text{ r/min}$$

a. 2 Slip is:

$$s = (n_s - n)/n_s \qquad \text{Eq. 17-2}$$
$$= (900 - 700)/900 = 0.222$$

a. 3 Mechanical power is:

$$P_m = 800 \text{ kW} \quad (= 1072 \text{ hp})$$

but

$$P_m = P_r (1 - s) \qquad \text{Eq. 17-8}$$
$$800 = P_r (1 - 0.222)$$

Power supplied to rotor is:

$$P_r = 1028 \text{ kW}$$

Power "dissipated" by rotor:

$$P_{jr} = sP_r \qquad \text{Eq. 17-7}$$
$$= 0.222 \times 1028 = 228 \text{ kW}$$

a. 4 Consequently, 228 kW are fed back to the ac line.

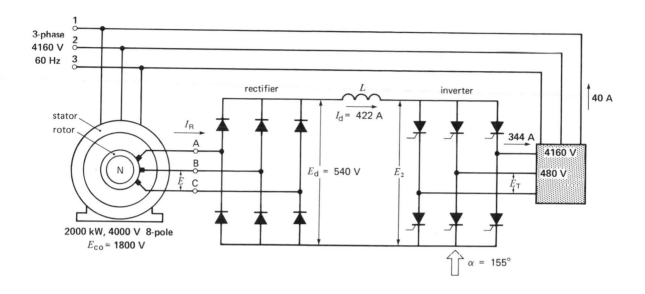

Figure 25-20   See Example 25-5.

b. 1   Rotor voltage $E = sE_{oc}$ Eq. 17-4
$$E = 0.222 \times 1800$$
$$= 400 \text{ V}$$

c. 1   Link voltage $E_d = 1.35\,E$
$$= 1.35 \times 400 = 540 \text{ V}$$

c. 2   Link current $I_d = 228\,000/540 = 422$ A

c. 3   The effective value of the rotor current is:
$$I_r = 0.816\,I_d = 0.817 \times 422 \quad \text{Eq. 23-15}$$
$$= 344 \text{ A}$$

d. $E_2 = 1.35\,E_T \cos \alpha$
$$540 = 1.35 \times 480 \cos \alpha$$
$$\alpha = 33.5°$$

The firing angle is actually $(180 - 33.5) = 146.5°$ because the converter acts as an inverter.

e. 1   The current in each phase of the 480 V line flows during $120°$ intervals and has a peak value of 422 A. The effective value is given by Eq. 23-15:
$$I = 344 \text{ A}$$

e. 2   Effective line current on the 4000 V side is:
$$I = (480/4000) \times 344 = 40 \text{ A}$$

25-12   Pulse width modulation

The frequency converters discussed so far create substantial harmonic voltages and currents. When these harmonics flow in the windings, they produce torque pulsations that are superimposed on the main driving torque. The pulsations are damped out at moderate and at high speeds owing to mechanical inertia. However, at low speeds, they may produce considerable vibration. Such torque fluctuations are unacceptable in some industrial applications, where fine speed control down to zero speed is required. Under these circumstances, the motor may be driven by *pulse width modulation* techniques.

To understand the principle, consider the voltage-fed frequency converter system shown in Fig. 25-21. A 3-phase bridge rectifier 1 produces a

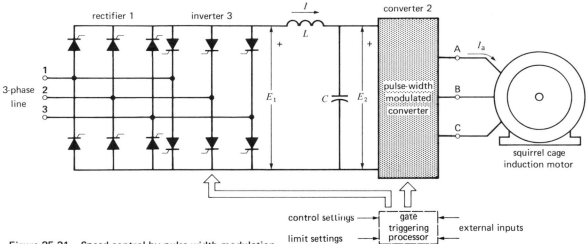

Figure 25-21   Speed control by pulse width modulation.

fixed voltage E_2 at the input to the self-commutated inverter 2. The inverter is triggered in a special way so that the output voltage is composed of a series of short positive pulses of constant amplitude, followed by an equal number of short negative pulses (Fig. 25-22a). The pulse width and pulse spacing are arranged so that the weighted average approaches a sine wave. The pulses as shown all have the same width, but in practice, the ones near the middle of the sine wave are made broader than those near the edges. By increasing the number of pulses per half cycle, we can make the output frequency as low as we please. Thus, to reduce the output frequency of Fig. 25-22a by a factor of 10, we increase the pulses per half-cycle from 5 to 50.

The pulse width and pulse spacing are specially designed so as to eliminate the low-frequency voltage harmonics, such as the 3rd, 5th, and 7th harmonics. The higher harmonics, such as the 17th, 19th, etc., are unimportant because they are damped out, both mechanically and electrically. Such pulse width modulation produces output currents having very low harmonic distorsion. Consequently, torque vibrations at low speeds and at standstill are eliminated.

In some cases, the output voltage has to be reduced while maintaining the same output frequen-

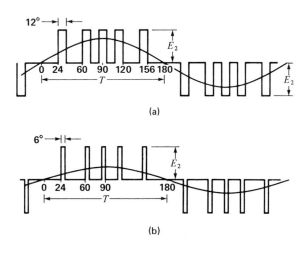

Figure 25-22   a. Waveform across one phase.
b. Waveform yielding the same frequency but half the voltage.

cy. This is done by reducing all the pulse widths in proportion to the desired reduction in output voltage. Thus, in Fig. 25-22b, the pulses are half as wide as in Fig. 25-22a, yielding an output voltage half as great, but having the same frequency. We can therefore vary both the output frequency and output voltage using a fixed dc input voltage. As a result, a simple diode bridge rectifier can be used to supply the fixed dc link voltage. The power factor of the 3-phase supply line is therefore high.

Four-quadrant operation can be achieved, but during power reversal, current I reverses while the polarity of E_2 remains the same. Consequently, an extra inverter 3 has to be placed in reverse parallel with rectifier 1 in order to feed power back to the line (Fig. 25-21). Rectifier 1 is automatically blocked while inverter 3 is in operation, and vice versa.

Pulse width modulation is effected by computer control of the gate triggering. The number of pulses per second is usually limited to a maximum of 400 and the pulse width is seldom less than 0.5 ms. Pulse width modulation is used to control induction motors up to several hundred horsepower.

QUESTIONS AND PROBLEMS

Practical level

25-1 Referring to Fig. 25-1, in which quadrants do the following torque-speed operating points occur?
   a. + 1650 r/min, + 100 N·m;
   b. + 3150 r/min, – 100 N·m.

25-2 Referring to Fig. 25-1, calculate the mechanical power [hp] of the motor when it runs at 450 r/min.

25-3 A standard 3-phase, 4-pole squirrel-cage induction motor is rated at 208 V, 60 Hz. We want the motor to run at a no-load speed of about 225 r/min while maintaining the same flux in the air gap. Calculate the required voltage and frequency to be applied to the stator.

25-4 Referring to Fig. 25-4, what is the current in the stator under the following conditions, knowing that the stator is energized at 460 V, 60 Hz?
   a. machine running as a motor at 1650 r/min and developing a torque of 100 N·m;
   b. machine running as a brake at 300 r/min;

   c. machine driven as an asynchronous generator at a torque of 120 N·m.

25-5 A 3-phase, 6-pole induction motor is driven by cycloconverter fed from a 60 Hz line. What is the approximate maximum speed that can be attained with this arrangement?

25-6 What is the basic difference between a naturally commutated and a self-commutated inverter?

25-7 A naturally commutated inverter can be used to drive a 3-phase synchronous motor but not a 3-phase induction motor. Explain.

25-8 In comparing the physical arrangement of the bridge rectifiers in one phase of the cycloconverter of Fig. 25-7, is there any difference with the bridge rectifier arrangement of Fig. 25-23?

25-9 A large induction motor has to run at a very low, steady speed. If electronic control is required, what type would be most appropriate?

Intermediate level

25-10 Referring to Fig. 25-1, and neglecting windage and friction losses, calculate the power P_r supplied to the rotor when the machine runs:
   a. as a motor at 1650 r/min;
   b. as a brake at 750 r/min;
   c. as a generator at 2550 r/min.

25-11 In Problem 25-10, calculate the value of the rotor I^2R losses in each case.

25-12 Referring to Fig. 25-1, calculate the voltage and frequency that must be applied to the machine so that it runs as a high-efficiency motor:
   a. At a speed of 1200 r/min developing a torque of 100 N·m;
   b. At a speed of 2400 r/min developing a torque of 60 N·m.

25-13 Referring to Fig. 25-4, calculate the voltage and frequency to be applied to the stator so that the locked rotor torque is 100 N·m at a current of 40 A.

25-14 The motor having the *T-n* characteristic given in Fig. 25-1 is running at a no-load speed of 1800 r/min. The total moment of inertia of the rotor and its load is 90 lb·ft². The speed has to be reduced to a no-load value of 1200 r/min, by suddenly changing the voltage and frequency applied to the stator. Calculate:
 a. the voltage and frequency required;
 b. the initial kinetic energy stored in the moving parts;
 c. the final kinetic energy in the moving parts;
 d. Is all the "lost" kinetic energy returned to the 3-phase line? Explain.

25-15 A 15 hp, 460 V, 3-phase, 60 Hz induction motor has the torque-speed characteristic given in Fig. 25-1.
 a. What is the new shape of the curve if we apply 230 V, 60 Hz to the stator?
 b. Calculate the new breakdown torque [ft·lb].

25-16 In Problem 25-15, calculate the stator voltage needed to reduce the breakdown torque to 60 N·m.

25-17 The blower motor in Fig. 25-8 has a nominal rating of 1/4 hp, 1620 r/min, 3-phase, 460 V. The respective torque-speed characteristics of the motor and blower are given in Fig. 25-9. Calculate the rotor I^2R losses (with the motor coupled to the blower):
 a. when the motor runs at rated voltage;
 b. when the stator voltage is reduced to 230 V;
 c. Is the rotor hotter in (a) or (b)?

25-18 In Problem 25-17, calculate the stator voltage required so that the blower will run at a speed of 810 r/min.

25-19 A 30 hp, 208 V, 3-phase, 3500 r/min, 60 Hz, wound-rotor induction motor produces an open-circuit rotor line voltage of 250 V. We wish to limit the locked-rotor torque to a maximum value of 40 N·m so as

to ensure a small starting current. A 3-phase bridge rectifier composed of six diodes is connected to the three slip rings. A single manual rheostat is connected across the dc output of the rectifier. Calculate:
 a. the synchronous speed of the motor;
 b. the power that is dissipated in the rotor circuit under locked-rotor conditions;
 c. the approximate dc output voltage;
 d. the approximate resistance of the rheostat and its power-handling capacity.

25-20 In Problem 25-19, a dc chopper is connected between the dc output of the rectifier and a 0.2 Ω resistor. If the chopper operates at a fixed frequency of 500 Hz, calculate the duration of the on-time T_a under locked-rotor conditions.

Advanced level

25-21 a. The squirrel-cage induction motor shown in Fig. 25-7 has a nominal rating of 50 hp, 460 V per phase, 60 Hz, 1100 r/min. The 3-phase line voltage is 208 V, 60 Hz. If we want the motor to run at a speed of about 200 r/min, while developing full-load torque, calculate the approximate voltage and frequency to be applied to the stator windings.
 b. If current I_a has an effective value of 60 A, calculate the approximate value of the peak current carried by each SCR.

25-22 The self-commutated inverter in Fig. 25-14a furnishes a motor current having an effective value of 26 A. What is the value of the dc link current?

25-23 A standard 50 hp, 1750 r/min, 3-phase, 200 V, 60 Hz squirrel-cage induction motor is driven by a current-fed self-commutated inverter shown in Fig. 25-14a. Calculate the voltage and frequency to be applied to the stator so that the motor develops its rated torque at 400 r/min. Assume the flux in the machine is constant.

PART VII

VII

ELECTRICAL ENERGY

26

BATTERIES AND FUEL CELLS

The invention of the storage cell in **1800** by Italian professor Alessandro Volta is probably one of the most important discoveries ever made in electricity. It enabled scientists, for the first time, to work with a continuous source of direct current. Previously, they could only study the momentary discharges produced by static electricity, which were far too brief to detect magnetic fields and other phenomena associated with current flow. The simple batteries Volta discovered paved the way to the fundamental discoveries of Oersted, Faraday, Henry and all the other scientists of the 19th century.

26-1 Principle of a cell

Now that we know how, it is very easy to build an electric cell. All we need to do is dip two dissimilar conductors into a jar of water and add a little acid. A difference of potential will spontaneously appear between the two electrodes. If we connect a resistance between the positive and negative terminals, a current will flow in the circuit (Fig. 26-1). The current produces a gradual change in the chemical composition of the electrolyte and of the two electrodes. Indeed, it is this chemical change that produces the electrical energy. When one of the electrodes (or the electrolyte) is more or less completely transformed, the difference of potential disappears and current ceases to flow. The cell is then said to be discharged.

In *primary cells*, the chemical reaction produces a progressive destruction of one or both electrodes and so the cell can no longer be used after it is discharged.

On the other hand, in *secondary cells*, the

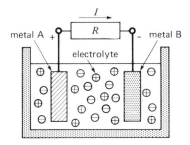

Figure 26-1 Two dissimilar electrodes immersed in an electrolyte produce an electric cell.

chemical change occurring during discharge is electrically reversible. We can rejuvenate such cells by passing a current through them in the opposite direction. This operation restores the electrodes and the electrolyte so that the entire cell reverts to its original chemical state.

Although the construction of a cell is quite simple, there is an ongoing search for electrodes and electrolytes that can deliver large quantities of energy, that last a long time, and that are light and inexpensive. Electrochemists have developed dozens of primary and secondary cells; some of the more common ones are listed in Table 26A. Individual cells develop potentials between 1.3 V and 2 V. The specific energy ranges from 40 kJ/kg to 300 kJ/kg depending on the type of cell and its duty cycle.

26-2 Theory of operation

In order to understand how a cell produces electricity, consider an electrolyte composed of a liquid solution of acid and water (Fig. 26-2a). The acid dissociates into positive and negative ions, as explained in Sec. 5-20. If we dip two dissimilar electrodes into the solution, one of them tends to capture positive ions while the other attracts negative ions. This affinity for one or the other of the two types of ions makes one electrode positive and the other negative (Fig. 26-2b).

If we connect a resistor between the two electrodes, an electric current will flow. The positive ions in solution gradually move towards the positive electrode while the negative ions gravitate towards the negative electrode.

When the positive ions touch the positive electrode, they *capture* electrons and immediately become neutral atoms. Conversely, when the negative ions touch *their* electrode, they *lose* electrons. It is precisely this exchange of electrons which produces the current in the circuit. The capture or loss of electrons also produces a chemical change. Thus, in Fig. 26-2c, the cross-hatched lines represent that portion of the positive electrode that has been chemically transformed. Similarly, the dotted

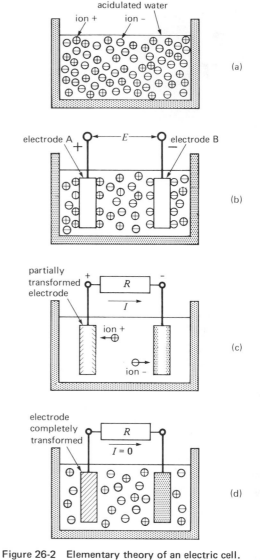

Figure 26-2   Elementary theory of an electric cell.
a. Electrolyte dissociates into positive and negative ions.
b. Electrode A has an affinity for positive ions.
c. Current flow produces a chemical change.
d. Current flow ceases when the chemical change is complete.

part of the negative electrode represents material that has undergone a chemical change. When one of the two plates is completely transformed, the current ceases to flow (Fig. 26-2d).

Despite their distinctive individual characteristics, primary and secondary cells share many common properties. We shall first study the similarities, leaving until later the special features that distinguish one cell from another.

26-3 Internal resistance

When a load is connected to a cell, the terminal voltage decreases. The voltage drop is due to the internal resistance of the cell. We can therefore represent a cell by an equivalent circuit composed of an internal voltage E_0, in series with a resistance r (Fig. 26-3). The internal voltage is equal to the

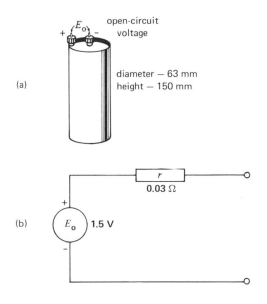

Figure 26-3   a.   No. 6 zinc-carbon dry cell.
b.   Equivalent circuit of the cell.

open-circuit voltage across the battery terminals. The internal resistance depends upon the capacity of the cell, its state of charge, its age, its temperature, and its chemical composition. For example, a new No. 6 carbon-zinc dry cell (diameter 63 mm, height 152 mm) has an internal resistance of about 0.03 Ω and an internal voltage of about 1.5 V.

For normal discharge currents, the internal IR

drop in a cell is about 10 percent of the open-circuit voltage.

Example 26-1:

A cell has an open-circuit voltage of 1.5 V and an internal resistance of 0.2 Ω. Calculate the current and the terminal voltage when the cell is connected to a resistance of 1 Ω (Fig. 26-4a).

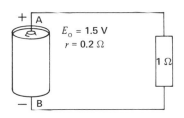

Figure 26-4   a.   See Example 26-1.

Solution:

a. 1   We can replace the cell by an equivalent circuit composed of a source of 1.5 V in series with a resistance of 0.2 Ω (Fig. 26-4b).

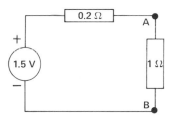

Figure 26-4   b.   Equivalent circuit.

a. 2   The total resistance of the circuit is 1.2 Ω
a. 3   Current is:
   $$I = 1.5/1.2 = 1.25 \text{ A}$$
a.4   The internal voltage drop is:
   $$IR = 1.25 \times 0.2 = 0.25 \text{ V}$$
a. 5   The voltage between terminals A and B is:
   $$E_{AB} = (1.5 - 0.25) = 1.25 \text{ V}$$

26-4 Discharge of a cell

When a cell is connected to a load, its terminal voltage remains steady for a long time and falls

TABLE 26A   TYPICAL PROPERTIES OF PRIMARY AND SECONDARY CELLS

Item	Unit	Primary cells					Secondary cells			
		carbon-zinc	zinc-mercury	alkaline-manganese	zinc-silver	zinc-air	lead-acid	nickel-cadmium	sodium-sulfur	lithium-iron disulfide
open-circuit voltage	V	1.5	1.35	1.5	1.6	1.45	2.0	1.3	2.1	2.3
min. operating voltage	V	0.8	0.9	0.8	0.9	1.1	1.7	1.0	1.5	1.5
specific energy	kJ/kg	150	300	200	300	650	40 to 80	70 to 120	225	550
energy per unit volume	kJ/dm^3	300	1200	450 to 700	1600	850	150 to 300	150 to 350	400	1200
permitted discharge rate	—	low	low	low	low	very low	high	very high	high	high
positive electrode	—	MnO_2 + C	Zn	Zn	Zn	O_2	PbO_2	NiOOH	S	FeS_2
negative electrode	—	Zn	HgO + graphite	MnO_2	Ag_2O	Zn	Pb	Cd	Na	Li
electrolyte	—	NH_4Cl $ZnCl$ + H_2O	KOH + ZnO + H_2O	KOH + H_2O	KOH + H_2O	KOH + H_2O	H_2SO_4 + H_2O	KOH + H_2O	Al_2O_3 + etc.	LiCl KCl
temperature range	°C	0 to + 50	0 to + 50	– 30 to + 50	0 to + 50	– 40 to + 40	– 40 to + 50	– 60 to + 40	+ 300	475
maximum storage time	years	1 - 3	5 - 7	4 - 5	4 - 5	3 - 4	2 - 4 month	4 - 6 month	—	—
useful life	years	2 - 3	4 - 5	3 - 4	4 - 5	2 - 3	5 - 20	10 - 20	—	—

steeply only when the cell is almost completely discharged (Fig. 26-5). The cell is considered to be discharged when the voltage reaches an arbitrary terminating value E_F (usually specified by the manufacturer). A primary cell may be drained beyond this point because it has to be discarded in any case. However, the emf of a secondary cell must not be allowed to fall below E_F because it may impair the cell's ability to recharge.

26-5 Capacity of a cell

The capacity of a cell is the amount of electricity it can deliver before the terminal voltage reaches

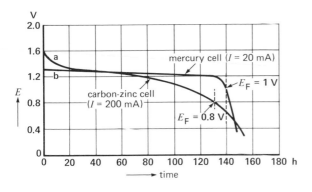

Figure 26-5   Typical discharge curves of a mercury cell and a carbon zinc cell.

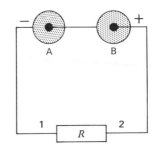

Figure 26-6   Cells connected in series.

the value E_F. The capacity is generally expressed in *ampere-hours* [A·h], although it can also be given in coulombs (1 A·h = 3600 C). A dry cell having a capacity of 30 A·h can deliver a current of 1 ampere for 30 hours or 1/10 of an ampere for 300 hours. However, it *cannot* deliver a current of 10 amperes for 3 hours (even though the product gives 30 A·h) because polarization becomes excessive and the terminal voltage falls off very quickly. The capacity of a cell is therefore not constant, but depends upon the current it delivers. The greater the current, the smaller the ampere-hour capacity will be. The capacity of a cell is usually specified by the manufacturer for a discharge period of 8 hours, 5 hours, or 1 hour.

Owing to the limited energy and the low voltage of dry cells, we often have to group them together to create medium voltage, high-power sources. Such a group of cells is called a battery.

26-6 Series grouping of cells

Cells are connected in series by successively joining the positive terminal of one cell to the negative terminal of the next (Fig. 26-6). The two remaining terminals are the terminals of the battery. The voltage of a group of cells connected in series is equal to the sum of the voltages of the cells. The internal resistance of the battery is equal to the sum of the internal resistances of the cells.

Example 26-2:
Three dry cells are connected in series with a flashlight bulb of 10 Ω. Each cell possesses an internal resistance of 0.3 Ω and an emf of 1.5 V. Calculate the power delivered to the lamp.

Solution:
a. 1   The emf of the battery is:
    $E = 3 \times 1.5 = 4.5$ V
a. 2   The internal resistance of the battery is:
    $r = 3 \times 0.3 = 0.9$ Ω
a. 3   The total resistance of the circuit is:
    $R = 0.9 + 10.0 = 10.9$ Ω
a. 4   The current is:
    $I = 4.5/10.9 = 0.412$ A
a. 5   The power delivered to the lamp is:
    $P = I^2R = (0.412)^2 \times 10 = 1.7$ W

26-7 Parallel grouping of cells

When a device draws a current greater than a single cell can deliver, we have to connect two or more cells in parallel. In this case, all the positive terminals are connected together and all the negative terminals are connected together (Fig. 26-7). The emf of the battery is the same as that of the individual cells. Cells connected in parallel must be of the same type.

If the cells have the same internal resistance, the current delivered by each cell is equal to the

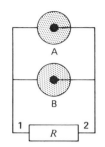

Figure 26-7   Cells connected in parallel.

total current divided by the number of cells. Should the resistance of one cell be higher than that of its neighbours, it will deliver a proportionately smaller current. It is particularly important to observe the polarities. In effect, if the polarity of one cell is accidentally reversed, the cell is destroyed in a few minutes.

26-8 Series-parallel grouping

When a load requires both a higher voltage and current than a single cell can provide, we must employ a series-parallel grouping. For example, the battery in Fig. 26-8 is composed of two parallel groups of two cells in series.

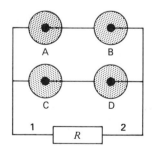

Figure 26-8   Cells in series-parallel.

Example 26-3:
The 6-volt coil of a relay has a resistance of 50 Ω. Dry cells having a capacity of 30 A·h and a potential of 1.5 V are available. How should the cells be connected in order that the relay may function for

a minimum period of 400 hours?

Solution:
a. 1   To function for 400 h, the current delivered by each cell must not exceed:
$$I = 30 \text{ A·h}/400 \text{ h} = 0.075 \text{ A}$$
a. 2   The current drawn by the coil is:
$$I = 6/50 = 0.12 \text{ A}$$
We must therefore connect two cells in parallel to furnish the current.
a. 3   However, two cells in parallel will provide only 1.5 V. Consequently, we must connect four cells in series to obtain the required 6 V. The mixed grouping is shown in Fig. 26-9. Note that each cell could easily supply a current of 0.12 A without diminishing its ampere-hour capacity. We could therefore use only four cells connected in series. However, with such an arrangement, the battery would be dead after about:
$$t = 30 \text{ A·h}/0.12 = 250 \text{ h}$$

26-9 Primary cells

Most primary cells are "dry" cells. Dry cells are so named because the liquid electrolyte is retained in an absorbant substance and the container is sealed. Dry cells may be carried about and set in any position without spilling the electrolyte.

26-10 Polarization

When a cell delivers current, hydrogen is usually given off at one of the electrodes, surrounding it with bubbles of gas. These bubbles act as insulators inhibiting the current flow. This obstructive phenomenon is called *polarization*. To prevent polarization, we add a depolarizing substance around the electrode to absorb the hydrogen as quickly as it is formed.

26-11 Carbon-zinc cell

The primary carbon-zinc cell, widely used for pocket lamps, has a zinc container that also acts as

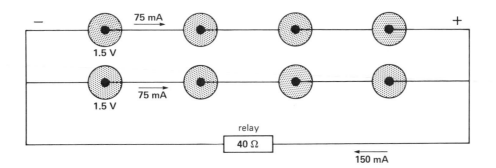

Figure 26-9   Electric battery to deliver 150 mA for 400 h (Example 26-3).

the negative electrode (Fig. 26-10). The positive electrode is made of a mixture of crushed carbon and maganese dioxide. The electrolyte is a solution of ammonium chloride and zinc chloride dissolved in water. The resulting paste surrounds a carbon rod that simply acts as a conductor to bring the current to the outer, positive terminal. The zinc container is separated from the paste by a layer of porous paper.

During discharge, the chemical reaction consumes the zinc at a rate of about 1 gram per ampere-hour.

26-12 Mercury cell

A primary mercury cell (Fig. 26-11) is small, robust, and has a very long shelf life. Its internal resistance is low and the terminal voltage remains very constant during the discharge period. The open-circuit terminal voltage is so stable that we sometimes use the cell as a reference source. For a given volume, a mercury cell contains four times as much energy as a carbon-zinc cell does. All these features explain its popularity in portable radios, missiles, instruments, electronic watches, and in

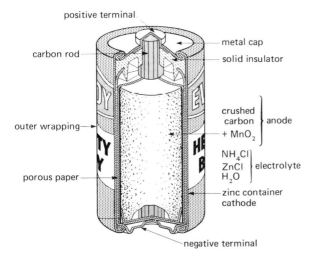

Figure 26-10   Carbon-zinc cell. *(Union Carbide)*

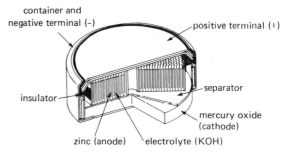

Figure 26-11   Mercury button cell. *(Duracell Inc., Mallory, Duracell Products Co.)*

pace-makers to stimulate the heart.

26-13 Alkaline-manganese cell

Primary alkaline-manganese cells may be stored for very long periods without deteriorating appreciably

(Fig. 26-12). Although their internal resistance is not as low as that of mercury cells, the voltage is very stable. Alkaline-manganese cells are superior to carbon-zinc cells for driving small motors, photographic apparatus, and other devices that draw substantial currents for short periods.

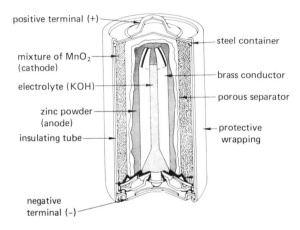

positive terminal (+)

steel container

mixture of MnO_2 (cathode)

electrolyte (KOH)

brass conductor

porous separator

zinc powder (anode)

insulating tube

protective wrapping

negative terminal (−)

Figure 26-12   Alkaline-manganese cell. *(Union Carbide)*

26-14 Life of a primary cell

Owing to secondary chemical effects, a cell deteriorates even when not in use. Thus, a new carbon-zinc cell, stored at normal room temperature, will lose most of its capacity by the end of 4 years. Temperature also influences the life of a cell: warm surroundings accelerate chemical reaction and appreciably diminish cell life.

26-15 Secondary cells

Lead cells and nickel-cadmium cells are the two most common types of secondary cells. They are mainly used as sources of reserve power during electric outages and to drive equipment that cannot be connected to a distribution system. These main applications enable us to distinguish two types of cells: those that provide intermittent, stand-by service (emergency lighting of buildings, auxiliary energy sources in electric substations) and those installed on mobile equipment (automo-

biles, lift-trucks, submarines, airplanes, etc.). The first application requires a very reliable cell which should last from 15 to 20 years. The second application requires a cell having a shorter life, but that can provide great energy in relation to its size.

Owing to its relatively low cost, the lead-acid cell is still the most common. However, nickel-cadmium cells are often preferred where high power must be furnished for short periods, or when periodic maintenance by qualified personnel is not feasible.

26-16 Efficiency of a secondary cell

When a secondary cell is being charged, most of the electricity it receives (coulombs) is recovered during the discharge period. Coulombic efficiencies of the order of 80 to 90 percent are common, depending upon the type of battery and its condition. Thus, a battery that receives 100 A·h of electricity during the charge period can deliver 80 to 90 A·h during the discharge interval.

On the other hand, the *energy* efficiency is only 50 to 70 percent because the charging voltage is considerably higher than the discharge voltage. Energy efficiency is defined as the ratio of the discharging energy in joules to the charging energy.

26-17 Production of hydrogen

All secondary cells contain water. Its molecules are composed of 2 atoms of hydrogen and 1 atom of oxygen. When we recharge a cell, and particularly if we exceed its normal charge level, the water gradually decomposes into hydrogen and oxygen. We can actually see the gases escape by the bubbling of the electrolyte. If sealed secondary cells are overcharged, the gases released may raise the internal pressure to as much as 400 kPa before the safety vent blows.

In the case of unsealed batteries, the hydrogen mixes with the surrounding air to create a potentially dangerous explosive mixture. The hydrogen content should never exceed 3% of the volume of air. This is why battery rooms are well ventilated and smoking is prohibited.

If we continue charging a battery that is already fully charged, the hydrogen released is given by the approximate equation:

$$V = 0.42 \frac{EQ}{e} \qquad (26\text{-}1)$$

where

V = volume of hydrogen released [L]
E = battery voltage [V]
Q = overcharge [A·h]
e = voltage per cell [V]
0.42 = constant, to take care of units

Each liter of hydrogen corresponds to the electrolysis of 0.8 mL of water; consequently, we must add water periodically to replenish the battery. In this regard, the reader should remember that electrolytes are *very* corrosive and accidental contact with the eyes must be strictly avoided.

Example 26-4:

A 12 V lead-acid automobile battery that is already fully charged, is inadvertently left connected to its charger. If the charging current is 2 A, calculate:

a. the amount of hydrogen released per day;
b. the amount of water that is lost.

Solution:

a. 1   The voltage per cell of a lead-acid battery is:
         $e = 2$ V
a. 2   The overcharge is:
         $Q = 2$ A x 24 h = 48 A·h
a. 3 $V = 0.42\,EQ/e$ Eq. 26-1
         $= 0.42 \times (12 \times 48/2)$
         $= 121$ L
b.   Amount of water lost by electrolysis:
         $V_w = 121 \times 0.8$
             $= 97$ mL, or nearly 1/4 pint

26-18 Lead-acid cell - theory of operation

Lead-acid cells have two sets of lead plates, composed respectively of spongy lead and of lead dioxide, immersed in a solution of sulfuric acid (Fig. 26-13). The chemical reaction that takes place

may be explained by the following elementary theory.

Figure 26-13a represents the chemical state of a lead cell that is fully charged. The positive plate is composed of lead dioxide (PbO_2) and the negative plate is of spongy lead. The plates are dipped in a solution of sulfuric acid (H_2SO_4).

As the cell discharges, both the lead dioxide and the spongy lead gradually change into lead sulfate ($PbSO_4$) (Fig. 26-13b). When the two plates attain the same chemical composition, the potential difference between them drops to zero, and the current stops (Fig. 26-13c).

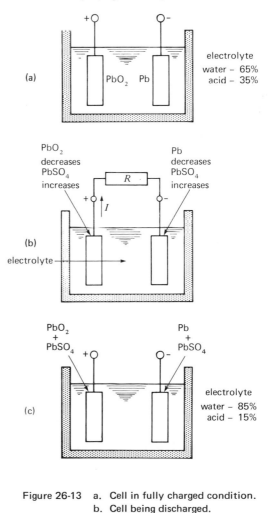

Figure 26-13   a. Cell in fully charged condition.
                      b. Cell being discharged.
                      c. Cell fully discharged.

We can restore the cell to its original state by connecting the terminals to a source of dc current as shown in Fig. 26-13d. The positive terminal of the charging source is connected to the positive terminal of the battery. In comparing Figs. 26-13b and 26-13d, we note that the charging current flows opposite to the discharge current.

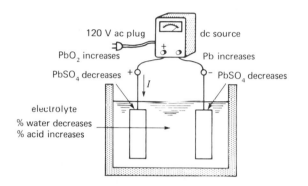

Figure 23-13   d. Cell being recharged.

26-19 Characteristics of a lead cell

Figure 26-14 shows the terminal voltage of a lead

cell and the corresponding density of the electrolyte during a charge and discharge cycle. The current is held at its rated value, based upon a discharged period of 8 hours. Thus, a 160 A·h battery has a rated current of 160 ÷ 8 = 20 amperes.

The internal resistance of a 12 V automobile battery (Fig. 26-15) is about 12 mΩ. Such a battery, composed of 6 cells connected in series, may supply between 200 and 400 amperes during the short period required to start a car.

Table 26A shows that the specific energy of a lead-acid cell ranges from 40 kJ/kg to 80 kJ/kg. The value depends mainly upon the density of the electrolyte. Thus, for long-term, high-reliability batteries, the relative density is kept at 1.21, compared to a density of 1.28 for an automobile battery. A lower density requires larger electrodes, but it ensures a longer life.

26-20 Maintenance of a battery

The maintenance a battery receives often depends upon its use. Thus, an automobile battery receives only minimal attention, whereas the emergency

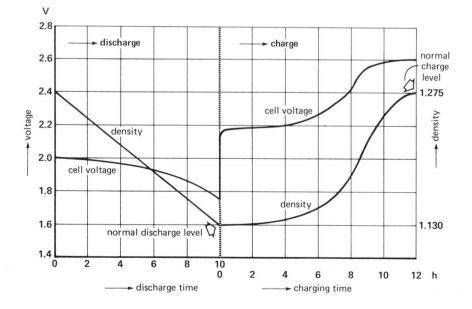

Figure 26-14   Electrolyte density and terminal voltage during the charge and discharge cycle of a lead-acid battery.

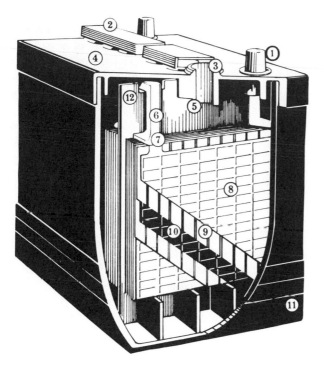

1. terminal
2. vent plug
3. filling hole
4. cover
5. electrolyte level indicator
6, 7. inter-cell connectors
8. negative plate
9. separator
10. positivo plate
11. container

Figure 26-15   Cutaway view of a 12 V automobile battery. *(Exide)*

batteries in a substation are continually monitored under a systematic maintenance program. The average life of the first battery is about 3 years while that of the second is 20 years.

The density of the electrolyte gives an indication of the state of charge of the battery. Pure water has a relative density of 1.00 while that of pure sulfuric acid is 1.85. The electrolyte in a fully charged battery has a density of about 1.28, as compared with 1.12 for a battery that is completely discharged. We can measure the density with a hydrometer. The hydrometer withdraws a sample of the electrolyte, and a graduated tubular floater gives the density reading. The electrolyte level should be checked frequently and, if necessary, distilled water should be added.

We should never leave a battery in a partially discharged condition because a lengthy storage period produces *sulfation* of the plates. This white, powdery deposit of lead sulfate hardens, and be-

comes insoluble in the acid. The active surface of the plates is consequently reduced, which raises the internal resistance and drops the capacity of the battery. If the sulfation is slight, we can eliminate it by over-charging the battery for several days, using a low current.

We must also exercise care when recharging a battery. Thus, the charging current must not be too great, otherwise the gaseous discharge may break up the plates. If we wish to accelerate the charging rate, we can use a higher initial current, but it must be limited to a value no greater than the capacity in ampere-hours. For example, the initial current should not exceed 160 amperes for a battery having a capacity of 160 A·h. The current must be gradually reduced as the battery charges up.

Partially discharged car batteries do not perform well in cold weather. Consequently, they should be kept well charged during winter, both to improve starting and to prevent the water from freezing and splitting the container. A fully charged battery can withstand temperatures as low as $-40°C$.

26-21 Nickel-cadmium cell

The nickel-cadmium secondary cell is composed of a positive electrode made of nickel hydroxide, and a negative electrode made of cadmium. Both are immersed in a solution of potassium hydroxide (20% KOH, 80% H_2O). The nickel-cadmium cell has the following advantages:
* It can deliver very high power (kW) for short periods;
* It may be completely discharged without affecting its characteristics;
* It discharges very, very slowly, during inactive periods;
* It releases no toxic gases;
* It is easy to maintain.

The high discharge rate of Ni-Cd cells explains why they are used to start large internal combustion engines. Furthermore, because they can be charged and discharged hundreds of times, Ni-Cd cells are very useful in stationary emergency-power installations. For long-term stand-by applications (Fig. 26-16), the specific energy (J/kg) is similar to that of lead-acid cells.

The density of the electrolyte does not change during charge and discharge. Consequently, we cannot determine the state of a Ni-Cd cell by means of a hydrometer. As in a lead-acid cell, the water in the electrolyte decomposes when the Ni-Cd cell is overcharged, releasing hydrogen and oxygen.

In summary, Ni-Cd batteries are particularly recommended for high-power discharge applications or where periodic maintenance is not feasible.

26-22 Special secondary cells

High-power, high energy-density secondary cells are presently under development as future power sources for electric cars. We mention only two of the possible candidates for this application:
1. the sodium-sulfur cell
2. the lithium-iron disulfide cell.

The properties of these cells are summarized in Table 26A. Note that both operate at high temperature. Figure 26-17 shows the construction of a large sodium-sulfur cell.

26-23 Fuel cells

Whenever a fuel such as wood, coal or natural gas burns, it combines with oxygen to create a new substance. The chemical reaction is called *oxida-*

Figure 26-16

Large battery installation to provide immediate standby emergency power for the lighting in an airport. The center rack contains 240 nickel-cadmium cells. The output of all the racks is connected to a 180 kW electronic inverter which converts the dc power to ac. A standby diesel-generator set is set in operation in the event of a prolonged outage. *(Nife-Jungner)*

A fuel cell basically consists of two electrodes, A and B, in contact with an electrolyte (Fig. 26-18). The fuel to be oxidized is brought in contact with electrode A, while oxygen is brought in contact with electrode B. An electrical load is con-

Figure 26-17   Large sodium-sulfur battery for electric vehicle. *(General Electric)*

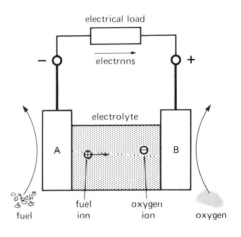

Figure 26-18   Basic principle of a fuel cell.

tion and the fuel is said to be oxidized. Oxidation is accompanied by the release of a large amount of energy that appears in the form of heat. The heat may be used to produce steam, which in turn can drive a turbogenerator set. Alternatively, the fuel can be burned in an internal combustion engine coupled to a generator. Unfortunately, the efficiency is very low when heat is converted into electrical energy in this way. Thus, for electrical outputs of 1 kW, the overall efficiency is only about 20%. It reaches a maximum of about 40% for outputs in the 500 MW range. The higher efficiencies are obtained by raising the temperature to the highest possible value that the metals can still safely withstand.

In the fuel cell, we get around the problem of low efficiency and high temperatures by causing the fuel to combine with oxygen in a different way. In essence, the fuel cell is a device that permits oxidation to take place without actually burning the fuel. The same amount of energy is released, but it appears in the form of electrical energy, rather than in the form of heat. How is this done?

nected across the terminals.

When the fuel touches its electrode, a special reaction takes place, causing the fuel to break up into positive ions and electrons. The ions move into the electrolyte and slowly migrate towards electrode B. The electrons, on the other hand, are captured by electrode A, flow through the load and on towards electrode B. The oxygen molecules that touch electrode B capture these electrons, thereby becoming negative oxygen ions. These ions diffuse into the electrolyte where they combine with the positive fuel ions to create a new substance that is electrically neutral. This substance accumulates with time, and must be removed to prevent contaminating the fuel cell.

Oxidation, therefore, takes place inside the electrolyte where the fuel and oxygen ions combine. However, no energy is released during this process: all the energy of oxidation appears as electrical energy at the electrodes. In effect, the electrons released at the "fuel" terminal are recaptured at the "oxygen" terminal with the result that a steady current flows through the load.

The electrode in contact with the fuel is always negative, while that in contact with the oxidizing agent is positive. The difference of potential across the terminals is typically between 0.5 V and 3 V, depending on the fuel used.

Ideally, the electric power supplied to the load is equal to the thermal power that would be released if the fuel were burned. There are, of course, some losses, but the efficiency of even small fuel cells is of the order of 40 percent. Clearly, this is an improvement, compared to the efficiency obtained by burning the fuel.

26-24 Hydrogen - oxygen fuel cell

Fuel cells are very complex devices; consequently, we shall limit our remarks to a simplified description of the hydrogen-oxygen fuel cell.

When hydrogen burns, it always combines in definite proportions with oxygen. In effect, for complete combustion, 1 kg of hydrogen consumes 8 kg of oxygen. In the process, about 120 MJ of heat is released, and the resulting product is ordinary water.

In a simple hydrogen fuel cell, the electrodes are made of platinum and the electrolyte is a solution of sulfuric acid. Hydrogen gas is continuously supplied to electrode A, and oxygen gas to electrode B (Fig. 26-19). Suppose that 1 kg of hydrogen moves through the cell *per hour* and that every molecule dissociates into 2 electrons and 2 positive hydrogen ions.

In order to neutralize the hydrogen ions, exactly 8 kg of oxygen must be furnished to electrode B, per hour. The hydrogen and oxygen ions combine in solution, and ordinary water is formed. Simultaneously, an electric current flows through the load, releasing 120 MJ of electrical energy. The power output of the cell corresponds therefore to 120 MJ/h, or 33.3 kW. This ideal output is not attained in practice because there are always losses in a fuel cell. Nevertheless, assuming an efficiency of 40%, the power output is still about 13 kW.

What is the approximate voltage developed by the fuel cell? We know that every molecule of hy-

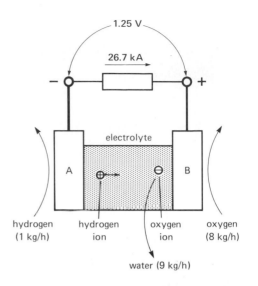

Figure 26-19   Simple hydrogen fuel cell (idealized conditions).

drogen releases 2 electrons when it comes in contact with the positive electrode. We also know that 1 kg of hydrogen contains very nearly 3×10^{26} molecules.[*] Consequently, the fuel cell produces about 6×10^{26} electrons per hour. Because each electron carries a charge of 1.6×10^{-19} coulombs, it follows that the total charge released is:

$Q = 6 \times 10^{26} \times 1.6 \times 10^{-19}$
$\quad = 96 \times 10^6$ C

The current flowing through the load is therefore:

$I = Q/t$
$\quad = 96 \times 10^6/3600$ s
$\quad = 26.7$ kA

Knowing that the power output is 33.3 kW, the terminal voltage is:

$E = P/I = 33.3/26.7$
$\qquad\quad = 1.25$ V

Thus, the maximum theoretical terminal voltage is 1.25 V. In practice, it is closer to about 0.8 V.

This example indicates that fuel cells, like bat-

* Deduced from the fact that Avogadro's number is 6.022×10^{26} per kilomole.

Figure 26-20   This modular fuel cell assembly contains a stack of 456 individual cells, connected in series. Twenty such units will be connected in series-parallel to make up a 4.5 MW installation. This dc power plant will be installed in downtown Manhattan and integrated into the network of the Consolidated Edison Co. of New York. Towards this end, the dc power will be converted to ac using a self-commuted inverter (See Chapter 23). Other details — electrolyte: phosphoric acid; operating temperature: 190°C; voltage per cell: about 0.7 V; current density per cell: 2500 A/m²; heat rate: 9500 Btu/kW·h; start-up time: 4 h from 21°C; load response time: 0.5 s from 35 percent power to rated output. *(Electric Power Research Institute)*

teries, are basically high-current, low-voltage dc devices. Higher voltages are obtained by connecting cells in series.

26-25 Types of fuel cells

Fuel cells is far more complex than the simple description given above might lead us to believe. The fuel may be a solid, liquid, or gas, and the electrolyte may be either a liquid or solid. Finally, oxygen may be provided in its pure state, or combined with other chemicals.

The operating temperature depends upon design; some fuel cells function at temperatures below 60°C while others perform best at 1000°C.

The reader has probably perceived that a fuel cell is basically a primary cell in which the reacting agents are fed continuously into the enclosure, while the unwanted byproducts are removed. Such a cell never discharges because the active materials are replenished as quickly as they are consumed. Nevertheless, the term "fuel cell" is apt because the electrical energy it produces can be related directly to the thermal energy that would be released if the fuel were burned.

In summary, the fuel cell is a unique power source because:

1. it is not a thermal machine, like a steam generator; it can therefore yield much higher efficiencies, in both small and large sizes;
2. it can be built in modular form and units can be added as the need arises;
3. it is noise-free and essentially pollution-free; consequently it can be installed in the heart of urban centers.

Figure 26-20 shows a power stack composed of over 400 individual fuel cells yielding a dc power output of 240 kW.

QUESTIONS AND PROBLEMS

Practical level

26-1 What is the difference between a primary cell and a secondary cell?

26-2 Name two types of primary cells and two types of secondary cells.

26-3 Why is smoking prohibited in battery rooms?

26-4 An electric starter for a 5000 hp diesel engine is battery-powered. Would a carbon-zinc battery be satisfactory? If not, what kind of battery do you propose?

26-5 What is the advantage of using sealed batteries?

26-6 A 12-volt lead-acid truck battery has a nominal capacity of 300 A·h.
   a. What is the rated current on an 8-hour basis?
   b. For how long could it deliver a current of 10 A?
   c. If the battery is discharged, what is the highest initial charging current that may be used?

26-7 A 12-volt nickel-cadmium battery has a capacity of 100 A·h based on a 5-hour discharge period. Calculate the terminal voltage when it delivers a current of 50 A knowing that the open-circuit voltage is 13 V and the internal resistance is 2.4 mΩ.

26-8 We wish to construct a battery using either lead-acid or Ni-Cd batteries. If the desired open-circuit voltage is 120 V, how many cells have to be connected in series in each case?

26-9 A hydrometer indicates an electrolyte density of 1.100 in a lead-acid storage battery. Should the battery be recharged? What should the density be when the battery is fully charged?

26-10 Describe the principle of operation of a fuel cell.

Intermediate level

26-11 A dry cell has an open-circuit voltage of 1.5 V. The voltage across the terminals is 1.2 V with a connected load of 3 Ω.
   a. Calculate the internal resistance of the cell.

b. If the terminals were shorted, what is the initial value of the discharge current?

26-12 A relay coil having a resistance of 20 Ω must be excited for about 250 h at a current of 300 mA. Individual cells having a capacity of 30 A·h and a voltage of 1.5 V are available. How many cells are needed and how must they be connected?

26-13 A large battery room contains 480 lead-acid cells operating at 120 V.

a. If the batteries are overcharged for 4 h at a current of 75 A, how much hydrogen is released [ft^3] ?

b. If each cell has a capacity of 150 A·h, what is the ampere-hour capacity of the battery?

26-14 A 6-volt carbon-zinc battery has the following dimensions: 5.5 in x 2.5 in x 4 in. Using the information given in Table 26A, calculate the approximate value of:

a. the available energy [W·h] ;

b. the capacity of the battery [A·h] .

c. For how long can it keep a 3 W lamp burning?

26-15 A mercury cell for a wristwatch has a diameter of 0.5 inches, a thickness of 3/16 inch.

a. If it has a capacity of 220 mA·h, for how many months can it keep the watch operating, if the circuit draws a current of 15 μA?

b. How much energy does the cell furnish during its lifetime [W·h] ?

Advanced level

26-16 a. Describe the principle of operation of a fuel cell.

b. Discuss the relative advantages of a generating station composed of fuel cells, compared to a thermal generating station.

26-17 A horse weighing 1600 lb can develop an average of 1 hp over an 8-hour period. Calculate the mass [lb] of a sodium-sulfur battery that can yield the same amount of energy before it has to be recharged.

26-18 A 120-volt battery driving a mining cart has to produce an average power of 21 kW over a 6-hour period. If lead-acid cells are used, calculate:

a. the approximate mass [lb] of the battery (in Table 26A, use 80 kJ/kg);

b. the mass if lithium-iron disulfide cells were used.

26-19 It is proposed that a lithium-iron disulfide battery be installed in conjunction with an electronic converter to store energy during off-peak hours of a power system. If the battery has to deliver 500 MW during a 2-hour peak, calculate:

a. the total weight of the battery;

b. the volume of the building if the battery occupies 30 percent of the space.

26-20 Calculate the efficiency of the fuel cell module shown in Fig. 26-20, and estimate its voltage and current output.

27
THE COST OF
ELECTRICITY

In 1978, the electric power utilities of North America supplied a total of 2.1×10^{12} kW·h of electrical energy to their industrial, commercial and residential customers (Table 27A). This enormous amount of energy represents 1 kW of power continually at the service of every man, woman and child, 24 hours per day. The production, transmission and distribution of this energy involves important costs that may be divided into two main categories - fixed costs and operating costs.

Fixed costs comprise the depreciation charges against buildings, dams, turbines, alternators, circuit breakers, transformers, transmission lines and any other equipment used in the production, transmission and distribution of electrical energy. These investments represent enormous sums: in 1975, they were about $350 000 per employee, compared to $60 000 per employee in the automobile industry.

Operating costs include salaries, fuel costs, administration, and any other daily or weekly expense.

Bearing in mind the relative importance of these two types of costs, utility companies have established rate structures that attempt to be as equitable as possible for their customers. The rates are based upon the following guidelines:
1. the amount of energy consumed [kW·h] or [J];
2. the *demand* or rate at which energy is consumed [kW];
3. the power factor of the load.

27-1 Tariff based upon energy

The cost of electricity depends, first, upon the amount of energy consumed. However, even if a customer uses no energy at all, he has to pay a minimum service charge, because it costs money to keep him connected to the line.

As consumption increases, the cost per kilowatthour drops, usually on a sliding scale. Thus, the domestic tariff may start at 10 cents per kilowatthour for the first one hundred kW·h, fall to 5 cent/kW·h for the next two hundred kW·h, and bottom out at 4 cent/kW·h for the rest of the energy consumed. The same general principle applies to medium-power and large-power users of electrical energy.

TABLE 27A   CONSUMPTION OF ELECTRICAL ENERGY IN NORTH AMERICA*

Type of customer	Number of customers	Total annual consumption [kW·h]	Monthly consumption per customer [kW·h]
Industrial	510 000	8.6×10^{11}	141 000
Commercial	10 100 000	5.3×10^{11}	4 400
Residential	85 600 000	7.5×10^{11}	730

* These statistics for 1978 were extrapolated from information supplied by the Edison Electric Institute. It is expected that the yearly increase in energy consumption will be about 4%. Thus, in 1990, the total energy will amount to 3.4×10^{12} kW·h.

27-2 Tariff based upon demand

The monthly cost of electricity supplied to a large customer depends not only upon the energy he consumes, but also upon the rate at which he uses it. In other words, the cost also depends upon the active *power* drawn from the line. To understand the reason for this, consider the following example.

Two factories A and B are respectively connected to a high-tension line by transformers T_A and T_B (Fig. 27-1). Factory A operates at full load, night and day, saturdays and sundays, constantly drawing 1000 kW of active power. At the end of the month (720 h), it has consumed a total of 1000 kW x 720 = 720 000 kW·h of energy.

Factory B consumes the same amount of energy, but its load is continually changing. Thus, power fluctuates between 50 kW and 3000 kW, sometimes reaching peaks as high as 4000 kW when big motors are started up. Obviously, the capacity of the transformer and transmission line supplying factory B must be greater than those supplying factory A. The electric utility must therefore invest more capital to service factory B; consequently, it is reasonable that B should pay more for its energy.

It is advantageous, both to the customer and the utility company that energy be consumed at a constant rate. The more regular the power flow, the less the cost.

27-3 Demand meter

The high power drawn by a motor during start-up does not last long enough to warrant the installation of correspondingly large equipment by the utility company. The question then arises: How long does the power surge have to last, in order to be considered significant? The answer depends upon several factors, but the period is usually taken to be 10, 15 or 30 minutes. For very large power users, such as municipalities, the averaging period may be as long as 60 minutes. It is called the *demand* interval.

To monitor the power flowing into the plant, a special meter is installed at the customer's service entrance. It automatically measures the average power during successive demand intervals (15 minutes, say). The average power measured during each interval is called the *demand*. As time goes by, the meter faithfully records the demand every 15 minutes and a pointer moves up and down a calibrated scale as the demand changes. In order to record the maximum demand, the meter carries a second pointer that is pushed upscale by the first.

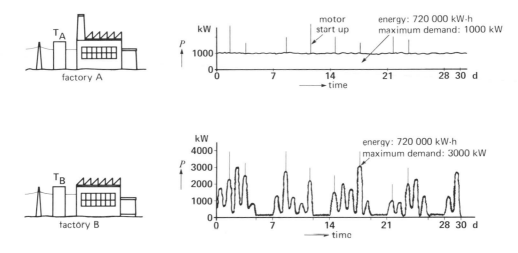

Figure 27-1   Comparison between two factories consuming the same energy but having different demands.

The second pointer simply sits at the highest position to which it is pushed. At the end of the month, a meterman takes the maximum demand reading and resets the pointer to zero. This special meter is called a demand meter, and it is installed at the service entrance of most industries and commercial establishments (Fig. 27-2). Figure 27-3 shows a printing demand meter for metering large industrial loads.

Figure 27-2   Combined   energy   and   demand   meter. *(Sangamo)*

Figure 27-3   Printing demand meter. *(General Electric)*

Example 27-1:

The graph in Fig. 27-4 represents the active power drawn by a large factory between 7:00 and 9:00 in the morning. The demand meter has a 30 min demand interval. Let us assume that at 7:00 the first pointer reads 2 MW while the second (pushed) pointer indicates 3 MW. What are the meter readings at the following times:

a. 7:30; b. 8:00;

c. 8:30; d. 9:00.

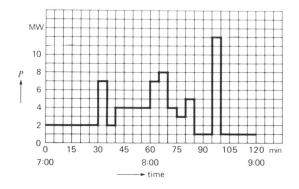

Figure 27-4 Active power absorbed by a plant.

Solution:

a. The average power (or demand) between 7:00 and 7:30 is obviously 2 MW. Consequently, pointer 1 indicates 2 MW at 7:30 and pointer 2 stays where it was at 3 MW.

b. The average power (or demand) between 7:30 and 8:00 is equal to the energy divided by time:

$$P_d = (7 \text{ MW} \times 5 \text{ min} + 2 \text{ MW} \times 5 \text{ min} + 4 \text{ MW} \times 20 \text{ min})/30 \text{ min}$$
$$= 4.17 \text{ MW}$$

During this interval, pointer 1 gradually moves from 2 MW (at 7:30) to 4.17 MW (at 8:00), pushing pointer 2 to 4.17 MW. Consequently, at 8:00, both pointers indicate 4.17 MW. Note that the demand reading is considerably less than the 7 MW peak which occurred during this interval.

c. The demand between 8:00 and 8:30 is:

$$P_d = (7 \times 5 + 8 \times 5 + 4 \times 5 + 3 \times 5 + 5 \times 5 + 1 \times 5)/30$$
$$= 4.67 \text{ MW}$$

Thus, at 8:30, both pointers have moved up to 4.67 MW.

d. The demand between 8:30 and 9:00 is:

$$P_d = (1 \times 5 + 12 \times 5 + 1 \times 20)/30$$
$$= 2.83 \text{ MW}$$

During this interval, pointer 1 drops from 4.67 MW to 2.83 MW, but pointer 2 sits at 4.67 MW, the previous maximum demand.

27-4 Tariff based upon power factor

Many alternating current machines, such as induction motors and transformers, absorb reactive power to produce their magnetic fields. The power factor of these machines is therefore less than unity and so, too, is the power factor of the factory where they are installed. A low power factor increases the cost of electrical energy, as the following example shows.

Consider two factories X and Y that consume energy at a uniform rate and have the same maximum demand. However, the power factor of X is unity while that of Y is 50 percent (Fig. 27-5).

The energy and demand being the same, the watthourmeters and demand meters will show the same reading at the end of the month. At first glance, it would appear that both users should pay the same bill. However, we must not overlook the apparent power drawn by each plant:

Apparent power drawn by factory X:

$$S = P/\cos \theta = 1000/1.0 \qquad \text{Eq. 8-5}$$
$$= 1000 \text{ kVA}$$

Apparent power drawn by factory Y:

$$S = P/\cos \theta = 1000/0.5 = 2000 \text{ kVA}$$

Because the line current is proportional to apparent power, factory Y draws twice as much current as factory X. The line conductors feeding Y must therefore be twice as big. Worse still, the transformers, circuit breakers, disconnect switches, and other devices supplying energy to Y, must have twice the rating of those supplying X.

The utility company must therefore invest more capital to service factory Y; consequently, it is logical that it should pay more for its energy, even though it consumes the same amount. In practice, the rate structure is designed to automatically in-

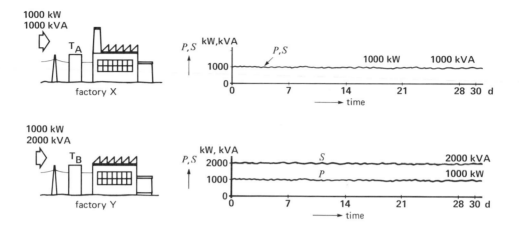

Figure 27-5   A low plant power factor requires larger utility company lines and equipment.

crease the billing whenever the power factor is low. Most electrical utilities require that the power factor of their industrial clients be 90 percent or more, in order to benefit from the minimum rate. When the power factor is too low, it is usually to the customer's advantage to improve it, rather than pay the higher monthly bill. This is usually done by installing capacitors at the service entrance to the plant, on the load side of the metering equipment. These capacitors may supply part of, or all the reactive power required by the plant. Industrial capacitors for power factor correction are made in single-phase and three-phase units rated from 5 kvar to 200 kvar.

27-5 Typical rate structures

Electrical power utility rates vary greatly from one area to another, and so we can only give a general overview of the subject. Most companies divide their customers into categories, according to their power demand. For example, one utility company distinguishes four power categories:
1. *Domestic power* - power corresponding to the needs of houses and rented apartments;
2. *Small power* - power of less than 100 kilowatts;
3. *Medium power* - power of 100 kilowatts and more, but less than 5000 kilowatts;
4. *Large power* - power in excess of 5000 kilo-

watts.

Table 27B shows typical rate structures that apply to each of these categories. In addition, a contract is usually drawn up between the electrical utility and the medium- or large-power customer. The contract may stipulate a minimum monthly demand, a minimum power factor, the voltage regulation, and various other clauses concerning firm power, growth rate, liability, off-peak energy, seasonal energy, price increases, and so forth. However, in most cases the rate schedule is quite straightforward as given in Table 27B.

27-6 Demand controllers

The maximum demand often plays an important role in compiling the electricity bill. Substantial savings can be made by keeping the maximum demand as low as possible. Thus, an alarm can be installed to sound a warning whenever the demand is about to exceed a pre-established maximum. Loads that are not absolutely essential can then be temporarily switched off until the peak has passed. This procedure can be carried out automatically by a demand controller that connects and disconnects individual loads so as to stay within the prescribed maximum demand (Fig. 27-7). Such a device can easily save thousands of dollars per year for a medium-power customer.

TABLE 27B   TYPICAL RATE STRUCTURES

Residential Rate Structure

Typical clauses:

1. "... This rate shall apply to electric service in a single private dwelling ...

2. ... This rate applies to single-phase alternating current at 60 Hz ..."

Rate Schedule

Minimum monthly charge - $5.00 plus
First 100 kW·h per month at 5 cent/kW·h
Next 200 kW·h per month at 3 cent/kW·h
Excess over 200 kW·h per month at 2 cent/kW·h

General Power Rate Structure (medium power)

Typical contract clauses:

1. "... The customer's maximum demand for the month, or its contract demand, is at least 50 kW but not more than 5000 kW ...

2. ... Utility Company shall make available to the customer 1000 kW of firm power during the term of this contract ...

3. ... The power shall be delivered at a nominal 3-phase line voltage of 480 V, 60 Hz ...

4. ... The power taken under this contract shall not be used to cause unusual disturbances on the Utility Company's system. In the event that unreasonable disturbances, including harmonic currents, produce interference with communication systems, the customer shall at his expense correct such disturbances ...

5. ... Voltage variations shall not exceed 7 percent up or down from the nominal line voltage ...

6. ... Utility Company shall make periodic tests of its metering equipment so as to maintain a high standard of accuracy ...

7. ... Customer shall use power so that current is reasonably balanced on the three phases. Customer agrees to take corrective measures if the current on the more heavily loaded phase exceeds the current in either of the two other phases by more than 20 percent. If said unbalance is not corrected, Utility Company may meter the load on individual phases and compute the billing demand as being equal to 3 times the maximum demand on any phase ...

TABLE 27B	TYPICAL RATE STRUCTURES

8. . . . The maximum demand for any month shall be the greatest of the demands measured in kilowatts during any 30 minute period of the month . . .

9. . . . If 90 percent of the highest average kVA measured during any 30 minute period is higher than the maximum demand, such amount will be used as the billing demand . . ."

Rate Schedule

Demand charge: $3.00 per month per kW of billing demand
Energy charge: 4 cent/kW·h for the first 100 hours of billing demand
2 cent/kW·h for the next 50 000 kW·h per month
1.2 cent/kW·h for the remaining energy

General Power Rate Structure (large power)

Typical contract clauses:

1. ". . . Customer's maximum demand for the month or its contract demand is greater than 5000 kW . . ."

2. Clauses similar to clauses 2 to 8 listed in the General Power (medium power) contract given above.

3. . . . Contract shall be for a duration of 10 years . . .

4. . . . The minimum bill for any one month shall equal 70 percent of the highest maximum demand charge during the previous 36 months . . .

5. . . . Utility Company shall not be obligated to furnish power in greater amounts than the customer's contract demand . . ."

Rate Schedule

Demand charge: First 75 000 kW of demand per month at $2.50 per kW
Excess over 75 000 kW of demand per month at $2.00 per kW
Additional charge for any demand in excess of customer's contract demand at $2.20 per month per kW
Energy charge: First 20 million kW·h per month at 6.1 mills per kW·h*
Next 30 million kW·h per month at 6.0 mills per kW·h
Additional energy at 5.9 mills per kW·h

* 1 mill = one thousandth of a dollar or one-tenth of a cent.

Example 27-2:

Billing of a domestic customer

A homeowner consumes 900 kW·h during the month of August. Calculate his electricity bill using the residential rate schedule given in Table 27B.

Solution:

Minimum charge	=	$ 5.00
First 100 kW·h @ 5 cent/kW·h	=	5.00
Next 200 kW·h @ 3 cent/kW·h	=	6.00
Remaining energy consumed (900 – 300) = 600 kW·h	=	12.00

    Total bill for the month: $28.00

This represents an average cost of 1800/900 = 3.11 cent/kW·h.

Demand meters are not usually installed in homes because the maximum demand seldom exceeds 10 kW.

Table 27C shows the energy consumed by various electrical appliances found in a home. Figure 27-6 is an example of an all-electric home, heated by baseboard heaters.

Figure 27-6   All-electric home that consumes a maximum of 9400 kW·h in January, and a minimum of 2100 kW·h in July.

TABLE 27C	AVERAGE MONTHLY CONSUMPTION OF HOUSEHOLD APPLIANCES

Average monthly consumption of a family of 5 persons in a modern house equipped with an automatic washing machine and a diswasher

appliance	kW·h consumed	appliance	kW·h consumed
Hot water heater (2000 gallons/month)	500	Automatic washing machine	100
Freezer	100	Coffee maker	9
Stove	100	Stereo system	9
Lighting	100	Radio	7
Dryer	70	Lawn mower	7
Dishwasher	30	Vacuum cleaner	4
Electric kettle	20	Toaster	4
Electric skillet	15	Clock	2
Electric iron	12		

Example 27-3:

Billing for a medium-power customer

A small industry operating night and day, 7 days a week, consumes 260 000 kW·h per month. The maximum demand is 1200 kW, and the maximum kVA demand is 1700 kVA. Calculate the electricity bill using the medium-power rate schedule.

Solution:

a. Clause 9 is important here because 0.90 x 1700 = 1530 kVA which is greater than the maximum demand of 1200 kW. Consequently, the demand for billing purposes is 1530 kW and not 1200 kW. In effect, the power factor of the plant is low; consequently, the "billing demand" is higher than the metered demand.

Applying the rate schedule, the demand charge is:

b. 1530 kW @ $3.00/kW = $ 4 590

The energy charge is:

c. 1 1530 kW x 100 hours

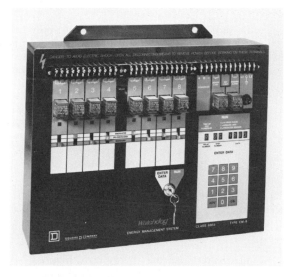

Figure 27-7 Automatic demand controller that sheds nonessential loads whenever the demand reaches a preset level. *(Square D)*

= 153 000 kW·h @ 4 cent/kW·h
153 000 x 0.04 = 6 120

c. 2 50 000 kW·h @ 2 cent/kW·h = 1 000

c. 3 The remainder of the energy is:
(260 000 – 153 000 – 50 000)
= 57 000 kW·h

c. 4 57 000 kW·h @ 1.2 cent/kW·h = 684

Total bill for the month: $12 394
The average cost of energy is:
unit cost = 12 394/260 000
= 4.77 cent/kW·h

Example 27-4:

Referring to Example 27-3, the plant engineering department makes a study of the situation and decides to install a demand controller to disconnect non-essential loads during peak load periods. The following results were obtained after the device was installed:

	new value	old value
energy consumed:	280 000 kW·h	260 000
maximum demand:	480 kW	1 200
maximum kVA demand:	700 kVA	1 700

Calculate the new bill and the new energy rate.

Solution:

a. The billing demand is the greater of 480 kW or 700 x 0.9 = 630 kVA. The demand for billing purposes is therefore 630 kW, again indicating a low plant power factor. The demand charge is:

b. 630 kW @ $3.00/kW = $1 890

The energy charge is:

c. 1 630 kW x 100 h
= 63 000 kW·h @ 4 cent/kW·h
63 000 x 0.04 = 2 520

c. 2 50 000 kW·h @ 2 cent/kW·h = 1 000

c. 3 The remainder of the energy is:
(280 000 – 63 000 – 50 000)
= 167 000 kW·h

c. 4 167 000 kW·h @ 1.2 cent/kW·h = 2 004

Total bill for the month: $7 414
The average energy cost is:

unit cost = 7 414/280 000

= 2.65 cent/kW·h

Note that the cost per kilowatthour has dropped very significantly by installing the demand controller. We can improve matters still more by installing capacitors to raise the power factor. The result is given in the next example.

Example 27-5:

Referring to Example 27-4, capacitors are installed at the service entrance to the plant, so as to raise the power factor to at least 90 percent. The following results were obtained some months after the capacitors were installed:

	new value	old value
energy consumed:	270 000 kW·h	280 000
maximum demand:	490 kW	480
maximum kVA demand:	520 kVA	700

Calculate the new bill and the corresponding energy rate.

metered demand. The power factor of the plant is now satisfactory, because even if more capacitors were added, the billing would not be reduced any more. The demand charge is:

b.   490 kW @ $3.00/kW   = $1 470

The energy charge is:

c. 1   490 kW x 100 h

= 49 000 kW·h @ 4 cent/kW·h

= 49 000 x 0.04   =   1 960

c. 2   50 000 kW·h @ 2 cent/kW·h   =   1 000

c. 3   The remainder of the energy is:

(270 000 − 49 000 − 50 000)

= 171 000 kW·h

c. 4   171 000 kW·h @ 1.2 cent/kW·h   =   2 052

Total bill for the month: $6 482

The average energy cost is:

unit cost = 6 482/270 000

= 2.4 cent/kW·h

It is clear that the capacitors have signif-

Figure 27-8

All-electric industry covering an area of 1300 m². It is heated by passing current through the reinforcing wire mesh embedded in the concrete floor. A load controller connects and disconnects the heating sections (nonpriority loads) depending on the level of production (priority loads). The demand is thereby kept below the desired preset level. Annual energy consumption: 375 000 kW·h; maximum demand in winter: 92 kW; maximum demand during the summer: 87 kW. *(Lab-Volt)*

Solution:

a.   The billing demand is the greater of 490 kW or 520 x 0.9 = 468 kVA. The billing demand is therefore 490 kW which is the same as the

icantly reduced the energy cost. Figure 27-8 shows a plant equipped with both a demand controller and power-factor correcting capacitors.

Example 27-6:

Billing of a large-power customer

A paper mill consumes 28 million kilowatthours of energy per month. The demand meter registers a peak demand of 43 000 kW. Calculate the monthly bill using the large-power rate schedule given in Table 27B.

Solution:

a. The demand charge is:

    43 000 kW @ $2.50/kW   = $107 500

b. The energy charge is:

    20 million kW·h @ 6.1 mill/kW·h

    $20 \times 10^6 \times 6.1/1000$ =   122 000

    8 million kW·h @ 6.0 mill/kW·h

    $8 \times 10^6 \times 6.0/1000$ =   48 000

    Total bill for the month: $277 500

Average cost per kW·h = $277\ 500/28 \times 10^6$

    = 9.9×10^{-3} = 9.9 mill

    or about 1 cent/kW·h

A monthly bill of nearly $300 000 may appear high, but we must remember that it probably represents less than 5 percent of the selling price of the finished product.

Figure 27-9 gives an idea of the power and energy consumed by a large city.

27-7 Power factor correction

Power factor correction (or improvement) is economically feasible when the monthly decrease in the cost of electric power exceeds the amortized cost of installing the required capacitors. In some cases, the customer has no choice, but must comply with the minimum power factor specified by the utility company.

The power factor may be improved by installing capacitors at the service entrance to the plant or commercial enterprise. In other cases, it may be desirable to correct the power factor of an individual device, if its power factor is particularly low.

Figure 27-9 In 1976, the City of Montreal with 1.84 million inhabitants consumed 16 763 GW·h of electric energy. Maximum demand during the winter, 3377 MW; during the summer, 1652 MW. *(Service de la C.I.D.E.M., Ville de Montréal)*

Example 27-7:

A factory draws an apparent power of 300 kVA at a power factor of 65% (lagging). What capacity in kvar must be installed at the service entrance to bring the overall power factor to:

a. unity;   b. 85 percent lagging.

Solution:

a. 1   Apparent power absorbed by the plant is:

$$S = 300 \text{ kVA}$$

a. 2   Active power absorbed by the plant is:

$$P = S \cos \theta \qquad \text{Eq. 8-5}$$
$$= 300 \times 0.65 = 195 \text{ kW}$$

a. 3   Reactive power absorbed by the plant is:

$$Q = \sqrt{S^2 - P^2} \qquad \text{Eq. 8-4}$$
$$= \sqrt{300^2 - 195^2} = 228 \text{ kvar}$$

To raise the power factor to unity, we have to supply all the reactive power absorbed by the load (228 kvar). The capacitors must therefore have a capacity of 228 kvar, or about 76 kvar per phase on a 3-phase line. Figure 27-10a shows the active and reactive power flow.

b. 1   The factory always draws the same amount of active power (195 kW) because the mechanical and thermal loads are fixed. If the new overall power factor is to be 0.85 lagging, the apparent power drawn from the line must be:

$$S = P/\cos \theta$$
$$= 195/0.85 = 230 \text{ kVA}$$

The new reactive power supplied by the line is:

$$Q = \sqrt{230^2 - 195^2} = 121 \text{ kvar}$$

Because the plant still needs 228 kvar and the line furnishes only 121 kvar, the difference must come from the capacitors. The capacity of these units is:

$$Q = (228 - 121) = 107 \text{ kvar}$$

Thus, if we can accept a power factor of 0.85 (instead of unity), we can reduce the size of the capacitor bank, and hence the cost. Figure 27-10b shows the power flow in the transmission line and the factory. Note that the factory draws the same active and reactive power, irrespective of the size of the capacitor installation.

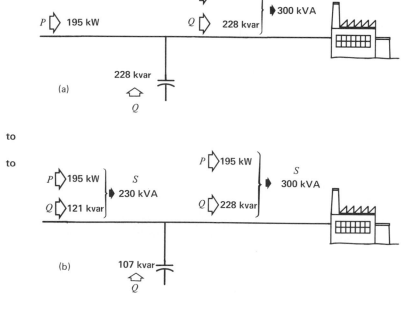

Figure 27-10

a. Overall power factor corrected to unity (Example 27-7).
b. Overall power factor corrected to 0.85.

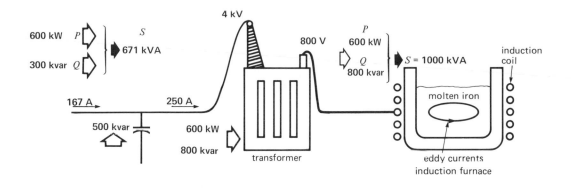

Figure 27-11   Individual power factor correction (Example 27-8).

Example 27-8:
A 600 kW induction furnace, connected to a 800 V single-phase line, operates at a power factor of 0.6 lagging. It is supplied by a 4 kV line and a step-down transformer (Fig. 27-11).
a. Calculate the current in the 4000 V line.
b. If we install a 500 kvar capacitor on the HV side of the transformer, calculate the new power factor and the new line current.

Solution:
This is an interesting example of a situation where individual power factor correction must be applied. The reason is that the induction furnace is a single-phase device whereas the plant is certainly energized by a 3-phase line. We cannot correct the power factor of single-phase equipment by adding balanced 3-phase capacitors at the service entrance.

a.1 Active power absorbed by the furnace is:
  $P = 600$ kW
a.2 Apparent power absorbed is:
  $S = P/\cos\theta = 600/0.6 = 1000$ kVA
a.3 Current in the 4 kV line is:
  $I = S/E = 1000/4 = 250$ A
b.1 Reactive power absorbed by the furnace is:
  $Q = \sqrt{S^2 - P^2} = \sqrt{1000^2 - 600^2}$
    $= 800$ kvar
b.2 Reactive power supplied by the capacitor is:
  $Q_c = 500$ kvar

b.3 Reactive power supplied by the line is:
  $Q_L = Q - Q_c = 800 - 500 = 300$ kvar
b.4 Active power drawn from the line is:
  $P_L = 600$ kW
b.5 Apparent power drawn from the line is:
  $S_L = \sqrt{P_L^2 + Q_L^2} = \sqrt{600^2 + 300^2}$
    $= 671$ kVA
b.6 New power factor is:
  $\cos\theta = P_L/S_L = 600/671 = 0.89$
b.7 The new line current is:
  $I = S_L/E = 671/4 = 167$ A

By installing a single-phase capacitor bank, the line current drops from 250 A to 167 A, which represents a decrease of 33%. It follows that the I^2R loss, and voltage drop on the supply line, will be greatly reduced. Furthermore, the power factor rises from 60% to 89% which will significantly reduce the monthly power bill. Finally, the 3-phase line currents are more likely to be reasonably balanced at the service entrance.

QUESTIONS AND PROBLEMS

Practical level

27-1 Explain what is meant by the following terms:

demand — billing demand
maximum demand — fixed cost
mill — demand interval

27-2 Using the rate schedule given in Table 27B calculate the power bill for a homeowner who consumes 920 kW·h in one month.

27-3 Explain why a low power factor in a factory results in a higher power bill.

27-4 Explain the behavior of a demand meter.

27-5 Using the rate schedule given in Table 27B for a medium power customer, calculate the monthly power bill under the following conditions:

demand meter reading	= 120 kW
billing demand	= 150 kW
energy consumed	= 36 000 kW·h

Intermediate level

27-6 The demand meter in a factory registers a maximum demand of 4300 kW during the month of May. The power factor is known to be less than 70 percent.
  a. If capacitors had been installed so as to raise the power factor to 0.9, would the maximum demand have been affected?
  b. Would the billing demand have been affected?

27-7 According to Example 27-1, the maximum demand registered at 8:00 is 4.17 MW. If the demand meter were replaced by another one having a demand interval of 15 min, calculate the new value of the maximum demand at 8:00.

27-8 a. Give an estimate of the energy consumed in one year by a modern city of 300 000 inhabitants in North America (Refer to Fig. 27-9).
  b. If the average rate is 40 mill/kW·h, calculate the annual cost of servicing the city.

27-9 A motor draws 75 kW from a 3-phase line, at a $\cos \theta = 0.72$ lagging.
  a. Calculate the value of Q and S absorbed by the motor.
  b. If a 20 kvar 3-phase capacitor is connected in parallel with the motor, what is the new value of P and Q supplied by the line?
  c. Calculate the percent drop in line current after the capacitor is installed.

27-10 A plant draws 160 kW at a lagging power factor of 0.55.
  a. Calculate the capacitors [kvar] required to raise the power factor to unity.
  b. If the power factor is only raised to 0.9 lagging, how much less would the capacitors cost (in percent)?

27-11 a. Assuming that power in a large industry can be purchased at 15 mill/kW·h, estimate the hourly cost of running a 4000 hp motor having an efficiency of 96 percent.
  b. If the motor runs night and day, 365 days per year, what would the annual saving be if the motor were redesigned to give an efficiency of 97%.

27-12 a. Referring to the residential rate schedule given in Table 27B, calculate the cost per kW·h if only 20 kW·h are consumed during a given month.
  b. The heating element on an electric stove is rated at 1200 W. Using the same rate schedule, what is the least possible cost of running it for one hour?

27-13 A barrel of oil costing 32 dollars contains 42 gal (U.S.) having a heating value of 115 000 Btu/gal. When the fuel is burned in a thermal generating station to produce electricity, the overall efficiency is typically 35%. Calculate the minimum cost per kilowatthour considering only the price of the fuel.

28

GENERATION
OF ELECTRICAL
ENERGY

Now that we are familiar with the principal machines and power devices, we are in a position to see how they are used in a large electrical system. Such a system comprises all apparatus used in the generation, transmission and distribution of electrical energy, starting from the generating station and ending up in the most remote summer home in the country. The next three chapters are therefore devoted to the following major topics:

- the generation of electrical energy
- the transmission of electrical energy
- the distribution of electrical energy

28-1 Demand of an electrical system

The total power drawn by the customers of a large utility system fluctuates between wide limits, depending on the seasons and time of day. Figure 28-1 shows how the system demand varies during a typical summer and winter day. The 15 GW winter peak occurs around 17:00 (5 p.m.) because increased domestic activity coincides with industrial and commercial centers that are still operating at full capacity.

The load curve of Fig. 28-2 shows the *seasonal* variations for the same system. Note that the peak demand during winter (15 GW) is more than twice

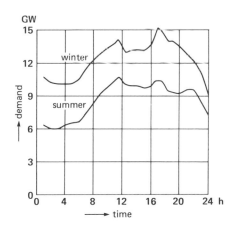

Figure 28-1   Demand curve of a large system during a summer and winter day.

the minimum demand during summer (6 GW).

In examining the curve, we note that the de-

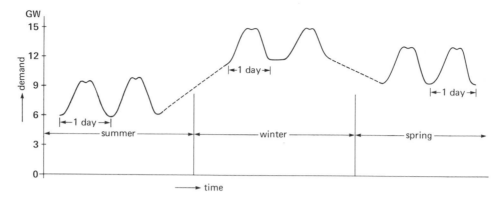

Figure 28-2   Demand curve of a large system during one year.

mand throughout the year never falls below 6 GW. This is the *base load* of the system. We also see that the annual *peak load* is 15 GW. The base load has to be fed 100 percent of the time but the peak load may be on for only 0.1 percent of the time. Between these two extremes, we have *intermediate loads* that have to be fed for less than 100 percent of the time.

If we plot the duration of each demand on an annual base, we obtain the *load duration curve* of Fig. 28-3. For example, the curve shows that a demand of 9 GW lasts 70 percent of the time while a demand of 12 GW lasts for only 15 percent of the time. The graph is divided into base, intermediate and peak load sections. The peak load portion usually includes demands that last for less than 15 percent of the time. On this basis, the system has to deliver 6 GW of base power, another 6 GW of intermediate power and 3 GW of peak power.

These power blocks give rise to three types of generating stations:

a. *Base power stations* that deliver full power at all times. Nuclear stations and coal-fired stations are particularly well adapted to furnish base demand;

b. *Intermediate power stations* that can respond relatively quickly to changes in demand, usually be adding or removing one or more generating units. Hydropower stations are well adapted for this purpose;

c. *Peak generating stations* that deliver power for brief intervals through the day. Such stations must be put in service very quickly. Consequently, they are equipped with prime movers such as diesel engines, gas turbines, compressed-air motors or pumped-storage turbines that can be started up in a few minutes. In this regard, it is worth mentioning that thermal generating stations, using gas or coal, take from 4 to 8 hours to start up, while nuclear stations may take several days. Obviously, such generating stations cannot be used to supply peak power.

Referring to Fig. 28-3, the dotted and cross-hatched areas indicate the relative amount of energy (kW·h) associated with the base, intermediate and peak loads. Thus, the base power stations supply 58 percent of the total annual energy requirements, while the peak load stations contribute only 1.3 percent. The peak load stations are in service for an average of only 1 hour per day. Consequently, peak power is very expensive because the stations that produce it are idle most of the time.

28-2 Location of the generating station

The physical location of the generating station, transmission lines and substations must be carefully studied to arrive at an acceptable, economic solution. We can sometimes locate a generating station next to the primary source of energy and use

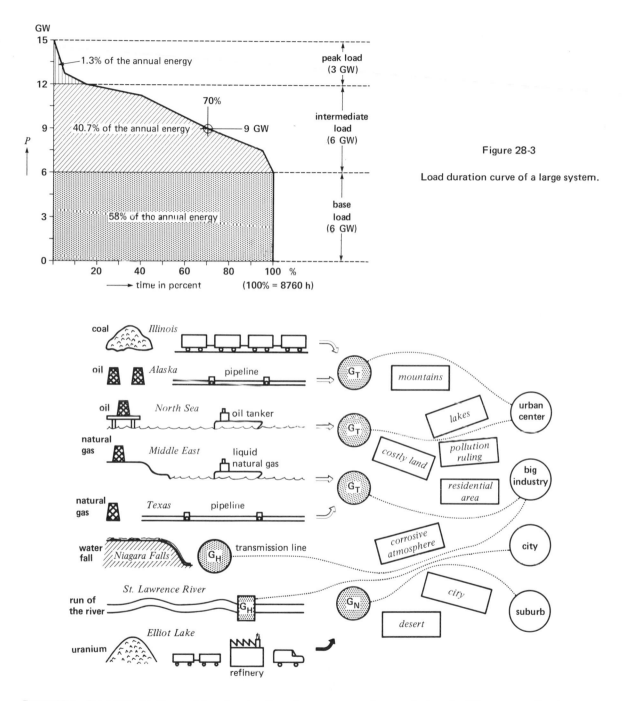

Figure 28-3

Load duration curve of a large system.

Figure 28-4 Extracting, hauling and transforming the primary sources of energy is done in different ways. Furthermore, the dotted transmission lines connecting the generating stations G with the consumers must go around various obstacles.

transmission lines to carry the electrical energy where it is needed. When this is neither practical or economical, we have to transport the primary energy (coal, gas, oil) by ship, train, or pipeline to the generating station. The generating station may therefore be near to, or far away from, the user of electrical energy. Figure 28-4 shows some of the obstacles that prevent transmission lines from following the shortest route. Owing to these obstacles, both physical and legal, transmission lines often follow a zig-zag path between the generating station and the ultimate user.

28-3 Types of generating stations

There are three main types of generating stations:
a. thermal generating station;
b. hydropower generating stations;
c. nuclear generating stations.

Although we can harness the wind, tides, and solar energy, these energy sources represent only a very small part of the total energy we need. It appears we are gradually moving towards an era of nuclear and solar energy, because the fossil-fuel resources of the world are being depleted at an accelerating rate.

28-4 Controlling the power balance between generator and load

The electrical energy consumed by a customer must immediately be supplied by the alternators because electrical energy cannot be stored. How do we maintain this almost instantaneous balance between the customer's requirements and the generated power? To answer the question, let us consider a single hydropower station supplying regional load R_1 (Fig. 28-5). Water behind the dam flows through the turbine runner, causing the turbine and alternator to rotate.

The mechanical power P_T developed by the turbine depends exclusively on the opening of the wicket gates that control the water flow. The greater the opening, the more water is admitted to the turbine and the increased power is immediate-ly transmitted to the alternator.

On the other hand, the electric power P_L drawn from the alternator depends exclusively on the load. When the mechanical power P_T supplied to the rotor is equal to the electrical power P_L consumed by the load, the alternator is in dynamic equilibrium and its speed remains constant. The electrical system is said to be *stable*.

However, we have just seen that the system demand fluctuates continually, so that P_L is sometimes greater, and sometimes less than P_T. If P_L is greater than P_T, the generating unit (turbine and alternator) begins to slow down. Conversely, if P_L is less than P_T, the unit speeds up.

The speed variation of the alternator is therefore an excellent indicator of the equilibrium between P_L and P_T, and hence of the stability of the system. If the speed falls, the wicket gates must open, and if it rises, they must close so as to maintain a continuous state of equilibrium between P_T and P_L. Although we could adjust the gates manually by observing the speed, we always use an automatic speed regulator.

Speed regulators, or governors, are extremely sensitive devices. They can detect speed changes as small as 0.02 percent. Thus, if the speed of an alternator increases from 1800 r/min to 1800.36 r/min, the governor begins to act on the wicket gate mechanism. If the load should suddenly increase, the speed will drop momentarily, but the governor will quickly bring it back to rated speed. The same corrective action takes place when the load is suddenly removed.

Clearly, any speed change produces a corresponding change in the system frequency. The frequency is therefore an excellent indicator of the stability of a system. *The system is stable so long as the frequency is constant.*

The governors of thermal and nuclear stations operate the same way, except that they regulate the steam valves, allowing more or less steam to flow through the turbines. The resulting change in steam flow has to be accompanied by a change in the rate of combustion. Thus, in the case of a coal-burning boiler, we have to reduce combustion as

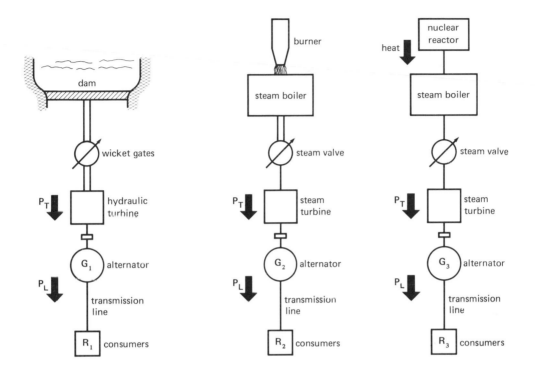

Figure 28-5   Power supplied to three independent regions.

soon as the valves are closed off, otherwise the boiler pressure will quickly exceed the safety limits.

28-5 Advantage of interconnected systems

Consider the three generating stations of Fig. 28-5, connected to their respective regional loads R_1, R_2, and R_3. Because the three systems are not connected, each can operate at its own frequency, and a disturbance on one does not affect the others. However, it is preferable to interconnect the systems because 1) it improves the overall stability, 2) it provides better continuity of service and 3) it is more economical. Figure 28-6 shows four interconnecting transmission lines, tying together both the generating stations and the regions being serviced. We now discuss the advantages of such a network.

1. **Stability.** Systems that are interconnected have greater reserve power than a system working alone. In effect, a large system is better able to withstand a large disturbance and, consequently, it is inherently more stable. For example, if the load suddenly increases in region R_1, energy immediately flows from G_2 and G_3 and over the interconnecting tie lines. The heavy load is therefore shared by all three stations instead of being carried by one alone.

2. **Continuity of service.** If a generating station should break down, or if it has to be shut down for annual inspection and repair, the customers it serves can temporarily be supplied by the two remaining stations. Energy flowing over the tie lines is automatically metered, and credited to the station that supplies it.

3. **Economy.** When several regions are intercon-

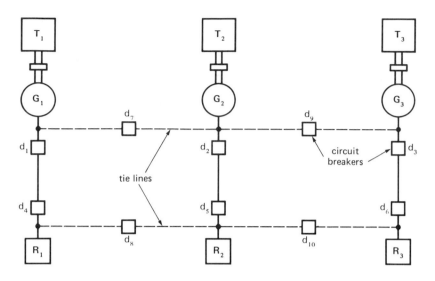

Figure 28-6   Three networks connected by 4 tie lines.

nected, the load can be shared between the various generating stations so that the overall operating cost is minimized. For example, instead of operating all three stations at reduced capacity during the night when demand is low, we can shut down one station completely and let the others carry the load. In this way, we reduce the operating cost of one station to "zero" while improving the efficiency of the other stations, because they now run closer to their rated capacity.

Electrical utility companies are therefore interested in grouping their resources by a grid of interconnecting transmission lines. A central dispatching office (control center) distributes the load among the various companies and generating stations so as to minimize the costs (Fig. 28-7). Owing to the complexity of some systems, control decisions are often made with the aid of a computer. The dispatching office also has to predict daily and seasonal load changes and direct the start-up and shut-down of generating units so as to maintain good stability of the immense and complicated network.

For example, the New England Power Exchange (NEPEX) coordinates the resources of 13 electrical utility companies serving Connecticut, Rhode Is-

land, Main and New Hamshire. It also supervises power flow between this huge network and the State of New York and Canada.

Although such interconnected systems must necessarily operate at the same frequency, we can still allocate the load among the various generating stations, according to a set program. Thus, if a generating unit has to deliver more power, we change its governor setting slightly so that more power is delivered to the alternator. The increased electrical output from this unit produces a corresponding decrease in the total power supplied by all the other generating stations.

28-6 Conditions during an outage

A major disturbance on a system creates a state of emergency and immediate steps must be taken so that it does not spread to other regions. The sudden loss of an important load or a permanent short-circuit of a transmission line constitutes a major disturbance.

If a big load is suddenly lost, the turbines begin to speed up and the frequency increases everywhere on the system. On the other hand, if a generator is disconnected, the speed of the remaining

Figure 28-7   Technicians in the control rooms of two generating stations communicate with each other or with a central dispatching office, while supervising the operation of their respective generating units.

generators decreases because they suddenly have to carry the entire load. The frequency starts to decrease - sometimes at the rate of 5 Hz per second if the power loss is big. Under these conditions, no time must be lost and, if conventional methods are unable to bring the frequency back to normal, one or more loads must be dropped. Such load shedding is done by frequency-sensitive relays that open selected circuit breakers as the frequency falls. For example, the relays may be set to shed 15 percent of the system load when the frequency reaches 59.3 Hz, another 15 percent when it reaches 58.9 Hz and a final 30 percent when the frequency is 58 Hz. Load shedding must be done in less than one second to "save" the loads judged to be of prime importance. As far as the disconnected customers are concerned, such an outage creates serious problems. Elevators stop between floors, arc furnaces start to cool down, paper tears as it moves through a paper mill, traffic lights stop functioning, and so forth. Clearly, it is in everyone's interest to provide uninterrupted service.

Experience over many years has shown that most system short-circuits are very brief. They may be caused by lightning, by polluted insulators, by falling trees, or by overvoltages created when circuit breakers open and close. Such disturbances usually produce a short-circuit between two phases

or between one phase and ground. Three-phase short-circuits are very rare. When we open a short-circuited line, the arc extinguishes almost immediately; consequently, we can reclose the circuit without fear that the arc will restrike. Owing to this feature, we can usually prevent a major outage by simply opening a short-circuited line and reclosing it very quickly. Naturally, such fast switching of circuit breakers is done automatically because it all happens in a matter of a few cycles.

28-7   Frequency and electric clocks

The frequency of a system fluctuates as the load varies, but the turbine governors always bring it back to 60 Hz. Owing to these fluctuations, the system gains or loses a few cycles throughout the day. When the accumulated loss or gain is about 180 cycles, the error is corrected by making all the alternators turn either faster or slower for a brief interval. The frequency correction is effected according to instructions from the dispatching center. In this way, a 60 Hz network generates exactly 5 184 000 cycles in a 24-hour period. Electric clocks connected to the network are therefore very accurate, because the position of the seconds hand is directly related to the number of elapsed cycles.

HYDROPOWER GENERATING STATIONS

Hydropower generating stations convert the energy of moving water into electrical energy by means of a hydraulic turbine coupled to an alternator.

28-8 Available power

The power that can be extracted from a waterfall depends on its height and rate of flow. The size and physical location of a hydropower station depends therefore on these two factors. The available power can be calculated by the equation:

$$P = 9.8\,qh \qquad (28-1)$$

where

P = available water power [kW]
q = water flow [m^3/s]
h = head of water [m]

Owing to friction losses in the water conduits, turbine casing, and the turbine itself, the mechanical power output of the turbine is somewhat less than that calculated by Eq. 28-1. However, the efficiency of large hydraulic turbines is between 90 and 94 percent. The alternator efficiency is even higher, ranging from 95 to 98 percent.

Example 28-1:
A large hydropower station has a head of 324 m and an average flow of 1370 m^3/s. The reservoir is composed of a series of lakes covering an area of 6400 km^2. Calculate:
a. the available hydraulic power;
b. the number of days this power could be sustained if the level of the impounded water were allowed to drop by 1 m (assume no precipitation or evaporation and neglect water brought in by surrounding rivers and streams).

Solution:
a. $P = 9.8\,qh$
    $= 9.8 \times 1370 \times 324$
    $= 4\,350\,000$ kW $= 4350$ MW

b. A drop of 1 m in the water level corresponds to 6400 $\times$ 10^6 m^3 of water. Because the flow is 1370 m^3/s, the time for all this water to flow through the turbines is:
    $t = 6400 \times 10^6/1370 = 4.67 \times 10^6$ s
    $= 1298$ h $= 54$ days
As a matter of interest, a flow of 1370 m^3/s is about 10 times the amount of water consumed by the city of New York, including all its suburbs.

28-9 Types of hydropower stations

Hydropower stations may be divided into three groups depending on the head of water:
1. high-head development;
2. medium-head development;
3. low-head development.
   High-head developments have heads in excess of 300 m, and high-speed Pelton turbines are used. Such generating stations are found in the Alps and other mountainous regions. The amount of impounded water is usually small.
   Medium-head developments have heads between 30 m and 300 m and medium-speed Francis turbines are used. The generating station is fed by water held back by a dam. The dam is usually built across a river bed in a relatively mountainous region. A great deal of water is impounded behind the dam (Fig. 28-8).
   Low-head developments have heads under 30 m and low-speed Kaplan or Francis turbines are used. These generating stations often extract the energy from flowing rivers. The turbines are designed to handle large volumes of water at low pressure. No reservoir is provided (Fig. 28-9).

28-10 Makeup of a hydropower plant

A hydropower installation consists of dams, waterways and conduits that impound and channel water towards the turbines. These, and other items described below, enable us to understand some of the basic features and components of a hydropower plant (see Fig. 28-10).

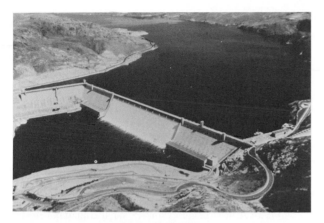

Figure 28-8

Grand Coulee Dam on the Columbia River, in the State of Washington, is 108 m high and 1270 m wide. It is the largest hydropower plant in the world, having 18 generating units of 125 MW each and 12 generating units of 600 MW each, for a total of 9450 MW of installed capacity. The spillway can be seen in the middle of the dam. *(General Electric)*

Figure 28-9

The Beauharnois generating station, on the St.Lawrence River, contains 26 3-phase alternators rated 50 MVA, 13.2 kV, 75 r/min, 60 Hz at a power factor of 0.8 lagging. An additional 10 units rated 65 MVA, 94.7 r/min, make up the complete installation. The output ranges between 1000 MW and 1575 MW depending upon the water flow. *(Hydro-Québec)*

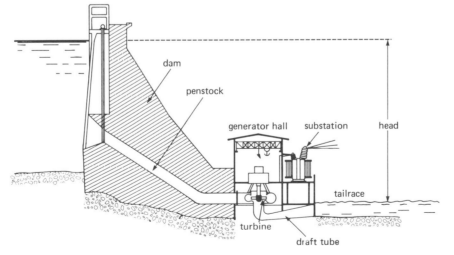

Figure 28-10   Cross section view of a medium-head hydropower plant.

1. Dams. Dams made of earth or concrete are built across river beds to channel the water towards the powerhouse and to create storage reservoirs. Reservoirs can compensate for the reduced precipitation during dry seasons, and for the abnormal flows that accompany heavy rains and melting snow. In effect, dams permit us to regulate the water flow throughout the year, so that the powerhouse may run at close to full capacity.

Spillways are provided next to the dam to discharge water whenever the reservoir level is too high. In effect, we have seen that the demand varies considerably throughout the day and is usually quite small at night. Consequently, we cannot always use the available water to supply energy to the system. If the water reservoir is small or almost nonexistant (such as in run-of-river stations), we unfortunately have to let the water through the spillway, without using it.

Dams often serve a dual purpose, providing irrigation and navigation facilities, in addition to their power-generating role.

2. Conduits penstocks and scroll-case. In large installations, conduits lead the water from the dam site to the hydro plant. They may be open canals, or tunnels carved through rock. They feed one or more penstocks (huge steel pipes) which bring the water to the individual turbines. Enormous valves, sometimes several meters in diameter, enable the water supply to be shut off in the conduits.

The penstocks channel the water into a scroll-case that surrounds the runner so that water is evenly distributed around its circumference. Guide vanes and wicket gates control the water flow so that it flows smoothly into the runner blades (see Figs. 28-11, 28-12 and 28-13). The wicket gates open and close in response to a powerful hydraulic mechanism that is controlled by the respective turbine governors.

3. Draft tube and tailrace. Water that has passed through the runner moves next through a carefully designed vertical channel, called draft tube. The draft tube improves the efficiency of the turbine.

Figure 28-11   Spiral case feeds water around the circumference of a 483 MW turbine. *(Marine Industrie)*

Figure 28-12   Inside the case a set of adjustable wicket gates control the amount of water flowing into the turbine. *(Marine Industrie)*

It leads out to the tailrace bringing the water into the downstream river bed.

4. Powerhouse. The powerhouse contains the alternators, transformers, circuit breakers, etc., and associated control apparatus. Instruments, relays, and meters are contained in a central control room where the entire station can be monitored and controlled. Finally, many other devices, too numerous to mention here, make up the complete hydropower station.

Figure 28-13

Runner of a Francis-type turbine being lowered into position at the Grand Coulee Dam. The turbine is rated at 620 MW, 72 r/min and operates on a nominal head of 87 m. Other details: runner diameter: 10 m; runner mass: 500 t; maximum head: 108 m; minimum head: 67 m; turbine efficiency: 93 percent; number of wicket gates: 32; mass per wicket gate: 6.3 t; turbine shaft length: 6.7 m; mass of shaft: 175 t. *(Les Ateliers d'Ingénierie Dominion)*

28-11 Pumped storage installations

We have already seen that peak power stations are needed to meet the variable system demand. To understand the different types of peaking systems used, consider a simple network in which the daily demand varies between 100 MW and 160 MW, as shown in Fig. 28-14. One obvious solution is to install a 100 MW base-power station and a peak-power unit of 60 MW, driven by a gas turbine.

However, another solution is to install a larger base-power unit of 130 MW and a smaller peaking

station of 30 MW. The peaking station must be able to both deliver and absorb 30 MW of electric power. During lightly loaded periods (indicated by a minus in Fig. 28-15), the peaking station receives and stores energy provided by the base-power generating plant. Then, during periods of heavy de-

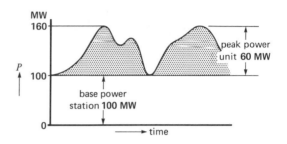

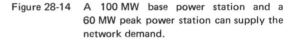

Figure 28-14   A 100 MW base power station and a 60 MW peak power station can supply the network demand.

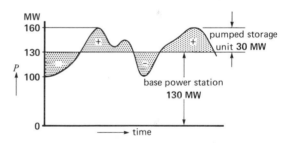

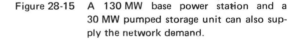

Figure 28-15   A 130 MW base power station and a 30 MW pumped storage unit can also supply the network demand.

mand (shown by a plus), the peaking station returns the energy it had previously stored.

This method has two advantages:

1. the base-power station is larger and, consequently, more efficient;
2. the peak-power station is much smaller and, therefore, less costly.

Large blocks of energy can only be stored mechanically, and this is why we often use a hydraulic pumped-storage system. Such a system consists of an upper and a lower reservoir connected by a penstock and an associated generating/pumping sta-

Figure 28-16

This pumped storage station, in Tennessee, pumps water from Lake Nickajack to the top of Raccoon Mountain, where it is stored in a 2 km² (≈ 500 acres) reservoir, giving a 316 m head. The four alternator/pump units can each deliver 425 MVA, during the system peaks. The units can be changed over from generators to pumps in a few minutes. *(Tennessee Valley Authority)*

tion. During system peaks, the station acts like an ordinary hydropower source, delivering electrical energy as water flows from the upper to the lower reservoir. However, during light load periods, the process is reversed. The alternator then operates as a synchronous motor, driving the turbine as an enormous pump. Water now flows from the lower to the upper reservoir, thereby storing energy in preparation for the next system peak (Fig. 28-16).

This generating/pumping cycle is repeated once or twice per day, depending on the nature of the system load. Peak-power alternators have ratings between 50 MW and 500 MW. They are reversible because the direction of rotation has to be changed when the turbine operates as a pump. Repeated starting of such big synchronous motors puts a heavy load on the transmission line, and special methods must be used to bring them up to speed. Pony motors are often used, but static electronic frequency converters are also gaining ground.

Pumped storage installations operating in conjunction with nuclear plants make a very attractive combination because the latter give best efficiency when operating at constant load.

THERMAL GENERATING STATIONS

Thermal generating stations produce electricity from the heat released by the combustion of coal, oil, or natural gas. Most stations have ratings between 200 MW and 1500 MW so as to attain the high efficiency and economy of a large installation. Such a station has to be seen to make one realize its enormous complexity and size.

Thermal stations are usually located near a river or lake because large quantities of cooling water are needed to condense the steam as it exhausts from the turbines. The hydraulic resources of most modern countries are already fully developed. Consequently, we have to rely on thermal and nuclear stations to supply the growing need for electrical energy.

28-12 The process of combustion

In every chemical reaction, the molecules of two substances combine to create a new substance. Thus, the chemical reaction between an atom of sodium (Na) and an atom of chlorine (Cl) produces a molecule of table salt (NaCl). Some chemical reactions, especially those involving oxygen atoms, produce not only a new substance, but also release energy in the form of heat. In some cases, the heat released is so great that the resulting high temperature causes the reacting elements to glow, creating what we call a fire.

Oxygen in the air reacts actively with carbon (C), hydrogen (H), sulfur (S), and every substance containing these atoms. This explains why coal, wood, oil, and natural gas burn in the presence of air.

28-13 Fuels

Atoms of oxygen combine in precise known proportions with atoms of carbon, hydrogen and sulfur. The amount of heat released and the resulting products of combustion can therefore be predicted in advance when we know the makeup of the fuel. Table 28A gives some of these details. For instance, during the complete combustion of 1 kg of carbon, 2.67 kg of oxygen is required and 33.8 MJ of heat

is released. The chemical reaction produces carbon dioxide, a nontoxic gas that has the same composition as the air we exhale. Because dry air contains 0.23 parts of oxygen per unit mass, we multiply the mass of oxygen by $1/0.23 = 4.3$ to obtain the mass of air needed to burn a given quantity of fuel.

By using the values given in Table 28A, we can calculate the heat released by a fuel if we know its composition. Tables 28B and 28C show the heat released by typical samples of coal, oil and natural gas.

Example 28-2:
The coal from a particular mine has the following composition, by weight:

carbon	60%	sulfur	3%
hydrogen	2%	other elements	35%

Calculate:

a. the approximate heat released per tonne (1000 kg) of coal;

b. the mass of air required to ensure complete combustion.

TABLE 28A   FUELS, HEAT, AND PRODUCTS OF COMBUSTION

Fuel	mass of fuel	mass of oxygen	heat released *	products of combustion	mass of air	volume of air†
—	(kg)	(kg)	MJ	—	(kg)	m³
carbon C	1	2.67	33.8	CO_2	11.5	9.6
hydrogen H	1	8	120*	H_2O	34.5	28.8
sulfur S	1	1	9.3	SO_2	4.3	3.6
methane CH_4	1	4	50*	$CO_2 + H_2O$	17.2	14.3
ethane C_2H_6	1	3.73	47.5*	$CO_2 + H_2O$	16.1	13.4
propane C_3H_8	1	3.64	46.5*	$CO_2 + H_2O$	15.6	13

* Energy that is available after having substracted the latent heat of vaporization of water (not recuperable in a boiler).

† At a temperature of 20°C and pressure of 101 kPa.

Solution:

a. 1 One tonne of coal contains the following masses of C, H_2 and S, which respectively release the heat indicated below:

carbon: 600 kg @ 33.8 MJ/kg   = 20.3 GJ
hydrogen: 20 kg @ 120 MJ/kg   = 2.4
sulfur: 30 kg @ 9.3 MJ/kg   = 0.3
other (noncombustible) elements:
    350 kg   = 0.0

    Total heat released   = 23 GJ

b. Referring to Table 28A, the mass of air required for each substance is:

carbon: 600 kg x 11.5   = 6900
hydrogen: 20 kg x 34.5   = 690
sulfur: 30 kg x 4.3   = 129

    Total air required   = 7719 kg
                              = 7.7 t

Note: The above calculations are very approximate, but they do indicate the order of magnitude of the quantities involved.

28-14 Products of combustion

Carbon dioxide (CO_2), sulfur dioxide (SO_2), and water are the main products of combustion when oil, coal, or gas are burned. Carbon dioxide and water produce no serious environmental effects, but sulfur dioxide creates substances that irritate the respiratory tracts. Dust and fly ash are other pollutants that may reach the atmosphere. Natural gas produces only water and CO_2. This explains why gas is used (rather than coal or oil), when atmospheric pollution must be reduced to a minimum.

28-15 Makeup of a thermal generating station

The basic structure and principal components of a thermal generating station are shown in Fig. 28-18. They are itemized and described below.

- A huge boiler (1) acts as a furnace, transferring heat from the burning fuel to row upon row of water tubes S_1, which entirely surround the flames. Water is kept circulating through the tubes by a pump P_1.

- A drum (2) containing water and steam under high pressure, produces the steam required by the turbines. It also receives the water delivered by boiler feed pump P_3. Steam races towards high-pressure turbine HP after having passed through superheater S_2. The superheater, composed of a series of tubes surrounding the flames, raises the steam temperature by about $200°C$. This ensures that the steam is absolutely dry and raises the overall efficiency of the station.

- High-pressure turbine (3) converts thermal energy into mechanical energy by letting the steam expand as it moves through the turbine blades. The temperature and pressure at the output of the turbine are therefore less than at the input. In order to raise the thermal efficiency and to prevent premature condensation, the steam passes through a reheater S_3, composed of a third set of heated tubes.

- Medium-pressure turbine (4) is similar to the high-pressure turbine, except that it is bigger so that the steam may expand still more.

- Low-pressure turbine LP (5), made in two identical sections, removes the remaining available energy from the steam. The steam flowing out of LP expands into an almost perfect vacuum created by the condenser.

- Condenser (6) causes the steam to condense by letting it flow over cooling pipes S_4. Cold water from an outside source flows through the pipes, thus carrying away the heat. A condensate pump P_2 removes the lukewarm condensed water and drives it through a reheater (7) towards feedwater pump (8).

- Reheater (7) is a heat exchanger. It receives hot steam, bled off from high-pressure turbine HP, to raise the temperature of the feedwater. Thermodynamic studies show that the overall thermal efficiency is improved when some steam is bled off this way, rather than letting it follow

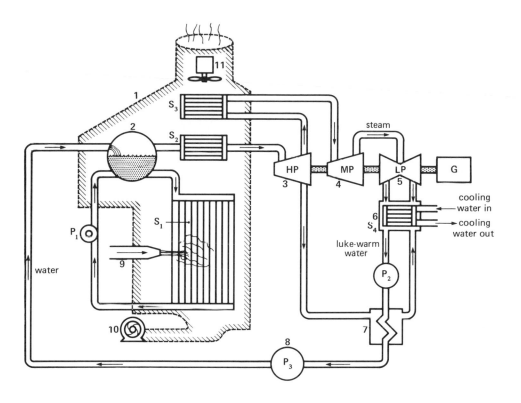

Figure 28-18   Principal components of a thermal power plant.

its normal course through all three turbines.

• Boiler feed pump (8) forces the feedwater into the high-pressure drum, thus completing the thermal cycle.

• The burners (9) supply and control the amount of gas, oil or coal injected into the boiler. Coal is pulverized before it is injected. Similarly, heavy bunker oil is preheated and injected as an atomized jet to improve surface contact (and combustion) with the surrounding air.

• A forced-draft fan (10) furnishes the enormous quantities of air needed for combustion (see Fig. 28-19).

• An induced-draft fan (11) carries the gases and other products of combustion towards cleans-

ing apparatus and from there to the stack and the outside air.

• Generator G, directly coupled to all three turbines, converts the mechanical energy into electrical energy.

In practice, a steam station has hundreds of other components and accessories to ensure high efficiency, safety, and economy. For example, control valves regulate the amount of steam flowing to the turbines; complex water purifiers maintain the required cleanliness and chemical composition of the feedwater; oil pumps keep the bearings properly lubricated, and so on. However, the basic components we have just described enable us to understand the operation and some of the basic problems of a thermal station.

TABLE 28B   TYPICAL COMPOSITION OF SAMPLES OF COAL AND OIL

substance in the sample	UNIT	mass of substance as a percent of mass of sample			
		coal sample A	coal sample B	oil sample X	oil sample Y
carbon	%	55	85	85.5	85
hydrogen	%	2	5	13.5	10.5
sulfur	%	1	4	1	4
waste	%	42	6	—	0.5
heat content in sample	MJ/kg	21	35	45	42

TABLE 28C   TYPICAL COMPOSITION OF NATURAL GAS

substance in the sample	Symbol	Unit	mass of substance as a percent of mass of sample	
			sample 1	sample 2
methane	CH_4	%	63	96
ethane	C_2H_6	%	3	0.1
propane	C_3H_8	%	2	0
nitrogen	N_2	%	28	1
waste		%	4	3
heat content in sample		MJ/kg	33.8	48

Figure 28-19   This forced-draft fan provides 455 m³/s of air at a pressure difference of 5.8 kPa for a thermal power station. It is driven by a 3-phase induction motor rated 12 000 hp (8955 kW), 60 Hz, 890 r/min. *(Novenco Inc.)*

28-16 Turbines

The low-, medium-, and high-pressure turbines possess a series of blades mounted on the drive shaft (Fig. 28-20). The steam is deflected by the blades, producing a powerful torque. The blades are made of special steel to withstand the high temperature and intense centrifugal forces.

Most turbines are coupled together to drive a common alternator. However, in some large installations, the HP turbine drives one alternator while the MP and LP turbines drive another having the same rating.

Figure 28-20   Low-pressure section of a 375 MW, 3600 r/min turbogenerator set, showing the radial blades. *(General Electric)*

28-17 Condenser

About one-half the energy produced in the boiler is removed from the steam when it exhausts into the condenser. Consequently, enormous quantities of water are needed to carry away the heat. The temperature of the cooling water increases typically by 5°C to 10°C as it flows through the condenser tubes. The condensed steam (condensate) usually has a temperature between 27°C and 33°C and the corresponding absolute pressure is a near-vacuum of about 5 kPa. The cooling water temperature is only a few degrees below the condensate temperature (see Fig. 28-21).

28-18 Cooling towers

If the thermal station is located in a dry region, or far away from a river or lake, we still have to cool the condenser, one way or another. We often use *evaporation* to produce a cooling effect. To understand the principle, consider a lake which exposes a large surface to the surrounding air. A lake evaporates continually, even at low temperatures, and tests have shown that for every kilogram of water that evaporates, the lake loses 2.4 MJ of heat. Con-

Figure 28-21   Condenser rated at 220 MW. Note the large pipes feeding cooling water into and out of the condenser. The condenser is as important as the boiler in thermal and nuclear power stations. *(Foster-Wheeler Energy Corporation)*

sequently, evaporation causes the lake to cool down.

Consider now a tub containing 100 kg of water at a certain temperature. If we can somehow cause 1 kg of water to evaporate, the temperature of the remaining 99 kg will drop by 5°C.* We conclude that whenever 1 percent of a body of water evaporates, the temperature of the remaining water drops by 5°C. Evaporation is therefore a very effective cooling process.

But how can we produce evaporation? Surprisingly, all that is needed is to expose a large surface of water to the surrounding air. The simplest way to do this is to break the water up into small droplets, and blow air through the resulting artificial rain.

In the case of a steam station, the warm cooling water from the condenser is broken up into small droplets inside a cooling tower (Fig. 28-22). Evaporation takes place, and the droplets are chilled as they fall towards an open reservoir below. The cool water is pumped from the reservoir and re-

*   The exact value is 5.8°C.

turned to the condenser where it again removes heat from the condensing steam. The cycle then repeats. Approximately 2 percent of the cooling water that flows through the condenser is lost by evaporation. This loss can be made up by an underground spring, a stream, or small lake.

28-19 Boiler feed pump

The boiler feed pump drives the feedwater into the high-pressure drum. The high pressure and large volume of water flowing through the pump requires a very powerful motor to drive it. In modern steam stations, the pumping power represents about 1 percent of the alternator output. Although this appears to be a significant loss, we must remember that the energy expended is later recovered when the high-pressure steam flows through the turbines. Consequently, the energy supplied to the feed pump motor is not really lost, except for the small portion consumed by the losses in the motor and pump themselves.

28-20 Energy flow diagram for a steam plant

Modern thermal generating stations are very similar throughout the world because all designers strive for high efficiency at lowest cost. This means that materials are strained to the limits of safety as far as temperature, pressure, and centrifugal forces are concerned. Because the same materials are available to all, the resulting systems are necessarily similar. Figure 28-23 shows a typical 540 MW turbine-generator set and Fig. 28-24 is a view of the control room.

Most modern boilers furnish steam at a temperature of 550°C and a pressure of 16.5 MPa. The overall efficiency (electrical output/thermal input) is then about 40 percent. The relative amounts of energy, steam flow, losses, and so forth, do not change very much, provided the temperature and pressure are about as indicated above. This enables us to draw a simple flow diagram showing the energy situation in a reduced-scale model of a typical steam station. Figure 28-25 shows such a model,

Figure 28-22

Cooling tower installed in a nuclear power station in Oregon. The generator output is 1280 MVA at a power factor of 0.88. Tower characteristics: height: 152 m; diameter at the base: 117 m; diameter at the top: 76 m; cooling water: 27 m³/s; water loss by evaporation: 0.7 m³/s. The temperature of the cooling water drops from 44.5°C to 24°C as it passes through the tower. *(Portland General Electric Company)*

Figure 28-23

This 540 MW turbogenerator set runs at 3600 r/min, generating a frequency of 60 Hz. The low-pressure turbine and alternator are in the background. *(General Electric)*

Figure 28-24

Control room of the 540 MW turbogenerator set. *(General Electric)*

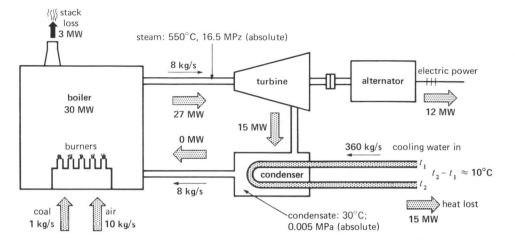

Figure 28-25   Scale model of a typical thermal generating station.

producing 30 MW of thermal power and generating 12 MW of electrical power.

For example, a 480 MW station (40 times more powerful than the model) has the following approximate characteristics:

Electric power output: 480 MW
Coal consumption: 40 kg/s
Air intake: 400 kg/s
Boiler thermal power: 1200 MW
Steam output: 320 kg/s
Cooling water requirement
   (with $\Delta t = 10°C$): 14 400 kg/s $= 14.4 \text{ m}^3/\text{s}$
Heat carried away by the cooling water: 600 MW

If a cooling tower is required, it would have to evaporate:

$q = 2\% \times 14.4 = 0.29 \text{ m}^3/\text{s}$

of cooling water. This loss has to be made up by a local source of water.

NUCLEAR GENERATING STATIONS

Nuclear stations produce electricity from the heat released by a *nuclear reaction*. When the nucleus of an atom splits in two (a process called atomic fission), a considerable amount of energy is re-

leased. Note that a chemical reaction, such as the combustion of coal, produces only a rearrangement of the atoms, without in any way affecting their nuclei.

A nuclear station is identical to a thermal station except that the boiler is replaced by a nuclear reactor. The reactor contains the fissile material that generates the heat. A nuclear station therefore contains an alternator, steam turbine, condenser, etc., similar to those found in a conventional steam station. The overall efficiency is also similar (between 30 and 40 percent) and a cooling system must be provided for. Consequently, nuclear stations are also located close to rivers and lakes. In dry areas, cooling towers are installed. Owing to these similarities, we will only examine the operating principle of the reactor itself.

28-21 Composition of an atomic nucleus

The nucleus of an atom contains two types of particles - protons and neutrons. The proton carries a positive charge, equal to the negative charge on an electron. The neutron, as its name implies, has no electric charge. Neutrons are therefore neither attracted nor repelled by protons and electrons.

Protons and neutrons have about the same

mass, and both weigh 1840 times as much as electrons do. The mass of an atom is therefore concentrated in its nucleus.

The number of protons and neutrons in the nucleus depends upon the element. Furthermore, because an atom is electrically neutral, the number of electrons is equal to the number of protons. Table 28D gives the atomic structure of a few important elements used in nuclear reactors. For example, there are three types of hydrogen atoms that can be distinguished from each other only by the makeup of the nucleus. First, there is ordinary hydrogen (H), whose nucleus contains 1 proton and no neutrons. Next, there are two rare forms, deuterium (D) and tritium (^{3}H) whose nuclei respectively contain 1 and 2 neutrons, in addition to the usual proton. These rare forms are called *isotopes* of hydrogen.

When two atoms of ordinary hydrogen unite with one atom of oxygen, we obtain ordinary water (H_2O) called *light* water. On the other hand, if 2 atoms of deuterium unite with 1 atom of oxygen, we obtain a molecule of *heavy* water (D_2O). The oceans contain about 1 kg of heavy water for every 7000 kg of sea water.

In the same way, two types of uranium atoms are found in nature: uranium 238 (^{238}U) and uranium 235 (^{235}U). Each contains 92 protons, but differing numbers of neutrons. Uranium 238 is very common, but the isotope ^{235}U is rare.*

Uranium 235 and heavy water deserve our attention because both are essential to the operation of the nuclear reactors we are about to describe.

28-22 Energy released by atomic fission

When the nucleus of an atom fissions, it splits in two. The total mass of the two atoms formed in this way is usually less than that of the original

---

* The ore from uranium mines contains the compound U_3O_8, very roughly in the proportion of 1000 to 1. Furthermore, U_3O_8 is actually composed of $^{238}U_3O_8$ and $^{235}U_3O_8$ in the relatively precise ratio of 139.8 to 1. It is very difficult to separate these two substances because ^{238}U and ^{235}U possess identical chemical properties (see Fig. 28-27).

TABLE 28D   ATOMIC STRUCTURE OF SOME ELEMENTS

Element	Symbol	Protons	Electrons	Neutrons	mass number (neutrons + protons)
hydrogen	H	1	1	0	1
deuterium	D	1	1	1	2
tritium	^{3}H	1	1	2	3
helium	He	2	2	2	4
carbon	C	6	6	6	12
iron	Fe	26	26	30	52
uranium 235	^{235}U	92	92	143	235
uranium 238	^{238}U	92	92	146	238

atom. If there is a loss in mass, energy is released according to Einstein's equation:

$$E = mc^2 \qquad (28\text{-}2)$$

where

E = energy released [J]
m = loss of mass [kg]
c = speed of light [3×10^8 m/s]

An enormous amount of energy is released because a loss in mass of only 1 μg produces 9 x x 10^{10} J, which is equivalent to the heat given off by burning 3 tonnes of coal. Uranium is one of those elements that loses mass when it fissions. However, uranium 235 is more "fissionable" than uranium 238 is, and so large separating plants have been built to isolate molecules containing ^{235}U from those containing ^{238}U.*

28-23 Chain reaction

How can we provoke the fission of a uranium atom? One way is to bombard its nucleus with neutrons. A neutron makes an excellent projectile because it is not repelled as it approaches the nucleus and, if its speed is not too great, it has a good chance of scoring a hit. If the impact is strong enough, the nucleus will split in two, releasing energy. The fission of one atom of ^{235}U frees 218 MeV of energy, mainly in the form of heat. Fission is a very violent reaction on an atomic scale, and it produces a second important effect: it ejects 2 or 3 neutrons that move at high speed way from the broken nucleus. These neutrons collide with other uranium atoms and so a chain reaction quickly takes place, releasing a tremendous amount of heat.

This is the principle that causes atomic bombs to explode. A mere 300 g of UO_2 (type ^{235}U) is

---

* For example, in uranium dioxide (UO_2) enrichment plants the ratio of $^{235}UO_2$ to $^{238}UO_2$ is typically raised to 0.03, compared to the natural, nonenriched ratio of 1/139.8 or 0.007 15.

enough to produce a violent explosion. Although a uranium mine also releases neutrons, the concentration of ^{235}U atoms is too low to produce a chain reaction.

In the case of a nuclear reactor, we have to slow down the neutrons to increase their chances of striking other uranium nuclei. Towards this end, small fissionable masses of a uranium compound (such as UO_2) are sunk within a *moderator*. The moderator may be ordinary water, heavy water, graphite, or any other material that can slow down neutrons without absorbing them. By using an appropriate geometrical distribution of the uranium fuel within the moderator, we can reduce the speed of the neutrons so they have the required velocity to initiate other fisions. Only then will a chain reaction take place, causing the reactor to "go critical".

As soon as the chain reaction starts, the temperature rises rapidly. To keep it at an acceptable level, a liquid or gas has to flow through the reactor to carry away the heat. The coolant may be heavy water, ordinary water, liquid sodium or a gas like helium or carbon dioxide. The hot coolant moves in a closed circuit which includes a heat exchanger. The latter transfers the heat to a steam generator that drives the turbines (Fig. 28-26).

28-24 Types of nuclear reactors

There are several types of reactors, but the following are the most important:

1. **Pressure water reactor (PWR).** Water is used as a coolant and it is kept under such high pressure that it cannot boil off into steam. We can use ordinary water, as in light-water reactors, or heavy water as in CANDU* reactors.

2. **Boiling water reactors (BWR).** The coolant in this reactor is ordinary water boiling under high pressure and releasing steam. This eliminates the

---

* CANDU: Canada Deuterium Uranium, developed by the Atomic Energy Commission of Canada.

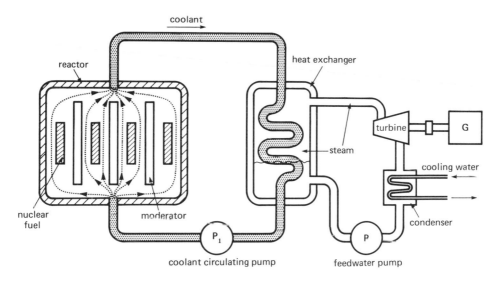

Figure 28-26   Schematic diagram of a nuclear power station.

need for a heat exchanger, because the steam circulates directly through the turbines. However, as in any light-water reactor, we must use enriched uranium dioxide containing about 3 percent ^{235}U.

3. High temperature gas reactor (HTGR).

This reactor uses an inert gas coolant such as helium or carbon dioxide. Owing to the high operating temperature (typically $750°C$), graphite is used as a moderator. The steam created by the heat exchanger is as hot as that produced in a conventional steam boiler. Consequently, the overall efficiency of these nuclear stations is about 40 percent.

4. Fast breeder reactor (FBR).

This reactor has the remarkable ability to both generate heat and create additional nuclear fuel while it is operating.

28-25   Example of a light-water reactor

Reactors that use ordinary water as a moderator are similar to those using heavy water, but the uranium-dioxide fuel has to be enriched. Enrichment means that the fuel bundles contain between 3 and 4 percent of ^{235}U, the remainder being ^{238}U. This enables us to reduce the size of the reactor for a given power output. On the other hand, the reactor has to be shut down about once a year to replace the expended bundles.

The generated heat, created mainly by the fission of uranium 235, is carried away by a coolant such as ordinary water, liquid sodium or a gas such as CO_2. As it flows through the heat exchanger, the coolant creates the steam that drives the turbine.

A typical nuclear power station located in Connecticut (Figs. 28-28 and 28-29) possesses a light-water reactor that drives a 3-phase, 667 MVA, 90 percent power factor, 19 kV, 60 Hz, 1800 r/min alternator. The reactor is composed of a vertical steel tank (2700 mm thick) having an external diameter of 4.5 m and a height of 12.5 m. The tank contains 157 vertical tubes which can lodge 157 large fuel assemblies. Each assembly is 3 m long and groups 204 fuel rods containing a total of 477 kg of enriched UO_2. The nuclear reaction is kept under control by 45 special-alloy control rods. When these rods are gradually lowered into the moderator, they absorb more and more neutrons. Consequently, they control the rate of the nuclear reaction and hence the amount of heat released by the reactor.

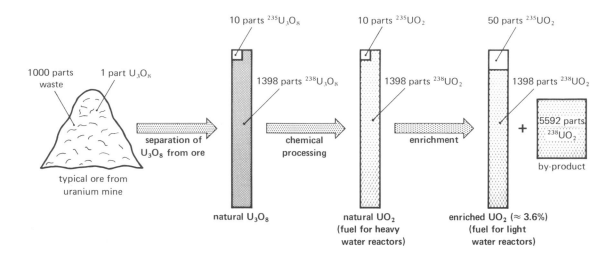

10 parts $^{235}U_3O_8$

10 parts $^{235}UO_2$

50 parts $^{235}UO_2$

1000 parts waste

1 part U_3O_8

1398 parts $^{238}U_3O_8$

1398 parts $^{238}UO_2$

1398 parts $^{238}UO_2$

typical ore from uranium mine

separation of U_3O_8 from ore

chemical processing

enrichment

+

5592 parts $^{238}UO_2$

by-product

natural U_3O_8

natural UO_2 (fuel for heavy water reactors)

enriched UO_2 ($\approx 3.6\%$) (fuel for light water reactors)

Figure 28-27   Various steps in the manufacture of nuclear fuel for heavy-water and light-water reactors. This extremely simplified diagram shows that in the process of enriching uranium dioxide, it is inevitable that large amounts of ^{238}U remain as a byproduct.

28-26 Example of a heavy-water reactor

The CANDU reactor uses heavy water, both as moderator and as a coolant. It differs from all other reactors in that is uses *natural* uranium dioxide as a fuel (see Fig. 28-30). One of the biggest installations of its kind is located at Pickering, a few kilometers east of Toronto, Canada. The nuclear station has 4 reactors and 4 turbines, each of which drives a 3-phase, 635 MVA, 85 percent power factor, 24 kV, 1800 r/min, 60 Hz alternator. Each reactor is coupled to 12 heat exchangers that provide the interface between the heavy water coolant and the ordinary steam that drives the tur-

Figure 28-28

Aerial view of a light-water nuclear generating station. The large rectangular building in the foreground houses a 667 MVA, 90 percent power factor, 19 kV, 60 Hz, 1800 r/min turbo-generator set; the circular building surrounds the reactor. *(Connecticut Yankee Atomic Power Company; photo by Georges Betancourt)*

Figure 28-29 Looking down into the water-filled refuelling cavity of the reactor. *(Connecticut Yankee Atomic Power Company; photo by Georges Betancourt)*

bines (Fig. 28-30).

Each reactor is enclosed in a large horizontal vessel (calandria) having a diameter of 8 m and a length of 8.25 m. The calandria possesses 390 horizontal tubes each housing 12 fuel bundles containing 22.2 kg of UO_2. Each bundle releases about 372.5 kW while it is in operation. Because there is a total of 4680 bundles, the reactor delivers 1740 MW of thermal power.

Twelve primary heat transfer pumps, each driven by an 1100 kW motor, push the heavy water coolant through the reactor and the heat exchangers. The heat exchangers produce the steam to drive the turbine. The steam exhausts into a condenser which is cooled by water from Lake Ontario.

The fuel bundles are inserted at one end of the calandria and, after a 19-month stay in the tubes, they are withdrawn from the other end. The bundles are inserted and removed on a continuous basis - an average of 9 bundles per day.

Table 28E gives typical characteristics of light- and heavy-water reactors.

28-27 Principle of a fast breeder reactor

A fast breeder reactor differs from other reactors because it can extract more of the available energy in the nuclear fuel. It possesses a central core consisting of a substance containing fissionable plutonium 239 (^{239}Pu). The core is surrounded by a blanket composed of substances containing nonfissionable uranium 238 (^{238}U). No moderator is used; consequently, the high-speed ("fast") neutrons generated by the fissioning ^{239}Pu bombard the nonfissionable atoms of ^{238}U. This nuclear reaction produces two important results:

a. the heat released by the fissioning core can be used to drive a steam-turbine;
b. some atoms of ^{238}U, in the surrounding blanket, capture the flying neutrons, thereby becoming fissionable ^{239}Pu. In other words, passive atoms of uranium 238 are transmuted into fissionable atoms of plutonium 239.

As time goes by, the blanket of nonfissionable ^{238}U is gradually transmuted to fissionable ^{239}Pu and waste products. The blanket is eventually removed and the materials are processed to recover the substances containing ^{239}Pu. The nuclear fuel recovered is placed in the central core to generate heat and to produce still more fuel in a newly relined blanket of substances containing uranium 238.

This process can be repeated until nearly 80 percent of the available energy in the uranium is extracted. This is much more efficient than the 2 percent now being extracted by conventional reactors.

The breeder reactor is particularly well adapted to complement existing light water reactors. The reason is that a great deal of ^{238}U is available as a byproduct in the manufacture of enriched ^{235}U (see Fig. 28-27). This otherwise useless material (now being stored) could be used to surround the core of a fast breeder reactor. By capturing fast neutrons, it could be rejuvenated as explained above, until most of the potential energy in the uranium is used up.

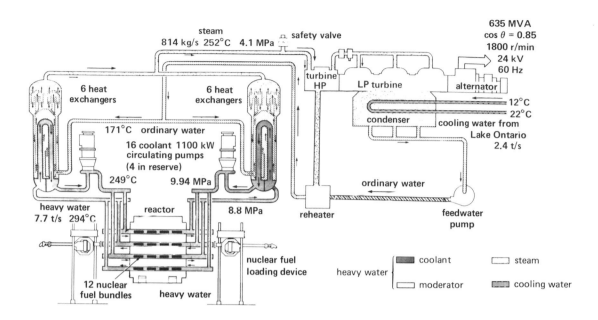

Figure 28-30 Simplified schematic diagram of a CANDU nuclear generating unit composed of one heavy-water reactor driving one alternator. *(Atomic Energy of Canada)*

28-28 Nuclear fusion

We have seen that splitting the nucleus of a heavy element such as uranium results in a decrease in mass and the release of energy. We can also produce energy by *combining* the nuclei of two light elements in a process called *nuclear fusion*. For example, energy is released by the fusion of an atom of deuterium with an atom of tritium. However, owing to the strong repulsion between the two nuclei (both are positive), they only unite (fuse) when they approach each other at high speed. The required velocity is close to the speed of light and corresponds to a thermodynamic temperature of several million degrees. If both the atomic concentration and speed are high enough, a self-sustaining chain reaction will result.

We can therefore produce heat by the fusion of two elements, and the hydrogen bomb is a good example of this principle. Unfortunately, we run into almost insurmountable problems when we try to *control* the fusion reaction, as we must do in a nuclear reactor. Basically, scientists have not yet succeeded in confining and controlling high-speed particles without at the same time slowing them down.

A major world-wide research effort is being devoted to solve this problem. If scientists succeed in domesticating nuclear fusion, it could mean the end of the energy shortage because hydrogen is the most common element on earth.

QUESTIONS AND PROBLEMS

Practical level

28-1 Explain the difference between a base load and peak load generating plant.

28-2 Why are nuclear power stations not suited to supply peak loads?

TABLE 28E   TYPICAL LIGHT-WATER AND HEAVY-WATER REACTORS

	Light-Water Reactor	Heavy-Water Reactor
Reactor Vessel		
external diameter	4.5 m	8 m
length	12.5 m	8.25 m
vessel thickness	274 mm	25.4 mm
weight empty	416 t	604 t
position	vertical	horizontal
number of fuel canals	157	390
type of fuel	enriched UO_2 (3.3%)	natural UO_2
total mass of fuel	75 t	104 t
Moderator		
type	light-water	heavy-water
volume	13.3 m^3	242 m^3
Reactor Cooling		
Heat produced in reactor	1825 MW	1661 MW
Coolant	light-water	heavy-water
Volume	249 m^3	130 m^3
Flow rate	12.8 t/s	7.73 t/s
Coolant temperature entering the reactor	285°C	249°C
Coolant temperature leaving the reactor	306°C	294°C
Coolant pumps	4	12
Total pump power	12 MW	14 MW
Electrical output		
3-phase, 1800 r/min, 60 Hz alternator	600 MW	540 MW

28-3 Referring to the coal-mine in Fig. 28-4, we have the choice of hauling the coal to a generating plant or installing the generating plant next to the mine mouth. What factors come into play in determining the best solution?

28-4 What is the best indicator of stability (or instability) of a network?

28-5 What is meant by the term network?

28-6 Give two reasons why systems are interconnected.

28-7 The river flow in Fig. 28-9 is 5000 m^3/s at a head of 24 m. Calculate the available hydraulic power.

28-8 Explain the operating principle of a thermal plant; a hydropower plant; a nuclear plant.

28-9 Name two basic differences between a light-water reactor and a heavy-water reactor?

28-10 Explain what is meant by: moderator, fission, fusion, neutron, heavy-water.

Intermediate level

28-11 The Zaïre River, in Africa, discharges, at a constant rate, 1300 km^3 of water per year. It has been proposed to build a series of dams in the region of Inga, where the river drops by 100 m. Calculate:
a. the water flow [m^3/s] ;
b. the power that could be harnessed [MW] ;
c. the discharge in cubic miles per year.

28-12 For how long does a 1500 MW alternator have to run to produce the same quantity of energy as that released by a 20 kilotonne atomic bomb? (See conversion charts in Appendix).

28-13 The demand of a municipality regularly varies between 60 MW and 110 MW in the course of one day, the average power being 80 MW. To produce the required energy, we have the following options:
a. install a base power generating unit and a diesel engine peaking plant;
b. install a base power generating unit and a pumped storage unit.

What are the respective capacities of the base power and peaking power plants in each case?

28-14 On a particular day, the head of Grand Coulee dam is 280 feet and the generators deliver 6000 MVA at a power factor of 0.9 lagging. Assuming the average turbine efficiency is 0.92 and the average alternator efficiency is 0.98, calculate:
a. the active power output [MW] ;
b. the reactive power supplied to the system [Mvar] ;
c. the amount of water flowing through the turbines [yd^3/s] .

28-15 Explain the principle of operation of a cooling tower.

28-16 A modern coal-burning thermal station produces an electrical output of 720 MW. Calculate the approximate value of:
a. the amount of coal consumed [tons (not tonnes) per day] ;
b. the amount of smoke, gas, and fly ash released [tons per day] ;
c. the cooling water flowing through the condenser, assuming a temperature rise of 10°C [m^3/s] .

28-17 In Problem 28-16, if a cooling tower is required, how much water must be drawn from a local stream [m^3/s] ? Can this water be recycled?

28-18 A fuel bundle of natural uranium dioxide has a mass of 22.2 kg when first inserted into a heavy-water reactor. If it releases an average of 372.5 kW of thermal energy during its 19-month stay in the reactor, calculate:
a. the total amount of heat released [J] and [Btu] ;
b. the reduction in weight of the bundle, due to the energy released [g] .

Advanced level

28-19a. Calculate the annual energy consumption [TW·h] of the system having the load duration curve given in Fig. 28-3.

b. If this energy were consumed at an absolutely uniform rate, what would the peak load be [GW] ?

28-20 A sample of coal has the following composition:

carbon 70%   sulfur 2%

hydrogen 3%   waste 25%

a. Calculate the heat released [Btu] in burning 1 ton (2000 lb) of this type of coal, assuming complete combustion;

b. what is the minimum weight of air required? [lb]

c. calculate the approximate amount of electrical energy [kW·h] that could be obtained, per ton.

28-21 Referring to Fig. 28-30, the temperature of the heavy water coolant drops from 294°C to 249°C in passing through the heat exchangers. Knowing that the reactor is cooled at the rate of 7.7 t/s of heavy-water, calculate the heat [MW] transmitted to the heat exchangers (specific heat of heavy-water is 4560 J/kg).

29

THE TRANSMISSION OF ELECTRICAL ENERGY

The transmission of electrical energy does not usually raise as much interest as does its generation and utilization: consequently, we often tend to neglect this important subject. This is unfortunate because the human and material resources involved in transmission are much greater than those devoted to generation.

Electrical energy is carried by conductors such as overhead transmission lines and underground cable. Although these conductors appear very ordinary, they hide important properties that greatly affect the transmission of electrical energy. In this chapter, we shall study these properties for every type of transmission line: high-voltage, low-voltage, high power, low power, aerial lines, underground lines, and so forth.

29-1 Principal components of a transmission system

In order to provide electrical energy in usable form, a transmission and distribution system must satisfy some basic requirements. The system must:
1. Provide, at all times, the power that consumers need;

2. Maintain a stable, nominal voltage that does not vary by more than ± 10%;
3. Maintain a stable frequency that does not vary by more than ± 0.1 Hz;
4. Supply energy at an acceptable price;
5. Observe standards of safety;
6. Protect the environment.

Figure 29-1 shows an elementary diagram of a transmission and distribution system. It consists of two generating stations G_1 and G_2, a few substations, an interconnecting substation and, finally, some commercial, residential and industrial loads. The energy is carried over lines designated *extra high* voltage (EHV), *high* voltage (HV), *medium* voltage (MV) and *low* voltage (LV), according to a scale of standardized voltages whose limiting values are given in Table 29A.

Substations (Fig. 29-1) serve to change the voltage by means of step-up and step-down transformers and to regulate it by means of synchronous condensers, static var compensators, or transformers with variable taps. Substations also contain circuit breakers, fuses and lightning arrestors, to protect expensive apparatus and to provide for quick isolation of critical lines from the system.

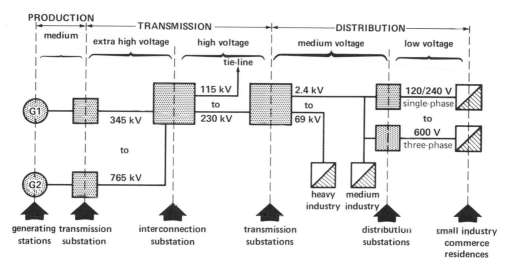

Figure 29-1   Single-line diagram of a generation, transmission and distribution system.

Interconnecting substations serve to tie different systems together, to enable power exchanges between them, and to increase the stability of the overall network.*

Electrical power utilities divide their systems into two major categories:

1. **transmission** systems in which the line voltage is roughly between 115 kV and 800 kV;
2. **distribution** systems in which the voltage generally lies between 120 V and 69 kV. Distribution systems are divided into medium-voltage distribution systems (2.4 kV to 69 kV) and low-voltage distribution systems (120 V to 600 V).

29-2 Types of power lines

The design of a power line depends upon:
1. the amount of active power it has to transmit;
2. the distance over which the energy must be carried;
3. the cost;
4. esthetic considerations, urban congestion, ease of installation, and so forth.

---

* A network is "an aggregation of interconnected conductors consisting of feeders, mains and services" (ref. IEEE Standard Dictionary of Electrical and Electronic Terms).

As mentioned previously, we distinguish four types of power lines, according to their voltage class:

1. **Low-voltage (LV) lines** are installed inside buildings, factories and houses to supply power to motors, electric stoves, lights, and so on. The service entrance panel constitutes the source and the lines are made of insulated cable or bus-bars operating at voltages below 600 V.

In some metropolitan areas, the distribution system consists of a grid of underground cables operating at 600 V or less. Such a network provides dependable service, because even the outage of one or several cables will not interrupt customer service. Today, however, we prefer to install medium-voltage radial distribution systems in the larger cities. In radial systems, the transmission lines spread out like fingers from one or more substations to feed power to various load centers.

2. **Medium-voltage (MV) lines** tie the load centers to the main substation of the utility company. The voltage is usually between 2.4 kV and 69 kV.

3. **High-voltage (HV) lines** connect the main substations to the generating stations. The lines are composed of aerial wire or underground cable op-

	VOLTAGE CLASSES AS APPLIED		
TABLE 29A	TO INDUSTRIAL AND COMMERCIAL POWER		

voltage class	nominal system voltage		
	two-wire	three-wire	four-wire
low	120	120/240 □	—
voltage	single phase	single phase	120/208 □
		480 V □	277/480 □
LV		600 V	347/600
medium		2 400	
voltage		4 160 □	
		4 800	
MV		6 900	
		13 800 □	7 200/12 470 □
		23 000	7 620/13 200 □
		34 500	7 970/13 800
		46 000	14 400/24 940 □
		69 000 □	19 920/34 500 □
high		115 000 □	
voltage		138 000 □	
		161 000	
HV		230 000 □	
extra high		345 000 □	
voltage		500 000 □	
EHV		735 000 - 765 000 □	

All voltages are 3-phase unless indicated otherwise.
Voltages designated by the symbol □ are preferred voltages.

*Note: Voltage class designations were approved for use by IEEE Standards Board
(September 4, 1975)*

erating at voltages below 230 kV. In this category, we also find lines that transmit energy between two large systems, to increase the stability of the network.

4. **Extra high voltage (EHV) lines** are used when generating stations are very far from the load centers. We put them in a separate class because of their special properties. Such lines operate at voltages up to 800 kV and may be as long as 1000 km. High-tension direct current lines, covered in Chapter 31, are also included in this group.

29-3 Standard voltages

To reduce the cost of distribution apparatus and to facilitate its protection, standards-setting organizations have established a number of standard voltages for transmission lines. These standards, given in Table 29A, reflect the various voltages presently used in North America. Voltages that bear the symbol □ are *preferred* voltages. Unless otherwise indicated, all voltages are three-phase.

29-4 Components of a transmission line

A transmission line is composed of conductors, insulators and supporting structures.

1. **Conductors.** Conductors for high-tension lines are always bare. We use stranded copper conductors or steel-reinforced aluminum cable (ACSR). ACSR conductors are usually preferred, because they result in a lighter and more economical line. Conductors obviously have to be spliced when a line is very long. Special care must be taken so that the joints have low resistance and great mechanical strength.

Just like the plates of a capacitor, the parallel conductors of a transmission line remain charged after the voltage source is de-energized. To prevent fatal shocks, we must always discharge the conductors to ground before beginning work on a de-energized line. Most high-tension lines are equipped with appropriate grounding switches, but, in some rural installations, a grounded conducting chain is thrown over the line.

2. **Insulators.** Insulators serve to support and anchor the conductors and to insulate them from ground. Insulators are usually made of porcelain, but glass and other synthetic insulating materials are also used.

From an electrical standpoint, insulators must offer a high resistance to surface leakage currents and they must be sufficiently thick to prevent puncture under the high voltage stresses they have to withstand. To increase the leakage path (and hence the leakage resistance), the insulators are molded in the shape of a skirt. From a mechanical standpoint, they must be strong enough to resist the pull due to the weight of the conductors.

There are two main types of insulators: *pin-type* insulators and *suspension-type* insulators (Figs. 29-2 and 29-3). The pin-type insulator has several porcelain skirts and the conductor is fixed at the top. A steel pin screws into the insulator so it can be bolted to a support.

For voltages above 70 kV, we always use suspension-type insulators, strung together by their ball and socket metallic parts. The number of insulators depends upon the voltage: for 110 kV, we generally use from 4 to 7; for 230 kV, from 13 to 16. Figure 29-4 shows an insulator arrangement for a 735 kV line. It is composed of 4 strings of 35 insulators each, to provide both mechanical and electrical strength.

3. **Supporting structures.** The supporting structure must keep the conductors at a safe height from the ground and at an adequate distance from each other. For voltages below 70 kV, we can use single wooden poles equipped with cross-arms, but for higher voltages, a wooden H-frame must be used. The wood is treated with creosote or special metallic salts to prevent it from rotting. For very high voltage lines, we always use steel towers made of galvanized angle-iron, bolted together.

The spacing between conductors must be sufficient to prevent arc-over under gusty wind conditions. The spacing has to be increased as we increase the distance between towers and as the line

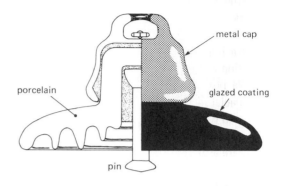

Figure 29-2   Sectional view of a 69 kV pin-type insulator. BIL: 270 kV; 60 Hz flashover voltage under wet conditions: 125 kV. *(Canadian Ohio Brass Co. Ltd.)*

Figure 29-3   Sectional view of a suspension insulator. Diameter: 254 mm; BIL: 125 kV, 60 Hz flashover voltage, under wet conditions: 50 kV. *(Canadian Ohio Brass Co. Ltd.)*

Figure 29-4

Lineman working "bare-handed" on a 735 kV line. He is wearing a special conductive suit so that his body is not subjected to high differences of potential. In the position shown, his potential with respect to ground is about 200 kV. *(Hydro-Québec)*

voltage becomes higher.

29-5 Construction of a line

Once we know the conductor size, the height of the poles and the distance between them (span), we can direct our attention to stringing the conductors. A wire supported between two points (Fig. 29-5) does not remain horizontal, but loops down at the middle. The vertical distance between

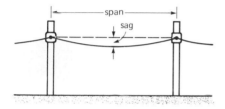

Figure 29-5   Span and sag of a line.

the straight line joining the points of support and the lowest point of the conductor is called *sag*. The tighter the wire, the smaller the sag will be.

Before undertaking the actual construction of a line, it is important to calculate the permissible sag and the corresponding mechanical pull. Among other things, temperature must be taken into account at the time the stringing takes place. On the one hand, the sag must not be too great, otherwise the wire will stretch even more during the summer heat, with the result that the clearance to ground may no longer be safe. On the other hand, if installed in the summer, the sag must not be too small otherwise the wire, contracting in winter, may become dangerously tight. Wind and sleet will add even more to the tractive force, and may ultimately cause the wire to break (Fig. 29-6).

29-6 Galloping lines

If a coating of sleet is deposited on a line during windy conditions, the line may begin to oscillate. Under certain conditions, the oscillations may become so large that the line is seen to actually "gal-

lop". Galloping lines can produce short-circuits between phases or snap the conductors. To eliminate the problem, we sometimes load the line with special mechanical weights, to dampen oscillations or to prevent them from building up.

29-7 Corona effect - radio interference

The very high voltages in use today produce a continual electrical discharge around the conductors, owing to local ionization of the air. This discharge, or corona effect, produces losses over the entire length of the transmission line. In addition, corona emits high frequency noise which interferes with nearby radio receivers and TV sets. To diminish corona, we must reduce the electric field (V/m) around the conductors, either by increasing their diameter or by arranging them in sets of two, three or more bundled conductors per phase (see Figs.

Figure 29-6   During winter, steel towers must carry the weight of both conductors and accumulated ice. *(Hydro-Québec)*

29-7a and 29-7b). This *bundling* arrangement also reduces the inductance of the line, enabling it to carry more power. This constitutes an additional benefit.

29-8 Pollution

Dust, acids, salts, and other pollutants in the atmosphere settle on insulators and reduce their insulating properties. Insulator pollution may produce short-circuits during storms or under momentary overvoltage conditions. Service interruption and the necessity to clean insulators periodically is therefore a constant concern, especially where pollution exists.

29-9 Ground wires

In Fig. 29-6, we see two bare conductors supported at the very top of the transmission-line towers. These conductors, called ground wires, serve to shield the line and intercept lightning strokes before they hit the current-carrying conductors below. Grounding wires normally do not carry current; consequently, they are often made of steel. They are solidly connected to ground at each tower.

29-10 Tower grounding

Transmission-line towers are always solidly connected to ground. Great care is taken to ensure that the ground resistance is low. In effect, when lightning hits a line, it creates a sudden voltage rise across the insulators as the lightning current discharges to ground. Such a voltage rise may produce a flashover across the insulators and a consequent line outage, as shown by the following example.

Example 29-1:
A 3-phase 69 kV transmission line with a BIL of 350 kV is supported on steel towers and protected by a circuit-breaker (Fig. 29-8). The ground resistance at each tower is 20 Ω whereas the neutral of the transmission line is *solidly* grounded at the

Figure 29-7   a. Four bundled conductors make up this phase of a 3-phase, 735 kV line.

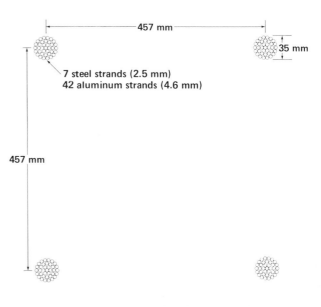

457 mm

35 mm

7 steel strands (2.5 mm)
42 aluminum strands (4.6 mm)

457 mm

Figure 29-7   b. Details of the bundled conductor.

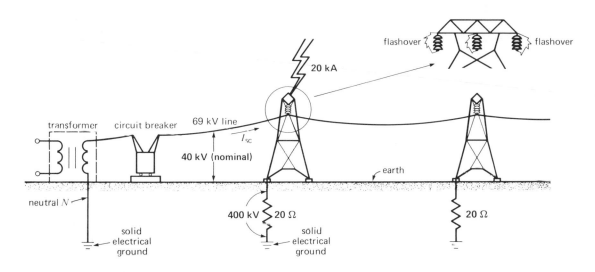

Figure 29-8   Flashover produced by a lightning as it flows to ground.

transformer just ahead of the circuit-breaker. During an electrical storm, one of the towers is hit by a lightning stroke of 20 kA.

a. Calculate the voltage across each insulator string under normal conditions;

b. Describe the sequence of events during and after the lightning stroke.

Solution:

a. Under normal conditions, the line-to-neutral voltage is 69 kV/1.73 = 40 kV and the current flowing in the tower ground resistance is zero. The steel tower is therefore at the same potential as the solid ground. It follows that the voltage across each insulator string (line to tower) is 40 kV (RMS).

b. When lightning strikes the tower, the voltage across the ground resistance suddenly leaps to 20 kA x 20 Ω = 400 kV. The voltage between the tower and *solid* ground is therefore 400 kV, and so the potential difference across all three insulator strings jumps to the same value. Because this impulse exceeds the insulator BIL of 350 kV, a flashover immediately occurs across the insulators, short-circuiting all three lines to the steel cross-arm. The resulting 3-phase short initiated by the lightning stroke will continue

to be fed and sustained by a heavy follow-through current from the 3-phase source. This short-circuit current I_{sc} will trip the circuit breaker, producing a line outage.

In view of the many customers affected by such a load interruption, we try to limit the number of outages by ensuring a low resistance between the towers and ground. In the preceding example, if the tower resistance had been 10 Ω instead of 20 Ω, the impulse voltage across the insulators would have risen to 200 kV and no flashover would have occured.

Note that lightning currents of 20 kA are quite frequent, and they last for only a few microseconds (see Sec. 4-12).

29-11 Electrical properties of transmission lines

The *fundamental* purpose of a transmission line is to carry *active* power. If it also has to carry reactive power, the latter should be small unless the line is very short. In addition, a power line should possess the following basic characteristics:

1. The voltage should remain as constant as possible over the entire length of the line, from source to load, and for all loads between zero and full load;

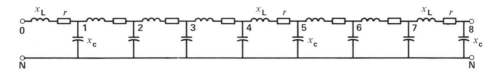

Figure 29-9 Distributed impedance of a transmission line.

2. The losses must be small so as to attain a high transmission efficiency;
3. The I^2R losses must not overheat the conductors.

If the line alone cannot satisfy the above requirements, supplementary equipment must be added until they are met.

29-12 Equivalent circuit of a line

In spite of their great variety, transmission lines possess similar electrical properties. In effect, an ac line possesses a resistance R, an inductive reactance X_L, and a capacitive reactance X_c. These impedances are uniformly distributed over the entire length of the line; consequently, we can represent the line by a series of identical sections, as shown in Fig. 29-9. Each section represents a portion of the line (1 km, for example), and elements r, x_L, x_c represent the impedances corresponding to this unit length.

We can simplify the circuit of Fig. 29-9 by lumping the individual resistances together to yield a total resistance R. In the same way, we obtain a total inductive reactance X_L and a total capacitive reactance X_C. It is convenient to assume that X_C is composed of two parts, each having a value $2X_C$ located at each end of the line. The resulting equivalent circuit of Fig. 29-10 is a good approximation of any 50 Hz or 60 Hz power line, provided its length is less than 500 km. Note that R and X_L increase as the length of the line increases, whereas X_C decreases with increasing length.

In the case of a 3-phase line, the above equivalent circuit represents one phase. Referring to Fig. 29-10, current I corresponds to the actual current flowing in one conductor and E is the voltage between the same conductor and neutral.

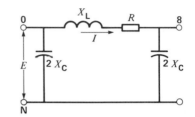

Figure 29-10 Equivalent lumped circuit of a transmission line.

29-13 Typical impedance values

Table 29B gives typical values of the inductive and capacitive reactances (x_L and x_c) for practical transmission lines operating at 60 Hz. Surprisingly, the respective impedances per unit length are reasonably constant for all aerial and underground lines. This is true whether the transmission line voltage is high or low, or the power great or small.

Note that the capacitive reactance of three-phase cables is about one hundred times smaller than that of aerial lines. This fact has a direct bearing on the distance that ac power can be transmitted by cable.

The resistance per unit length varies greatly with conductor size and so it is impossible to give a "typical" value. Table 29C gives the resistance as well as the ampacity for several aerial conductors.

Example 29-2:
A 3-phase 230 kV transmission line having a length of 50 km is composed of three ACSR conductors having a cross-section of 1000 MCM.
a. Determine the equivalent circuit of the 3-phase line;
b. Give the equivalent circuit, per phase.

TABLE 29C   RESISTANCE AND AMPACITY OF SOME BARE AERIAL CONDUCTORS

conductor size		resistance per conductor at 75°C		ampacity in free air*	
AWG	cross section mm²	copper Ω/km	ACSR Ω/km	copper A	ACSR A
10	5.3	3.9	6.7	70	—
7	10.6	2.0	3.3	110	—
4	21.1	0.91	1.7	180	140
1	42.4	0.50	0.90	270	200
3/0	85	0.25	0.47	420	300
300 MCM	152	0.14	0.22	600	500
600 MCM	304	0.072	0.11	950	750
1000 MCM	507	0.045	0.065	1300	1050

*   The ampacity indicated is the maximum that may be used without weakening the conductor by overheating. In practice, the actual line current may be only 25% of the indicated value.

Solution:

a. 1   Referring to Tables 29B and 29C, the approximate line impedances are:

r = 0.065 Ω/km;
x_L = 0.5 Ω/km;
x_c = 300 kΩ/m

a. 2   The line impedances *per phase* are:

R = 0.065 x 50 = 3.25 Ω
X_L = 0.5 x 50 = 25 Ω
X_C = 300 000/50 = 6000 Ω

a. 3   The capacitive reactance at each end of the line is:

$2X_C$ = 2 x 6000 = 12 kΩ

a. 4   The equivalent circuit of the 3-phase line is shown in Fig. 29-11. Note that the line ca-

TABLE 29B	TYPICAL IMPEDANCE VALUES PER KILOMETER FOR 3-PHASE, 60 Hz LINES	
type of line	x_L Ω	x_c Ω
aerial line;	0.5	300 000
underground cable	0.1	3 000

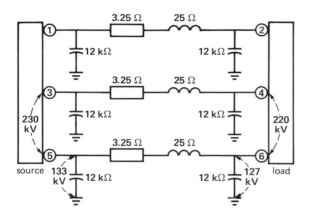

Figure 29-11   Equivalent circuit of a 3-phase line.

pacitance acts as if it were composed of six capacitors connected between the lines and ground. This arrangement holds true even when the neutrals of the source and load are not grounded. The result is that the voltage across each capacitor is given by $E_c = E/\sqrt{3}$ where E is the respective line voltage at the source or load. Thus, assuming the line voltage is 230 kV at the source and 220 kV at the load, the voltages across the respective capacitors are:

$$E_{CS} = 230/\sqrt{3} = 133 \text{ kV}$$
$$E_{CL} = 220/\sqrt{3} = 127 \text{ kV}$$

b. 1   The equivalent circuit per phase is given in Fig. 29-12.

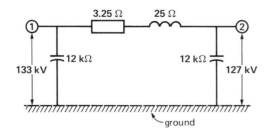

Figure 29-12   Equivalent circuit of one phase.

29-14   Simplifying the equivalent circuit

We can sometimes simplify the transmission line circuit even more by eliminating one, two or all the elements shown in Fig. 29-10. The validity of this simplification depends upon the relative magnitude of the active and reactive powers P_J, Q_L, Q_C associated with the line, compared to the active power P absorbed by the load. Referring to Fig. 29-13, these powers are:

P = active power absorbed by the load;

P_J = I^2R, active power dissipated in the line;

Q_L = I^2X_L, reactive power absorbed by the line;

Q_C = E^2/X_C, reactive power generated by the line.

If any one of these powers is negligible compared with the active power P, we can neglect the corresponding circuit element. For example, low-voltage lines are always short; consequently, X_C is high and so E^2/X_C is negligible. This permits us to represent such lines by the circuit shown in Fig. 29-14. On the other hand, long, high-voltage lines can be represented by the circuit of Fig. 29-15 because the I^2R losses are relatively small whereas the reactive powers Q_L and Q_C are not.

Example 29-3:
The transmission line shown in Fig. 29-11 delivers 300 MW to the 3-phase load. If the line voltage at the sending and receiving end is 230 kV, calculate:

a. the active and reactive powers associated with the line;

b. the approximate equivalent circuit, per phase.

Solution:
Referring to Fig. 29-16a, we have:

a. 1   The line-to-neutral voltage at the load is:
$$E = 230/1.73 = 133 \text{ kV}$$

a. 2   The active power transmitted to the load per phase is:
$$P = 300/3 = 100 \text{ MW}$$

a. 3   The load current is:
$$I = 100 \text{ MW}/133 \text{ kV} = 750 \text{ A}$$

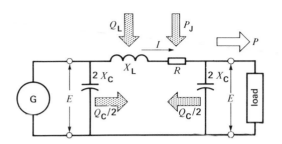

Figure 29-13   Active and reactive powers of a transmission line.

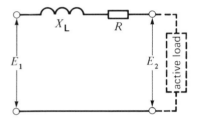

Figure 29-14   Equivalent circuit of a short LV line.

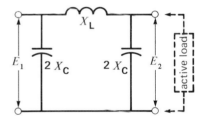

Figure 29-15   Equivalent circuit of a long HV line.

a. 4   If we temporarily neglect the presence of the 12 kΩ capacitor in parallel with the load, the line current is equal to the load current (Fig. 29-16b).

a. 5 I^2R losses in the line:

$$P_J = I^2R = 3.25 \times 750^2$$
$$= 1.83 \text{ MW} \quad (1.8 \text{ percent of } P)$$

a. 6   Reactive power absorbed by the line:

$$Q_L = I^2X_L = 25 \times 750^2$$
$$= 14.1 \text{ Mvar} \quad (14 \text{ percent of } P)$$

a. 7   Reactive power generated by the line:

$$Q_C = E^2/X_C = 133\,000^2/6000$$
$$= 3 \text{ Mvar} \quad (3 \text{ percent of } P)$$

b. 1   Comparing the relative values of P_J, Q_L and P, it is clear that we can neglect the resistance and the capacitance of the line. They represent less than 3 percent of the power transmitted. The resulting equivalent circuit is a simple inductive reactance of 25 Ω (Fig. 29-16c).

29-15   Voltage regulation and maximum power

The voltage of a transmission line should remain as constant as possible even under variable load conditions. Ordinarily, the voltage regulation from zero to full load should not exceed ± 5% of the nominal voltage (though we can sometimes accept a regulation as high as ± 10%).

Another important consideration is the behavior of the line under temporary, abnormal overloads. Consequently, in order to determine the voltage regulation and to establish their power-

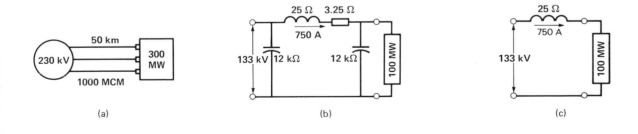

Figure 29-16   Progressive simplification of a 735 kV line (Example 29-3).

handling capability, we now examine four types of lines:
1. resistive line;
2. inductive line;
3. inductive line with compensation;
4. inductive line connecting two large systems.

In our analysis, the lines connect a load (or receiver) E_R to a source (or sender) E_S. The load is assumed to have all possible values, ranging from no-load to a short-circuit. However, we assume the load power factor to be unity because we are only interested in the *active* power the line can transmit. Consequently, the load can be represented by a variable resistance absorbing a power P.

29-16 Resistive line

The transmission line of Fig. 29-17a possesses a resistance R. Starting from an open circuit, we grad-

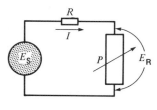

(a)

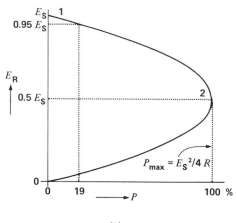

(b)

Figure 29-17   Characteristics of a resistive line.

ually reduce the load resistance until it becomes zero. During this process, we observe the terminal voltage E_R across the load, as well as the active power P it receives. If numerical values were given, a few simple calculations would enable us to draw a graph of E_R as a function of P. However, we prefer to use a generalized curve that shows the relationship between E_R and P for *any* transmission line having an arbitrary resistance R.

The generalized shape of this graph is given in Fig. 29-17b. It reveals the following information:
a. There is an upper limit to the power the line can transmit to the load. In effect, $P_{max} = E_S^2/4$ and this maximum is reached when $E_R = 0.5 E_S$.
b. If we permit a maximum regulation of 5 percent ($E_R = 0.95 E_S$), the line can carry a load that is 19% of P_{max}. The line could transmit more power, but the customer voltage E_R would then be too low.

Note that the sender must furnish the power P absorbed by the load, plus the I^2R losses in the line.

Example 29-4:
A single-phase transmission line having a resistance of 1 Ω is connected to a sender voltage of 100 V. Calculate:
a. the maximum power the line can transmit;
b. the receiver power for a receiver voltage of 95 V.

Solution:
a.   The maximum power that can be transmitted is:
$$P_{max} = E_S^2/4R = 100^2/(4 \times 1)$$
$$= 2500 \text{ W}$$
b. 1 The voltage drop in the line resistance is:
$$E_S - E_R = 100 - 95 = 5 \text{ V}$$
b. 2 The line current is therefore:
$$I = (E_S - E_0)/R = 5/1$$
$$= 5 \text{ A}$$
b. 3 The receiver power is:
$$P = E_S I = 95 \times 5 = 475 \text{ W}$$
(Note that 475/2500 = 0.19 or 19%).

29-17 Inductive line

Let us now consider a line having negligible resistance but possessing an inductive reactance X (Fig. 29-18a). The receiver again operates at unity power factor, and so it can be represented by a variable resistance absorbing a power P. As in the case of a resistive line, voltage E_R diminishes as the load increases, but the regulation curve has a different shape (Fig. 29-18b). In effect, the generalized graph of E_R as a function of P reveals the following information:

a. The line can transmit a maximum power $P_{max} = E_S^2/2X$. The corresponding terminal voltage $E_R = 0.707 E_S$. For a given line impedance and sender voltage, the reactive line can therefore deliver twice as much power as a resistive line can (compare $P = E_S^2/2X$ and $P = E_S^2/4R$).

b. If we again allow a maximum regulation of 5

percent, we discover the line can carry a load that is 60% of P_{max}. Thus, for a given line impedance, and a regulation of 5 percent, the inductive line can transmit *six* times as much active power as a resistive line can.

The sender has to supply the active power P consumed by the load plus the reactive power I^2X absorbed by the line.

29-18 Compensated inductive line

We can improve the regulation and power-handling capacity of an inductive line by adding a capacitive reactance X_C across the load (Fig. 29-19a). All we have to do is to adjust the value of X_C so that the reactive power E_S^2/X_C supplied by the capacitor is at all times equal to *one half* the reactive power I^2X absorbed by the line. For such a compensated line the receiver voltage E_R will always be equal to

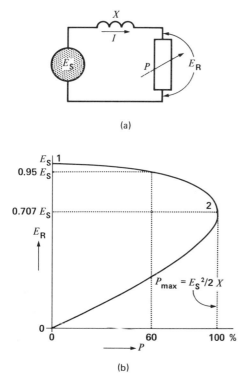

(a)

(b)

Figure 29-18   Characteristics of an inductive line.

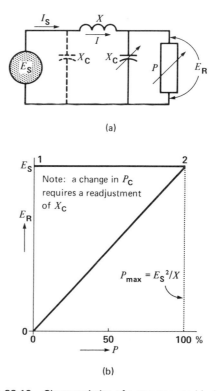

(a)

(b)

Figure 29-19   Characteristics of a compensated inductive line.

the sender voltage E_S, irrespective of the active power P absorbed by the load.

However, there still is an upper limit to the power the line can transmit. A detailed analysis shows that we can maintain a constant load voltage up to a maximum of $P_{max} = E_S{}^2/X$. Beyond this limit, E_R gradually decreases to zero in a diagonal straight line, as shown by the graph of Fig. 29-19b. Note that:

a. The voltage regulation is perfect until the load power reaches the limiting value $P_{max} = E_S{}^2/X$.

b. The compensated inductive line can deliver twice as much power (P_{max}) as an uncompensated line can. Moreover, it has the advantage of maintaining a constant load voltage.

Capacitor X_C supplies half the reactive power absorbed by the line; the remaining half is supplied by the sender E_S. If necessary, we can add a second capacitor X_C (shown dotted) at the input to the line. The source has then only to supply the active power P, while the reactive power is supplied by the capacitors at both ends.

29-19 Inductive line connecting two systems

Large cities and other regional users of electrical energy are always interconnected by one or more transmission lines. Such a network improves the stability of the system and enables it to better endure momentary short-circuits and other disturbances. Interconnecting lines also permit energy exchanges between electrical utility companies.

On such lines, the voltages at each end remain essentially independent of each other, both in value and in phase. In effect, because of their power and size, both areas act as independent, infinite buses.* Figure 29-20 shows the equivalent circuit of an inductive line connecting two such power areas S and R. We assume the terminal voltages E_S and E_R are fixed, each possessing the same magnitude E. Regarding the exchange of active power between the two areas, we examine three distinct cases:

* See Sec. 19-13.

1. E_S and E_R in phase;
2. E_S leading E_R by an angle δ;
3. E_S lagging E_R by an angle δ.

1. E_S and E_R in phase. In this case, the line current is zero and no power is transmitted.

2. E_S leads E_R by an angle δ (Fig. 29-20). Region S supplies power to region R and, from the phasor diagram, we can prove (Sec. 19-17) that the active power transmitted is given by:

$$P = \frac{E^2}{X} \sin \delta \qquad (29\text{-}1)$$

where

P = active power transmitted per phase [MW] †

E = line-to-neutral voltage [kV]

X = inductive reactance of the line, per phase [Ω]

δ = phase angle between the voltages at each end of the line [°]

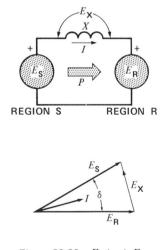

Figure 29-20 E_S leads E_R.

† In this equation, if E represents the line voltage, P is the *total* power transmitted by the three phases.

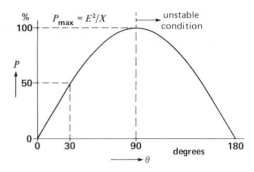

Figure 29-21   a. Power versus angle characteristic.

Figure 29-21a shows the active power transmitted as a function of the voltage phase angle between the two regions. Note that the power increases progressively and attains a maximum value of E^2/X when the phase angle is 90°. In effect, just as in other transmission lines we have studied, a line connecting two power centers can transmit only so much power and no more. The limit is the same as that of a compensated inductive line. Although we can still transmit power when the phase angle exceeds 90°, we avoid this condition because it corresponds to an unstable mode of operation. When δ approaches 90°, the two regions are at the point of "pulling apart" and the line circuit-breakers will trip.

Figure 29-21b shows the load voltage E_R as a function of the active power transmitted. It is sim-

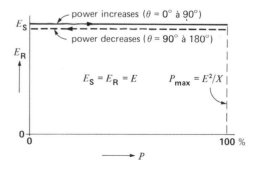

Figure 29-21   b. Voltage versus power characteristic.

ply a horizontal line that stretches to a maximum value $P_{max} = E^2/X$ before falling back again to zero (dotted line). This regulation curve should be compared with that of Fig. 29-19b.

Note that the line voltage drop E_X is quite large, even though the terminal voltages E_S and E_R are equal. Referring to Fig. 29-20, it is clear that the line voltage "drop" increases as the phase angle between E_S and E_R increases.

3. E_S **lags behind** E_R **by an angle** δ (Fig. 29-22). The active power has the same value as before, but it now flows in the opposite direction, from region R towards region S. The graph of active power ver-

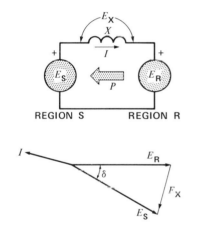

Figure 29-22 E_R leads E_S.

sus phase angle is identical to that shown in Fig. 29-21a.

If we compare Figs. 29-20 and 29-22, we note that the direction of power flow does not depend upon the relative magnitudes of E_S and E_R (they are equal), but *only upon the phase angle between them*. On inductive lines, active power always flows from the leading to the lagging voltage side.

29-20 Review of power transmission

To sum up, there is always a limit to the amount of power a line can transmit. The maximum power

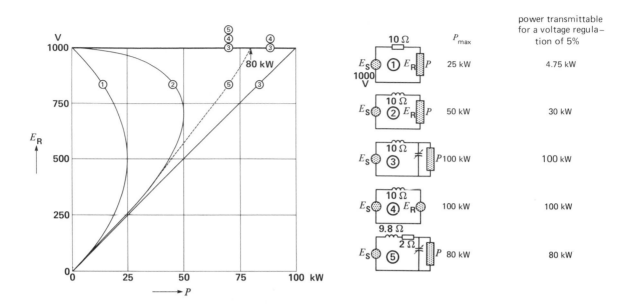

Figure 29-23   Comparison of the power-handling properties of various transmission lines.

is proportional to the square of the voltage and inversely proportional to the impedance of the line. Figure 29-23 enables us to compare actual values of power and voltage for the four types of transmission lines we have studied. Each line possesses an impedance of 10 Ω and the sender furnishes a voltage E_S of 1000 V. It is clear that the E_R versus P curves become flatter and flatter as we progress from a resistive to an inductive to a compensated line.

The table next to the graph shows the maximum power that can be transmitted, assuming a regulation of 5% or better. Thus, the resistive line can transmit 4.75 kW, whereas the inductive line can transmit 30 kW.

Because all lines possess some resistance, we also show the voltage-power curve of a compensated line having a reactance of 9.8 Ω and a resistance of 2 Ω (curve 5). This line also has an impedance of 10 Ω, but the maximum power it can transmit drops to 80 kW, compared to 100 kW for a line possessing no resistance.

29-21 Choosing the line voltage

We have seen that for a given transmission line and for a given voltage regulation, the power that can be transmitted is proportional to E^2/Z, where E is the voltage of the line and Z, its impedance. Because Z is proportional to the length of the line, we deduce that the line voltage is given by:

$$E^2 = kPl$$

that is:

$$\boxed{E = k\sqrt{Pl}} \qquad (29\text{-}2)$$

where

E = 3-phase line voltage [kV]

P = power to be transmitted [kW]

l = length of the transmission line [km]

k = coefficient that depends on the type of line and the allowable voltage regulation. Typical values are:

　　k = 0.1 for an uncompensated line having a regulation of 5 percent

　　k = 0.06 for a compensated line

Equation 29-2 is not exact, but it does give us

an idea of the magnitude of the line voltage E. The value finally chosen depends upon economic factors as well as technical considerations; in general, the actual voltage will lie between $0.6\,E$ and $1.5\,E$.

Example 29-5:

Power has to be carried over a distance of 20 km to feed a 10 MW unity power factor load. If the line is uncompensated:

a. Determine the line voltage;

b. Select an appropriate wire size;

c. Calculate the voltage regulation.

Solution:

a. 1 Because the line is not compensated, we assume $k = 0.1$:

$$E = k\sqrt{Pl} \qquad \text{Eq. 29-2}$$
$$= 0.1\sqrt{10\,000 \times 20}$$
$$= 44.7 \text{ kV}$$

a. 2 Any voltage between $0.6 \times 44.7\,\text{kV}$ (= 27 kV) and $1.5 \times 44.7\,\text{kV}$ (= 67 kV) is acceptable. We shall use a standard line voltage of 34.5 kV.

a. 3 The line-to-neutral voltage is:

$$E = 34.5/\sqrt{3} = 19.9 \text{ kV}$$

b. 1 The conductor size depends, first, upon the current to be carried. The line current is:

$$I = S/(1.73\,E) \qquad \text{Eq. 9-9}$$
$$= 10 \times 10^{6}/(1.73 \times 34\,500)$$
$$= 167 \text{ A}$$

b. 2 According to Table 29C, we can use a No. 1 ACSR conductor:

ampacity = 200 A
$$R = 0.9 \ \Omega/\text{km} \times 20 \text{ km} = 18 \ \Omega$$
$$X_L = 0.5 \ \Omega/\text{km} \times 20 \text{ km} = 10 \ \Omega$$
(Table 29B)

c. 1 The IR drop in the line is:
$$IR = 167 \times 18 = 3006 \text{ V}$$

c. 2 The IX_L drop in the line is:
$$IX_L = 167 \times 10 = 1670 \text{ V}$$

c. 3 The line-to-neutral load voltage is 19 900 V. The complete circuit diagram per phase is given in Fig. 29-24a. The corresponding

phasor diagram is given in Fig. 29-24b. The sender voltage may be calculated as follows:

$$E_S = \sqrt{(19\,900 + 3006)^2 + 1670^2}$$
$$= 22\,967 \text{ V}$$

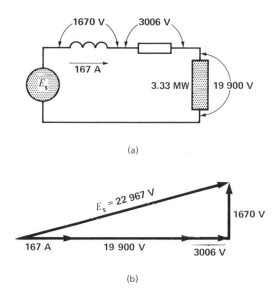

(a)

(b)

Figure 29-24   a. Transmission line under load.
                  b. Corresponding phasor diagram.

c. 4 If the load were removed, E_R would rise to 22 967 V. The voltage regulation is therefore:

regulation = (22 967 – 19 000)/22 967
$$= 3067/22\,967$$
$$= 0.133 \text{ or } 13.3 \text{ percent}$$

29-22 Methods of increasing the power capacity

High-voltage lines are mainly inductive and they possess a reactance of about 0.5 Ω/km. This creates problems when we have to transmit large blocks of power over great distances. Suppose, for example, that we have to carry 4000 MW over a distance of 400 km. The reactance of the line is 400 km $\times$ 0.5 Ω/km = 200 Ω per phase. Since the highest practical voltage is about 800 kV, the line can transmit no more than:

$$P_{max} = E^2/X = 800^2/200 \qquad \text{Eq. 29-1}$$
$$= 3200 \text{ MW}$$

To transmit 4000 MW, the only solution is to use *two lines in parallel*, one beside the other. Note that doubling the size of the conductors would not help, because it is the *reactance* and not the resistance of the conductors that determines the maximum power. To carry large blocks of power, we sometimes erect two, three and even four transmission lines in parallel, which follow the same corridor across the countryside (Fig. 29-25). In addition to high cost, the use of parallel lines tends to

Figure 29-25   Two 735 kV transmission lines in parallel carrying electrical energy to a large city. Each phase is composed of 4 bundled conductors (see Fig. 29-7). *(Hydro-Québec)*

create serious problems of land expropriation. Consequently, we sometimes use special methods to increase the maximum power of a line. In effect, when we can no longer increase the line voltage, we try to reduce the line reactance X_L. This is done by using several conductors per phase, kept apart by spacers. Such *bundled conductors* can reduce the reactance by as much as 40 percent permitting an increase of 67 percent in the power-handling capability of the line.

Another method uses capacitors in series with the three lines to artificially reduce the value of X_L. With this arrangement, the maximum power is:

$$P_{max} = E^2/(X_L - X_{cs})$$

where X_{cs} is the reactance of the series capacitors per phase. Such series compensation is also used to regulate the voltage of medium-voltage lines when the load fluctuates rapidly.

29-23 Extra high voltage lines

When electrical energy is transmitted at extra high voltages, special problems arise that require the installation of large compensating devices to regulate the voltage and to guarantee stability. Among these devices are synchronous capacitors, inductive reactors, static var compensators, and shunt and series capacitors.

To understand the need for such devices, and to appreciate the magnitude of the powers involved, consider a three-phase, 727 kV, 60 Hz line, having a length of 600 km. The inductive and capacitive reactances are respectively 0.5 Ω and 300 kΩ for each kilometer of length. We shall first determine the equivalent circuit of the transmission line per phase:

Sender voltage per phase (line-to-neutral) is:

E_S = 727/1.73 = 420 kV

Inductive reactance per phase is:

X_L = 0.5 x 600 = 300 Ω

Capacitive reactance per phase is:

X_C = 300 kΩ/600 = 500 Ω

Equivalent capacitive reactance at each end of the line is:

X_{C1} = X_{C2} = 2 x 500 Ω = 1000 Ω

The equivalent circuit per phase is shown in Figure 29-26. Let us now study the behavior of the line under no-load and full-load conditions.

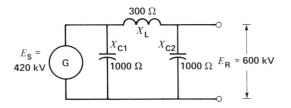

Figure 29-26   EHV transmission line at no-load.

No-load operation. At no-load, the circuit formed by X_L in series with X_{C2} produces a partial resonance and the terminal voltage E_R rises to 600 kV. This represents an increase of 43 percent above the nominal voltage of 420 kV (Fig. 29-26). Such an abnormally high voltage is unacceptable. The only feasible way to reduce it is to connect an inductive reactance X_{L2} at the end of the line (Fig. 29-27). If we make X_{L2} equal to X_{C2}, the resulting parallel

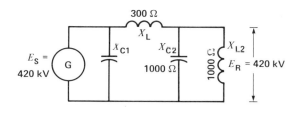

Figure 29-27   EHV reactor compensation.

resonance brings voltage E_R back to 420 kV. In effect, the reactive power generated by X_{C2} ($420^2/1000 = 176$ Mvar) is entirely absorbed by X_{L2}. The latter must, therefore, have a capacity of 176 Mvar *per phase*.

Despite this inductive compensation, we still have a reactive power of 176 Mvar, generated by X_{C1}, which has to be absorbed by alternator G.

However, a capacitive load at the terminals of an alternator creates overvoltages, unless we reduce the alternator exciting current (Sec. 19-11). But under-excitation is not recommended because it leads to instability. Consequently, we must install a second inductive reactance of 176 Mvar close to the generating station. In the case of long transmission lines, we often connect several inductive reactances along the line to distribute the compensation evenly over its length.

Inductive reactors (fixed or variable) are composed of a large coil placed inside a tank and immersed in oil (Fig. 29-28). A laminated steel core split up into a series of short air-gaps carries the magnetic flux. Very intense magnetic forces are developed which, on a 60 Hz system, continuously oscillate between zero and several tons, at a frequency of 120 Hz. The core laminations and all metallic parts must be firmly secured to reduce vibration and to limit the noise to an acceptable level.

Operation under load, characteristic impedance. Returning again to the original uncompensated line (Fig. 29-26), let us connect a small unity power factor load across the open terminals. If we progressively increase the load, terminal voltage E_R

Figure 29-28

Three large 110 Mvar, single-phase reactors, installed in a substation to compensate the line capacitance of a very long 3-phase 735 kV transmission line. *(Hydro-Québec)*

will gradually decrease from its open-circuit value of 600 kV and, for one particular load, it will exactly equal the voltage E_S of the source (Fig. 29-29). This particular load is called the *surge-impedance load*. For most aerial lines, it corresponds

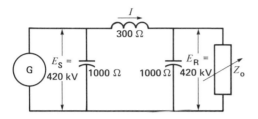

Figure 29-29   Surge impedance loading of a line.

to a line-to-neutral load resistance of about 400 Ω per phase. This critical load resistance is independent of the system frequency. The surge impedance load (SIL) of a transmission line is therefore given by the approximate equation:

$$\boxed{\text{SIL} = E^2/400} \qquad (29\text{-}3)$$

where

SIL = surge impedance load [MW]
E = 3-phase line voltage [kV]

In our example, the total surge-impedance load is approximately $727^2/400 = 1320$ MW.

When a transmission line delivers active power corresponding to its surge-impedance load, the reactive power generated by the capacitance of the line is equal to that absorbed by its inductance. The line, in effect, compensates itself. If the load exceeds the surge-impedance load, we can keep E_R at 420 kV by adding extra capacitors at the end of the line. However, the maximum power is still limited to

$P = 3 \times (E^2/X) = 3 \times (420^2/300)$
  $= 1764$ MW.

On the other hand, if the load is less than the surge-impedance load, we must add inductive reactance at the end of the line to maintain a constant voltage. Because the load continually changes throughout the day, we must continually vary the magnitude of the capacitive and inductive reactance to

keep a steady voltage. This is done by means of static var compensators (Fig. 29-32), or rotating synchronous machines. The latter can deliver or absorb reactive power according to whether they are over- or under-excited (Sec. 20-10).

29-24 Power exchange

We sometimes have to install an additional transmission line on systems that are already tightly interconnected. Such a line may be required to meet the energy needs of a rapidly-growing area or to improve the overall stability of the network. In such cases, we use special methods so that the additional line will transmit the required power.

Consider, for example, two major power centers A and B that are already interconnected by a grid of transmission lines (Fig. 29-30). The respective voltages E_a and E_b are equal, but E_a leads E_b by an angle δ. If we decide to connect these two centers

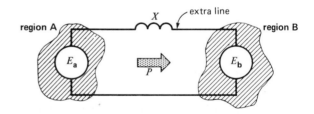

Figure 29-30   Power flow between two regions.

by an extra transmission line having a reactance X, the active power P will automatically flow from A to B because E_a leads E_b (see Sec. 29-19). Furthermore, phase angle δ and reactance X will inherently dictate the magnitude of the power transmitted because $P = (E^2/X) \sin \delta$.

Unfortunately, the value and direction of P may not correspond at all to what we want to achieve. For example, if we wish to transmit energy from region B to region A, the installation of a simple line will not do, for reasons we have just explained.

However, we can *force* an energy exchange in

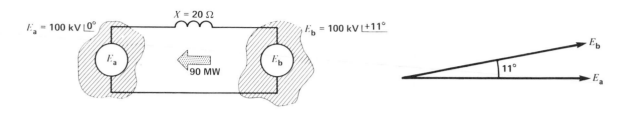

Figure 29-31   a. An ordinary line causes power to flow in the wrong direction.

one direction or the other by artificially modifying the phase angle between the two regions. All we have to do is to introduce a phase-shift autotransformer (Sec. 13-11) at one end of the line; by varying the phase angle of this transformer, we can completely control the power flow between the two centers.

Example 29-6:
Figure 29-31a shows the voltages and phase angle between two regions A and B that already form part of a network (not shown). Voltage E_b leads E_a by 11° and both voltages have a value of 100 kV. A new tie line having a reactance $X = 20\ \Omega$ connects the two regions. Calculate:

a. the power transmitted by the line and the direction of power flow, if no phase-shift transformer is employed;

b. the required phase-shift of the transformer so that the line will transmit 70 MW from A to B.

Solution:
a. The power is given by:

$$P = (E^2/X) \sin \delta \qquad \text{Eq. 29-1}$$
$$= (100^2/20) \sin 11° = 95\ \text{MW}$$

Because E_b leads E_a, the 95 MW will flow from B to A.

b. Let us first calculate the phase angle δ_1 required between opposite ends of the line, so that it will transmit 70 MW. We have:

$$P = (E^2/X) \sin \delta_1 \qquad \text{Eq. 29-1}$$
$$70 = (100^2/20) \sin \delta_1$$
$$\therefore \quad \sin \delta_1 = 0.14$$
from which $\delta_1 = 8°$

Consequently, voltage E_d must *lead* E_b by 8° in order that 70 MW may flow from A to B (Fig. 29-31b). Referring to the phasor diagram, and noting that E_b already leads E_a by 11°, it follows that E_d must be 19° ahead of E_a. The autotransformer T must therefore produce a phase-shift of 19° between its primary and secondary windings and the secondary voltage must lead the primary voltage. We can put the autotransformer at either end of the line or, if necessary, in the middle.

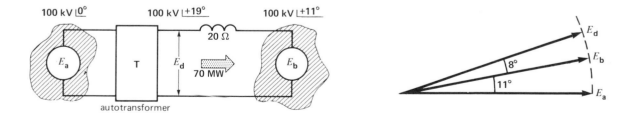

Figure 29-31   b. A phase-shift autotransformer can force power to flow in the desired direction (Example 29-6).

Figure 29-32   Static var compensator for HV line. *(General Electric)*

We now consider a practical application of a phase-shift transformer. To facilitate power exchange between the State of Connecticut and Long Island, New York, six 138 kV submarine cables (2 per phase) were installed between Norwalk and Northport, at the bottom of Long Island Sound (Fig. 29-33). Because the two regions were already interconnected by a grid of transmission lines above ground, it was decided to install a

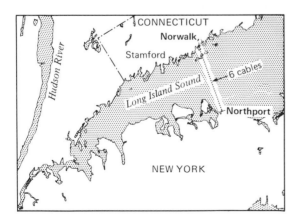

Figure 29-33   Six single-phase submarine cables join Connecticut to Long Island.

phase-shift autotransformer at Northport to control a maximum power flow of 300 MW. Variable taps enable a maximum phase-shift of ± 25 degrees. On the other side, at Norwalk, a 300 MVA variable-voltage autotransformer was installed to provide voltage control of up to ± 10 percent, without phase-shift. By varying the phase-shift and the voltage at either end of the 19 km line, it is possible to control the power flow between the two regions, in one direction or the other, depending on the need. The following technical details point out the magnitude of the power involved in such a cable transmission system:

Each of the six single-phase cables possesses a resistance of 1.3 Ω, and inductive reactance of 1.1 Ω and a capacitive reactance of 375 Ω (see Fig. 5-25). The latter can be represented by two reactances of 750 Ω at each end of the cable. The full-load current is 630 A and the line-to-neutral voltage is 80 kV. Referring to the equivalent circuit diagram for one cable (Fig. 29-34), we can readily calculate the value of the various powers.

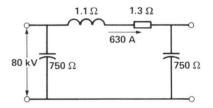

Figure 29-34   Equivalent circuit of each submarine cable.

In comparing the powers, we can see that the capacitance of the cable is far more important than its resistance or inductance are. A cable acts like an enormous capacitor, contrary to an aerial line which is mainly inductive. It is precisely the inherent large capacitance that prevents us from using cables to transmit energy over long distances. This restriction does not apply when direct current is used because capacitance then has no effect. Direct current transmission lines are covered in Chapter 31.

Submarine cable installation	Power per cable	Total power
I^2R losses:		
$P_J = 630^2 \times 1.1$		
$= 0.516$ MW	0.516 MW	3.1 MW
Reactive power generated:		
$Q_C = (80\ 000)^2/375$		
$= 17.06$ Mvar	17 Mvar	102 Mvar
Reactive power absorbed:		
$Q_L = 630^2 \times 1.1$		
$= 0.436$ Mvar	0.44 Mvar	2.6 Mvar
Active power transmitted:		
$P = 630 \times 80\ 000$		
$= 50$ MW	50 MW	300 MW

QUESTIONS AND PROBLEMS

Practical level

29-1 Standard voltages are grouped into four main classes. Name them and state the approximate voltage range of each.

29-2 Explain what is meant by:
— suspension-type insulator
— ground wire
— corona effect
— sag of a transmission line
— galloping line
— reactance of a line

29-3 Why must transmission line towers be solidly connected to ground?

29-4 The 735 kV transmission line, 745 miles long, transmits a power of 800 MW.
a. Is there an appreciable voltage difference between the two ends of the line, measured line to neutral?
b. Is there a phase angle between corresponding voltages?

29-5 In some areas, two identical three-phase lines are installed side by side, supported by sepa-

rate towers. Could we replace these two lines by a single line by simply doubling the size of the conductors? Explain.

29-6 Why do we seldom install underground cable (instead of aerial transmission lines) between generating stations and distant load centers?

29-7 In Problem 29-4, the average line span is 480 m. How many towers are erected between the source and the load?

29-8 A 20 km transmission line operating at 13.2 kV has just been disconnected from the source. A lineman may receive a fatal shock if he does not first connect the line to ground before touching it. Explain.

29-9 What is the ampacity of a 600 MCM ACSR cable suspended in free air? Why can a copper conductor having the same cross-section carry a current that considerably greater?

Intermediate level

29-10 Each phase of the two 735 kV lines shown in Fig. 29-25 is composed of 4 bundled subconductors. The current per phase is 2000 A and the resistance of each subconductor is 0.045 Ω/km. Calculate:
a. the total power transmitted by both lines at unity power factor load;
b. the total I^2R loss, knowing the lines are 350 miles long;
c. the I^2R loss as a percent of the total power transmitted.

29-11 A single-phase transmission line possesses a resistance R of 15 Ω (Fig. 29-17a). The source E_s is 6000 V and the impedance of the unity power factor load varies between 285 Ω and 5 Ω.
a. Calculate the terminal voltage E_R and the power P absorbed by the load when the impedance is successively 285 Ω, 45 Ω, 15 Ω, and 5 Ω.
b. Draw the graph of the terminal voltage E_R as a function of the power P.

29-12 In Problem 29-11, what is the phase angle between E_R and E_s when the load is 45 Ω?

29-13 The transmission line in Problem 29-11 is replaced by another having an inductive reactance of 15 Ω (Fig. 29-18a).
   a. Calculate the voltage E_R at the terminals of the load and the power P it absorbs for the same impedance values.
   b. Draw the graph of E_R as a function of P.

29-14 In Problem 29-13, what is the phase angle between E_R and E_s when the load impedance is 45 Ω? Does E_R lead or lag behind E_s?

29-15 A single-phase transmission line possesses an inductive reactance of 15 Ω. It is supplied by a source E_s of 6000 V.
   a. Calculate the voltage E_R at the end of the line for the following capacitive loads: 285 Ω, 45 Ω.
   b. Calculate the phase angle between E_R and E_s when the load is 45 Ω.

29-16 The line in Problem 29-15 possesses a resistance of 15 Ω (instead of a reactance of 5 Ω).
   a. Calculate the voltage E_R at the end of the line for a capacitive load of 45 Ω.
   b. If a line is purely resistive, can we raise the voltage at the end of the line by connecting a capacitor across it?

29-17 The following information is given for Fig. 29-19a:
   — terminal voltage E_R: 6000 V
   — equivalent load resistance: 45 Ω
   — inductive reactance X_L of the line: 15 Ω
   — capacitive reactance X_c
      in parallel with the load: 150 Ω
   Neglecting the dotted reactance X_c in parallel with the source, calculate:
   a. the reactive power supplied by the capacitor;
   b. the line current I;
   c. the reactive power absorbed by the line;
   d. the reactive power supplied by the source;
   e. the apparent power supplied by the source;
   f. the voltage E_s of the source;
   g. the voltage E_R and the power P when E_s is 6 kV.

29-18 In Problem 29-17 what is the phase angle between E_R and E_s?

29-19 Referring to Fig. 29-9, each section of the circuit represents a transmission line length of 1 km in which the impedances are x_L = 0.5 Ω; r = 0.25 Ω; x_c = 300 kΩ. Calculate the values of X_L, R and X_c if the circuit is reduced to that shown in Fig. 29-10.

29-20 What is meant by the term surge impedance load?

Advanced level

29-21 A three-phase 230 kV transmission line having a reactance of 43 Ω per phase connects two regions that are 50 miles apart. The phase angle between the voltages at the two ends of the line is 20°. Calculate:
   a. the active power transmitted by the line;
   b. the current in each conductor;
   c. the total reactive power absorbed by the line;
   d. the reactive power supplied to the line by each region.

29-22 A three-phase aerial line connected to a 115 kV 3-phase source has a length of 200 km. It is composed of three 600 MCM type ACSR conductors. Referring to Fig. 29-13 and Tables 29B and 29C, if there is no load on the line, calculate:
   a. the value of R, X_L and X_c, per phase;
   b. the voltage between conductors at the "load" (open end);
   c. the current drawn from the source, per phase;
   d. the total reactive power received by the source;
   e. the total I^2R loss in the line.

29-23 a. In Problem 29-22 calculate the current drawn from the source if a 3-phase short occurs across the end of the line.
   b. Compare this current with the ampacity of the conductors.

30
DISTRIBUTION OF ELECTRICAL ENERGY

In Chapter 29, we mentioned that an electrical power system is composed of high-voltage transmission lines that feed power to a medium-voltage (MV) network by means of substations. In North America, these MV networks generally operate at voltages between 2.4 kV and 69 kV. In turn, they supply millions of independent low-voltage systems that function between 120 V and 600 V.

In this chapter, we cover the following main topics:

1. Substations;
2. Protection of medium-voltage distribution systems;
3. Low-voltage distribution;
4. Electrical installation in buildings.

SUBSTATIONS

Substations are used throughout the electrical system. Starting with the generating station, a substation raises the medium-voltage generated by the alternators to the high-voltage needed to transmit the energy economically.

The high transmission-line voltage is again re-duced in other substations located close to the power-consuming centers. The electrical equipment in such distribution substations is similar to that found in substations associated with generating plants.

30-1 Substation equipment

A substation usually contains the following major apparatus:

transformers	grounding switches
circuit-breakers	surge arresters
horn-gap switches	current-limiting reactors
disconnect switches	instrument transformers

In the description which follows, we study the basic principles of this equipment. Furthermore, to understand how it all fits together, we conclude our study with a typical substation that provides power to a large suburb.

30-2 Circuit-breakers

Circuit-breakers are designed to interrupt either

normal or short-circuit currents. The behave like big switches that may be opened or closed by local push-buttons or by distant radio signals. Furthermore, circuit-breakers will automatically open a circuit whenever the line current exceeds a preset limit. They can be set more accurately than fuses can, and they do not have to be replaced after each fault.

The most important types of circuit-breakers are:
1. Oil circuit-breakers (OCB's);
2. Air-blast circuit-breakers;
3. SF_6 circuit-breakers;
4. Vacuum circuit-breakers.

The nameplate on a circuit breaker usually indicates (1) the maximum steady-state current it can carry, (2) the maximum interrupting current, (3) the maximum line voltage, and (4) the interrupting time in cycles. The interrupting time may last from 2 to 8 cycles on a 60 Hz system. To cut large currents this quickly, we have to ensure rapid deionization of the arc, combined with rapid cooling. High-speed interruption limits the damage to transmission lines and equipment and, equally important, it improves the stability of the system.

The tripping action is usually produced by means of an overload relay than can detect abnormal line conditions. For example, the relay coil in Fig. 30-1 is connected to the secondary of a current transformer. The primary carries the line current of the phase that has to be protected. If the line current exceeds a preset limit, the secondary current will cause relay contacts C_1C_2 to close. As soon as they close, the tripping coil is energized by an auxiliary dc source. This causes the main line contacts to open, thus interrupting the circuit.

1. Oil circuit-breakers. Oil circuit-breakers are composed of a steel tank filled with insulating oil. In one version (Fig. 30-2), a group of porcelain bushings channels the 3-phase line currents to a set of fixed contacts. A group of movable contacts, actuated by an insulated rod, opens and closes the circuit. When the circuit-breaker is closed, the line

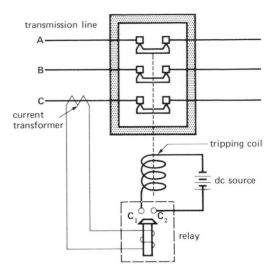

Figure 30-1 Elementary circuit-breaker tripping circuit.

current of each phase penetrates the tank by way of one porcelain bushing, flows through the first fixed contact, the movable contact, the second fixed contact, and then on out by a second bushing.

If an overload occurs, the tripping coil releases a powerful spring which pulls on the insulated rod, causing the contacts to open. As soon as the contacts separate, a violent arc is created, which volatilizes the surrounding oil. The pressure of the hot gases creates turbulence around the contacts. This causes cool oil to swirl around the arc, thus extinguishing it.

In modern high-power breakers, the arc is confined to an explosion chamber so that the pressure of the hot gases produces a powerful jet of oil. The jet is made to flow across the path of the arc, to extinguish it. Other types of circuit-breakers are designed so that the arc is deflected and lengthened by a self-created magnetic field. The arc is blown against a series of insulating plates that break up the arc and cool it down. Figures 30-3 and 30-4 show the appearance of two typical OCBs.

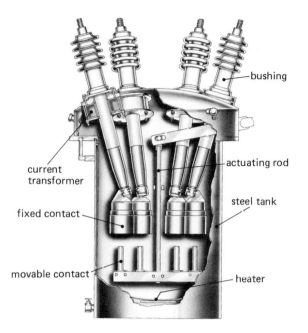

bushing

current transformer

actuating rod

fixed contact

steel tank

movable contact

heater

Figure 30-2

Cross section of an oil circuit-breaker. The diagram shows 4 of the six bushings; the heater keeps the oil at a satisfactory temperature during cold periods. *(Canadian General Electric)*

Figure 30-3

Three-phase oil circuit-breaker rated at 1200 A at 115 kV. It can interrupt a current of 50 kA in 3 cycles on a 60 Hz system. Other characteristics: height: 3660 mm; diameter: 3050 mm; mass: 21 t; BIL 550 kV. *(General Electric)*

Figure 30-4

Minimum oil circuit-breaker installed in a 420 kV, 50 Hz substation. Rated current: 2000 A; rupturing capacity: 25 kA; height (less support): 5400 mm; length: 6200 mm; 4 circuit breaking modules in series. *(Brown Boveri)*

Figure 30-5

Air blast circuit-breaker rated 2000 A at 362 kV. It can interrupt a current of 40 kA in 3 cycles on a 60 Hz system. It consists of 3 identical modules connected in series, each rated for a nominal voltage of 121 kV. The compressed air reservoir can be seen at the left. Other characteristics: height: 5640 mm; overall length: 9150 mm; BIL 1300 kV. *(General Electric)*

2. Air-blast circuit-breakers. These circuit-breakers interrupt the circuit by blowing compressed air at supersonic speed across the opening contacts. Compressed air is stored in reservoirs at a pressure of about 3 MPa ($\approx$ 435 psi) and is replenished by a compressor located in the substation. The most powerful circuit-breakers can typically open short-circuit currents of 40 kA at a line voltage of 765 kV in a matter of 2 cycles on a 60 Hz line. The noise accompanying the air blast is so loud that noise-suppression methods must be used when the breakers are installed near residential areas. Figure 30-5 shows a typical air-blast circuit-breakers, and Fig. 30-6 is a cross section of the contact mechanism.

3. SF$_6$ circuit-breakers. These totally enclosed circuit-breakers, insulated with SF$_6$ gas,[*] are used whenever space is at a premium, such as in downtown substations (Fig. 30-7). They are much smaller than any other type of circuit-breaker of equivalent power and are far less noisy than air-circuit-breakers.

4. Vacuum circuit-breakers. These circuit-breakers operate on a different principle from other breakers because there is no gas to ionize when the contacts open. They are hermetically sealed; consequently, they are silent and never become polluted

* Sulfur hexafluoride.

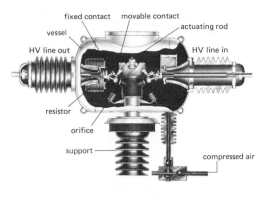

Figure 30-6

Cross section of one module of an air blast circuit breaker. When the breaker trips, the rod is driven upwards, separating the fixed and movable contacts. The intense arc is immediately blown out by a jet of compressed air coming from the orifice. The resistor dampens the overvoltages that occur when the breaker opens. *(General Electric)*

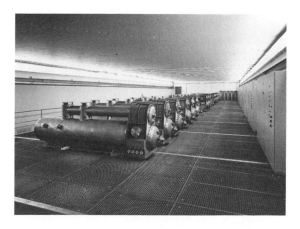

Figure 30-7

Group of 15 totally enclosed SF_6 circuit breakers installed in an underground substation of a large city. Rated current: 1600 A; rupturing current: 34 kA; normal operating pressure: 265 kPa ($\approx$ 38 psi); pressure during arc extinction: 1250 kPa ($\approx$ 180 psi). These SF_6 circuit-breakers take up only 1/16 of the volume of conventional circuit breakers having the same rupturing capacity. *(Brown Boveri)*

(Fig. 30-8). Their interrupting capacity is limited to about 30 kV. For higher voltages, several circuit-breakers are connected in series.

Figure 30-8

Three-phase vacuum circuit-breaker having a rating of 1200 A at 25.8 kV. It can interrupt a current of 25 kA in 3 cycles on a 60 Hz system. Other characteristics: height: 2515 mm; mass: 645 kg; BIL: 125 kV. *(General Electric)*

30-3 Air-break switches

Air-break switches can interrupt the exciting currents of transformers, or the moderate capacitive currents of unloaded transmission lines. They cannot interrupt normal load currents.

Air-break switches are composed of a movable blade that engages a fixed contact; both are mounted on insulating supports (Figs. 30-9, 30-10). Two arcing horns are attached to the fixed and movable contacts, so that when the main contact is broken, an arc is set up between the arcing-horns. The arc moves upward due to the combined

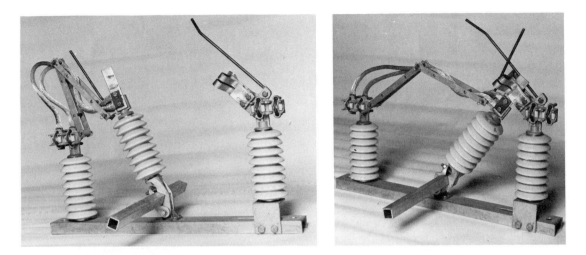

Figure 30-9   One pole of a horn gap disconnecting switch rated 600 A, 27 kV, 60 Hz; (left) open position, (right) closed position. *(Dominion Cutout)*

Figure 30-10   One pole of a 3-phase 3000 A, 735 kV, 60 Hz horn gap disconnecting switch (left) in the open position, (right) in the closed position. The switch can be operated manually by turning a hand wheel or remotely by means of a motorized drive immediately below the hand wheel. Other characteristics: height when closed: 12 400 mm; length: 7560 mm; mass: 3 t; maximum current carrying capacity for 10 cycles: 120 kA; BIL: 2200 kV. *(Kearney)*

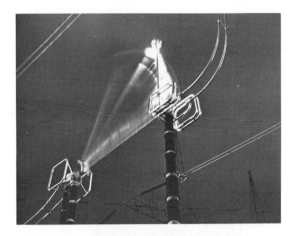

Figure 30-11   The arc produced between the horns of a disconnecting switch as it cuts the exciting current of a HV transformer provides the light to take this night picture. *(Hydro-Québec)*

action of the magnetic field and the hot air currents it produces. As the arc rises, it becomes longer until it eventually blows out (Fig. 30-11). Although the horns become pitted and gradually wear out, they can easily be replaced.

30-4 Disconnecting switches

Unlike air-break switches, disconnecting switches are unable to interrupt any current at all. They must only be opened and closed when the current is zero. They are basically isolating switches, enabling us to isolate oil circuit-breakers, transformers, transmission lines, and so forth, from a live network. Disconnecting switches are essential to carry out maintenance work and to reroute power flow.

Figure 30-12 shows a 2000 A, 15 kV disconnecting switch. It is equipped with a latch to prevent the switch from opening under the strong electromagnetic forces that accompany short-circuits. The latch is disengaged by inserting a hookstick into the ring and pulling the movable blade out of the fixed contact.

Figure 30-13 shows another so-called "disconnect" that carries a larger current, but at a much lower voltage. It, too, is opened by means of a manual hookstick. Figure 30-14 shows another type of disconnecting switch and Fig. 30-15 shows how the fixed and movable contacts engage. Fig-

Figure 30-12   This hookstick-operated disconnecting switch is rated 2000 A, 15 kV and has a BIL of 95 kV. *(Dominion Cutout)*

Figure 30-13   Disconnecting switch rated 10 kA, 1 kV for indoor use. *(Montel/Sprecher and Schuh)*

Figure 30-14   Disconnecting switch rated 600 A, 46 kV for sidewise operation. *(Kearney)*

ure 30-16 shows maintenance personnel working on a large disconnecting switch in a HV substation.

30-5 Grounding switches

Grounding switches are safety switches that ensure a transmission line is definitely grounded while repairs are being carried out. Figure 30-17 shows three such switches in the open, horizontal position. To short-circuit the line to ground, all three grounding switches swing up to engage the stationary contact connected to each phase. Grounding switches are opened and closed only after the lines are de-energized.

30-6 Lightning arresters

The purpose of a lightning arrester* is to limit the overvoltages that may occur across transformers and other electrical apparatus due either to lightning or switching surges. The upper end of the arrester is connected to the line or terminal that has to be protected, and the lower end is solidly connected to ground.

---

*   Also called surge arrester or surge diverter.

Figure 30-15

The blade of a vertical motion disconnecting switch is in tight contact with two fixed contacts, thanks to the pressure exerted by two powerful springs. When the switch opens, the blade twists on its axis as it moves upward. During switch closure, the reverse rotary movement exerts a wiping action against the fixed contacts, thus ensuring excellent contact. *(Kearney)*

The arrester is composed of an external porcelain tube containing an ingenious arrangement of stacked discs, air gaps, ionizers, and coils. The discs (or valve blocks) are composed of a silicon carbide material known by trade names such as thyrite®, autovalve®, etc. As we saw in Sec. 5-16, this material has a resistance that decreases dramatically with increasing voltage. The typical E-I characteristic of a surge arrester is given in Fig. 30-18.

Under normal voltage conditions, spark gaps prevent any current from flowing through the tubular column. If an overvoltage occurs, the spark gaps break down and the surge discharges to

Figure 30-16

Like all electric equipment, disconnecting switches have to be overhauled and inspected at regular intervals. During such operations, the current has to be diverted by way of auxiliary tie lines within the substation. The picture shows one pole of a 3-phase disconnect rated 2000 A, 345 kV. *(Hydro-Québec)*

Figure 30-17

Combined disconnecting switch and grounding switch rated at 115 kV. The grounding switch blades are shown in the open, horizontal position. These blades pivot upwards to engage three fixed contacts at the same time, as the line is opened. *(Kearney)*

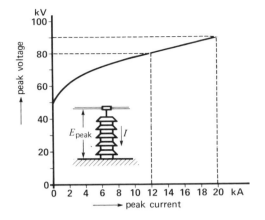

Figure 39-18   Voltage-current characteristic of a lightning arrester having a nominal rating of 30 kV (42.4 kV peak), used on a 34.5 kV line.

ground. The 60 Hz follow-through current is limited by the resistance of the valve blocks and the arc is simultaneously stretched and cooled in a whole series of arc chambers. The arc is quickly snuffed out and the arrester is then ready to protect the line against the next voltage surge. The discharge period is very short, rarely lasting more than a fraction of a millisecond.

Lightning arresters also enable us to reduce the BIL requirements of apparatus installed in substations (Sec. 4-15). On HV and EHV systems, the reduction in BIL significantly reduces the cost of the installed apparatus. Figure 30-19 shows a lightning arrester installed in an EHV substation.

30-7 Current-limiting reactors

The MV bus in a substation usually energizes several feeders which carry power to various load centers around the station. It so happens that the output impedance of the MV bus is usually very low. Consequently, if a short-circuit should occur on one of the feeders, the resulting short-circuit current can be disastrous.

Consider, for example, a 3-phase 69 MVA, 220 kV/24.9 kV transformer having an impedance of 8% and a nominal secondary current of 1600 A. It supplies power to eight 200 A feeders connected to the common MV bus (Fig. 30-20). Each feeder is protected by a 24.9 kV, 200 A circuit-breaker having a rupturing capacity of 4000 A. Because the transformer impedance is 8%, it can deliver a secondary short-circuit current of:

$$I = 1600 \times (100/8)$$
$$= 20\,000 \text{ A}$$

This creates a problem because if a feeder becomes short-circuited, the resulting current flow (20 000 A) is 100 times greater than the normal rating (200 A) of the feeder and its associated circuit-breaker. The circuit-breaker could be damaged

Figure 30-19   Lightning arresters protect this EHV transformer. *(General Electric)*

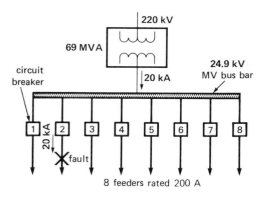

Figure 30-20   MV bus bar feeding 8 lines.

in attempting to interrupt the circuit. Furthermore, the feeder might burn over its entire length, from the circuit-breaker to the fault. Finally, a violent explosion would take place at the fault itself owing to the tremendous amount of thermal energy released by the burning arc.

To prevent this from happening, a current-limiting reactor is connected in series with each phase of the feeder (Fig. 30-21). The reactance must be high enough to keep the current below the interrupting capacity of the circuit-breaker, but not so high as to produce a large voltage drop under normal full-load conditions. Figure 30-22 shows three current-limiting reactors in series with a HV line.

Figure 30-22   Three 2.2 Ω reactors rated 500 A connected in series with a 120 kV, 60 Hz line. They are insulated from ground by four insulating columns and each is protected by a lightning arrester. *(Hydro-Québec)*

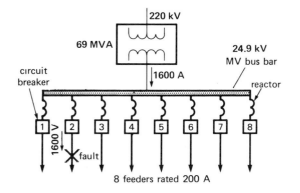

Figure 30-21   Current-limiting reactors reduce the short-circuit current.

30-8 Grounding transformer

We sometimes have to create a neutral on a 3-phase, 3-wire system, to change it into a 3-phase, 4-wire system. This can be done by means of a *grounding* transformer. It is basically a 3-phase autotransformer in which identical primary and secondary windings are connected in series in zig-

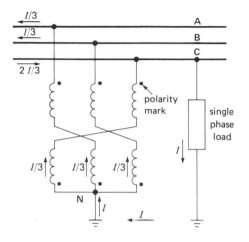

Figure 30-23 Grounding transformer to create a neutral.

zag fashion on a 3-legged core (Fig. 30-23).

If we connect a single-phase load between one line and neutral, load current I divides into three equal currents $I/3$ in each winding. Because the currents are equal, the neutral point stays fixed, and the line-to-neutral voltages remain balanced as they would be on a regular 4-wire system. In practice, the single-phase loads are distributed as evenly as possible between the three phases and neutral so that the unbalanced load current I is relatively small.

30-9 Example of a substation

Figure 30-24 shows the principal elements of a typical modern substation providing power to a large suburb. Power is fed into the substation at 220 kV and is distributed at 24.9 kV to various load centers within about a 5 km radius.

The substation is fed by 3 different lines, all operating at 220 kV (Fig. 30-25). It contains six 3-phase transformers rated at 36/48/60 MVA, 220 kV/24.9 kV. The windings are connected in wye-delta and automatic tap-changers regulate the secondary voltage.

A neutral is established on the MV side by means of 3-phase grounding transformers. Consequently, single-phase power can be provided at $24.9/\sqrt{3}$ = 14.4 kV.

Minimum-oil circuit-breakers having an interrupting capacity of 32 kA protect the HV side. Conventional oil circuit-breakers having an interrupting capacity of 25 kA are used on the MV side. Furthermore, all the outgoing feeders are protected by circuit-breakers having an interrupting capacity of 12 kA.

This completely automatic and unattended substation covers an area of 235 m x 170 m. However, line switching and other operations can be carried out by radio command from a dispatching center.

Figure 30-24

Aerial view of a substation serving a large suburb. The 220 kV lines (1) enter the substation and move through disconnecting switches (2) and circuit-breakers (3) to energize the primaries of the transformers (4). The secondaries are connected to a MV bus (5) operating at 24.9 kV. Grounding transformers (6) and MV circuit-breakers (7) feed power through current limiting reactors (8). The power is carried away by 36 aerial and underground feeders to energize the suburb (9).

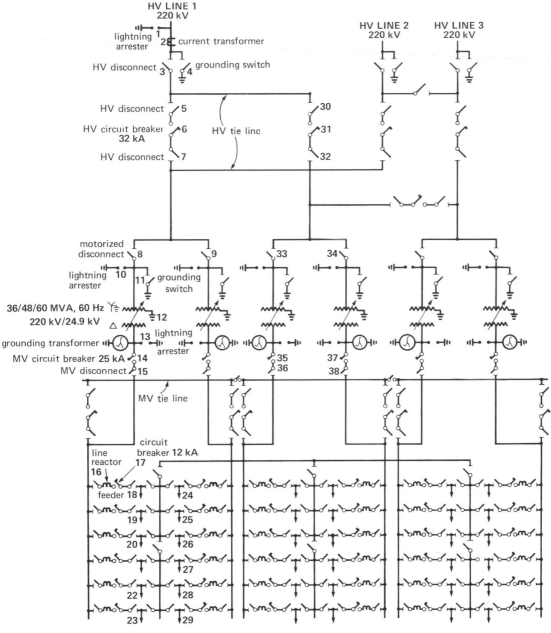

Figure 30-25   Schematic diagram of the substation in Fig. 30-24.

1

2

3

4

5

6

7

8

9

10

11

12

The substation provides service to hundreds of single-family homes, dozens of apartment buildings, several business and shopping centers, a large university and some industries. Figure 30-26 shows the basic layout and components of the substation.

30-10 Medium-voltage distribution

Thirty-six 3-phase feeders (30 active and 6 spares), rated at 24.9 kV, 400 A lead outwards from the substation. Each feeder is equipped with three current-limiting reactors that limit the line-to-ground short-circuit currents to a maximum of 12 kA. Some feeders are underground, others overhead and still others are underground/overhead.

Underground feeders are composed of three single-phase stranded aluminum cables insulated with polyethylene. The insulation is in turn surrounded by a spiral wrapping of tinned copper conductors which act as the ground. The cable is pulled through underground concrete duct (Fig. 30-27) or simply buried in the ground. Spare cables are invariably buried along with active cables to provide alternate service in case of a fault.

The 24.9 kV aerial lines are supported on wooden poles. The latter also carry the LV circuits, and telephone cable. The 24.9 kV lines are tapped

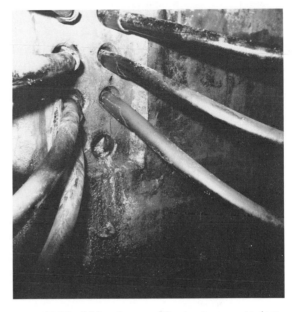

Figure 30-27   MV underground feeders in concrete duct.

Figure 30-26

This sequence of photos enables us to see how energy flows through the substation, starting from the 220 kV lines until it leaves by the 24.9 kV feeders.

1. 220 kV incoming line.
2. The line passes through 3 CT's (left) and the substation apparatus is protected by 3 lightning arresters (right).
3. Three HV disconnecting switches are placed ahead of the circuit-breakers.
4. Minimum volume oil circuit-breakers composed of 3 modules in series permit the line to be opened and closed under load.
5. Three-phase transformer bank steps down the voltage from 220 kV to 24.9 kV. Lightning arresters on the right protect the HV windings.
6. MV line from the transformer feeds the 24.9 kV bus.
7. Grounding transformer and its associated oil circuit-breaker having a rupturing capacity of 25 kA.
8. Current-limiting reactors.
9. Three-phase circuit-breaker having a rupturing capacity of 12 kA.
10, 11 MV underground feeder rated 400 A, 24.9 kV/14.4 kV leads into the ground towards a load center in the suburb.
12. All steel supports are solidly grounded by bare copper conductors to prevent overvoltages across equipment due to lightning strokes and other disturbances. Typical station ground resistance: 0.1 Ω.

off at various points to supply 3-phase and single-phase power to residences, commercial establishments, recreation centers, and so forth (Fig. 30-28).

For nearby areas, the 24.9 kV line voltage is regulated within acceptable limits by the tap-changing transformers at the substation. In more remote districts, special measures have to be taken to keep the voltage reasonably stable with changing load. Thus, self-regulating autotransformers (Fig. 30-29) are often installed.

30-11 Low-voltage distribution

Two systems are provided on the low-voltage side of this typical suburban network:
1. Single-phase 120/240 V with grounded neutral;
2. Three-phase 600/347 V with grounded neutral.

The first system is mainly used in individual dwellings and for single-phase power up to 150 kVA. The second is used in industry, large buildings and commercial centers where the power requirement is under 2000 kVA.

For single-phase service, the transformers are usually rated between 10 kVA and 167 kVA and they are pole-mounted. The voltage rating is 14 400 V/240-120 V. The transformers possess a single high-voltage bushing connected to one side of the HV winding. The other side of the winding is connected to the steel enclosure which, in turn, is connected by a neutral conductor to ground (Fig. 30-30).

In the case of 3-phase installations, 3 single-phase transformers rated at 14 400 V/347 V are used. The units are connected in wye-wye and the neutral on the primary side is solidly grounded.

Figure 30-28   MV aerial feeder serving a residential district.

Figure 30-29   Automatic tap-changing autotransformer maintains steady voltages on long rural lines. *(General Electric)*

The secondary side provides a line voltage of 600 V, and it may or may not be grounded. Such standard distribution transformers have no taps, and no circuit-breakers or fuses are used on the secondary side. The primary HV terminal is, however, protected by a cut-out in order to prevent excessive damage to equipment in case of a fault (see Figs. 30-30 and 30-31).

Protection of Medium-Voltage Distribution Systems

Medium-voltage lines must be adequately protected against short-circuits so as to limit damage to equipment and to restrict the outage to as small an area as possible. Such line faults can occur in various ways: falling branches, icing, defective equipment, lines that touch, and so forth. According to statistics, 85 percent of the short-circuits are temporary, lasting only a fraction of a second. The same studies reveal that 70 percent take place between one line and ground. Finally, short-circuits involving all 3 phases of a transmission line are rare. The methods of protection are based upon these statistics, and on the necessity to provide continuity of service to the customers.

30-12 Coordination of the protective devices

When a fault occurs, the current increases sharply, not only on the faulted line, but on all lines that directly or indirectly lead to the short-circuit. To prevent the overload current from simultaneously tripping all the associated protective devices, we

Figure 30-30 A fused cutout (top left) and lightning arrester (top right) protect a single-phase transformer rated 25 kVA, 14.4 kV/240 V - 120 V, 60 Hz.

Figure 30-31 Expulsion type fused cutout rated 7.5 kV, 300 A. *(Dominion Cutout)*

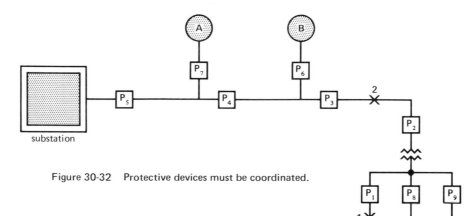

Figure 30-32   Protective devices must be coordinated.

must design the system so that the devices trip selectively.

A well-coordinated system will cause only those devices next to the short-circuit to open, leaving the others intact. To achieve this, the tripping current and tripping time of each device is set to protect the line and associated apparatus, while restricting the outage to the smallest number or customers.

Consider, for example, the simple distribution system of Fig. 30-32 composed of a main feeder from a substation, supplying a group of subfeeders. The subfeeders deliver power to loads A, B, C, D and E. A protective device is installed at the input to each subfeeder so that if a short-circuit occurs, it alone will be disconnected from the system. For instance, a short-circuit at point (1) will trip device P_1 but not P_2. Similarly, a fault at point (2) will open P_3 but not P_4, and so on. A short-circuit must be cleared in a matter of a few cycles. Consequently, the coordination between the protective devices involves delays that are measured in milliseconds. We must therefore know the value of the potential short-circuit currents everywhere on the network. We must also know the tripping characteristics of the fuses and circuit-breakers throughout the system. The most important protective devices used on MV lines are:

1. Fused cutouts;
2. Reclosers;
3. Sectionalizers.

30-13 Fused cutouts

A *fused cutout* is essentially a fused disconnecting switch. The fuse pivots about one end and the circuit can be opened by pulling on the other end with a hookstick (Fig. 30-31). Cutouts are relatively inexpensive and are used to protect transformers and small single-phase feeders against overloads. They are designed to that when the fuse blows, it automatically swings down, indicating that a fault has occured on the line.

Fused cutouts possess a fuse link that is kept taut by a spring. The fuse link assembly is placed inside a porcelain or glass tube filled with boracic acid, oil, or carbon tetrachloride. The fuse link must be replaced every time it blows, which often produces a relatively long outage. To ensure good coordination, the current/time characteristics are selected very carefully for each cutout. A burned out fuse link must always be replaced by another having exactly the same rating.

30-14 Reclosers

A *recloser* is a circuit-breaker that opens on short-circuit and automatically recloses after a brief time delay. The delay may range from a fraction of a

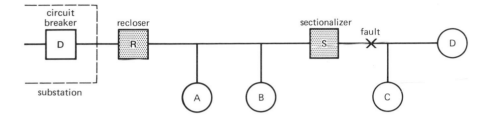

Figure 30-34   Recloser/sectionalizer protective system.

second to several seconds. The open/close sequence may be repeated 2 or 3 times, depending on the internal control setting of the recloser. If the short-circuit does not clear itself after 2 or 3 attempts to reclose the line, the recloser opens the circuit permanently. A repair crew must then locate the fault, remove it, and reset the recloser.

Reclosers rated at 24.9 kV can interrupt fault currents up to 12 000 A. They are made for either single-phase or 3-phase lines, and are usually pole-mounted (Fig. 30-33). Reclosers are self-powered, drawing their energy from the line and storing it in powerful actuating springs and electromagnets.

30-15 Sectionalizers

When a main feeder is protected by several fuses spotted over the length of the line, it is often difficult to obtain satisfactory coordination between them, based on fuse-blowing time alone. Under these circumstances, we resort to *sectionalizers*. A sectionalizer is a special circuit-breaker that trips depending on the number of times a recloser has tripped further up along the line. In other words, a sectionalizer works according to the "instructions" of a recloser.

For example, consider a recloser R and a sectionalizer S protecting an important main feeder (Fig. 30-34). If a fault occurs at the point shown the recloser will automatically open and reclose the circuit, according to a predetermined program. A recorder inside the *sectionalizer* counts the number of times the recloser has tripped and, just before it trips for the last time, the sectionalizer itself trips - but permanently. In so doing, it deprives customers C and D of power but it also isolates the fault. Consequently, when the recloser closes for the last time, it will stay closed and customers A and B will continue to receive service. Unlike reclosers, sectionalizers are not designed to interrupt line currents. Consequently, they must trip during the interval when the line current is zero, which coincides with the time when the recloser itself is open.

Sectionalizers are available for single-phase and

Figure 30-33   Automatic recloser protecting a 3-phase feeder.

3-phase transmission lines. They offer several advantages over fused cutouts. They can be reclosed on a dead short-circuit without fear of exploding, and there is no delay in looking for a fuse link having the correct caliber.

30-16 Review of MV protection

If we examine the single line diagram of a typical distribution system, we find it contains dozens of automatic reclosers, sectionalizers and fused cutouts. The reclosure of circuit-breakers at the substation may be coordonated with reclosers and sectionalizers of the system. The variety of devices available makes it possible to provide adequate protection of MV lines by using combinations such as:
1. circuit-breaker — fuse;
2. circuit-breaker — fuse — fuse;
3. circuit-breaker — recloser — fuse;
4. circuit-breaker — recloser — sectionalizer;
5. circuit-breaker — sectionalizer — recloser — sectionalizer — fuse, etc.

In urban areas, the lines are relatively short and the possibility of faults is rather small. Such lines are subdivided into 3 or 4 sections protected by single-pole fused cutouts. Reclosers and sectionalizers are not required. On the other hand, in outlying districts, a 24.9 kV line may be quite long and consequently more exposed to faults. In such cases, the line is subdivided into sections and protected by reclosers and sectionalizers to provide satisfactory service.

LOW-VOLTAGE DISTRIBUTION

We have seen that electrical energy is delivered to the customer via HV substations through MV networks and finally by LV circuits. In this section, we briefly cover the organization of a LV distribution system.

30-17 LV distribution system

The most common LV systems used in North America are:

1. Single-phase, 2-wire, 120 V;
2. Single-phase, 3-wire, 240/120 V;
3. Three-phase, 4-wire, 208/120 V;
4. Three-phase, 3-wire, 480 V;
5. Three-phase, 4-wire, 480/277 V;
6. Three-phase, 3-wire, 600 V;
7. Three-phase, 4-wire, 600/347 V.

In Europe and other countries of the world, 3-phase 380/220 V, 50 Hz systems are widely used. Despite the different voltages employed, the basic principles of LV distribution are the same.

Single-phase, 2-wire, 120 V system. This simple distribution system is only used for very small loads. When heavier loads have to be serviced, the 120 V system is not satisfactory because large conductors are required. Furthermore, the voltage drop becomes significant even over short distances.

Single-phase, 3-wire, 240/120 V system. In order to reduce the current and hence conductor size, the voltage is raised to 240 V. However, because the 120 V level is still very useful, the 240 V/120 V 3-wire system was developed (Fig. 30-35). This type of distribution system is widely used. It is produced by a distribution transformer having a double secondary winding (Sec. 12-1). The common wire, called neutral, is solidly connected to ground. When the "live" lines A and B are equally loaded, the current in the neutral is zero. When the loading is unequal, the neutral current is equal to the difference between the line currents $I_A - I_B$ (Fig. 30-35). We try to distribute the 120 V loads as equally as possible between the two live wires.

What are the advantages of such a 3-wire system?
1. The line to ground voltage is only 120 V, which is reasonably safe for use in a home;
2. Lighting and small motor loads can be energized at 120 V, while larger loads such as electric stoves, large motors, and so forth, can be fed from the 240 V line.

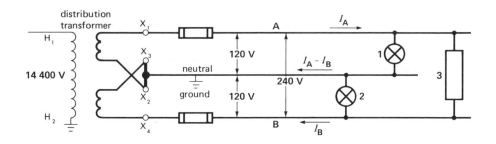

Figure 30-35   Single-phase 240 V/120 V distribution system.

Both live lines are protected by fuses or circuit-breakers. However, such protective devices must *never* be placed in series with the neutral conductor. The reason is that if the device trips, the line-to-neutral voltages become unbalanced. The voltage across the lightly loaded 120 V line goes up, while that across the more heavily loaded side goes down. This means that some lights are dimmer than others and, moreover, the intensity will vary as refrigerator motors, electric stove elements, and so forth, are switched on and off.

Three-phase, 4-wire, 208/120 V system. We can create a 3-phase, 4-wire system by using 3 single-phase transformers connected in delta-wye. The neutral of the secondary is grounded, as shown in Fig. 30-36.

This system is used in commercial buildings and small industries because the 208 V line voltage can be used for electric motors or other large loads, while the 120 V lines can be used for lighting circuits and convenience outlets. The loads between the three "live" lines and neutral are balanced as much as possible. When the loads are perfectly balanced, the current in the neutral wire is zero.

Three-phase, 3-wire, 600 V system. A 600 V, 3-phase, 3-wire system is used for power circuits where fairly large motors are installed, ranging up to 300 hp (Fig. 30-37). Separate 600 V/240-120 V

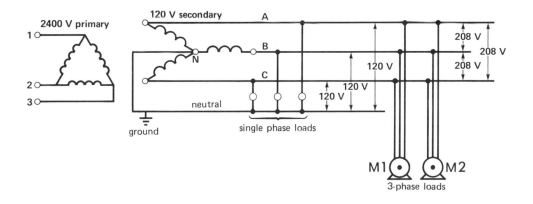

Figure 30-36   Three-phase 4-wire 208 V/120 V distribution system.

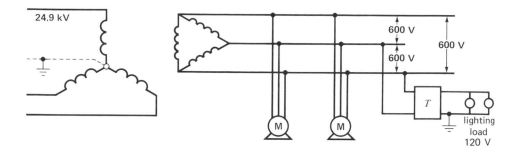

Figure 30-37   Three-phase 3-wire 600 V distribution system.

step-down transformers, spotted throughout the premises, are used to service the lighting loads and convenience outlets.

Three-phase, 4-wire, 600/347 V system. In large buildings and commercial centres, a 600 V, 4-wire distribution system is used because it enables operating motors at 600 V and fluorescent lights at 347 V. For 120 V convenience outlets, separate transformers are required, usually fed from the 600 V line.

The same remarks apply to 480/277 V, 4-wire systems.

30-18 Grounding electrical installations

The grounding of electrical systems is probably one of the less well understood aspects of electricity. Nevertheless, it is a very important subject and an effective way of preventing accidents.

As we have seen, most electrical distribution systems in buildings are grounded, usually by connecting the neutral to a water pipe or the massive steel structure. On low-voltage systems, the purpose of grounding is mainly to reduce the danger of electric shock. On the other hand, as far as *equipment* grounding is concerned, it is systematically employed on HV, MV and LV systems.

30-19 Electric shock

It is difficult to specify whether a voltage is dan-

gerous or not because electric shock is actually caused by *the current* that flows through the human body. The current depends mainly upon the skin contact resistance because, by comparison, the resistance of the body itself is negligible. The contact resistance varies with the thickness, wetness, and resistivity of the skin.

It is generally claimed that currents below 5 mA are not dangerous. Between 10 mA and 20 mA, the current is potentially dangerous because the victim loses muscular control and so may not be able to let go; above 50 mA, the consequences may be fatal.

The resistance of the human body, taken between two hands, or between one hand and a leg, ranges from 500 Ω to 50 kΩ. If the resistance of a dry hand is, say, 50 kΩ, then momentary contact with a 600 V line may not be fatal (I = 600 V/50 kΩ = 12 mA). But the resistance of a clammy hand is much lower, so that an ac voltage as low as 25 V may prove fatal (I = 25 V/500 Ω = 50 mA).

When ac current flows through the body, the muscular contractions prevent the victim from letting go. The current is particularly dangerous when it flows in the region of the heart. It induces temporary paralysis and, if it flows long enough, fibrillation may result. Fibrillation is a very rapid and uncoordinated heart beat that is not synchronized with the pulse beat. In such cases, the person can be revived by applying artificial respiration.

Statistical investigations have shown that a cur-

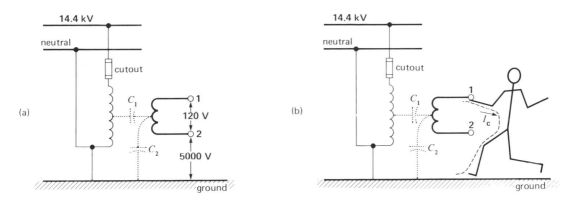

Figure 30-38   Transformer capacitance may produce high voltages on the LV side of a transformer.

rent may cause death if it satisfies the following equation:

$$I = 116/\sqrt{t}$$ * (30-1)

where

I = current flow through the body [mA]

t = time of current flow [s]

116 = an empirical constant, expressing the probability of a fatal outcome

According to tests, a person has 1 chance in 20 of dying if this equation is satisfied. For example, a current of 116 mA flowing for 1 s is fatal in one case out of 20.

30-20   Grounding of 120 V and 240 V/120 V systems

Suppose the primary winding of a distribution transformer is connected between the line and neutral of a 14.4 kV line (Fig. 30-38a). If the secondary conductors are ungrounded, it would appear that a person could touch either one without harm because there is no ground return. However, this is not true.

* Dalziel, C.F., 1968 "Reevaluation of Lethal Electric Currents", *IEEE, Transactions on Industry and General Application*, Vol. IGA-4, No. 5, pages 467 to 475.

First, the capacitive coupling C_1, C_2 between the primary, secondary, and ground, can produce a high voltage between the secondary lines and ground. Depending upon the relative magnitude of C_1 and C_2, it may be as high as 20 to 40 percent of the primary voltage. If a person touches either one of the secondary wires, the resulting capacitive current I_c could be dangerous, even in the case of small transformers. For example, if I_c is only 20 mA, the person may no longer be able to let go (Fig. 30-38b).

Even more serious, suppose that a high-voltage wire accidentally touches a 120 V conductor. This could be caused by an internal fault in the transformer, or by a branch or tree falling across the MV and LV lines. Under these circumstances, the much higher than normal 14.4 kV between the secondary conductors and ground would immediately puncture the 120 V insulation, causing a massive flashover. The flashover could occur anywhere on the secondary network, possible inside a home or factory. Consequently, an ungrounded secondary system is a potential fire hazard and may produce grave accidents under abnormal conditions.

On the other hand, if one of the secondary lines is grounded, the accidental contact between a HV and a LV conductor produces a dead short. The short-circuit current follows the dotted path shown in Fig. 30-39. The high current will blow

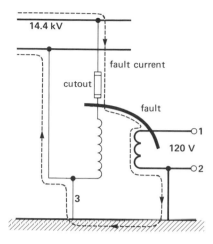

Figure 30-39   A HV to LV fault is not dangerous if the secondary is solidly grounded.

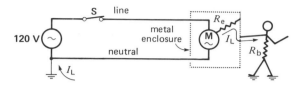

Figure 30-40   Ungrounded metallic enclosures are potentially dangerous.

the fuse on the MV side, thus effectively disconnecting the transformer and secondary distribution system from the MV line. In conclusion, if the neutral is *solidly* grounded, the potential difference between the ground and live conductor 1 will never exceed 120 V. However, if the ground electrode has a high resistance (Sec. 5-19), the voltage produced by a MV-LV short-circuit current may still be large, and potentially dangerous.

30-21 Equipment grounding

The consumer is constantly touching electrical equipment of all kinds, ranging from domestic appliances and hand-held tools to industrial motors, switchgear, heating equipment, and so forth. As we have seen the voltages and currents associated with this equipment far exceeds those the human body can stand. Consequently, special precautions are taken to ensure that the equipment is safe.

In order to understand the safety features of modern distribution systems, let us begin with a simple single-phase circuit composed of a 120 V source connected to a motor M (Fig. 30-40). The motor is inside a piece of equipment that is sur-

rounded by an ungrounded metal enclosure. The neutral is solidly grounded at the service entrance. If a person touches the metal enclosure, nothing will happen if the equipment is functioning correctly. But if the winding insulation becomes faulty, the resistance R_e between the motor and the enclosure may drop from several megohms to only a few hundred ohms or less. A person having a body resistance R_b would complete the current path to ground as shown in Fig. 30-40. If R_e is small (as it could be), the leakage current I_L could be dangerously high. This system is unsafe.

It appears that we could remedy the situation by bonding the enclosure to the grounded neutral wire (Fig. 30-41). The leakage current now flows

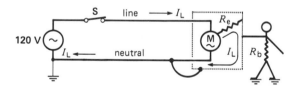

Figure 30-41   Bonding enclosure to the neutral wire appears safe.

from the motor, through the enclosure, and straight back to the neutral wire, but the enclosure remains at ground potential. Consequently, the operator would not experience any shock. The trouble with this solution is that the neutral wire may become open either accidentally or due to a faulty installation. For example, if the switch is inadvertently connected in series with the neutral rather than the live wire, the motor can still be shut on and off. However, if someone touched the enclosure while the motor is *off*, he could receive a bad shock (Fig. 30-42). The reason is that when the motor is off, the potential of the enclosure rises to that of the live conductor.

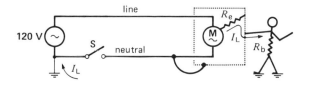

Figure 30-42   Enclosure-to-neutral bonding can still be dangerous.

To get around this problem, we install a third wire, called *ground* wire, between the enclosure and the system ground (Fig. 30-43). With this arrangement, the neutral wire must never be connected to the enclosure, except for special exceptions allowed by the NEC.

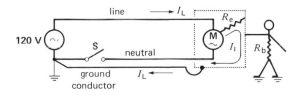

Figure 30-43   A ground conductor bonded to the enclosure is safe.

The ground wire may be bare, or, if insulated, it is colored green. In armored-cable and conduit installations, the armor and conduit serve as the ground conductor. The locknuts, squeeze-connectors, threads and bushings must be tight so as to make a good electrical connection between the service ground and the hundreds of outlets that sometimes make up a large installation.

Many electrical outlets are now provided with receptacles having three contacts - one live, one neutral, and one ground (Fig. 30-44). Portable hand-tools such as electric drills are equipped with

a 3-conductor cord and a 3-prong plug. So-called double-insulated tools having plastic enclosures may dispense with the ground wire requirement.

30-22 Ground-fault circuit-breaker

The grounding schemes we have covered are usually adequate, but further safety measures are needed in some cases. Suppose for example, that a person sticks his finger into a lamp socket (Fig. 30-45). Although the metal enclosure is securely grounded, the person will still receive a painful shock. Again, suppose a 120 V electric toaster tumbles into a swimming pool. The wet heating elements and contacts will produce a hazardous leakage current throughout the pool, even if the frame is securely grounded. Devices have been developed that will cut the source of power as soon

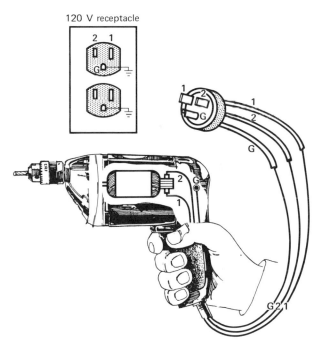

Figure 30-44   The metal housing of hand tools must be grounded.

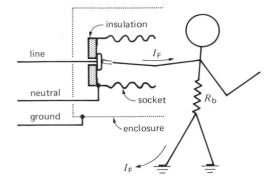

Figure 30-45   Special case where grounding does not afford protection.

as such accidents take place. These *ground-fault circuit-breakers* will typically trip in 25 ms if the leakage current exceeds 5 mA. How do these protective devices operate?

A small current transformer surrounds the live and neutral wires as shown in Fig. 30-46. The secondary winding is connected to a sensitive electronic detector that can trigger a circuit-breaker CB in series with the 120 V line. Under normal conditions, the currents flowing in the line conductor and the neutral are equal, and so the net current $(I_W - I_N)$ flowing through the hole in the toroidal core is zero. Consequently, no flux is produced in the core, the induced voltage E_F is zero, and breaker CB does not trip.

Suppose now that a fault current I_F leaks directly from the line wire to ground. This could happen if someone touched a live terminal, or if the motor M fell into the water (Fig. 30-46). A fault current would also be produced if the insulation broke down between the motor and its grounded enclosure. Under any of these conditions, the net current flowing through the hole of the CT is no longer zero but equal to I_F. A flux is set up and a voltage E_F is induced which trips CB. Because an unbalance of only 5 mA has to be detected, the core material must be extremely permeable at low flux densities. The relative permeability is typically **40 000** at a density of 4 mT and Supermalloy is used for this purpose.

30-23   Electrical installation in buildings

The electrical distribution system in a building is the final link between the consumer and the original source of electrical energy. All such "in-house" distribution systems, be they large or small, must meet certain basic requirements. They relate to:
1. Safety:
   a. protection against electric shock;
   b. protection of conductors against physical damage;
   c. protection against overloads;
   d. protection against hostile environments.
2. Conductor voltage drop:
      It should not exceed 1 or 2 percent.

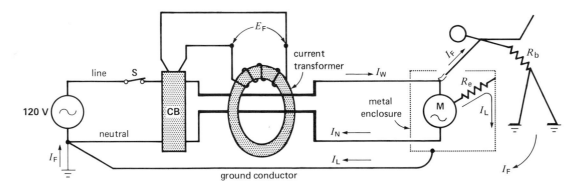

Figure 30-46   Ground fault circuit-breaker trips when external leakage currents flow.

3. Life expectancy:

The distribution system should last a minimum of 50 years.

4. Economy:

The cost of the installation should be minimized while observing the pertinent standards.

Standards are set by the National Electrical Code* and every electrical installation must be approved by an inspector before it can be put into service.

30-24 Principal components of an electrical installation

Many components are used in the makeup of an electrical installation. The block diagrams of Figs. 30-47 and 30-48, together with the following definitions,† will help the reader understand the purpose of some of the major items.

1. **Service conductors.** These are the conductors that extend from the street main or from transformers, to the service equipment of the premises supplied.

2. **Service equipment.** The necessary equipment, usually consisting of a circuit-breaker or switch and fuses, and their accessories, located near the point of entrance of supply conductors to a building or other structure, or an otherwise defined area, and intended to constitute the main control and means of cutoff of the supply.

3. **Metering equipment.** Various meters and recorders to indicate the electrical energy consumed on the premises.

4. **Panel board.** A single panel or group of panel units designed for assembly in the form of a single panel; including buses, automatic overcurrent de-

---

* In Canada, by the Canadian Electrical Code.

† Some taken from the National Electrical Code.

vices, and with or without switches for the control of light, heat or power circuits; designed to be placed in a cabinet or cutout box placed in or against a wall or partition and accessible only from the front.

5. **Switchboard.** A large single panel, frame, or assembly of panels on which are mounted, on the face, or back, or both, switches, overcurrent and other protective devices, buses, and usually instruments. Switchboards are generally accessible from the rear as well as from the front and are not intended to be installed in cabinets.

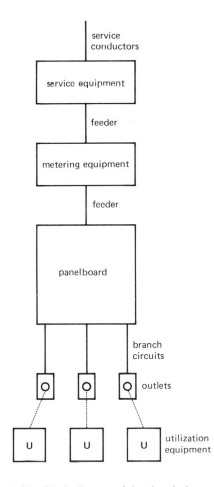

Figure 30-47   Block diagram of the electrical system in a residence.

6. Feeder. All circuit conductors between the service equipment, or the generator switchboard of an isolated plant, and the final branch-circuit overcurrent device.

7. Branch circuit. The circuit conductors between the final overcurrent device protecting the circuit and the outlet(s).

8. Outlet. A point on the wiring circuit at which current is taken to supply utilization equipment.

9. Receptacle. A contact device installed at the outlet for the connection of a single attachment plug.

10. Utilization equipment. Equipment which utilizes electrical energy for mechanical, chemical, heating, lighting, or similar services.

The greatly simplified diagrams of Fig. 30-47 and 30-48 indicate the type of distribution systems used respectively in a home and in an industrial or commercial establishment.

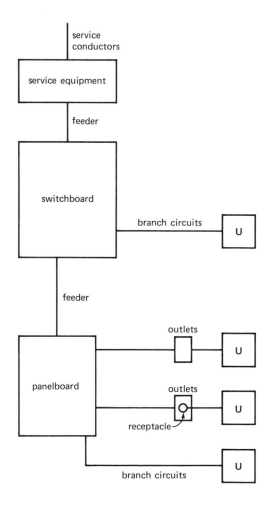

Figure 30-48   Block diagram of the electrical system in an industrial or commercial establishment.

QUESTIONS AND PROBLEMS

Practical level

30-1 What is the difference between a circuit-breaker and a disconnecting switch?

30-2 Name 4 types of circuit breakers.

30-3 The disconnector shown in Fig. 30-13 can dissipate a maximum of 200 W. Calculate the maximum value of the contact resistance.

30-4 What is the purpose of a grounding switch?

30-5 Name some of the main components of a substation.

30-6 The ground resistance of a substation is 0.35 Ω. Calculate the rise in potential of the steel structure if the station is hit by a 50 kA lightning stroke.

30-7 What is the purpose of the following equipment:

fuse cutout   receptacle
recloser   ground wire
current-limiting reactor surge arrester

Intermediate level

30-8 A lightning arrester having the characteristics shown in Fig. 30-18 is connected to a line having a line-to-neutral voltage of 34.5 kV. Calculate:
   a. the peak voltage between line and neutral;
   b. the current flowing in the arrester under

these conditions.

30-9 In Problem 30-8, an 80 kV surge appears between line and neutral. Calculate:
 a. the peak current in the arrester;
 b. the peak power dissipated in the arrester;
 c. the energy dissipated in the arrester if the surge effectively lasts for 5 μs.

30-10 Figure 30-25 is a schematic diagram of a substation and Fig. 30-26 shows the actual equipment. Can you correlate the symbols of the schematic diagram with the equipment?

30-11 Repairs have to be carried out on HV circuit breaker No. 6 shown in Fig. 30-25. If the three 220 kV lines must be kept in service, which disconnecting switches must be kept open?

30-12 The current-limiting reactors (8) shown in Fig. 30-26 limit the short-circuit current to 12 kA on the 24.9 kV feeders. Calculate the reactance and inductance of each coil.

30-13 In Fig. 30-35, resistive loads 1, 2 and 3 respectively absorb 1200 W, 2400 W and 3600 W. Calculate the current:
 a. in lines A and B;
 b. in the neutral conductors;
 c. in the HV line.

30-14 In Fig. 30-37, the lighting circuit is off and the two motors together draw 420 kVA from the 600 V line. Calculate the current in the MV lines.

30-15 Draw a graph (I versus t) of Equation 30-1 for currents between 10 mA and 2 A. Cross-hatch the potentially mortal regions. State whether the following conditions are dangerous:
 a. 300 mA for 10 ms;
 b. 30 mA for 2 min.

30-16 Explain the operation of a ground fault circuit-breaker.

Advanced level

30-17 The following loads are connected to the 240 V/120 V line shown in Fig. 30-35.

load 1: 6 kW, cos θ = 1.0
load 2: 4.8 kW, cos θ = 0.8 lagging
load 3: 18 kVA, cos θ = 0.7 lagging
 a. Calculate the currents in lines A and B, and the neutral;
 b. What is the current in the MV line?
 c. What is the power factor on the MV side?

30-18 Referring to Fig. 30-36, the connected loads have the following ratings:
single phase loads : 30 kW each
motor M1 : 50 kVA, cos θ = 0.5 lagging
motor M2 : 160 kVA, cos θ = 0.80 lagging
 a. Calculate the currents in the secondary windings.
 b. Calculate the line currents and power factor on the 2400 V side.

30-19 In Fig. 30-37, a sensitive voltmeter reads 300 V between one 600 V line and ground, even though the 600 V system is not grounded. Can you explain this phenomenon?

30-20 The oil in the big power transformer shown on the extreme left-hand side of Fig. 30-25 has to be filtered and cleaned. Without interrupting the power flow from the three 220 kV incoming lines, state which circuit breakers and which disconnecting switches have to be opened, and in what order?

30-21 Referring to Fig. 30-26, the three aluminum conductors that make up the 3-phase 24.9 kV feeder (item 10.11) each have a cross-section of 500 MCM. The cable possesses the following characteristics, per phase, and per kilometer of length:
resistance : 0.13 Ω
inductive reactance : 0.1 Ω
capacitive reactance : 3000 Ω
 a. Draw the equivalent circuit of one phase if the line length is 5 km.
 b. If no current-limiting line reactors are used, calculate the short-circuit current if a fault occurs at the end of the line.
 c. Given the 12 kA rating of the circuit breakers, is a line reactor needed in this special case?

31

DIRECT
CURRENT
TRANSMISSION

The development of high-power, high voltage converters has made it possible to transmit and control large blocks of power using direct current. Direct current transmission offers unique features that complement the characteristics of existing ac networks. We cover here some of the various ways it is being adapted and used, both in North America and throughout the world. However, before undertaking this chapter, the reader should first review the principles of power electronics covered in Chapter 23.

31-1 Features of dc transmission

What are the advantages of transmitting power by dc rather than by ac? They may be listed as follows:

1. DC power can be controlled much more quickly. For example, power in the megawatt range can be reversed in a dc line in less than one second. This feature makes it useful to operate dc transmission lines in parallel with existing ac networks. When instability is about to occur (due to a disturbance on the ac system), the dc

power can be changed in amplitude to counteract and dampen out the power oscillations. Quick power control also means that dc short-circuit currents can be limited to much smaller values than those encountered on ac networks.

2. DC power can be transmitted in cables over great distances. We have seen that the capacitance of a cable limits ac power transmission to a few tens of kilometers (Sec. 29-24). Beyond this limit, the reactive power generated by cable capacitance exceeds the rating of the cable itself. Because capacitance does not come into play under steady-state dc conditions, there is theoretically no limit to the distance that power may be carried this way. As a result, power can be transmitted by cable under large bodies of water, where the use of ac cables is unthinkable. Furthermore, underground dc cable may be used to deliver power into large urban centers. Unlike overhead lines, underground cable is invisible, free from atmospheric pollution, and solves the problem of securing rights of way.

3. We have seen that ac power can only be trans-

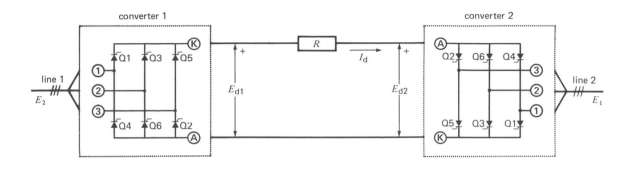

Figure 31-1 Elementary dc transmission system.

mitted between centers operating at the same frequency. Furthermore, the power transmitted depends upon line reactance and the phase angle between the voltages at each end of the line (Sec. 29-19). The big advantage of dc is that systems having different frequencies can now be tied together. In effect, when power is transmitted by dc, frequencies and phase angles make no difference, and line reactance does not limit the steady-state power flow. If anything, it is only the resistance of the line that limits the flow. This also means that power can be transmitted over greater distances by using dc. However, this is a marginal benefit because large blocks of ac power are already being carried over distances exceeding 1000 km.

4. Overhead dc transmission lines become economically competitive with ac lines when the length of the line exceeds several hundred kilometers. The width of the power corridor is less, and experience to date has shown that outages due to lightning are somewhat reduced. Consequently, dc transmission lines are being used to carry electric power directly from a generating station located near a coal mine or waterfall, to the load center.

If dc transmission offers such important advantages, why is it not more widely used? The reason is (1) reliable HV converters have only recently been developed, (2) they are still relatively expensive, and (3) it is not easy to tap power off at dif-

ferent points along a dc line. In effect, dc lines are usually point-to-point systems, tying one large generating station to one large power-consuming center. Electronic converters are installed at each end of the transmission line, but none in between.

31-2 Basic dc transmission system

A dc transmission system consists basically of a dc transmission line connecting two ac systems. A converter at one end of the line converts ac power into dc power while a similar converter at the other end reconverts the dc power into ac power. One converter acts therefore as a rectifier, the other as an inverter.

Stripped of everything but the bare essentials, the transmission system may be represented by the circuit of Fig. 31-1. Converter 1 is a 3-phase 6-pulse rectifier which converts the ac power of line 1 into dc power. The dc power is carried over a 2-conductor transmission line and reconverted to ac power by means of converter 2, acting as an inverter. Both the rectifier and inverter are naturally commutated by the respective line voltages to which they are connected (Secs. 23-8, 23-17, and 23-25). Consequently, the networks can function at entirely different frequencies, without affecting power transmission between them.

Power flow may be reversed by changing the firing angles α_1 and α_2, so that converter 1 becomes an inverter and converter 2 a rectifier. Changing

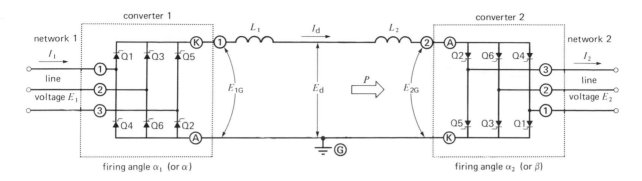

Figure 31-2   Smoothing inductors L_1 and L_2 are required.

the angles reverses the polarity of the conductors, but the direction of current flow remains the same. The reason is that thyristors can only conduct in one direction.

The dc voltages E_{d1} and E_{d2} at each converter station are identical, except for the IR drop in the line. The drop is usually so small that we can neglect it, except insofar as it affects losses, efficiency and conductor heating.

Owing to the high voltages and currents encountered in transmission lines, each thyristor shown in Fig. 31-1 is actually composed of several thyristors connected in series-parallel. Such a group of thyristors is often called a valve. Thus, a valve for a 50 kV, 1000 A converter would typically be composed of 2 parallel groups of 50 thyristors connected in series, or 100 SCRs per bridge arm. Each converter in Fig. 31-1 would therefore contain 600 thyristors. The 100 thyristors in each bridge arm are triggered simultaneously so they together act like a super-SCR.

31-3 Voltage, current and power relationships

In a practical transmission line, smoothing inductors L_1 and L_2 (Fig. 31-2) must be used as a buffer between the ripple-free dc voltage E_d and the undulating output of the converters (Secs. 23-9, 23-25, and 23-28). They also reduce the ac harmonic currents flowing in the transmission line to an acceptable level.

Typical waveshapes of E_{1G} and E_{2G} on the dc side of the rectifier and inverter, respectively, are shown in Figs. 31-3a and 31-3b. We have assumed firing angles of 15° and 165°, respectively, for the rectifier and inverter. Consequently, the dc line voltage is given by:

$$E_d = 1.35\,E\cos\alpha \qquad \text{Eq. 23-14}$$
$$= 1.35\,E\cos 15°$$
$$= 1.304\,E$$

The potential difference between E_{1G} and E_d appears across inductor L_1. Similarly, the difference between E_{2G} and E_d appears across L_2.

In the case of dc transmission lines, the rectifier angle α_1 is simply designated α. Furthermore, the inverter firing angle is not considered as a delay (α_2) with respect to the rectifier zero firing point, but as an angle of advance β. Thus, as regards the inverter, instead of stating that $\alpha_2 = 165°$ (as we have done in all previous inverter circuits, Chapters 23, 24, and 25), we now refer to it as an angle of advance $\beta = 15°$ (Fig. 31-3b). The value of β is related to α_2 by the simple equation:

$$\beta = 180 - \alpha_2$$

Note that the inverter voltage is zero when β is 90°, and maximum when $\beta = 0°$.

The voltage, current, and power relationships of a dc transmission system are the same as those for any circuit containing ac/dc power converters. Re-

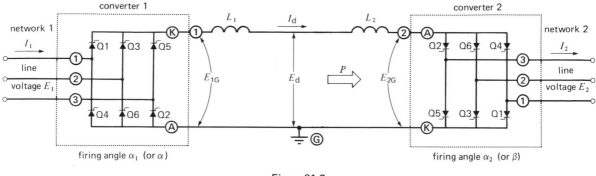

Figure 31-2

ferring to Fig. 31-2, and based upon equations we have already seen, the relationships may be stated as follows:

$$P = E_d I_d$$

$$E_d = 1.35\, E_1 \cos \alpha \qquad \text{Eq. 23-14}$$

$$E_d = 1.35\, E_2 \cos \beta$$

$$I_1 = I_2 = 0.816\, I_d \qquad \text{Eq. 23-15}$$

$$Q_1 = P \tan \alpha \qquad \text{Eq. 23-16}$$

$$Q_2 = P \tan \beta$$

where

P = active power transmitted [W]

E_d = dc line voltage [V]

I_d = dc line current [A]

I_1, I_2 = effective values of the currents in ac lines 1 and 2 [A]

E_1, E_2 = effective values of the respective ac line voltages [V]

Q_1, Q_2 = reactive powers absorbed by converters 1 and 2 [var]

α = rectifier angle of delay [°]

β = inverter angle of advance [°]

In order to keep the reactive powers Q_1 and Q_2 as low as possible, we attempt to make α and β approach 0°. However, for practical reasons, both are kept at around 30°. Using these values, we can calculate the relative magnitudes of the voltages and currents in a typical transmission line delivering, say, 100 A at a potential of 1000 V (Fig. 31-4). Thus, we have:

$$E_d = 1.35\, E_1 \cos \alpha$$
$$1000 = 1.35\, E_1 \cos 30$$
$$\therefore\ E_1 \approx 850 \text{ V}$$

Furthermore,

$$E_d = 1.35\, E_2 \cos \beta$$
$$1000 = 1.35\, E_2 \cos 150$$
$$\therefore\ E_2 \approx 850 \text{ V}$$

$$I_1 = I_2 = 0.816\, I_d$$
$$= 0.816 \times 100$$
$$\approx 82 \text{ A}$$

$$P = E_d I_d = 1000 \times 100$$
$$= 100 \text{ kW}$$

$$Q_1 = P \tan \alpha$$
$$= 100 \tan 30$$
$$= 58 \text{ kvar}$$

$$Q_2 = P \tan \beta$$
$$= 100 \tan 150$$
$$= 58 \text{ kvar}$$

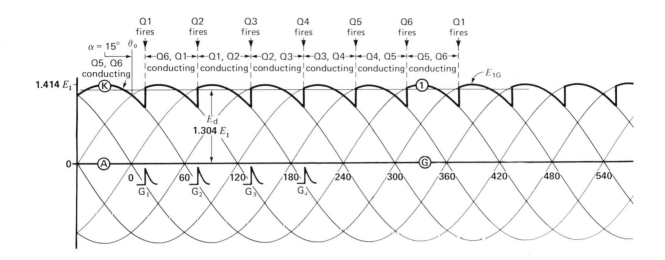

Figure 31-3   a.   Rectifier voltage waveshape for $\alpha = 15°$.

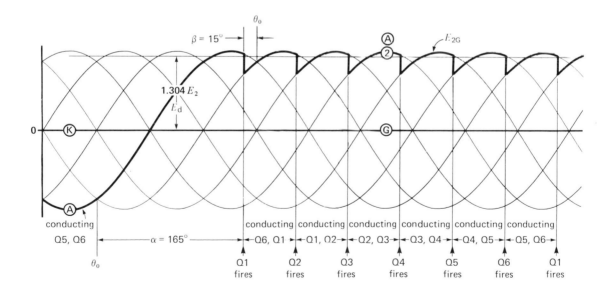

Figure 31-3   b.   Inverter voltage waveshape for $\beta = 15°$.

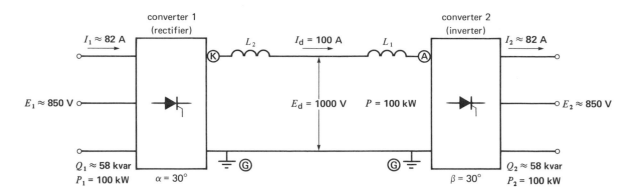

Figure 31-4   Scale model of a simple dc transmission system.

Figure 31-4 may thus be used as a scale model to determine the order of magnitude of the voltages, currents, and power in *any* dc transmission system.

Example 31-1:
A dc transmission line operating at 150 kV carries a current of 400 A. Calculate the approximate value of:
a. the ac line voltage at each converter station;
b. the ac line current;
c. the active power absorbed by the rectifier;
d. the reactive power absorbed by each converter;
e. the apparent power supplied to the rectifier.

Solution:
Using the scale model of Fig. 31-4 we find:
a. The ac line voltages are:

$E_1 = E_2 =$ (150 kV/1000 V) x 850 = 127.5 kV

b. The effective value of the line current is:

$I_1 = I_2 =$ (400/100) x 82 = 328 A

c. The active power absorbed by the rectifier:

P_1 = 150 kV x 400 A = 60 000 kW = 60 MW

d. The reactive power absorbed by each converter:

$Q_1 = Q_2 =$ (60 000/100) x 58 = 34 800 kvar
  = 34.8 Mvar

e. Apparent power supplied to the rectifier:

$S = \sqrt{P_1{}^2 + Q_1{}^2} = \sqrt{60^2 + 34.8^2}$
  = 69.4 MVA

31-4 Power fluctuations on a dc line

In order to ensure stability in transmitting dc power, the rectifier and inverter must have special voltage-current characteristics. These characteristics are shaped by computer-controlled gate firing circuits. We can best understand the need for such controls by studying the behavior of the system when the controls are absent.

Figure 31-5 shows a dc transmission line having a resistance R. The converters produce voltages E_{d1} and E_{d2}, and the resulting dc line current is obviously given by:

$$I_d = (E_{d1} - E_{d2})/R \qquad (31-1)$$

The line resistance is always small; consequently, a very small difference between E_{d1} and E_{d2} can produce full-load current in the line. Furthermore, a small change in either E_{d1} or E_{d2} can produce a very big change in I_d. For example, if E_{d1} increases by only a few percent, the line current can easily double. Conversely, if E_{d2} increases by only a few percent, the line current can fall to zero.

Unfortunately, both E_{d1} and E_{d2} are subject to sudden changes because the associated ac line voltages E_1 and E_2 may fluctuate. The fluctuations may be due to sudden load changes on the ac networks or to any number of other system disturbances than can occur. Owing to the almost instantaneous response of the converters and transmis-

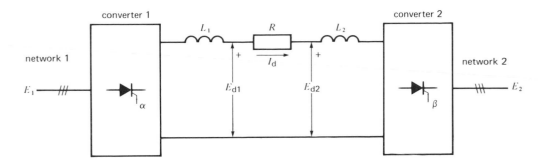

Figure 31-5 A small change in either E_{d1} or E_{d2} produces a very big change in I_d.

sion line, the dc current swings wildly under those conditions, producing erratic power swings between the two networks. Such power surges are unacceptable because they tend to destabilize the ac networks at each end, and because they produce misfiring of the SCRs.

It is true that the firing angles α and β could be modulated to counteract the ac line voltage fluctuations. However, it is preferable to design the system so that large, unpredictable dc power surges are inherently impossible. We now show how this is done.

31-5 Typical rectifier and inverter characteristic

In a practical dc transmission system, the computer-controlled rectifier circuit is designed to yield the E-I curve shown in Fig. 31-6a. Assuming a fixed ac line voltage E_1, the dc output voltage E_d is kept constant until the line current I_d reaches a value I_1. Beyond this point, E_d drops sharply as can be seen on the curve. This E-I characteristic is obtained by keeping α constant at about $30°$ until current I_d approaches the desired value I_1. The firing angle then increases (automatically) at a very rapid rate, so that I_d is equal to I_1, when $E_{d1} = 0$. In other words, if a short-circuit were to occur across the dc side of the rectifier, the resulting dc current would be equal to I_1.

As regards the inverter, it is designed to give the E-I curve shown in Fig. 31-6b. Assuming a fixed ac line voltage E_2, voltage E_{d2} is maintained at zero until the dc line current reaches a value I_2. This

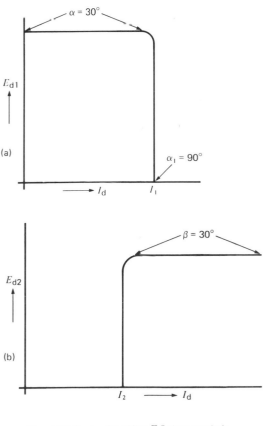

Figure 31-6 a. Rectifier E-I characteristic.
b. Inverter E-I characteristic.

means that from zero to I_2, the firing angle $\beta = 90°$. As soon as the dc current approaches the desired value I_2, the firing angle α_2 increases (automatically) to a fixed limiting value of about $30°$.

Under normal operating conditions, the inverter voltage level is made to be slightly below the rectifier voltage level. Furthermore, limiting current I_2 is made slightly smaller than I_1. The effect of these constraints can best be seen by superposing the rectifier and inverter characteristics (Fig. 31-7a). The actual transmission-line voltage and current corresponds to the point of intersection of the two

31-6 Power control

To vary the power flow over the dc line, the rectifier and inverter E-I characteristics are modified simultaneously. Voltages E_{d1} and E_{d2} are kept constant but I_1 and I_2 are varied simultaneously while keeping the current margin fixed. Thus, Fig. 31-7b shows the new E-I characteristics for a transmis-

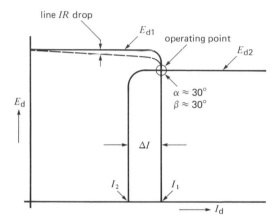

Figure 31-7 a. Operating point when the transmission line delivers rated power.

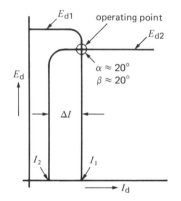

Figure 31-7 b. Operating point when the line delivers 20 percent rated power.

curves. It is obvious that the line current I_d is equal to I_1 (determined by the rectifier characteristic) while the line voltage E_d is equal to E_{d2} (determined by the inverter characteristic). The difference between I_1 and I_2 is called the current margin ΔI. It is made to equal about 10 percent of the rated line current.

If the line has appreciable resistance, the IR drop modifies the effective rectifier characteristic so that it follows the dotted line shown in Fig. 31-7a. This, however, does not affect the operating point under normal conditions. The effective power input to the inverter is therefore given by the product $E_{d2}I_1$.

sion line current I_d smaller than that in Fig. 31-7a. By thus shifting the E-I characteristics back and forth, we can cause the dc power to vary from zero to maximum. Note that the line voltage E_d is constant, and that it is always determined by the inverter. On the other hand, the magnitude of the line current is determined by the rectifier.

At this point, the reader may wonder why the E-I characteristics have been given such odd shapes to attain such a simple result. The reason is that the dc system must be able to accommodate serious ac voltage fluctuations at either end of the line without affecting the dc power flow too much. It must also limit the magnitude of the fault currents, should a short-circuit occur on the dc line. We now explain how this is achieved.

31-7 Effect of voltage fluctuations

Referring to Fig. 31-7a, let us assume that the dc line carries full-load current I_1. If the ac voltage of line 1 increases suddenly, E_{d1} rises in proportion,

in current is not excessive. Consequently, the power flow is again not affected too much. As soon as the disturbance is cleared, the E-I characteristics return to the original curves given in Fig. 31-7a.

Finally, one of the worst conditions that can arise is short-circuit on the dc line. Here again, the rectifier supplies a maximum current I_1 while the

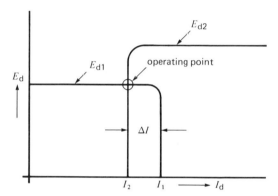

Figure 31-8   Change in operating point when E_1 falls drastically.

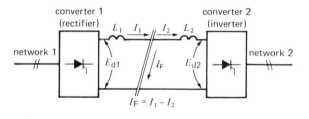

Figure 31-9   The short-circuit current at the fault cannot exceed 10 percent of the rated line current.

but this has no effect on I_1 or E_{d2}. Consequently, the power flow over the line is unaffected.

On the other hand, if line voltage E_2 decreases, E_{d2} decreases in proportion. The dc line current is unaffected, but, because E_{d2} is smaller than before, the dc power is also less. However, the percent change in power cannot exceed the percent change in ac voltage E_2.

Next, if a large disturbance occurs on line 1, E_{d1} may fall drastically. This produces a new operating point, shown in Fig. 31-8. The dc line current decreases suddenly from I_1 to I_2, while the dc voltage decreases equally suddenly from E_{d2} to E_{d1}. With a current margin of 10 percent, the drop

rectifier draws a maximum current I_2. Consequently, the fault current

$$I_F = (I_1 - I_2)$$

is only 10 percent of the normal line current (Fig. 31-9). Fault currents are therefore much smaller than on ac transmission lines. In addition, because E_{d1} and E_{d2} are close to zero, the power delivered to the fault is small.

It is now clear that the special shape of the E-I characteristics prevents large power fluctuations on the line, and limits the short-circuit currents. In practice, the actual E-I characteristics differ slightly from those shown in Fig. 31-7. However, the basic principle remains the same.

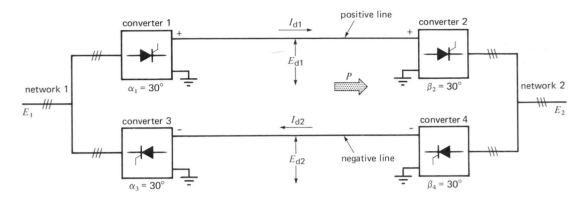

Figure 31-10   a. Bipolar line transmitting power from network 1 to network 2.

31-8 Bipolar transmission line

Most dc transmission lines are *bipolar*. They possess a positive line and a negative line and a common ground return (Fig. 31-10a). A converter is installed at the end of each line, and the line currents I_{d1} and I_{d2} flow in the directions shown. Converters 1 and 3 act as rectifiers while converters 2 and 4 are inverters. Power obviously flows over both lines from ac network 1 to ac network 2. The ground current is $I_{d1} - I_{d2}$. It is usually small because we try to maintain equal currents in the positive and negative lines.

The bipolar arrangement has three advantages. First, the ground current is small, under normal conditions. Consequently, corrosion of underground pipes, structures, and so forth is minimized. Second, the same transmission-line towers can carry two lines, thus doubling the power, with a relatively small increase in cost. Third, if one line is knocked out of service, the other can continue to function, delivering half the normal power between the ac networks.

31-9 Power reversal

To reverse power flow in a bipolar line, we change the firing angles, so that all the rectifiers become inverters and vice versa. This reverses the polarity of the transmission lines, but the line currents I_{d1}

and I_{d2} continue to flow in the same direction as before (Fig. 31-10b).

31-10 Components of a dc transmission line

In order to function properly, a dc transmission system must have auxiliary components, in addition to the basic converters. Referring to Fig. 31-11, the most important components are:
1. dc line inductors (L)
2. harmonic filters on the dc side (F_{dc})
3. converter transformers (T)
4. reactive power source (Q)
5. harmonic filters on the ac side (F_{ac})
6. microwave communications link between the converter stations
7. ground electrodes (Gd)

The need for these components is explained in the following sections.

31-11 Direct current inductors and harmonic filters

Voltage harmonics are produced on the dc side of both the rectifier and inverter (Sec. 23-9). They give rise to 6th and 12th harmonic currents, and such currents, if allowed to flow over the dc line, could produce serious noise on neighboring telephone lines. Consequently, filters are required to prevent the currents from flowing over the line.

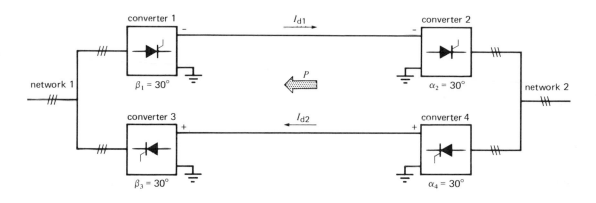

Figure 31-10 b. Power reversal is obtained by reversing the line polarities.

The filters consist of two inductors L and a shunt filter F_{dc}. The latter is composed of two series LC circuits, each tuned to respectively short-circuit the 6th and 12th harmonic currents to ground.

The inductors L also prevent the dc line current from increasing too rapidly if a line fault should occur. This enables the thyristors to establish control before the current becomes too large to handle electronically.

31-12 Converter transformers

The basic purpose of the converter transformer is to transform the network voltage E_{L1} to yield the ac voltage E_1 required by the converter. Three-phase transformers, connected in either wye-wye or wye-delta, are used. A tertiary winding (Sec. 13-5) is often added to supply local power needs. It may also be designed for direct connection to a source of reactive power (Q).

As we have seen, the dc line voltage E_d is kept essentially constant from no-load to full-load. Furthermore, to reduce the reactive power absorbed by the converters, firing angles α and β should be kept small. This means that the ratio between ac voltage input and dc voltage output of the convert-

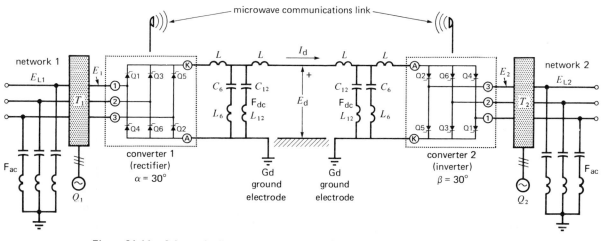

Figure 31-11 Schematic diagram showing some of the more important components of a HVDC transmission system.

ers is essentially fixed. Because E_d is constant, it follows that E_1 must also be constant.

Unfortunately, the network voltage E_{L1} may vary significantly throughout the day. Consequently, the converter transformers on the rectifier side must be equipped with taps so that the variable input voltage E_{L1} will give a reasonably constant output voltage E_1. The taps are switched automatically by a motorized tap changer whenever the network voltage E_{L1} changes for a significant length of time. For the same reasons, taps are needed on the converter transformers on the inverter side.

31-13   Reactive power source

The reactive power absorbed by the converters must be supplied by the ac network or by a local reactive power source. Because the active power transmitted varies throughout the day, the reactive source must also be varied. Consequently, either variable static capacitors or a synchronous capacitor are required (Sec. 20-10).

31-14   Harmonic filters on the ac side

Three-phase 6-pulse converters produce 5th, 7th, 11th, 13th (and higher) current harmonics on the ac side. These harmonics are a direct result of the choppy current waveforms (Sec. 23-29). Again for reasons of telephone interference, these currents must not be allowed to flow over the ac lines (Sec. 23-8). Consequently, they are bypassed through low-impedance filters F_{ac} connected between the three-phase lines. The filters for each frequency are connected in wye, and the neutral point is grounded. On a 60 Hz network, each 3-phase filter is composed of a set of series-resonant LC circuits respectively tuned to 300, 420, 660, and 780 Hz.

31-15   Communications link

In order to control the converters at both ends of the line, a communications link between them is essential. For example, to maintain the current margin ΔI (Fig. 31-7), the inverter at one end of the line must "know" what the rectifier current setting I_1 is. This information must continually be relayed by a high-speed communications link between the two converters.

31-16   Ground electrode

Particular attention is paid to the ground electrode at each end of the dc line. Direct currents in the ground are much more corrosive than ac currents. Consequently, the actual ground electrode is usually situated several kilometers from the converter station, to ensure that dc leakage currents create no local problem. The dc ground wire between the station and the actual grounding site is either pole-mounted or enclosed in a shielding cable. At the site, special means are used to minimize electrode resistance. This is particularly important in monopolar systems. In such installations, the ground current may be several hundred amperes, and the heat generated may eventually dry out the grounding bed, causing the ground resistance to increase.

The best grounds are obtained next to, or in, large bodies of water. But even in this case, elaborate grounding methods must be used.

31-17   Typical converter station

Figure 31-12 shows the elementary circuit diagram of a typical monopolar inverter station. The incoming dc line operates at 150 kV, and power is fed into a 230 kV, 3-phase, 60 Hz power line. Two smoothing inductors, each having an inductance of 0.5 H, are connected in series with the dc line. The two LC filters effectively short-circuit the 6th and 12th harmonic voltages generated on the dc side of the converter. The 9 Ω and 11 Ω resistors make the filters less sensitive to slight frequence changes.

Three single-phase transformers connected wye-wye (with a tertiary winding) are connected to the ac side of the converter. A 160 Mvar synchronous capacitor, connected to the tertiary winding, provides the reactive power for the converter.

Filters for the 5th, 7th, 11th, and 13th harmon-

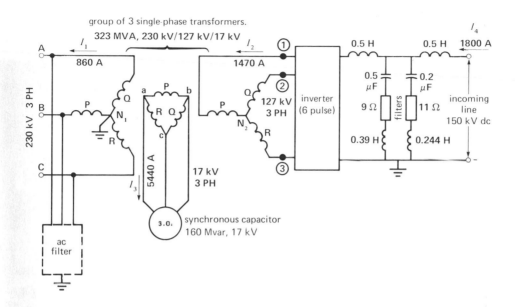

Figure 31-12 Simplified circuit of a 150 kV, 1800 A inverter. See components illustrated in Figs. 31-13 to 31-18.

ic currents are connected between the three ac lines and neutral of the 230 kV system. As previously explained, the filters shunt the ac harmonic currents so that they do not enter the 230 kV line.

Figure 31-13 through 31-17 give us an idea of the size of these various components, and of the immense switchyard needed to accommodate them.

Figure 31-13

These 12 single-phase harmonic filters at an inverter station are tuned for 300 Hz, 420 Hz, 660 Hz and 780 Hz. They are connected between the three lines and neutral of the outgoing 230 kV, 60 Hz transmission line. The filter in the foreground is tuned to 720 Hz. It is series-circuit composed of a 2 Ω resistor, a group of capacitors having a total capacitance of 0.938 μF and an oil-filled inductor of 44.4 mH. The 720 Hz reactive power associated with the LC circuit amounts to 18.8 Mvar. *(GEC Power Engineering Limited, England)*

Figure 31-14   Three-phase converter transformer bank rated 230 kV/127 kV/17 kV composed of 3 single-phase transformers each rated 323 MVA. *(Manitoba Hydro)*

Figure 31-16   Three-phase 6-pulse inverter rated 270 MW, 150 kV. It is insulated from ground by means of 4 porcelain columns; it corresponds to inverter 2 in Fig. 31-12. *(Manitoba Hydro)*

Figure 31-15   View of one 0.5 H smoothing inductor on the 450 kV dc side of the inverter station. The second inductor can be seen in the distance (lower right - hand corner). The space between the two units permits installing the filters on the dc side. *(Manitoba Hydro)*

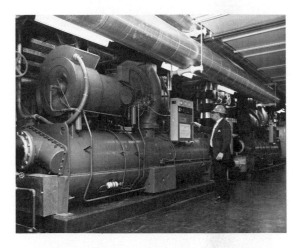

Figure 31-17   Partial view of the refrigeration unit needed to cool the inverters. *(Manitoba Hydro)*

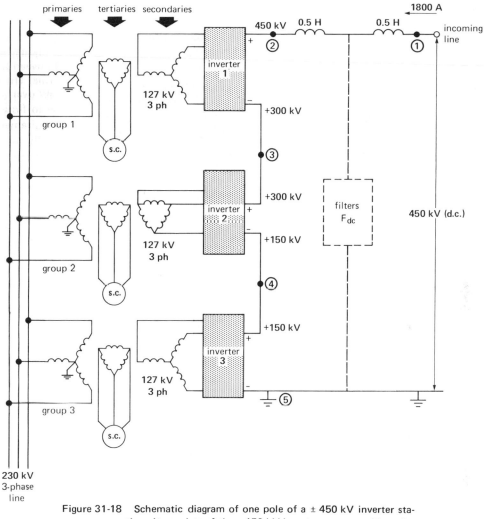

Figure 31-18   Schematic diagram of one pole of a ± 450 kV inverter station. It consists of three 150 kV inverters connected in series on the dc side.

31-18 Cascaded converter station

In some bipolar lines, the converters are cascaded (connected in series) to produce a higher dc voltage. Figure 31-18 shows how three 150 kV converters are cascaded to produce a 450 kV dc output. The dc sides are connected in series, while the ac sides are essentially connected in parallel to the 3-phase line. This means that converter 2 (and its transformer) functions at a dc potential of 150 kV above ground. Converter 1 (and its transformer) operates at an even higher dc potential of 300 kV. As a result, these converters have to be mounted on insulating columns (Fig. 31-16).

Note that only the 127 kV secondary windings of converter transformers 1 and 2 operate at dc potentials of 150 kV and 300 kV respectively. The windings must be especially well insulated to withstand these dc voltages, in addition to the 127 kV ac voltage.

The 127 kV windings of transformer No. 1 are wye-connected, while those of transformer No. 2 are delta-connected. This produces a 30° phase shift in the voltages and currents on the 127 kV side. The result is that the 5th and 7th harmonics tend to cancel each other and do not therefore appear in the 230 kV line. Consequently, the filtering equipment for these frequencies is substantially reduced.

31-19 Typical installations

Power transmission by direct current is being used in many parts of the world. It will no doubt grow in importance as the reliability, speed and versatility of power converters is realized. The following installations give the reader an idea of the various types of systems that have been built, and the particular problem they were designed to solve.

1. **Schenectady.** Of historical interest is the 17 mile, 5.25 MW, 30 kV transmission line installed between Mechanicville and Schenectady, New York, in 1936. Using mercury arc converters, it tied together a 40 Hz and 60 Hz system.

2. **Gotland.** The first important dc transmission line was installed in Sweden, in 1954. It connected the Island of Gotland (in the middle of the Baltic Sea) to the mainland, by a 96 km submarine cable. The single-conductor cable operates at 100 kV and transmits 20 MW. The ground current returns by the sea.

3. **English Channel.** In 1961, a bipolar submarine link was laid in the English Channel between England and France. Two cables, one operating at + 100 kV and the other at – 100 kV, laid side by side, together carry 160 MW of power in one direction or the other. The power exchange between the two countries was found to be economical because the time zones are different, and consequently, the system peaks do not occur at the same time. Furthermore, France has excess hydro generating capacity during the spring, thus making the export of power attractive.

4. **Pacific Intertie.** In 1970, a bipolar link operating at ± 400 kV was installed between the Dalles, Oregon, and Los Angeles, California. The overhead line transmits a total of 1440 MW over a distance of 1370 km. Power can be made to flow in either direction, depending upon the requirements of the respective NW and SW regions. The dc link also helps stabilize the 3-phase ac transmission system connecting the two regions.

5. **Nelson River.** The hydropower generated by the Nelson River, situated 890 km north of Winnipeg, Canada, is transmitted by means of two bipolar lines operating at ± 450 kV. Each bipolar line carries 1620 MW, which is converted and fed into the ac system near Winnipeg. According to studies made, it was slightly more economical to transmit power by dc rather than by ac over this considerable distance.

6. **Eel River.** The converter station at Eel River, Canada, provides an asynchronous intertie between the 230 kV electrical systems of Quebec and New Brunswick. Although both systems operate at a nominal frequency of 60 Hz, it was not feasible to connect them directly, owing to stability considerations. In this application, the dc "transmission line" is only a few meters long, representing the length of the conductors needed to connect the rectifiers and inverters. Power may flow in either direction, up to a maximum of 320 MW (See Fig. 31-20).

7. **CU Project.** The power output of a generating station situated next to the lignite coal mines near Underwood, North Dakota, is converted to dc and transmitted 436 miles eastwards to a terminal near Minneapolis, Minnesota, where it is reconverted to ac. The bipolar line transmits 1000 MW at 1250 A, ± 400 kV. A metallic ground return is provided, in the event that one line should be out for a prolonged period (see Fig. 31-21).

8. EPRI Compact Substation. This project aims at compacting HVDC converter stations so they may be installed in dense metropolitan areas where land costs are high. All components are gas insulated with SF_6 and the thyristors are cooled with liquid freon. This Electric Power Research Institute development project is located in Queen's, New York (Fig. 31-22). It provides a bipolar dc link having a capacity of 100 MW at $\pm$ 50 kV. It connects two large Consolidated Edison substations that are about 700 m apart. The input and output ac feeders are composed of two 138 kV underground cables. Their shunt capacitance is so large that no ac filters are needed and it also furnishes all the reactive power absorbed by the converters.

Figure 31-19

This converter station and switchyard at Eel River connects the ac networks of Quebec and New Brunswick by means of a dc link. The rectifier and inverter are both housed in the large building in the center. It pioneered the commercial use of solid-state thyristors in HVDC applications. *(New Brunswick Electric Power Commission)*

Figure 31-20

View of one 6-pulse thyristor valve housed in its rectangular steel cubicle. It is fed by a 3-phase converter transformer and yields an output of 2000 A at 40 kV. The hundreds of individual thyristors it contains are triggered by a reliable, interference-free fiber-optic control system.

Eight such cubicles, together with three synchronous capacitors and four converter transformers make up the entire converter terminal. *(General Electric)*

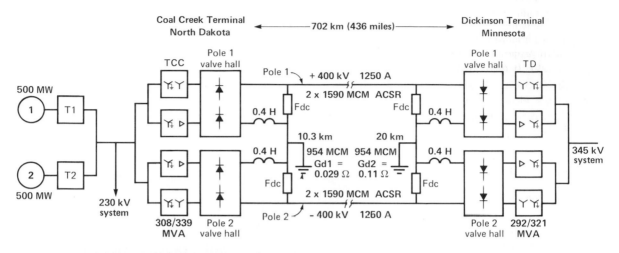

Figure 31-21 Simplified schematic diagram of the bipolar HVDC transmission system delivering 1000 MW over a distance of 702 km. The output from two 500 MW turboalternators is stepped up to 230 kV and transmitted to the Coal Creek Terminal where the power is converted to dc. The on-load tap-changing converter transformers TCC are connected wye-wye and wye-delta for 12-pulse converter operation.

The 0.4 H smoothing inductors are in series with the grounded lines, thus significantly reducing the insulation requirements. The dc filters Fdc, each composed of a 48.8 mH inductor in series with a 1 μF capacitor bank, prevent the 12th harmonic voltage from reaching the dc lines.

The positive and negative lines consist of two bundled conductors. The dc grounds are situated at 10.3 km and 20 km from the respective terminals. Under normal conditions the line currents are controlled automatically so that the ground current is 20 A or less. However, if one pole is out of service for short periods, the ground current can be as high as 1375 A.

The 12-pulse inverter station (Dickinson Terminal) feeds into a 345 kV, 60 Hz system, and tap-changing converter transformers TD are used to regulate the inverter voltage level.

The control system is arranged to operate each terminal unmanned from a control center located in Minnesota.

QUESTIONS AND PROBLEMS

Practical level

31-1 Give three examples where dc power transmission is particularly useful.

31-2 Name the principal components making up a dc transmission system.

31-3 Which harmonics occur on the ac side of a converter? On the dc side?

31-4 What is the purpose of the large dc line inductors?

31-5 A dc transmission line operating at 50 kV, carries a current of 600 A. The terminal contains a single 3-phase, 6-pulse converter.

a. Calculate the approximate value of the secondary ac line voltage of the converter transformer.

b. What is the effective value of the secondary line current?

31-6 The bipolar line shown in Fig. 31-10a operates at a potential of ± 150 kV. If the dc line currents are respectively 600 A and 400 A, calculate:

a. the power transmitted between the two ac networks;

b. the value of the ground current.

31-7 The transmission line shown in Fig. 31-5 possesses a resistance of 10 Ω. The rectifier (converter 1) produces a dc voltage of

Figure 31-22   This compact HVDC converter terminal in Queens, New York, is being used to obtain performance, reliability, and operating data for HVDC installations in urban areas. This, prototype installation will also be used in
the development of compact cables, filters, and related equipment. *(Electric Power Research Institute)*

102 kV while the inverter generates 96 kV.

a. Calculate the dc line current and the power transmitted to network 2.

b. If the gates of the inverter are fired a little earlier in the cycle, will the dc line current increase or decrease? Explain.

c. If the inverter gates are fired so that the inverter generates 110 kV, will the power flow reverse? Explain.

31-8 When a short-circuit occurs on a dc line, the current in the fault itself is less than the full load current. Explain.

31-9 Why is a communications link needed between the converter stations of a dc line?

Intermediate level

31-10 The converters shown in Fig. 31-2 are identical 3-phase 6-pulse units, producing a voltage E_d of 50 kV and a current I_d of 1200 A.

a. Calculate the dc current per valve (bridge arm).

b. What is the approximate peak inverse voltage across each valve?

31-11 Referring to Section 31-19, part 4, calculate the line current per pole on the Pacific Intertie, and estimate the total reactive power absorbed by each converter station.

31-12 The ground electrode of a bipolar converter station is located 15 km from the station, and possesses a ground resistance of 0.5 Ω. If the line currents in each pole are respectively 1700 A and 1400 A, calculate the power loss at the electrode.

31-13 Referring to Fig. 31-11, it is given that E_d = 450 kV, I_d = 1800 A and the two dc smoothing inductors L each have an inductance of 0.5 H. If a short-circuit occurs between line and ground close to the rectifier station, calculate the magnitude of the rectifier current after 5 ms, assuming the gate triggering remains unaltered.

31-14 Each pole of the bipolar Nelson project (Fig. 31-18) is composed of two conductors (2-conductor bundle) of ACSR cable. Each conductor is composed of 72 strands of aluminum (diameter 0.16 in) and a central 7-strand (diameter 0.1067 in) steel core. Each 2-conductor bundle carries a nominal current of 1800 A over a distance of 550 miles. The voltage at the rectifier terminal is 450 kV. Neglecting the presence of the steel core, calculate:
  a. the effective cross section of the 2-conductor bundle [in^2] ;
  b. the line resistance of the 2-conductor bundle at a temperature of 20°C;
  c. the corresponding I^2R loss;
  d. the dc voltage at the inverter terminal;
  e. the efficiency of the line (neglecting corona losses).

Advanced level

31-15 Referring to Fig. 31-12, calculate:
  a. the resonant frequency of the two dc filters;
  b. the value of the respective series impedances.
  c. What is the dc voltage across the capacitors?

31-16 a. In Problem 31-15, if the sixth harmonic voltage generated on the dc side of the inverter is 20 kV, calculate the value of the corresponding harmonic current.
  b. Calculate the value of the harmonic voltage across the 0.5 H inductor.
  c. What is the value of the 360 Hz voltage at the input to the second 0.5 H line inductor?

31-17 In a dc transmission system, the rectifier and inverter stations have the typical E-I characteristics given in Figure 31-6. The ac line voltage at the inverter is 68 kV and the firing angle β is 28°. The dc transmission line possesses a resistance of 2 Ω and carries a current of 1200 A. If the ac line voltage at the rectifier station is 75 kV, calculate:
  a. firing angle α;
  b. the reactive power absorbed by the rectifier station.
  c. Would a tap-changing converter transformer be useful in this case? Explain.

REFERENCES

BOOKS

Allegheny, 1961. *Electrical materials handbook.* Pittsburgh: Allegheny Ludlum Steel Corp.

ANSI, 1971. *Schedules of preferred ratings and related required capabilities for ac high-voltage circuit breakers rated on a symmetrical current basis (ANSI C 37.06-1971).* New York: American National Standards Institute.

——, 1976. *Metric Practice* (ANSI Z210.1-1976)/(IEEE Std 268-1976)/(ASTM E380-76). New York: Institute of Electrical and Electronic Engineers.

ASME, 1968. *Letter symbols for quantities used in electrical science and electrical engineering.* New York: American Soc. of Mech. Eng.

ASTM, 1979. *Annual Book of ASTM standards: Part 6 Copper and copper alloys.* Philadelphia: American Society for Testing and Materials.

Bean, R.L.; Chackan, N.; Moore, H.R. and Wentz, E.C. 1959. *Transformers for the electric power industry.* New York: McGraw.

Bedford, B.D. and Hoft, R.G. 1964. *Principles of inverter circuits.* New York: Wiley.

Beeman, D. 1955. *Industrial power systems handbook.* New York: McGraw.

Biddle, 1967. *"Getting down to earth . . . " A manual on earth resistance testing for the practical man.* Plymouth Meeting: James G. Biddle Co.

Bloomquist, W.C. (ed). 1950. *Capacitors for industry.* New York: Wiley.

Boll, R. 1978. *Soft magnetic materials.* Philadelphia: Heyden.

Bonneville Power Administration (1976). *Transmission line reference book HVDC to $\pm$ 600 kV.* Palo Alto: Electric Power Research Institute.

Bosich, J.F. 1970. *Corrosion prevention for practicing engineers.* New York: Barnes and Noble.

Boylestad, R.L. 1977. *Introductory circuit analysis.* Columbus: Merrill.

Bruins, P.F. 1968. *Plastics for electrical insulation.* New York: Wiley.

Cairns, E.J. and McBreen, J. 1975. *Advanced batteries: candidates for vehicle propulsion.* GMR-1871. Warren, Mich.: Research Labs, Gen. Motors.

Canadian Standards Association, 1979. *Canadian metric practice guide Z 234.1-79.* Toronto: CSA.

Considine, D.M. (ed). 1977. *Energy technology handbook.* New York: McGraw.

Dunning, J.S.; Bradley, T.G. and Zeitner, E.J. 1976. *Development of compact lithium/iron disulfide electrochemical cells.* GMR-2168. Warren, Mich.: Research Labs, Gen. Motors.

Edison Electric Institute, 1973. *Questions and answers about the electric utility industry.* New York: Edison Electric Institute.

Elgerd, O. 1971. *Electric energy systems theory: an introduction.* New York: McGraw.

Federal Power Commission, 1978. *Electric power statistics.* Washington: Superintendent of Documents.

Fink, D.G. and Beaty, H.W. (eds). 1978. *Standard handbook for electrical engineers.* New York: McGraw.

General Electric, 1975. *Transmission line reference book* 345 kV *and above.* New York: Fred Weidner and Son Printers Inc.

——, 1971. *Nickel cadmium battery application engineering handbook (GET 3148).* Gainesville: General Electric Company Battery Products Section.

Gingrich, H.W. 1979. *Electrical machinery, transformers, and controls.* Englewood Cliffs: Prentice-Hall.

Gyugyi, L. and Pelly, B.R. 1976. *Static power frequency changers.* New York: Wiley.

Hogerton, J.F. 1963. *Atomic fuel.* Oak Ridge: U.S. Atomic Energy Commission.

Holden, A. 1965. *The nature of solids.* New York: Columbia University Press.

IEEE, 1972. *IEEE standard dictionary of electrical and electronic terms.* New York: Wiley.

——, 1972. *IEEE recommended practice for grounding of industrial and commercial power systems.* New York: Inst. Electrical and Electronic Eng.

——, 1973. *IEEE standard and general requirements for distribution, power, and regulating transformers.* (IEEE Std 462-1973/ANSI C57. 12.00). New York: Inst. Electrical and Electronic Eng.

——, 1974. *IEEE Standard practices and requirements for thyristor converters for motor drives.* New York: Inst. Electrical and Electronic Eng.

——, 1979. *IEEE Standard metric practice.* New York: Inst. Electrical and Electronic Eng.

——, 1968. *IEEE guide for transformer impulse tests.* New York: Inst. Electrical and Electronic Eng.

——, 1971. *Graphic symbols for electrical and electronics diagrams.* New York: Inst. Electrical and Electronic Eng.

Indiana, 1963. *Design and application of permanent magnets, Manual 7.* Valparaiso: Indiana General Corp.

Jackson, H.W. 1976. *Introduction to electric circuits.* Englewood Cliffs: Prentice-Hall.

Johnson, R.C. 1971. *Electrical wiring.* Englewood Cliffs: Prentice-Hall.

Kimbark, E.W. 1971. *Direct current transmission.* New York: Wiley.

——, 1956. *Power system stability.* New York: Dover.

Kip, A.F. 1969. *Fundamentals of electricity and magnetism.* New York: McGraw.

Kosow, I.L. 1972. *Electric machinery and transformers.* Englewood Cliffs: Prentice-Hall.

——, 1973. *Control of electric machines.* Englewood Cliffs: Prentice-Hall.

Kurtz, E.B. and Showmaker, T.J. 1976. *The lineman's and cableman's handbook.* New York: McGraw.

Levine, J.N. 1961. *Selected papers on new techniques for energy conversion.* New York: Dover.

Liptai, R.G. and Pearson, J.W. 1975. *Metrication - managing the industrial transition.* Philadelphia: American Society for Testing and Materials.

Lye, R.W. (ed). 1976. *Power converter handbook.* Peterborough: Can. Gen. Electric Co. Ltd.

Marbury, R.E. 1949. *Power capacitors.* New York: McGraw.

Matsch, L.W. 1977. *Electromagnetic and electromechanical machines.* New York: Dun-Donnelley.

Morganite, 1961. *Carbon brushes for electrical machines.* London: Morganite Carbon Limited.

Moskowitz, L.R. 1976. *Permanent magnet design and application handbook.* Boston: Cahners.

NEMA, 1970. *Electrical insulation terms and definitions.* New York: Insulating materials division, National Elect. Manufacturers Ass'n.

———, 1978. *Motors and Generators* (ANSI/NEMA No. MG1-1978). Washington: National Electrical Manufacturers Association.

Nesbitt, E.A. 1962. *Ferromagnetic domains, a basic approach to the study of magnetism.* Baltimore: Bell Telephone Labs, Inc.

OECD, 1974. *Twenty-seventh survey of electric power equipment.* Paris: Organization for Economic Co-operation and Development.

Ogorkiewicz, R.M. 1970. *Engineering properties of thermoplastics.* London: Wiley.

Oklahoma State University, 1977. *Power factor improvement.* Stillwater: OSU Elect. Power Technology Dept.

Olson, E. 1966. *Applied Magnetism.* New York: Springer-Verlag.

Pearman, R.A. 1980. *Power electronics, solid state motor control.* Reston: Reston Publ. Co.

Pelly, B.R. 1971. *Thyristor phase-controlled converters and cycloconverters.* New York: Wiley.

Ramshaw, R.S. 1973. *Power electronics thyristor controlled power for electric motors.* London: Chapman and Hall/Halsted Pr.

Richardson, D.V. 1978. *Rotating electric machinery and transformer technology.* Reston: Reston Publ. Co.

Russel, C.R. 1967. *Elements of energy conversion.* New York: Pergamon Press.

Sanford, R.L. and Cooter, I.L. 1962. *Basic magnetic quantities and the measurement of the magnetic properties of materials.* Washington: Nat. Bur. Stds., U.S. Dept. Comm.

Schaefer, J. 1965. *Rectifier circuits.* New York: Wiley.

Schonland, B. 1964. *The flight of thunderbolts.* London: Clarendon Press.

Seel, F. 1963. *Atomic structure and chemical bonding.* New York: Wiley.

Stevenson, W.D. Jr. 1975. *Elements of power system analysis.* New York: McGraw.

Stigant, S.A. and Franklin, A.C. 1973. *The J & P transformer book.* Boston Newnes-Butterworths.

Tarnawecky, M.Z. 1971. *Manitoba power conference EHV-DC.* Winnipeg: University of Manitoba.

TVA, 1975. *Annual report of the Tennessee Valley Authority, Volume II - appendices.* Knoxville: Tennessee Valley Authority.

Uhlmann, E. 1975. *Power transmission by direct current.* New York: Springer-Verlag.

Uman, M.A. 1969. *Lightning.* New York: McGraw.

Underwriters Laboratories, 1978. *Systems of insulating materials* (UL 1446). Northbrook, I11: Underwriters Laboratories Inc.

———, 1977. *Ground-fault circuit interrupters* (UL 943). Northbrook, I11: Underwriters Laboratories Inc.

Veinott, C.G. 1972. *Computer-aided design of Electrical Machinery.* Cambridge: MIT Pr.

Vielstich, W. 1970. *Fuel cells.* London: Wiley.

Vinal, G.W. 1950. *Primary batteries.* New York: Wiley.

———, 1955. *Storage batteries.* New York: Wiley.

Werninck, E.H. (ed). 1978. *Electric motor handbook.* New York: McGraw.

Waddicor, H. 1935. *The principles of electric power transmission by alternating currents.* New York: Wiley.

Westinghouse, 1964. *Electrical transmission and distribution reference book.* Pittsburgh: Westinghouse Electric Corp.

——, 1965. *Electrical utility engineering reference book.* Pittsburgh: Westinghouse Electric Corp.

Wildi, T. 1973. *Understanding Units.* Québec: Sperika Enterprises Ltd.

REFERENCES

ARTICLES AND TECHNICAL REPORTS

Allan, R. (ed). 1975. Power semiconductors. *Spectrum.* Nov.: 37-44.

Allan, J.A. et al 1975. Electrical aspects of the 8750 hp gearless ball-mill drive at St. Lawrence Cement Company. *IEEE Trans. Ind. Gen. Appl.* 1A-11, No. 6: 681-687.

Apprill, M.R. 1978. Capacitors reduce voltage flicker. *Electrical World* 189, No. 4: 55-56.

American National Standard, 1979. Standard nominal diameters and cross sectional areas of AWG sizes of solid round wires used as electrical conductors. *ANSI/ASTM* B 258-65 (Part 6): 536-542.

Bachmann, K. 1972. Permanent magnets. *Brown Boveri Rev.* 9: 464-468.

Balzhiser, R.E. 1977. Capturing a star: controlled fusion power. *EPRI Journal* Dec.: 6-13.

Beaty, H.W. (ed). 1978. Underground distribution. *Electrical World* 189, No. 9: 51-66.

Beaty, H.W. 1978. Charts determine substation grounds. *Electrical World* 189, No. 2: 56-58.

Beutler, A.J.; Staats, G. 1969. A 100 000-Joule system for charging permanent magnets. *IEEE Trans. Ind. and Gen. Appl.* 1GA-5, No. 1: 95-100.

Creek, F.R.L. 1976. Large 1200 MW four-pole generators for nuclear power stations in the USA. *GEC J. Sc. and Tech.* 43, No. 2: 68-76.

Dalziel, C.F. 1968. Reevaluation of lethal electric currents. *IEEE Trans. Ind. and Gen. Appl.* 1GA-4, No. 5: 467-475.

Drake, W.D. and Sarris, A.E. 1968. Lightning protection for cement plants. *IEEE Trans. Ind. Gen. Appl.* 1GA-4, No. 1: pp. 57-67.

Duff, D.L. and Ludbrook, A. 1968. Semiconverter rectifier go high power. *IEEE Trans. Ind. Gen. Appl.* 1GA-4, No. 2: 185-192.

DuPont, 1972. "Tefzel" ETFE fluoropolymer: an exciting new resin. *Journal of "Teflon" 13: No. 1.*

Electrical World 1978. Statistical Report. *Electrical World* 189, No. 6: 75-106.

Elliott, T.C. (ed). 1976. Demand control of industry power cuts utility bills, points to energy savings. *Power* 120, No. 6: 19-26.

Engstrom, L.; Mutanda, N.M. Adams, N.G. and Flisberg, G. 1975. Refining copper with HVDC. *Spectrum* 12, No. 12: 40-45.

Fagenbaum, J. 1980. Cogeneration: an energy saver. *Spectrum* 17, No. 8: 30-34.

Fickett, A. 1976. Fuel cels - versatile power generators. *EPRI Journal*, April: 14-19.

Friedlander, G.D. 1977. UHV: onward and upward. *Spectrum* 14, No. 2: pp. 57-65.

Gazzana-Priaroggia, P.; Piscioneri, J.H. and Margolin, S.W. 1971. The Long Island Sound submarine cable interconnection. *Spectrum* 8, No. 10: pp. 63-71.

Goodbrand, W. and Ross, C.A. 1972. Load-rejection testing of steam turbine generators. *Ontario Hydro Rsrch Qtly.* 24, No. 2: 1-7.

Graf, K. 1973. AC pump drives with static-converter slip-power recovery system for the Lake Constance water supply scheme. *Siemens Rev.*, XL: 539-542.

Hand, C. (ed). 1977. Batteries. *Canadian Electronics Engineering*, May: 19-28.

Harris, S.W. 1960. Compensating dc motors for fast response. *Control Enginnering* Oct. 1960: 115-118.

Herbst, W.; Käuferle, F.P. and Reichert, K. 1974. Controllable static reactive power compensation. *Brown Boveri Rev.* 61: 433-439.

Hoffmann, A.H. 1969. Brushless synchronous motors for large industrial drives. *IEEE Trans. Ind. and Gen. Appl.* 1GA-5, No. 2: 158-162.

Holburn, W.W. 1970. Brushless excitation of 660 MW generators. *Journal of Science and Technology.* 37, No. 2: 85-90.

Hossli, W. (ed). 1976. Large steam turbines. *Brown Boveri Rev.* 63: 84-147.

Iwanusiw, O.W. 1970. Remanent flux in current transformers. *Ont. Hydro Rsrch Qtly.* 3rd Qtr: 18-21.

Jacobs, A.P. and Walsh, G.W. 1968. Application considerations for SCR dc drives and associated power systems. *IEEE Trans. Ind. Gen. Appl.* 1GA-4, No. 4: pp. 396-404.

Javetski, J. 1978. Cogeneration. *Power.* 122, No. 4: 35-40.

Joos, G. and Barton, H.B. 1975. Four-quadrant DC variable-speed drives - design considerations. *Proc. IEEE,* 63, No. 12: 1660-1668.

Joyce, J.S. 1974. Factors influencing reliability of large generators for nuclear power plants. *IEEE Trans. Pwr App. Syst.,* PAS-93, No. 1: 210-219.

Kolm, H.H. and Thornton, R.D. 1973. Electromagnetic flight. *Scientific American.* 229, No. 4: pp. 17-25.

Krick, N. and Noser, R. 1976. The growth of turbo-generators. *Brown Boveri Rev.* 63: 148-155.

Kronenberg, K.J. and Bohlmann, M.A. 1968. Stability of permanent magnets. *Applied Magnetics; Indiana Current Corp.* 15, No. 1.

Lee, R.H. 1975. The effect of color on temperature of electrical enclosures subject to solar radiation. *IEEE Trans. Ind. Appl.* IA-11, No. 6: 646-653.

Lengyel, G. 1962. Arrhenius theorem aids interpretation of accelerated life test results. *Can. Electronics Engineering.* Nov.: 35-39.

Lyman, J. (ed). 1975. Battery technology. *Electronics.* April: 75-82.

Machine Design, 1979. Electrical and electronics reference issue. *Machine Design Magazine* 51, No. 11: entire issue.

Manian, V.S. 1971. Electric water heaters for high-usage residences. *Ontario Hydro Rsrch Quarterly,* 3rd quarter: 7-13.

McCormick, L.S. and Hedding, R.A. 1974. Phase-angle regulators optimize transmission line power flows. *Westinghouse Engineer.* July: 87-91.

McLaughlin, M.H. and Vonzastrow, E.E. 1975. Power semiconductor equipment cooling methods and application criteria. *IEEE Trans. Ind. Appl.* IA-11, No. 5: 546-555.

Matta, U. (ed). 1975. Industrial electro-heat. *Brown Boveri Rev.* 62: 4-67.

Methé, M. 1971. The distribution of electrical power in large cities. *Canadian Engineering Journal.* October: 15-18.

Minnick, L. and Murphy, M. 1976. The breeder - when and why. *EPRI Journal* March: 6-11.

Moor, J.C. 1977. Electric drives for large compressors. *IEEE Trans. Ind. Gen. Appl.* 1A-13, No. 5: pp. 441-449.

Patterson, W.A. 1977. The Eel River HVDC scheme. *Can. Elec. Eng. J.* 2, No. 1: 9-16.

Quinn, G.C. 1977. Plant primary substation trends. *Power* 121, No. 3: 29-35.

Rieder, W. 1971. Circuit breakers. *Scientific American* 224, No. 1: 76-84.

Robinson, E.R. 1968. Redesign of dc motors for applications with thyristors power supplies. *IEEE Trans. Ind. Gen. Appl.* 1GA-4, No. 5: 508-514.

Robichaud, G.G. and Tulenko, J.S. 1978. Core design: fuel management as a balance between in-core efficiency and economics. *Power* 122, No. 5: 52-57.

Schwieger, R. (ed). 1978. Industrial Boilers - what's happening to-day. *Power* 122, No. 2: s1-s24.

Sebesta, D. 1978. Responsive ties avert system breakup. *Electrical World* 180, No. 7: 54-55.

Selzer, A. 1971. Switching in vacuum: a review. *Spectrum* 8, No. 6: pp. 26-37.

Sullivan, R.P. 1978. Uranium: key to the nuclear fuel cycle. *Power* 122, No. 4: 58-63.

Summers, C.M. 1971. The conversion of energy *Scientific American* 224, No. 3: 148-160.

Vendryes, G.A. 1977. Superphénix: a full-scale breeder reactor. *Scientific American* 236, No. 3: pp. 26-35.

Waldinger, H. 1971. Converter-fed drives. *Siemens Rev.* 38: 387-390.

Woll, R.F. 1964, High temperature insulation in ac motors. *Westinghouse Engineer* Mar. 1964: 1-5.

Woll, R.F. 1975. Effect of unbalanced voltage on operation of polyphase induction motors. *IEEE Trans. Ind. Appl.* IA-11, No. 1: 38-42.

Woll, R.F. 1977. Electric motor drives for oil well walking beam pumps. *IEEE Trans. Ind. Gen. Appl.* 1A-13, No. 5: 437-441.

Woodson, R.D. 1971. Cooling towers. *Scientific American* 224, No. 5: 70-78.

Zimmermann, J.A. 1969. Starting requirements and effects of large synchronous motors. *IEEE Trans. Ind. and Gen. Appl.* 1GA-5, No. 2: 169-175.

APPENDIX

The conversion charts listed in this Appendix make it very easy to convert from one unit to any other. Their use is explained in Chapter 1, Section 10.

Quantities such as AREA, MASS, VOLUME, and so forth, are listed in alphabetical order for quick reference. Multipliers between units are either exact, or accurate to ± 0.1 percent.

Examples at the bottom of each page are intended to assist the reader in applying the conversion rule which basically states:

WITH THE ARROW — MULTIPLY

AGAINST THE ARROW — DIVIDE

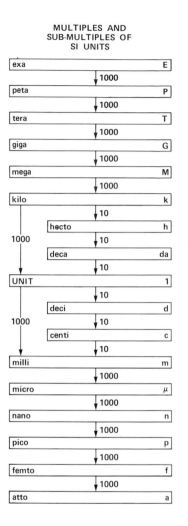

MULTIPLES AND SUB-MULTIPLES OF SI UNITS

exa   E
↓ 1000
peta   P
↓ 1000
tera   T
↓ 1000
giga   G
↓ 1000
mega   M
↓ 1000
kilo   k
↓ 10
hecto   h
↓ 10
deca   da
↓ 10
UNIT   1
↓ 10
deci   d
↓ 10
centi   c
↓ 10
milli   m
↓ 1000
micro   μ
↓ 1000
nano   n
↓ 1000
pico   p
↓ 1000
femto   f
↓ 1000
atto   a

(1000 from kilo to milli; 1000 from UNIT to milli)

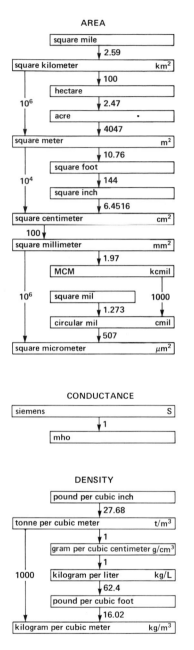

AREA

square mile
↓ 2.59
square kilometer   km²
↓ 100
hectare
↓ 2.47
acre
↓ 4047
square meter   m²
↓ 10.76
square foot
↓ 144
square inch
↓ 6.4516
square centimeter   cm²
↓ 100
square millimeter   mm²
↓ 1.97
MCM   kcmil
↓ (1000)
square mil
↓ 1.273
circular mil   cmil
↓ 507
square micrometer   μm²

(10^6 from square kilometer to square meter; 10^4 from square meter to square centimeter; 10^6 from square centimeter to square micrometer)

CONDUCTANCE

siemens   S
↓ 1
mho

DENSITY

pound per cubic inch
↓ 27.68
tonne per cubic meter   t/m³
↓ 1
gram per cubic centimeter   g/cm³
↓ 1
kilogram per liter   kg/L
↓ 62.4
pound per cubic foot
↓ 16.02
kilogram per cubic meter   kg/m³

(1000 from tonne per cubic meter to kilogram per cubic meter)

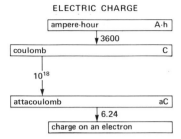

ELECTRIC CHARGE

ampere-hour   A·h
↓ 3600
coulomb   C
↓ 10^{18}
attacoulomb   aC
↓ 6.24
charge on an electron

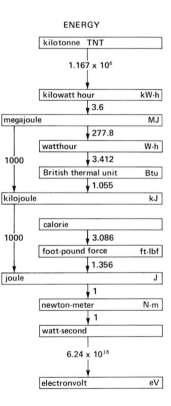

ENERGY

kilotonne TNT
↓ 1.167×10^6
kilowatt hour   kW·h
↓ 3.6
megajoule   MJ
↓ 277.8
watthour   W·h
↓ 3.412
British thermal unit   Btu
↓ 1.055
kilojoule   kJ
↓ (1000)
calorie
↓ 3.086
foot-pound force   ft·lbf
↓ 1.356
joule   J
↓ 1
newton-meter   N·m
↓ 1
watt-second
↓ 6.24×10^{18}
electronvolt   eV

(1000 from megajoule to kilojoule; 1000 from kilojoule to joule)

Example: Convert 1590 MCM to square inches.

Solution: 1590 MCM = 1590 (÷ 1.97) (÷ 100) (÷ 6.4516) in² = 1.25 in².

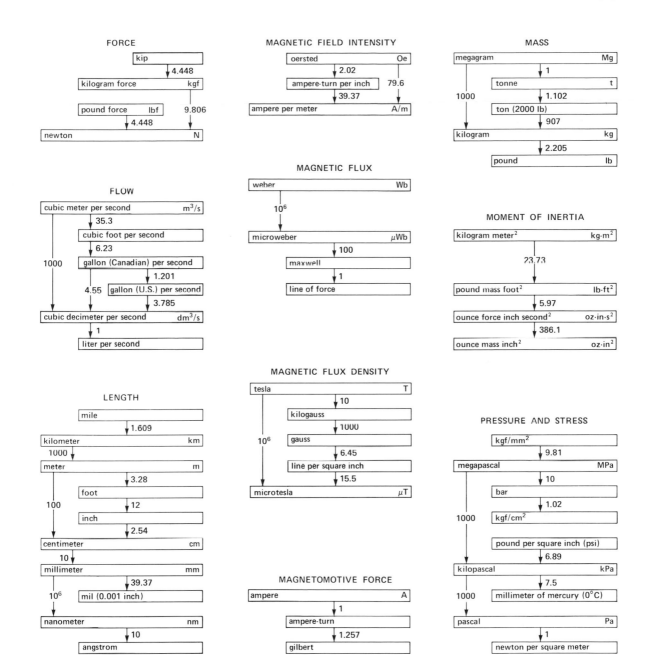

FORCE

kip
↓ 4.448
kilogram force   kgf
pound force   lbf   9.806
↓ 4.448
newton   N

FLOW

cubic meter per second m^3/s
↓ 35.3
cubic foot per second
↓ 6.23
gallon (Canadian) per second
↓ 1.201
1000   4.55   gallon (U.S.) per second
↓ 3.785
cubic decimeter per second dm^3/s
↓ 1
liter per second

LENGTH

mile
↓ 1.609
kilometer   km
1000 ↓
meter   m
↓ 3.28
foot
100   ↓ 12
inch
↓ 2.54
centimeter   cm
10 ↓
millimeter   mm
↓ 39.37
10^6 mil (0.001 inch)
nanometer   nm
↓ 10
angstrom

MAGNETIC FIELD INTENSITY

oersted   Oe
↓ 2.02
ampere-turn per inch   79.6
↓ 39.37
ampere per meter   A/m

MAGNETIC FLUX

weber   Wb
↓ 10^6
microweber μWb
↓ 100
maxwell
↓ 1
line of force

MAGNETIC FLUX DENSITY

tesla   T
↓ 10
kilogauss
↓ 1000
10^6 gauss
↓ 6.45
line per square inch
↓ 15.5
microtesla μT

MAGNETOMOTIVE FORCE

ampere   A
↓ 1
ampere-turn
↓ 1.257
gilbert

MASS

megagram   Mg
↓ 1
tonne   t
1000   1.102
ton (2000 lb)
↓ 907
kilogram   kg
↓ 2.205
pound   lb

MOMENT OF INERTIA

kilogram meter^2 kg·m^2
↓ 23.73
pound mass foot^2 lb·ft^2
↓ 5.97
ounce force inch second^2 oz·in·s^2
↓ 386.1
ounce mass inch^2 oz·in^2

PRESSURE AND STRESS

kgf/mm^2
↓ 9.81
megapascal   MPa
↓ 10
bar
↓ 1.02
1000   kgf/cm^2
pound per square inch (psi)
↓ 6.89
kilopascal   kPa
↓ 7.5
1000   millimeter of mercury (0°C)
pascal   Pa
↓ 1
newton per square meter

Example: Convert 580 psi to megapascals.
Solution: 580 psi = 580 (x 6.89) (÷ 1000) MPa = 4 MPa.

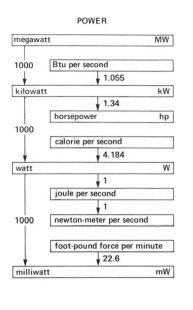

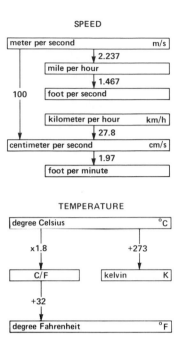

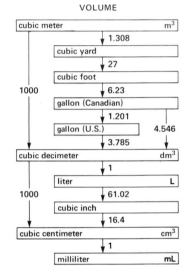

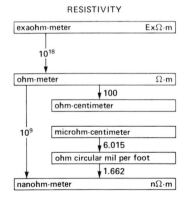

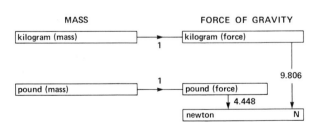

Example: Calculate the force of gravity (in newtons) that acts on a mass of 9 lb.

Solution: 9 lbm → 9 (x 1) (x 4.448) N = 40 N.

Note: The following numerical answers are usually rounded off to an accuracy of ± 1%. Answers that are preceded by the symbol ≈ are accurate to ± 5%.

CHAPTER 1. 7) MW 8) TJ 9) mP 10) kHz 11) GJ
12) mA 13) mWb 14) cm 15) L 16) mg 17) μs
18) mK 19) mrad 21) mT 22) mm 23) r 24) MΩ
25) MP 26) ms 27) pF 28) kV 29) MA 30) kA
31) km 32) nm 33) mL 34) L/s or m^3/s 35) Hz
36) radian, (rad) 37) weber, (Wb) 38) kg/m^3 39) watt, (W)
40) kelvin (K) or °C 41) kg 62) 11.95 yd^2 63) 126.9 mm^2
64) 2.549 in^2 65) 6.591 mm^2 66) 2.59 km^2 67) 76.77 Btu/s
68) 0.746 kW 69) 7.079 m^3 70) 13.56 x 10^6 μJ
71) 4.536 kgf 72) 0.93 J 73) 12 kilogauss 74) 1.417 kg
75) 6049.6 A/m 76) 3.107 mile 77) 288 000 C 78) 111.2 N
79) 11.34 kg 80) 6614 lb 81) 0.001 Wb 82) 8304 kg/m^3
83) 67.7 mbar 84) 1.378 x 10^6 Pa 85) 482.3 x 10^3 N/m^2
86) 1.57 rad/s 87) 393 K 88) 366.3 K 89) 120°C

CHAPTER 2. 2a) E_{16} = - 80 V 2b) E_{25} = + 80 V 2c) E_{52} = 0
3) + 20 V, 0 V, - 30 V, + 30 V 6) 20 V 7) 480 V, 100 Hz;
120 V, 25 Hz; 480 V, 50 Hz 8) 30 V 9) 1.8 V 11) 320, 160,
112 12) 480 kA/m 13a) 960 N 13b) 960 N

CHAPTER 3. 1) 392 N 2) 735 N, 2940 J 4) 60 N·m
5) 336 hp 6) 14.5 hp 7a) 134 hp 7b) 83.3%
7c) 68 256 Btu/h 8) 415.7 J 9a) 1096 J, 4384 J 9b) 3288 J
9c) 242 kJ 11c) 50 N·m 13) zero 14) 209 W 15) 209 W
16) 15.54 N·m, 2.53 hp 17a) 13.1 N·m 17b) 88.8 kJ
17c) 2195 W 17d) 2400 W 18a) 1222 N·m 18b) 1332 N·m
19a) 475 N·m 19b) 4455 N 20) 3.04 MJ 21) 7.68 MJ
22) 239 r/min 23) 47.8 r/min, 1156 ft·lb

CHAPTER 4. 8) 180°C 12) 1 year 13) 32 years
14) 560 V 15a) 300 kV 15b) 75 kV 15c) 5 kV 16) 7 mm
20) 750 kV 21a) 30 kA, 6 x 10^6 MW 21b) 47.5 h
22) 1.94°C 23) 216°C

CHAPTER 5. 14) No. 9 15) No. 10 16) 1.77 Ω
17) 254 mm^2 18) 25% 19) No. 4 21) 296 cmil
22) 7.39 km, 21.3 kg 23) 4.8 cmil, 1.6 omil 24) 35 Ω
25a) 1.33 Ω 25b) 19.1 kW 26) 3.65 Ω 27) 0.17 N, 1.07 N
28) 2.22 MW, 13.2 MW, 10.3 MW 29) 200 Ω 30) 13.3 kN,
83.6 kN 31) 86.1°C 32) 100 Ω 33a) 1.63 mm
33b) 51.5 kW 34) 424 A

CHAPTER 6. 1a) 4800 1b) 336 4) 333 000, 10 000, 10
6) 0.51 J 7) 2228 Btu/h 8) 980 W 9) 2000 W, 250 W
10) 44 11) 20 loops per second 12) 315 J 13a) 2000 A
13b) 2 mWb 13c) 2J 14a) 18 kJ/m^3 14b) 36 kJ/m^3
14c) 40 kJ/m^3 15a) 203 A 15b) 264 A 17a) 2.5 in
17b) 4.5 in^2 17c) 11.25 in^3 18a) 0.17 T 18b) 22 in^3

CHAPTER 7. 1b) 31.3 V, 141 V, - 200 V 2) 70.7 V
3a) 12 A 3b) 169.7 V 3c) 1440 W 3d) 2880 W 4) 23 Hz
5) 36°, 1.67 ms 6a) I_1 lags I_3 by 60° 6b) I_3 lags I_2 by 90°
6c) E lags I_1 by 150° 7a) 7 A 7b) 6.7 A 7c) 4.36 A
7d) 4.36 A 8a) I_a lags behind I_2 by 128° 8b) I_2 leads I_b 26°
8c) I_c leads I_2 173.4° 8d) I_2 leads I_d 6.6° 9a) 5.66 V at + 45°
9b) 16.5 V at + 76° 9c) 46.8 V at + 160° 9d) 16 V at - 90°
10a) 12 V·s 10b) 3 A, 6 A 11b) 2400 W 12) 7.07 A
13a) 120 V 13b) 72.1 V 14a) 3200 J 14b) 0 W, 3200 J

CHAPTER 8. 2) 100 kVA 6) 64.3% 7) 667 kVA, 291 kvar
8) 4.34 kvar 9a) 1440 W 9b) 1440 VA 9c) 2880 W
9d) 1/120 s 10a) 1440 var 10b) 1440 VA 10c) 1440 W
10d) 1440 W 10e) 1/240 s 11a) A is an active source
11b) C is the active source 11c) F is the reactive source
11d) J is the reactive source 11e) K is the active source
11f) M is the reactive source 12) 7.2 A, 9.6 A, 12 A 13) 2665 var
14a) 2765 W 14b) 745 var 14c) 2864 var 14d) 11.9 A
14e) 96.5% 15a) 3.71 Ω 15b) 6.32 Ω 15c) 12.16 Ω
16a) 320 kW 16b) 240 kvar 17a) 1200 W 17b) 500 var
17c) 1300 VA 17d) 0,923 18a) 2 kW 18b) zero

CHAPTER 9. 1) 4157 V 2a) 100, 0, 50, 50, 86.6 V
2b) +, 0, +, +, - 2c) + 50, - 86.6, - 100, + 50, + 86.6 V
3) yes 4a) 358 V 4b) 23.87 A 4c) 25.6 kW 5a) 694 A
5b) 13.2 kV 5c) 9.16 MW 5d) 27.5 MW 5e) 19 Ω
6) a - b - c 7) 26 kVA 8) 120 V 9a) 13 kW 9b) 6.5 kW
10) single-phase 11a) 41.6 A 11b) 41.6 A 11c) 2.89 Ω
12a) 160 Ω 12b) 480 Ω 13) 0.9 Ω 14) 8 Ω 15a) 89.5%
15b) 62.3 kVA 15c) 37.2 kvar 15d) 80.2% 16a) 14.6 kW,
7.78 kvar, 16.5 kVA 16b) 270 V 17a) XZY 18a) 5 A
18b) 19.9 kvar 19) lamp connected to Z 20) 20 kW
21a) 18 A 21b) 130 A 21c) 72.1 A 22) 18.3 A
23a) 4.63 Ω, 3.44 Ω 23b) 36.7° 24a) 9.6 Ω 24b) 7.72 Ω,
5.79 Ω

CHAPTER 10. 2) 1190 7) 63.2% 8) 0.20 hp
9a) 6097 VA 9b) 82% 10a) 21.32 kW 10b) 24.3 hp
10c) 64 kW·h 11) 1800 W 13) ± 5.6 kW·h 14a) 15.5%
14b) 88.4 A 16) 0.9 mA 17a) 67 450 Ω 17b) 50 V
17c) 111 V, 0 V 18) 27.8 V 19) 10 r/min 20a) meter reads
1 percent high 20) no change

CHAPTER 11. 11) 120 V 12) 60 kV 13a) 110.4 V
13b) 0.353 A, 22.08 A 14) 2.42 kW, 11 A, 22 A
15) 50 A, 1250 A 16) 1.88 mWb 18a) 360 V
18b) 2.25 mWb 18c) 0.9 mWb 19a) zero 19b) 520 V
19c) additive 20a) short-circuit 23a) 72 Ω 23b) 0.115 Ω
24a) 126.4 V 24b) 95 A, 5.22 A 25a) 466.9 kW 25b) 99.1%
27) 9 mWb 29a) 84.1 Ω 29b) 3.04% 29c) 23.36 mΩ
29d) 3.04% 29e) 976 W 29f) 1.3%, 2.75% 30a) 1.9 Ω,
36.6 Ω 30b) 453.6 Ω 30c) 8.4% 31) 99.08% 32) 19.4 t
33a) 28.9 W/kg 33b) 156 W/kg

CHAPTER 12. 5) 300 turns 6) 0.2 V 7) 1300 kVA
8) H1 - X2 or H2 - X1 together 9) 92 kVA 10) 13 mA
11a) 0.45 V 11b) 2.25 mV 11c) 250 A/5 A

CHAPTER 13. 2a) 20.8 A, 433 A 2b) 20.8 A, 250 A
3) 1506 A, 65 kA 4) 17.7 kA, 2175 A 5a) 1278 A 5b) yes
6a) delta-delta 6b) 577 A, 48.1 A 6c) 333 A, 27.8 A
7a) 236 kVA 7b) 1.89 kV 8a) no 8b) 433 kVA
9a) 347, 600, 347 V

CHAPTER 14. 9a) 103 V 9b) 288 W 9c) 10.4 N·m
10a) 138 V 10b) polarity reverses 10c) voltage increases by less
than 10% 11a) 5000 A 11b) 7800 A 12) 2 A 13) A = 0 V,
B = 18 V, C = 0 V 14) at 90° E_A = 20 V, at 120° E_A = 18 V
16) 276.5 V 17a) 12 brush sets, 150 A 19) E_{XY} is (+)
20a) E_{34} is (-) 21a) 292 V 21b) 0.436 T 21c) 530 μs
22) 1.67 mV

CHAPTER 15. 9a) 221 V 9b) 13.8 kW 9c) 13.26 kW,
17.8 hp 10a) 1533 A 10b) 1.85 Ω 11a) 2975 A
11b) 2400 A 12) 144 V, 9.6 V 13a) 11.7 kW, 94%
13b) 10 A 13c) 8 mΩ, 223 V 14a) 0.48 Ω 14b) 30 kW,
1.9 kW, 0 kW 15a) 0.96 Ω 15b) 45 kW, 8.43 kW, 0 kW
16a) 17.8 kg·m^2 16b) 140 kJ 16c) 70 kJ 19) 2113 ft·lb, 0;
2113 ft·lb, 1056 ft·lb 20) 50 hp, 180 r/min

CHAPTER 16. 7) 131.4 kW, 547 A 8) 22.8 hp 9) 882 A
10a) 28°C 10b) 88°C too hot 12) 82.9% 13) 10.4 kW,
14 hp 14) < 155°C, < 145°C, < 129°C 15a) 139°C
15b) 108°C 16a) 133°C 16b) 105°C 17a) 2.28 A/mm^2
17b) 13.4 W/kg 18a) 58.9 W/kg 18b) 985 cmil/A
19) 26.2 kW, 35 hp 20a) 60 Hz 20b) ≈ 400 W 21) ≈ 384 h
22) 67%

CHAPTER 17. 10a) 360 r/min 10b) no 10c) 20
12) 936 A, 156 A, 55 A 15a) 600 r/min 15b) 564 r/min
16a) 3 V, 45 Hz 16b) 0.67 V, 10 Hz 16c) 1 V, 15 Hz
17a) 15 A, 90 A, 5 A 17b) 882 r/min, 812 N·m 18) 97.2 A
19a) 120 V, 30 Hz 19b) 60 V, 15 Hz 19c) 960 V, 240 Hz
20a) − 8.66 A, + 8.66 A, 0 20b) 86.6 A 21) 3, slot 1 to slot 8
22a) 23.56 m/s 22b) 3.3 V 22c) 196.35 mm 23a) 87.9%
23b) 2298 kW 23c) 34.5 kW 23d) 2251 kW, 30.3 kN·m, 96%
26) 1500 A, 75 N 27) 20 N 28) 38 kW 29a) 68.9 mΩ,
4.49 mΩ 29b) 1067 V, 40 Hz; 16 V, 0.6 Hz 29c) 1035 kvar
29d) 2.07 kW 29e) 50 kW 30a) 18.6 Mvar 30b) 670 kW
30c) 1498 kW 30d) zero 30e) 23.8 kN·m 31a) 400 V
31b) 508 mΩ, 314 kW 31c) 455 A 32) 264.5 mm
34) 100 mm, 120 kW

CHAPTER 18. 10) 78% 12a) 2.25 kW or 3 hp 12b) 54 or
56 13) 438, 280, 315 ft·lb; zero, 675, 765 r/min 14a) 3.8 kW
14b) 586 r/min 14c) 222.4 kW 14d) 5.3 kW 15a) 140 Hz
15b) 54 kW 15c) 41.2 kvar 15d) 75.4 A 15e) 100 hp
16a) 1109 A 16e) 434 kW 16c) 90.2 kW 16d) 5212 A,
83.5 kN·m 16e) 43.3 kN·m 17) 4.7°C 18a) 39.4 s
18b) 105 000 Btu 19a) 456 V, 1740 r/min 19b) ≈ 16 hp
20a) design D 20b) same 21a) 54.5 mΩ 21b) 2.23 Ω
22a) ≈ 1100 N·m 23a) 11.4 kg·m² 23b) 270 kJ 24a) 795 kg,
1509 lb·ft² 24b) 810 r/min, 260 ft·lb 24c) 675 ft·lb
25a) 916 N·m 25b) 1.3 s 25c) 11.3 kJ 25d) 6.7 s
26a) 134 r/min 26b) 11:1 26c) 358 A 26d) 50 t
26e) 728 MJ 26f) 39 min 26g) 91 kW·h

CHAPTER 19. 3a) 20 poles 3b) 360 r/min 6) 12 poles
7) 4 9a) 7500 V, 50 Hz 9b) 37.5 V, 0.25 Hz 12a) 150 A
12b) 50 A 13) 2400 V 14b) 353, 600, 750, 600, 0 kW
15) 807 mm 16a) 145 Ω 16b) 192 Ω 16c) 144 Ω
16d) 10 A 16e) 1750 V 16f) 3031 V 16g) 57.6 kW
16h) 36.9° 17a) 16 kV 17b) 250 A 18a) 44.56°
18b) 2.475° 18c) 7.87 in 19a) 0.45 Ω 19b) 0.412 Ω
19c) 0.916 19d) 1.09 20a) 8130 kW 20b) 720 kW
20c) 24.23 MN·m 20d) 22.7°C 21) 0.409 T 22a) 126 MW
22c) 85.7% leading 23a) 228.5 MW 23b) 7230 A
24) delivers 63.3 Mvar to the bus 25a) 236 r/min 25b) 108°,
1944°

CHAPTER 20. 7) 2300 hp 9a) 2217 kVA 9b) 90.2%
9c) 956 kvar 9d) 32 poles 11a) 333 A 11b) 3 Ω
12) 500 r/min 14a) 36.9° 14b) 2.16 MW 14c) 100%
14d) zero 15a) 300 A, zero degrees 15d) 720 kvar delivered
17a) 3457 kVA 17b) 289 A 17c) 5889 V 17d) 1.44°
17e) 1569 kvar 17f) 9150 hp 18a) 4741 V 18b) 32.8°
19a) 14.6 Ω, 1594 kV 19b) 6807 ft·lb 19c) 142 A
20a) 4269 kW, 45.3 kN·m 20b) 1704 kW, 36.2 kN·m
20c) 40.7 kN·m 20d) ≈ 83 s

CHAPTER 21. 9a) 14 A, 14 A 9b) 33.8° 9c) 26.8 A
9d) 70.7% 10a) no 10b) probably not 12a) 50 Hz
12c) 4.26 J 13a) 8 mhp 13b) 55% 13c) 0.28
13d) 0.79 p.u., 1.06 p.u. 14a) 4.42 ft·lb 14b) 4.44 p.u.
14c) zero 14d) 2.5 15a) 157.5 V 15b) 93° 16a) 0.577 A
16b) 86.6% lagging 16c) 30 W 16d) 37% 17a) 1.94 Ω
17b) 21.2 A, 84 V 17c) 3.2 N·m

CHAPTER 22. 10a) 2 min 10b) 5 s 12) 208 days
16) 28 s 17a) 834 ft·lb, 358 ft·lb 17b) 149 ft·lb

CHAPTER 23. 6a) 1620 V 6b) 200 A 6c) 600 A
7a) 3240 V 7b) 200 A 8a) 540 V 8b) 540 V 8c) 18 A
8d) 9 A 8e) 9720 W 9a) 150 V 9b) 3.75 A 10) 45 kW
11a) 0.72 kW 11b) 99.96% 12a) negative 12b) increasing
13a) 36 A 13b) 170 V 13c) 65 J 13d) 0.1 H 13e) 170 V
14a) 324 V 14b) 243 kW 14c) 750 A 14d) 5.55 ms
14e) 612 V 14f) zero 14g) 45.4 V 15a) 24 kW
15b) 24 kW 15c) 12 A 15d) 400 A 15e) 300 Hz
15f) 167 Ω 16a) zero 16b) 77.7° 16c) 102.3°
17a) 14 kV 17b) 6.3 MW 17c) 367 A 18a) 23.5 Mvar
18a) 40 kV 18b) − 14.64 kV 18c) 54.6 kV 19) 6.25 mH
20a) 340 V 20b) 147° 20c) 30.4 A 21) 1/24 s

22a) 3.29 V·s 22b) 0.47 H 23a) 120 A 23b) 48 kJ
23c) 4 kW 24a) 120 A 24b) 80 A 24c) 200 A
24d) 160 A 25a) 135.2° 25b) 57.5 kW 25c) 57 kvar

CHAPTER 24. 3) 91 Hz 4a) 626 V 4b) 169 kW 4c) 90 A
4d) 220 hp 9a) 30 A 9b) 4.29 hp 9c) 63.7 N·m
9d) 56.4 ft·lb 10b) 5 Hz 11a) 200 V 11b) 283 V
11e) 2 A 12a) 31.3° 12b) 5.1 kvar 12c) 28.6 A
12d) 113 V 13a) 85° 13b) 16.8 kvar 14a) 343 A
14b) 146° 15a) 300 V 15b) 315 V 16a) 150 A
16b) 75 V 16c) 60 V 17a) zero 17b) 150 A 17c) 15 V
18a) 100.7° 18b) 87.3° 19a) 232 A, 240 A 19b) 700 V
20a) 192 V 20b) 250 A 22a) 500 Hz 22b) − 563 V
22c) 43.5° 22d) 73.7 A 22e) 360 Hz 22f) 3000 Hz
22g) 72.5% 22h) 60 A

CHAPTER 25. 1a) quadrant 1 1b) quadrant 4 2) 6.3 hp
3) 26 V, 7.5 Hz 4a) 40 A 4b) 120 A 10a) 18.85 kW
10b) 11.3 kW 10c) 30.15 kW 11a) 1.57, 16, 12.6 kW
12a) 345 V, 45 Hz 12b) 633 V, 82.5 Hz 13) 38 V, 5 Hz
14a) 307 V, 40 Hz 14b) 67.3 kJ 14c) 29.9 kJ 15b) 29.5 ft·lb
16) 166 V 17a) 20.7 W 17h) 36.3 W 18) 184 V
19a) 3600 r/min 19b) 15 kW 19c) 337 V 19d) 7.6 Ω
20) 325 μs 21a) 115 V, 15 Hz 21b) 85 A 22) 31.8 A
23) 50 V, 15 Hz

CHAPTER 26. 6a) 37.5 A 6b) 30 h 6c) 300 A
7) 12.88 V 8) 60, 92 11a) 0.75 Ω 11b) 2 A 12) 12 cells,
4 in series and 3 such groups in parallel 13a) 267 ft³
13b) 1200 A·h 14a) 75 W·h 14b) 12.5 A·h 14c) 25 h
15a) 20 months 15b) 0.2 W·h 17) 210 lb 18a) 12 500 lb
18b) 1818 lb 19a) ≈ 6500 t 19b) ≈ 10 000 m³ 20) 36%,
≈ 320 V, ≈ 700 A

CHAPTER 27. 2) $33.80 5) $1470 7) 4.33 kW
8) 109 million dollars 9a) 72 kvar, 104 kVA 9b) 75 kW,
52 kvar 9c) 12.5% 10a) 243 kvar 10b) 34% 11a) $46.64/h
11b) $4205 12a) 30 cent/kW·h 12b) 2.4 cents
13) 6.45 cent/kW·h

CHAPTER 28. 7) 1176 MW 11a) 41 200 m³/s
11b) 40 400 MW 11c) 312 mi³ 12) 15.6 h 13a) 60 MW,
50 MW 13b) 80 MW, 30 MW 14a) 5400 MW 14b) 2615 Mvar
14c) 9364 yd³/s 16a) 5715 tons per day 16b) ≈ 57 000 tons
per day 16c) 21.6 m³/s 17) 0.43 m³/s 18a) 1.86 x 10¹³ J,
1.76 x 10¹⁰ Btu 18b) 0.207 g 19a) 90.6 TW·h 19b) 10.34 GW
20a) 23.6 x 10⁶ Btu 20b) 9.17 tons 20c) 6916 kW·h per ton
21) 1580 MW

CHAPTER 29. 7) ≈ 2500 9a) 750 A 10a) 5092 MW
10b) 151.8 MW 10c) ≈ 3% 11a) 5700 V, 114 kW; 4500 V,
450 kW; 3000 V, 600 kW; 1500 V, 450 kW 13a) 5992 V, 126 kW;
5692 V, 720 kW; 4243 V, 1200 kW; 1897 V, 720 kW 14) 18.4°
E_R lags behind E_S 15a) 6577 V, 9000 V 15b) in phase
16a) 5692 V 16b) no 17a) 240 kvar 17b) 139.2 A
17c) 290.6 kvar 17d) 50.6 kvar 17e) 802 kVA 17f) 5759 V
17g) 6251 V, 868 kW 18) 20.3° 19) 4 Ω, 37.5 kΩ, 2 Ω
21a) 421 MW 21b) 1074 A 21c) 148.8 Mvar 21d) 74.3 Mvar
22a) 22 Ω, 100 Ω, 1500 Ω 22b) 119 kVar 22c) 45 A
22d) 9 Mvar 22e) 34.6 kW 23a) 648.5 A

CHAPTER 30. 3) 2 μΩ 6) 17.5 kΩ 8a) 48.8 kV 8b) zero
9a) 12 000 A 9b) 960 MW 9c) 4800 J 12) 1.2 Ω, 3.18 mH
13a) 25 A, 35 A 13b) 60 A 13c) 0.5 A 14) 9.74 kA
15a) 300 mA for 10 ms is not dangerous 15b) 30 mA for 2 min is
hazardous 17a) 115.6 A, 124.7 A, 31.6 A 17b) 2 A
17c) 81.8% lagging 18a) 777 A 18b) 86.8% 21b) 17.5 kA

CHAPTER 31. 5a) 42.5 kV 5b) 492 A 6a) 150 MW
6b) 200 A 7a) 600 A, 57.6 MW 10a) 400 A 10b) 600 kV
11) 1800 A, 835 Mvar 12) 45 kW 13) 4050 A 14a) 2.9 in²
14b) 13.38 Ω 14c) 43.4 MW 14d) 426 kV 14e) 94.7%
15a) 360 Hz, 720 Hz 15b) 9 Ω, 11 Ω 15c) 150 kV
16a) 17.7 A 16b) 15.6 kV 16c) 159 V 17a) 45.9°
17b) 87.2 Mvar